INTERNATIONAL SERIES OF MONOGRAPHS ON PHYSICS

T0202393

INTERNATIONAL SERIES OF MONOGRAPHS ON PHYSICS

CP Violation

GUSTAVO CASTELO BRANCO

Instituto Superior Técnico
and
Centro de Física Teórica de Partículas

LUÍS LAVOURA

Universidade de Lisboa
and
Centro de Física Teórica de Partículas

JOÃO PAULO SILVA

Instituto Superior de Engenharia de Lisboa
and
Centro de Física Teórica de Partículas

CLARENDON PRESS · OXFORD

UNIVERSITY PRESS

Great Clarendon Street, Oxford, OX2 6DP,
United Kingdom

Oxford University Press is a department of the University of Oxford.
It furthers the University's objective of excellence in research, scholarship,
and education by publishing worldwide. Oxford is a registered trade mark of
Oxford University Press in the UK and in certain other countries

© G. C. Branco, L. Lavoura, and J. P. Silva 1999

The moral rights of the author have been asserted

First published 1999
First published in paperback 2014

Impression: 1

Published in the United States of America by Oxford University Press
198 Madison Avenue, New York, NY 10016, United States of America

British Library Cataloguing in Publication Data

Data Available

ISBN 978-0-19-850399-6 (hbk.)
ISBN 978-0-19-871675-4 (pbk.)

Printed and bound in Great Britain by
Clays Ltd, St Ives plc

Links to third party websites are provided by Oxford in good faith and
for information only. Oxford disclaims any responsibility for the materials
contained in any third party website referenced in this work.

PREFACE

CP violation is an intriguing, elusive subject and our current knowledge of it is rather limited, both at the experimental and theoretical levels. On the one hand, CP violation has only been observed in the neutral-kaon system; on the other hand, in that system CP violation is very solidly established. From the theoretical standpoint, CP violation can be incorporated in the three-generation standard model (SM), which easily leads to the right order of magnitude for that effect, after one takes into account the experimental values for the quark mixing angles. However, we lack a fundamental understanding of the origin of CP violation. This is all the more important, because CP violation is one of the crucial ingredients necessary to generate the observed matter–antimatter asymmetry of the Universe. It is now believed that it is not possible to generate a baryon asymmetry of the observed size exclusively with the CP violation present in the SM. New sources of CP violation in models beyond the SM can play an important role in the explanation of the observed size of this asymmetry.

In spite of the importance of the phenomenon of CP violation, at present there is no self-contained textbook on the subject, covering both its phenomenological and theoretical aspects. It is this lack that we have aimed at eliminating. We have tried to write a text which, starting from basic and well known concepts, can lead graduate students and professional physicists alike into a reasonable understanding of the intricacies of CP violation. We have been particularly keen about adopting a consistent notation, and about self-containedness: we only assume knowledge of ordinary quantum mechanics, in the first part of the book, and of the standard model of electroweak interactions, from the second part onwards. We have also not hesitated in providing a detailed derivation of many results which remain poorly or only superficially explained in the literature.

The book is divided in four parts, aiming at an increasingly specialized group of readers. Most of the topics in the first two parts of the book might be included in a standard particle-physics course discussing electroweak interactions. The intended readership for the first part is very broad, including any student or physicist wishing to learn the basics of CP violation; this part is accessible to anyone familiar with ordinary quantum mechanics, and only little knowledge of particle physics and field theory is assumed. We explain what CP violation is and what are its basic observed features. Special emphasis is given to the phenomenology of CP violation in neutral-meson systems, considering the specific cases of the neutral kaons and neutral-B mesons and the approximations relevant for each of them. We discuss various ways to measure CP violation, especially in the neutral-B systems. Throughout, we use quantities which are invariant under arbitrary rephasings of the state vectors, using that property to identify the physical, measurable quantities.

Part II deals with the Kobayashi–Maskawa mechanism of CP violation in the SM. The readers are assumed to have some familiarity with gauge theories in general and the SM in particular. We study the unitarity triangles and their relevance for CP violation, describe various parametrizations of the Cabibbo–Kobayashi–Maskawa matrix, and discuss the experimental constraints on that matrix; we then review the computation of the CP-violating parameters ϵ and ϵ'/ϵ, being careful to present the analysis in such a way that it can easily incorporate new experimental data.

The third and fourth parts are narrower in scope. Part III is devoted to the model-building subtleties related to CP violation, and to various possible CP-violation mechanisms. Specific models are considered, the intention being to illustrate particular mechanisms of CP violation within minimal extensions of the SM. Thus, each model should be taken as representative of a whole set of possibilities. We work out models with, in turn, an extended scalar sector, fermion sector, and gauge sector. We also discuss the strong CP problem and describe some of its possible solutions.

We repeatedly emphasize the fact that CP violation arises as a clash between the CP-transformation properties of different terms in the Lagrangian. Although CP violation is due to the presence of complex phases in field theory, physical CP-violating quantities should not depend on the particular basis that one chooses to work in. This philosophy naturally leads to the construction of weak-basis-invariant CP-violating quantities; those quantities automatically eliminate the spurious phases which may always be brought in and out of the Lagrangian by means of rephasings of the fundamental fields.

It is generally believed that a deeper understanding of CP violation will require its experimental observation outside the neutral-kaon complex. This lacuna will be partially filled by the upcoming experimental studies at B factories; various tests of the SM, and the corresponding searches for new physics, will be conducted at those machines. These exciting prospects have provided further motivation for writing this book, which we hope will prove to be a timely publication. Thus, Part IV is specifically dedicated to the possibilities for the study of CP violation, in particular through the observation of CP asymmetries, at B factories. Our analysis is mostly model-independent, and we try to distinguish between theoretical expectations and the actual measuring capabilities.

It is not possible to cover all aspects of CP violation in a book of this size and, of course, our experience and interests have influenced the choice of topics. Three important subjects which are not dealt with in this book are electric dipole moments, baryogenesis, and supersymmetric models. These are very specialized areas of research which would require considerable space for a thorough and pedagogical introduction. However, the new sources of CP violation which arise in models beyond the SM, presented in detail in Chapters 22–26, will have an impact on baryogenesis. Furthermore, the techniques introduced in those chapters can be readily extended to the case of supersymmetric models. We may refer the interested reader to the existing monographs on baryogenesis (Cohen *et al.* 1993; Turok 1993; Rubakov and Shaposhnikov 1996; Trodden 1998) and on electric dipole moments (Khriplovich and Lamoreaux 1997).

There are many chapters and sections in the book which, having been included for the sake of completeness, may be skipped without undue loss of continuity or understanding. We have marked those chapters and sections with an asterisk in the Contents, and we have usually also called attention to this fact in the beginning of the chapter or section.

Whenever using experimental data, we have used the values given in the 1996 edition of the Review of Particle Properties (Particle Data Group 1996). The 1998 edition (Particle Data Group 1998) was not used because it appeared only shortly before completion of the manuscript; moreover, the physics in this book does not rely heavily on any precise experimental values.

In our bibliography we have made an effort to cite the original relevant literature on each topic which appeared up to the summer of 1998. But, in a field which evolves as rapidly as CP violation, it is impossible to keep track of all the relevant articles in the literature. The fact that many topics have been studied for a long time only makes things worse. We apologize for any omissions, which should not be interpreted as reflecting any negative opinion on our part.

Lisbon, Portugal   G. C. Branco
November 1998   L. Lavoura
                                                                            J. P. Silva

ACKNOWLEDGEMENTS

This book benefited greatly from the generous help of many colleagues and friends. We are especially grateful to Lincoln Wolfenstein for his generous support and advice over the years, for countless conversations, for the critical reading of various parts of the draft, and for enlightening discussions during his visit to Lisbon. Special thanks also go to Boris Kayser for the many conversations that we had during his visit to Lisbon.

Our colleagues Evgeny Akhmedov, Augusto Barroso, Stefano Bertolini, Francisco Botella, Jean-Marc Gérard, Walter Grimus, Hans-Günther Moser, Mário Santos, and Helmut Vogel have read and criticized parts of the draft. We thank them for their many comments and suggestions. Needless to say, ours is the blame for any inadequacies that remain.

Gustavo Castelo Branco is thankful to Rabindra Mohapatra for having introduced him to the subject of this book, two decades ago. We are also grateful to António Amorim, Andrea Brignole, John Donoghue, Ricardo González Felipe, Paulo Parada, Margarida Nesbitt Rebelo, Jorge Romão, Luca Silvestrini, João Soares, and Alfred Stadler for discussions.

We are indebted to Marc Baillargeon and Paulo Nogueira for allowing us to use, before public release, their excellent software 'Scribble', with which most Feynman graphs have been drawn.

João Paulo Silva thanks the *Instituto Superior de Engenharia de Lisboa* for relieving him from his teaching duties during the Fall semester of 1997. We also want to thank many of our colleagues and the staff at the *Centro de Física das Interacções Fundamentais* and at the *Centro de Física Nuclear da Universidade de Lisboa*, in particular Ana Eiró, who helped us on various occasions. Ricardo González Felipe, Paulo Nogueira, Pedro Sacramento, and Vítor Rocha Vieira have helped solving various computer problems. Eduardo Lopes rescued portions of the book from a crashed computer hard disk.

Our editor, Soenke Adlung, has provided us with friendly and valuable advice along the way. We are also indebted to the staff of Oxford University Press, especially to Kim Roberts and to Julia Tompson, for their help in the preparation of the manuscript.

Finally, we wish to thank our families for their forbearance and relentless support. To some extent, this book is as much theirs as it is ours. Luís Lavoura wants to dedicate this book to the memory of his mother, who died shortly before completion of the manuscript.

CONTENTS

Part I

CP in quantum mechanics

1

THE MEANING OF THE DISCRETE SYMMETRIES

1.1 Parity and time reversal in classical physics

Left–right symmetry—also called space-inversion or parity symmetry—and time-reversal symmetry are two invariances of classical physics—of classical mechanics and of classical gravitational and electromagnetic interactions—which were recognized long before the advent of quantum mechanics and of quantum field theory. We shall review the meaning of those symmetries in classical physics before implementing them in a quantum-mechanical context.

1.1.1 *Parity*

Parity symmetry, usually called P, consists in the invariance of physics under a discrete transformation which changes the sign of the space coordinates x, y, and z. This corresponds to the inversion of the three coordinate axes through the origin, a transformation which changes the handedness of the system of axes. A right-handed system becomes left-handed upon the parity transformation (see Fig. 1.1).

Parity symmetry is sometimes called mirror symmetry, because the inversion of the coordinate axes may be achieved in two steps, through a mirror reflection on a coordinate plane followed by a rotation by an angle π around the axis perpendicular to that plane (see Fig. 1.2). From the basic assumption of isotropy of space it follows that physics is invariant under a rotation. Therefore, the relevant point is whether physics is invariant under the mirror reflection too. Thus, P symmetry is in practice equivalent to symmetry under a mirror reflection. As the mirror interchanges left and right—for instance, our right arm is the left arm

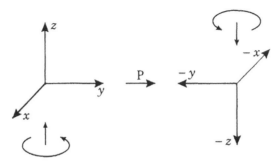

FIG. 1.1. A right-handed coordinate system becomes left-handed under the parity transformation.

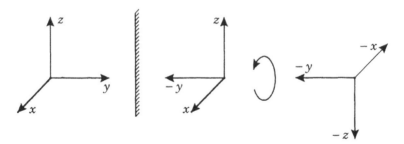

FIG. 1.2. The parity transformation as a reflection on a mirror followed by a
    rotation.

of our mirror image—parity symmetry is also called left–right symmetry.

Applying two parity transformations in succession is equivalent to no transformation at all. The square of the parity transformation is the identity transformation.

The parity transformation changes the sign of the position vector of a particle:
$\vec{r} \rightarrow -\vec{r}$. As a consequence, the velocity of the particle,

$$\vec{v} = \frac{d\vec{r}}{dt},$$ (1.1)

also changes sign under P. The same happens with the momentum

$$\vec{p} = m\vec{v}.$$ (1.2)

The angular momentum,

$$\vec{J} = \vec{r} \times \vec{p},$$ (1.3)

is invariant under P, because both $\vec{r}$ and $\vec{p}$ change sign. According to Newton's
law, the force which acts on a particle is equal to the rate of change of its
momentum,

$$\vec{F} = \frac{d\vec{p}}{dt}.$$ (1.4)

Therefore, under parity $\vec{F} \rightarrow -\vec{F}$.

The Lorentz force acting on a particle with electric charge q is given by

$$\vec{F}_{\text{Lorentz}} = q\left(\vec{E} + \vec{v} \times \vec{B}\right),$$ (1.5)

where $\vec{E}$ is the electric-field strength and $\vec{B}$ is the magnetic-field strength. Under
parity $\vec{F}_{\text{Lorentz}}$ and $\vec{v}$ change sign. Therefore, P must transform $\vec{E} \rightarrow -\vec{E}$ and
$\vec{B} \rightarrow \vec{B}$.

The scalar potential V and the vector potential $\vec{A}$ are defined through

$$\vec{E} = -\vec{\nabla}V - \frac{\partial \vec{A}}{\partial t},$$
$$\vec{B} = \vec{\nabla} \times \vec{A}.$$ (1.6)

The operator $\vec{\nabla} = \partial/\partial\vec{r}$ changes sign under P. Therefore, parity transforms $\vec{A} \to -\vec{A}$ while V is left invariant.

We shall not demonstrate the invariance of the whole body of classical mechanics and electromagnetism under the parity transformation delineated above. That invariance is established by surveying all the equations of classical physics and checking that they are invariant under parity.

Vectors are generically defined as three-component objects which transform in the same way as $\vec{r}$ under a rotation of the system of coordinate axes. This prescription does not tell us how vectors should transform under parity. As the square of the parity transformation is the identity transformation, there are two types of vectors: those which change sign under P and those which do not. Vectors which change sign under parity, like $\vec{p}$ and $\vec{E}$, are called polar vectors, or simply vectors. Entities like $\vec{B}$ and $\vec{J}$, which do not change their sign under a mirror reflection, are called axial vectors, or pseudovectors.

Analogously, there are quantities which are invariant under a space rotation but change sign under parity. This is the case in particular of the scalar product of a vector and a pseudovector, e.g. $\vec{E} \cdot \vec{B}$. Those quantities are called pseudoscalars, as opposed to (proper) scalars, which are mirror-invariant.

1.1.2 Time reversal and $\hat{T}$

Let us now consider the time-reversal transformation, usually called T. This consists of changing the sign of the time coordinate t. From eqn (1.1) we see that, when $t \to -t$, the velocity $\vec{v} \to -\vec{v}$. The momentum $\vec{p}$ also changes sign. The angular momentum $\vec{J} \to -\vec{J}$. On the other hand, from eqn (1.4), as both $\vec{p}$ and t change sign under time reversal, the force $\vec{F}$ remains invariant.

As $\vec{F}_{\text{Lorentz}}$ in eqn (1.5) must be invariant while $\vec{v}$ changes sign, $\vec{E} \to \vec{E}$ but $\vec{B} \to -\vec{B}$ under time reversal. From eqns (1.6) we see that $V \to V$ but $\vec{A} \to -\vec{A}$ under T.

The mathematical transformation delineated above, under which t, $\vec{p}$, and other entities change sign, may be called $\hat{T}$. The genuine time-reversal transformation T goes beyond $\hat{T}$, since it also interchanges final states and initial states. Time reversal is related to the following fundamental question that one may ask about the laws of Nature: let us consider the final state of some process, invert the velocities of all particles in that state, and let it evolve; shall we obtain the former initial state with all velocities reversed?

1.1.3 Spin, dipole moments, and helicity

Spin is a concept extraneous to classical physics. However, there is no problem in integrating it as an *ad hoc* quantity. Some particles are postulated to have associated with them an intrinsic angular momentum $\vec{s}$, which is in everything identical to a classical angular momentum $\vec{J}$. As such, $\vec{s} \to \vec{s}$ under parity and $\vec{s} \to -\vec{s}$ under time reversal.

The spin $\vec{s}$ is observable through interactions which are proportional to it. If a particle with spin $\vec{s}$ moves in an electromagnetic field with field strengths $\vec{E}$ and $\vec{B}$, there may exist in the Hamiltonian terms of the form

Table 1.1 *P and $\hat{T}$ transformations in classical physics.*

Name	Symbol	P	$\hat{T}$
Time	t	+	−
Position	$\vec{r}$	−	+
Energy	E	+	+
Momentum	$\vec{p}$	−	−
Spin	$\vec{s}$	+	−
Helicity	h	−	+
Electric-field strength	$\vec{E}$	−	+
Magnetic-field strength	$\vec{B}$	+	−
Magnetic dipole moment	d_m	+	+
Electric dipole moment	d_e	−	−

$$-d_e\, \vec{s} \cdot \vec{E} \tag{1.7}$$

and

$$-d_m\, \vec{s} \cdot \vec{B}, \tag{1.8}$$

the numerical coefficients d_e and d_m having the appropriate dimensions. If the interaction in eqn (1.7) exists the particle is said to possess an electric dipole moment d_e. If the interaction in eqn (1.8) is present the particle has a magnetic dipole moment d_m.

Checking the transformation rules of $\vec{E}$ and of $\vec{B}$ under parity and under time reversal, we conclude that $d_m \to d_m$ under any of those transformations, while $d_e \to -d_e$ under any of them. Therefore, electric dipole moments violate both P and T. On the other hand, magnetic dipole moments violate neither P nor T and, indeed, they provide a practical way of measuring the spin of a particle.

An important quantity is the sign of the projection of a particle's spin $\vec{s}$ along the direction of its momentum $\vec{p}$. This is called the helicity h of the particle:

$$h = \frac{\vec{s} \cdot \vec{p}}{|\vec{s}|\,|\vec{p}|}. \tag{1.9}$$

Helicity is a pseudoscalar, because it is the dot product of a polar vector ($\vec{p}$) and an axial vector ($\vec{s}$). On the other hand, $h \to h$ under T.

1.1.4 *Summary*

The preceding results are summarized in Table 1.1, in which we have indicated whether the relevant quantities are invariant (denoted by a '+' sign) or change their sign (denoted by a '−' sign) under the P and $\hat{T}$ transformations.

1.1.5 *Relativistic mechanics*

When one makes the transition to relativistic mechanics, the time and the position vector get united in the position four-vector[1] $x^\mu = (t, \vec{r})$, while the en-

---

[1] We use units such that $c = \hbar = 1$.

Table 1.2 *P and $\hat{T}$ transformations in relativistic physics.*

Name	Symbol	P	$\hat{T}$
Position	x^μ	x_μ	$-x_\mu$
Derivative	∂^μ	∂_μ	$-\partial_\mu$
Momentum	p^μ	p_μ	p_μ
Potential	A^μ	A_μ	A_μ
Field tensor	$F^{\mu\nu}$	$F_{\mu\nu}$	$-F_{\mu\nu}$

ergy and three-momentum become components of the momentum four-vector $p^\mu = (E, \vec{p})$. The derivative four-vector is $\partial^\mu = (\partial/\partial t, -\vec{\nabla})$. The scalar and vector potentials of electromagnetism are united in the four-vector $A^\mu = (V, \vec{A})$. The angular momentum $\vec{J}$ becomes part of an antisymmetric tensor $M^{\mu\nu} = x^\mu p^\nu - p^\mu x^\nu$. In the same way, $\vec{E}$ and $\vec{B}$ are united in an electromagnetic-field tensor $F^{\mu\nu} = \partial^\mu A^\nu - \partial^\nu A^\mu$.

All four-vectors behave in the same way under parity: their time component is left unchanged, while their space components change sign. We denote this by $x^\mu \to x_\mu$, $\partial^\mu \to \partial_\mu$, $p^\mu \to p_\mu$, and $A^\mu \to A_\mu$. Tensors also have their indices lowered by P: $M^{\mu\nu} \to M_{\mu\nu}$ and $F^{\mu\nu} \to F_{\mu\nu}$. Parity invariance may thus be extended to relativistic mechanics.

The transformation properties under time reversal are more complex, since not all four-vectors behave in the same way. The four-vectors $x^\mu \to -x_\mu$ and $\partial^\mu \to -\partial_\mu$ behave in a different way[2] from the four-vectors $p^\mu \to p_\mu$ and $A^\mu \to A_\mu$. The electromagnetic-field tensor $F^{\mu\nu} \to -F_{\mu\nu}$, and the angular-momentum tensor $M^{\mu\nu} \to -M_{\mu\nu}$ under $\hat{T}$. All equations of relativistic mechanics are invariant under the time-reversal transformation delineated above.

In Table 1.2 we have indicated how relativistic tensors of interest transform under P and under $\hat{T}$.

1.2   The meaning of P and of T

Parity and time reversal are closely related to some of the most basic questions that one may ask about the laws of Nature.

Suppose that one watches some physical event in a mirror. Does the event that one sees there look real? Does the event seen in the mirror correspond to something allowed by the laws of Nature? This is the basic question that P symmetry addresses.

Now suppose instead that one has the physical event filmed and then watches the film running backwards. Will the events seen in the backward-running film look possible and realistic, or will they be at odds with the laws of Nature? This is the issue raised by T symmetry.

---

[2]Thus, $\hat{T}$ differs from the discrete transformation of the Poincaré group under which all four-vectors $V^\mu \to -V_\mu$. The latter discrete transformation is sometimes misleadingly called T.

1.2.1   *P- and T-asymmetry of the observed events*

In a certain sense, it is an obvious fact that left and right are distinct in Nature. They represent more than mere conventions. Most people display greater skilfulness with their right than with their left hand. Our liver is located in the right side, our heart in the left side of our body. Therefore, no one would confuse the mirror image of a human being with a real person! At a more fundamental level, the aminoacids in life's chemistry are not identical with their mirror images. Many organic molecules have a right-handed and a left-handed version, and one of the versions occurs much more often in the biosphere than the other one. However, the above asymmetries are in general considered to be accidents of life's evolution on Earth, and not the consequence of a fundamental left–right asymmetry in the laws of Nature.

As for T, a 'time arrow' seems to exist in observed events, not only in biology, but also in the more fundamental realms of physics. A piece of wood burns down to ashes and smoke, but ashes and smoke have never been seen to absorb heat from their surroundings and generate a piece of wood. Naively one might think that this asymmetry in the time evolution of physical systems is in contradiction with the laws of classical physics, since Newton's law is invariant under time reversal, in the sense that, if $\vec{r}(t)$ is a possible trajectory, then $\vec{r}(-t)$ is also an allowed trajectory. The asymmetry in the time evolution is one of the postulates of classical thermodynamics, which states that, if a system at a certain instant is in a non-equilibrium macroscopic state, it will evolve into another state with higher entropy. This is the second law of thermodynamics, for which an explanation is given within the framework of statistical mechanics. Systems evolve in the time direction that they do because the final macroscopic configuration is microscopically more probable than the initial one. This in no way implies a time-reversal asymmetry in the fundamental laws of microscopic physics.

1.2.2   *A thought experiment about T*

The fundamental laws of classical physics are T invariant but, because of the large number of individual particles and collisions involved, macroscopic systems display T-asymmetry. In order to understand how this comes about, a simple thought experiment (Lee 1990) may help.

Suppose that, in a large country with lots of intersecting roads, one thousand drivers start from the same place and drive one thousand kilometres each. Suppose that no directions are marked on the roads, and each driver meets multiple road crossings, each time choosing at will, with no outside help, which new road he shall take.

Now consider the time-reversed situation. Each of the thousand drivers starts from the final point that he has reached in the previous journey, and drives once again one thousand kilometres, once again choosing, more or less at random, which new road he will take at each crossing. One asks oneself, will all drivers, at the end of their second journey, meet at the starting point of the first one? Clearly, the probability that this happens is extremely small.

Although the individual motion of each driver has obeyed time-reversal-invariant laws—they have driven one thousand kilometers in each journey, and at each crossing they have chosen, according to the same chance rules, which new road to take—the observed motion of the total system was time-reversal-asymmetric. This was due to two reasons: first, the numerous road crossings at which each driver had to make a choice; second, the large number of drivers involved.

In the same way, in classical mechanics, the large number of particles and the large number of collisions among them render the time-reversed motion of a macroscopic system extremely improbable.

1.2.3   *A thought experiment about P*

One may wonder, where does the observed difference between right and left (see § 1.2.1) come from? Is it built into the fundamental equations of physics? Or is it just a chance consequence of the particular development that life took on Earth?

One may translate this question into a thought experiment. Suppose that we were able to build a live being, for instance a dog or a fly, completely made up of organic molecules of the wrong handedness. The question then is, would this artificial being be able to live and function properly? Would it be competitive in a Darwinian sense with the existing forms of life, or would it suffer from some intrinsic disadvantage because of the opposite handedness of its biochemistry?

For a mechanical analogue of this question (Lee 1990), consider two mirror-symmetric cars. They are of the same model, but each of them is the mirror image of the other one.[3] One asks oneself, will these cars run in the same way? If the two cars are accelerated with the gas pedal tilted at the same angle, will they move forward at the same speed? Or might one of them, for instance, stay stuck or even move backwards?

1.2.4   *Summary*

In spite of the observed reality that life is mirror-asymmetric, the equations of classical physics are left–right symmetric. In spite of the obvious arrow of time in real physical events, classical mechanics and classical gravitational and electromagnetic interactions do not have a preferred time direction. The question then is whether left–right symmetry and time-reversal symmetry carry over to the microscopic world. Is there somewhere a fundamental P asymmetry which might explain the observed left–right asymmetry of the biosphere? Or should that asymmetry be assigned to fortuitous initial conditions? And is there, somewhere in the fundamental interactions beyond classical physics' realm, a T asymmetry? Could such an asymmetry help explain the observed time arrow of events?

---

[3]This must not be confused with two cars of the same model with the driving wheels on opposite sides. Such cars do not have mirror-symmetric engines.

1.3   Charge conjugation

Contrary to P and T, charge-conjugation symmetry C does not have an analogue in classical physics. This symmetry is related to the existence of an antiparticle for every particle. This is a prediction of relativistic quantum theory which has been brilliantly confirmed by experiment, in particular through the discovery of the positron (Anderson 1933) and of the antiproton (Chamberlain *et al.* 1955). It should be emphasized that the notion of antiparticle exists neither in classical physics nor in non-relativistic quantum mechanics.

In relativistic quantum field theory, one can associate both positively and negatively charged particles with each (complex) field ϕ. Moreover, there is a C transformation which transforms ϕ into a related field, e.g. $\phi^\dagger$, which has opposite U(1) charges—electric charge, baryon and lepton number, and flavour quantum numbers such as strangeness, the third component of isospin, and so on. The transformed field obeys the same relativistic equation of motion as the original one. It has the same mass, but its interaction with an electromagnetic potential is characterized by opposite electric charge.

C symmetry asserts that antiparticles behave in exactly the same way as the corresponding particles, and that it is a mere matter of convention which of them we call 'particles' and which we call 'antiparticles'.

1.3.1   *A thought experiment about C*

Why should C symmetry be important? Experimentally it is known that, when a particle and its antiparticle collide, they have a high probability of annihilating. Let us then consider the following thought experiment (Lee 1990).

Suppose that our civilization came into contact with another civilization on a distant planet. The contact might take place via exchange of electromagnetic messages, without any charged particle ever being exchanged. After years of friendly correspondence, the two civilizations might want to physically meet, for instance through the sending of a space vessel. The problem would then be to know whether the other civilization is made out of matter or of antimatter. Indeed, if it were made out of antimatter, physical contact would be impossible, lest annihilation destroys both meeting parties. How could civilizations communicate to each other whether each one's 'matter' is the same as the other one's, or whether they are made out of the antimatter of each other?

In order to communicate this, one needs some absolute way of distinguishing matter from antimatter. If C symmetry holds, matter and antimatter are distinguishable only by practical example, i.e., they are a convention. C symmetry must be violated, some physical event must occur differently with matter and antimatter, in order that an explanation to our distant partners of how that event happens in our world lets them know what 'matter' means to us.

1.4   Violation of C, P, and CP

It turns out that the whole body of weak interactions works differently for matter and antimatter. Moreover, weak interactions also are left–right asymmetric. On

the other hand, after simultaneous C and P transformations, (most) weak interactions remain identical to themselves—cross sections and decay rates remain
unchanged.

The composite transformation CP, made out of simultaneous C and P transformations, then acquires relevance. Namely, the conceptual problem of distinguishing matter from antimatter (see § 1.3.1) can only acquire a solution if we
are able to eliminate the convention of what is 'left' and what is 'right' from the
game. It is not enough that C be violated, CP must be violated too in order that
matter may be distinguished from antimatter.

1.4.1   *The experiment of Wu et al.*

Historically, the possibility that weak interactions violate parity was first suggested by Lee and Yang (1956). They examined the experimental evidence then
available and concluded that parity invariance of the weak interactions was 'only
an extrapolated hypothesis unsupported by experimental evidence', i.e., it had
not yet been probed. They went on to suggest experiments which might test
whether parity is conserved. They observed that, in order to test parity violation, one should try and measure some pseudoscalar quantity, like for instance
the helicity of some particle, or more generally the scalar product of the spin of
a particle and the momentum of some other particle. If the expectation value of
a pseudoscalar observable is found to be non-zero, then there is parity violation.

Parity violation in the β decay of ^{60}Co was discovered soon afterwards (Wu
et al. 1957). The nuclide ^{60}Co decays through β^- emission to an excited state of
^{60}Ni, which then decays to its fundamental state through the emission of two successive gammas. In the experiment of Wu *et al.*, a ^{60}Co source was incorporated
into a crystal of cerium magnesium. When a small magnetic field (~ 0.05 T) is
applied to the crystal, there is an alignment of the electronic spins, generating
inside the crystal a strong magnetic field (~ 10–100 T). This in turn polarizes
the ^{60}Co nuclei through the hyperfine coupling, provided the whole system is
at a sufficiently low temperature (~ 0.01 K). The low temperature was achieved
through a process of adiabatic demagnetization.

A scintillation counter was used to measure the intensity of the β^- emission
relative to the orientation of the polarizing field and as a function of the temperature. It was found that the electrons are emitted preferentially in the direction
opposite to the one of the applied magnetic field, and that that preference disappears when the crystal warms up. This means that the scalar product of the
spin of the ^{60}Co nuclei and the momentum of the emitted electron has a non-zero
expectation value. As that scalar product changes sign under P, parity violation
in β decay was established.

1.4.2   *The helicity of the electron neutrino*

After the experiment of Wu *et al.* (1957), another experiment of great importance in testing the nature of the weak interactions was the one of Goldhaber
et al. (1958). There, the helicity of the neutrino emitted in the electron capture
by ^{152}Eu was measured. The experiment was particularly ingenious because, as

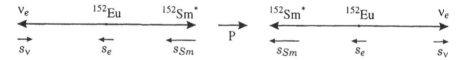

FIG. 1.3. The electron capture by ^{152}Eu, yielding ^{152}Sm* and ν_e, and its P-transformed process.

the neutrino hardly interacts with matter, some way of indirectly measuring its helicity had to be devised.

The nuclide ^{152}Eu has zero spin. It captures an electron from the K shell—with zero orbital angular momentum—of the atom, giving rise to an excited state with spin 1 of ^{152}Sm, and emitting a neutrino ν_e. The excited state then decays to the fundamental state of ^{152}Sm, which is spinless, through emission of a photon.

Conservation of angular momentum along the direction of flight of the neutrino and of the ^{152}Sm* nucleus, in the rest frame of ^{152}Eu, implies that the handedness[4] of the neutrino is the same as the handedness of ^{152}Sm*—see Fig. 1.3. (This is because the angular momentum of the initial state, constituted by the ^{152}Eu nucleus and by the K-shell electron, is just the spin of the electron.) Again, it can easily be shown that, if the photon from the decay of that excited state is emitted in the same direction as the neutrino, that photon has the same handedness as the excited state and, therefore, as the neutrino. Thus, the measurement of the helicity of the neutrino is reduced to a selection of the events in which the gamma is emitted in the same direction as the neutrino, and to a measurement of the helicity of the gamma in those events. It was found that the gammas have helicity -1. Thus, the ν_e emitted in the electron capture must have helicity -1, i.e., be left-handed.

The fact that the neutrino emitted in electron capture always has helicity -1 constitutes a violation of parity, as can be seen in Fig. 1.3.

After the original experiment of Goldhaber et al. (1958) on the electron neutrino, many other experiments have attempted to measure the helicity of the neutrinos. In particular, the helicity of the muonic neutrino has been directly measured in a nice experiment by Roesch et al. (1982). It has always been found that neutrinos have helicity -1, while antineutrinos have helicity $+1$.

1.4.3 CP

In general, one finds that C is violated together with P in the weak interactions. Let us give an example of this. The charged pion π^+ decays predominantly to $\mu^+ \nu_\mu$. The muon neutrino from the decay is left-handed (Garwin et al. 1957; Friedman and Telegdi 1957), see Fig. 1.4 (a). The P-conjugate process, in which the ν_μ would be right-handed, never occurs—see Fig. 1.4 (b). The C-conjugate

---

[4] Handedness is equivalent to the helicity of a particle. If a particle has helicity $+1$ it is said to be right-handed. The particle is left-handed if it has helicity -1.

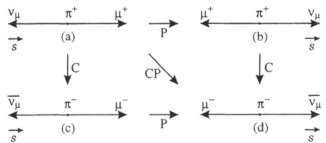

FIG. 1.4. C, P, and CP transformation of the decay $\pi^+ \to \mu^+ \nu_\mu$.

process is the decay of π^- to $\mu^- \bar{\nu}_\mu$, the $\bar{\nu}_\mu$ being left-handed—Fig. 1.4 (c). It turns out that this process never occurs either. Instead, $\pi^- \to \mu^- \bar{\nu}_\mu$ occurs at the same rate as $\pi^+ \to \mu^+ \nu_\mu$, only the $\bar{\nu}_\mu$ is right-handed, as in Fig. 1.4 (d).

This fact represents a simultaneous violation of C and of P. P is violated because the neutrinos have a definite handedness. If P were not violated the neutrinos should, with equal probability, be either left-handed or right-handed, both in $\pi^+ \to \mu^+ \nu_\mu$ and in $\pi^- \to \mu^- \bar{\nu}_\mu$. C is violated because the neutrinos in the two observed decays have opposite handedness, and therefore the decays are not C-conjugates of each other.

On the other hand, the combined symmetry CP is preserved. Indeed, when we simultaneously interchange $\pi^+ \leftrightarrow \pi^-$, $\mu^+ \leftrightarrow \mu^-$, and $\nu_\mu \leftrightarrow \bar{\nu}_\mu$ (C transformation), and also interchange left- and right-handedness of the neutrinos and of the muons (P transformation), the decay rates are equal. C and P are violated, but the combined symmetry CP is not.

Clearly, $\pi^\pm$ decays cannot provide a solution to the communication problem of § 1.3.1. The distinction of matter from antimatter using these decays requires a previous convention about the handedness of the coordinate system. We are unable to explain to our far-away partners what we mean by 'a positively charged pion' because, if we try and tell them that 'it decays into a neutral particle with negative helicity' they will ask us how do we define the sign of the helicity, and such a definition is equivalent to a *convention* for the handedness of the coordinate system.

1.4.4 *CP violation*

The fact that CP symmetry is preserved even while C and P symmetries are violated was first pointed out by Landau (1957), who suggested that neutrinos are always left-handed and antineutrinos are always right-handed.[5] Only much later was CP discovered to be violated too (Christenson *et al.* 1964).

The clearest evidence for CP violation is the charge asymmetry in K_{l3} decays. The kaon K_L is a neutral particle with well-defined mass and decay width.

---

[5] Actually, the first suggestion that C and P might be separately violated, while CP would be conserved, was due to Wick *et al.* (1952). They considered that possibility 'remote at the moment'.

There is no other particle with equal mass. Therefore, K_L must be its own antiparticle. It decays both to $\pi^+ e^- \bar{\nu}_e$ and to the C-conjugate mode $\pi^- e^+ \nu_e$. However, it decays slightly less often to the first than to the second mode. This fact unequivocally establishes both C violation and CP violation.

Indeed, when we consider the total decay rates, we have already performed an integration over the momenta of all the particles resulting from the decay, as well as a sum over their spin states. These integrations and sums eliminate parity from consideration. Then, if the decay rates to $\pi^+ e^- \bar{\nu}_e$ and to $\pi^- e^+ \nu_e$ are different, there is violation of CP, not only of C.

We now have the solution to the thought experiment of § 1.3.1. We should tell our partners on a distant planet to observe K_{l3} decays: the decay which occurs less often gives rise to a pion with the same electric charge as the proton that we are made of. They would thus learn what we mean by 'matter'. No reference to right-handed and left-handed coordinate systems would be needed. Indeed, we might afterwards use pion decays to explain to our distant friends what we mean by 'left' and 'right'.

1.4.5 *Theoretical importance of CP violation*

Besides being a fascinating effect because of its elusiveness at both the experimental and theoretical levels, CP violation might also play an important role in our understanding of cosmology. This is because the observed baryon asymmetry of the Universe, i.e., the fact that there is much more matter than antimatter in the observed Universe, could only be generated from an initial situation in which the amounts of matter and antimatter would be equal if there is CP violation. This fact was first shown by Sakharov (1967), who pointed out that baryon-number violation, C and CP violation, and a departure from thermal equilibrium, are all necessary in order for it to be possible to generate a net baryon asymmetry in the early Universe.

Another interesting consequence of CP violation would be the possibility that elementary particles have electric dipole moments. We have seen in § 1.1.3 that electric dipole moments violate both P and T. Composite particles like nuclei or atoms may display a net electric dipole moment because of the existence of degenerate states with different properties under parity and time reversal; on the contrary, any electric dipole moment of an elementary particle would in a downright manner violate both P and T. Now, T violation is connected through the CPT theorem to CP violation—this means that it would be very difficult to conceive of a theory which would violate T without simultaneously violating CP (see § 2.5 below). Thus, although not strictly necessary for the existence of electric dipole moments of elementary particles, theoreticians certainly do not expect them to exist unless CP is violated.

THE DISCRETE SYMMETRIES IN QUANTUM PHYSICS

2.1 Introduction

In quantum theory the transformations P, T, and C are represented by operators $\mathcal{P}$, $\mathcal{T}$, and $\mathcal{C}$, respectively. In this chapter we introduce the main properties of those operators. We shall not construct them explicitly in formal quantum field theory; for this reason, some of the properties will remain undemonstrated.

The symmetry T requires a mathematical formalism distinct from the one used for P and C symmetries. We shall treat C and P jointly, leaving T to a separate section, which may be skipped without loss of continuity.

In quantum theory, the fundamental operator is the Hamiltonian $\mathcal{H}$, which is Hermitian and generates time translations $\exp\left(-i\mathcal{H}\Delta t\right)$. The time translation from $t' = -\infty$ to $t = +\infty$ is the scattering matrix

$$S = \lim_{t \to +\infty} \lim_{t' \to -\infty} \exp\left[-i\mathcal{H}\left(t - t'\right)\right]. \tag{2.1}$$

The matrix S is unitary. It is usually separated into a trivial part, the unit matrix, and a non-trivial transition matrix T:

$$S = 1 + iT. \tag{2.2}$$

Unitarity then yields

$$T - T^{\dagger} = iT^{\dagger}T. \tag{2.3}$$

2.2 Definition of $\mathcal{C}$ and of $\mathcal{P}$

2.2.1 *How to define the operators $\mathcal{C}$ and $\mathcal{P}$*

In classical physics, the parity transformation does not affect the time coordinate. In particular, it commutes with a time translation $t \to t + \Delta t$. Following the principle of correspondence for passing from classical mechanics to quantum mechanics (Dirac 1958), we require the quantum-mechanical representations of parity through the operator $\mathcal{P}$, and of the time translation through the operator $\exp\left(-i\mathcal{H}\Delta t\right)$, to reproduce these features. The operator $\mathcal{P}$ must therefore commute with $\mathcal{H}$. Now, in quantum theory, $[\mathcal{P}, \mathcal{H}] = 0$ means that parity is a good symmetry of Nature. This contradicts what we have seen in the previous chapter, that parity is *not* a symmetry of the weak interactions. We conclude that there is no operator $\mathcal{P}$ satisfying the requisites that an operator must have in order to be a good quantum-mechanical representation of the parity transformation—in particular, commuting with the weak Hamiltonian $\mathcal{H}_{\mathrm{W}}$. There is no operator $\mathcal{P}$

which adequately represents the parity transformation in quantum mechanics (Lee and Wick 1966).

The same of course applies to the C transformation. Our picture of it implies that it must commute with time translations. Therefore, the operator C must satisfy $[C, \mathcal{H}] = 0$. But we know that C is not a good symmetry of Nature. Therefore, there is no operator C which represents correctly the C transformation.

In this book, however, we shall often be dealing with operators $\mathcal{P}$ and C, and with their composite operator $C\mathcal{P}$. How do we define them? We start from the realization that electromagnetic interactions are C and P invariant. We then assume that strong interactions are C and P invariant too; all available experimental data warrant this assumption. Thus, when weak interactions are switched off and only the strong and electromagnetic interactions are taken into account, we can define legitimate C and $\mathcal{P}$ operators. We then use these operators as probes for the (non-)invariance of the various interactions beyond the strong and electromagnetic ones.

It is important to understand from the very beginning this strategy for defining $\mathcal{P}$ and C. We restrict ourselves to an appropriate part of the complete Lagrangian, namely, the kinetic Lagrangian and the electromagnetic-interaction Lagrangian. This part of the Lagrangian is C and P invariant. We define operators C and $\mathcal{P}$ suitable to this part of the Lagrangian. We then gradually proceed to other parts of the Lagrangian, at each step including new interactions. We probe whether these parts of the Lagrangian are invariant under the action of the operators C and $\mathcal{P}$ previously defined. If they are not, then there is C or P violation.

2.2.2 *Internal symmetries*

There is in general some arbitrariness in the definitions of $\mathcal{P}$ and C in a quantum theory. This is because the Lagrangian has internal symmetries. The simplest internal symmetry corresponds to the possibility of rephasing (changing the phase of) each individual quantum field. More generally, whenever there are various quantum fields with the same quantum numbers, there is an internal symmetry mixing those fields. That symmetry is unitary, because we want the kinetic terms of the Lagrangian to preserve their normalization.

Now, the action of the operators C and $\mathcal{P}$ may include an arbitrary reshuffling of some internal indices, i.e., an internal-symmetry transformation may be juxtaposed on the basic transformation effected by either C or $\mathcal{P}$. The quantum fields are mathematical constructs and do not have an immediate correspondence with experimental observables. Therefore, an operator $\mathcal{P}$ which mixes some fields occupying an equivalent position in the Lagrangian is just as good a representation of parity as an operator $\mathcal{P}$ which does not mix them. This is the origin of a large degree of arbitrariness in the definition of $\mathcal{P}$, and analogously also of C and $C\mathcal{P}$.

2.3   Properties of C and of $\mathcal{P}$

The operators C and $\mathcal{P}$ are unitary,

$$C^\dagger = C^{-1},$$
$$P^\dagger = P^{-1}. \tag{2.4}$$

When defining C and P we assume that C and P are good symmetries, as we have explained in the previous section. Therefore,

$$[C, \mathcal{H}] = 0,$$
$$[P, \mathcal{H}] = 0. \tag{2.5}$$

For the scattering matrix, one has

$$CSC^\dagger = S,$$
$$PSP^\dagger = S. \tag{2.6}$$

In quantum field theory each field may be written as a linear combination of creation and destruction operators for particles and for antiparticles. For instance, the electron field ψ is a linear combination of operators a which destroy electrons and of operators $b^\dagger$ which create positrons; its C-conjugate field ψ^c is a linear combination of operators $a^\dagger$, which create electrons, and operators b, which destroy positrons. If $|e^-\rangle$ represents an electron with momentum $\vec{p}$ and helicity h, and $|e^+\rangle$ represents a positron with the same momentum and helicity, then these states may be written as

$$|e^-\rangle = a^\dagger(\vec{p}, h)\,|0\rangle,$$
$$|e^+\rangle = b^\dagger(\vec{p}, h)\,|0\rangle, \tag{2.7}$$

where $|0\rangle$ is the vacuum state. The operator C transforms electrons into positrons and vice versa, without affecting the momentum and the spin. Thus,

$$Ca^\dagger(\vec{p}, h)\,C^\dagger = \exp(i\vartheta)\,b^\dagger(\vec{p}, h),$$
$$Cb^\dagger(\vec{p}, h)\,C^\dagger = \exp(-i\vartheta)\,a^\dagger(\vec{p}, h). \tag{2.8}$$

The phase ϑ in eqns (2.8) is arbitrary. The vacuum state does not change under the action of C:

$$C|0\rangle = |0\rangle. \tag{2.9}$$

Combining eqns (2.7)–(2.9) one obtains

$$C|e^-\rangle = \exp(i\vartheta)\,|e^+\rangle,$$
$$C|e^+\rangle = \exp(-i\vartheta)\,|e^-\rangle. \tag{2.10}$$

The operator C does not commute with the charge operator Q. For instance,

$$Q|e^-\rangle = -e|e^-\rangle,$$
$$Q|e^+\rangle = e|e^+\rangle. \tag{2.11}$$

(We denote the positron charge by e.) Equations (2.10) and (2.11) show that C does not commute with Q.

Particles, which are simultaneous eigenstates of $\mathcal{H}$ and of $\mathcal{Q}$, cannot be eigen-states of $\mathcal{C}$ if they have non-zero charge. Moreover, not all particles with zero charge are eigenstates of $\mathcal{C}$. Indeed, what we have said about $\mathcal{Q}$ is valid not only for electric charge, but also for any other conserved charge, such as baryon number and lepton number—and strangeness and the other flavour quantum numbers, as long as the strong and electromagnetic interactions conserve them. None of those operators commutes with $\mathcal{C}$. Therefore, a particle must have zero value of all those operators if it is to be an eigenstate of $\mathcal{C}$. For instance, the neutral pion π^0 and the photon γ are eigenstates of $\mathcal{C}$, but the neutron n and the neutral kaon K^0 are not, because the former has non-zero baryon number and the latter has non-zero strangeness.

The operator $\mathcal{P}$ is associated with an observable, a quantum number, which is conserved as long as interactions are P invariant. That quantum number is named the parity of the state. In particular, as long as $\mathcal{P}$ commutes with the Hamil-tonian, each non-degenerate particle must be an eigenstate of $\mathcal{P}$. As quantum-mechanical state vectors are defined up to an arbitrary phase, the fact that the square of the classical P transformation is the identity transformation only im-plies that $\mathcal{P}^2$ applied on any quantum state vector should multiply that vector by a phase. Thus, $\mathcal{P}$ does not in general satisfy $\mathcal{P}^2 = 1$. The parities of quantum states may in general be any complex number with modulus 1.

We may however choose a convention for $\mathcal{P}$ in which a basic set of fields all have parity +1. Thus, in nuclear physics the intrinsic parities of the proton, the neutron, and the Λ baryon are chosen to be +1. Once this is done, all other particles and nuclei have intrinsic parities which are either +1 or −1.

It must be stressed that this phase choice for the action of $\mathcal{P}$ on hadron states is really just a convention, which does not have to be adopted in order to reach the physical conclusions that follow from parity conservation by the strong interactions. In this book we shall not be using any phase convention when we deal with the operator that interests us most, $\mathcal{CP}$. We find this practice advisable because it automatically prevents us from attributing physical significance to phases which have none.

The operator $\mathcal{C}$ is also associated with a quantum number, which is named C-parity. One finds from eqns (2.10) that $\mathcal{C}^2 = 1$; as a consequence, C-parity can only take the values +1 and −1. In order for a quantum state to be an eigenstate of $\mathcal{C}$ it must have zero charge, as we have seen. Therefore, only neutral particles may be assigned a C-parity. For charged particles, the eigenstates of C are artificial constructs, the superposition of a particle and the corresponding antiparticle in the same state of motion (i.e., with the same momentum and spin).

The discrete symmetries C and P are different in one important respect from the ordinary continuous symmetries of quantum mechanics, like translations and rotations. While the conserved quantities associated with the latter (momentum and angular momentum, respectively) are additive, the quantum numbers associ-ated with the former are multiplicative. For instance, translations are generated by a differential operator, which acts on the product of two functions by acting successively on each of them:

$$\vec{\nabla}(fg) = \left(\vec{\nabla}f\right)g + f\left(\vec{\nabla}g\right). \qquad (2.12)$$

As a consequence, the momentum associated with the product wave function is the sum of the momenta associated with the factor wave functions:

$$\left.\begin{array}{c} -i\vec{\nabla}f = \vec{p}_f f \\ -i\vec{\nabla}g = \vec{p}_g g \end{array}\right\} \Rightarrow -i\vec{\nabla}(fg) = (\vec{p}_f + \vec{p}_g)(fg). \qquad (2.13)$$

On the other hand, the parity transformation changes $\vec{r} \to -\vec{r}$ simultaneously in all functions that it is applied to. Now,

$$\left.\begin{array}{c} f(-\vec{r}) = (-1)^{P_f} f(\vec{r}) \\ g(-\vec{r}) = (-1)^{P_g} g(\vec{r}) \end{array}\right\} \Rightarrow f(-\vec{r})g(-\vec{r}) = (-1)^{P_f+P_g} f(\vec{r})g(\vec{r}). \qquad (2.14)$$

We see that the parities of f and of g must be multiplied in order to obtain the parity of (fg). Parity is a multiplicative quantum number, just as C-parity. This is the opposite of what happens with linear and angular momentum (and baryon number, and electric charge...), which are additive quantum numbers. The difference has its root in the fact that the additive quantum numbers are the eigenvalues *of the generators* of some continuous transformations, while parity and C-parity are the eigenvalues *of (discrete) transformations themselves.*

2.4 The operator $\mathcal{T}$

The implementation of time-reversal symmetry in a quantum-mechanical context is subtle. This is because the relevant operator, $\mathcal{T}$, is not unitary but rather antiunitary.

2.4.1 *Antiunitary operators*

In order to understand what antiunitarity means, let us start by considering which requirements we should impose on a quantum-mechanical symmetry transformation. Suppose that an operator θ transforms $|\psi\rangle$ into $|\psi_\theta\rangle$ and $|\chi\rangle$ into $|\chi_\theta\rangle$:

$$\begin{aligned} |\psi_\theta\rangle &= \theta|\psi\rangle, \\ |\chi_\theta\rangle &= \theta|\chi\rangle. \end{aligned} \qquad (2.15)$$

In order for θ to represent a symmetry, we might be tempted to require that it preserves the quantum-mechanical bracket:

$$\langle\chi_\theta|\psi_\theta\rangle = \langle\chi|\theta^\dagger\theta|\psi\rangle = \langle\chi|\psi\rangle. \qquad (2.16)$$

Equation (2.16) is, however, too restrictive. Indeed, quantum-mechanical probabilities are not given by the brackets themselves, but rather by their moduli. Therefore, in order for θ to be a symmetry, it should be enough to impose

$$|\langle \chi_\theta | \psi_\theta \rangle| = |\langle \chi | \psi \rangle|. \tag{2.17}$$

Equation (2.17) allows for another possibility, besides the one in eqn (2.16):

$$\langle \chi_\theta | \psi_\theta \rangle = \langle \chi | \psi \rangle^* = \langle \psi | \chi \rangle. \tag{2.18}$$

It may be proved that any transformation θ satisfying eqn (2.17) must follow either the rule in eqn (2.16) or the one in eqn (2.18), apart from trivial phase changes. The proof of this theorem (Wigner 1932; Gottfried 1966; Weinberg 1995) is rather involved, and we shall not deal with it here. It is enough to assert that any quantum-mechanical symmetry θ must either preserve the values of all quantum-mechanical brackets, as in eqn (2.16), or else transform the brackets into their complex conjugates, as in eqn (2.18). Symmetries of the first kind are represented by unitary operators. This is the case, in particular, for $\mathcal{P}$ and $\mathcal{C}$. Symmetries of the second kind are represented by antiunitary operators. This is the case of $\mathcal{T}$, and also of the combined operator $\mathcal{CPT}$, to be discussed in the next section.

2.4.2 *Operating rules for antiunitary transformations*

Antiunitary operators θ may be interpreted as consisting of the product of an unitary operator, U_θ, by an operator K which complex-conjugates all the c-numbers to the right of it (Wigner 1932): $\theta = U_\theta K$. When applied to a ket $|\psi\rangle$, understood as a column matrix of complex numbers, θ acts in the following manner:

$$|\psi_\theta\rangle = U_\theta K |\psi\rangle = U_\theta |\psi^*\rangle. \tag{2.19}$$

Obviously then,

$$\theta \left(c_1 |\psi\rangle + c_2 |\chi\rangle \right) = U_\theta \left(c_1^* |\psi^*\rangle + c_2^* |\chi^*\rangle \right)$$
$$= c_1^* |\psi_\theta\rangle + c_2^* |\chi_\theta\rangle, \tag{2.20}$$

where c_1 and c_2 are arbitrary complex coefficients. The property in eqn (2.20) is called antilinearity.

As K just complex-conjugates all c-numbers to its right, $K^2 = 1$. Therefore, $\theta^{-1} = K U_\theta^\dagger$. Thus, when applied on an operator $\mathcal{O}$, understood as a square matrix of complex numbers, θ acts in the following way:

$$\mathcal{O}_\theta \equiv \theta \mathcal{O} \theta^{-1}$$
$$= U_\theta K \mathcal{O} K U_\theta^\dagger$$
$$= U_\theta \mathcal{O}^* U_\theta^\dagger. \tag{2.21}$$

The operator $\theta^\dagger = K^\dagger U_\theta^\dagger$ is not identical with θ^{-1}, contrary to what happens with unitary operators. Indeed, in order to reproduce eqn (2.18), the operator $K^\dagger$ must be interpreted as complex-conjugating all c-numbers to its *left*:

$$\langle \chi_\theta | \psi_\theta \rangle = \langle \chi | \theta^\dagger \theta | \psi \rangle = \langle \chi | K^\dagger U_\theta^\dagger U_\theta K | \psi \rangle = \langle \chi^* | \psi^* \rangle. \tag{2.22}$$

Now consider $\langle \chi_\theta | \mathcal{O}_\theta | \psi_\theta \rangle$. This should be understood as the product of the bra $\langle \chi_\theta |$ by the ket $\mathcal{O}_\theta | \psi_\theta \rangle$. Thus,

$$
\begin{aligned}
\langle \chi_\theta | \mathcal{O}_\theta | \psi_\theta \rangle &= \left(\langle \chi | \theta^\dagger \right) \left(\theta \mathcal{O} \theta^{-1} \theta | \psi \rangle \right) \\
&= \left(\langle \chi | K^\dagger U_\theta^\dagger \right) \left(U_\theta K \mathcal{O} K U_\theta^\dagger U_\theta K | \psi \rangle \right) \\
&= \langle \chi^* | \mathcal{O}^* | \psi^* \rangle \\
&= \langle \psi | \mathcal{O}^\dagger | \chi \rangle .
\end{aligned} \tag{2.23}
$$

2.4.3 Basis-dependence of K

The complex-conjugation operator K is not well defined, contrary to what one might grasp from the previous section. In order to understand this, let us consider two different bases for the quantum-mechanical Hilbert space, $\{|a_k\rangle\}$ and $\{|b_k\rangle\}$. In the first basis, the ket $|a_1\rangle$ is given by a column matrix in which the first entry is unity and all other entries are zero. As that column matrix is real, we find that $K|a_1\rangle = |a_1\rangle$. On the other hand, in the second basis,

$$
|a_1\rangle = \sum_k \langle b_k | a_1 \rangle | b_k \rangle . \tag{2.24}
$$

The kets $|b_k\rangle$ are now the basis vectors, which are represented by real column matrices with all entries equal to zero, except for the kth entry which is unity. They are therefore invariant under the action of K. Hence,

$$
K|a_1\rangle = \sum_k \langle b_k | a_1 \rangle^* | b_k \rangle , \tag{2.25}
$$

which is in general different from $|a_1\rangle$ in eqn (2.24). We conclude that in the first basis $K|a_1\rangle = |a_1\rangle$ while in the second basis $K|a_1\rangle \neq |a_1\rangle$. The action of K depends on the basis that one uses for the Hilbert space.

Of course, we would like the operator θ, which corresponds to a physical symmetry transformation, not to depend on the basis. This means that U_θ must be basis-dependent, and its basis-dependence must offset the one of K. In the words of Gottfried (1966), 'if the basis is changed, the work of U_θ and K has to be reapportioned'.

2.4.4 $\mathcal{T}$ as an antiunitary operator

A straightforward way to realize that the operator $\mathcal{T}$, representing the time-reversal transformation in quantum mechanics, must be antiunitary, is to consider the basic quantum commutator

$$
[r_j, p_k] = i \delta_{jk} , \tag{2.26}
$$

where r_j and p_k are Cartesian components of the position vector $\vec{r}$ and the momentum vector $\vec{p}$, respectively, of a particle. Under time reversal p_k changes sign but r_j does not, as we know from Chapter 1. This is in contradiction with eqn (2.26) unless we assume that $\mathcal{T}$ also changes i into $-i$.

One further indication is the behaviour of the Schrödinger equation for a free particle,

$$i\frac{\partial \psi}{\partial t} = -\frac{\vec{\nabla}^2 \psi}{2m},$$ (2.27)

under time reversal. In classical mechanics, a free particle has a time-reversal invariant motion, and we would like the same to happen in quantum mechanics. But the operator $\partial/\partial t$ is T-odd while the operator $\vec{\nabla}$ is T-even. This is impossible to reconcile with eqn (2.27) unless $\mathcal{T}$ changes $i \rightarrow -i$ and $\psi \rightarrow \psi^*$.

The operator $\mathcal{T}$ therefore has to be antiunitary. As the classical time-reversal transformation leaves the energy invariant, we require

$$\mathcal{T}\mathcal{H}\mathcal{T}^{-1} = \mathcal{H}.$$ (2.28)

From eqn (2.1) and from antiunitarity it then follows that

$$\mathcal{T}S\mathcal{T}^{-1} = S^\dagger.$$ (2.29)

Then, from eqn (2.2),

$$\mathcal{T}T\mathcal{T}^{-1} = T^\dagger.$$ (2.30)

In order to obtain correspondence with classical mechanics, $\mathcal{T}$ must be defined in such a way that

$$\mathcal{T}|\vec{p}_a, \vec{s}_a\rangle = \exp\left(i\theta_a\right)|-\vec{p}_a, -\vec{s}_a\rangle,$$ (2.31)

where $|\vec{p}_a, \vec{s}_a\rangle$ is a state in which particles have momenta $\vec{p}_a$ and spins $\vec{s}_a$. Equation (2.31) is the quantum version of the classical rule according to which time reversal inverts the momenta and spins of all particles. It follows from eqns (2.23), (2.29), and (2.31), that

$$\langle\vec{p}_a, \vec{s}_a|S|\vec{p}_b, \vec{s}_b\rangle = \exp\left[i\left(\theta_a - \theta_b\right)\right]\langle-\vec{p}_b, -\vec{s}_b|S|-\vec{p}_a, -\vec{s}_a\rangle.$$ (2.32)

Therefore, T invariance implies

$$|\langle\vec{p}_a, \vec{s}_a|S|\vec{p}_b, \vec{s}_b\rangle| = |\langle-\vec{p}_b, -\vec{s}_b|S|-\vec{p}_a, -\vec{s}_a\rangle|.$$ (2.33)

Equation (2.33) is known as the reciprocity relation, or as the 'principle of detailed balance'. It states that the probability of an initial state b being scattered into a final state a is the same as the probability that an initial state identical to a, but with all the momenta and spins reversed, scatters into the final state b with all the momenta and spins reversed. This is, of course, just the quantum version of the classical picture of T symmetry.

2.4.5 T invariance and 'T conservation'

The operator $\mathcal{T}$ does not have meaningful eigenvalues, contrary to what happens with $\mathcal{C}$ and $\mathcal{P}$. To see why this is so, suppose that

$$\mathcal{T}|\psi\rangle = k|\psi\rangle,$$ (2.34)

with k some complex number. Then, because of antiunitarity,

$$\mathcal{T}\left[\exp\left(i\zeta\right)|\psi\rangle\right] = \exp\left(-i\zeta\right)\mathcal{T}|\psi\rangle = k\exp\left(-2i\zeta\right)\left[\exp\left(i\zeta\right)|\psi\rangle\right]. \qquad (2.35)$$

Quantum mechanics states that the kets $|\psi\rangle$ and $\exp\left(i\zeta\right)|\psi\rangle$ correspond to the same physical state. Equations (2.34) and (2.35) mean that that physical state sometimes has $\mathcal{T}$ eigenvalue k, and sometimes has $\mathcal{T}$ eigenvalue $k\exp\left(-2i\zeta\right)$. This implies that $\mathcal{T}$ cannot be associated with a quantum number.

Because of this fact, it is incorrect to refer to T invariance as 'T conservation', or to say that 'T is conserved'. As there is no quantum number associated with $\mathcal{T}$, nothing is conserved when there is T invariance.

Another way to see this is by supposing that at $t' \to -\infty$ we had the initial state prepared to be an eigenstate of $\mathcal{T}$, say

$$\mathcal{T}|\psi(t')\rangle = k|\psi(t')\rangle.$$

This quantum state would evolve and, at $t \to +\infty$, we would obtain

$$|\psi(t)\rangle = S|\psi(t')\rangle.$$

Applying $\mathcal{T}$ to this state, and using eqn (2.29), we obtain

$$\mathcal{T}|\psi(t)\rangle = \mathcal{T}S\mathcal{T}^{-1}\mathcal{T}|\psi(t')\rangle = kS^\dagger|\psi(t')\rangle,$$

which in general is not proportional to $|\psi(t)\rangle$. Thus, at $t \to +\infty$ we no longer have an eigenstate of $\mathcal{T}$. This means that there is no T conservation.

2.4.6 Kramers' degeneracy

While $\mathcal{T}$ is antiunitary and cannot be associated with a quantum number, $\mathcal{T}^2 = U_T K U_T K = U_T U_T^*$ is unitary and can be associated with a quantum number. However, there are restrictions on that quantum number, which is not allowed to be an arbitrary number of modulus one. In order to see this, let us suppose that

$$\mathcal{T}^2|\psi\rangle = k|\psi\rangle \qquad (2.36)$$

for some state $|\psi\rangle$, where k is a number of modulus one. Then,

$$\mathcal{T}^2\left(\mathcal{T}|\psi\rangle\right) = \mathcal{T}^3|\psi\rangle = \mathcal{T}k|\psi\rangle = k^*\left(\mathcal{T}|\psi\rangle\right). \qquad (2.37)$$

This means that the eigenvalues of $\mathcal{T}^2$ for the states $|\psi\rangle$ and $\mathcal{T}|\psi\rangle$ are complex-conjugates of each other. If one makes the reasonable assumption that those two states should have the same value of $\mathcal{T}^2$, one concludes that that value can only be either $+1$ or -1.

By explicit construction of the operator $\mathcal{T}$ it may be shown (see for instance Sakurai 1994) that bosonic systems (systems with integer angular momentum) have $\lambda = +1$, while fermionic systems (with half-integer angular momentum) have $\lambda = -1$. This has an interesting consequence:

Theorem 2.1 *In a T invariant fermionic system, all energy eigenstates are (at least) doubly degenerate.*

(This is called Kramers' degeneracy after its discoveror.)

Proof Consider an energy eigenstate $|E\rangle$ with $\mathcal{H}|E\rangle = E|E\rangle$. Using eqn (2.28) we find that $|E_{\mathcal{T}}\rangle \equiv \mathcal{T}|E\rangle$ is also an eigenstate of $\mathcal{H}$ with energy E. On the other hand, $|E_{\mathcal{T}\mathcal{T}}\rangle \equiv \mathcal{T}^2|E\rangle = -|E\rangle$ because the system has half-integer angular momentum. Therefore,

$$-\langle E_{\mathcal{T}}|E\rangle = \langle E_{\mathcal{T}}|E_{\mathcal{T}\mathcal{T}}\rangle = \langle E_{\mathcal{T}}|E\rangle. \qquad (2.38)$$

(In the second step we have used eqn 2.18.) We conclude that $\langle E_{\mathcal{T}}|E\rangle = 0$. Thus, $|E_{\mathcal{T}}\rangle$ cannot be identical with $|E\rangle$, i.e., there is degeneracy.   □

Kramers' degeneracy occurs in particular in any fermionic system immersed in a static electric field $\vec{E}$, no matter how complicated that field is, provided there is no magnetic field $\vec{B}$.

2.5   CPT

At this point, it is useful to introduce the combined transformation CPT, which simultaneously performs a reflection of the space axes through the origin, inverts the time evolution, and interchanges particles and antiparticles. This operation is represented in quantum mechanics by the operator $\mathcal{CPT}$. Being the product of two unitary operators and an antiunitary one, $\mathcal{CPT}$ is antiunitary.

Any of the discrete symmetries C, P, or T, or any combination thereof, may be violated in Nature. However, there is a strong theoretical prejudice against the possibility that CPT is violated. This prejudice is based on the so-called 'CPT theorem' (Lueders 1954; Pauli 1955; Jost 1957, 1963; Streater and Wightman 1964) which, starting from very general properties of quantum field theory—in particular, Lorentz invariance and local (anti-)commutation relations obeying the spin-statistics connection—asserts that any such theory is CPT invariant. Because of this theorem, it is very difficult to conceive a realistic, sensible relativistic quantum theory in which CPT is violated, and usually one just assumes that Nature is CPT invariant.

String theory is a non-local theory and may also present spontaneous breaking of Lorentz invariance. Then, it would not satisfy the assumptions of the CPT theorem, and might display CPT violation. Some theoretical work has been done in this direction (Kostelecký and Potting 1991), but the issue is far from clear.

A simple consequence of the CPT theorem is that violation of one of the discrete symmetries implies violation of the complementary one. For instance, if CP is violated, then T must be violated too. However, the CPT theorem applies to quantum field theories, it does not apply to specific observables, which may violate a symmetry without violating the complementary one. For instance, if the electron turned out to have an electric dipole moment, this would represent a violation of T (and of P), but it would not represent a violation of CP. However, the Lagrangian responsible for the electric dipole moment of the electron must violate both T and CP.

Table 2.1 *Some experimental bounds on the differences of masses and lifetimes in particle–antiparticle pairs.*

ψ	$\left(m_\psi - m_{\bar\psi}\right)/m_{\text{average}}$	$\left(\tau_\psi - \tau_{\bar\psi}\right)/\tau_{\text{average}}$
e^+ (positron)	$< 4 \times 10^{-8}$	—
μ^+ (antimuon)	—	$(2 \pm 8) \times 10^{-5}$
π^+ (charged pion)	$(2 \pm 5) \times 10^{-4}$	$(5.5 \pm 7.1) \times 10^{-4}$
n (neutron)	$(9 \pm 5) \times 10^{-5}$	—

The theoretical bias in favour of CPT invariance should not hinder experimentalists in an effort towards observing CPT violation, or setting upper bounds on CPT-violating quantities. Most equalities following from CPT symmetry also follow from either C or CP symmetries; as we already know the latter symmetries to be violated in Nature, we must have recourse to CPT to enforce those equalities.[6]

The most basic consequence of CPT symmetry is the equality of the masses of a particle and its antiparticle. The lifetimes of a particle and its antiparticle should also be equal—which is not surprising, because decay widths are equivalent to imaginary parts of masses. These equalities have been experimentally tested for some particles, as seen in Table 2.1.

Another consequence of CPT invariance is the fact that the electric charges q of a particle and its antiparticle should be exactly symmetric. Also, the magnetic dipole moments μ of a particle and its antiparticle should be opposite; for the leptons, which are point-like particles, it is usual to state this in terms of the gyromagnetic ratios, denoted by the letter g. The corresponding experimental bounds are given in Table 2.2.

Table 2.2 *Experimental bounds on the sums of electric charges, and on the differences of magnetic dipole moments, in particle–antiparticle pairs.*

ψ	$\left|q_\psi + q_{\bar\psi}\right|/e$	$\left(\mu_\psi - \left|\mu_{\bar\psi}\right|\right)/\left|\mu_{\text{average}}\right|$	$\left(g_\psi - g_{\bar\psi}\right)/g_{\text{average}}$
e^+	$< 4 \times 10^{-8}$	—	$(-0.5 \pm 2.1) \times 10^{-12}$
μ^+	—	—	$(-2.6 \pm 1.6) \times 10^{-8}$
p (proton)	$< 2 \times 10^{-5}$	$(-2.6 \pm 2.9) \times 10^{-3}$	—

These experimental values and bounds are those given by the Particle Data Group (1996, pp. 249, 250, 320, 321, 561, and 567). In any case, the best tests of CPT symmetry come from the neutral-kaon system, specifically, from the agreement of experiment with the predictions of the CPT invariant phenomenology. We shall deal with this point at some length in Chapter 8. The value given by the Particle Data Group (1996, p. 59) is

$$\left|\frac{m_{K^0} - m_{\overline{K^0}}}{m_{\text{average}}}\right| < 9 \times 10^{-19}, \tag{2.39}$$

[6]In any case, we may expect the violation of the equalities to be at least as suppressed as CP violation normally is. Therefore, we should expect those violations to be extremely small.

which looks like a very impressive bound. However, the derivation of the bound in eqn (2.39) has subtleties which are far from trivial, partly because various CPT-violating parameters might be cancelling out in the final result (Lavoura 1992*a*).

3

P, T, AND C INVARIANCE OF QED

3.1 Introduction

In this chapter we make an overview of the charged Klein–Gordon and Dirac fields in interaction with the electromagnetic field—of quantum electrodynamics (QED) of spin-0 and spin-1/2 particles. We assume the reader already to have some acquaintance with the subject, and we do not pretend our overview to be very pedagogical. Our emphasis is on the P, T, and C transformations of the fields. We fix some notation, and present various formulae which will be used later in the book.

3.2 The photon field

The Lagrangian for the photon field is

$$\mathcal{L}_A = -\tfrac{1}{4} F_{\mu\nu} F^{\mu\nu}$$
$$= -\tfrac{1}{2} \left(\partial_\mu A_\nu\right)\left(\partial^\mu A^\nu\right) + \tfrac{1}{2} \left(\partial_\mu A_\nu\right)\left(\partial^\nu A^\mu\right). \tag{3.1}$$

The field $A^\mu(t, \vec{r})$ is, after second quantization, a Hermitian operator. It transforms under P in the following way (see Table 1.2):

$$\mathcal{P} A_\mu(t, \vec{r}) \mathcal{P}^\dagger = A^\mu(t, -\vec{r}). \tag{3.2}$$

Taking into account that $\vec{r} \to -\vec{r}$ implies $\partial^\mu \to \partial_\mu$, we see that

$$\mathcal{P} \mathcal{L}_A(t, \vec{r}) \mathcal{P}^\dagger = \mathcal{L}_A(t, -\vec{r}). \tag{3.3}$$

Similarly,

$$\mathcal{T} A_\mu(t, \vec{r}) \mathcal{T}^{-1} = A^\mu(-t, \vec{r}). \tag{3.4}$$

As $t \to -t$ implies $\partial^\mu \to -\partial_\mu$, we see that

$$\mathcal{T} \mathcal{L}_A(t, \vec{r}) \mathcal{T}^{-1} = \mathcal{L}_A(-t, \vec{r}). \tag{3.5}$$

Physical intuition tells us that the charged current $j^\mu = (\rho, \vec{j})$ must change sign under C—both the charge density and the density of current change sign. Since $\mathcal{C} j^\mu \mathcal{C}^\dagger = -j^\mu$, and in order for the electromagnetic interaction $A_\mu j^\mu$ to be C invariant, we postulate

$$\mathcal{C} A_\mu(t, \vec{r}) \mathcal{C}^\dagger = -A_\mu(t, \vec{r}), \tag{3.6}$$

and therefore

$$\mathcal{C}\mathcal{L}_{\mathrm{A}}\left(t,\vec{r}\right)\mathcal{C}^{\dagger}=\mathcal{L}_{\mathrm{A}}\left(t,\vec{r}\right). \tag{3.7}$$

Equations (3.3), (3.5), and (3.7) introduce the P, T, and C transformation rules for the Lagrangian or any part thereof; for consistency with electromagnetism, we must have

$$\mathcal{P}\mathcal{L}_{\mathrm{part}}\left(t,\vec{r}\right)\mathcal{P}^{\dagger}=\mathcal{L}_{\mathrm{part}}\left(t,-\vec{r}\right), \tag{3.8}$$
$$\mathcal{T}\mathcal{L}_{\mathrm{part}}\left(t,\vec{r}\right)\mathcal{T}^{-1}=\mathcal{L}_{\mathrm{part}}\left(-t,\vec{r}\right), \tag{3.9}$$
$$\mathcal{C}\mathcal{L}_{\mathrm{part}}\left(t,\vec{r}\right)\mathcal{C}^{\dagger}=\mathcal{L}_{\mathrm{part}}\left(t,\vec{r}\right), \tag{3.10}$$

for any Hermitian operator $\mathcal{L}_{\mathrm{part}}$ which is part of the total Lagrangian $\mathcal{L}$.[7]

3.3   The Klein–Gordon field

The Klein–Gordon Lagrangian for a spin-0 field ϕ with mass m and electric charge q, moving in the electromagnetic field given by the potential A^{μ}, is

$$\mathcal{L}_{\mathrm{KG}}=\left(\partial_{\mu}\phi^{\dagger}-iqA_{\mu}\phi^{\dagger}\right)\left(\partial^{\mu}\phi+iqA^{\mu}\phi\right)-m^{2}\phi^{\dagger}\phi. \tag{3.11}$$

After second-quantization ϕ is an operator field. In this context we remind that Hermitian conjugation both complex-conjugates all c-numbers and transforms all operators into their Hermitian conjugates, while complex conjugation—involved, for instance, in the operation of $\mathcal{T}$—only complex-conjugates the c-numbers, leaving the operators unaffected. Thus, $\phi^{*}\neq\phi^{\dagger}$.

Under parity,

$$\mathcal{P}\phi\left(t,\vec{r}\right)\mathcal{P}^{\dagger}=\exp\left(i\alpha_{p}\right)\phi\left(t,-\vec{r}\right). \tag{3.12}$$

The phase α_{p} is arbitrary. Taking the Hermitian conjugate of eqn (3.12) we obtain

$$\mathcal{P}\phi^{\dagger}\left(t,\vec{r}\right)\mathcal{P}^{\dagger}=\exp\left(-i\alpha_{p}\right)\phi^{\dagger}\left(t,-\vec{r}\right). \tag{3.13}$$

Together with eqn (3.2) and the change $\partial^{\mu}\to\partial_{\mu}$, we find

$$\mathcal{P}\mathcal{L}_{\mathrm{KG}}\left(t,\vec{r}\right)\mathcal{P}^{\dagger}=\mathcal{L}_{\mathrm{KG}}\left(t,-\vec{r}\right), \tag{3.14}$$

as it should, cf. eqn (3.8). Notice that

$$\mathcal{P}^{2}\phi\left(t,\vec{r}\right)\mathcal{P}^{\dagger^{2}}=\exp\left(2i\alpha_{p}\right)\phi\left(t,\vec{r}\right). \tag{3.15}$$

$\mathcal{P}^{2}$ multiplies the Klein–Gordon field by a phase.

---

[7]If, for instance, $\mathcal{L}_{\mathrm{part}}$ is not P invariant, then there is no operator $\mathcal{P}$ satisfying eqn (3.8). That equation only informs us how a parity invariant part of the Lagrangian should behave under parity.

Under time-reversal we have

$$\mathcal{T}\phi\left(t,\vec{r}\right)\mathcal{T}^{-1} = \exp\left(i\alpha_t\right)\phi\left(-t,\vec{r}\right), \tag{3.16}$$

with an arbitrary phase α_t. From eqn (3.16) we derive

$$\mathcal{T}\phi^\dagger\left(t,\vec{r}\right)\mathcal{T}^{-1} = \exp\left(-i\alpha_t\right)\phi^\dagger\left(-t,\vec{r}\right). \tag{3.17}$$

Then, for instance,

$$\mathcal{T}\left(\partial^\mu\phi + iqA^\mu\phi\right)\mathcal{T}^{-1} = e^{i\alpha_t}\left(\partial^\mu\phi - iqA_\mu\phi\right),$$

in which we have omitted writing down the space–time coordinates, as we shall sometimes be doing. As $\partial^\mu \to -\partial_\mu$ when $t \to -t$, we find:

$$\mathcal{T}\mathcal{L}_{\mathrm{KG}}\left(t,\vec{r}\right)\mathcal{T}^{-1} = \mathcal{L}_{\mathrm{KG}}\left(-t,\vec{r}\right). \tag{3.18}$$

As $\mathcal{T}$ complex-conjugates all complex numbers, eqn (3.16) also leads to

$$\begin{aligned}
\mathcal{T}^2\phi\left(t,\vec{r}\right)\mathcal{T}^{-2} &= \mathcal{T}\exp\left(i\alpha_t\right)\phi\left(-t,\vec{r}\right)\mathcal{T}^{-1} \\
&= \exp\left(-i\alpha_t\right)U_{\mathcal{T}}\phi^*\left(-t,\vec{r}\right)U_{\mathcal{T}}^\dagger \\
&= \phi\left(t,\vec{r}\right).
\end{aligned} \tag{3.19}$$

Thus, $\mathcal{T}^2 = 1$ on the Klein–Gordon field, as on the photon field.

Under charge conjugation,

$$\mathcal{C}\phi\left(t,\vec{r}\right)\mathcal{C}^\dagger = \exp\left(i\alpha_c\right)\phi^\dagger\left(t,\vec{r}\right). \tag{3.20}$$

Therefore,

$$\mathcal{C}\phi^\dagger\left(t,\vec{r}\right)\mathcal{C}^\dagger = \exp\left(-i\alpha_c\right)\phi\left(t,\vec{r}\right). \tag{3.21}$$

Together with eqn (3.6), we obtain

$$\begin{aligned}
\mathcal{C}\mathcal{L}_{\mathrm{KG}}\left(t,\vec{r}\right)\mathcal{C}^\dagger &= \left(\partial_\mu\phi + iqA_\mu\phi\right)\left(\partial^\mu\phi^\dagger - iqA^\mu\phi^\dagger\right) - m^2\phi\phi^\dagger \\
&= \mathcal{L}_{\mathrm{KG}}\left(t,\vec{r}\right),
\end{aligned} \tag{3.22}$$

because ϕ and $\phi^\dagger$ commute. Also,

$$\mathcal{C}^2\phi\left(t,\vec{r}\right)\mathcal{C}^{\dagger^2} = \phi\left(t,\vec{r}\right). \tag{3.23}$$

Thus, $\mathcal{C}^2 = 1$ on the Klein–Gordon field.

3.4   The Dirac field

3.4.1   *Dirac matrices*

The Dirac matrices γ^μ are four 4×4 matrices which obey the anti-commutation algebra (Clifford algebra)

$$\{\gamma^\mu, \gamma^\nu\} = 2g^{\mu\nu}. \tag{3.24}$$

There is an infinite number of sets of four matrices obeying this algebra. Each of these sets is called a representation of the algebra, or simply a representation of the Dirac matrices.[8]

---

[8] For a good introduction to Clifford algebras and Dirac matrices in a space–time of arbitrary dimension, see Sohnius (1985).

The matrix

$$\gamma_5 \equiv i\gamma^0\gamma^1\gamma^2\gamma^3 \tag{3.25}$$

obeys a Clifford algebra together with the γ^μ. Indeed,

$$\{\gamma_5, \gamma^\mu\} = 0, \tag{3.26}$$
$$(\gamma_5)^2 = 1. \tag{3.27}$$

The eigenvalue of γ_5 is called chirality. From eqn (3.27), chirality may be either $+1$ or -1. We define the projection matrices of chirality,

$$\gamma_R \equiv \frac{1+\gamma_5}{2},$$
$$\gamma_L \equiv \frac{1-\gamma_5}{2}. \tag{3.28}$$

They satisfy the usual properties of projection matrices:

$$\gamma_R + \gamma_L = 1,$$
$$(\gamma_R)^2 = \gamma_R,$$
$$(\gamma_L)^2 = \gamma_L, \tag{3.29}$$
$$\gamma_R\gamma_L = \gamma_L\gamma_R = 0.$$

In the massless limit, a fermion of chirality $+1$ is right-handed, and a fermion of chirality -1 is left-handed. This is the reason for the choice of the subscripts 'R' and 'L' for the projection matrices of chirality.

From the commutators of pairs of Dirac matrices we define matrices $\sigma^{\mu\nu}$:

$$\sigma^{\mu\nu} \equiv \tfrac{i}{2}[\gamma^\mu, \gamma^\nu]. \tag{3.30}$$

For every representation of the Dirac matrices there are non-singular 4×4 matrices A and C such that

$$A\gamma_\mu = \gamma_\mu^\dagger A, \tag{3.31}$$
$$\gamma_\mu C = -C\gamma_\mu^T. \tag{3.32}$$

Then,

$$A\gamma_5 = -\gamma_5^\dagger A, \tag{3.33}$$
$$\gamma_5 C = C\gamma_5^T, \tag{3.34}$$
$$A\sigma_{\mu\nu} = \sigma_{\mu\nu}^\dagger A, \tag{3.35}$$
$$\sigma_{\mu\nu} C = -C\sigma_{\mu\nu}^T. \tag{3.36}$$

A and C also have the following properties:

$$A^\dagger = A, \tag{3.37}$$

$$C^T = -C, \tag{3.38}$$

$$CA^*C^*A = 1. \tag{3.39}$$

The usual representations of the Dirac matrices (Dirac–Pauli, Weyl, and Majorana representations) have the extra feature that

$$\gamma_0^\dagger = \gamma_0,$$
$$\gamma_k^\dagger = -\gamma_k, \tag{3.40}$$

for $k = 1$, 2, and 3. For those representations

$$A = \gamma^0. \tag{3.41}$$

However, all the relevant properties of the Dirac matrices are independent of their representation, and we do not need to assume eqns (3.40). Therefore, in general eqn (3.41) does not hold.

3.4.2 *Dirac spinors*

A Lorentz transformation is a linear coordinate transformation

$$x^\mu \to x'^\mu = \Lambda^\mu{}_\nu x^\nu, \tag{3.42}$$

given by a real 4×4 matrix Λ such that

$$g_{\alpha\beta} = \Lambda^\mu{}_\alpha \Lambda^\nu{}_\beta g_{\mu\nu}. \tag{3.43}$$

In particular, an infinitesimal Lorentz transformation

$$\Lambda^\mu{}_\nu = \delta^\mu_\nu + \lambda^\mu{}_\nu \tag{3.44}$$

has real infinitesimal transformation parameters $\lambda^\mu{}_\nu$ satisfying $\lambda_{\mu\nu} = -\lambda_{\nu\mu}$.

A Dirac spinor is a 4×1 column matrix of fields $\psi(x)$ transforming under a Lorentz transformation as

$$\psi(x) \to \psi'(x') = S(\Lambda)\psi(x'), \tag{3.45}$$

where $S(\Lambda)$ is a non-singular 4×4 matrix such that

$$S(\Lambda)^{-1}\gamma^\mu S(\Lambda) = \Lambda^\mu{}_\nu \gamma^\nu. \tag{3.46}$$

In particular, for the infinitesimal Lorentz transformation in eqn (3.44),

$$S(\Lambda) = 1 + \tfrac{1}{8}\lambda_{\mu\nu}\left[\gamma^\mu, \gamma^\nu\right]. \tag{3.47}$$

The matrix in eqn (3.47) has the following properties:

$$S(\Lambda)^\dagger A = AS(\Lambda)^{-1},$$
$$C\left[S(\Lambda)^{-1}\right]^T = S(\Lambda)C. \tag{3.48}$$

These properties of $S(\Lambda)$ hold for any Lorentz transformation, and not only for infinitesimal ones. This results from the fact that any finite Lorentz transformation may be constructed as a product of infinitesimal ones and, furthermore, if eqns (3.48) hold for any two matrices $S(\Lambda_1)$ and $S(\Lambda_2)$, they also hold for their product $S(\Lambda_1) S(\Lambda_2)$.

For every Dirac spinor ψ, we define

$$\overline{\psi} \equiv \psi^\dagger A, \tag{3.49}$$

$$\psi^c \equiv C\overline{\psi}^T$$
$$= CA^T \psi^{\dagger T}. \tag{3.50}$$

Using eqns (3.37)–(3.39) we find

$$(\psi^c)^c = \psi \tag{3.51}$$
$$\overline{\psi^c} = -\psi^T C^{-1}. \tag{3.52}$$

From eqns (3.45) and (3.48) it follows that

$$\overline{\psi'} = \overline{\psi} S(\Lambda)^{-1}, \tag{3.53}$$
$$\psi'^c = S(\Lambda) \psi^c. \tag{3.54}$$

Thus, ψ^c transforms under a Lorentz transformation in the same way as ψ.

A Majorana spinor is a Dirac spinor such that

$$\psi^c = e^{i\zeta} \psi, \tag{3.55}$$

where ζ is an arbitrary phase.

A Weyl spinor is a Dirac spinor which is an eigenstate of γ_5. We easily derive

$$\gamma_5 \psi = \pm \psi \Leftrightarrow \overline{\psi} \gamma_5 = \mp \overline{\psi} \Leftrightarrow \gamma_5 \psi^c = \mp \psi^c. \tag{3.56}$$

Thus, if ψ is a Weyl spinor, then ψ^c is also a Weyl spinor but with opposite chirality. As a corollary, a spinor cannot simultaneously have Majorana and Weyl character.

For every spinor ψ, we define its chiral components

$$\psi_R \equiv \gamma_R \psi,$$
$$\psi_L \equiv \gamma_L \psi, \tag{3.57}$$

which are Weyl spinors. Obviously, $\psi = \psi_R + \psi_L$. Moreover,

$$\overline{\psi_R} = \overline{\psi} \gamma_L,$$
$$\overline{\psi_L} = \overline{\psi} \gamma_R. \tag{3.58}$$

Therefore,

$$(\psi_R)^c = (\psi^c)_L,$$
$$(\psi_L)^c = (\psi^c)_R. \tag{3.59}$$

3.4.3 The Dirac Lagrangian

The Lagrangian for a Dirac field ψ with mass m and electric charge q in interaction with the electromagnetic field A^μ is

$$\mathcal{L}_{\mathrm{D}} = \overline{\psi}\left[\gamma_\mu\left(i\partial^\mu - qA^\mu\right) - m\right]\psi. \tag{3.60}$$

This Lagrangian is Hermitian. Indeed, the total derivative $\partial^\mu\left(\overline{\psi}\gamma_\mu\psi\right)$ vanishes because of current conservation. Therefore,

$$\mathcal{L}_{\mathrm{D}} = \psi^\dagger A\left[\frac{i}{2}\gamma_\mu\left(\partial^\mu - \overleftarrow{\partial^\mu}\right) - q\gamma_\mu A^\mu - m\right]\psi, \tag{3.61}$$

which is obviously Hermitian.

3.4.4 Parity

From the Dirac algebra we know that

$$\gamma^0\gamma_\mu\gamma^0 = \gamma^\mu. \tag{3.62}$$

This suggests[9]

$$\mathcal{P}\psi\left(t,\vec{r}\right)\mathcal{P}^\dagger = \exp\left(i\beta_p\right)\gamma^0\psi\left(t,-\vec{r}\right). \tag{3.63}$$

Then,

$$\begin{aligned}\mathcal{P}\overline{\psi}\left(t,\vec{r}\right)\mathcal{P}^\dagger &= \mathcal{P}\psi^\dagger\left(t,\vec{r}\right)\mathcal{P}^\dagger A \\ &= \exp\left(-i\beta_p\right)\overline{\psi}\left(t,-\vec{r}\right)\gamma^0.\end{aligned} \tag{3.64}$$

Also,

$$\mathcal{P}^2\psi\left(t,\vec{r}\right)\mathcal{P}^{\dagger^2} = \exp\left(2i\beta_p\right)\psi\left(t,\vec{r}\right). \tag{3.65}$$

The Dirac action is P invariant. Indeed,

$$\begin{aligned}\mathcal{P}\mathcal{L}_{\mathrm{D}}\left(t,\vec{r}\right)\mathcal{P}^\dagger &= \overline{\psi}\left(t,-\vec{r}\right)\gamma^0\left\{\gamma_\mu\left[i\partial^\mu - qA_\mu\left(t,-\vec{r}\right)\right] - m\right\}\gamma^0\psi\left(t,-\vec{r}\right) \\ &= \overline{\psi}\left(t,-\vec{r}\right)\left\{\gamma^\mu\left[i\partial^\mu - qA_\mu\left(t,-\vec{r}\right)\right] - m\right\}\psi\left(t,-\vec{r}\right) \\ &= \mathcal{L}_{\mathrm{D}}\left(t,-\vec{r}\right).\end{aligned}$$

In the last line of this derivation we have taken into account that $\partial_\mu \to \partial^\mu$ when $\vec{r} \to -\vec{r}$.

3.4.5 Time reversal

Under time reversal, we have

$$\mathcal{T}\psi\left(t,\vec{r}\right)\mathcal{T}^{-1} = U_T\psi^*\left(t,\vec{r}\right)U_T^\dagger = \exp\left(i\beta_t\right)\gamma_0^*\gamma_5^*C^*A\psi\left(-t,\vec{r}\right). \tag{3.66}$$

Then,

$$\mathcal{T}\overline{\psi}\left(t,\vec{r}\right)\mathcal{T}^{-1} = U_T\overline{\psi}^*\left(t,\vec{r}\right)U_T^\dagger$$

[9]Equation (3.62) is analogous to eqn (3.46), and eqn (3.63) is analogous to eqn (3.45), when $\Lambda^\mu{}_\nu = \mathrm{diag}\left(1,-1,-1,-1\right)$ and $S\left(\Lambda\right) = \exp\left(i\beta_p\right)\gamma^0$.

$$= \left[U_T \psi^* \left(t, \vec{r} \right) U_T^\dagger \right]^\dagger A^*$$
$$= -\exp\left(-i\beta_t\right) \psi^\dagger \left(-t, \vec{r}\right) AC\gamma_5^T \gamma_0^T A^*$$
$$= \exp\left(-i\beta_t\right) \psi^\dagger \left(-t, \vec{r}\right) \left(C^{-1}\right)^* \gamma_5^* \gamma_0^*. \tag{3.67}$$

This transformation leaves the Dirac action invariant. Indeed,

$$\mathcal{T} \mathcal{L}_\mathrm{D} \left(t, \vec{r}\right) \mathcal{T}^{-1}$$
$$= U_T \mathcal{L}_\mathrm{D}^* \left(t, \vec{r}\right) U_T^\dagger$$
$$= U_T \overline{\psi}^* \left(t, \vec{r}\right) U_T^\dagger \left\{ \gamma_\mu^* \left[-i\partial^\mu - q U_T A^{\mu*} \left(t, \vec{r}\right) U_T^\dagger \right] - m \right\} U_T \psi^* \left(t, \vec{r}\right) U_T^\dagger$$
$$= \psi^\dagger \left(-t, \vec{r}\right) \left(C^{-1}\right)^* \gamma_5^* \gamma_0^* \left\{ \gamma_\mu^* \left[-i\partial^\mu - q A_\mu \left(-t, \vec{r}\right) \right] - m \right\} \gamma_0^* \gamma_5^* C^* A\psi \left(-t, \vec{r}\right)$$
$$= \overline{\psi} \left(-t, \vec{r}\right) \left\{ \gamma^\mu \left[-i\partial^\mu - q A_\mu \left(-t, \vec{r}\right) \right] - m \right\} \psi \left(-t, \vec{r}\right)$$
$$= \mathcal{L}_\mathrm{D} \left(-t, \vec{r}\right),$$

where we have taken into account that $t \to -t$ implies $\partial^\mu \to -\partial_\mu$. Also notice that

$$\mathcal{T}^2 \psi \left(t, \vec{r}\right) \mathcal{T}^{-2} = \exp\left(-i\beta_t\right) \gamma_0 \gamma_5 C A^* U_T \psi^* \left(-t, \vec{r}\right) U_T^\dagger$$
$$= \gamma_0 \gamma_5 C A^* \gamma_0^* \gamma_5^* C^* A\psi \left(t, \vec{r}\right)$$
$$= -\psi \left(t, \vec{r}\right). \tag{3.68}$$

Thus, $\mathcal{T}^2$ acting on the Dirac field is equal to -1, while it is equal to $+1$ when acting either on the Klein–Gordon or on the photon field. In general, $\mathcal{T}^2$ is -1 for fermionic fields, $+1$ for bosonic fields, as anticipated in § 2.4.6.

3.4.6 Charge conjugation

Let us now consider the action of the charge-conjugation operator. It is given by

$$\mathcal{C}\psi \mathcal{C}^\dagger = \exp\left(i\beta_c\right) \psi^c. \tag{3.69}$$

Then,

$$\mathcal{C}\overline{\psi} \mathcal{C}^\dagger = \exp\left(-i\beta_c\right) \overline{\psi^c}. \tag{3.70}$$

Moreover, on account of eqn (3.51),

$$\mathcal{C}^2 \psi \mathcal{C}^{\dagger^2} = \psi. \tag{3.71}$$

The C transformation leaves the Dirac Lagrangian invariant. Indeed,

$$\mathcal{C} \mathcal{L}_\mathrm{D} \mathcal{C}^\dagger = -\psi^T C^{-1} \left[\gamma_\mu \left(i\partial^\mu + q A^\mu \right) - m \right] C \overline{\psi}^T$$
$$= \psi^T \left[\gamma_\mu^T \left(i\partial^\mu + q A^\mu \right) + m \right] \overline{\psi}^T$$
$$= -\overline{\psi} \left[\gamma_\mu \left(i\overleftarrow{\partial^\mu} + q A^\mu \right) + m \right] \psi$$
$$= \mathcal{L}_\mathrm{D}.$$

We have passed from the second to the third line by transposition. When doing this we have assumed ψ and $\psi^\dagger$ to anti-commute, because they are fermionic fields.

3.4.7 *CP*

Putting together the parity and the charge-conjugation transformations, we obtain the CP transformation, effected by the operator $\mathcal{CP}$. For two arbitrary spinors ψ and χ, we have

$$
\begin{aligned}
(\mathcal{CP})\,\psi\,(\mathcal{CP})^{\dagger} &= \exp\left(i\xi_{\psi}\right)\gamma^{0}C\overline{\psi}^{T}, \\
(\mathcal{CP})\,\chi\,(\mathcal{CP})^{\dagger} &= \exp\left(i\xi_{\chi}\right)\gamma^{0}C\overline{\chi}^{T}.
\end{aligned}
\tag{3.72}
$$

Then,

$$
\begin{aligned}
(\mathcal{CP})\,\overline{\psi}\,(\mathcal{CP})^{\dagger} &= -\exp\left(-i\xi_{\psi}\right)\psi^{T}C^{-1}\gamma^{0}, \\
(\mathcal{CP})\,\overline{\chi}\,(\mathcal{CP})^{\dagger} &= -\exp\left(-i\xi_{\chi}\right)\chi^{T}C^{-1}\gamma^{0}.
\end{aligned}
\tag{3.73}
$$

For simplicity of notation we have omitted an explicit reference to the space–time coordinates $(t,\vec{r})$ and $(t,-\vec{r})$. It will later be useful to have the CP transformation properties of various field bilinears. We easily obtain:

$$
(\mathcal{CP})\left(\overline{\psi}\chi\right)(\mathcal{CP})^{\dagger} = \exp\left[i\left(\xi_{\chi}-\xi_{\psi}\right)\right]\left(\overline{\chi}\psi\right),
\tag{3.74}
$$

$$
(\mathcal{CP})\left(\overline{\psi}\gamma_{5}\chi\right)(\mathcal{CP})^{\dagger} = -\exp\left[i\left(\xi_{\chi}-\xi_{\psi}\right)\right]\left(\overline{\chi}\gamma_{5}\psi\right),
\tag{3.75}
$$

$$
(\mathcal{CP})\left(\overline{\psi}\gamma^{\mu}\chi\right)(\mathcal{CP})^{\dagger} = -\exp\left[i\left(\xi_{\chi}-\xi_{\psi}\right)\right]\left(\overline{\chi}\gamma_{\mu}\psi\right),
\tag{3.76}
$$

$$
(\mathcal{CP})\left(\overline{\psi}\gamma^{\mu}\gamma_{5}\chi\right)(\mathcal{CP})^{\dagger} = -\exp\left[i\left(\xi_{\chi}-\xi_{\psi}\right)\right]\left(\overline{\chi}\gamma_{\mu}\gamma_{5}\psi\right).
\tag{3.77}
$$

Taking appropriate linear combinations of these equations we find the equivalent equations

$$
(\mathcal{CP})\left(\overline{\psi}\gamma_{L}\chi\right)(\mathcal{CP})^{\dagger} = \exp\left[i\left(\xi_{\chi}-\xi_{\psi}\right)\right]\left(\overline{\chi}\gamma_{R}\psi\right),
\tag{3.78}
$$

$$
(\mathcal{CP})\left(\overline{\psi}\gamma_{R}\chi\right)(\mathcal{CP})^{\dagger} = \exp\left[i\left(\xi_{\chi}-\xi_{\psi}\right)\right]\left(\overline{\chi}\gamma_{L}\psi\right),
\tag{3.79}
$$

$$
(\mathcal{CP})\left(\overline{\psi}\gamma^{\mu}\gamma_{L}\chi\right)(\mathcal{CP})^{\dagger} = -\exp\left[i\left(\xi_{\chi}-\xi_{\psi}\right)\right]\left(\overline{\chi}\gamma_{\mu}\gamma_{L}\psi\right),
\tag{3.80}
$$

$$
(\mathcal{CP})\left(\overline{\psi}\gamma^{\mu}\gamma_{R}\chi\right)(\mathcal{CP})^{\dagger} = -\exp\left[i\left(\xi_{\chi}-\xi_{\psi}\right)\right]\left(\overline{\chi}\gamma_{\mu}\gamma_{R}\psi\right).
\tag{3.81}
$$

3.4.8 *Field bilinears and the discrete transformations*

We may now elaborate Table 3.1, which gives the result of the application of the various discrete transformations to field bilinears. In writing down that table, we have ommitted reference to the transformation of the coordinate variables. We have also ommitted all the free phases present in each discrete transformation. As a consequence, in general only the *relative* transformation rules of two different field bilinears under each discrete symmetry is relevant.

3.5   Relative parities of a particle and its antiparticle

Consider again the Klein–Gordon field ϕ and its charge-conjugated field. It is clear that if parity transforms

$$
\mathcal{P}\phi\left(t,\vec{r}\right)\mathcal{P}^{\dagger} = \exp\left(i\alpha_{p}\right)\phi\left(t,-\vec{r}\right),
\tag{3.82}
$$

Table 3.1 *Action of some discrete transformations on Dirac-field bilinears.*

	P	T	C	CP	CPT
$\bar{\psi}\chi$	$\bar{\psi}\chi$	$\bar{\psi}\chi$	$\bar{\chi}\psi$	$\bar{\chi}\psi$	$\bar{\chi}\psi$
$\bar{\psi}\gamma_5\chi$	$-\bar{\psi}\gamma_5\chi$	$\bar{\psi}\gamma_5\chi$	$\bar{\chi}\gamma_5\psi$	$-\bar{\chi}\gamma_5\psi$	$-\bar{\chi}\gamma_5\psi$
$\bar{\psi}\gamma_L\chi$	$\bar{\psi}\gamma_R\chi$	$\bar{\psi}\gamma_L\chi$	$\bar{\chi}\gamma_L\psi$	$\bar{\chi}\gamma_R\psi$	$\bar{\chi}\gamma_R\psi$
$\bar{\psi}\gamma_R\chi$	$\bar{\psi}\gamma_L\chi$	$\bar{\psi}\gamma_R\chi$	$\bar{\chi}\gamma_R\psi$	$\bar{\chi}\gamma_L\psi$	$\bar{\chi}\gamma_L\psi$
$\bar{\psi}\gamma^\mu\chi$	$\bar{\psi}\gamma_\mu\chi$	$\bar{\psi}\gamma_\mu\chi$	$-\bar{\chi}\gamma^\mu\psi$	$-\bar{\chi}\gamma_\mu\psi$	$-\bar{\chi}\gamma^\mu\psi$
$\bar{\psi}\gamma^\mu\gamma_5\chi$	$-\bar{\psi}\gamma_\mu\gamma_5\chi$	$\bar{\psi}\gamma_\mu\gamma_5\chi$	$\bar{\chi}\gamma^\mu\gamma_5\psi$	$-\bar{\chi}\gamma_\mu\gamma_5\psi$	$-\bar{\chi}\gamma^\mu\gamma_5\psi$
$\bar{\psi}\gamma^\mu\gamma_L\chi$	$\bar{\psi}\gamma_\mu\gamma_R\chi$	$\bar{\psi}\gamma_\mu\gamma_L\chi$	$-\bar{\chi}\gamma^\mu\gamma_R\psi$	$-\bar{\chi}\gamma_\mu\gamma_L\psi$	$-\bar{\chi}\gamma^\mu\gamma_L\psi$
$\bar{\psi}\gamma^\mu\gamma_R\chi$	$\bar{\psi}\gamma_\mu\gamma_L\chi$	$\bar{\psi}\gamma_\mu\gamma_R\chi$	$-\bar{\chi}\gamma^\mu\gamma_L\psi$	$-\bar{\chi}\gamma_\mu\gamma_R\psi$	$-\bar{\chi}\gamma^\mu\gamma_R\psi$
$\bar{\psi}\sigma^{\mu\nu}\chi$	$\bar{\psi}\sigma_{\mu\nu}\chi$	$-\bar{\psi}\sigma_{\mu\nu}\chi$	$-\bar{\chi}\sigma^{\mu\nu}\psi$	$-\bar{\chi}\sigma_{\mu\nu}\psi$	$\bar{\chi}\sigma^{\mu\nu}\psi$

then

$$\mathcal{P}\left[\mathcal{C}\phi\left(t,\vec{r}\right)\mathcal{C}^\dagger\right]\mathcal{P}^\dagger = \exp\left(-i\alpha_p\right)\left[\mathcal{C}\phi\left(t,-\vec{r}\right)\mathcal{C}^\dagger\right].\qquad(3.83)$$

Now consider instead the Dirac field ψ and its charge-conjugated field. Clearly, if

$$\mathcal{P}\psi\left(t,\vec{r}\right)\mathcal{P}^\dagger = \exp\left(i\beta_p\right)\gamma^0\psi\left(t,-\vec{r}\right),\qquad(3.84)$$

then

$$\mathcal{P}\left[\mathcal{C}\psi\left(t,\vec{r}\right)\mathcal{C}^\dagger\right]\mathcal{P}^\dagger = -\exp\left(-i\beta_p\right)\gamma^0\left[\mathcal{C}\psi\left(t,-\vec{r}\right)\mathcal{C}^\dagger\right],\qquad(3.85)$$

where we have used $CA^T\gamma^{0*} = C\gamma^{0T}A^T = -\gamma^0 CA^T$. Now if we compare eqns (3.82) and (3.83) with their analogous eqns (3.84) and (3.85), we find that for the Dirac field we have a relative minus sign in the transformation laws for ψ and $\mathcal{C}\psi\mathcal{C}^\dagger$ while, for the Klein–Gordon field, the sign in the transformation laws of ϕ and $\mathcal{C}\phi\mathcal{C}^\dagger$ is the same.

In reality, what matters is the parity of a particle–antiparticle pair. The parity of a charged particle is arbitrary—the phases α_p and β_p are arbitrary—but the parity of a particle–antiparticle pair may be experimentally probed—see the next chapter for examples. In the case of Klein–Gordon particles, the intrinsic parities of a particle and its antiparticle cancel each other ($e^{i\alpha_p}e^{-i\alpha_p} = 1$), but in the case of Dirac fields they give rise by themselves alone to a minus sign. Observable consequences of this minus sign will be studied in the next chapter.

3.6   Electric and magnetic dipole moments

In a non-relativistic approximation—see for instance Bjorken and Drell (1964)—the four-component Dirac spinor ψ may be separated into two two-component spinors φ and χ, the components of φ being much larger than the ones of χ. The two-component spinor φ obeys the equation

$$\left(i\frac{\partial}{\partial t} - qA^0\right)\varphi = \frac{1}{2m}\left(-i\vec{\nabla} - q\vec{A}\right)^2\varphi - \frac{q}{m}\vec{s}\cdot\vec{B}\varphi.\qquad(3.86)$$

Here, $\vec{s} = \vec{\sigma}/2$ are the Pauli matrices divided by two, representing the spin. One sees in eqn (3.86) that φ, besides the minimal coupling to the electromagnetic field given by the transformations $E \to E - qA^0$ and $\vec{p} \to \vec{p} - q\vec{A}$, also has a magnetic dipole moment q/m.

Now suppose that we modify the Dirac Lagrangian and use, instead of eqn (3.60), the non-renormalizable Lagrangian

$$\mathcal{L} = \overline{\psi} \left[\gamma_\mu \left(i\partial^\mu - qA^\mu \right) - m + \frac{q}{2m} \sigma^{\mu\nu} \left(F + iG\gamma_5 \right) \left(\partial_\nu A_\mu \right) \right] \psi, \qquad (3.87)$$

with real numbers F and G. It may easily be checked that the Lagrangian in eqn (3.87) is Hermitian. It is also C invariant. Indeed,

$$\mathcal{C} \left[\overline{\psi} \sigma^{\mu\nu} \left(F + iG\gamma_5 \right) \psi \right] \mathcal{C}^\dagger = -\psi^T C^{-1} \sigma^{\mu\nu} \left(F + iG\gamma_5 \right) C \overline{\psi}^T$$
$$= -\overline{\psi} \sigma^{\mu\nu} \left(F + iG\gamma_5 \right) \psi.$$

As A^μ changes sign under C, the new terms in the Lagrangian of eqn (3.87) are C invariant. As for P invariance,

$$\mathcal{P} \left[\overline{\psi} \sigma^{\mu\nu} \left(F + iG\gamma_5 \right) \psi \right] \mathcal{P}^\dagger = \overline{\psi} \gamma^0 \sigma^{\mu\nu} \left(F + iG\gamma_5 \right) \gamma^0 \psi$$
$$= \overline{\psi} \sigma_{\mu\nu} \left(F - iG\gamma_5 \right) \psi;$$

for T invariance,

$$\mathcal{T} \left[\overline{\psi} \sigma^{\mu\nu} \left(F + iG\gamma_5 \right) \psi \right] \mathcal{T}^{-1} = \psi^\dagger \left(C^{-1} \right)^* \gamma_5^* \gamma_0^* \left(\sigma^{\mu\nu} \right)^* \left(F - iG\gamma_5^* \right) \gamma_0^* \gamma_5^* C^* A\psi$$
$$= -\overline{\psi} \sigma_{\mu\nu} \left(F - iG\gamma_5 \right) \psi.$$

Therefore, the F-term is both P and T invariant, while the G-term violates both P and T. Thus, G has the same transformation properties as an electric dipole moment. Indeed, making the same non-relativistic approximation as before, one finds that φ now satisfies the equation

$$\left(i\frac{\partial}{\partial t} - qA^0 \right) \varphi = \frac{1}{2m} \left(-i\vec{\nabla} - q\vec{A} \right)^2 \varphi - \frac{q}{m} \left[(1 + F)\, \vec{s} \cdot \vec{B} + G\vec{s} \cdot \vec{E} \right] \varphi. \quad (3.88)$$

The spinor φ now has an electric dipole moment $(q/m)\, G$ and a magnetic dipole moment $(q/m)\, (1 + F)$. Both dipole moments are proportional to q, and therefore they have opposite sign for a particle and its antiparticle. F is usually called the anomalous magnetic dipole moment of ψ.

As magnetic and electric dipole moments have opposite signs for a particle and its antiparticle, one concludes that a Majorana field cannot have any of those moments. Indeed,

$$\overline{\psi^c} \sigma^{\mu\nu} \left(F + iG\gamma_5 \right) \psi^c = -\overline{\psi} \sigma^{\mu\nu} \left(F + iG\gamma_5 \right) \psi. \qquad (3.89)$$

The Lagrangian in eqn (3.87) may arise at higher order in perturbation theory, with coefficients F and G depending on the momentum transfer to the photon. The G-term only appears if the underlying theory is both P- and T-violating.

3.7 P, T, and C invariance of the strong interactions

We have seen that the quantum electrodynamics of scalars and of spin-1/2 particles is P, T, and C invariant. In the case of P and T invariances, this is not surprising, because these are symmetries of classical electromagnetism—see Chapter 1—which we would want to recover as invariances of quantum electromagnetism too.

It is important to emphasize the crucial role that the electromagnetic interaction plays in defining, in particular, what form the C transformation should take. A particle and its antiparticle must have electromagnetic interactions characterized by the opposite sign of the electric charge. If a field has zero electric charge, then its C- and CP-transformed fields are largely arbitrary. We shall encounter this problem in § 23.6.

Once it is found that the electromagnetic interaction is C, P, and T invariant, it is natural to assume that other interactions enjoy these invariances too. We may assume, in particular, that the strong interaction has these invariances. Indeed, up to now there is no experimental evidence for C, P, or T violation by the strong interaction. In the next chapter we shall focus on some consequences of the C, P, and T invariance of the strong and electromagnetic interactions.

The quantum-chromodynamics (QCD) (strong-interaction) Lagrangian is

$$\mathcal{L}_{\text{QCD}} = -\tfrac{1}{4}F_{\mu\nu}^a F^{a\mu\nu} + \overline{q_x}\left[\delta_{xy}\left(i\gamma^\mu\partial_\mu - m_q\right) + g_s\gamma^\mu G_\mu^a \frac{\lambda_{xy}^a}{2}\right]q_y. \tag{3.90}$$

Here, g_s is the strong coupling constant; x and y are colour indices; q_x is a quark field; m_q is its mass; G_μ^a (with a from 1 to 8) are the gluon fields; the λ^a are the Gell-Mann matrices:

$$\lambda^1 = \begin{pmatrix} 0 & 1 & 0 \\ 1 & 0 & 0 \\ 0 & 0 & 0 \end{pmatrix}, \quad \lambda^2 = \begin{pmatrix} 0 & -i & 0 \\ i & 0 & 0 \\ 0 & 0 & 0 \end{pmatrix}, \quad \lambda^3 = \begin{pmatrix} 1 & 0 & 0 \\ 0 & -1 & 0 \\ 0 & 0 & 0 \end{pmatrix}, \quad \lambda^4 = \begin{pmatrix} 0 & 0 & 1 \\ 0 & 0 & 0 \\ 1 & 0 & 0 \end{pmatrix},$$

$$\lambda^5 = \begin{pmatrix} 0 & 0 & -i \\ 0 & 0 & 0 \\ i & 0 & 0 \end{pmatrix}, \quad \lambda^6 = \begin{pmatrix} 0 & 0 & 0 \\ 0 & 0 & 1 \\ 0 & 1 & 0 \end{pmatrix}, \quad \lambda^7 = \begin{pmatrix} 0 & 0 & 0 \\ 0 & 0 & -i \\ 0 & i & 0 \end{pmatrix}, \quad \lambda^8 = \tfrac{1}{\sqrt{3}}\begin{pmatrix} 1 & 0 & 0 \\ 0 & 1 & 0 \\ 0 & 0 & -2 \end{pmatrix}. \tag{3.91}$$

The field-strength tensor $F_{\mu\nu}^a = -F_{\nu\mu}^a$ is given by

$$F_{\mu\nu}^a \equiv \partial_\mu G_\nu^a - \partial_\nu G_\mu^a + g_s f^{abc} G_\mu^b G_\nu^c, \tag{3.92}$$

where the f^{abc} are the structure constants of SU(3), defined by

$$\left[\frac{\lambda^a}{2}, \frac{\lambda^b}{2}\right] = if^{abc}\frac{\lambda^c}{2}. \tag{3.93}$$

They are given by

$$\begin{aligned} f^{123} &= 1, \quad f^{147} = \tfrac{1}{2}, \quad f^{156} = -\tfrac{1}{2}, \quad f^{246} = \tfrac{1}{2}, \quad f^{257} = \tfrac{1}{2}, \\ f^{345} &= \tfrac{1}{2}, \quad f^{367} = -\tfrac{1}{2}, \quad f^{458} = \tfrac{3}{2}, \quad f^{678} = \tfrac{3}{2}, \end{aligned} \tag{3.94}$$

and by the fact that they are antisymmetric in all three indices.

Notice that the Gell-Mann matrices are Hermitian and they are either symmetric or antisymmetric: $\lambda^{aT} = \lambda^{a*} = s^a \lambda^a$, with $s^a = +1$ for $a = 1, 3, 4, 6$, and 8, and $s^a = -1$ for $a = 2, 5$, and 7. Also notice that $f^{abc} \neq 0 \Rightarrow s^b s^c = -s^a$.

The QCD Lagrangian already incorporates the assumption of P, C, and T invariance of the strong interactions. The quark fields transform under the discrete symmetries as studied in § 3.4.4–3.4.6; the phases β_p, β_t, and β_c are independent of the colour index. The gluon fields transform as

$$\mathcal{P} G_\mu^a (t, \vec{r}) \, \mathcal{P}^\dagger = G^{a\mu} (t, -\vec{r}),$$
$$\mathcal{T} G_\mu^a (t, \vec{r}) \, \mathcal{T}^{-1} = s^a G^{a\mu} (-t, \vec{r}), \qquad (3.95)$$
$$\mathcal{C} G_\mu^a (t, \vec{r}) \, \mathcal{C}^\dagger = -s^a G_\mu^a (t, \vec{r}).$$

Therefore, the field-strength tensor transforms as

$$\mathcal{P} F_{\mu\nu}^a (t, \vec{r}) \, \mathcal{P}^\dagger = F^{a\mu\nu} (t, -\vec{r}),$$
$$\mathcal{T} F_{\mu\nu}^a (t, \vec{r}) \, \mathcal{T}^{-1} = -s^a F^{a\mu\nu} (-t, \vec{r}), \qquad (3.96)$$
$$\mathcal{C} F_{\mu\nu}^a (t, \vec{r}) \, \mathcal{C}^\dagger = -s^a F_{\mu\nu}^a (t, \vec{r}).$$

Given these transformation laws, the P, T, and C invariance of $\mathcal{L}_{\text{QCD}}$ is obvious.

However, in order to obtain invariance of the strong interaction under the discrete transformations, it is not sufficient to prove the invariance of its Lagrangian. QCD has a non-trivial vacuum structure, and that vacuum must be invariant if the strong interaction is to be invariant. In general, the vacuum of QCD gives rise to both P and T violation, as we shall see in Chapter 27. For most of this book, however, we shall neglect this awkward 'strong CP problem', and assume that the vacuum of QCD is invariant under the discrete transformations. Only then can our theoretical understanding of the strong interaction incorporate the invariance under the discrete symmetries that its experimental exploration exhibits.

3.8 Conclusions

Besides deriving many formulae—in particular the ones relating to the Dirac theory—which will be used later in the book, we have seen in this chapter how QED fixes the CP-transformation properties of the electromagnetic, Klein–Gordon, and Dirac fields:

$$(\mathcal{CP}) A^\mu (t, \vec{r}) \, (\mathcal{CP})^\dagger = -A_\mu (t, -\vec{r}); \qquad (3.97)$$
$$(\mathcal{CP}) \phi (t, \vec{r}) \, (\mathcal{CP})^\dagger = \exp(i\alpha) \, \phi^\dagger (t, -\vec{r}); \qquad (3.98)$$
$$(\mathcal{CP}) \psi (t, \vec{r}) \, (\mathcal{CP})^\dagger = \exp(i\beta) \, \gamma_0 C A^T {\psi^\dagger}^T (t, -\vec{r}); \qquad (3.99)$$
$$(\mathcal{CP}) \overline{\psi} (t, \vec{r}) \, (\mathcal{CP})^\dagger = -\exp(-i\beta) \, \psi^T (t, -\vec{r}) \, C^{-1} \gamma_0. \qquad (3.100)$$

QED provides the model that must be followed by the CP transformation of any other interactions, $\mathcal{L}_{\text{part}}$, in the complete Lagrangian:

$$(\mathcal{CP})\, \mathcal{L}_{\text{part}}\,(t,\vec{r})\,(\mathcal{CP})^{\dagger} = \mathcal{L}_{\text{part}}\,(t,-\vec{r})\,. \qquad (3.101)$$

In particular, we have also shown, in § 3.7, that CP invariance is also a property of the strong interaction.

4

APPLICATIONS OF THE DISCRETE SYMMETRIES

4.1 Introduction

We present in this chapter some elementary applications of the discrete symmetries P, C, and T. These examples are extracted from the realms of nuclear physics and of low-energy particle physics. For more details, the reader is advised to consult the books that we have used in preparing this brief account (Sakurai 1964; Perkins 1987; Lee 1990; Burcham and Jobes 1995).

In a first reading, this chapter may be skipped without inconvenience.

4.2 C invariance: Furry's theorem

We have seen in the previous chapter that C transforms A^μ into $-A^\mu$. The creation operator $a^\dagger(\vec{p}, h)$ for a photon of momentum $\vec{p}$ and helicity h transforms according to

$$\mathcal{C} a^\dagger(\vec{p}, h) \mathcal{C}^\dagger = -a^\dagger(\vec{p}, h). \tag{4.1}$$

Therefore, a one-photon state $|\gamma\rangle = a^\dagger(\vec{p}, h)|0\rangle$ has C-parity -1: $\mathcal{C}|\gamma\rangle = -|\gamma\rangle$. The C-parity of an n-photon state $|n\gamma\rangle$ is the product of the C-parities of the n photons:

$$\mathcal{C}|n\gamma\rangle = (-1)^n |n\gamma\rangle. \tag{4.2}$$

If there is C invariance,

$$\langle n'\gamma|S|n\gamma\rangle = \langle n'\gamma|\mathcal{C}^\dagger \mathcal{C} S \mathcal{C}^\dagger \mathcal{C}|n\gamma\rangle$$
$$= (-1)^{(n+n')} \langle n'\gamma|S|n\gamma\rangle. \tag{4.3}$$

Therefore, if $n + n'$ is odd then the matrix element must vanish. This is Furry's theorem (Furry 1937): a state with an odd number of photons (and no other particles) cannot scatter into a state with an even number of photons, or vice versa. An elementary example is the three-photon vertex at one loop in the quantum electrodynamics of spin-1/2 fermions. The sum of the two relevant graphs, depicted in Fig. 4.1, vanishes (Feynman 1949).

The neutral pion π^0 decays mostly (branching ratio 98.8%) to two photons. This is an electromagnetic decay, and the electromagnetic interaction is C invariant. From this fact we infer that

$$\mathcal{C}|\pi^0\rangle = +|\pi^0\rangle. \tag{4.4}$$

Similarly, the meson η, which decays with probability 39.3% to two photons, is C-even. C conservation implies that none of these mesons may decay to an odd number of photons. Indeed, it is found experimentally that

$$\text{BR}\left(\pi^0 \to 3\gamma\right) < 3.1 \times 10^{-8},$$
$$\text{BR}\left(\eta \to 3\gamma\right) < 5 \times 10^{-4}. \tag{4.5}$$

4.3 T invariance: the spin of the pion

4.3.1 $2 \to 2$ scattering

Let us apply the reciprocity relation to the scattering of two particles 1 and 2, constituting the initial state i, to two particles 3 and 4, constituting the final state f. In the centre-of-momentum frame, the particles 1 and 2 have momenta $\vec{p}_i$ and $-\vec{p}_i$, respectively, and the particles 3 and 4 have momenta $\vec{p}_f$ and $-\vec{p}_f$, respectively. The energy in that frame is $s = (p_1 + p_2)^2 = (p_3 + p_4)^2$, where p_k is the four-momentum of the particle k, for $k = 1, 2, 3,$ or 4. Also, we denote by j_k the spin of the particle k.

Let S_{fi} be the relevant S-matrix element. The differential cross-section relative to an element of solid angle $d\Omega_f$ for the final-state momentum $\vec{p}_f$ is

$$\frac{d\sigma}{d\Omega_f} = \frac{|S_{fi}|^2}{48\pi^2 s} \frac{p_f}{p_i}, \tag{4.6}$$

where $p_f = |\vec{p}_f|$ and $p_i = |\vec{p}_i|$. In writing eqn (4.6) it has been assumed that the particle wave functions are normalized in such a way that the density of each particle species is $2E$ particles per unit volume, where E is the energy of the particle. This normalization affects the normalization of the matrix S; our $|S_{fi}|^2$ is equal to $16E_1 E_2 E_3 E_4$ times what it would have been had we chosen the normalization in which the density of particles is one particle per unit volume.

If we measure the cross-section without polarizing the initial-state particles 1 and 2, and without observing the polarizations of the final-state particles 3 and 4, the relevant differential cross-section is the sum over the final spins, averaged over the initial spins, of the expression in eqn (4.6), i.e.,

$$\frac{d\sigma}{d\Omega_f} = \frac{\sum_{s_k, k=1}^{4} |S_{fi}|^2}{48\pi^2 s (2j_1 + 1)(2j_2 + 1)} \frac{p_f}{p_i}. \tag{4.7}$$

For the inverse reaction, in which particles 3 and 4 are scattered into particles 1 and 2, with the same energy s in the centre-of-momentum frame, we have, because of the reciprocity relation,

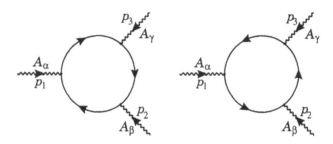

FIG. 4.1. One-loop Feynman diagrams for the three-photon vertex in QED.

$$\frac{d\sigma}{d\Omega_i} = \frac{\sum_{s_k,k=1}^{4} |S_{fi}|^2}{48\pi^2 s\,(2j_3+1)\,(2j_4+1)\,p_f}\,p_i,$$ (4.8)

with the same $|S_{fi}|$. Taking the ratio of eqns (4.7) and (4.8), we obtain

$$\frac{d\sigma\,(1+2\to3+4)}{d\sigma\,(3+4\to1+2)} = \frac{p_f^2\,(2j_3+1)\,(2j_4+1)\,d\Omega_f}{p_i^2\,(2j_1+1)\,(2j_2+1)\,d\Omega_i}.$$ (4.9)

Integrating the differential cross-sections over the solid angles for their respective final-state momenta, we obtain

$$\frac{\sigma\,(1+2\to3+4)}{\sigma\,(3+4\to1+2)} = \frac{p_f^2\,(2j_3+1)\,(2j_4+1)}{p_i^2\,(2j_1+1)\,(2j_2+1)}\,c,$$ (4.10)

where c is a factor which accounts for the possibility that the particles 1 and 2, or the particles 3 and 4, are identical, in which case the integrations over the final-state solid angle span only 2π steradians. Thus, $c = 1/2$ if particles 3 and 4 are identical but particles 1 and 2 are distinct; $c = 2$ if particles 1 and 2 are identical but particles 3 and 4 are distinct; and $c = 1$ otherwise.

4.3.2 The spin of the pion

Equation (4.10) was used (Marshak 1951; Cheston 1951) to determine the spin of the charged pion π^+ by using the reactions, assumed to be T invariant, $d+\pi^+ \leftrightarrow p+p$, where d is the deuteron and p is the proton. Knowing that the proton spin is $1/2$ and the deuteron spin is 1, we obtain

$$\frac{\sigma\,(d+\pi^+\to p+p)}{\sigma\,(p+p\to d+\pi^+)} = \frac{2p_p^2}{3p_d^2\,(2j_\pi+1)}.$$ (4.11)

The comparison of the cross sections, measured at similar energies in the centre-of-momentum frame (Durbin et al. 1951; Clark et al. 1951, 1952; Cartwright et al. 1953), gave $j_\pi = 0$.

   In high-energy nucleon–nucleon collisions, the yields of neutral and charged pions are equal, indicating that the spin multiplicities of those particles are the same; hence, the spin of the neutral pion must be zero too.

4.4 P invariance: two-photon decay of a spin-0 particle

Consider a spin-0 particle decaying to two photons. In the rest frame of the decaying particle one of the photons has momentum $\vec{p}$ and the other one has momentum $-\vec{p}$. Besides the direction of $\vec{p}$, there are two other relevant directions in the problem: the directions of the electric fields of the two photons, $\vec{E}_1$ and $\vec{E}_2$. Indeed, the magnetic fields, $\vec{B}_1$ and $\vec{B}_2$, are perpendicular to the electric fields and to $\vec{p}$, i.e., $\vec{B}_1\cdot\vec{E}_1 = \vec{B}_1\cdot\vec{p} = \vec{B}_2\cdot\vec{E}_2 = \vec{B}_2\cdot\vec{p} = 0$. Therefore, they do not constitute independent directions. Also, $\vec{E}_1\cdot\vec{p} = \vec{E}_2\cdot\vec{p} = 0$.

The decay amplitude must be of the form

$$A\vec{E}_1 \cdot \vec{E}_2 + B\left(\vec{E}_1 \times \vec{E}_2\right) \cdot \vec{p}, \tag{4.12}$$

where A and B are numbers which depend only on the mass of the decaying particle. Notice that there is symmetry under the interchange of the two photons—under $\vec{E}_1 \leftrightarrow \vec{E}_2$ and $\vec{p} \rightarrow -\vec{p}$—as Bose–Einstein statistics requires.

Under parity a spin-zero field gets multiplied by $\exp{(i\alpha_p)}$ and its charge-conjugated field gets multiplied by $\exp{(-i\alpha_p)}$. If the particle is identical with its antiparticle—if the Klein–Gordon operator field is Hermitian—then $\exp{(i\alpha_p)}$ must be either $+1$ or -1. The particle may thus be a scalar (if it has parity $+1$) or a pseudoscalar (if it has parity -1). As for the three vectors $\vec{p}$, $\vec{E}_1$, and $\vec{E}_2$, they all change sign under parity. Therefore, under parity B changes sign but A does not.

Suppose that parity is conserved in the decay of a spin-0 particle which is identical with its antiparticle. It follows that, if the particle is a scalar then $B = 0$, and the polarizations of the two photons in the final state are predominantly parallel—the probability that the two photons have polarizations at an angle θ between themselves is proportional to $\cos^2\theta$, where θ is the angle between the directions of $\vec{E}_1$ and $\vec{E}_2$. If the decaying particle is a pseudoscalar then $A = 0$, and the probability law for the polarizations of the two photons is $\sin^2\theta$.

This argument (Yang 1950) led to the experimental determination of the parity of the neutral pion. That meson decays to $e^+e^+e^-e^-$ with a branching ratio of $(3.14 \pm 0.30) \times 10^{-5}$. The electrons and positrons result from the internal conversion of the two virtual photons of the main decay mode, $\pi^0 \rightarrow 2\gamma$, into electron–positron pairs. It was shown by Kroll and Wada (1955) that the electron and the positron are preferentially aligned with the electric field of the photon which originated them, and the correlation between the photon polarizations persists as a correlation between the planes defined by the momenta of the two electron–positron pairs. If θ is the angle between those planes, then the probability distribution of θ should be proportional to $1 + K\cos 2\theta$ for a scalar π^0, or to $1 - K\cos 2\theta$ for a pseudoscalar π^0, with $K = 0.47$ (Kroll and Wada 1955). By comparing the orientation of the two planes, Plano et al. (1959) found that they are mostly perpendicular. Therefore, π^0 has parity -1.

4.5   C- and P-parities of positronium

Let us now consider the C and P quantum numbers of positronium, i.e., of a state consisting of a positron and an electron bound by the Coulomb interaction.

As the electromagnetic interaction is P and C invariant, we may adiabatically switch it off without altering the C and P quantum numbers of the state. We are thus able to eliminate the spin–orbit interaction as well as the presence of virtual photons. The C- and P-parities of positronium are therefore equal to the products of the C- and P-parities of the separate spin and orbital wave functions, in their non-relativistic limit, and of the intrinsic C- and P-parities.

The orbital wave function is given by a spherical harmonics $Y_l^m(\vec{p}/p)$ times a function of p. Here, l is the total orbital angular momentum, m is the projection of the angular momentum along the z direction, $\vec{p}$ is the relative momentum of electron and positron, and $p = |\vec{p}|$. A fundamental property of Y_l^m is that it acquires a sign $(-1)^l$ under the parity transformation $\vec{p} \to -\vec{p}$. If the vector $\vec{p}$ has a direction defined by the polar angles θ and ϕ, that transformation is equivalent to $\theta \to \pi - \theta$ and $\phi \to \pi + \phi$. Thus, $Y_l^m(\pi - \theta, \pi + \phi) = (-1)^l Y_l^m(\theta, \phi)$. Parity does not affect the spins of the electron and positron. If the electron field gets multiplied by $\exp(i\beta_p)$ upon a parity transformation, the positron field gets multiplied by $-\exp(-i\beta_p)$, as we have seen in § 3.5. Therefore, the overall parity of positronium is

$$P = (-1)^{l+1}. \tag{4.13}$$

Charge conjugation interchanges the creation operators of the positron and of the electron. When one brings back the operators to their original position a minus sign arises, because fermionic operators anticommute. Besides, when interchanging the positron and the electron one changes the sign of their relative momentum $\vec{p}$, thus generating a sign $(-1)^l$ in the wave function. As for the spin wave function, the total spin may be either 0 or 1. The combination of two one-half spins to form an $S = 1$ state is symmetric under the interchange of the spins; the $S = 0$ state is antisymmetric. A sign $(-1)^{S+1}$ is thus generated by the interchange of the spins of the electron and positron. Overall, we have for positronium

$$C = (-1)^{l+S}. \tag{4.14}$$

Comparing this result with the one in § 4.2, we arrive at the following important conclusion: a positronium state with even $l + S$ cannot decay to an odd number of photons; a state with odd $l + S$ cannot decay to an even number of photons. Thus, for $l = 0$,[10] positronium may decay to two photons if $S = J = 0$, but it must decay to three photons if $S = J = 1$, where J is the total angular momentum. The decay rates may be computed and were first measured by Deutsch (1953), who found excellent agreement with theory.

In the case $l = S = J = 0$, the positronium parity is negative and the two resulting photons must have preferentially orthogonal polarizations, as was shown in the previous section. This was checked by looking at the Compton scattering of the photons. Compton scattering depends strongly on the polarization of the photons, being more likely to occur on a plane normal to the electric vector of the incoming photon. Using this method, Wu and Shaknov (1950) have experimentally demonstrated that the two photons from positronium decay have mostly orthogonal polarizations, thus implicitly demonstrating the correctness of the theoretical prediction that the electron and the positron have opposite intrinsic parities.

---

[10]When $l = 0$ there is a spatial superposition of the wave functions of the electron and positron, allowing annihilation to take place.

It is worth pointing out that the result in eqn (4.14) is valid in general for any system of a particle and its antiparticle, of whatever intrinsic spin s, bound by the Coulomb interaction. Indeed, the interchange of the particle and antiparticle by the C transformation yields a factor $(-1)^l$ from the orbital wave function, a factor $(-1)^{2s}$ from the fermionic or bosonic nature of the fields, and a factor $(-1)^{S+2s}$ from the spin wave function, where S is the total spin of the bound state. The latter factor is a particular case of the general rule for Clebsch–Gordan coefficients

$$\langle j_1 j_2 m_1 m_2 | j_1 j_2 JM \rangle = (-1)^{J-j_1-j_2} \langle j_2 j_1 m_2 m_1 | j_2 j_1 JM \rangle. \qquad (4.15)$$

The two factors $(-1)^{2s}$ cancel out, and the C-parity of the particle-antiparticle bound state always ends up being given by eqn (4.14).

4.6   The intrinsic parities of mesons and baryons

4.6.1   *Flavour and intrinsic parities*

Strong and electromagnetic interactions separately conserve each flavour—u, d, s, and so on. Therefore, for each single flavour we may set by convention the intrinsic parity of one hadron with that flavour to be $+1$ (or we may set it to be any other arbitrary phase)—for more details, see Weinberg (1995, p. 124). The standard convention is to assume the intrinsic parities of the proton, the neutron, and the Λ, to be equal to $+1$. As these particles are in the same SU(3) multiplet, they are characterized by the same state of inner motion—they are three-quark states in which the relative orbital angular momentum of each pair of quarks is zero. This means that, as a matter of fact, we are making the convention that the u, d, and s quarks all have the same intrinsic parity. Thus, if

$$\mathcal{P}u\mathcal{P}^\dagger = \exp(i\beta_u)\,\gamma^0 u,$$
$$\mathcal{P}d\mathcal{P}^\dagger = \exp(i\beta_d)\,\gamma^0 d,$$
$$\mathcal{P}s\mathcal{P}^\dagger = \exp(i\beta_s)\,\gamma^0 s,$$

we are making the convention $\beta_u = \beta_d = \beta_s$.

From the convention that the proton, the neutron, and the Λ all have parity $+1$, we may derive step-by-step the intrinsic parities of all non-charmed, non-beautiful mesons, baryons, and nuclei. For instance, the deuteron is composed of a neutron and a proton, in a state of relative orbital angular momentum $L = 0$, with a small (about 7% in amplitude) admixture of an $L = 2$ component. This orbital angular momentum gives rise to a parity $(-1)^L = 1$. The intrinsic parities of neutron and proton are both 1 by convention; the deuteron therefore has parity 1 too.

4.6.2 The parities of pions and kaons

Most known mesons are a bound state of a quark q and an antiquark $\bar{q}'$.[11] We denote by L the orbital angular momentum of the quark–antiquark system, and by S its total spin, which may be either 0 or 1; the meson spin is denoted by J.

If $q' = q$, as happens in the case of the π^0, then the meson is identical with its antimeson, and it has a C-parity equal to $(-1)^{L+S}$, as in the case of positronium. Moreover, its parity does not depend on any convention for the relative parities of the various quark flavours. Thus, we were able in the previous sections to find experimental evidence for the parity of the π^0 being -1, and its C-parity being $+1$; the neutral pion, which is spinless, must therefore have $L = S = 0$.

If, on the other hand, the flavours q and q' are distinct, the meson is not an eigenstate of C and, moreover, its intrinsic parity is dependent on the convention that we made in the previous subsection for the relative parities of proton, neutron, and Λ. However, as long as all quark flavours have the same parity by convention, the parities of all mesons are given by the same rules as the parities of positronium states: the parity of any meson is thus given by $(-1)^{L+1}$.

In general, $L = 0$ states may have either $J = S = 0$ and then they are pseudoscalars, or they may have $J = S = 1$ and they are pseudovectors. These are the stable mesons, and they all have negative parity. In the same way, the stable baryons are three-quark states in which the orbital angular momentum of any two quarks is zero; all stable baryons have parity $+1$, just as the nucleons and the Λ.

The parity of the charged pion may be experimentally derived from observation (Panofsky *et al.* 1951; Chinowsky and Steinberger 1954) of the capture of slow pions by deuterium,

$$\pi^- + d \to n + n. \tag{4.16}$$

The pion and the deuteron must be in an s-wave, which has a non-zero probability of the two particles coinciding at the same point of space. As the pion is spinless and the deuteron has spin 1, it follows that the total angular momentum on each side of eqn (4.16) is 1. The system of two neutrons has orbital angular momentum L and total spin S. The symmetry of its spatial wave function is $(-1)^L$ and that of its spin wave function is $(-1)^{S+1}$. The total symmetry must be negative because of Fermi–Dirac statistics. Therefore, $L + S$ is even. On the other hand, L and S must combine to give a total angular momentum $J = 1$. The only values satisfying these requirements are $L = S = 1$.

The intrinsic parities of the nucleons cancel on both sides of the reaction in eqn (4.16), because of the convention that the proton and the neutron have the same intrinsic parity. The two-neutron state has $(-1)^L = -1$, while the deuteron has $(-1)^L = +1$. Therefore, π^- has odd parity.

If π^- has parity $\exp(i\alpha_p) = -1$, then π^+ has parity $\exp(-i\alpha_p) = -1$ too. We conclude that the π^+ is P-odd too.

---

[11] Some observed mesons appear not to fit this paradigm. They might be either multiquark states, or else have an important gluonic component—be either glueballs or hybrid quark-antiquark-gluon mesons. See for instance Particle Data Group (1996, p. 557).

The intrinsic parities of the kaons may be found by studying strong processes in which they are produced in association with the Λ (associated production). It is found that the kaons have negative parity.

4.7   The relative phase of G_V and G_A

Consider the β decay of the neutron n to the proton p,

$$n \to pe^- \bar{\nu}_e. \tag{4.17}$$

According to the Fermi theory, modified to include maximal parity violation in the leptonic sector, the effective Hamiltonian for this decay must be the product of a leptonic and a hadronic current, in which the leptonic current is of the $V-A$ form,

$$J_\mu^{\text{leptonic}} = \bar{e}\gamma_\mu \gamma_L \nu_e, \tag{4.18}$$

while the hadronic current should be

$$J_\mu^{\text{hadronic}} = \bar{p}\gamma_\mu \left(G_V + G_A\gamma_5\right) n. \tag{4.19}$$

Thus, as the Hamiltonian is Hermitian,

$$\mathcal{H}_{\text{eff}} = \left(\bar{e}\gamma_\mu\gamma_L\nu_e\right)\left[\bar{p}\gamma^\mu \left(G_V + G_A\gamma_5\right)n\right] + \left(\overline{\nu_e}\gamma_\mu\gamma_L e\right)\left[\bar{n}\gamma^\mu \left(G_V^* + G_A^*\gamma_5\right)p\right]. \tag{4.20}$$

4.7.1   T invariance

Now consider the T transformation:

$$\mathcal{T}n\mathcal{T}^{-1} = \gamma_0^*\gamma_5^* C^* An, \tag{4.21}$$

$$\mathcal{T}\nu_e\mathcal{T}^{-1} = \gamma_0^*\gamma_5^* C^* A\nu_e, \tag{4.22}$$

$$\mathcal{T}\bar{p}\mathcal{T}^{-1} = p^\dagger \left(C^{-1}\right)^* \gamma_5^*\gamma_0^*, \tag{4.23}$$

$$\mathcal{T}\bar{e}\mathcal{T}^{-1} = e^\dagger \left(C^{-1}\right)^* \gamma_5^*\gamma_0^*. \tag{4.24}$$

The arbitrary phases in the T transformation have been omitted for simplicity. Explicit mention of the coordinate change $t \to -t$ was omitted too. Clearly,

$$\mathcal{T}\left(\bar{e}\gamma_\mu\gamma_L\nu_e\right)\left[\bar{p}\gamma^\mu \left(G_V + G_A\gamma_5\right)n\right]\mathcal{T}^{-1}$$

$$= \left[e^\dagger \left(C^{-1}\right)^* \gamma_5^*\gamma_0^*\gamma_\mu^*\gamma_L^*\gamma_0^*\gamma_5^* C^* A\nu_e\right]$$

$$\times \left\{p^\dagger \left(C^{-1}\right)^* \gamma_5^*\gamma_0^* \left[\gamma^\mu \left(G_V + G_A\gamma_5\right)\right]^* \gamma_0^*\gamma_5^* C^* An\right\}$$

$$= \left(\bar{e}\gamma^\mu\gamma_L\nu_e\right)\left[\bar{p}\gamma_\mu \left(G_V^* + G_A^*\gamma_5\right)n\right],$$

in which, we emphasize once again, arbitrary phases have been omitted. It is now clear that T invariance of the Hamiltonian means

$$G_V + G_A\gamma_5 = \exp\left(i\zeta\right)\left(G_V^* + G_A^*\gamma_5\right), \tag{4.25}$$

i.e., G_V/G_A must be real in order for T invariance to hold. Experimentally (Particle Data Group 1996, p. 48), it is found that the phase of G_V/G_A is $(180.07 \pm 0.18)°$, confirming T invariance in neutron decay.

4.7.2 CP invariance

The CPT theorem tells us that, if the effective Hamiltonian in eqn (4.20) violates T, then it should also violate CP. This is indeed true. Using eqn (3.80), we see that

$$(\mathcal{CP}) \left(\bar{e} \gamma_\mu \gamma_L \nu_e \right) (\mathcal{CP})^\dagger = - \exp\left[i \left(\xi_{\nu_e} - \xi_e \right) \right] \left(\overline{\nu_e} \gamma^\mu \gamma_L e \right) ; \qquad (4.26)$$

while, using eqns (3.76) and (3.77), we have

$$(\mathcal{CP}) \left[\bar{p} \gamma^\mu \left(G_V + G_A \gamma_5 \right) n \right] (\mathcal{CP})^\dagger = - \exp\left[i \left(\xi_n - \xi_p \right) \right] \left[\bar{n} \gamma_\mu \left(G_V + G_A \gamma_5 \right) p \right] . \qquad (4.27)$$

Thus,

$$\begin{aligned}
(\mathcal{CP}) \mathcal{H}_{\text{eff}} (\mathcal{CP})^\dagger = {} & \exp\left(-i\zeta \right) \left(\overline{\nu_e} \gamma^\mu \gamma_L e \right) \left[\bar{n} \gamma_\mu \left(G_V + G_A \gamma_5 \right) p \right] \\
& + \exp\left(i\zeta \right) \left(\bar{e} \gamma^\mu \gamma_L \nu_e \right) \left[\bar{p} \gamma_\mu \left(G_V^* + G_A^* \gamma_5 \right) n \right], \qquad (4.28)
\end{aligned}$$

where $\zeta = \xi_e + \xi_p - \xi_{\nu_e} - \xi_n$ is an arbitrary phase. CP invariance requires

$$\begin{aligned}
(\mathcal{CP}) \mathcal{H}_{\text{eff}} (\mathcal{CP})^\dagger &= \mathcal{H}_{\text{eff}} \\
\Rightarrow G_V + G_A \gamma_5 &= \exp\left(i\zeta \right) \left(G_V^* + G_A^* \gamma_5 \right) . \qquad (4.29)
\end{aligned}$$

This is the same as eqn (4.25), as it should be.

An important point should be called to the reader's attention in this example. CP invariance of an effective Hamiltonian (or Lagrangian) in practice requires that the coupling constants in that effective Hamiltonian (in this case, G_V and G_A) be relatively real. It is not really each individual coupling constant that must be real in order for CP invariance to hold. Indeed, there are arbitrary phases in the CP transformation. Those phases may be adjusted in order to obtain CP invariance of each individual term. Rather, it is the *relative* phase of the coupling constants which must vanish (or be π) in order for CP invariance to prevail.

5

WEAK AND STRONG PHASES

5.1 Complex CP conditions

Let us consider the transitions of the CP-conjugate initial states i and $\bar{i}$ to the final states f and g, and to their CP-conjugate states $\bar{f}$ and $\bar{g}$, respectively.

We call the reader's attention to the following: whenever we write 'f' and '$\bar{f}$', for instance, we mean by this two CP-conjugate states in a *physical* sense. Thus, if f is composed of the particles $a, b, \ldots$, with momenta $\vec{p}_a, \vec{p}_b, \ldots$, and spins $\vec{s}_a, \vec{s}_b, \ldots$, respectively, then $\bar{f}$ is composed of the antiparticles $\bar{a}, \bar{b}, \ldots$, with momenta $-\vec{p}_a, -\vec{p}_b, \ldots$, and spins $\vec{s}_a, \vec{s}_b, \ldots$, respectively. This must be contrasted with the kets $|f\rangle$ and $|\bar{f}\rangle$, which are *mathematical* entities belonging to the formal structure of quantum mechanics. Those kets may be rephased at will, and they are related to each other by a *mathematical* operator $\mathcal{CP}$, which transforms each of them into the other one up to an arbitrary phase—see eqns (5.2) below. Thus, the fact that $\bar{f}$ is the CP-conjugate state of f does *not* mean that a relation $\mathcal{CP}|f\rangle = |\bar{f}\rangle$ holds between the kets. Rather, we have $\mathcal{CP}|f\rangle = e^{i\xi_f}|\bar{f}\rangle$.

First we deal with the general case $i \neq \bar{i}$, $f \neq \bar{f}$, and $g \neq \bar{g}$. We want to find the conditions on the transition amplitudes imposed by CP invariance. CP invariance implies, for the transition matrix T,

$$(\mathcal{CP}) \, T \, (\mathcal{CP})^{\dagger} = T. \tag{5.1}$$

CP may or may not be an invariance of Nature. On the other hand, the square of the CP transformation is, in classical physics, identical with the identity transformation, and therefore $(\mathcal{CP})^2$ corresponds to a conserved quantum number. The value of $(\mathcal{CP})^2$ for initial and final states must be identical, and it is an arbitrary, purely conventional phase. Without loss of generality we shall always assume $(\mathcal{CP})^2 = 1$. This assumption simplifies the algebra.

The CP transformation of the kets reads

$$\begin{aligned}
\mathcal{CP}|i\rangle &= e^{i\xi_i}|\bar{i}\rangle, & \mathcal{CP}|\bar{i}\rangle &= e^{-i\xi_i}|i\rangle, \\
\mathcal{CP}|f\rangle &= e^{i\xi_f}|\bar{f}\rangle, & \mathcal{CP}|\bar{f}\rangle &= e^{-i\xi_f}|f\rangle, \\
\mathcal{CP}|g\rangle &= e^{i\xi_g}|\bar{g}\rangle, & \mathcal{CP}|\bar{g}\rangle &= e^{-i\xi_g}|g\rangle.
\end{aligned} \tag{5.2}$$

The phases ξ_i, ξ_f, and ξ_g are arbitrary. We want to see whether it is possible to choose them so that CP invariance of the phenomenology is obtained.

From eqns (5.1) and (5.2) we derive the CP constraints on the transition amplitudes:

$$\langle f|T|i\rangle = e^{i(\xi_i - \xi_f)}\langle \bar{f}|T|\bar{i}\rangle, \tag{5.3}$$

$$\langle \bar{f}|T|i\rangle = e^{i(\xi_i + \xi_f)}\langle f|T|\bar{i}\rangle, \tag{5.4}$$

$$\langle g|T|i\rangle = e^{i(\xi_i - \xi_g)}\langle \bar{g}|T|\bar{i}\rangle, \tag{5.5}$$

$$\langle \bar{g}|T|i\rangle = e^{i(\xi_i + \xi_g)}\langle g|T|\bar{i}\rangle. \tag{5.6}$$

From these equations it is clear that the modulus of each transition amplitude must equal the modulus of the amplitude for the CP-conjugate transition. The quantities

$$\left|\langle f|T|i\rangle\right| - \left|\langle \bar{f}|T|\bar{i}\rangle\right|, \tag{5.7}$$

$$\left|\langle \bar{f}|T|i\rangle\right| - \left|\langle f|T|\bar{i}\rangle\right|, \tag{5.8}$$

$$\left|\langle g|T|i\rangle\right| - \left|\langle \bar{g}|T|\bar{i}\rangle\right|, \tag{5.9}$$

$$\left|\langle \bar{g}|T|i\rangle\right| - \left|\langle g|T|\bar{i}\rangle\right| \tag{5.10}$$

violate CP.

If we were considering solely the decays to f and to $\bar{f}$ we would have only eqns (5.3) and (5.4), which are two complex equations involving two arbitrary phases ξ_i and ξ_f. Then, there would be no CP-violating quantities beyond the ones in eqns (5.7) and (5.8). Similarly, if we were considering only the decays to g and to $\bar{g}$ the only physical consequences of CP invariance would be the vanishing of the quantities in eqns (5.9) and (5.10). However, when we consider simultaneously the decays to f and to $\bar{f}$ and also to g and to $\bar{g}$, we see that eqns (5.3)–(5.6) correspond to four complex equations with only three arbitrary phases. Then, a physical CP condition on the phases of the decay amplitudes must remain. Indeed, we easily find that the complex quantity

$$\langle f|T|i\rangle\langle \bar{f}|T|i\rangle\langle g|T|\bar{i}\rangle\langle \bar{g}|T|\bar{i}\rangle - \langle g|T|i\rangle\langle \bar{g}|T|i\rangle\langle f|T|\bar{i}\rangle\langle \bar{f}|T|\bar{i}\rangle \tag{5.11}$$

must vanish if CP invariance holds.

Let us now consider the case in which $\bar{f} = f$ and $\bar{g} = g$. Then, $\exp(i\xi_f) = \eta_f$ and $\exp(i\xi_g) = \eta_g$, where $\eta_f = \pm 1$ and $\eta_g = \pm 1$ are the CP-parities of f and of g, respectively. The conditions in eqns (5.3)–(5.6) reduce to

$$\langle f|T|i\rangle = \eta_f e^{i\xi_i}\langle f|T|\bar{i}\rangle, \tag{5.12}$$

$$\langle g|T|i\rangle = \eta_g e^{i\xi_i}\langle g|T|\bar{i}\rangle, \tag{5.13}$$

from which we derive the complex CP condition

$$\frac{\langle g|T|\bar{i}\rangle}{\langle g|T|i\rangle} = \eta_f \eta_g \frac{\langle f|T|\bar{i}\rangle}{\langle f|T|i\rangle}. \tag{5.14}$$

Thus, the complex quantity

$$\langle f|T|i\rangle\langle g|T|\bar{i}\rangle - \eta_f \eta_g \langle g|T|i\rangle\langle f|T|\bar{i}\rangle \tag{5.15}$$

violates CP.

5.2   Weak phases and strong phases

In the course of this book we shall be discovering that the presence of complex phases in the transition amplitudes is closely related to CP violation. A cavalier argument for this is the following: as a result of the CPT theorem, CP violation is equivalent to T violation; T transforms numbers into their complex conjugates; therefore, T and CP violation must arise from some numbers being different from their complex conjugates, i.e., from their being non-real.

It is important to stress from the very beginning that the phase of a transition amplitude is arbitrary and non-physical, because in quantum theory kets and bras may be rephased at will. As both $|i\rangle$ and $\langle f|$ may be rephased at will, the phase of $\langle f|T|i\rangle$ is arbitrary. Only phases which are rephasing-invariant, i.e., which do not change when the state vectors are rephased, may have a physical meaning and, in particular, lead to CP violation. Those phases are in general the *relative* phases of the various coherent contributions to a particular transition amplitude. The phase of each partial amplitude may be changed at will and is meaningless, but the relative phase of two partial amplitudes is rephasing-invariant and in general has observable consequences.

Three kinds of phases may arise in transition amplitudes:

- 'weak' or CP-odd phases,

- 'strong' or CP-even phases,

- 'spurious' CP-transformation phases.

The designations 'weak' and 'strong' do *not* mean that the origins of the phases are in weak and in strong interactions, respectively. A weak phase is defined to be one which has opposite signs in the transition amplitude for a process and in the transition amplitude for its CP-conjugate process. A strong phase has the same sign in the amplitudes for two CP-conjugate processes. Spurious phases are global, purely conventional relative phases between an amplitude and the amplitude for the CP-conjugate process; they do not originate in any dynamics, they just come from the assumed CP transformation of the field operators and of the kets and bras they act upon.

Weak phases usually originate from complex couplings in the Lagrangian. We have given one example in § 4.7: the phases of the two complex coupling constants G_V and G_A are CP-odd. The couplings G_V and G_A appear in the transition amplitude for neutron decay $n \rightarrow pe^-\overline{\nu_e}$, while G_V^* and G_A^* appear in the transition amplitude for the CP-conjugate antineutron decay $\bar{n} \rightarrow \bar{p}e^+\nu_e$. While the absolute phases of G_V and of G_A are irrelevant, the relative phase $\arg G_V - \arg G_A$ leads to T and CP violation if it is neither 0 nor π.

Another example of CP-odd phases are the phases of the matrix elements of the Cabibbo–Kobayashi–Maskawa matrix in the charged-current weak Lagrangian. As the Lagrangian is Hermitian, those complex phases change sign when one passes to the CP-conjugate process.

Strong phases may have two different origins. First, they may arise from the traces of products of an even number n of γ matrices together with γ_5. If $n = 0$

or $n = 2$ those traces vanish, $\mathrm{tr}\,\gamma_5 = 0$ and $\mathrm{tr}\,(\gamma^\mu\gamma^\nu\gamma_5) = 0$, but if n is four or larger the trace does not in general vanish, and is imaginary:

$$\mathrm{tr}\,\left(\gamma^\mu\gamma^\nu\gamma^\chi\gamma^\xi\gamma_5\right) = -4i\epsilon^{\mu\nu\chi\xi}. \tag{5.16}$$

Here, $\epsilon^{\mu\nu\chi\xi}$ is the completely antisymmetric tensor. Its overall sign is fixed by $\epsilon^{0123} = 1$. The tensor $\epsilon^{\mu\nu\chi\xi}$ is the Minkowski-space extension of the completely antisymmetric tensor ϵ^{ijk} of Euclidean space, which arises in triple cross products $\vec{a}\cdot\left(\vec{b}\times\vec{c}\right)$ of three vectors $\vec{a}$, $\vec{b}$, and $\vec{c}$. Indeed, traces like the one in eqn (5.16) yield terms in cross-sections proportional to the triple cross product of three (spin or momentum) vectors. Those terms are odd under the transformation $\hat{T}$ defined in § 1.1.2.

A second possible origin for strong phases are final-state-interaction (FSI) scatterings from on-shell states. These include the well-known final-state phase shifts of nuclear physics. Those phase shifts are equal for two CP-conjugate processes, and therefore they constitute strong phases.

The FSI allow the various final states of the weak decay to scatter elastically or inelastically via non-weak interactions. Thus, the total amplitude for a particular decay $i \to f$ includes contributions from processes $i \to f' \to f$, where the decay $i \to f'$ is weak, and the state f' subsequently scatters into f via the strong (or electromagnetic) interaction. If the intermediate state f' is on mass shell this generates an absorptive part in the amplitude. This is the origin of the CP-even phase. Whereas the CP-odd (weak) phase originates in the weak decay $i \to f'$, the CP-even (strong) phase arises in the $f' \to f$ scattering, and is dominated by the strong interaction.

In perturbation theory, strong phases appear as absorptive parts in Feynman loop integrals, which may be computed with the help of Cutkosky cuts. The computation of the absorptive parts may be tiresome, because it requires that one goes beyond Born approximation, and there may be many diagrams. Some authors like to use a trick in which they avoid computing all those absorptive parts: they only consider Born-approximation diagrams, but use Breit–Wigner propagators for the internal particles in those diagrams. Indeed, the imaginary part of the Breit–Wigner propagator equals the absorptive part that would arise from allowing the corresponding particle to be on shell. However, using a Breit–Wigner propagator for a gauge particle may be inconsistent with gauge invariance. Also, the Breit–Wigner propagator only accounts reasonably well for the absorptive parts when the propagating particle is almost on shell. Besides, in the complete computation there might be other intermediate states which might be put on shell beyond the ones accounted for by the Breit–Wigner propagator.

In general, experimental information on the FSI for heavy-meson systems is lacking, and it would anyway be hard to interpret, due to the large number of available intermediate states f'. One must then rely on theoretical modelling. For lighter mesons the situation is more favourable, as we shall see in particular in Chapter 8.

5.3   CP violation and interfering amplitudes

We next demonstrate that, if one wants to have CP violation in the transitions of i and $\bar{i}$ to f and $\bar{f}$ only—omitting g and $\bar{g}$ for the moment—then the transition amplitudes must be the sum of two or more interfering amplitudes with different weak phases and different strong phases.

Weak phases change sign under CP, and we might think that, once there is a CP-odd phase in an amplitude, CP violation automatically arises. To see that this is not so, consider for instance that

$$\begin{aligned}
\langle f|T|i\rangle &= A e^{i(\delta+\phi)}, \\
\langle \bar{f}|T|\bar{i}\rangle &= A e^{i(\delta-\phi+\theta)}.
\end{aligned} \tag{5.17}$$

Here, A is a real positive number, ϕ is a weak (CP-odd) phase, δ is a strong (CP-even) phase, and θ is a spurious phase, an arbitrary CP-transformation phase which in general arises but has no bearing on the question of whether CP will be violated or not. One immediately notices that the arrangement in eqns (5.17) does not correspond to CP violation, since the phases of the amplitudes are irrelevant, only their moduli matter. The quantity $|\langle f|T|i\rangle| - |\langle \bar{f}|T|\bar{i}\rangle| = A - A$ vanishes, and therefore CP is conserved.

An alternative way to arrive at the same conclusion is the following. We know from eqn (5.3) that CP is conserved once phases ξ_i and ξ_f such that

$$\langle f|T|i\rangle = e^{i(\xi_i-\xi_f)}\langle \bar{f}|T|\bar{i}\rangle \tag{5.18}$$

exist. Now, it is obvious that if we choose

$$\xi_i - \xi_f = 2\phi - \theta, \tag{5.19}$$

then eqn (5.18) is satisfied. It is important to stress that ξ_i and ξ_f are *free*; we have the right to choose them in the effort to obtain CP invariance of the phenomenology.

As CP invariance holds when the amplitudes are as in eqns (5.17), let us instead suppose that

$$\begin{aligned}
\langle f|T|i\rangle &= A_1 e^{i(\delta_1+\phi_1)} + A_2 e^{i(\delta_2+\phi_2)}, \\
\langle \bar{f}|T|\bar{i}\rangle &= A_1 e^{i(\delta_1-\phi_1+\theta)} + A_2 e^{i(\delta_2-\phi_2+\theta)}.
\end{aligned} \tag{5.20}$$

There are now two interfering amplitudes; they have moduli A_1 and A_2, CP-even phases δ_1 and δ_2, and CP-odd phases ϕ_1 and ϕ_2, respectively. These two interfering amplitudes may arise for instance from two different Feynman diagrams for the processes. The vertices of those diagrams would involve different CP-odd factors, while absorptive parts or traces with γ_5 should also be present. A spurious phase θ is common in both partial amplitudes. CP violation is now possible, because the moduli of the total amplitudes differ. Indeed,

$$|\langle f|T|i\rangle|^2 - |\langle \bar{f}|T|\bar{i}\rangle|^2 = -4A_1 A_2 \sin(\delta_1 - \delta_2) \sin(\phi_1 - \phi_2). \tag{5.21}$$

In order to obtain CP violation, the interfering amplitudes must have different strong phases ($\delta_1 \neq \delta_2$) and different weak phases ($\phi_1 \neq \phi_2$). CP-even phases are just as necessary as CP-odd phases in order to obtain CP violation.

Instead of the quantity in eqn (5.21), a more relevant quantity is the asymmetry

$$\frac{|\langle f|T|i\rangle|^2 - |\langle \bar{f}|T|\bar{i}\rangle|^2}{|\langle f|T|i\rangle|^2 + |\langle \bar{f}|T|\bar{i}\rangle|^2} = \frac{-2A_1 A_2 \sin(\delta_1 - \delta_2) \sin(\phi_1 - \phi_2)}{A_1^2 + A_2^2 + 2A_1 A_2 \cos(\delta_1 - \delta_2) \cos(\phi_1 - \phi_2)}. \quad (5.22)$$

In order for the asymmetry to be large three conditions must be met:

1. the difference between the weak phases of the two interfering amplitudes should be close to $\pi/2$, i.e., $\cos(\phi_1 - \phi_2) \approx 0$;
2. the difference between the strong phases of the two interfering amplitudes should be close to $\pi/2$, i.e., $\cos(\delta_1 - \delta_2) \approx 0$;
3. the difference between the moduli of the two interfering amplitudes should be small, i.e., $A_1 \approx A_2$.

In the limiting case $|\phi_1 - \phi_2| = |\delta_1 - \delta_2| = \pi/2$ and $A_1 = A_2$, the absolute value of the asymmetry attains its maximum value 1.

5.4   CP violation without strong phases

If we consider simultaneously the transitions to two different final states f and g, and to their CP-conjugate states $\bar{f}$ and $\bar{g}$, then CP violation may be observed even in the absence of strong phases and of interfering amplitudes. Here we study only the simple case in which f and g are two eigenstates of CP with the same CP-parity. Suppose for instance that $\eta_f = \eta_g = +1$, and that

$$\begin{aligned}
\langle f|T|i\rangle &= A_1 e^{i(\delta_1 + \phi_1)}, \\
\langle f|T|\bar{i}\rangle &= A_1 e^{i(\delta_1 - \phi_1 + \theta)}, \\
\langle g|T|i\rangle &= A_2 e^{i(\delta_2 + \phi_2)}, \\
\langle g|T|\bar{i}\rangle &= A_2 e^{i(\delta_2 - \phi_2 + \theta)}.
\end{aligned} \quad (5.23)$$

The quantity in eqn (5.15), which we already know to violate CP, then is

$$\langle f|T|i\rangle\langle g|T|\bar{i}\rangle - \langle g|T|i\rangle\langle f|T|\bar{i}\rangle = 2i A_1 A_2 e^{i(\delta_1 + \delta_2 + \theta)} \sin(\phi_1 - \phi_2). \quad (5.24)$$

In this case, the strong phases δ_1 and δ_2 are basically irrelevant for CP violation. They might be absent and CP violation would still exist. Only the weak phases ϕ_1 and ϕ_2 are necessary.

In general, the quantity in eqn (5.24) will not be observable, because f and g are in principle unconnected final states. Some relationship between these two states must exist in order that a physical decay involves both of them simultaneously. This is precisely what happens in the decays of the neutral kaons to $\pi^+\pi^-$ and $\pi^0\pi^0$. Indeed, these physical states are superpositions of two eigenstates of CP with CP=+1—the state of two pions with isospin 2 and the state of two pions with isospin 0. Those are the states f and g in that case. A quantity analogous to the one in eqn (5.24) is then defined to be the CP-violating parameter ϵ', as will be seen in Chapter 8.

5.5   Hermiticity of the transition matrix

We shall be dividing the Hamiltonian $\mathcal{H}$ into two parts: the strong and electromagnetic Hamiltonian, which is dominant and is P, T, and C invariant, and the weak Hamiltonian $\mathcal{H}_W$, which violates the discrete symmetries and is treated as a perturbation. Correspondingly, the transition matrix is written

$$T = T_{\text{strong}} + T_{\text{weak}}, \tag{5.25}$$

where T_{strong} includes both the strong and electromagnetic interactions, and is of order 0 in any weak coupling constant g.[12] Equation (2.3) then yields

$$T_{\text{strong}} - T_{\text{strong}}^{\dagger} = iT_{\text{strong}}^{\dagger} T_{\text{strong}}, \tag{5.26}$$

$$T_{\text{weak}} - T_{\text{weak}}^{\dagger} = i\left(T_{\text{strong}}^{\dagger} T_{\text{weak}} + T_{\text{weak}}^{\dagger} T_{\text{strong}} + T_{\text{weak}}^{\dagger} T_{\text{weak}}\right). \tag{5.27}$$

It may happen that, for some set of weak transitions $i \to f$, $i \to f'$, and so on, $T_{\text{strong}} = 0$, i.e., $S_{\text{strong}} = 1$. This is the case when both the initial and the final states of the transition are not scattered by the strong and electromagnetic interactions In this case,

$$T_{\text{weak}} - T_{\text{weak}}^{\dagger} = iT_{\text{weak}}^{\dagger} T_{\text{weak}}. \tag{5.28}$$

Then, to first order in the weak coupling constants, T_{weak} is Hermitian. This is because the right-hand side of eqn (5.28) is $\sim g^2$. We shall consider instances of this case later in the book, when we treat semileptonic decays of neutral-meson systems in which there is only one hadron in the final state; in those cases, there can be no final-state scattering due to the strong interaction, and $T_{\text{strong}} = 0$.[13]

We now re-formulate, following Wolfenstein (1991), the reasoning above. Suppose that strong and electromagnetic interactions do not cause the state i either to decay or to mix with other states. The state i decays only via the weak interaction. We expand the S matrix as a power series in the weak coupling g:

$$S = S_0 + iT_1 + iT_2 + \cdots, \tag{5.29}$$

where $T_k \propto g^k$. In this expansion, S_0 is the S matrix for the strong and electromagnetic interactions. Unitarity of S implies $S_0^{\dagger} = S_0^{-1}$ and $T_1^{\dagger} = S_0^{\dagger} T_1 S_0^{\dagger}$. Therefore,

$$\langle f|T_1^{\dagger}|i\rangle = \sum_{i'}\sum_{f'}\langle f|S_0^{\dagger}|f'\rangle\langle f'|T_1|i'\rangle\langle i'|S_0^{\dagger}|i\rangle. \tag{5.30}$$

As $|i\rangle$ is an eigenstate of S_0,

---

[12]In general, various weak interactions may exist. Those interactions may have different coupling constants g. Then, T_{strong} is that part of the transition matrix which does not depend on any of the weak coupling constants.

[13]As a matter of fact, T_{strong} is not zero even in this case, because of the presence of an electromagnetic (Coulomb) final-state scattering. Only when this scattering is neglected, and we only treat the weak interactions to order g, can we assert that T_{weak} is Hermitian.

$$\langle i'|S_0^\dagger|i\rangle = \delta_{i'i}, \tag{5.31}$$

i.e., i is stable under the strong and electromagnetic interactions, it decays only weakly. The final state $\langle f|$ may also be an eigenstate of S_0,

$$\langle f|S_0^\dagger|f'\rangle = \delta_{ff'}, \tag{5.32}$$

when there are no final-state interactions. Under these assumptions,

$$\langle f|T_1^\dagger|i\rangle = \langle f|T_1|i\rangle, \tag{5.33}$$

which means that T_1 is Hermitian for these particular initial and final states.

5.6 Consequences of CPT invariance

Now consider the consequences of CPT invariance. If

$$\begin{aligned} |i\rangle &= \mathcal{CPT}|\tilde{i}\rangle, \\ |f\rangle &= \mathcal{CPT}|\tilde{f}\rangle, \end{aligned} \tag{5.34}$$

and as (from CPT invariance) $(\mathcal{CPT})\, T_1\, (\mathcal{CPT})^{-1} = T_1^\dagger$, we have

$$\begin{aligned} \langle f|T_1^\dagger|i\rangle &= \langle \tilde{f}|T_1|\tilde{i}\rangle^* \\ &= \sum_{i'}\sum_{f'}\langle f|S_0^\dagger|f'\rangle\langle f'|T_1|i'\rangle\langle i'|S_0^\dagger|i\rangle. \end{aligned} \tag{5.35}$$

As $|i\rangle$ is an eigenstate of S_0, we obtain

$$\langle \tilde{f}|T_1|\tilde{i}\rangle^* = \sum_{f'}\langle f|S_0^\dagger|f'\rangle\langle f'|T_1|i\rangle. \tag{5.36}$$

Then, using the unitarity of S_0,

$$\begin{aligned} \sum_f \left|\langle \tilde{f}|T_1|\tilde{i}\rangle\right|^2 &= \sum_f\sum_{f'}\sum_{f''}\langle f|S_0^\dagger|f'\rangle\langle f'|T_1|i\rangle\langle f|S_0^\dagger|f''\rangle^*\langle f''|T_1|i\rangle^* \\ &= \sum_{f'}\langle f'|T_1|i\rangle\sum_{f''}\langle f''|T_1|i\rangle^*\sum_f\langle f|S_0^\dagger|f'\rangle\langle f|S_0^\dagger|f''\rangle^* \\ &= \sum_{f'}\langle f'|T_1|i\rangle\sum_{f''}\langle f''|T_1|i\rangle^*\delta_{f''f'} \\ &= \sum_{f'}|\langle f'|T_1|i\rangle|^2 \,. \end{aligned} \tag{5.37}$$

The sum over states eliminates from consideration the inversion of the momenta and spins originating in the T transformation. Therefore,

$$\sum_f \left|\langle \tilde{f}|T_1|\tilde{i}\rangle\right|^2 = \sum_f |\langle \bar{f}|T_1|\bar{i}\rangle|^2 \,, \tag{5.38}$$

where $\bar{i}$ and $\bar{f}$ are the CP-conjugate states of i and f, respectively. We conclude that the total decay widths of i and $\bar{i}$ are equal as a consequence of CPT invariance. This precludes observation of a CP-violating difference between the total decay widths of a particle and its antiparticle, and constitutes an obstacle in the study of CP violation.

Equation (5.37) is even more powerful because it applies separately to each set of final states which is not connected by the FSI (by the action of S_0) to the other possible final states. Weak interactions cause i to decay to a multitude of final states $f, f', f'', \ldots$, which are connected among themselves by strong and/or electromagnetic FSI, but are not connected by the FSI to the other possible final states. Then, S_0 is unitary in the subspace spanned by $f, f', f'', \ldots$, and the reasoning leading to eqn (5.37) holds. The CP-conjugate state of i, $\bar{i}$, decays to the CP-conjugate final states $\bar{f}, \bar{f}', \bar{f}'', \ldots$. CPT invariance forces the total widths to be identical:

$$
\begin{aligned}
&|\langle f|T_1|i\rangle|^2 + |\langle f'|T_1|i\rangle|^2 + |\langle f''|T_1|i\rangle|^2 + \cdots \\
&= |\langle \bar{f}|T_1|\bar{i}\rangle|^2 + |\langle \bar{f}'|T_1|\bar{i}\rangle|^2 + |\langle \bar{f}''|T_1|\bar{i}\rangle|^2 + \cdots .
\end{aligned}
\tag{5.39}
$$

However, the fact that there are final-state interactions connecting $f, f', f'', \ldots$ among themselves allows for CP violation to occur, embodied in different partial decay widths:

$$
\begin{aligned}
|\langle f|T_1|i\rangle|^2 &\neq |\langle \bar{f}|T_1|\bar{i}\rangle|^2, \\
|\langle f'|T_1|i\rangle|^2 &\neq |\langle \bar{f}'|T_1|\bar{i}\rangle|^2, \\
|\langle f''|T_1|i\rangle|^2 &\neq |\langle \bar{f}''|T_1|\bar{i}\rangle|^2.
\end{aligned}
\tag{5.40}
$$

Pais and Treiman (1975) have put forward a number of interesting instances of this. For instance, let X_S denote a sum over all possible hadron states with total strangeness S. Then, CPT invariance implies

$$
\Gamma\left(D^+ \to l^+\nu_l X_{-1}\right) = \Gamma\left(D^- \to l^-\bar{\nu}_l X_1\right),
\tag{5.41}
$$
$$
\Gamma\left(D^+ \to l^+\nu_l X_0\right) = \Gamma\left(D^- \to l^-\bar{\nu}_l X_0\right).
\tag{5.42}
$$

As another example, consider the two-pion and three-pion decays of the charged kaons. Because of G-parity, which is a particular combination of C symmetry and isospin symmetry, and as such is preserved by the strong interaction—but not by electromagnetism—, a state with an even number of pions cannot scatter into a state with an odd number of pions, and vice-versa. Two-pion final states are thus disconnected from three-pion final states. Thus, when the electromagnetic FSI are neglected,

$$
\Gamma\left(K^+ \to \pi^+\pi^0\right) = \Gamma\left(K^- \to \pi^-\pi^0\right),
\tag{5.43}
$$

preventing CP violation in the two-pion decays. On the other hand, CP violation in the three-pion decays may occur, but CPT still imposes a condition on it:

$$
\begin{aligned}
&\Gamma\left(K^+ \to \pi^+\pi^0\pi^0\right) - \Gamma\left(K^- \to \pi^-\pi^0\pi^0\right) \\
&= \Gamma\left(K^- \to \pi^+\pi^-\pi^-\right) - \Gamma\left(K^+ \to \pi^-\pi^+\pi^+\right).
\end{aligned}
\tag{5.44}
$$

In order to have CP violation in the partial decay rates, as in eqns (5.40), final-state-interaction phases are essential. The picture that one should keep in mind is a process $i \to f' \to f$, where the decay $i \to f'$ is weak and then there is a final-state scattering $f' \to f$; if it is possible to put the intermediate state f' on shell, CP violation may arise.

The consistency with the requirements of CPT invariance—that total semi-inclusive widths should be equal for particles and antiparticles—should always be carefully checked in any explicit computation. It is easy to run into pitfalls, as was shown in particular examples by Gérard and Hou (1988, 1989) and Soares (1992).

5.7   T violation and $\hat{\text{T}}$ violation

CP-violating (CP-odd) observables may be either $\hat{\text{T}}$-odd, and then they are CP$\hat{\text{T}}$-even, or $\hat{\text{T}}$-even, and then they are CP$\hat{\text{T}}$-odd. Observables of the first kind typically are triple cross products constructed out of the momentum and/or spin vectors of the particles in some process; say, $\vec{p}_a \cdot (\vec{p}_b \times \vec{p}_c)$. What violates CP is the difference between the expectation value of such an observable for a process and for the CP-conjugate process, viz., $\langle \vec{p}_a \cdot (\vec{p}_b \times \vec{p}_c) \rangle - \langle \vec{p}_{\bar{a}} \cdot (\vec{p}_{\bar{b}} \times \vec{p}_{\bar{c}}) \rangle$. Observables of the second kind typically are partial-width asymmetries, as in the previous section.

The difference between T and $\hat{\text{T}}$ must be stressed. T involves an interchange of initial and final states which, in a quantum-mechanical problem, is impossible to reproduce in the laboratory; the final state of a scattering or decay is a coherent superposition of outgoing quantum-mechanical spherical waves, and setting up an apparatus which would produce the T-reversed state, a coherent superposition of incoming spherical waves, is impossible (Lee 1990). This is the reason why one cannot directly test T symmetry in the laboratory, and must have recourse to consequences of that symmetry, like the principle of detailed balance—see eqn (2.33).

If final-state interactions may be neglected, then violation of $\hat{\text{T}}$ implies violation of T. This is because, in that case, the transition matrix T is, to first order in the weak interactions, Hermitian. Now, T invariance implies $|\langle f|T|i\rangle| = |\langle i_T|T|f_T\rangle|$, where f_T and i_T are the T-transformed states of f and i, respectively. If T is Hermitian, we then have $|\langle f|T|i\rangle| = |\langle f_T|T|i_T\rangle|$, which means precisely $\hat{\text{T}}$ invariance. Thus, when FSI are absent, T invariance implies $\hat{\text{T}}$ invariance, and therefore violation of $\hat{\text{T}}$ implies violation of T.

Conversely, FSI may lead to $\hat{\text{T}}$ violation without the occurrence of any T violation (and, from the CPT theorem, CP violation). Thus, $\hat{\text{T}}$ violation does not have to correspond to CP violation; a non-zero value for the triple product of momenta and/or spins should be confronted with the analogous observable in the CP-conjugate process in order to ascertain CP violation (Rindani 1995; Yuan 1995).

6

NEUTRAL-MESON SYSTEMS: MIXING

6.1 Introduction

In this and the next chapter we discuss the mixing and decays of the P^0 and $\overline{P^0}$ mesons. In our notation, P^0 may refer to either K^0, D^0, B_d^0, or B_s^0. We deal here with the general phenomenology relevant for any neutral-meson system. The specifics of particular systems will be discussed later.

If only the strong and electromagnetic interactions existed, P^0 and $\overline{P^0}$ would be stable and form a particle–antiparticle pair with common mass m_0. Because of the weak interactions, P^0 and $\overline{P^0}$ decay. Moreover, neither electric-charge conservation nor any other conservation law respected by the weak interactions prevent P^0 and $\overline{P^0}$ from having both real and virtual transitions to common states n. As a consequence, P^0 and $\overline{P^0}$ mix, i.e., they oscillate between themselves before decaying. Similarly, there are theoretical speculations about the possible mixing of neutrinos and, if baryon number is not conserved, of the neutron and antineutron.

Thus, $|P^0\rangle$ and $|\overline{P^0}\rangle$ are eigenstates of the strong and electromagnetic interactions with common mass m_0 and opposite flavour content. Since flavour is conserved in the strong and electromagnetic interactions, $\langle P^0|\overline{P^0}\rangle = 0$. When the weak-interaction Hamiltonian $\mathcal{H}_W$ is switched on, $|P^0\rangle$ and $|\overline{P^0}\rangle$ both mix and decay to other states.

6.2 Mixing

In principle, we would like to consider the evolution of a state of the general form

$$a(t)|P^0\rangle + b(t)|\overline{P^0}\rangle + c_1(t)|n_1\rangle + c_2(t)|n_2\rangle + c_3(t)|n_3\rangle + \cdots,$$

where n_1, n_2, and so on, are states to which either P^0 or $\overline{P^0}$ may decay, and t is the time measured in the P^0–$\overline{P^0}$ rest frame. The evolution of such a state is in general very complicated. If however

- for $t = 0$ only $a(t)$ and $b(t)$ are non-zero, while all $c_i(0) = 0$,
- we want to compute only the values of $a(t)$ and $b(t)$, not the values of the $c_i(t)$,
- the times t in which we are interested are much larger than the typical strong-interaction scale,

then a great simplification is achieved, as was first shown by Weisskopf and Wigner (1930 a,b). In the Wigner–Weisskopf approximation, which we shall use

throughout, a beam of oscillating and decaying neutral mesons is described, in
its rest frame, by the two-component wave function

$$|\psi(t)\rangle = \psi_1(t)|P^0\rangle + \psi_2(t)|\overline{P^0}\rangle, \qquad (6.1)$$

where t is the proper time. The wave function evolves according to a Schrödinger-
like equation:

$$i\frac{d}{dt}\begin{pmatrix} \psi_1 \\ \psi_2 \end{pmatrix} = \begin{pmatrix} R_{11} & R_{12} \\ R_{21} & R_{22} \end{pmatrix} \begin{pmatrix} \psi_1 \\ \psi_2 \end{pmatrix}. \qquad (6.2)$$

The matrix R is not Hermitian, else the mesons would just oscillate, they would
not decay—see eqn (6.9) below. It may be written

$$R = M - \tfrac{i}{2}\Gamma, \qquad (6.3)$$

with

$$M = M^\dagger, \\ \Gamma = \Gamma^\dagger. \qquad (6.4)$$

Clearly,

$$M = \tfrac{1}{2}(R + R^\dagger), \\ \Gamma = i(R - R^\dagger). \qquad (6.5)$$

The matrices M and Γ are given, in second-order perturbation theory, by sums
over intermediate states n:

$$M_{ij} = m_0\delta_{ij} + \langle i|\mathcal{H}_W|j\rangle + \sum_n P\frac{\langle i|\mathcal{H}_W|n\rangle\langle n|\mathcal{H}_W|j\rangle}{m_0 - E_n}, \\ \Gamma_{ij} = 2\pi\sum_n \delta(m_0 - E_n)\langle i|\mathcal{H}_W|n\rangle\langle n|\mathcal{H}_W|j\rangle. \qquad (6.6)$$

The operator P projects out the principal part. The intermediate states contributing to M are virtual, while the ones contributing to Γ are physical states
to which both P^0 and $\overline{P^0}$ decay. The latter states may be grouped together in
decay channels c for which, by definition, the matrix elements of $\mathcal{H}_W$ are the
same. For instance, decay channels group together states with the same quantum numbers, but with the decay particles flying off in different directions. The
density of states of each channel is

$$\rho_c = \sum_{n_c} \delta(m_0 - E_{n_c}). \qquad (6.7)$$

Therefore, another way to write Γ_{ij} is

$$\Gamma_{ij} = 2\pi\sum_c \rho_c\langle i|\mathcal{H}_W|c\rangle\langle c|\mathcal{H}_W|j\rangle. \qquad (6.8)$$

From eqns (6.2)–(6.4) it follows that

$$\frac{d}{dt}\left(|\psi_1|^2 + |\psi_2|^2\right) = -\left(\psi_1^* \ \psi_2^*\right)\Gamma\begin{pmatrix}\psi_1\\\psi_2\end{pmatrix}. \tag{6.9}$$

The mesons P^0 and $\overline{P^0}$ decay, therefore the left-hand side of the above equation must be negative for any values of ψ_1 and ψ_2. Hence, Γ is positive definite, i.e., Γ_{11}, Γ_{22}, and $\det\Gamma$ are positive.

In quantum mechanics, all kets may be rephased at will, no measurable consequences following from the rephasing of a ket. We shall be careful to keep the phenomenology of neutral-meson mixing and decay invariant under the rephasings

$$\begin{aligned}|P^0\rangle &\to e^{i\gamma}|P^0\rangle,\\|\overline{P^0}\rangle &\to e^{i\bar{\gamma}}|\overline{P^0}\rangle.\end{aligned} \tag{6.10}$$

The diagonal matrix elements of R are invariant under this rephasing, but the off-diagonal ones are not:

$$\begin{aligned}M_{12} &\to e^{i(\bar{\gamma}-\gamma)}M_{12},\\\Gamma_{12} &\to e^{i(\bar{\gamma}-\gamma)}\Gamma_{12},\\M_{21} &\to e^{i(\gamma-\bar{\gamma})}M_{21},\\\Gamma_{21} &\to e^{i(\gamma-\bar{\gamma})}\Gamma_{21},\end{aligned} \tag{6.11}$$

as can be seen from eqns (6.6). As a consequence, from the eight real numbers (four moduli and four phases) in R, only seven have physical significance.

The two eigenstates of R may be distinguished by labels a and b. Since R is not Hermitian, its eigenvalues are complex and we write them $\mu_a = m_a - (i/2)\Gamma_a$ and $\mu_b = m_b - (i/2)\Gamma_b$, where m_a and m_b are the masses of P_a and P_b, respectively, while Γ_a and Γ_b are their decay widths. Let us denote $\Delta m = m_a - m_b$ and $\Delta\Gamma = \Gamma_a - \Gamma_b$. At this stage the labels a and b do not have any physical meaning. Hence, the signs of Δm and of $\Delta\Gamma$ are arbitrary. However, their relative sign has physical significance: it indicates whether it is the heaviest or the lightest state which lives longer.

In fact, there are three distinct questions begging an answer (Azimov 1990): Which, P_a or P_b, is the heavier eigenstate? Which eigenstate lives longer? And, which eigenstate decays most often to a particular CP-even (or CP-odd) final state? Neither of these questions has physical meaning in itself, because a and b are meaningless labels, i.e., P_a and P_b are a priori equivalent and we have not yet defined a way to distinguish them. However, the relative answers to those questions are physically meaningful.

For the B^0–$\overline{B^0}$ systems it has become customary to choose the mass of the eigenstates as label: $a = H$ and $b = L$ for the heavy and light eigenstate, respectively. Then, Δm is positive by definition, while both the sign of $\Delta\Gamma$ and the dominant CP content of the eigenstates are physically meaningful. A different nomenclature is used in the K^0–$\overline{K^0}$ system. There, the lifetimes of the eigenstates are widely different and one uses them to label the eigenstates: $a = L$

refers to the long-lived neutral kaon K_L, and $b = S$ refers to the short-lived neutral kaon K_S. Then, $\Delta\Gamma < 0$ by definition, and the experimental determination of the sign of Δm is necessary. It turns out that K_L is heavier than K_S: $\Delta m > 0$. Moreover, K_S is found to decay predominantly into CP-even states.

In this book we label the states by their mass: P_H is the heaviest eigenstate and P_L is the lightest one. Henceforth,

$$m \equiv \frac{m_H + m_L}{2},$$
$$\Gamma \equiv \frac{\Gamma_H + \Gamma_L}{2}, \qquad (6.12)$$

while

$$\Delta\mu \equiv \mu_H - \mu_L$$
$$= \Delta m - \tfrac{i}{2}\Delta\Gamma,$$
$$\Delta m \equiv m_H - m_L, \qquad (6.13)$$
$$\Delta\Gamma \equiv \Gamma_H - \Gamma_L.$$

The mass difference Δm is positive by definition. The sign of $\Delta\Gamma$ is physically meaningful.[14]

In the different neutral-meson systems these observables have different orders of magnitude and, therefore, different approximations are justified. Of course, any given expression may be written using either the quantities on the left-hand side or those on the right-hand side of eqns (6.12) and (6.13), but the physical interpretation is typically more transparent with one choice than with the other one. In the kaon case the lifetimes are widely different and writing the decay widths in terms of Γ_S and Γ_L highlights the fact that the K_S component disappears much earlier than the K_L one: after a while we have an almost pure K_L beam. On the other hand, in the B and D systems the decay widths are expected to be very similar and it makes more sense to use Γ together with a small modulation dependent on $\Delta\Gamma$. It is then useful to introduce the dimensionless parameters

$$x \equiv \frac{\Delta m}{\Gamma}, \qquad y \equiv \frac{\Delta\Gamma}{2\Gamma}. \qquad (6.15)$$

The range of y is from -1 to 1, approaching these limiting values when one decay width is much larger than the other one, as in the kaon case. On the other hand, x is positive by definition. We find it useful to define another parameter:

$$u \equiv -\frac{\Delta\Gamma}{2\Delta m} = -\frac{y}{x}. \qquad (6.16)$$

---

[14]Be careful to note that m_H and m_L are not the eigenvalues of M, and Γ_H and Γ_L are not the eigenvalues of Γ. Still,

$$\operatorname{tr} M = m_H + m_L = 2m,$$
$$\operatorname{tr} \Gamma = \Gamma_H + \Gamma_L = 2\Gamma. \qquad (6.14)$$

The parameter u happens to be very close to 1 in the kaon system. In the other neutral-meson systems its value has not yet been experimentally determined.

6.3 Discrete symmetries

The CP transformation interchanges P^0 and $\overline{P^0}$. Choosing $(\mathcal{CP})^2 = 1$,

$$\begin{aligned}
\mathcal{CP}|P^0\rangle &= e^{i\xi}|\overline{P^0}\rangle, \\
\mathcal{CP}|\overline{P^0}\rangle &= e^{-i\xi}|P^0\rangle.
\end{aligned} \tag{6.17}$$

Analogously,

$$\begin{aligned}
\mathcal{CPT}|P^0\rangle &= e^{i\nu}|\overline{P^0}\rangle, \\
\mathcal{CPT}|\overline{P^0}\rangle &= e^{i\nu}|P^0\rangle.
\end{aligned} \tag{6.18}$$

Here we have taken into account the fact that $\mathcal{CPT}$ is antiunitary, contrary to $\mathcal{CP}$ which is unitary. From eqns (6.17) and (6.18) we get

$$\begin{aligned}
\mathcal{T}|P^0\rangle &= e^{i(\nu-\xi)}|P^0\rangle, \\
\mathcal{T}|\overline{P^0}\rangle &= e^{i(\nu+\xi)}|\overline{P^0}\rangle.
\end{aligned} \tag{6.19}$$

Both $\mathcal{T}$ and $\mathcal{CPT}$ interchange outgoing with ingoing states. However, as P^0 and $\overline{P^0}$ are taken to be in their rest frame, we do not have to concern ourselves with that point.

The phases ξ and ν are not invariant under the rephasing in eqn (6.10), rather

$$\begin{aligned}
\xi &\to \xi + \gamma - \bar{\gamma}, \\
\nu &\to \nu - \gamma - \bar{\gamma}.
\end{aligned} \tag{6.20}$$

We emphasize that the phases ξ and ν must not be seen as defined *a priori*. Rather, CP invariance exists if there is any phase ξ such that the phenomenology is left invariant by the transformation in eqns (6.17). Analogously, there is CPT invariance if one can find a phase ν such that the transformation in eqns (6.18) leaves the phenomenology invariant. Strong and electromagnetic interactions are invariant under the CP and CPT transformations *for any choice* of the phases ξ and ν, in the same way that quantum electrodynamics is invariant under the P, T, and C transformations of the field ψ, for any choice of the transformation phases β_p in eqn (3.63), β_t in eqn (3.66), and β_c in eqn (3.69), respectively.

We may define eigenstates of CP

$$|P_\pm\rangle \equiv \frac{1}{\sqrt{2}}\left(|P^0\rangle \pm e^{i\xi}|\overline{P^0}\rangle\right). \tag{6.21}$$

The factor $2^{-1/2}$ is for normalization. The kets $|P_\pm\rangle$ are the eigenstates of the CP transformation in eqns (6.17) corresponding to the eigenvalues ± 1. Notice the following two points:

1. A choice of the relative phase of $|P_+\rangle$ and $|P_-\rangle$ was implicitly done when writing down eqns (6.21). Indeed, we chose $\langle P^0|P_+\rangle$ to have the same phase as $\langle P^0|P_-\rangle$. Whenever using eqns (6.21), one should be careful not to attribute physical significance to any phase which would vary if the phases of $|P_+\rangle$ and of $|P_-\rangle$ were to be independently changed.

2. The phase ξ of the CP transformation in eqns (6.17) is arbitrary and devoid of physical meaning, as we emphasized in the previous paragraph. This fact is a further source of arbitrariness in the definition of $|P_\pm\rangle$.

We shall not be using the states $|P_\pm\rangle$ in the rest of this book.

Let us consider the CP-, CPT-, and T-invariance conditions on the matrix elements of M and Γ. Defining $\mathcal{H}_{\text{CP}} \equiv (CP)\, \mathcal{H}_{\text{W}}\, (CP)^\dagger$, and similarly $\mathcal{H}_{\text{CPT}} \equiv (CPT)\, \mathcal{H}_{\text{W}}\, (CPT)^{-1}$ and $\mathcal{H}_{\text{T}} \equiv T\mathcal{H}_{\text{W}}T^{-1}$, we derive for instance

$$\Gamma_{11} = 2\pi \sum_n \delta(m_0 - E_n)\langle P^0|\mathcal{H}_{\text{W}}|n\rangle\langle n|\mathcal{H}_{\text{W}}|P^0\rangle$$

$$= 2\pi \sum_n \delta(m_0 - E_n)\left(e^{-i\xi}\langle \overline{P^0}|\mathcal{H}_{\text{CP}}|\overline{n}\rangle\right)\left(e^{i\xi}\langle \overline{n}|\mathcal{H}_{\text{CP}}|\overline{P^0}\rangle\right)$$

$$= 2\pi \sum_n \delta(m_0 - E_n)\langle \overline{P^0}|\mathcal{H}_{\text{CP}}|n\rangle\langle n|\mathcal{H}_{\text{CP}}|\overline{P^0}\rangle$$

$$= 2\pi \sum_n \delta(m_0 - E_n)\left(e^{-i\nu}\langle \overline{P^0}|\mathcal{H}_{\text{CPT}}|\overline{n}\rangle\right)^*\left(e^{i\nu}\langle \overline{n}|\mathcal{H}_{\text{CPT}}|\overline{P^0}\rangle\right)^*$$

$$= 2\pi \sum_n \delta(m_0 - E_n)\langle \overline{P^0}|\mathcal{H}_{\text{CPT}}|n\rangle\langle n|\mathcal{H}_{\text{CPT}}|\overline{P^0}\rangle$$

$$= 2\pi \sum_n \delta(m_0 - E_n)\left[e^{i(\xi-\nu)}\langle P^0|\mathcal{H}_{\text{T}}|n\rangle\right]^*\left[e^{i(\nu-\xi)}\langle n|\mathcal{H}_{\text{T}}|P^0\rangle\right]^*$$

$$= 2\pi \sum_n \delta(m_0 - E_n)\langle P^0|\mathcal{H}_{\text{T}}|n\rangle\langle n|\mathcal{H}_{\text{T}}|P^0\rangle.$$

We conclude that CPT and CP invariance ($\mathcal{H}_{\text{CPT}} = \mathcal{H}_{\text{W}}$ and $\mathcal{H}_{\text{CP}} = \mathcal{H}_{\text{W}}$, respectively) imply $\Gamma_{11} = \Gamma_{22}$, while T invariance ($\mathcal{H}_{\text{T}} = \mathcal{H}_{\text{W}}$) does not have any consequence for Γ_{11}. In a similar way we derive all the results in Table 6.1. We discover that CP symmetry is equivalent, in the case of M and Γ, to the simultaneous existence of CPT and T symmetry.

Since the phase ξ is arbitrary, and since M and Γ are Hermitian, the equations $M_{21} = \exp(2i\xi)M_{12}$ and $\Gamma_{21} = \exp(2i\xi)\Gamma_{12}$, which follow from either T or CP symmetry, are not a constraint when taken separately. They only acquire a meaning when taken together. They mean that

Table 6.1 *Effects of the discrete symmetries.*

Symmetry	Diagonal elements	Off-diagonal elements
CPT	$M_{11} = M_{22}$, $\Gamma_{11} = \Gamma_{22}$	no effect
T	no effect	$M_{21} = \exp(2i\xi)M_{12}$, $\Gamma_{21} = \exp(2i\xi)\Gamma_{12}$
CP	$M_{11} = M_{22}$, $\Gamma_{11} = \Gamma_{22}$	$M_{21} = \exp(2i\xi)M_{12}$, $\Gamma_{21} = \exp(2i\xi)\Gamma_{12}$

$$\mathrm{Im}\left(M_{12}^{*}\Gamma_{12}\right) = 0 \tag{6.22}$$

or, equivalently, that

$$|R_{12}| = |R_{21}| \tag{6.23}$$

if either T or CP symmetry hold.

It is convenient to introduce the real dimensionless T- and CP-violating parameter

$$\delta \equiv \frac{|R_{12}| - |R_{21}|}{|R_{12}| + |R_{21}|} \tag{6.24}$$

and the complex dimensionless CPT- and CP-violating parameter

$$\theta \equiv \frac{R_{22} - R_{11}}{\Delta\mu}. \tag{6.25}$$

Together with the two complex masses μ_H and μ_L, the parameters δ and θ may be taken to be the seven observables in R.

Notice that $-1 \leq \delta \leq +1$ by definition. There is however a stronger bound on $|\delta|$ when $2|M_{12}| \neq |\Gamma_{12}|$. We define the phase

$$\varpi \equiv \arg\left(M_{12}^{*}\Gamma_{12}\right), \tag{6.26}$$

which is manifestly rephasing-invariant—cf. eqn (6.11). Then,

$$\frac{\delta}{1+\delta^2} = \frac{2|M_{12}\Gamma_{12}|\sin\varpi}{4|M_{12}|^2 + |\Gamma_{12}|^2}. \tag{6.27}$$

From this equation it follows that

$$\begin{aligned}
|\delta| &\leq \frac{|\Gamma_{12}|}{2|M_{12}|}, \\
|\delta| &\leq \frac{2|M_{12}|}{|\Gamma_{12}|}.
\end{aligned} \tag{6.28}$$

Thus, $|\delta|$ can only reach 1 if $|\Gamma_{12}| = 2|M_{12}|$.

6.4 The mass eigenstates

The eigenvalue equation for R yields

$$\Delta\mu = \sqrt{4R_{12}R_{21} + (R_{22} - R_{11})^2}, \tag{6.29}$$

or equivalently

$$(\Delta\mu)^2 \left(1 - \theta^2\right) = 4R_{12}R_{21}. \tag{6.30}$$

The eigenvectors of R may be written

$$\begin{aligned}
|P_H\rangle &= p_H|P^0\rangle + q_H|\overline{P^0}\rangle, \\
|P_L\rangle &= p_L|P^0\rangle - q_L|\overline{P^0}\rangle.
\end{aligned} \tag{6.31}$$

The signs in front of q_H and q_L are just a convention, which may differ among different authors, or even from one neutral-meson system to another within the same paper. The normalization conditions are $|p_H|^2 + |q_H|^2 = |p_L|^2 + |q_L|^2 = 1$. The diagonalization of R fixes the ratios

$$\frac{q_H}{p_H} = \frac{\Delta\mu(1+\theta)}{2R_{12}} = \frac{2R_{21}}{\Delta\mu(1-\theta)},$$
$$\frac{q_L}{p_L} = \frac{\Delta\mu(1-\theta)}{2R_{12}} = \frac{2R_{21}}{\Delta\mu(1+\theta)}. \tag{6.32}$$

Their magnitudes, $|q_H/p_H|$ and $|q_L/p_L|$, are measurable. On the other hand, their phases do not have any physical significance. Indeed, under independent rephasings of the flavour eigenstates, see eqns (6.10), and of the mass eigenstates,

$$|P_H\rangle \rightarrow e^{i\gamma_H}|P_H\rangle,$$
$$|P_L\rangle \rightarrow e^{i\gamma_L}|P_L\rangle, \tag{6.33}$$

the coefficients get transformed as

$$q_H \rightarrow e^{i(\gamma_H - \bar{\gamma})}q_H,$$
$$q_L \rightarrow e^{i(\gamma_L - \bar{\gamma})}q_L,$$
$$p_H \rightarrow e^{i(\gamma_H - \gamma)}p_H,$$
$$p_L \rightarrow e^{i(\gamma_L - \gamma)}p_L. \tag{6.34}$$

Therefore, the only relevant phase is that of the ratio

$$\zeta = \frac{q_H/p_H}{q_L/p_L}. \tag{6.35}$$

We may use eqns (6.32) to show that

$$\delta = \frac{|p_L/q_L| - |q_H/p_H|}{|p_L/q_L| + |q_H/p_H|}, \tag{6.36}$$

$$\theta = \frac{q_H/p_H - q_L/p_L}{q_H/p_H + q_L/p_L}$$
$$= \frac{\zeta - 1}{\zeta + 1}. \tag{6.37}$$

Hence, CPT is violated to the extent that q_H/p_H differs from q_L/p_L, while T is violated if $|p_L/q_L| \neq |q_H/p_H|$.

Contrary to what happens with P^0 and $\overline{P^0}$, P_H and P_L have exponential evolution laws with well-defined masses and decay widths. Thus,

$$|P_H(t)\rangle = e^{-i\mu_H t}|P_H\rangle = e^{-im_H t}e^{-\Gamma_H t/2}|P_H\rangle,$$
$$|P_L(t)\rangle = e^{-i\mu_L t}|P_L\rangle = e^{-im_L t}e^{-\Gamma_L t/2}|P_L\rangle. \tag{6.38}$$

The symbol t always refers to the time measured *in the rest frame of the decaying particle*.[15] From eqns (6.38),

---

[15] As we have seen, P_H and P_L have different masses and, hence, different rest frames. Still, in the Wigner–Weisskopf approximation t is the time measured in the rest frame given by the common mass m_0 from the strong and electromagnetic interactions.

$$\left|\langle P^0|P_H(t)\rangle\right|^2 = e^{-\Gamma_H t}\left|\langle P^0|P_H\rangle\right|^2,$$

$$\left|\langle \overline{P^0}|P_H(t)\rangle\right|^2 = e^{-\Gamma_H t}\left|\langle \overline{P^0}|P_H\rangle\right|^2, \tag{6.39}$$

which displays an exponential fall-off in the probabilities to observe a P^0 or a $\overline{P^0}$, identifying Γ_H as the decay width of P_H. Similarly, Γ_L is the decay width of P_L.

The bracket $\langle P_H|P_L\rangle$ has an ill-defined phase, because the relative phase of $|P_H\rangle$ and $|P_L\rangle$ has not been fixed, see eqns (6.33). But

$$\begin{aligned}
|\langle P_H|P_L\rangle|^2 &= \frac{1 + |q_H/p_H|^2|q_L/p_L|^2 - 2\mathrm{Re}\left[(q_H/p_H)(q_L/p_L)^*\right]}{(1 + |q_H/p_H|^2)(1 + |q_L/p_L|^2)} \\
&= \frac{2|R_{12}|^2 + 2|R_{21}|^2 - |\Delta\mu|^2(1 - |\theta|^2)}{2|R_{12}|^2 + 2|R_{21}|^2 + |\Delta\mu|^2(1 + |\theta|^2)} \\
&= \frac{(1 + \delta^2)|1 - \theta^2| - (1 - \delta^2)(1 - |\theta|^2)}{(1 + \delta^2)|1 - \theta^2| + (1 - \delta^2)(1 + |\theta|^2)}.
\end{aligned} \tag{6.40}$$

Therefore, $\langle P_H|P_L\rangle = 0$ if and only if both δ and $\mathrm{Im}\,\theta$ vanish. This means that CP invariance in mixing ($\delta = \theta = 0$) implies $\langle P_H|P_L\rangle = 0$, but the converse is not true: $\langle P_H|P_L\rangle$ may vanish while $\mathrm{Re}\,\theta$ does not, with CP and CPT thus being violated.

6.5 Unitarity

At any instant t, the state $|\psi(t)\rangle = \psi_1(t)|P^0\rangle + \psi_2(t)|\overline{P^0}\rangle$ decays to a state f with a probability proportional to

$$\begin{aligned}
|\langle f|T|\psi(t)\rangle|^2 &= |\psi_1|^2|\langle f|T|P^0\rangle|^2 + |\psi_2|^2|\langle f|T|\overline{P^0}\rangle|^2 \\
&\quad + 2\mathrm{Re}\left(\psi_1\psi_2^*\langle f|T|P^0\rangle\langle f|T|\overline{P^0}\rangle^*\right).
\end{aligned} \tag{6.41}$$

We assume that the final-state kinematical factors and integrations are already incorporated in the definition of T, in such a way that the probability that ψ decays to f between instants t and $t + dt$ is given by $|\langle f|T|\psi(t)\rangle|^2\,dt$. Thus, the matrix elements of T have dimension square-root of mass.

Because of the unitarity of the evolution, the norm of the total state vector must be conserved. Thus, the decrease in the norm of $|\psi(t)\rangle$ must be compensated by the increases in the norms of all decay products:

$$\sum_f |\langle f|T|\psi(t)\rangle|^2 = -\frac{d\langle\psi(t)|\psi(t)\rangle}{dt}$$

$$= -\frac{d}{dt}\left(|\psi_1|^2 + |\psi_2|^2\right). \tag{6.42}$$

Remembering eqn (6.9), we see that, as ψ_1 and ψ_2 are arbitrary,

$$\Gamma_{11} = \sum_f |\langle f|T|P^0\rangle|^2, \tag{6.43}$$

$$\Gamma_{22} = \sum_f |\langle f|T|\overline{P^0}\rangle|^2, \tag{6.44}$$

$$\Gamma_{12} = \sum_f \langle f|T|P^0\rangle^* \langle f|T|\overline{P^0}\rangle. \tag{6.45}$$

Equations (6.43)–(6.45) are the unitarity relations (Bell and Steinberger 1966). Notice their similarity with eqn (6.8).

Let us define

$$z \equiv \frac{|\Gamma_{12}|}{\Gamma}. \tag{6.46}$$

From the second eqn (6.14) we know that $2\Gamma = \Gamma_{11} + \Gamma_{22}$. Therefore,

$$z = \frac{2\left|\sum_f \langle f|T|P^0\rangle^* \langle f|T|\overline{P^0}\rangle\right|}{\sum_f \left(|\langle f|T|P^0\rangle|^2 + |\langle f|T|\overline{P^0}\rangle|^2\right)}. \tag{6.47}$$

It follows that $z \le 1$.

We may derive other unitarity relations, equivalent to the above ones, by repeating the reasoning in a slightly different form (Bell and Steinberger 1966). We assume that at instant $t = 0$ the state $|\psi(t)\rangle$ was equal to $\chi_H|P_H\rangle + \chi_L|P_L\rangle$. Thus,

$$|\psi(t)\rangle = \chi_H e^{-i\mu_H t}|P_H\rangle + \chi_L e^{-i\mu_L t}|P_L\rangle.$$

The norm of this state is

$$\langle \psi(t)|\psi(t)\rangle = |\chi_H|^2 e^{-\Gamma_H t} + |\chi_L|^2 e^{-\Gamma_L t} + 2\mathrm{Re}\left[\chi_H^* \chi_L e^{(-\Gamma + i\Delta m)t}\langle P_H|P_L\rangle\right].$$

The rate of decrease of the norm at instant $t = 0$ is

$$-\frac{d\langle\psi(t)|\psi(t)\rangle}{dt}\bigg|_{t=0} = \Gamma_H|\chi_H|^2 + \Gamma_L|\chi_L|^2 + 2\mathrm{Re}\left[(\Gamma - i\Delta m)\chi_H^* \chi_L\langle P_H|P_L\rangle\right].$$

This rate of decrease must equal

$$\sum_f |\langle f|T|\psi(0)\rangle|^2 = \sum_f |\chi_H\langle f|T|P_H\rangle + \chi_L\langle f|T|P_L\rangle|^2.$$

Therefore, as χ_H and χ_L are arbitrary,

$$\Gamma_H = \sum_f |\langle f|T|P_H\rangle|^2 = \sum_f \Gamma_{fH}, \tag{6.48}$$

$$\Gamma_L = \sum_f |\langle f|T|P_L\rangle|^2 = \sum_f \Gamma_{fL}, \tag{6.49}$$

$$(\Gamma - i\Delta m)\langle P_H|P_L\rangle = \sum_f \langle f|T|P_H\rangle^* \langle f|T|P_L\rangle. \tag{6.50}$$

We have introduced here the partial decay widths of P_H and P_L to the channel f, which we have named Γ_{fH} and Γ_{fL}, respectively.

Defining

$$\phi_f \equiv \arg \frac{\langle f|T|P_H\rangle}{\langle f|T|P_L\rangle}, \tag{6.51}$$

one may write eqn (6.50) as

$$(\Gamma + i\Delta m)\,\langle P_L|P_H\rangle = \sum_f \sqrt{\Gamma_{fH}\Gamma_{fL}}\, e^{i\phi_f}. \tag{6.52}$$

From this unitarity relation we derive the Bell–Steinberger (1966) inequalities

$$|\langle P_L|P_H\rangle|^2 \le \frac{\left(\sum_f \sqrt{\Gamma_{fH}\Gamma_{fL}}\right)^2}{\Gamma^2 + (\Delta m)^2} \tag{6.53}$$

$$\le \frac{\Gamma_H \Gamma_L}{\Gamma^2 + (\Delta m)^2} = \frac{1-y^2}{1+x^2}. \tag{6.54}$$

This upper bound on the overlap of $|P_H\rangle$ and $|P_L\rangle$ is called the unitarity bound. The overlap is a CP-violating quantity, as we know from the previous section. Still, remarkably, it is possible to put an upper bound on it exclusively from the knowledge of the non-CP-violating observables x and y.

6.6 CPT-invariant case

From now on we shall, unless otherwise explicitly stated, assume CPT not to be violated. If CPT is a good symmetry the matrix $\boldsymbol{R}$, instead of having seven real observables, only has five, because θ vanishes. From eqns (6.32) it is clear that, when CPT invariance is assumed, $q_H/p_H = q_L/p_L$. As $|p_H|^2 + |q_H|^2 = |p_L|^2 + |q_L|^2 = 1$, we conclude that p_H and p_L then have the same modulus. It is convenient to fix the relative phase of $|P_H\rangle$ and $|P_L\rangle$ in such a way that p_H and p_L also have the same phase and are therefore equal. Once this is done, q_H and q_L become equal too. We thus write simply[16]

$$\begin{aligned}|P_H\rangle &= p|P^0\rangle + q|\overline{P^0}\rangle,\\ |P_L\rangle &= p|P^0\rangle - q|\overline{P^0}\rangle.\end{aligned} \tag{6.55}$$

Equivalently,

$$\begin{aligned}|P^0\rangle &= \frac{1}{2p}\left(|P_H\rangle + |P_L\rangle\right),\\ |\overline{P^0}\rangle &= \frac{1}{2q}\left(|P_H\rangle - |P_L\rangle\right).\end{aligned} \tag{6.56}$$

It is important to stress that, once CPT is violated, this phase convention for the relative phase of $|P_H\rangle$ and $|P_L\rangle$ becomes ill-defined. When $q_H/p_H \ne q_L/p_L$,

---

[16]Note that some authors use $|P_H\rangle = p|P^0\rangle - q|\overline{P^0}\rangle$ and $|P_L\rangle = p|P^0\rangle + q|\overline{P^0}\rangle$ instead of eqns (6.55). This is a simple matter of convention, but it leads to formulas that differ from ours by $q \to -q$.

fixing the phases of p_H and p_L to be equal does not lead to q_H and q_L with the same phase, and vice versa. Therefore, in the CPT-violating case there is no advantage in making any particular phase convention for p_H/p_L or for q_H/q_L.

It is important to note that, if we interchange $|P_H\rangle$ and $|P_L\rangle$ in eqns (6.55), we have $q/p \to -q/p$, together with $\Delta m \to -\Delta m$ and $\Delta\Gamma \to -\Delta\Gamma$. Thus, the sign of q/p, just as that of $\Delta\Gamma$, acquires a meaning only relative to the sign of Δm. In this book we choose $\Delta m > 0$ by convention.

We have

$$\Delta\mu = \Delta m - \tfrac{i}{2}\Delta\Gamma = 2\sqrt{R_{12}R_{21}}. \tag{6.57}$$

The relative signs of $\sqrt{R_{12}}$ and $\sqrt{R_{21}}$ are chosen so that $\Delta m > 0$. One derives from eqns (6.32) that

$$\frac{q}{p} = \frac{\Delta\mu}{2M_{12} - i\Gamma_{12}} = \frac{2M_{12}^* - i\Gamma_{12}^*}{\Delta\mu} = \sqrt{\frac{2M_{12}^* - i\Gamma_{12}^*}{2M_{12} - i\Gamma_{12}}}. \tag{6.58}$$

By definition $|p|^2 + |q|^2 = 1$. Notice that q, p, and q/p are not invariant under a rephasing of the kets and, therefore, their phases cannot be measured.

Once we have assumed the phase convention for the relative phase of $|P_H\rangle$ and $|P_L\rangle$ embodied in eqns (6.55), the bracket $\langle P_H|P_L\rangle$ becomes real. Indeed, we have, for the T- and CP-violating parameter δ introduced in eqn (6.24),

$$\delta = |p|^2 - |q|^2 = \langle P_L|P_H\rangle. \tag{6.59}$$

Thus,

$$|p|^2 = \frac{1+\delta}{2},$$
$$|q|^2 = \frac{1-\delta}{2}. \tag{6.60}$$

Squaring eqn (6.57) and separating the real and imaginary parts we obtain

$$(\Delta m)^2 - \tfrac{1}{4}(\Delta\Gamma)^2 = 4|M_{12}|^2 - |\Gamma_{12}|^2, \tag{6.61}$$
$$(\Delta m)(\Delta\Gamma) = 4\text{Re}\left(M_{12}^*\Gamma_{12}\right). \tag{6.62}$$

On the other hand, from eqn (6.24),

$$\delta = \frac{2\text{Im}\left(M_{12}^*\Gamma_{12}\right)}{(\Delta m)^2 + |\Gamma_{12}|^2}. \tag{6.63}$$

It is easy to invert the system of eqns (6.61)–(6.63) to find

$$|M_{12}|^2 = \frac{4(\Delta m)^2 + \delta^2(\Delta\Gamma)^2}{16(1-\delta^2)}, \tag{6.64}$$

$$|\Gamma_{12}|^2 = \frac{(\Delta\Gamma)^2 + 4\delta^2(\Delta m)^2}{4(1-\delta^2)}. \tag{6.65}$$

From eqn (6.65) it follows that $|\Gamma_{12}|^2 \geq (\Delta\Gamma)^2/4$. Therefore, $z^2 \geq y^2$.

Equations (6.64) and (6.65) may be inverted:

$$(\Delta m)^2 = \frac{4|M_{12}|^2 - \delta^2|\Gamma_{12}|^2}{1 + \delta^2}, \tag{6.66}$$

$$(\Delta \Gamma)^2 = \frac{4|\Gamma_{12}|^2 - 16\delta^2|M_{12}|^2}{1 + \delta^2}. \tag{6.67}$$

It is useful to derive the identities

$$\frac{q}{p}\Gamma_{12} = \frac{y + i\delta x}{1 + \delta}\Gamma, \tag{6.68}$$

$$\frac{q}{p}M_{12} = \frac{x - i\delta y}{2(1 + \delta)}\Gamma, \tag{6.69}$$

and

$$\frac{q}{p} = \sqrt{\frac{1 - \delta}{1 + \delta}} \exp\left(i \arg \Gamma_{12}^*\right) \frac{-u + i\delta}{\sqrt{u^2 + \delta^2}}, \tag{6.70}$$

$$= \sqrt{\frac{1 - \delta}{1 + \delta}} \exp\left(i \arg M_{12}^*\right) \frac{1 + i\delta u}{\sqrt{1 + \delta^2 u^2}}. \tag{6.71}$$

We have used the quantities x, y, and u defined in eqns (6.15) and (6.16). The square roots in eqns (6.70) and (6.71) are positive by definition. Indeed, in deriving those equations we have used the convention that Δm is positive.

From eqn (6.65) we derive

$$\delta^2 = \frac{z^2 - y^2}{z^2 + x^2}. \tag{6.72}$$

Therefore, δ^2 is a monotonically increasing function of z^2. The bound $z^2 \leq 1$ corresponds to the unitarity bound of eqn (6.54),

$$\delta^2 \leq \frac{1 - y^2}{1 + x^2}. \tag{6.73}$$

Sometimes eqn (6.45) enables us to place a bound on z way beyond $z^2 \leq 1$. If this can be done, we obtain from eqn (6.72) a bound on $|\delta|$ better than the one in eqn (6.73).

6.7   The case of CP conservation

From Table 6.1 we see that, if CP is conserved, then

$$M_{12}^* = e^{2i\xi} M_{12}, \tag{6.74}$$
$$\Gamma_{12}^* = e^{2i\xi} \Gamma_{12}.$$

Then, from eqn (6.58),

$$\frac{q}{p} = \pm e^{i\xi}. \tag{6.75}$$

The phase ξ appears in the CP transformation in eqns (6.17). Therefore, CP invariance implies

$$\begin{aligned} CP|P_H\rangle &= \pm|P_H\rangle, \\ CP|P_L\rangle &= \mp|P_L\rangle, \end{aligned} \tag{6.76}$$

i.e., $|P_H\rangle = |P_\pm\rangle$ and $|P_L\rangle = |P_\mp\rangle$. This means that the eigenstates of the Hamiltonian are also eigenstates of CP and have opposite eigenvalues.

The sign in eqns (6.75) and (6.76) must be taken from experiment. Only experiment can determine whether the heaviest eigenstate is mainly CP-even and the lightest eigenstate is mainly CP-odd, or the opposite situation occurs. This applies for each neutral-meson system separately.

6.8 The reciprocal basis

An important consequence of CP conservation is the unitarity of the transformation which relates the flavour eigenstates with the mass eigenstates. Indeed, when CP is conserved the matrix R, although not Hermitian, commutes with its Hermitian conjugate. Matrices R satisfying $\left[R, R^\dagger\right] = 0$ are called 'normal' matrices. It can be shown that

$$\left[R, R^\dagger\right] = 0 \Leftrightarrow [M, \Gamma] = 0 \Leftrightarrow \delta = \text{Im}\,\theta = 0 \Leftrightarrow \langle P_H|P_L\rangle = 0. \tag{6.77}$$

Moreover, there is a theorem stating that a matrix is normal if and only if it can be diagonalized by a unitary transformation.

Notice that the crucial point is not whether the matrix R is Hermitian or not but, rather, whether R commutes with its Hermitian conjugate or not. Indeed, if both M and Γ are non-zero then R is not Hermitian; still, R can be diagonalized by a unitary transformation as long as $[M, \Gamma] = 0$.

On the other hand, we can see already from eqn (6.40) that $|P_H\rangle$ and $|P_L\rangle$ are not orthogonal if either $\delta \neq 0$ or $\text{Im}\,\theta \neq 0$. This is due to the fact that the transformation in eqns (6.31), given by the matrix

$$X = \begin{pmatrix} p_H & p_L \\ q_H & -q_L \end{pmatrix}, \tag{6.78}$$

is not unitary. In order to see this, consider for simplicity the CPT-invariant case:

$$X = \begin{pmatrix} p & p \\ q & -q \end{pmatrix}. \tag{6.79}$$

Then,

$$X^\dagger X = \begin{pmatrix} 1 & \delta \\ \delta & 1 \end{pmatrix}, \quad XX^\dagger = \begin{pmatrix} 1+\delta & 0 \\ 0 & 1-\delta \end{pmatrix}, \tag{6.80}$$

where use was made of eqns (6.60).

In fact, if R is not normal then it is not diagonalized by a unitary transformation $X^\dagger R X$, but rather by a general similarity transformation:

$$X^{-1} R X = \text{diag}\left(\mu_H, \mu_L\right). \tag{6.81}$$

Here,

$$X^{-1} = \frac{1}{p_H q_L + p_L q_H} \begin{pmatrix} q_L & p_L \\ q_H & -p_H \end{pmatrix} \tag{6.82}$$

is the matrix inverse of X. Since X is not unitary, the left-eigenvectors of R, $\langle \tilde{P}_H|$ and $\langle \tilde{P}_L|$, are not the Hermitian conjugates of its right-eigenvectors $|P_H\rangle$ and $|P_L\rangle$. Indeed, $\langle \tilde{P}_H|$ and $\langle \tilde{P}_L|$ are obtained from X^{-1}:

$$\begin{aligned}
\langle \tilde{P}_H| &= \frac{1}{p_H q_L + p_L q_H} \left(q_L \langle P^0| + p_L \langle \overline{P^0}| \right), \\
\langle \tilde{P}_L| &= \frac{1}{p_H q_L + p_L q_H} \left(q_H \langle P^0| - p_H \langle \overline{P^0}| \right).
\end{aligned} \tag{6.83}$$

The vectors $|\tilde{P}_H\rangle$ and $|\tilde{P}_L\rangle$ form the 'reciprocal basis' (Sachs 1963, 1964; Enz and Lewis 1965; see also Alvarez-Gaumé et al. 1998) of the basis given by $|P_H\rangle$ and $|P_L\rangle$. The reciprocal basis may alternatively be defined through the equations

$$\begin{aligned}
\langle \tilde{P}_H|P_L\rangle &= \langle \tilde{P}_L|P_H\rangle = 0, \\
\langle \tilde{P}_H|P_H\rangle &= \langle \tilde{P}_L|P_L\rangle = 1.
\end{aligned} \tag{6.84}$$

The need for a reciprocal basis is common to all quantum-mechanical problems in which the effective Hamiltonian yields a matrix which is not normal—see for instance Löwdin (1998).

Equation (6.81) means that

$$\begin{aligned}
R &= \mu_H |P_H\rangle\langle \tilde{P}_H| + \mu_L |P_L\rangle\langle \tilde{P}_L| \\
&= \frac{1}{p_H q_L + p_L q_H} \left[\mu_H \begin{pmatrix} p_H \\ q_H \end{pmatrix} \begin{pmatrix} q_L & p_L \end{pmatrix} + \mu_L \begin{pmatrix} p_L \\ -q_L \end{pmatrix} \begin{pmatrix} q_H & -p_H \end{pmatrix} \right].
\end{aligned} \tag{6.85}$$

Moreover, it follows from eqns (6.84) that $|P_H\rangle\langle \tilde{P}_H|$ and $|P_L\rangle\langle \tilde{P}_L|$ are projection operators, and in particular

$$|P_H\rangle\langle \tilde{P}_H| + |P_L\rangle\langle \tilde{P}_L| = 1, \tag{6.86}$$

i.e., $|P_H\rangle\langle \tilde{P}_H| + |P_L\rangle\langle \tilde{P}_L|$ constitutes a partition of unity, just as $|P^0\rangle\langle P^0| + |\overline{P^0}\rangle\langle \overline{P^0}|$. As a result, the time-evolution operator for the neutral-meson system is

$$\exp\left(-i\,R\,t\right) = e^{-i\mu_H t} |P_H\rangle\langle \tilde{P}_H| + e^{-i\mu_L t} |P_L\rangle\langle \tilde{P}_L|. \tag{6.87}$$

As an application, let us consider the decay chain $i \to X\{P_H, P_L\} \to Xf$ in which an initial state i decays into an intermediate state XP_H or XP_L, which after a time t decays into the final state Xf. The complete amplitude for this process involves the amplitude for the initial decay into XP_H or XP_L, the time-evolution amplitude for this state, given by eqn (6.87), and finally the amplitude for the decay into Xf. Suppressing the reference to X, we find

$$A\left(i \to P_{H,L} \to f\right) = \langle f|T|P_H\rangle\, e^{-i\mu_H t}\, \langle \tilde{P}_H|T|i\rangle + \langle f|T|P_L\rangle\, e^{-i\mu_L t}\, \langle \tilde{P}_L|T|i\rangle. \tag{6.88}$$

This is an exact expression. However, sometimes it is possible to choose a final state f and to set the experimental conditions in such a way as to maximize the importance of $i \to XP_H \to Xf$ relative to $i \to XP_L \to Xf$. In that case we may make the approximation

$$\begin{aligned}
A\left(i \to P_{H,L} \to f\right) &\approx A\left(i \to P_H \to f\right)\\
&= \langle f|T|P_H\rangle\, e^{-i\mu_H t}\, \langle \tilde{P}_H|T|i\rangle\\
&= \langle f|T|P_H\rangle\, e^{-i\mu_H t}\, \left[\langle \tilde{P}_H|P^0\rangle\langle P^0|T|i\rangle + \langle \tilde{P}_H|\overline{P^0}\rangle\langle \overline{P^0}|T|i\rangle\right],
\end{aligned} \tag{6.89}$$

where we have used the partition of unity $1 = |P^0\rangle\langle P^0| + |\overline{P^0}\rangle\langle \overline{P^0}|$ to derive the last line. When one uses the approximation in eqn (6.89), one talks about 'the decay $i \to XP_H$',[17] and writes

$$\begin{aligned}
A\left(i \to XP_H\right) &= \langle \tilde{P}_H|P^0\rangle\, A(i \to XP^0) + \langle \tilde{P}_H|\overline{P^0}\rangle\, A(i \to X\overline{P^0})\\
&= \tfrac{1}{2}\left[p^{-1}A(i \to XP^0) + q^{-1}A(i \to X\overline{P^0})\right],
\end{aligned} \tag{6.90}$$

where, in the last line, we have assumed the CPT-invariant case:

$$\begin{aligned}
\langle \tilde{P}_H| &= \tfrac{1}{2}\left(p^{-1}\langle P^0| + q^{-1}\langle \overline{P^0}|\right),\\
\langle \tilde{P}_L| &= \tfrac{1}{2}\left(p^{-1}\langle P^0| - q^{-1}\langle \overline{P^0}|\right).
\end{aligned} \tag{6.91}$$

Therefore, the ratio of the two component amplitudes in eqn (6.90) is given by $q^{-1}/p^{-1} = p/q$, and not by q^*/p^*—as would have been the case if we had used $\langle P_H|$ instead of $\langle \tilde{P}_H|$. The difference between q^{-1}/p^{-1} and q^*/p^* only disappears in the limit $|q/p| = 1$. This will be important in § 33.1.2 where we study the decay $B_d^0 \to J/\psi K_S$.

---

[17]Nevertheless, strictly speaking, it is eqn (6.88) which expresses the correct way to think about decays into neutral-meson eigenstates (Enz and Lewis 1965; Kayser, private communication). The point is that, since CP is violated, there is no final state f that can be obtained only from P_H and not from P_L. There will always be a non-zero amplitude for the decay path $i \to XP_L \to Xf$. We shall come back to this point when we discuss cascade decays in Chapter 34.

7

NEUTRAL-MESON SYSTEMS: DECAYS

7.1 The parameters λ_f

Consider the decays of P^0 and $\overline{P^0}$ into a final state f. Phenomenologically, there are two independent decay amplitudes,[18]

$$
\begin{aligned}
A_f &\equiv \langle f|T|P^0 \rangle, \\
\bar{A}_f &\equiv \langle f|T|\overline{P^0} \rangle,
\end{aligned}
\tag{7.1}
$$

entering in the description of those decays. Physics must be invariant under the phase redefinitions in eqns (6.10) and also under

$$
|f\rangle \to e^{i\gamma_f}|f\rangle.
\tag{7.2}
$$

The phases γ, $\bar{\gamma}$, and γ_f are independent. Under these rephasings,

$$
\begin{aligned}
A_f &\to e^{i(\gamma-\gamma_f)}A_f, \\
\bar{A}_f &\to e^{i(\bar{\gamma}-\gamma_f)}\bar{A}_f, \\
\frac{q}{p} &\to e^{i(\gamma-\bar{\gamma})}\frac{q}{p}.
\end{aligned}
\tag{7.3}
$$

We see that the quantities which are rephasing-invariant, and therefore have a chance to be observable, are the magnitudes

$$
\left|\frac{q}{p}\right|, \ |A_f|, \ |\bar{A}_f|,
\tag{7.4}
$$

and the complex parameter

$$
\lambda_f \equiv \frac{q}{p}\frac{\bar{A}_f}{A_f}.
\tag{7.5}
$$

From the quantities in eqns (7.4) and (7.5) only four real numbers are independent: three moduli and one phase. However, in the following discussion the question of independence is not important.

It is sometimes convenient to use

$$
\bar{\lambda}_f \equiv \frac{1}{\lambda_f}.
\tag{7.6}
$$

Some authors use different definitions for $\bar{\lambda}$. We use the notation in eqn (7.6) for all decay modes f, avoiding any confusion.

---

[18] We shall implicitly assume the squared amplitudes to incorporate the relevant phase-space factors and integrations. Hence, all $|A|^2$ have mass dimension.

If there were CPT violation in the mixing, we should construct two parameters λ_f, namely

$$\lambda_f^H \equiv \frac{q_H}{p_H} \frac{\bar{A}_f}{A_f},$$

$$\lambda_f^L \equiv \frac{q_L}{p_L} \frac{\bar{A}_f}{A_f}, \tag{7.7}$$

which are invariant under rephasings of both the flavour and the mass eigenstates. Obviously,

$$\frac{\lambda_f^H}{\lambda_f^L} = \frac{q_H/p_H}{q_L/p_L} = \zeta, \tag{7.8}$$

so that λ_f^H and λ_f^L are equal in the CPT-invariant case.

It is also useful to consider the decay amplitudes for the mass eigenstates,

$$A_f^H \equiv \langle f|T|P_H \rangle = p_H A_f + q_H \bar{A}_f,$$
$$A_f^L \equiv \langle f|T|P_L \rangle = p_L A_f - q_L \bar{A}_f. \tag{7.9}$$

Then, the relevant invariant quantities are

$$\frac{p_L}{p_H} \frac{A_f^H}{A_f^L} = \frac{1 + \lambda_f^H}{1 - \lambda_f^L}. \tag{7.10}$$

Alternatively, one may use

$$\frac{q_L}{q_H} \frac{A_f^H}{A_f^L} = \frac{1 + \bar{\lambda}_f^H}{-1 + \bar{\lambda}_f^L}. \tag{7.11}$$

The quantities in eqns (7.10) and (7.11) are related through ζ. They are equal in the CPT-invariant case.

7.2 CP-violating observables

7.2.1 *Final states which are not CP eigenstates*

We consider the decays of P^0 and $\overline{P^0}$ to f and to the CP-conjugate decay channel $\bar{f}$. We assume f and $\bar{f}$ to be distinct.

If CP is conserved in the mixing, we know from eqns (6.74) and (6.75) that there is a phase ξ such that

$$M_{12}^* = e^{2i\xi} M_{12},$$
$$\Gamma_{12}^* = e^{2i\xi} \Gamma_{12},$$
$$\frac{q^2}{p^2} = e^{2i\xi}. \tag{7.12}$$

From eqns (7.12) follows the condition of CP conservation in the mixing:

$$\left| \frac{q}{p} \right| = 1 \Leftrightarrow \sin \varpi = 0, \tag{7.13}$$

as we saw in the previous chapter.

CP transforms

$$CP|f\rangle = e^{i\xi_f}|\bar{f}\rangle,$$
$$CP|\bar{f}\rangle = e^{-i\xi_f}|f\rangle. \tag{7.14}$$

Together with eqn (6.17), this leads to the CP-invariance conditions for the decay amplitudes

$$\bar{A}_{\bar{f}} = e^{i(\xi_f - \xi)}A_f,$$
$$A_{\bar{f}} = e^{i(\xi_f + \xi)}\bar{A}_f. \tag{7.15}$$

From eqns (7.15), after elimination of the arbitrary phases ξ and ξ_f, the conditions for CP conservation in the decay amplitudes follow:

$$|A_f| = |\bar{A}_{\bar{f}}|,$$
$$|A_{\bar{f}}| = |\bar{A}_f|. \tag{7.16}$$

These conditions are just what one would expect: the probabilities of the decays of P^0 to f and of $\overline{P^0}$ to $\bar{f}$ must be equal. CP-invariance conditions analogous to eqns (7.16) also hold in the case of the decays of charged particles, which do not mix.

From eqns (7.15) one derives

$$A_f A_{\bar{f}} = e^{2i\xi}\bar{A}_{\bar{f}}\bar{A}_f. \tag{7.17}$$

We may combine eqn (7.17) with eqns (7.12) and obtain extra conditions for CP conservation, involving the phases of combinations of mixing matrix elements and decay matrix elements:

$$\arg\left(M_{12}^2 A_f \bar{A}_{\bar{f}}^* A_{\bar{f}} \bar{A}_{\bar{f}}^*\right) = 0, \tag{7.18}$$

$$\arg\left(\Gamma_{12}^2 A_f \bar{A}_{\bar{f}}^* A_{\bar{f}} \bar{A}_{\bar{f}}^*\right) = 0, \tag{7.19}$$

$$\arg\left(\frac{p^2}{q^2} A_f \bar{A}_{\bar{f}}^* A_{\bar{f}} \bar{A}_{\bar{f}}^*\right) = 0. \tag{7.20}$$

In particular, we find that CP conservation implies

$$\lambda_f = \frac{1}{\lambda_{\bar{f}}}. \tag{7.21}$$

The moduli are equal because of eqns (7.13) and (7.16). The phases are equal because of eqn (7.20).

7.2.2 Classification of CP violation

There may be three different types of CP violation:

- CP violation in the mixing (this is called by some authors 'indirect CP violation'), when eqn (7.13) does not hold;
- CP violation in the decay amplitudes (which is usually called 'direct CP violation'), when eqns (7.16) do not hold;

- CP violation in a phase mismatch between the mixing parameters and the decay amplitudes (we propose to call this 'interference CP violation', although other authors use different names), when eqn (7.20) does not hold.

Direct CP violation is not specific to systems of neutral mesons. If both direct and indirect CP violation are absent, then $|\lambda_f| = 1/|\lambda_{\bar{f}}|$; still, interference CP violation may be present when

$$\arg \lambda_f + \arg \lambda_{\bar{f}} \neq 0. \qquad (7.22)$$

Since we are free to rephase all kets and bras, CP violation may only arise from the clash between two phases, and never from only one phase. Indirect CP violation arises from a clash between the phases of M_{12} and of Γ_{12}. Direct CP violation arises from a clash between the phases of two interfering decay amplitudes in the total decay amplitude—see Chapter 5. Interference CP violation arises from a clash between the phase of q/p and the phases of the decay amplitudes.

The definition that we adopt for interference CP violation, in eqn (7.22), is rather arbitrary. That definition follows from eqn (7.20). However, we might just as well have adopted either eqn (7.18) or eqn (7.19) as the definition for interference CP conservation. Unfortunately, as soon as there is CP violation in the mixing the phases of M_{12}, of Γ_{12}, and of p/q are all different; thus, mixing CP violation implies the existence of interference CP violation for at least two of the three alternative definitions.

7.2.3   Final states which are CP eigenstates

Let us now consider the case of decays to a CP eigenstate, i.e., the case $\bar{f} = f$. Then, the two eqns (7.16) become identical, i.e., we only have one direct-CP-invariance condition:

$$|A_f| = |\bar{A}_f|. \qquad (7.23)$$

Equation (7.22) becomes $2 \arg \lambda_f \neq 0$. Thus, the condition for the absence of interference CP violation is

$$\operatorname{Im} \lambda_f = 0. \qquad (7.24)$$

From eqns (7.13), (7.23), and (7.24), we conclude that CP invariance requires

$$\lambda_f = \pm 1 \qquad (7.25)$$

if f is a CP eigenstate.

7.2.4   Different decay channels

At this point, we should remember Chapter 5 and note that, if we consider two different decay modes f and g (with $g \neq f$ and $g \neq \bar{f}$), the conditions for direct CP invariance involve not only the moduli of the decay amplitudes, they also involve their phases. In particular, CP invariance requires the quantity in eqn (5.11) to vanish, and therefore

$$A_f A_{\bar{f}} \bar{A}_g \bar{A}_{\bar{g}} = A_g A_{\bar{g}} \bar{A}_f \bar{A}_{\bar{f}}. \tag{7.26}$$

Hence,

$$\lambda_f \lambda_{\bar{f}} = \lambda_g \lambda_{\bar{g}}. \tag{7.27}$$

In the case $\bar{f} = f$ and $\bar{g} = g$, eqn (5.14) yields

$$\lambda_g = \eta_f \eta_g \lambda_f. \tag{7.28}$$

If f and g have the same CP-parity λ_f should be equal to λ_g; if they have opposite CP-parities, then CP conservation implies $\lambda_f = -\lambda_g$. In any case, λ_f must be either $+1$ or -1, as stated in eqn (7.25).

In the next chapter we shall see that, in the neutral-kaon decays to two pions, there is a CP-violating parameter ϵ' which measures a violation of eqn (7.28), in the sense that two parameters λ are different when $\epsilon' \neq 0$—see in particular eqn (8.89).

7.3   The superweak theory

The superweak theory of Wolfenstein (1964) was an attempt at a theoretical explanation of the CP violation observed that same year in the neutral-kaon system (Christenson et al. 1964). Amazingly, during more than thirty years of hard experimental work, that theory seemed to be able to account for all observed CP-violating phenomena. The situation was changed by the recent result of the KTeV Collaboration, which indicated a non-zero value for ϵ'/ϵ, thus confirming an earlier result of the NA31 Collaboration (1993).

The superweak theory is a purely phenomenological assumption, and it is difficult to ground it on a complete gauge theory of the electroweak interactions. However, a few gauge models in which CP violation seems to have effective superweak features are available in the literature (Lavoura 1994; Bowser-Chao et al. 1998; Georgi and Glashow 1998).

The superweak theory was originally assumed for the neutral-kaon system only. We may however extend it to any other neutral-meson system. Indeed, the superweak theory may constitute a good gauge to evaluate future observations of CP violation in, for instance, the B_d^0-$\overline{B_d^0}$ system. Exploratory attempts at such a comparison have been done (Winstein 1992; Soares and Wolfenstein 1992; Winstein and Wolfenstein 1993).

7.3.1   Basic assumption

The basic assumption of the superweak theory for the decays of the P^0-$\overline{P^0}$ system is the following: there is no CP violation in the decay amplitudes. This means that there is a phase ξ and, for each pair of CP-conjugate final states f and $\bar{f}$, there is a phase ξ_f, such that the decay amplitudes satisfy eqns (7.15). These conditions imply the absence of direct CP violation. In particular, eqns (7.16) follow; also, for any other pair of CP-conjugate decay channels g and $\bar{g}$, eqn (7.27) holds.[19]

If one considers instead a CP-eigenstate decay channel f, with CP-parity η_f, one has

$$\lambda_f = \frac{q}{p} e^{-i\xi} \eta_f. \tag{7.29}$$

Thus, if the superweak theory is valid then eqn (7.28) holds. The parameters λ are equal for the decays into two CP eigenstates with the same CP-parity. This means that *all interference CP violation is basically identical in the superweak theory.*

7.3.2 Γ_{12}

We recall that, according to eqn (6.45),

$$\Gamma_{12} = \sum_f A_f^* \bar{A}_f, \tag{7.30}$$

where the sum extends over all decay modes f; in particular, the sum includes all pairs of CP-conjugate decay modes. For instance,

$$\Gamma_{12} = A_f^* \bar{A}_f + A_{\bar{f}}^* \bar{A}_{\bar{f}} + A_g^* \bar{A}_g + A_{\bar{g}}^* \bar{A}_{\bar{g}} + \cdots.$$

From the CP-invariance conditions in eqns (7.15) it follows that

$$A_f \bar{A}_f^* = e^{2i\xi} \bar{A}_{\bar{f}} A_{\bar{f}}^*. \tag{7.31}$$

Therefore,

$$\Gamma_{12} = e^{-2i\xi} \Gamma_{12}^*. \tag{7.32}$$

We then obtain

$$\Gamma_{12} A_f \bar{A}_f^* = \Gamma_{12}^* A_{\bar{f}}^* \bar{A}_{\bar{f}}, \tag{7.33}$$

which is valid for any f and $\bar{f}$.

7.3.3   *Source of CP violation*

From the previous subsections we gather that there is no direct CP violation in the superweak theory. Besides, there is no CP violation from the clash between the phase of Γ_{12} and the phases of the decay amplitudes. It follows that in the superweak theory the only source of CP violation is the clash between the phase of M_{12} and the phases of the decay amplitudes or, equivalently, of Γ_{12}.

The basic and original idea of the superweak theory was that the decay amplitudes originate from the weak interaction, which is assumed to be CP-conserving. On the other hand, a new, much weaker—'superweak'—interaction is assumed to exist, which only contributes to M_{12}. This superweak contribution to M_{12} has a phase mismatch with the decay amplitudes and with Γ_{12}.

---

[19] At this juncture it is important to call attention to the fact that some authors use the term 'direct CP violation' to mean any CP-violating effect which disproves the superweak theory. For those authors, in particular Winstein and Wolfenstein (1993), direct CP violation exists not only when relations such as eqns (7.16) and (7.26) are violated; it exists whenever the superweak theory is violated.

7.3.4 *Other consequences*

Let us consider some other consequences of the superweak theory for the decays of the P^0-$\overline{P^0}$ system to a CP eigenstate f. From eqns (7.29), (6.70), and (7.32) one derives

$$\lambda_f = \pm\sqrt{\frac{1-\delta}{1+\delta}}\frac{-u+i\delta}{\sqrt{u^2+\delta^2}}. \tag{7.34}$$

Now define

$$\begin{aligned}
\epsilon_f &\equiv \frac{\langle f|T|P_H\rangle}{\langle f|T|P_L\rangle} \\
&= \frac{pA_f+q\bar{A}_f}{pA_f-q\bar{A}_f} \\
&= \frac{1+\lambda_f}{1-\lambda_f}. \tag{7.35}
\end{aligned}$$

Then, from eqn (7.34),

$$\begin{aligned}
\frac{2\mathrm{Re}\,\epsilon_f}{1+|\epsilon_f|^2} &= \frac{1-|\lambda_f|^2}{1+|\lambda_f|^2} \\
&= \delta \tag{7.36}
\end{aligned}$$

$$\begin{aligned}
\frac{\mathrm{Im}\,\epsilon_f}{\mathrm{Re}\,\epsilon_f} &= \frac{2\mathrm{Im}\,\lambda_f}{1-|\lambda_f|^2} \\
&= \pm\sqrt{\frac{1-\delta^2}{u^2+\delta^2}}. \tag{7.37}
\end{aligned}$$

If for simplicity we assume that δ is very small, so that δ^2 may be neglected, then either

$$|\epsilon_f| \approx \frac{|\delta|}{2}\sqrt{1+\frac{1}{u^2}}, \tag{7.38}$$

or

$$|\epsilon_f| \approx \frac{2}{|\delta|}\sqrt{\frac{u^2}{1+u^2}}, \tag{7.39}$$

i.e., either P_H decays to f much more often than P_L, or vice versa, depending on the CP parity of f. In any case, the ratio of one partial decay width to the other one is always the same. This is an important prediction of the superweak theory. For the phase of ϵ_f we have, in the same approximation,

$$\frac{\mathrm{Im}\,\epsilon_f}{\mathrm{Re}\,\epsilon_f} \approx \pm\frac{1}{|u|}. \tag{7.40}$$

Equations (7.38) and (7.40) are well verified in the case of the CP-violating parameter ϵ of the two-pion decays of the K^0-$\overline{K^0}$ system, as we shall see in the next chapter.

7.4 Main conclusions

We have made the following classification for CP violation in neutral-meson systems:

- CP violation in the mixing (indirect CP violation). This occurs whenever $\delta \neq 0$, i.e., whenever $|q/p| \neq 1$. Unitarity imposes an upper limit on δ—see eqn (6.73). This upper limit may be improved through eqn (6.72) if we are able to use eqn (6.45) to put a bound on $|\Gamma_{12}|$.

- CP violation in the decay amplitudes (direct CP violation). This occurs whenever $|A_f| \neq |\bar{A}_{\bar{f}}|$. It is the only type of CP violation possible in the case of charged mesons. As shown in Chapter 5, $|A_f| \neq |\bar{A}_{\bar{f}}|$ requires the presence of (at least) two interfering amplitudes with different weak phases and different strong phases.

- CP violation in the interference between the mixing and decay amplitudes (interference CP violation). It occurs whenever $\arg \lambda_f + \arg \lambda_{\bar{f}} \neq 0$. If f is a CP eigenstate, this is equivalent to $\operatorname{Im} \lambda_f \neq 0$.

When f is a CP eigenstate, CP invariance implies $\lambda_f = \pm 1$. Moreover, if f and g are CP eigenstates with CP eigenvalues η_f and η_g, respectively, then CP invariance requires $\lambda_f = \eta_f \eta_g \lambda_g$.

8

THE NEUTRAL-KAON SYSTEM

8.1 Introduction

The neutral kaons K^0 and $\overline{K^0}$ are two of the eight members of the octet of light spin-0 mesons with negative parity, which also includes the charged kaons $K^{\pm}$, the pions $\pi^{\pm}$ and π^0, and the η (see Fig. 8.1). The kaons are strange particles, the strangeness of K^0 and of K^+ being $+1$, while that of $\overline{K^0}$ and of K^- is -1. In the quark model, $K^0 \sim \bar{s}d$, $\overline{K^0} \sim s\bar{d}$, $K^+ \sim \bar{s}u$, and $K^- \sim s\bar{u}$.

The neutral kaons constitute the only system in which CP violation has been observed up to now. Indeed, their mixing makes them an excellent laboratory to look for very weak effects, like CP violation and CPT violation (Kostelecký 1998). The measurement of the tiny mass difference between the long-lived and the short-lived neutral kaons is one of the most precise measurements in particle physics.

We discuss in this chapter the specifics of neutral-kaon mixing and decays, relying on notation and formulae from Chapters 6 and 7. We assume CPT invariance.

8.2 Special features

The neutral-kaon system has two features which distinguish it from other neutral-meson systems.

First feature: the lifetimes of the two eigenstates of mixing are very different. As a consequence, it is usual to distinguish the eigenstates of mixing by their lifetimes instead of distinguishing them by their masses: K_S is the short-lived neutral kaon and K_L is the long-lived neutral kaon. The corresponding masses

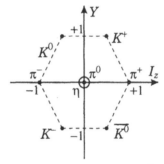

FIG. 8.1. The SU(3) octet of light $J^P = 0^-$ mesons.

and decay widths are given by $\mu_S = m_S - (i/2)\,\Gamma_S$ and $\mu_L = m_L - (i/2)\,\Gamma_L$, respectively. We define

$$\Delta m \equiv m_L - m_S,$$
$$\Delta\Gamma \equiv \Gamma_L - \Gamma_S. \tag{8.1}$$

Also, $\Gamma \equiv (\Gamma_S + \Gamma_L)/2$ as in eqn (6.12). Experimentally,

$$\tau_L \equiv \frac{1}{\Gamma_L} = (5.17 \pm 0.04) \times 10^{-8}\,\text{s},$$
$$\tau_S \equiv \frac{1}{\Gamma_S} = (8.927 \pm 0.009) \times 10^{-11}\,\text{s}, \tag{8.2}$$

while

$$\Delta m = (3.491 \pm 0.009) \times 10^{-12}\,\text{MeV} = (5.304 \pm 0.014) \times 10^{9}\,\text{s}^{-1}. \tag{8.3}$$

We see that $\Gamma_S \approx 579\Gamma_L$. Therefore, $\Delta\Gamma \approx -\Gamma_S$ and $\Gamma \approx \Gamma_S/2$. Moreover, $\Gamma_S \approx 2\Delta m$. From the present point of view, the latter approximate equality is just a coincidence. Remembering eqns (6.15), we see that in the neutral-kaon system $x \approx -y \approx u \approx 1$.

By definition, $\Delta\Gamma < 0$. The sign of Δm was experimentally determined by means of regeneration experiments; it was found that $\Delta m > 0$, as anticipated in eqn (8.3). The average mass of the neutral kaons is

$$m_K = 497.672 \pm 0.031\,\text{MeV}. \tag{8.4}$$

Notice that the mass difference Δm is fourteen orders of magnitude smaller than the average mass. This is a consequence of the fact that Δm arises at second order in the weak Hamiltonian, which has a typical strength 10^{-7} that of the strong Hamiltonian, responsible for m_K.

As K_L is heavier than K_S, we should identify P_H of Chapter 6 with K_L, while P_L is K_S. Accordingly, we write

$$|K_L\rangle = p_K|K^0\rangle + q_K|\overline{K^0}\rangle,$$
$$|K_S\rangle = p_K|K^0\rangle - q_K|\overline{K^0}\rangle, \tag{8.5}$$

or

$$|K^0\rangle = \frac{1}{2p_K}\left(|K_L\rangle + |K_S\rangle\right),$$
$$|\overline{K^0}\rangle = \frac{1}{2q_K}\left(|K_L\rangle - |K_S\rangle\right), \tag{8.6}$$

with

$$|p_K|^2 = \frac{1+\delta}{2},$$
$$|q_K|^2 = \frac{1-\delta}{2}. \tag{8.7}$$

Second feature: the kinematically allowed phase space for the two-pion decay channels $\pi^+\pi^-$ and $\pi^0\pi^0$ is much larger than the one for any other decay channel.

If it were not for CP symmetry, the decays to two pions would be dominant for both K_S and K_L. As a matter of fact, the two-pion decays are dominant for K_S, but not for K_L. Experimentally,

$$
\begin{aligned}
\mathrm{BR}\left(K_S \to \pi^+\pi^-\right) &= (6.861 \pm 0.028) \times 10^{-1}, \\
\mathrm{BR}\left(K_S \to \pi^0\pi^0\right) &= (3.139 \pm 0.028) \times 10^{-1}, \\
\mathrm{BR}\left(K_L \to \pi^+\pi^-\right) &= (2.067 \pm 0.035) \times 10^{-3}, \\
\mathrm{BR}\left(K_L \to \pi^0\pi^0\right) &= (9.36 \pm 0.20) \times 10^{-4}.
\end{aligned}
\tag{8.8}
$$

As both the kaons and the pions are spinless, the two-pion state resulting from the decay of a neutral kaon is in an s wave. That state then has $C = P = CP = +1$. If CP was conserved K_S and K_L would be eigenstates of CP, one of them with eigenvalue $+1$ and the other one with eigenvalue -1 (remember eqn 6.76). The eigenstate with eigenvalue $+1$ would decay to two pions, the one with eigenvalue -1 would not. Thus, the short-lived neutral kaon K_S, which decays predominantly to two pions, would be the CP-even superposition of $|K^0\rangle$ and $|\overline{K^0}\rangle$. On the other hand, K_L, being prevented by CP symmetry from decaying to the kinematically favoured two-pion states, would automatically have a lifetime much longer than that of K_S.

Small CP violation in the kaon system slightly disturbs this state of affairs. The original discovery of CP violation (Christenson *et al.* 1964) consisted in the observation of two-pion decays of K_L. Intrinsically and *a priori*, K_L is equivalent to K_S and there is no CP violation in the fact that K_L by itself alone decays to two pions. CP violation rather lies in the fact that both K_S and K_L, which are mixtures of two CP-conjugate states, decay to the same CP eigenstate (Sachs 1987).

Thus, the dominance of the two-pion decay modes is closely related to the large difference between the lifetimes of the two eigenstates of mixing. The eigenstate K_S, which is allowed by CP symmetry to decay to the kinematically favoured two-pion states, automatically has a much smaller lifetime than K_L, which only decays to two pions because of CP violation.

As CP violation in neutral-kaon mixing is very small, K^0 and $\overline{K^0}$ are approximately half K_S and half K_L. This means that the evolution of a neutral-kaon beam is characterized by two-pion decays, from the K_S component, at a short distance from the production vertex, followed at much larger distances by the decays of K_L, given by

$$
\begin{aligned}
\mathrm{BR}\left(K_L \to \pi^\pm e^\mp \nu_e\right) &= (3.878 \pm 0.027) \times 10^{-1}, \\
\mathrm{BR}\left(K_L \to \pi^\pm \mu^\mp \nu_\mu\right) &= (2.717 \pm 0.025) \times 10^{-1}, \\
\mathrm{BR}\left(K_L \to \pi^0\pi^0\pi^0\right) &= (2.112 \pm 0.027) \times 10^{-1}, \\
\mathrm{BR}\left(K_L \to \pi^+\pi^-\pi^0\right) &= (1.256 \pm 0.020) \times 10^{-1}.
\end{aligned}
\tag{8.9}
$$

8.3 Unitarity bound

Equations (6.53) and (6.54) read, in the case of neutral kaons,

$$|\langle K_S|K_L\rangle|^2 = \delta^2 \leq \frac{\Gamma_S\Gamma_L}{\Gamma^2 + (\Delta m)^2}\left[\sum_f \sqrt{BR(K_S \to f)BR(K_L \to f)}\right]^2 \quad (8.10)$$

$$\leq \frac{\Gamma_S\Gamma_L}{\Gamma^2 + (\Delta m)^2}. \quad (8.11)$$

From the experimental values in eqns (8.2) and (8.3), together with eqn (8.11), one obtains $|\delta| \leq 6 \times 10^{-2}$. This bound indicates that CP violation in neutral-kaon mixing is very small. The strength of this bound is a direct consequence of $|y| \approx 1$.

A better bound can be obtained if we use eqn (8.10), together with the experimental results in eqns (8.8) for the two-pion decay modes, which dominate the sum. We obtain

$$|\delta| \leq 3.4 \times 10^{-3}. \quad (8.12)$$

With such a small $|\delta|$, and Δm and $\Delta\Gamma$ being of the same order of magnitude, it is reasonable to approximate eqns (6.64) and (6.65) by

$$\Delta m \approx 2|M_{12}|$$
$$\Delta\Gamma \approx -2|\Gamma_{12}|. \quad (8.13)$$

From eqn (6.63), and as

$$\Delta m \approx 2\,|M_{12}| \approx -\tfrac{1}{2}\Delta\Gamma \approx |\Gamma_{12}|, \quad (8.14)$$

we derive

$$\varpi \equiv \arg\left(M_{12}^*\Gamma_{12}\right) \approx \pi - 2\delta. \quad (8.15)$$

The complex numbers M_{12} and Γ_{12} have a phase difference close to π, because of eqn (6.62) and $(\Delta m)(\Delta\Gamma) < 0$.

8.4 Leptonic asymmetry

The leptonic asymmetries are the clearest signs of CP violation in the neutral-kaon system. They are defined by

$$\delta_l \equiv \frac{\Gamma(K_L \to \pi^- l^+ \nu_l) - \Gamma(K_L \to \pi^+ l^- \bar{\nu}_l)}{\Gamma(K_L \to \pi^- l^+ \nu_l) + \Gamma(K_L \to \pi^+ l^- \bar{\nu}_l)}, \quad (8.16)$$

where l may be either the electron e or the muon μ. If CP is conserved then K_L, being a neutral particle with unique mass and decay width, must be an eigenstate of CP. If CP is conserved a CP eigenstate must decay with equal probability to two states which are CP-conjugates of each other. Therefore, $\delta_l \neq 0$ is an unmistakable signature of CP violation (see § 1.4.4).

Experiment indicates that δ_e and δ_μ are almost equal:

$$\delta_l = (3.27 \pm 0.12) \times 10^{-3}. \quad (8.17)$$

The equality of δ_e and δ_μ follows from the universality of the weak interaction.

One should remember the $\Delta S = \Delta Q$ rule for the decays of strange particles. This rule implies that K^0 decays to $\pi^- l^+ \nu_l$ but not to $\pi^+ l^- \bar{\nu}_l$, while $\overline{K^0}$ decays to $\pi^+ l^- \bar{\nu}_l$ but not to $\pi^- l^+ \nu_l$:

$$\langle \pi^+ l^- \bar{\nu}_l | T | K^0 \rangle = \langle \pi^- l^+ \nu_l | T | \overline{K^0} \rangle = 0. \tag{8.18}$$

Thus, the semileptonic decays tag the flavour of the neutral kaon. Moreover, there is only one hadron—the charged pion—in the semileptonic final states $\pi^\pm l^\mp \nu_l$, and therefore no final-state strong interactions may scatter those final states into or from other final states; as a consequence, CPT invariance leads to

$$|\langle \pi^- l^+ \nu_l | T | K^0 \rangle| = |\langle \pi^+ l^- \bar{\nu}_l | T | \overline{K^0} \rangle|. \tag{8.19}$$

Then, the leptonic asymmetry measures the difference between the probability of finding a K^0 and the probability of finding a $\overline{K^0}$ in K_L. Therefore, when the $\Delta S = \Delta Q$ rule is strictly valid,

$$\delta_e = \delta_\mu = \frac{|p_K|^2 - |q_K|^2}{|p_K|^2 + |q_K|^2} = \delta. \tag{8.20}$$

Notice the closeness between the value of δ in eqn (8.17) and the unitarity bound in eqn (8.12). From the derivation of the unitarity bound in Chapter 6 we learn what this means: the relevant phases

$$\begin{aligned}
\phi_{+-} &\equiv \arg \frac{\langle \pi^+ \pi^- | T | K_L \rangle}{\langle \pi^+ \pi^- | T | K_S \rangle}, \\
\phi_{00} &\equiv \arg \frac{\langle \pi^0 \pi^0 | T | K_L \rangle}{\langle \pi^0 \pi^0 | T | K_S \rangle},
\end{aligned} \tag{8.21}$$

must be very close to each other.

We may also define leptonic asymmetries for the semileptonic K_S decays,

$$\delta'_l \equiv \frac{\Gamma(K_S \to \pi^- l^+ \nu_l) - \Gamma(K_S \to \pi^+ l^- \bar{\nu}_l)}{\Gamma(K_S \to \pi^- l^+ \nu_l) + \Gamma(K_S \to \pi^+ l^- \bar{\nu}_l)}. \tag{8.22}$$

Their measurement may be possible at a ϕ factory (Buchanan *et al.* 1992). CPT invariance together with the $\Delta S = \Delta Q$ rule predict $\delta'_l = \delta_l$.

Violation of the $\Delta S = \Delta Q$ rule is parametrized by

$$\begin{aligned}
x_l &\equiv \frac{\langle \pi^- l^+ \nu_l | T | \overline{K^0} \rangle}{\langle \pi^- l^+ \nu_l | T | K^0 \rangle}, \\
\bar{x}_l &\equiv \frac{\langle \pi^+ l^- \bar{\nu}_l | T | K^0 \rangle^*}{\langle \pi^+ l^- \bar{\nu}_l | T | \overline{K^0} \rangle^*}.
\end{aligned} \tag{8.23}$$

As long as final-state interactions in the semileptonic decays may be neglected, the right-hand side of eqn (5.27) vanishes. Thus, $(CPT)\, T\, (CPT)^{-1} = T^\dagger = T$ and therefore $\langle f | T | i \rangle = \langle f_{CPT} | T | i_{CPT} \rangle^*$, where $|i_{CPT}\rangle \equiv CPT |i\rangle$ and $|f_{CPT}\rangle \equiv$

$\mathcal{CPT}|f\rangle$. Thus, CPT invariance implies $x_l = \bar{x}_l$. The parameter x_l is expected to be $\sim 10^{-7}$, basically because $\langle \pi^- l^+ \nu_l | T | K^0 \rangle$ and $\langle \pi^+ l^- \bar{\nu}_l | T | \overline{K^0} \rangle$ are first-order in $\mathcal{H}_W$, while $\langle \pi^- l^+ \nu_l | T | \overline{K^0} \rangle$ and $\langle \pi^+ l^- \bar{\nu}_l | T | K^0 \rangle$ are second-order. Experiment indicates that x_l is of order 10^{-2} or smaller.

The general expression for δ_l, assuming CPT invariance but allowing for violation of the $\Delta S = \Delta Q$ rule, is

$$
\begin{aligned}
\delta_l &= \frac{\left| \sqrt{R_{12}} + \sqrt{R_{21}} x_l \right|^2 - \left| \sqrt{R_{21}} + \sqrt{R_{12}} x_l^* \right|^2}{\left| \sqrt{R_{12}} + \sqrt{R_{21}} x_l \right|^2 + \left| \sqrt{R_{21}} + \sqrt{R_{12}} x_l^* \right|^2} \\
&= \frac{(|R_{12}| - |R_{21}|)\left(1 - |x_l|^2\right)}{(|R_{12}| + |R_{21}|)\left(1 + |x_l|^2\right) + 4\mathrm{Re}\left(\sqrt{R_{12}^* R_{21}}\, x_l\right)}.
\end{aligned}
\tag{8.24}
$$

Thus, δ_l must originate in mixing CP violation, $|R_{12}| \neq |R_{21}|$, even when the $\Delta S = \Delta Q$ rule is violated.

8.5 The parameters η

We define, for an arbitrary decay channel f, the parameters

$$
\eta_f = |\eta_f| e^{i\phi_f} \equiv \frac{\langle f | T | K_L \rangle}{\langle f | T | K_S \rangle} r.
\tag{8.25}
$$

The factor r is introduced in order to obtain rephasing-invariance. It is largely arbitrary, it must satisfy only two conditions. First, it must depend on the phases of the kets $|K_S\rangle$ and $|K_L\rangle$ in such a way as to offset the phase-convention dependence of the ratio $\langle f|T|K_L \rangle / \langle f|T|K_S \rangle$. Only then is the phase ϕ_f physical. Second, in the CPT invariant case, and with the phase convention of eqns (8.5), one must have $r = 1$. Thus, $\langle K^0 | K_L \rangle = \langle K^0 | K_S \rangle$ and $\langle \overline{K^0} | K_L \rangle = -\langle \overline{K^0} | K_S \rangle$ must imply $r = 1$. For instance, Kayser (1996) has suggested

$$
r = \frac{\langle K^0 | K_S \rangle}{\langle K^0 | K_L \rangle}
$$

and Lavoura (1991) has suggested

$$
r = \frac{2 \langle K^0 | K_S \rangle \langle \overline{K^0} | K_S \rangle}{\langle K^0 | K_L \rangle \langle \overline{K^0} | K_S \rangle - \langle K^0 | K_S \rangle \langle \overline{K^0} | K_L \rangle}.
$$

The exact definition of r is immaterial as long as we assume CPT invariance and the phase convention in eqns (8.5). It becomes important only when we want to study the CPT-violating case. We shall assume $r = 1$, but we write down r explicitly in the definitions of parameters, whenever necessary.

With $r = 1$,

$$
|\eta_f| = \sqrt{\frac{\Gamma_{fL}}{\Gamma_{fS}}} = \sqrt{\frac{\Gamma_L}{\Gamma_S} \frac{\mathrm{BR}(K_L \to f)}{\mathrm{BR}(K_S \to f)}}
\tag{8.26}
$$

is directly measurable.

If we define, as in eqn (7.5),

$$\lambda_f \equiv \frac{q_K}{p_K} \frac{\bar{A}_f}{A_f}, \tag{8.27}$$

where $A_f \equiv \langle f|T|K^0 \rangle$ and $\bar{A}_f \equiv \langle f|T|\overline{K^0} \rangle$, then

$$\eta_f = \frac{1 + \lambda_f}{1 - \lambda_f}. \tag{8.28}$$

The parameters η are measured by observing the time dependence of the decays of tagged neutral kaons. For instance, if at production time a neutral-kaon beam had strangeness $+1$, we would use

$$\Gamma[K^0(t) \to f] = \frac{|\langle f|T|K_S \rangle|^2}{2(1+\delta)} \left[e^{-\Gamma_S t} + |\eta_f|^2 e^{-\Gamma_L t} + 2|\eta_f| e^{-\Gamma t} \cos(\phi_f - \Delta m t) \right]. \tag{8.29}$$

If at production time the strangeness of the neutral-kaon beam was -1, we would use

$$\Gamma[\overline{K^0}(t) \to f] = \frac{|\langle f|T|K_S \rangle|^2}{2(1-\delta)} \left[e^{-\Gamma_S t} + |\eta_f|^2 e^{-\Gamma_L t} - 2|\eta_f| e^{-\Gamma t} \cos(\phi_f - \Delta m t) \right]. \tag{8.30}$$

The time t is measured in the rest frame of the neutral kaons. Notice the relative minus sign between the interference terms in eqns (8.29) and (8.30). The interference pattern in those equations, which is displayed in Fig. 8.2, is one of the best experimental demostrations of quantum mechanics in the realm of particle physics.

Summing eqns (8.29) and (8.30) over all decay modes f, by means of the unitarity eqns (6.48)–(6.50), one obtains

$$\sum_f \Gamma[K^0(t) \to f] = \frac{1}{2(1+\delta)} \left\{ \Gamma_S e^{-\Gamma_S t} + \Gamma_L e^{-\Gamma_L t} \right.$$
$$\left. + 2\delta e^{-\Gamma t} \left[\Gamma \cos(\Delta m t) + \Delta m \sin(\Delta m t) \right] \right\},$$
$$\sum_f \Gamma[\overline{K^0}(t) \to f] = \frac{1}{2(1-\delta)} \left\{ \Gamma_S e^{-\Gamma_S t} + \Gamma_L e^{-\Gamma_L t} \right.$$
$$\left. - 2\delta e^{-\Gamma t} \left[\Gamma \cos(\Delta m t) + \Delta m \sin(\Delta m t) \right] \right\}. \tag{8.31}$$

The difference between the two decay curves in eqns (8.31) depends only on the CP-violating parameter δ.

For the two main decay channels, $\pi^+\pi^-$ and $\pi^0\pi^0$, the parameters η are named η_{+-} and η_{00}, respectively. Their measured values are

$$|\eta_{+-}| = (2.285 \pm 0.019) \times 10^{-3},$$
$$|\eta_{00}| = (2.275 \pm 0.019) \times 10^{-3},$$
$$\phi_{+-} = (43.7 \pm 0.6)^\circ,$$
$$\phi_{00} = (43.5 \pm 1.0)^\circ. \tag{8.32}$$

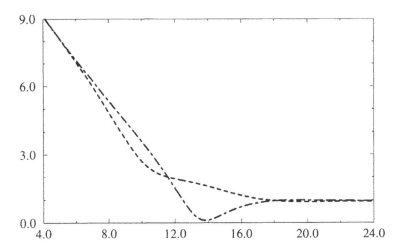

FIG. 8.2. The logarithms of $\Gamma[K^0(t) \to f]$ (dashed line) and of $\Gamma[\overline{K^0}(t) \to f]$ (dashed-dotted line) plotted against the time t measured in units τ_S. We have used the values of Γ_L/Γ_S, $\Delta m/\Gamma_S$, and δ in eqns (8.2)–(8.3) and (8.17). Also, $|\eta_f| = \delta/\sqrt{2}$ and $\phi_f = 43.49°$ are the (approximate) values relevant for the two-pion decay modes. For an appropriate scaling of the logarithms we have taken $|\langle f|T|K_S\rangle|^2 = 10^6$. For $t < 6\tau_S$ both curves approximately follow the simple exponential decay of K_S; for $t > 18\tau_S$ they both approximate the exponential decay of K_L. The interference between $K_L \to \pi\pi$ and $K_S \to \pi\pi$ is maximal for $t \sim 9\text{–}15\tau_S$, and has opposite sign in the decays of $K^0(t)$ and of $\overline{K^0}(t)$.

Particularly important are the ratio

$$\left|\frac{\eta_{00}}{\eta_{+-}}\right| = 0.9956 \pm 0.0023 \tag{8.33}$$

and the phase difference

$$\phi_{00} - \phi_{+-} = (-0.2 \pm 0.8)° . \tag{8.34}$$

If the $\Delta S = \Delta Q$ rule is valid, then $\eta_{\pi^- l^+ \nu_l} = +1$ and $\eta_{\pi^+ l^- \bar{\nu}_l} = -1$.

8.6 Regeneration

This section and the next one consider specific methods for the experimental study of the neutral-kaon system, and may be skipped without loss of continuity.

Suppose that we have a beam of neutral kaons and, after letting it evolve for a proper time much longer than τ_S but much shorter that τ_L, we have it incident on a thin slab of material, called a regenerator. When it is incident on the regenerator, the beam is almost exclusively K_L, because the K_S component has decayed away. Inside the regenerator kaons are scattered by strong interaction with the nuclei in the regenerator. The strong interaction distinguishes between K^0 and $\overline{K^0}$, which have different forward-scattering amplitudes off the regenerator. As a consequence, from the opposite side of the regenerator emerges a superposition of K^0 and $\overline{K^0}$ different from the incident one. Emerging from the regenerator we will not have a K_L beam, rather a beam of neutral kaons with a regenerated K_S component.

We may describe the process in the following way. We name $\langle K_L R|T|K_L R \rangle$ the amplitude for a K_L incident on the regenerator to emerge as K_L. The amplitude for the process in which a K_L incident on the regenerator emerges as K_S is $\langle K_S R|T|K_L R \rangle$. Thus, the neutral-kaon state which emerges from the regenerator is

$$|K_r\rangle = \langle K_S R|T|K_L R\rangle |K_S\rangle + \langle K_L R|T|K_L R\rangle |K_L\rangle. \tag{8.35}$$

We now observe the decays of K_r to the channel f as a function of the proper time t, measured with the initial time being the instant at which the kaon beam emerges from the regenerator. We get

$$\Gamma\left[K_r(t) \to f\right] = |\langle f|T|K_S\rangle|^2 \, |\langle K_S R|T|K_L R\rangle|^2$$
$$\times \left[e^{-\Gamma_S t} + |v_f|^2 \, e^{-\Gamma_L t} + 2 \, |v_f| \, e^{-\Gamma t} \cos\left(\theta_f - \Delta m t\right)\right], \tag{8.36}$$

where

$$v_f = |v_f| e^{i\theta_f} \equiv \frac{\langle K_L R|T|K_L R\rangle \, \langle f|T|K_L\rangle}{\langle K_S R|T|K_L R\rangle \, \langle f|T|K_S\rangle}. \tag{8.37}$$

Notice that v_f is a rephasing-invariant quantity, and therefore its phase θ_f is measurable.

Regeneration is a way of measuring the parameters η_f whenever we are able to make a reliable theoretical computation of $\langle K_L R|T|K_L R\rangle / \langle K_S R|T|K_L R\rangle$, so that we are able to extract η_f from v_f.

8.7   Correlated decays

We consider in this section the decays of $K^0 \overline{K^0}$ pairs in an antisymmetric correlated state. This is important because such correlated states will be copiously produced in the upcoming ϕ factories, in particular at DAΦNE. The resonance ϕ has spin 1, and upon its decay the resulting $K^0 \overline{K^0}$ pair is in a p wave. This correlated state is C- and P-odd and is written, in the rest frame of the decaying ϕ,

$$|\phi^-\rangle = \frac{1}{\sqrt{2}} \left[|K^0(\vec{k})\rangle \otimes |\overline{K^0}(-\vec{k})\rangle - |\overline{K^0}(\vec{k})\rangle \otimes |K^0(-\vec{k})\rangle\right]$$
$$= \frac{1}{2\sqrt{2}p_K q_K} \left[|K_S(\vec{k})\rangle \otimes |K_L(-\vec{k})\rangle - |K_L(\vec{k})\rangle \otimes |K_S(-\vec{k})\rangle\right]. \tag{8.38}$$

Notice the absence of $|K_L(\vec{k})\rangle \otimes |K_L(-\vec{k})\rangle$ and $|K_S(\vec{k})\rangle \otimes |K_S(-\vec{k})\rangle$ components. This is because the p wave is antisymmetric, and two identical bosons in an antisymmetric state would violate Bose symmetry. This fact holds not only for the initial instant at which ϕ decays into a kaon pair, but also for any later time, even after the neutral kaons have oscillated back and forth into each other. The antisymmetry of the wave function is preserved by the linearity of the oscillation. This holds even when CPT is violated in the mixing.

Let the kaon with momentum $\vec{k}$ decay to the state f at time t_1 and the kaon with momentum $-\vec{k}$ decay at time t_2 to the state g. The density of probability for this decay is

$$\left| \langle f, t_1; g, t_2 | T | \phi^- \rangle \right|^2 = \frac{|\langle f | T | K_S \rangle \langle g | T | K_S \rangle|^2}{2(1 - \delta^2)} \left\{ |\eta_f|^2 e^{-\Gamma_L t_1 - \Gamma_S t_2} \right.$$

$$\left. + |\eta_g|^2 e^{-\Gamma_S t_1 - \Gamma_L t_2} - 2|\eta_f \eta_g| e^{-\Gamma(t_1 + t_2)} \cos\left[\Delta m(t_2 - t_1) + \phi_f - \phi_g\right] \right\}. \quad (8.39)$$

If f and g are eigenstates of CP with the same CP-parity, this decay is forbidden by CP symmetry. Indeed, in that case the eigenstates of mass would coincide with the eigenstates of CP; K_S would have CP=+1 and K_L would have CP= -1. Equation (8.38) tells us that, if in one side of the detector we have K_S, in the opposite side of the detector we must have K_L. Thus, the occurrence of two final CP eigenstates with the same CP-parity in both sides of the detector is forbidden by CP symmetry.

Suppose that experimentally we do not observe t_2. Then, it is adequate to integrate eqn (8.39) from $t_2 = 0$ to $t_2 = \infty$ and obtain a distribution dependent only on t_1:

$$\left| \langle f, t_1; g | T | \phi^- \rangle \right|^2 = \frac{|\langle f | T | K_S \rangle \langle g | T | K_S \rangle|^2}{2\Gamma(1 - \delta^2)} \left[\frac{|\eta_f|^2}{1 - y} e^{-\Gamma_L t_1} + \frac{|\eta_g|^2}{1 + y} e^{-\Gamma_S t_1} \right.$$

$$\left. - 2|\eta_f \eta_g| \frac{e^{-\Gamma t_1}}{\sqrt{1 + x^2}} \cos\left(-\Delta m t_1 + \phi_f - \phi_g + \tilde{\phi}_{sw} \right) \right], \quad (8.40)$$

where $\tilde{\phi}_{sw} \equiv \arctan(\Delta m / \Gamma) \approx 43.39°$. If we do not observe t_1 either, we obtain the probability for the decays to f and g to occur at any time:

$$\left| \langle f; g | T | \phi^- \rangle \right|^2 = \frac{|\langle f | T | K_S \rangle \langle g | T | K_S \rangle|^2}{2\Gamma^2 (1 - \delta^2)} \left[\frac{|\eta_f|^2 + |\eta_g|^2}{1 - y^2} - \frac{2\text{Re}\left(\eta_f \eta_g^* \right)}{1 + x^2} \right]. \quad (8.41)$$

Summing this expression over all decay channels f and g we obtain 1 as we should, after using the unitarity relations.

We may also sum eqn (8.40) over all decay channels g and obtain the time distribution of the decays of the meson with momentum $\vec{k}$ to the channel f:

$$\left| \langle f, t_1 | T | \phi^- \rangle \right|^2 = \frac{|\langle f | T | K_S \rangle|^2}{2(1 - \delta^2)}$$

$$\times \left[e^{-\Gamma_S t_1} + |\eta_f|^2 e^{-\Gamma_L t_1} - 2\delta |\eta_f| e^{-\Gamma t_1} \cos(\phi_f - \Delta m t_1) \right]. \quad (8.42)$$

This distribution is the average of those in eqns (8.29) and (8.30). This is because at $t = 0$ the probabilities of having a K^0 or a $\overline{K^0}$ with momentum $\vec{k}$ are equal.

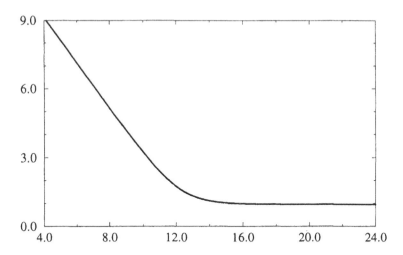

FIG. 8.3. The decay curve in eqn (8.42). The notation and the values used are the same as for drawing Fig. 8.2.

The interference term is suppressed by δ and is therefore very small (see Fig. 8.3).

We may also consider the situation in which only the relative time $\Delta t = t_1 - t_2$ is measured, while $t_1 + t_2$ remains unobserved. Then, we have to integrate the expression in eqn (8.39) over $t_1 + t_2$ from $|\Delta t|$ to $+\infty$, obtaining the probability distribution

$$|\langle f; g; \Delta t|T|\phi^-\rangle|^2 = \frac{|\langle f|T|K_S\rangle \langle g|T|K_S\rangle|^2}{4\Gamma(1-\delta^2)} \left[|\eta_f|^2 e^{-\Gamma_L|\Delta t|} + |\eta_g|^2 e^{-\Gamma_S|\Delta t|} \right.$$
$$\left. -2|\eta_f \eta_g| e^{-\Gamma|\Delta t|} \cos(\phi_f - \phi_g - \Delta m|\Delta t|) \right], \qquad (8.43)$$

valid for $\Delta t > 0$. For $\Delta t < 0$ we must use eqn (8.43) with f and g interchanged.

An interesting particular case is $f = g$. In that case the distribution in eqn (8.43) becomes symmetric in Δt:

$$|\langle f; f; \Delta t|T|\phi^-\rangle|^2 = \frac{|\langle f|T|K_S\rangle \langle f|T|K_L\rangle|^2}{4\Gamma(1-\delta^2)} \left[e^{-\Gamma_S|\Delta t|} + e^{-\Gamma_L|\Delta t|} \right.$$
$$\left. -2e^{-\Gamma|\Delta t|} \cos(\Delta m|\Delta t|) \right]. \qquad (8.44)$$

This distribution allows, with any decay channel f, measurements of Γ_S, Γ_L, and $|\Delta m|$. It vanishes at $\Delta t = 0$ as a result of Bose symmetry: the original state being antisymmetric, it cannot yield two simultaneous identical bosonic states f.

All the above decay distributions provide various possibilities for the measurement of the mixing and CP-violation parameters in the K^0–$\overline{K}^0$ system at a ϕ factory, depending on the decay channels and time distributions used (Buchanan *et al.* 1992). Some of the phenomenological formulas in this section are still valid in the case of CPT violation in neutral-kaon mixing, because they depend crucially on the antisymmetry of the original state ϕ^-, not on the assumption of CPT invariance.

8.8 Two-pion decays

8.8.1 *Parametrization*

Both the kaons and the pions are spinless particles. Therefore, when a neutral kaon decays to two pions, the latter must be in a state of zero angular momentum. The pions are bosons, therefore their total wave function must be symmetric. Being in a state of zero angular momentum, their isospin state must be even. Thus, the state $\langle 2\pi, I = 1| = (\langle \pi^+\pi^-| - \langle \pi^-\pi^+|)/\sqrt{2}$ must be discarded. The symmetric combination $(\langle \pi^+\pi^-| + \langle \pi^-\pi^+|)/\sqrt{2}$ may then be simply denoted $\langle \pi^+\pi^-|$. The isospin decomposition of the two-pion states is

$$\langle \pi^+\pi^-| = \sqrt{\tfrac{1}{3}}\langle 2\pi, I = 2| + \sqrt{\tfrac{2}{3}}\langle 2\pi, I = 0|,$$
$$\langle \pi^0\pi^0| = \sqrt{\tfrac{2}{3}}\langle 2\pi, I = 2| - \sqrt{\tfrac{1}{3}}\langle 2\pi, I = 0|,$$
(8.45)

or equivalently

$$\langle 2\pi, I = 0| = \sqrt{\tfrac{2}{3}}\langle \pi^+\pi^-| - \sqrt{\tfrac{1}{3}}\langle \pi^0\pi^0|,$$
$$\langle 2\pi, I = 2| = \sqrt{\tfrac{1}{3}}\langle \pi^+\pi^-| + \sqrt{\tfrac{2}{3}}\langle \pi^0\pi^0|.$$
(8.46)

We shall denote the state $\langle 2\pi, I = 0|$ by $\langle 0|$, and the state $\langle 2\pi, I = 2|$ by $\langle 2|$.

It is important to call the reader's attention to the normalization of the two-pion states that we are using. Namely, we are considering that the two neutral pions in $\langle \pi^0\pi^0|$ are identical particles, and the π^+ and π^- in $\langle \pi^+\pi^-|$ are identical particles too. Thus, for instance, we compute

$$\Gamma\left(K_S \to \pi^+\pi^-\right) = \frac{1}{32\pi m_K}\sqrt{1 - \frac{4m_\pi^2}{m_K^2}}\left|\sqrt{\tfrac{1}{3}}\langle 2|T|K_S\rangle + \sqrt{\tfrac{2}{3}}\langle 0|T|K_S\rangle\right|^2, \quad (8.47)$$

$$\Gamma\left(K_S \to \pi^0\pi^0\right) = \frac{1}{32\pi m_K}\sqrt{1 - \frac{4m_\pi^2}{m_K^2}}\left|\sqrt{\tfrac{2}{3}}\langle 2|T|K_S\rangle - \sqrt{\tfrac{1}{3}}\langle 0|T|K_S\rangle\right|^2. \quad (8.48)$$

Some authors use a normalization in which the matrix elements are $\sqrt{2}$ times smaller than ours. This is because they want to compute $\Gamma\left(K_S \to \pi^+\pi^-\right)$ taking π^+ and π^- to be *distinguishable* particles, while computing $\Gamma\left(K_S \to \pi^0\pi^0\right)$ with identical neutral pions. They write

$$\Gamma\left(K_S \to \pi^+\pi^-\right) = \frac{1}{16\pi m_K}\sqrt{1 - \frac{4m_\pi^2}{m_K^2}}\left|\sqrt{\tfrac{1}{3}}\langle 2|T|K_S\rangle + \sqrt{\tfrac{2}{3}}\langle 0|T|K_S\rangle\right|^2, \quad (8.49)$$

$$\Gamma\left(K_S \to \pi^0\pi^0\right) = \frac{1}{32\pi m_K}\sqrt{1 - \frac{4m_\pi^2}{m_K^2}}\left|\frac{2}{\sqrt{3}}\langle 2|T|K_S\rangle - \sqrt{\frac{2}{3}}\langle 0|T|K_S\rangle\right|^2. \quad (8.50)$$

Notice the difference between the denominators $(16\pi m_K)$ in eqn (8.49) and $(32\pi m_K)$ in eqn (8.50). We use for the main amplitude $\langle 0|T|K^0\rangle$ the value

$$\left|\langle 0|T|K^0\rangle\right| = 4.71 \times 10^{-4}\,\mathrm{MeV}, \quad (8.51)$$

while authors using the other normalization give $\left|\langle 0|T|K^0\rangle\right| = 3.33 \times 10^{-4}\,\mathrm{MeV}$.

We must take into account the $|\Delta I| = 1/2$ rule for kaon decays. The kaons have isospin $1/2$, and that rule tells us that they decay predominantly to $\langle 0|$, not so much to $\langle 2|$. It is convenient to normalize the four relevant decay amplitudes by the largest of them, which is $\langle 0|T|K_S\rangle$. We thus define (Chau 1983)

$$\omega \equiv \frac{\langle 2|T|K_S\rangle}{\langle 0|T|K_S\rangle}, \quad (8.52)$$

$$\epsilon \equiv \frac{\langle 0|T|K_L\rangle}{\langle 0|T|K_S\rangle}r, \quad (8.53)$$

$$\epsilon_2 \equiv \frac{\langle 2|T|K_L\rangle}{\langle 0|T|K_S\rangle}r. \quad (8.54)$$

Both ω and ϵ_2 violate the $|\Delta I| = 1/2$ rule. Both ϵ and ϵ_2 violate CP. Notice that ϵ is the parameter η for the decay to $2\pi, I = 0$—cf. eqns (8.25) and (8.53). However, as this two-pion state is not experimentally observed, rather it is a theoretical concoction, we use the notation 'ϵ' instead of, say, 'η_0'.

Instead of ϵ_2 it is convenient to use a different parameter, which also violates both CP and the $|\Delta I| = 1/2$ rule:

$$\epsilon' \equiv \frac{\epsilon_2 - \epsilon\omega}{\sqrt{2}} = \frac{\langle 2|T|K_L\rangle\langle 0|T|K_S\rangle - \langle 0|T|K_L\rangle\langle 2|T|K_S\rangle}{\sqrt{2}\langle 0|T|K_S\rangle^2}r. \quad (8.55)$$

We find that $\epsilon' \neq 0$ represents direct CP violation, i.e., CP violation in the decay amplitudes. Indeed, the two-pion states have CP=+1. And

$$\langle 2|T|K_L\rangle\langle 0|T|K_S\rangle - \langle 0|T|K_L\rangle\langle 2|T|K_S\rangle \propto \langle 2|T|K^0\rangle\langle 0|T|\overline{K^0}\rangle - \langle 2|T|\overline{K^0}\rangle\langle 0|T|K^0\rangle$$

is a directly CP-violating quantity, as we have seen in eqn (5.24).

From eqns (8.45) we find

$$\begin{aligned}
\eta_{+-} &= \epsilon + \frac{\epsilon'}{1 + \omega/\sqrt{2}} \\
&\approx \epsilon + \epsilon', \\
\eta_{00} &= \epsilon - \frac{2\epsilon'}{1 - \sqrt{2}\omega} \\
&\approx \epsilon - 2\epsilon'.
\end{aligned} \quad (8.56)$$

We have used $|\omega| \ll 1$, which is a consequence of the $|\Delta I| = 1/2$ rule. (We shall compute ω explicitly soon.)

Equations (8.53) and (8.55) constitute a 'theoretical' definition of ϵ and ϵ'. They have the advantage of an easy theoretical interpretation. Some authors however prefer an 'experimental' definition of those parameters, which directly connects them to the measured quantities η_{+-} and η_{00}. They define

$$\epsilon \equiv \frac{2\eta_{+-} + \eta_{00}}{3},$$

$$\epsilon' \equiv \frac{\eta_{+-} - \eta_{00}}{3}. \tag{8.57}$$

The 'theoretical' and 'experimental' definitions yield parameters ϵ and ϵ' which differ only slightly. Indeed, eqns (8.56) lead to eqns (8.57) when $|\omega| \ll 1$.

From eqns (8.56) or (8.57) it follows that

$$\frac{\eta_{00}}{\eta_{+-}} = \frac{1 - 2\epsilon'/\epsilon}{1 + \epsilon'/\epsilon}$$

$$\approx 1 - 3\frac{\epsilon'}{\epsilon}, \tag{8.58}$$

where we have anticipated that $|\epsilon'| \ll |\epsilon|$. On the other hand, we shall soon see that ϵ'/ϵ is predicted to be approximately real. Therefore,

$$\left| \frac{\eta_{00}}{\eta_{+-}} \right|^2 \approx 1 - 6\frac{\epsilon'}{\epsilon}. \tag{8.59}$$

It is in this context that the experimental result in eqn (8.33) becomes important. It displays a two standard deviation of ϵ'/ϵ from zero. In any case, it is clear that ϵ'/ϵ is at most $\sim 10^{-3}$. A large experimental effort is being continually invested in the experimental determination of

$$\left| \frac{\eta_{00}}{\eta_{+-}} \right|^2 = \frac{\mathrm{BR}(K_L \to \pi^0\pi^0)\mathrm{BR}(K_S \to \pi^+\pi^-)}{\mathrm{BR}(K_S \to \pi^0\pi^0)\mathrm{BR}(K_L \to \pi^+\pi^-)}, \tag{8.60}$$

in the hope of achieving a better determination of ϵ'/ϵ.

8.8.2 ω and possible $\Delta I = 5/2$ transitions

In this subsection, which some readers may prefer to skip, we make a detour and investigate how the value of ω is determined from experiment. From eqns (8.47) and (8.48), and from the definition of ω in eqn (8.52), we derive

$$\frac{\Gamma(K_S \to \pi^+\pi^-)}{\Gamma(K_S \to \pi^0\pi^0)} = 0.985 \left| \frac{\sqrt{2} + \omega}{1 - \sqrt{2}\omega} \right|^2. \tag{8.61}$$

The factor 0.985 accounts for two breakings of isospin symmetry: the different masses of the neutral and charged pions, and the Coulomb interaction in the final state $\pi^+\pi^-$. Assuming ω to be small, we have

$$\frac{\Gamma\left(K_S \to \pi^+\pi^-\right)}{\Gamma\left(K_S \to \pi^0\pi^0\right)} \approx 1.97\left(1 + 3\sqrt{2}\mathrm{Re}\,\omega\right).$$ (8.62)

From the experimental data in eqns (8.8) we obtain

$$\mathrm{Re}\,\omega \approx 0.026.$$ (8.63)

In order to compute $|\omega|$ we must compare the rate of $K_S \to 2\pi$ with the one of $K^+ \to \pi^+\pi^0$. For this purpose we first use isospin symmetry to relate $\langle 2|T|K^0\rangle$ and $\langle \pi^+\pi^0|T|K^+\rangle$. The initial states K^0 and K^+ form a doublet of isospin. The final states $2\pi, I = 2$ and $\pi^+\pi^0$ are components of a quintuplet of isospin. The transition matrix effecting the transition between an initial $I = 1/2$ and a final $I = 2$ state must be the sum of a $\Delta I = 3/2$ part and a $\Delta I = 5/2$ part, which we denote $T^{(3/2)}$ and $T^{(5/2)}$, respectively. Thus,

$$\langle 2|T|K^0\rangle = \langle 2|T^{(3/2)}|K^0\rangle + \langle 2|T^{(5/2)}|K^0\rangle,$$
$$\langle \pi^+\pi^0|T|K^+\rangle = \langle \pi^+\pi^0|T^{(3/2)}|K^+\rangle + \langle \pi^+\pi^0|T^{(5/2)}|K^+\rangle.$$ (8.64)

In order to parametrize the relative strength of $T^{(3/2)}$ and $T^{(5/2)}$, we introduce

$$a \equiv \frac{\langle 2|T^{(5/2)}|K^0\rangle}{\langle 2|T^{(3/2)}|K^0\rangle}.$$ (8.65)

Working out the Clebsch–Gordan coefficients, we find

$$\langle \pi^+\pi^0|T^{(3/2)}|K^+\rangle = \sqrt{\tfrac{3}{2}}\langle 2|T^{(3/2)}|K^0\rangle,$$
$$\langle \pi^+\pi^0|T^{(5/2)}|K^+\rangle = -\sqrt{\tfrac{2}{3}}\langle 2|T^{(5/2)}|K^0\rangle.$$ (8.66)

Therefore,

$$\begin{aligned}
\langle \pi^+\pi^0|T|K^+\rangle &= \sqrt{\tfrac{3}{2}}\langle 2|T^{(3/2)}|K^0\rangle - \sqrt{\tfrac{2}{3}}\langle 2|T^{(5/2)}|K^0\rangle \\
&= \left(\sqrt{\tfrac{3}{2}} - \sqrt{\tfrac{2}{3}}a\right)\langle 2|T^{(3/2)}|K^0\rangle \\
&= \frac{3 - 2a}{\sqrt{6}\left(1 + a\right)}\langle 2|T|K^0\rangle.
\end{aligned}$$ (8.67)

Now, from the first eqn (8.6), and from eqns (8.52)–(8.55), we have

$$\langle 2|T|K^0\rangle = \frac{\omega + \sqrt{2}\epsilon' + \epsilon\omega}{2\omega p_K}\langle 2|T|K_S\rangle.$$ (8.68)

Therefore,

$$|\langle \pi^+\pi^0|T|K^+\rangle|^2 = \left|\frac{3 - 2a}{1 + a}\right|^2\frac{|\omega + \sqrt{2}\epsilon' + \epsilon\omega|^2}{12\left(1 + \delta\right)|\omega|^2}|\langle 2|T|K_S\rangle|^2$$

$$\approx \frac{9 - 12\,\mathrm{Re}\,a}{1 + 2\,\mathrm{Re}\,a} \frac{1}{12\,(1 + \delta)} \, |\langle 2|T|K_S\rangle|^2$$

$$\approx \tfrac{3}{4}\left(1 - \tfrac{10}{3}\mathrm{Re}\,a\right) |\langle 2|T|K_S\rangle|^2 \,. \tag{8.69}$$

We have anticipated here the small values of a and of ω, and also used the experimental facts that $\epsilon \sim 10^{-3}$, $\epsilon' \sim 10^{-6}$, and $\delta \sim 10^{-3}$ are all small.

From eqn (8.69) we derive

$$\frac{\Gamma\left(K^+ \to \pi^+\pi^0\right)}{\Gamma\left(K_S \to 2\pi\right)} \approx \tfrac{3}{4}\left(1 - \tfrac{10}{3}\mathrm{Re}\,a\right) \frac{|\langle 2|T|K_S\rangle|^2}{|\langle 0|T|K_S\rangle|^2 + |\langle 2|T|K_S\rangle|^2}$$

$$\approx \tfrac{3}{4}\left(1 - \tfrac{10}{3}\mathrm{Re}\,a\right)|\omega|^2 \,. \tag{8.70}$$

Inserting the experimental values in the left-hand side of eqn (8.70), one gets

$$|\omega| \approx 0.045\left(1 + \tfrac{5}{3}\mathrm{Re}\,a\right). \tag{8.71}$$

As we shall see later in this chapter, theoretically one predicts the phase of ω to be close to $-\pi/4$. Let us assume this theoretical prediction to be correct. Then, from eqns (8.63) and (8.71), one obtains

$$\mathrm{Re}\,a \approx -0.11. \tag{8.72}$$

We have thus proved the self-consistency of our assumption that a is small.

8.8.3 Decay amplitudes

Let us consider the decay amplitudes

$$\begin{aligned}
\langle I_{\mathrm{out}}|T|K^0\rangle &= A_I e^{i\delta_I}, \\
\langle I_{\mathrm{out}}|T|\overline{K^0}\rangle &= \bar{A}_I e^{i\delta_I},
\end{aligned} \tag{8.73}$$

for $I = 0$ and $I = 2$. We have explicitly factored out the phases δ_I, which are the final-state-interaction (strong-interaction) phase shifts of the two-pion states with definite isospin, defined by

$$|I_{\mathrm{in}}\rangle = e^{2i\delta_I}|I_{\mathrm{out}}\rangle. \tag{8.74}$$

These strong-interaction phase shifts depend on the angular momentum and on the energy of the two-pion system in its centre-of-momentum frame. The relevant δ_I are for an energy equal to m_K and for zero angular momentum. The experimental result is

$$\delta_2 - \delta_0 = (-41.4 \pm 8.1)^\circ. \tag{8.75}$$

In this treatment of the FSI, only the strong interaction is taken into account, while the final-state electromagnetic interaction is neglected. The states $\langle 0|$ and $\langle 2|$ are eigenstates of the strong interaction, but they are mixed by the electromagnetic interaction, which does not conserve isospin.

Let us consider the consequences of CPT for A_I and $\bar{A}_I$. CPT transforms

$$
\begin{aligned}
\mathcal{CPT}|K^0\rangle &= e^{i\nu}|\overline{K^0}\rangle, \\
\mathcal{CPT}|\overline{K^0}\rangle &= e^{i\nu}|K^0\rangle, \\
\mathcal{CPT}|0_{\text{out}}\rangle &= e^{i\chi}|0_{\text{in}}\rangle, \\
\mathcal{CPT}|2_{\text{out}}\rangle &= e^{i\chi}|2_{\text{in}}\rangle.
\end{aligned}
\tag{8.76}
$$

The CPT-transformation phases for the $|0\rangle$ and $|2\rangle$ states are equal because we do not want CPT to mix $|\pi^+\pi^-\rangle$ with $|\pi^0\pi^0\rangle$, as would otherwise happen. CPT invariance of the transition matrix implies

$$
\begin{aligned}
A_I e^{i\delta_I} &= \langle I_{\text{out}}|T|K^0\rangle \\
&= \langle \mathcal{CPT}(I_{\text{out}})|T|\mathcal{CPT}(K^0)\rangle\rangle^* \\
&= \langle I_{\text{in}}|T|\overline{K^0}\rangle^* e^{i(\chi-\nu)} \\
&= \langle I_{\text{out}}|T|\overline{K^0}\rangle^* e^{i(\chi-\nu+2\delta_I)} \\
&= \bar{A}_I^* e^{i(\chi-\nu+\delta_I)}.
\end{aligned}
$$

Therefore, CPT symmetry implies

$$
\begin{aligned}
A_0 &= \bar{A}_0^* e^{i(\chi-\nu)} \\
A_2 &= \bar{A}_2^* e^{i(\chi-\nu)}.
\end{aligned}
\tag{8.77}
$$

The phases χ and ν are unphysical and meaningless. However, the following equations are physically meaningful, because they are χ- and ν-independent:

$$
\begin{aligned}
|A_0| &= |\bar{A}_0|, \\
|A_2| &= |\bar{A}_2|, \\
A_0^* A_2 &= \bar{A}_0 \bar{A}_2^*, \\
A_0 \bar{A}_2^* &= A_2 \bar{A}_0^*.
\end{aligned}
\tag{8.78}
$$

These are the consequences of CPT invariance.

8.8.4 ϵ

We define the parameters λ for the decays of the neutral kaons to two pions either with isospin zero or with isospin two:

$$
\lambda_I \equiv \frac{q_K}{p_K} \frac{\bar{A}_I}{A_I}.
\tag{8.79}
$$

Then,

$$
\epsilon = \frac{1+\lambda_0}{1-\lambda_0},
\tag{8.80}
$$

as in eqn (8.28). If CP was conserved then λ_0 would be -1 and ϵ would vanish.

Because of the first eqn (8.78),

$$\lambda_0 = \sqrt{\frac{1-\delta}{1+\delta}}\, e^{i\theta},$$ (8.81)

where $\theta \equiv \arg \lambda_0$. Then, from eqn (8.80) we may derive

$$\frac{2\mathrm{Re}\,\epsilon}{1+|\epsilon|^2} = \frac{1-|\lambda_0|^2}{1+|\lambda_0|^2} = \delta$$ (8.82)

and

$$\frac{\mathrm{Im}\,\epsilon}{\mathrm{Re}\,\epsilon} = \frac{\sqrt{1-\delta^2}}{\delta}\sin\theta.$$ (8.83)

It is clear that ϵ may be non-zero because of either CP violation in mixing ($\delta \neq 0$) or interference CP violation ($\sin\theta \neq 0$).[20] Thus, ϵ contains no direct CP violation, but it may originate either in mixing or interference CP violation.

We now define

$$\varsigma \equiv \arg(\Gamma_{12} A_0 \bar{A}_0^*),$$ (8.84)

which we shall use together with $\varpi = \arg\left(M_{12}^* \Gamma_{12}\right)$. From eqn (6.70) we have as $x > 0$,

$$\theta + \varsigma = \arg\left(\frac{q_K}{p_K}\Gamma_{12}\right) = \arg\left(-u + i\delta\right).$$

As a consequence, from eqn (8.83),

$$\frac{\mathrm{Im}\,\epsilon}{\mathrm{Re}\,\epsilon} = \sqrt{\frac{1-\delta^2}{u^2+\delta^2}}\left(\cos\varsigma + \frac{u\sin\varsigma}{\delta}\right).$$ (8.85)

All the above equations are exact.

8.8.4.1 *A note on phase conventions* From eqn (8.80) it follows that, in a phase convention in which $\bar{A}_0 = A_0$, $q_K/p_K = (\epsilon - 1)/(\epsilon + 1)$. In such a phase convention we may write

$$\begin{aligned} p_K &= \frac{1+\epsilon}{\sqrt{2(1+|\epsilon|^2)}}, \\ q_K &= -\frac{1-\epsilon}{\sqrt{2(1+|\epsilon|^2)}}. \end{aligned}$$ (8.86)

Alternatively, in the phase convention $\bar{A}_0 = -A_0$,

$$\begin{aligned} p_K &= \frac{1+\epsilon}{\sqrt{2(1+|\epsilon|^2)}}, \\ q_K &= \frac{1-\epsilon}{\sqrt{2(1+|\epsilon|^2)}}. \end{aligned}$$ (8.87)

Both eqns (8.86) and eqns (8.87) are used by many authors. It must be emphasized that the phase conventions $A_0 = \pm\bar{A}_0$ do not exhaust the freedom that

---

[20] When $\delta = 0$, $\epsilon = i\sin\theta/(1 - \cos\theta)$ must originate in $\sin\theta \neq 0$.

one has in rephasing the kaon kets; indeed, we may rephase both $|K^0\rangle$ and $|\overline{K^0}\rangle$ at will, which means that there are two rephasing degrees of freedom, while only one rephasing is needed in order to achieve either $\bar{A}_0 = A_0$ or $\bar{A}_0 = -A_0$. It must also be emphasized that these phase conventions have nothing to do with what is called 'a phase convention for the CP transformation', like fixing $CP|K^0\rangle = |\overline{K^0}\rangle$ or $CP|K^0\rangle = -|\overline{K^0}\rangle$. Indeed, such 'conventions' convey a wrong idea about the meaning of CP symmetry. The free phase ξ_K in the CP transformation $CP|K^0\rangle = \exp(i\xi_K)|\overline{K^0}\rangle$ is not to be fixed by any convention, rather it is a phase that must be kept free in an effort to find a phenomenology which is CP invariant. CP invariance exists if there is *any* phase ξ_K such that the phenomenology turns out to be invariant under that transformation; ξ_K should not be restricted by assuming *a priori* that $\exp(i\xi_K)$ must be either $+1$ or -1.

8.8.5 ϵ' and ω

For ϵ' and ω we derive

$$\omega = e^{i(\delta_2-\delta_0)} \frac{p_K A_2 - q_K \bar{A}_2}{p_K A_0 - q_K \bar{A}_0} = e^{i(\delta_2-\delta_0)} \frac{A_2}{A_0} \frac{1-\lambda_2}{1-\lambda_0}, \tag{8.88}$$

and

$$\epsilon' = e^{i(\delta_2-\delta_0)} \frac{2 p_K q_K \left(\bar{A}_2 A_0 - A_2 \bar{A}_0\right)}{\sqrt{2}\left(p_K A_0 - q_K \bar{A}_0\right)^2} = \sqrt{2} e^{i(\delta_2-\delta_0)} \frac{A_2}{A_0} \frac{\lambda_2 - \lambda_0}{(1-\lambda_0)^2}. \tag{8.89}$$

Direct CP violation in ϵ' lies in the difference between λ_2 and λ_0, cf. eqn (7.28).

8.8.6   *Approximations: ϵ*

We now recall eqn (7.30). The main decay channel is $|2\pi, I = 0\rangle$. This is overwhelmingly dominant, therefore

$$\Gamma_{12} \approx A_0^* \bar{A}_0. \tag{8.90}$$

This is the crucial approximation in the analysis of the two-pion decays of the neutral kaons. It leads to $\varsigma = 0$. Indeed, one may show (Lavoura 1992*a*) that the present experimental data are already good enough to exclude $|\varsigma| > 5 \times 10^{-5}$. This is important, because ς must be much smaller than δ if we want to neglect the second term in the right-hand side of eqn (8.85).

Equation (8.90) effectively reduces interference CP violation in the $2\pi, I = 0$ channel to mixing CP violation. Indeed, when $\varsigma = 0$, the phases of $A_0^* \bar{A}_0$ and of Γ_{12} are equal, and the only independent phase to cause CP violation is $\varpi = \arg(M_{12}^* \Gamma_{12})$. This is the reason why many authors talk about ϵ representing mixing CP violation in the kaon system. In all rigour, ϵ arises from both mixing CP violation and interference CP violation, but eqn (8.90) reduces the latter to the former.

Let us then assume $\varsigma = 0$. From eqn (8.85) we get the important prediction $\arg \epsilon \approx \phi_{sw}$, where

$$\phi_{sw} \equiv \arctan \frac{1}{u} \approx 43.49° \tag{8.91}$$

is the so-called 'superweak phase'. Taking into account eqn (8.82), we find

$$\epsilon \approx \frac{\delta}{\sqrt{2}} e^{i\pi/4}. \tag{8.92}$$

Equations (8.91) and (8.92) agree with the predictions of the superweak theory in eqns (7.40) and (7.38), respectively.

Now,

$$\varpi \equiv \arg\left(M_{12}^* \Gamma_{12}\right) \approx -\arg(M_{12} A_0 \bar{A}_0^*), \tag{8.93}$$

because of eqn (8.90). Therefore,

$$\sin \varpi \approx -\frac{\mathrm{Im}(M_{12} A_0 \bar{A}_0^*)}{|M_{12} A_0 \bar{A}_0|}. \tag{8.94}$$

But, from eqn (8.15), we find

$$\delta \approx \frac{\sin \varpi}{2}. \tag{8.95}$$

Therefore,

$$\delta \approx -\frac{\mathrm{Im}\left(M_{12} A_0 \bar{A}_0^*\right)}{(\Delta m)\left|A_0 \bar{A}_0\right|}. \tag{8.96}$$

The value given by the Particle Data Group (1996) is

$$\epsilon = (2.280 \pm 0.013) \times 10^{-3} e^{i\pi/4}; \tag{8.97}$$

this is a fit using eqns (8.32) and the first eqn (8.57). Comparing eqns (8.92) and (8.97), one has

$$(3.224 \pm 0.018) \times 10^{-3} \approx -\frac{\mathrm{Im}\left(M_{12} A_0 \bar{A}_0^*\right)}{(\Delta m)\left|A_0 \bar{A}_0\right|}. \tag{8.98}$$

Equation (8.98) is the starting point for the theoretical fits of $|\epsilon|$ or, equivalently, of δ.

8.8.7 *Approximations: ϵ' and ω*

As $|\epsilon|$ is very small, we may approximate

$$\lambda_0 \approx -1. \tag{8.99}$$

Then,

$$\frac{A_2}{A_0} \lambda_2 = \lambda_0 \frac{\bar{A}_2}{\bar{A}_0} \approx -\frac{A_2^*}{A_0^*}, \tag{8.100}$$

where we have used eqns (8.78). With these approximations, we get

$$\omega = e^{i(\delta_2 - \delta_0)} \left(\frac{A_2}{A_0} - \frac{A_2}{A_0} \lambda_2\right) \frac{1}{1 - \lambda_0}$$

$$\approx e^{i(\delta_2 - \delta_0)} \mathrm{Re}\frac{A_2}{A_0}. \tag{8.101}$$

$$\epsilon' = \sqrt{2}e^{i(\delta_2-\delta_0)}\left(\frac{A_2}{A_0}\lambda_2 - \frac{A_2}{A_0}\lambda_0\right)\frac{1}{(1-\lambda_0)^2}$$

$$\approx \frac{i}{\sqrt{2}}e^{i(\delta_2-\delta_0)}\mathrm{Im}\frac{A_2}{A_0}. \tag{8.102}$$

8.8.8 *Conclusions*

The phenomenological scheme includes two important approximations:

1. $u \equiv -\Delta\Gamma/(2\Delta m) \approx 1$;
2. $\Gamma_{12} \approx A_0^* \bar{A}_0$.

Approximation 1 is an experimental fact which, from the point of view of present theoretical knowledge, is just a coincidence—although a very useful one. Approximation 2 basically follows from the $|\Delta I| = 1/2$ rule. In practice, its important consequence is that the phase ς in eqn (8.84) is extremely close to zero.

Based on these approximations, the phenomenological scheme makes four predictions:

1. The values of ϵ and of δ are related by eqn (8.82);
2. The phase of ϵ is equal to the superweak phase;
3. Assuming $\mathrm{Im}\,(A_2/A_0) > 0$, the phase of ϵ' is $\delta_2 - \delta_0 + \pi/2 \approx \pi/4$;
4. Assuming $\mathrm{Re}\,(A_2/A_0) > 0$, the phase of ω is $\delta_2 - \delta_0 \approx -\pi/4$.

Predictions 1 and 2 are well verified experimentally. We do not yet have enough experimental information on the phases of ϵ' and ω, but there is no reason to suspect that predictions 3 and 4 do not hold, especially when we take into account the possible existence of $\Delta I = 5/2$ transitions. Deviations from the predictions 1–4 might signal CPT violation (Barmin *et al.* 1984; Lavoura 1991).

9

HEAVY NEUTRAL-MESON SYSTEMS

9.1 Introduction

In this chapter we derive some theoretical formulas needed in the study of the heavy-neutral-meson systems D^0-$\overline{D^0}$, B_d^0-$\overline{B_d^0}$, and B_s^0-$\overline{B_s^0}$. Their quark contents are $D^0 = c\bar{u}$, $B_d^0 = \bar{b}d$, and $B_s^0 = \bar{b}s$. We denote a generic heavy neutral meson by P^0. We denote by Q the heavy quark (or antiquark) and by q the light quark (or antiquark).

Much of the interest in B decays had its origin in the seminal articles by Carter and Sanda (1980, 1981) and by Bigi and Sanda (1981, 1987). In writing some of the sections in this chapter we have also profited from the reviews by Fridman (1988), Dunietz (1994), and Xing (1996).

The formulas that we shall be deriving can also be applied to the neutral-kaon system. However, in that case it is more convenient to write the decay rates in terms of the decay amplitudes of the mass eigenstates, $\langle f|T|K_S\rangle$ and $\langle f|T|K_L\rangle$, as we have done in the previous chapter. In the heavy-neutral-meson systems one uses the decay amplitudes of the flavour eigenstates, $A_f \equiv \langle f|T|P^0\rangle$ and $\bar{A}_f \equiv \langle f|T|\overline{P^0}\rangle$.

Many experiments on the heavy P^0-$\overline{P^0}$ systems involve determining the flavour of a neutral meson at its production time and/or when it decays. The flavour of the meson at the time of decay may be found by looking for flavour-specific decays. Analogously to the kaon decays into $\pi^\pm l^\mp \nu_l$, these are decays into final states which can be reached either from P^0 but not from $\overline{P^0}$, or the other way round. For example, B_d^0 has a probability of around 10% of decaying semileptonically into a positively charged lepton ($l^+ = e^+$ or μ^+), a neutrino ν_l and hadrons, through the quark subprocess $\bar{b} \to \bar{c}l^+\nu_l$. Similarly, $\overline{B_d^0}$ may decay semileptonically, through $b \to cl^-\bar{\nu}_l$, yielding a negatively charged lepton. In general, the charge of the lepton in the final state has the same sign as the charge of the decaying heavy quark. This is the $\Delta B = \Delta Q$ rule for semileptonic B decays, analogous to the $\Delta S = \Delta Q$ rule of kaon decays in § 8.4. Just as in the kaon case, this rule is expected to be almost exact in the standard model (SM) and in most of its extensions, and it is *assumed* in most experimental analysis of B decays.[21]

The determination of the initial flavour of a neutral meson is usually called 'tagging' and is done using the rule of associated production. The production of the mesons is dominated either by the strong interaction, as in $p\bar{p}$ collisions,

---

[21] The impact of possible $\Delta B = -\Delta Q$ amplitudes has been extensively discussed by Dass and Sarma (1994, 1996*a,b*).

or by the electromagnetic interaction, as in the process $e^+e^- \to \phi \to K^0\overline{K^0}$ and in the analogous process $e^+e^- \to \Upsilon(4S) \to B_d^0\overline{B_d^0}$. Both the strong and the electromagnetic interactions conserve flavour and, therefore, a quark Q is always produced in association with its antiquark $\bar{Q}$. Thus, the reasoning behind the tagging strategy is: if we detect a quark Q in one side of the detector, we know that the quark in the opposite side must be $\bar{Q}$. This strategy can only work when a charged meson containing the quark Q (or the antiquark $\bar{Q}$) is observed in one side of the detector. We then know that the neutral meson in the opposite side had the corresponding antiquark $\bar{Q}$ (or quark Q) *at production time*. Indeed, while charged mesons do not oscillate, neutral mesons do. The tagging of the neutral meson is performed by identifying the flavour of the charged meson it was produced together with, through the decay of the latter. For instance, this is done at LEP when states such as $B_d^0 B^- X^+$ are produced, where X^+ is a collection of particles with total charge $+1$. An analogous strategy was followed by the CPLEAR Collaboration, who used the processes $p\bar{p} \to K^0 K^- \pi^+$ and $p\bar{p} \to \overline{K^0} K^+ \pi^-$ for kaon production, and tagged the neutral kaons by the charged kaon they were produced together with.

On the other hand, when two neutral mesons are produced in the strong or electromagnetic process, both of them oscillate. Detecting a flavour-specific final state in one side of the detector informs us about the flavour of that meson at its decay time, but not about its flavour at production time.

Therefore, one must consider four classes of initial conditions (Bigi and Sanda 1981, 1987):

A) When a charged meson is produced in association with its antimeson, there is no mixing. Then, all we can have is direct CP violation.

B) When a charged meson is produced in association with a neutral meson, the decay of the charged meson tags the initial flavour of the neutral one. (One may also have the neutral meson produced in association with a baryon.) We then have a single tagged neutral meson. The decay-rate formulas relevant for this case are derived in § 9.2. Those formulas are applied in the ensuing sections to show how tagged decays can be used to extract information about mixing and CP violation.

C) When two neutral mesons, P^0 and $\overline{P^0}$, are produced, identifying the flavour of one of them at decay time does not identify the initial flavour of the other one. Three different cases may then be considered:

    1. The two neutral mesons are produced through an intermediate $Q\bar{Q}$ resonance with odd orbital angular momentum. The mesons then appear in a correlated, antisymmetric wave function, cf. § 8.7. This is the case for the B_d^0–$\overline{B_d^0}$ pair produced from the decay of the $\Upsilon(4S)$. Antisymmetry of the wave function ensures that the dependence on the time difference $t_- \equiv t_1 - t_2$ between the times t_1 and t_2 of the decays of the two mesons, is the same as the time dependence of a single tagged neutral meson. (We show this explicitly in § 9.8.1.) This is the reason why one often disregards the correlated nature of the

meson production and discusses instead tagged, single-meson decay rates, in the context of B-physics experiments at the $\Upsilon(4S)$.

2. The two neutral mesons are produced through an intermediate $Q\bar{Q}$ resonance with even angular momentum. The wave function is symmetric. This is the case for the B_s^0–$\overline{B_s^0}$ pair produced from the decay of the $\Upsilon(5S)$. The formulas relevant for this case are derived in § 9.7 together with the ones relevant for case 1.

3. The two neutral mesons are uncorrelated. This situation occurs, for instance, at LEP. Usually, one integrates over the time t_2 of the tagging decay, obtaining a formula for the time evolution t_1 of the meson in the opposite side of the detector. The corresponding formulas are derived in § 9.9.

Measurements of mixing and CP violation using the decays of two correlated or uncorrelated neutral mesons are discussed in § 9.10.

D) The two heavy neutral mesons produced may not be each other's antiparticle. For instance, we could produce B_s^0 together with $\overline{B_d^0}$ and a set of particles with total strangeness $S = +1$. Now, although B_s^0 and $\overline{B_d^0}$ cannot interfere along their respective evolutions, they can mix with their corresponding antiparticles. As in the previous case, identifying the flavour of one meson at decay time does not identify the initial flavour of the other meson. We shall mention this case only briefly, in connection with uncorrelated neutral-meson production, as at LEP. We refer the reader to the work by Bigi and Sanda (1981, 1987).

9.2   Tagged decays

In this section we assume that the initial flavour of the meson has been tagged, for instance, by the decay of an associated charged meson. We consider the time-dependent decay rates of P^0 and $\overline{P^0}$.

Suppose that a P^0 $(\overline{P^0})$ is created at time $t = 0$, and denote by $P^0(t)$ $(\overline{P^0}(t))$ the state that it evolves into after a time t, measured in its rest frame. To find out the time evolution we use eqns (6.56), (6.38), and (6.55) to obtain

$$|P^0(t)\rangle = g_+(t)|P^0\rangle + \frac{q}{p}g_-(t)|\overline{P^0}\rangle,$$
$$|\overline{P^0}(t)\rangle = \frac{p}{q}g_-(t)|P^0\rangle + g_+(t)|\overline{P^0}\rangle, \tag{9.1}$$

where

$$g_\pm(t) \equiv \tfrac{1}{2}\left(e^{-i\mu_H t} \pm e^{-i\mu_L t}\right). \tag{9.2}$$

Care must be exercised when comparing apparently similar formulas from different authors, since extra minus signs sometimes appear in the definition of $g_-(t)$, as well as in the definitions of $\Delta\Gamma$ and of q.

The following formulas are useful:

$$|g_\pm(t)|^2 = \tfrac{1}{4}\left[e^{-\Gamma_H t} + e^{-\Gamma_L t} \pm 2e^{-\Gamma t}\cos(\Delta m\, t)\right]$$

$$= \frac{e^{-\Gamma t}}{2}\left[\cosh\frac{\Delta\Gamma t}{2} \pm \cos\left(\Delta mt\right)\right],$$

$$g_+^*(t)g_-(t) = \tfrac{1}{4}\left[e^{-\Gamma_H t} - e^{-\Gamma_L t} - 2ie^{-\Gamma t}\sin(\Delta m\, t)\right] \tag{9.3}$$

$$= -\frac{e^{-\Gamma t}}{2}\left[\sinh\frac{\Delta\Gamma t}{2} + i\sin\left(\Delta mt\right)\right].$$

Integrating over time, we obtain

$$G_\pm \equiv \int_0^{+\infty} |g_\pm(t)|^2 dt$$

$$= \frac{1}{2\Gamma}\left(\frac{1}{1-y^2} \pm \frac{1}{1+x^2}\right),$$

$$G_{+-} \equiv \int_0^{+\infty} g_+^*(t)g_-(t)dt \tag{9.4}$$

$$= \frac{1}{2\Gamma}\left(\frac{-y}{1-y^2} + \frac{-ix}{1+x^2}\right).$$

Therefore,

$$F \equiv \frac{G_-}{G_+} = \frac{x^2 + y^2}{2 + x^2 - y^2}. \tag{9.5}$$

From eqns (9.1), the probability that a particle initially identified as a P^0 is again identified as a P^0 at time t, is equal to the probability that a particle which was $\overline{P^0}$ at time $t = 0$ is again $\overline{P^0}$ at time t:

$$\text{Prob}[P^0(t) = P^0] = \text{Prob}[\overline{P^0}(t) = \overline{P^0}] = |g_+(t)|^2. \tag{9.6}$$

On the other hand, the probabilities that a particle identified as a P^0 at time $t = 0$ becomes $\overline{P^0}$ at time t, and that a particle identified as a $\overline{P^0}$ at time $t = 0$ becomes P^0 at time t, are only equal if CP is conserved in the mixing:

$$\text{Prob}[\overline{P^0}(t) = P^0] = \left|\frac{p}{q}\right|^2 |g_-(t)|^2,$$

$$\text{Prob}[P^0(t) = \overline{P^0}] = \left|\frac{q}{p}\right|^2 |g_-(t)|^2. \tag{9.7}$$

In the CPT-violating case we have, instead of eqns (9.1),

$$|P^0(t)\rangle = [g_+(t) - \theta g_-(t)]\,|P^0\rangle + \frac{q_H}{p_H}(1-\theta)g_-(t)|\overline{P^0}\rangle,$$

$$|\overline{P^0}(t)\rangle = \frac{p_L}{q_L}(1-\theta)g_-(t)|P^0\rangle + [g_+(t) + \theta g_-(t)]\,|\overline{P^0}\rangle. \tag{9.8}$$

Therefore, once CPT is violated it is not true any more that

$$\text{Prob}[P^0(t) = P^0] = \text{Prob}[\overline{P^0}(t) = \overline{P^0}].$$

On the other hand,

$$\text{Prob}[P^0(t) = \overline{P^0}] = \text{Prob}[\overline{P^0}(t) = P^0]$$

means that there is T invariance in mixing, even when CPT is violated (Kabir 1970).

Using eqns (9.1), we find

$$\Gamma[P^0(t) \to f] = |A_f|^2 \left\{ |g_+(t)|^2 + |\lambda_f|^2 |g_-(t)|^2 + 2\text{Re} \left[\lambda_f g_+^*(t) g_-(t) \right] \right\},$$

$$\Gamma[P^0(t) \to \bar{f}] = |\bar{A}_{\bar{f}}|^2 \left| \frac{q}{p} \right|^2 \left\{ |g_-(t)|^2 + |\bar{\lambda}_{\bar{f}}|^2 |g_+(t)|^2 + 2\text{Re} \left[\bar{\lambda}_{\bar{f}} \, g_+(t) g_-^*(t) \right] \right\},$$

$$\Gamma[\overline{P^0}(t) \to f] = |A_f|^2 \left| \frac{p}{q} \right|^2 \left\{ |g_-(t)|^2 + |\lambda_f|^2 |g_+(t)|^2 + 2\text{Re} \left[\lambda_f g_+(t) g_-^*(t) \right] \right\},$$

$$\Gamma[\overline{P^0}(t) \to \bar{f}] = |\bar{A}_{\bar{f}}|^2 \left\{ |g_+(t)|^2 + |\bar{\lambda}_{\bar{f}}|^2 |g_-(t)|^2 + 2\text{Re} \left[\bar{\lambda}_{\bar{f}} g_+^*(t) g_-(t) \right] \right\}.$$

$$(9.9)$$

These expressions give us the probability, divided by dt, that the state which initially was P^0 (or $\overline{P^0}$) decays to the final state f (or $\bar{f}$) during the time interval $[t, t+dt]$.

Experimentally it may be impossible to measure the time dependence. In that case all we can measure are the total numbers of events. These numbers are proportional to

$$\Gamma[P^0 \to f] = |A_f|^2 \left[G_+ + |\lambda_f|^2 G_- + 2\text{Re}\left(\lambda_f G_{+-}\right) \right],$$

$$\Gamma[P^0 \to \bar{f}] = |\bar{A}_{\bar{f}}|^2 \left| \frac{q}{p} \right|^2 \left[G_- + |\bar{\lambda}_{\bar{f}}|^2 G_+ + 2\text{Re}\left(\bar{\lambda}_{\bar{f}} G_{+-}^*\right) \right],$$

$$\Gamma[\overline{P^0} \to f] = |A_f|^2 \left| \frac{p}{q} \right|^2 \left[G_- + |\lambda_f|^2 G_+ + 2\text{Re}\left(\lambda_f G_{+-}^*\right) \right],$$

$$\Gamma[\overline{P^0} \to \bar{f}] = |\bar{A}_{\bar{f}}|^2 \left[G_+ + |\bar{\lambda}_{\bar{f}}|^2 G_- + 2\text{Re}\left(\bar{\lambda}_{\bar{f}} G_{+-}\right) \right].$$

$$(9.10)$$

To determine the inclusive rates one uses the unitarity relations

$$\sum_f |A_f|^2 = \Gamma_{11} = \Gamma,$$

$$\sum_f |A_f|^2 |\lambda_f|^2 = \left| \frac{q}{p} \right|^2 \Gamma_{22} = \frac{1-\delta}{1+\delta}\Gamma,$$

$$\sum_f |A_f|^2 \lambda_f = \frac{q}{p}\Gamma_{12} = \frac{y+i\delta x}{1+\delta}\Gamma.$$

$$(9.11)$$

We find

$$\sum_f \Gamma[P^0(t) \to f] = \Gamma \left\{ |g_+(t)|^2 + \frac{1-\delta}{1+\delta} |g_-(t)|^2 + 2\text{Re} \left[\frac{y + i\delta x}{1 + \delta} g_+^*(t) g_-(t) \right] \right\},$$

$$\sum_f \Gamma[\overline{P^0}(t) \to f] = \Gamma \left\{ |g_+(t)|^2 + \frac{1+\delta}{1-\delta} |g_-(t)|^2 + 2\text{Re} \left[\frac{y - i\delta x}{1 - \delta} g_+^*(t) g_-(t) \right] \right\}.$$

$$(9.12)$$

One may check the correctness of these expressions by integrating over time to show that

$$\sum_f \Gamma[P^0 \to f] = \sum_f \Gamma[\overline{P^0} \to f] = 1. \tag{9.13}$$

Indeed, the meson created at time $t = 0$ must have probability 1 of decaying to any final state f at any later time.

Equations (9.9) may be written in the form

$$\Gamma[P^0(t) \to f] = |A_f|^2 \frac{e^{-\Gamma t}}{2} (H + I),$$

$$\Gamma[\overline{P^0}(t) \to f] = |A_f|^2 \frac{e^{-\Gamma t}}{2} \left| \frac{p}{q} \right|^2 (H - I),$$

$$(9.14)$$

where we have used eqns (9.3) and defined

$$H \equiv \left(1 + |\lambda_f|^2\right) \cosh \frac{\Delta \Gamma t}{2} - 2\text{Re}\lambda_f \sinh \frac{\Delta \Gamma t}{2},$$

$$I \equiv \left(1 - |\lambda_f|^2\right) \cos(\Delta m t) + 2\text{Im}\lambda_f \sin(\Delta m t).$$

$$(9.15)$$

The function H depends on exponentials and on $\Delta\Gamma$; the function I is oscillatory and depends on Δm. If f is a CP eigenstate, then H is CP-conserving, while I is CP-violating, because CP conservation then imposes $\lambda_f = \pm 1$.

9.3 Flavour-specific decays

Let us denote by o a final state to which only P^0 can decay, and by $\bar{o}$ its CP-conjugate state, to which only $\overline{P^0}$ can decay. (Typically, o is a semileptonic state.) Thus, the decays of P^0 into $\bar{o}$ and of $\overline{P^0}$ into o are forbidden. This corresponds to $A_{\bar{o}} = \bar{A}_o = 0$, and therefore $\lambda_o = \bar{\lambda}_{\bar{o}} = 0$. Equations (9.9) become

$$\Gamma[P^0(t) \to o] = |A_o|^2 |g_+(t)|^2,$$

$$\Gamma[P^0(t) \to \bar{o}] = |\bar{A}_{\bar{o}}|^2 \left| \frac{q}{p} \right|^2 |g_-(t)|^2,$$

$$\Gamma[\overline{P^0}(t) \to o] = |A_o|^2 \left| \frac{p}{q} \right|^2 |g_-(t)|^2,$$

$$\Gamma[\overline{P^0}(t) \to \bar{o}] = |\bar{A}_{\bar{o}}|^2 |g_+(t)|^2.$$

$$(9.16)$$

Clearly, $\Gamma[P^0(t) \to \bar{o}]$ and $\Gamma[\overline{P^0}(t) \to o]$ vanish at $t = 0$, but they are non-zero for $t \neq 0$ due to the mixing of the neutral mesons. It is also due to mixing that one can find two states o (or two states $\bar{o}$) as the result of the decay of a state which initially was a P^0–$\overline{P^0}$ pair. This effect will be discussed when we come to the associated production of two neutral mesons.

9.3.1 *Time-integrated probabilities*

Suppose that at $t = 0$ we start with N_0 mesons P^0. The number of final states o that we obtain in the time interval $[t, t + dt]$ is then

$$N_0 \, |A_o|^2 \, |g_+(t)|^2 \, dt.$$

The number of final states $\bar{o}$ obtained in the same time interval is

$$N_0 \, |\bar{A}_{\bar{o}}|^2 \, \left| \frac{q}{p} \right|^2 \, |g_-(t)|^2 \, dt.$$

If the typical decay time of the P mesons is very small we may be unable to observe the time dependence of the flavour transitions. In that case, all we have access to is the total number of events in a given final state. The total number of o and $\bar{o}$ decays obtained from N_0 mesons P^0 is

$$\begin{aligned}
N[P^0 \to o] &= N_0 \, |A_o|^2 \, \mathrm{TIP}[P^0 \to P^0], \\
N[P^0 \to \bar{o}] &= N_0 \, |\bar{A}_{\bar{o}}|^2 \, \mathrm{TIP}[P^0 \to \overline{P^0}],
\end{aligned} \tag{9.17}$$

respectively. Here we have introduced the time-integrated probabilities (TIP)

$$\begin{aligned}
\mathrm{TIP}[P^0 \to P^0] &= G_+, \\
\mathrm{TIP}[P^0 \to \overline{P^0}] &= \left| \frac{q}{p} \right|^2 G_-,
\end{aligned} \tag{9.18}$$

respectively.

Similarly, if we start out with N_0 particles $\overline{P^0}$, we obtain

$$N[\overline{P^0} \to o] = N_0 \, |A_o|^2 \, \mathrm{TIP}[\overline{P^0} \to P^0],$$

$$N[\overline{P^0} \to \bar{o}] = N_0 \, |\bar{A}_{\bar{o}}|^2 \, \mathrm{TIP}[\overline{P^0} \to \overline{P^0}]$$

final states o and $\bar{o}$, respectively. We have defined

$$\begin{aligned}
\mathrm{TIP}[\overline{P^0} \to \overline{P^0}] &= G_+, \\
\mathrm{TIP}[\overline{P^0} \to P^0] &= \left| \frac{p}{q} \right|^2 G_-.
\end{aligned} \tag{9.19}$$

These TIP have time dimension and are not probabilities. They become probabilities only when multiplied by the square of the decay amplitude into the relevant final state, which has mass dimension.

The notation of the TIP may be a bit misleading. For instance, one might be surprised by the fact that

$$\text{TIP}[P^0 \to \overline{P^0}] + \text{TIP}[P^0 \to P^0] = \frac{1}{\Gamma(1-y^2)}\left[1 + \left(\left|\frac{q}{p}\right|^2 - 1\right)\frac{x^2 + y^2}{2(1+x^2)}\right] \neq \frac{1}{\Gamma}.$$

The reason is simple: not all the P^0 mesons decay into flavour-specific final states. Many final states may be reached from both P^0 and $\overline{P^0}$. It is precisely these channels that contribute to $|\Gamma_{12}|$. If these common states did not exist, we would have $\Gamma_{12} = 0$, implying both $|q/p| = 1$ and $y = 0$. The sum of the TIPs would then be $1/\Gamma$.

9.3.2 Pais–Treiman parameters

It is usual to introduce the Pais–Treiman parameters (Pais and Treiman 1975) as measures of mixing:

$$
\begin{aligned}
r &\equiv \frac{\text{TIP}[P^0 \to \overline{P^0}]}{\text{TIP}[P^0 \to P^0]} = \left|\frac{q}{p}\right|^2 F, \\
\bar{r} &\equiv \frac{\text{TIP}[\overline{P^0} \to P^0]}{\text{TIP}[\overline{P^0} \to \overline{P^0}]} = \left|\frac{p}{q}\right|^2 F,
\end{aligned}
\tag{9.20}
$$

where F has been defined in eqn (9.5). CP violation in the mixing is probed by the difference between the Pais–Treiman parameters:

$$\delta = \frac{\sqrt{\bar{r}} - \sqrt{r}}{\sqrt{\bar{r}} + \sqrt{r}}. \tag{9.21}$$

The quantity $F = \sqrt{r\bar{r}}$ measures the amount of mixing. Mixing is maximal if $|y| = 1$ (Fridman 1988), corresponding to very different lifetimes, as in the K^0–$\overline{K^0}$ system:

$$y = \pm 1 \Rightarrow F = 1. \tag{9.22}$$

This happens because, if $|y| \to 1$, after an infinitesimally small time one of the eigenstates P_H or P_L has completely decayed away and we have only the other eigenstate which, under the assumption of CP conservation, is an equal admixture of P^0 and $\overline{P^0}$. Mixing is also maximal in the limit of very large x (Fridman 1988),

$$x = \infty \Rightarrow F = 1. \tag{9.23}$$

This is due to the fact that, when the typical oscillation time $1/\Delta m$ is much smaller than the typical decay time $1/\Gamma$, then the initial P^0 oscillates back and forth to and from $\overline{P^0}$ many times before decaying, thus appearing to be an equal admixture of P^0 and $\overline{P^0}$.

9.3.3   CP-violating asymmetries

One may test for CP violation through the asymmetry

$$A_M = \frac{N[\overline{P^0} \to o] - N[P^0 \to \bar{o}]}{N[\overline{P^0} \to o] + N[P^0 \to \bar{o}]}$$

$$= \frac{|p/q|^2 |A_o|^2 - |q/p|^2 |\bar{A}_{\bar{o}}|^2}{|p/q|^2 |A_o|^2 + |q/p|^2 |\bar{A}_{\bar{o}}|^2}. \tag{9.24}$$

This asymmetry, however, is unable to separate mixing CP violation from direct CP violation. One may get a clean measurement of direct CP violation from

$$A_D = \frac{N[P^0 \to o] - N[\overline{P^0} \to \bar{o}]}{N[P^0 \to o] + N[\overline{P^0} \to \bar{o}]}$$

$$= \frac{|A_o|^2 - |\bar{A}_{\bar{o}}|^2}{|A_o|^2 + |\bar{A}_{\bar{o}}|^2}. \tag{9.25}$$

As we have seen in Chapter 5, there is direct CP violation when at least two amplitudes, having different strong and weak phases, contribute coherently to the decay; the asymmetry may be sizeable if the two amplitudes have comparable magnitudes.

In some cases, the asymmetry in eqn (9.25) is expected to be small. As an example, consider the decay $B_d^0 \to h^- l^+ \nu_l$, where h^- is a single negatively charged hadron. As there is a single hadron, there can be no final-state-interaction (FSI) CP-even phase due to the strong interaction. There could be an FSI phase shift due to electroweak scattering, but this is very small (Dass and Sarma 1996a,b). Then, CPT invariance itself implies

$$|\langle h^- l^+ \nu_l |T| B_d^0 \rangle| = |\langle h^+ l^- \bar{\nu}_l |T| \overline{B_d^0} \rangle|,$$

i.e., there is no direct CP violation. This is the same that happens in the neutral-kaon decays to $\pi^{\pm} l^{\mp} \nu_l$, as we have seen in eqn (8.19).

Let us assume this particular case of no direct CP violation. If $|A_o| = |\bar{A}_{\bar{o}}|$, one is able to measure the Pais–Treiman parameters:

$$\frac{N[P^0 \to \bar{o}]}{N[P^0 \to o]} = r,$$

$$\frac{N[\overline{P^0} \to o]}{N[\overline{P^0} \to \bar{o}]} = \bar{r}. \tag{9.26}$$

The asymmetry A_M then becomes a measure of CP violation in mixing:

$$A_M = \frac{|p/q|^2 - |q/p|^2}{|p/q|^2 + |q/p|^2} = \frac{\bar{r} - r}{\bar{r} + r} = \frac{2\delta}{1 + \delta^2}$$

$$= \frac{|R_{12}|^2 - |R_{21}|^2}{|R_{12}|^2 + |R_{21}|^2} = \frac{4\mathrm{Im}\,(M_{12}^* \Gamma_{12})}{4|M_{12}|^2 + |\Gamma_{12}|^2}. \tag{9.27}$$

In the kaon system, this asymmetry is $\approx 6.5 \times 10^{-3}$.

9.4   The case of CP conservation in mixing

In the case of heavy neutral mesons, it is usually *assumed* that mixing CP violation $|q/p| - 1$ is small and can be neglected, at least when compared with interference CP violation. We shall soon discuss the experimental status of this assumption, and will later turn to its justification in a standard-model computation of $|q/p| - 1$. In this section, we keep to standard practice and assume that $|q/p| = 1$. Then, using eqns (9.14) and (9.15), we find

$$\frac{\Gamma[P^0(t) \to f] - \Gamma[\overline{P^0}(t) \to f]}{|A_f|^2 \exp(-\Gamma t)} = I$$

$$= \left(1 - |\lambda_f|^2\right) \cos(\Delta mt) + 2\mathrm{Im}\lambda_f \sin(\Delta mt),$$

$$(9.28)$$

$$\frac{\Gamma[P^0(t) \to f] + \Gamma[\overline{P^0}(t) \to f]}{|A_f|^2 \exp(-\Gamma t)} = H$$

$$= \left(1 + |\lambda_f|^2\right) \cosh\frac{\Delta\Gamma t}{2} - 2\mathrm{Re}\lambda_f \sinh\frac{\Delta\Gamma t}{2}.$$

$$(9.29)$$

9.4.1   *No direct CP violation: CP eigenstates*

In the case of the decays of the neutral mesons to a CP eigenstate f, there are only two independent decay amplitudes and decay rates. Indeed, $\bar{f} = f$ implies $A_{\bar{f}} = A_f$ and $\bar{A}_{\bar{f}} = \bar{A}_f$. Therefore

$$\lambda_f = \lambda_{\bar{f}} = \frac{1}{\bar{\lambda}_{\bar{f}}} = \frac{1}{\bar{\lambda}_f}, \qquad (9.30)$$

and the interference-CP-violation parameter is $\arg\lambda_f + \arg\lambda_{\bar{f}} = 2\arg\lambda_f$ or $\mathrm{Im}\lambda_f$, as stated in eqn (7.24). Moreover, as $(\mathcal{CP})^2 = 1$,

$$\mathcal{CP}|f\rangle = \eta_f|f\rangle, \qquad (9.31)$$

with $\eta_f = 1$ for a CP-even and $\eta_f = -1$ for a CP-odd final state f.

Equation (9.28) then exhibits two different sources of CP violation:

Direct CP violation: ($|A_f| \neq |\bar{A}_f|$) is probed by the first term in the right-hand side of eqn (9.28).

Interference CP violation: ($\mathrm{Im}\lambda_f \neq 0$) is probed by the second term in the right-hand side of eqn (9.28).

The theoretical estimate of direct-CP-violating quantities is usually plagued by hadronic uncertainties. For this reason, finding situations in which that type of CP violation can be neglected is crucial. We will now show that this occurs in decays into CP eigenstates that are dominated by a single weak phase. In that case, the calculation of λ_f is also free of hadronic uncertainties.

In order to simplify our presentation, we keep only the CP-transformation properties of the CP eigenstate final state f through η_f. For all the other (spurious) phases arising in the CP transformations of the kets and of the quark field operators of the underlying theory, we follow the usual practice and eliminate them from the game. Their impact is discussed in detail in Appendix A, where we show explicitly that our conclusions are not affected by this simplification.[22] We take

$$\frac{q}{p} = -e^{2i\phi_M} \tag{9.32}$$

and

$$\begin{aligned} A_f &= A e^{i\phi_A} e^{i\delta_A} \\ \bar{A}_f &= \eta_f A e^{-i\phi_A} e^{i\delta_A}. \end{aligned} \tag{9.33}$$

The minus sign in eqn (9.32) is introduced to ease the comparison with the SM calculation to be performed in Chapter 33. The decay amplitude has been parameterized in terms of its modulus A, weak phase ϕ_A, and strong phase δ_A. We obtain

$$\lambda_f = -\eta_f e^{2i(\phi_M - \phi_A)}. \tag{9.34}$$

The crucial feature of eqn (9.34) for the study of CP violation is the fact that both the moduli of the decay amplitudes A_f and $\bar{A}_f$, A, and the FSI effects that those amplitudes contain, δ_A, cancel out in λ_f. This effectively eliminates the hadronic uncertainties from the computation of the CP-violating parameter $\mathrm{Im}\lambda_f$, and, thus, from the CP-violating asymmetry. This is the reason behind the importance of looking for decays into CP eigenstates, with decay amplitudes dominated by a single weak phase: *one has a direct measurement of a weak phase in the Lagrangian*. If there are several weak phases contributing to the decay, we also have direct CP violation, hindering the extraction of the individual weak phases $\phi_A - \phi_M$.

We may take the ratio of eqns (9.28) and (9.29) to form the time-dependent asymmetry

$$\begin{aligned} A_{CP}(t) &= \frac{\Gamma[P^0(t) \to f] - \Gamma[\overline{P^0}(t) \to f]}{\Gamma[P^0(t) \to f] + \Gamma[\overline{P^0}(t) \to f]} \\ &= \frac{\mathrm{Im}\lambda_f \sin(\Delta m t)}{\cosh(\Delta\Gamma t/2) - \mathrm{Re}\lambda_f \sinh(\Delta\Gamma t/2)}. \end{aligned} \tag{9.35}$$

All hadronic uncertainties have cancelled out, since the result depends only on λ_f. Only the phase $2(\phi_M - \phi_A)$ is important.

---

[22] In this section we assume that there is no CP violation in the mixing. Therefore $q/p = \pm e^{i\xi}$, where ξ is the phase appearing in the CP transformation of the ket describing the neutral meson. This seems to be in contradiction with eqn (9.32). This problem is clarified and explained in detail in Appendix A.

One can also look for CP violation in the time-integrated rates. The relevant formulae are the same as in eqns (9.28) and (9.29), with $|g_\pm(t)|^2$ and $g_+^*(t)g_-(t)$ substituted by $G_\pm$ and G_{+-}, respectively. One obtains

$$
\begin{aligned}
A_{CP} &= \frac{\Gamma[P^0 \to f] - \Gamma[\overline{P^0} \to f]}{\Gamma[P^0 \to f] + \Gamma[\overline{P^0} \to f]} \\
&= \frac{1-y^2}{1+x^2} \frac{1 - |\lambda_f|^2 + 2x\mathrm{Im}\lambda_f}{1 + |\lambda_f|^2 - 2y\mathrm{Re}\lambda_f} \\
&= \frac{1-y^2}{1+x^2} \frac{x\mathrm{Im}\lambda_f}{1 - y\mathrm{Re}\lambda_f}.
\end{aligned}
\tag{9.36}
$$

These time-integrated asymmetries are useful only for systems in which x is not much larger than 1, and y^2 is not too close to unity; otherwise the suppression factor $(1-y^2)/(1+x^2)$ becomes very small. This is not the case for the B_s^0–$\overline{B_s^0}$ system, forcing us to look for time-dependent rates, which are experimentally more challenging.

9.4.2 Small direct CP violation: CP eigenstates

We now want to learn what happens if, besides the dominant amplitude A_1 with weak phase ϕ_{A1}, there is another interfering amplitude A_2 with a different weak phase ϕ_{A2}. In this case

$$
\begin{aligned}
A_f &= A_1 e^{i\phi_{A1}} e^{i\delta_1} + A_2 e^{i\phi_{A2}} e^{i\delta_2}, \\
\bar{A}_f &= \eta_f \left(A_1 e^{-i\phi_{A1}} e^{i\delta_1} + A_2 e^{-i\phi_{A2}} e^{i\delta_2} \right).
\end{aligned}
\tag{9.37}
$$

The weak phases ϕ_{A1}, ϕ_{A2}, and ϕ_M, and the strong phases δ_1 and δ_2 are not rephasing-invariant, but the differences $\phi_1 \equiv \phi_{A1} - \phi_M$, $\phi_2 \equiv \phi_{A2} - \phi_M$, and $\Delta = \delta_2 - \delta_1$ can be measured. Indeed,

$$
\lambda_f = -\eta_f e^{-2i\phi_1} \frac{1 + re^{i(\phi_1-\phi_2)}e^{i\Delta}}{1 + re^{-i(\phi_1-\phi_2)}e^{i\Delta}},
\tag{9.38}
$$

where $r = A_2/A_1$. Therefore,

$$
\begin{aligned}
\mathrm{Re}\lambda_f &= -\eta_f \frac{\cos 2\phi_1 + 2r\cos(\phi_1+\phi_2)\cos\Delta + r^2\cos 2\phi_2}{1 + r^2 + 2r\cos(\phi_1 - \phi_2 - \Delta)}, \\
\mathrm{Im}\lambda_f &= \eta_f \frac{\sin 2\phi_1 + 2r\sin(\phi_1+\phi_2)\cos\Delta + r^2\sin 2\phi_2}{1 + r^2 + 2r\cos(\phi_1 - \phi_2 - \Delta)}, \\
1 + |\lambda_f|^2 &= \frac{2 + 4r\cos(\phi_1 - \phi_2)\cos\Delta + 2r^2}{1 + r^2 + 2r\cos(\phi_1 - \phi_2 - \Delta)}, \\
1 - |\lambda_f|^2 &= \frac{4r\sin(\phi_1 - \phi_2)\sin\Delta}{1 + r^2 + 2r\cos(\phi_1 - \phi_2 - \Delta)}.
\end{aligned}
\tag{9.39}
$$

Equations (9.39) exhibit the symmetry $\phi_1 \leftrightarrow \phi_2$, $\Delta \leftrightarrow -\Delta$, and $r \leftrightarrow 1/r$, as they should. If A_2 is much smaller than A_1, we may expand in powers of r to obtain

$$\frac{2\mathrm{Im}\lambda_f}{1+|\lambda_f|^2} \approx \eta_f \left[\sin 2\phi_1 - 2r\cos 2\phi_1 \sin(\phi_1 - \phi_2)\cos\Delta\right], \tag{9.40}$$

$$\frac{-2\mathrm{Re}\lambda_f}{1+|\lambda_f|^2} \approx \eta_f \left[\cos 2\phi_1 + 2r\sin 2\phi_1 \sin(\phi_1 - \phi_2)\cos\Delta\right], \tag{9.41}$$

$$\frac{1-|\lambda_f|^2}{1+|\lambda_f|^2} \approx 2r\sin(\phi_1 - \phi_2)\sin\Delta. \tag{9.42}$$

The direct-CP-violation observable in eqn (9.42) vanishes when $\Delta = 0$. However, as pointed out by Gronau (1993), even if the FSI are very small and $\Delta \approx 0$, the interference-CP-violation observable in eqn (9.40) does not provide a clean measurement of a single weak phase. The presence of a second amplitude with a different weak phase may ruin the measurement of $\sin 2\phi_1$, even if it has the same strong phase. This occurs even for moderate values of r.

9.4.3 No direct CP violation: CP non-eigenstates

If the final states are not CP eigenstates, there is no relation between $\bar{A}_f$ and A_f, and the FSI phases do not cancel in their ratio and in λ_f. Let us take once more $q/p = -e^{2i\phi_M}$. Assuming the decays to be dominated by only one weak phase, one has one weak phase and one strong phase in the amplitudes A_f and $\bar{A}_f$ and another weak phase and strong phase in the amplitudes $\bar{A}_f$ and $A_{\bar{f}}$:

$$\begin{aligned} A_f &= A e^{i\phi_a} e^{i\delta_a}, \\ \bar{A}_{\bar{f}} &= A e^{-i\phi_a} e^{i\delta_a}, \end{aligned} \tag{9.43}$$

and

$$\begin{aligned} \bar{A}_f &= B e^{i\phi_b} e^{i\delta_b}, \\ A_{\bar{f}} &= B e^{-i\phi_b} e^{i\delta_b}, \end{aligned} \tag{9.44}$$

where we have discarded all the phases brought about by the CP transformation. A and B are real by definition.

Interference CP violation is related to

$$\begin{aligned} \lambda_f &= -\frac{B}{A} e^{2i\phi} e^{i\Delta}, \\ \bar{\lambda}_{\bar{f}} &= -\frac{B}{A} e^{-2i\phi} e^{i\Delta}, \end{aligned} \tag{9.45}$$

where $2\phi \equiv 2\phi_M + \phi_b - \phi_a$ and $\Delta \equiv \delta_b - \delta_a$. There is no direct CP violation, as $|\lambda_f| = |\bar{\lambda}_{\bar{f}}|$. Clearly, $|\lambda_f| = |\bar{\lambda}_{\bar{f}}| \neq 1$. If f were a CP eigenstate, then A_f and $\bar{A}_f$ would be related by a CP transformation such that $B = A$, $\delta_b = \delta_a$, and $\phi_b = -\phi_a$. We would then have $\Delta = 0$, $\bar{\lambda}_{\bar{f}} = \lambda_f^*$, $|\lambda_f| = 1$, thus recovering the case in § 9.4.1.

Equations (9.9) reduce to

$$\Gamma[P^0(t) \to f] = A^2 \left\{ |g_+(t)|^2 + |\lambda_f|^2 |g_-(t)|^2 + 2\mathrm{Re}\left[\lambda_f g_+^*(t)g_-(t)\right] \right\},$$

$$\Gamma[P^0(t) \to \bar{f}] = A^2 \left\{ |g_-(t)|^2 + |\lambda_f|^2 |g_+(t)|^2 + 2\mathrm{Re}\left[\bar{\lambda}_{\bar{f}}\, g_+(t)g_-^*(t)\right] \right\},$$

$$\Gamma[\overline{P^0}(t) \to f] = A^2 \left\{ |g_-(t)|^2 + |\lambda_f|^2 |g_+(t)|^2 + 2\mathrm{Re}\left[\lambda_f g_+(t)g_-^*(t)\right] \right\},$$

$$\Gamma[\overline{P^0}(t) \to \bar{f}] = A^2 \left\{ |g_+(t)|^2 + |\lambda_f|^2 |g_-(t)|^2 + 2\mathrm{Re}\left[\bar{\lambda}_{\bar{f}} g_+^*(t)g_-(t)\right] \right\}. \tag{9.46}$$

Fitting for the time dependence of the various decay curves allows, in principle, for the extraction of $|\lambda_f|$, $\arg(-\lambda_f) = 2\phi + \Delta$, and $\arg(-\bar{\lambda}_{\bar{f}}) = -2\phi + \Delta$. We may thus recover the weak phase 2ϕ as well as the strong phase Δ (Aleksan et al. 1991). The limitations of this method lie in the need for high experimental precision in the measurement of the decay curves. Moreover, in some cases we may not really be able to determine $\arg \lambda_f$ and $\arg \bar{\lambda}_{\bar{f}}$ experimentally. Rather, we may only be able to determine some trigonometric functions thereof. Then, discrete ambiguities arise in the determination of 2ϕ and of Δ.

We may define CP asymmetries like the ones in eqns (9.35) and (9.36), but they are not very illuminating. For instance, the time-integrated asymmetry

$$A_{CP} = \frac{\Gamma[P^0 \to f] - \Gamma[\overline{P^0} \to \bar{f}]}{\Gamma[P^0 \to f] + \Gamma[\overline{P^0} \to \bar{f}]}$$
$$= \frac{xD_M\mathrm{Im}\left(\lambda_f - \bar{\lambda}_{\bar{f}}\right) - y\mathrm{Re}\left(\lambda_f - \bar{\lambda}_{\bar{f}}\right)}{(1+D_M) + (1-D_M)|\lambda_f|^2 + xD_M\mathrm{Im}\left(\lambda_f + \bar{\lambda}_{\bar{f}}\right) - y\mathrm{Re}\left(\lambda_f + \bar{\lambda}_{\bar{f}}\right)}, \tag{9.47}$$

where $D_M \equiv (1-y^2)/(1+x^2)$ is known as the dilution factor.

9.5 Inclusive decays

Thus far we have concentrated on exclusive decays, i.e., decays into specific, fully reconstructed final states. Let us now see whether something can be learned from the rates obtained by summing over all final states. Those rates have been given in eqns (9.12). One sees that the two inclusive rates $\Gamma[P^0(t) \to \mathrm{all}]$ and $\Gamma[\overline{P^0}(t) \to \mathrm{all}]$ are equal if and only if there is no CP violation in the mixing, i.e., if $\delta = 0$—this had already been seen, in the neutral-kaon case, in eqns (8.31). Therefore, the rate difference yields a measure of mixing CP violation. Indeed, the asymmetry

$$A_M^{\mathrm{incl}}(t) \equiv \frac{\Gamma[P^0(t) \to \mathrm{all}] - \Gamma[\overline{P^0}(t) \to \mathrm{all}]}{\Gamma[P^0(t) \to \mathrm{all}] + \Gamma[\overline{P^0}(t) \to \mathrm{all}]}$$
$$= -2\delta \frac{|g_-(t)|^2 + \mathrm{Re}\left[(y - ix)g_+^*(t)g_-(t)\right]}{(1-\delta^2)|g_+(t)|^2 + (1+\delta^2)|g_-(t)|^2 + 2\mathrm{Re}\left[(y - i\delta^2 x)g_+^*(t)g_-(t)\right]}$$
$$= \frac{\delta\left[-\cosh(\Delta\Gamma t/2) + y\sinh(\Delta\Gamma t/2) + \cos(\Delta m t) + x\sin(\Delta m t)\right]}{\cosh(\Delta\Gamma t/2) - y\sinh(\Delta\Gamma t/2) - \delta^2\cos(\Delta m t) - \delta^2 x\sin(\Delta m t)}. \tag{9.48}$$

Equation (9.13) guarantees that the time-integrated version of this quantity vanishes, as required by CPT invariance. (Indeed, $G_- + \mathrm{Re}\left[(y - ix)G_{+-}\right] = 0$.) If one *assumes* that $y = 0$ and $\delta \ll 1$, then

$$A_M^{\mathrm{incl}}(t) \approx \delta\left[x\sin\left(\Delta mt\right) - 2\sin^2\left(\Delta mt/2\right)\right], \tag{9.49}$$

a result derived by Beneke *et al.* (1997). This inclusive asymmetry increases with x and may be important in the B_s^0–$\overline{B_s^0}$ system. The time dependence in eqn (9.49) has two important implications. First, it allows a determination of Δm without resorting to flavour-specific final states. Second, it provides an automatic discrimination between the B_d^0–$\overline{B_d^0}$ and B_s^0–$\overline{B_s^0}$ systems. Of course, observing this time dependence has a cost in statistics. Also, if y turns out to be large, one should use the full expression in eqn (9.48) instead of the approximate eqn (9.49).

One might also consider partially inclusive asymmetries, where one sums over all final states with a particular flavour content. These asymmetries depend on all sources of CP violation: direct, mixing, and interference. Beneke *et al.* (1997) have claimed that such asymmetries might be useful in B decays. Their analysis uses local quark–hadron duality, in which one assumes that the quark-diagram kinematics is not affected by hadronization. The advantage of working with partially inclusive rates is that they are much larger than exclusive rates. On the other hand, the sum over exclusive modes 'dilutes' possible CP-violating asymmetries. Within a given model, this dilution factor may be calculable, up to hadronic matrix elements. However, the correctness of the local duality assumption and the effect of specific experimental conditions on the calculation of the dilution factors may be hard to quantify.

9.6   Untagged decays

Experimentally, the tagging of a P^0 or $\overline{P^0}$ inevitably implies a loss of statistics. This problem is worse when the experiments are performed in the dirty environment of hadron colliders. We want to discuss what can be learned from the decays of untagged mesons. This case is all the more important because one may design experiments with P^0–$\overline{P^0}$ pairs that reproduce this untagged condition (Yamamoto 1997a).

The untagged decay rates are

$$\Gamma_f(t) \equiv \Gamma[P^0(t) \to f] + \Gamma[\overline{P^0}(t) \to f]$$

$$= |A_f|^2 \frac{e^{-\Gamma t}}{2}\left[\left(1 + \left|\frac{p}{q}\right|^2\right)H + \left(1 - \left|\frac{p}{q}\right|^2\right)I\right]$$

$$= |A_f|^2 \frac{e^{-\Gamma t}}{1 - \delta}(H - \delta I). \tag{9.50}$$

H and I have been defined in eqns (9.15).

Let us consider first untagged decays into flavour-specific final states o and $\bar{o}$. Then,

$$\Gamma_o(t) = |A_o|^2 \frac{e^{-\Gamma t}}{1 - \delta} \left[\cosh \frac{\Delta \Gamma t}{2} - \delta \cos(\Delta m t) \right],$$

$$\Gamma_{\bar{o}}(t) = |\bar{A}_{\bar{o}}|^2 \frac{e^{-\Gamma t}}{1 + \delta} \left[\cosh \frac{\Delta \Gamma t}{2} + \delta \cos(\Delta m t) \right]. \tag{9.51}$$

The corresponding time-integrated rates are

$$\Gamma_o = \frac{|A_o|^2}{\Gamma(1 - \delta)} \left(\frac{1}{1 - y^2} - \frac{\delta}{1 + x^2} \right),$$

$$\Gamma_{\bar{o}} = \frac{|\bar{A}_{\bar{o}}|^2}{\Gamma(1 + \delta)} \left(\frac{1}{1 - y^2} + \frac{\delta}{1 + x^2} \right). \tag{9.52}$$

One can learn about mixing CP violation through the asymmetries

$$A_M^U(t) \equiv \frac{\Gamma_o(t) - \Gamma_{\bar{o}}(t)}{\Gamma_o(t) + \Gamma_{\bar{o}}(t)}$$

$$= \delta \frac{\cosh(\Delta \Gamma t/2) - \cos(\Delta m t)}{\cosh(\Delta \Gamma t/2) - \delta^2 \cos(\Delta m t)}. \tag{9.53}$$

The time-integrated version is

$$A_M^U \equiv \frac{\Gamma_o - \Gamma_{\bar{o}}}{\Gamma_o + \Gamma_{\bar{o}}}$$

$$= \delta \frac{x^2 + y^2}{1 + x^2 - \delta^2(1 - y^2)}. \tag{9.54}$$

In writing eqns (9.53) and (9.54), we have assumed that there is no direct CP violation, i.e., that $|A_o| = |\bar{A}_{\bar{o}}|$. If one abandons this assumption the expressions get considerably more complicated. For instance, A_M^U becomes

$$A_M^U = \frac{\delta(x^2 + y^2) + \kappa\left[1 + x^2 - \delta^2(1 - y^2)\right]}{1 + x^2 - \delta^2(1 - y^2) + \kappa\delta(x^2 + y^2)}, \tag{9.55}$$

where

$$\kappa \equiv \frac{|A_o|^2 - |\bar{A}_{\bar{o}}|^2}{|A_o|^2 + |\bar{A}_{\bar{o}}|^2}. \tag{9.56}$$

One should compare the measure of mixing CP violation, A_M^U in eqn (9.54), obtained from untagged data samples, with the observable A_M in eqn (9.27), corresponding to tagged data samples. In eqn (9.54), the CP-conserving mixing parameters x and y are not disentangled from the CP-violating parameter δ. For example, if δ is small, A_M^U is approximately equal to the product of δ and the CP-conserving quantity $(x^2 + y^2)/(1 + x^2)$. The latter must be measured before δ may be extracted. In addition, this CP-conserving factor may suppress A_M^U.

Let us now return to eqn (9.50). In the rest of this section we assume that $|q/p| = 1$. Then, I and the oscillatory dependence on $\Delta m t$ drop out from the un-tagged decay rates, and the result is proportional to the function H in eqn (9.15).

We first consider the decay rate into a flavour-specific final state o,

$$\Gamma_o(t) = \frac{|A_o|^2}{2} \left(e^{-\Gamma_H t} + e^{-\Gamma_L t} \right). \tag{9.57}$$

Fitting the data to this theoretical curve enables a measurement of $\Delta\Gamma$ (Dunietz 1995).

Now consider untagged decays into CP eigenstates. When the decay amplitudes into these CP eigenstates are dominated by a single weak phase, $|\lambda_f| = 1$ and the only CP violation appears in the difference of $|\mathrm{Re}\lambda_f|$ from 1. However, $\mathrm{Re}\lambda_f$ appears multiplied by $\sinh(\Delta\Gamma t/2)$, and can only be measured if $\Delta\Gamma$ is large. Therefore, untagged decays are useful only when $\Delta\Gamma$ is sufficiently large. That is the case in the neutral-kaon system, and may also be the case in the B_s^0–$\overline{B_s^0}$ system (Beneke et $al.$ 1996). The time-integrated rates are enhanced when $|y|$ approaches 1.

In § 9.4.1 we have shown that, when y vanishes, tagged decays into CP eigenstates dominated by a single weak phase provide a clean measurement of the sine of the CP-violating phase $2(\phi_A - \phi_M)$. Here we measure exclusively the cosine of that angle. The use of untagged decays in the context of the B_s^0–$\overline{B_s^0}$ system has been extensively discussed by Dunietz (1995), who presents many candidate channels to search for CP violation.

9.7   Correlated mesons

In this and the following sections we consider the case in which a P^0–$\overline{P^0}$ pair is created. Depending on how that pair is produced, the two mesons may have either correlated or uncorrelated wave functions. The former case occurs, for instance, at the production threshold of $Q\bar{Q}$ bound states such as J/ψ or Υ, or at the production threshold of a P^0–$\overline{P^0}{}^*$ pair, with its subsequent decay to P^0–$\overline{P^0}$. Uncorrelated wave functions occur in $p\bar{p}$ collisions, in the decays of the Z boson, and when the production of the heavy mesons is done on the e^+e^- continuum.

We first consider the case in which the P^0–$\overline{P^0}$ pair is produced in a state Φ^c with definite parity and C-parity $\eta_c = \pm 1$. This happens with the B_d^0–$\overline{B_d^0}$ pairs produced at the $\Upsilon(4S)$ (as in the forthcoming Belle and BaBar experiments), in which case $\eta_c = -1$, and with the B_s^0–$\overline{B_s^0}$ pairs produced at the $\Upsilon(5S)$, in which case $\eta_c = +1$. The formalism is the same as for the K^0–$\overline{K^0}$ pairs produced in the decay of the ϕ, which was treated in the previous chapter, except that we now want to write all formulas in terms of A_f and $\bar{A}_f$, and besides we want to consider the case $\eta_c = +1$, instead of treating only the simpler case $\eta_c = -1$.

The relevant resonance is typically produced in e^+e^- collisions and has angular momentum l. It decays into a P^0–$\overline{P^0}$ pair, which must also have angular momentum l. Therefore, both the parity and the C-parity of the pair are $\eta_c = (-1)^l$. The CP-parity is always +1. The initial pair is in a state

$$|\Phi^c\rangle = \tfrac{1}{\sqrt{2}} \left[|P^0(\vec{k})\rangle \otimes |\overline{P^0}(-\vec{k})\rangle + \eta_c |\overline{P^0}(\vec{k})\rangle \otimes |P^0(-\vec{k})\rangle \right], \tag{9.58}$$

where $\vec{k}$ and $-\vec{k}$ are the three-momenta of the left-moving and right-moving meson, respectively, in the resonance's rest frame.

The meson with momentum $\vec{k}$ decays at time t_1 into the final state f, and the meson with momentum $-\vec{k}$ decays at time t_2 into the final state g. The amplitude for this process is

$$\langle f, t_1; g, t_2 | T | \Phi^c \rangle = \tfrac{1}{\sqrt{2}} \{ a_c \left[g_-(t_1) g_+(t_2) + \eta_c g_+(t_1) g_-(t_2) \right]$$
$$+ b_c \left[g_+(t_1) g_+(t_2) + \eta_c g_-(t_1) g_-(t_2) \right] \} , \qquad (9.59)$$

where

$$\begin{aligned}
a_c &\equiv \frac{q}{p} \bar{A}_f \bar{A}_g + \eta_c \frac{p}{q} A_f A_g \\
&= A_f \bar{A}_g \left(\lambda_f + \eta_c \bar{\lambda}_g \right) , \\
b_c &\equiv A_f \bar{A}_g + \eta_c \bar{A}_f A_g \\
&= A_f \bar{A}_g \left(1 + \eta_c \lambda_f \bar{\lambda}_g \right) .
\end{aligned} \qquad (9.60)$$

The decay rate is proportional to

$$|\langle f, t_1; g, t_2 | T | \Phi^c \rangle|^2 = e^{-\Gamma t_+} \left[\frac{|a_c + b_c|^2}{8} e^{-\Gamma y t_c} + \frac{|a_c - b_c|^2}{8} e^{\Gamma y t_c} \right.$$
$$\left. + \frac{|b_c|^2 - |a_c|^2}{4} \cos\left(\Gamma x t_c \right) + \frac{\mathrm{Im}\left(a_c b_c^* \right)}{2} \sin\left(\Gamma x t_c \right) \right] , \quad (9.61)$$

where $t_c \equiv t_1 + \eta_c t_2$. The domain $t_1 \in [0, +\infty]$ and $t_2 \in [0, +\infty]$ corresponds to $t_- \equiv t_1 - t_2 \in [-\infty, +\infty]$ and $t_+ \equiv t_1 + t_2 \in [|t_-|, +\infty]$. For $\eta_c = +1$ the rate in eqn (9.61) depends only on t_+. Given N P^0–$\overline{P^0}$ pairs,

$$N \, |\langle f, t_1; g, t_2 | T | \Phi^c \rangle|^2 \, dt_1 \, dt_2$$

is the number of events characterized by a decay into the final state f during the time interval $[t_1, t_1 + dt_1]$ and a decay into the final g during the time interval $[t_2, t_2 + dt_2]$.

The times t_1 and t_2 are measured in the respective meson's rest frame. In the case of the $\Upsilon(4S)$, the resulting B_d^0 and $\overline{B}_d^0$ mesons move slowly in the resonance's rest frame. It is then practically immaterial whether t_1 and t_2 are measured in the meson's rest frame or in the resonance's rest frame. However, the difference between the times measured in the two frames must be taken into account in the case of the creation of K^0–$\overline{K^0}$ pairs from the decay of the ϕ (Kayser 1996).

In practice, unless the bunch dimensions are much smaller than the paths of the decaying mesons, the position of the resonance will not be determined with sufficient accuracy to measure the individual decay times. The situation is worse when the two mesons are produced nearly at rest, for then the decay vertices and the time difference may not be measurable in practice. This is the case with B_d^0–$\overline{B}_d^0$ production at the $\Upsilon(4S)$. The solution is to build experiments with e^+ and e^- beams of different energies, so that the mesons have a significant boost in the laboratory frame. The decay rates for experiments in which only the time difference t_- between the two decays is measured are obtained by integrating eqn (9.61) over t_+,

$$|\langle f; g; t_-|T|\Phi^c\rangle|^2 \equiv \int_{|t_-|}^{+\infty} dt_+ \left|\frac{\partial(t_1, t_2)}{\partial(t_+, t_-)}\right| |\langle f, t_1; g, t_2|T|\Phi^c\rangle|^2, \qquad (9.62)$$

where the Jacobian factor is

$$\left|\frac{\partial(t_1, t_2)}{\partial(t_+, t_-)}\right| = \tfrac{1}{2}. \qquad (9.63)$$

For $\eta_c = +1$ one obtains (Bigi and Sanda 1981; Fridman 1988; Xing 1996)

$$
\begin{aligned}
|\langle f; g; t_-|T|\Phi^+\rangle|^2 &= \frac{e^{-\Gamma|t_-|}}{\Gamma} \left[\frac{|a_+ + b_+|^2}{16(1+y)} e^{-\Gamma y|t_-|} + \frac{|a_+ - b_+|^2}{16(1-y)} e^{\Gamma y|t_-|} \right. \\
&\quad + \frac{|b_+|^2 - |a_+|^2}{8} \frac{\cos(\Gamma x|t_-|) - x\sin(\Gamma x|t_-|)}{1+x^2} \\
&\quad \left. + \frac{\mathrm{Im}(a_+ b_+^*)}{4} \frac{\sin(\Gamma x|t_-|) + x\cos(\Gamma x|t_-|)}{1+x^2} \right]. \qquad (9.64)
\end{aligned}
$$

For $\eta_c = -1$ the result is (Bigi and Sanda 1981; Fridman 1988; Xing 1996)

$$
\begin{aligned}
|\langle f; g; t_-|T|\Phi^-\rangle|^2 &= \frac{e^{-\Gamma|t_-|}}{\Gamma} \left[\frac{|a_- + b_-|^2}{16} e^{-\Gamma y t_-} + \frac{|a_- - b_-|^2}{16} e^{\Gamma y t_-} \right. \\
&\quad \left. + \frac{|b_-|^2 - |a_-|^2}{8} \cos(\Gamma x t_-) + \frac{\mathrm{Im}(a_- b_-^*)}{4} \sin(\Gamma x t_-) \right]. (9.65)
\end{aligned}
$$

If we integrate over t_-, we obtain the completely time-integrated rate, which may be written for $\eta_c = \pm 1$ as

$$
\begin{aligned}
|\langle f; g|T|\Phi^c\rangle|^2 &= \frac{1}{4\Gamma^2} \left[(|b_c|^2 + |a_c|^2) \frac{1+\eta_c y^2}{(1-y^2)^2} - 2\mathrm{Re}(a_c b_c^*) \frac{(1+\eta_c)y}{(1-y^2)^2} \right. \\
&\quad \left. + (|b_c|^2 - |a_c|^2) \frac{1-\eta_c x^2}{(1+x^2)^2} + 2\mathrm{Im}(a_c b_c^*) \frac{(1+\eta_c)x}{(1+x^2)^2} \right]. \qquad (9.66)
\end{aligned}
$$

If we use the unitarity relations,

$$
\begin{aligned}
\sum_f \sum_g |a_c|^2 &= \frac{2\Gamma^2}{1-\delta^2} \left[1 + \eta_c y^2 + \delta^2(1 - \eta_c x^2) \right], \\
\sum_f \sum_g |b_c|^2 &= \frac{2\Gamma^2}{1-\delta^2} \left[1 + \eta_c y^2 - \delta^2(1 - \eta_c x^2) \right], \qquad (9.67) \\
\sum_f \sum_g a_c b_c^* &= \frac{2\Gamma^2}{1-\delta^2} (1 + \eta_c)(y - i\delta^2 x).
\end{aligned}
$$

we can show that, summing eqn (9.66) over f and g, one obtains 1 as one should.

9.8   Quantum-mechanical effects with correlated states

9.8.1   Quantum mechanics of parity-odd P^0–$\overline{P^0}$ pairs

We highlight here an important feature of correlated mesons in a parity-odd state, i.e., when $\eta_c = -1$, as occurs for the B_d^0–$\overline{B}_d^0$ pairs formed from the $\Upsilon(4S)$.

Let us assume that the meson decaying at time t_2 decays into the flavour-specific final state o. The decay identifies the meson as being P^0 at that time. In general, this would not tell us anything about the flavour of the other meson. However, if the two mesons have started out in the parity-odd state Φ^-, we have

$$a_- = -\frac{p}{q} A_f A_o,$$
$$b_- = -\bar{A}_f A_o,$$

(9.68)

and, from eqn (9.65) and the third eqn (9.9), we find that

$$|\langle f; o; t_-|T|\Phi^-\rangle|^2 = \frac{|A_o|^2}{4\Gamma} \frac{\exp(-\Gamma|t_-|)}{\exp(-\Gamma t_-)} \Gamma[\overline{P^0}(t_-) \to f].$$

(9.69)

Similarly, if the meson decaying at time t_2 decays to $\bar{o}$, we learn that the flavour of that meson is $\overline{P^0}$ at that time. Then,

$$a_- = \frac{q}{p} \bar{A}_f \bar{A}_{\bar{o}},$$
$$b_- = A_f \bar{A}_{\bar{o}},$$

(9.70)

and, from eqn (9.65) and the first eqn (9.9),

$$|\langle f; \bar{o}; t_-|T|\Phi^-\rangle|^2 = \frac{|\bar{A}_{\bar{o}}|^2}{4\Gamma} \frac{\exp(-\Gamma|t_-|)}{\exp(-\Gamma t_-)} \Gamma[P^0(t_-) \to f].$$

(9.71)

How should we interpret the results in eqns (9.69) and (9.71)? When the initial $\overline{P^0}$ meson evolves in time, it oscillates back and forth into and from P^0. The same occurs with the initial P^0. But, the antisymmetry of the correlated wave function under the change $\vec{k} \to -\vec{k}$ is preserved by the linearity of the evolution. Hence, if at some instant $t_2 \neq 0$ the right-moving meson is found—from its flavour-tagging decay—to be P^0, then the left-moving meson *at that instant* is certainly $\overline{P^0}$. That left-moving meson will *from that instant on* evolve as a tagged $\overline{P^0}$. Thus, time-dependent experiments starting from the state Φ^- and tagging the flavour of one meson automatically reproduce the situation discussed in § 9.2.

However, it is important to note that what one usually calls 'time-integrated measurements' are *not* the same in the two situations. In § 9.2 we have considered the time evolution of a neutral meson whose flavour content was determined at time $t = 0$. To obtain time-integrated observables we have integrated over the time variable from $t = 0$ to $t = +\infty$. In the case of a correlated initial state Φ^-, we determine the flavour of the meson in one side of the detector at time t_2. At that instant the meson in the opposite side of the detector has opposite flavour, and evolves thereafter as a tagged initial state with time variable $t_- = t_1 - t_2$. However, if we are unable to measure t_-—this is the case when the two mesons move very slowly in the laboratory frame—we must integrate t_- from $-\infty$ to $+\infty$. The problem, as we shall later show, is that the terms proportional to $\mathrm{Im}\lambda_f$ in the time-integrated asymmetries of § 9.2 vanish when the time integration is performed over the domain $t_- = [-\infty, +\infty]$ instead of being performed over the domain $t = [0, +\infty]$. Hence, the usual time-integrated asymmetries—in which one tags one meson and looks for the decay of the other meson into a CP eigenstate—cannot be used to test for interference CP violation at symmetric colliders producing $\Upsilon(4S)$, because, in that case, we are unable to measure t_-.

Of course, if we can measure the time difference, and in particular its sign, we may select the events in the positive-t_- interval. Only then are we able to reproduce the time-integrated results of § 9.2.

The tagging possibility described in this section does not occur for the Φ^+ state. If the mesons arise from Φ^+, tagging the flavour of the right-moving meson to be P^0 at some instant *does not* guarantee that the left-moving meson is $\overline{P^0}$ at the same instant.

9.8.2 *Other quantum-mechanical effects*

Using eqns (6.56) we may rewrite $|\Phi^+\rangle$ and $|\Phi^-\rangle$ as

$$|\Phi^+\rangle = \frac{1}{2\sqrt{2}pq} \left[|P_H(\vec{k})\rangle \otimes |P_H(-\vec{k})\rangle - |P_L(\vec{k})\rangle \otimes |P_L(-\vec{k})\rangle \right],$$

$$|\Phi^-\rangle = \frac{1}{2\sqrt{2}pq} \left[|P_L(\vec{k})\rangle \otimes |P_H(-\vec{k})\rangle - |P_H(\vec{k})\rangle \otimes |P_L(-\vec{k})\rangle \right]. \tag{9.72}$$

Thus, if we have Φ^-, and if P_H is found in one side of the detector, we are sure that there is P_L in the opposite side of the detector, and vice versa. On the other hand, with Φ^+, either we have P_L in both sides of the detector, or P_H in both sides.

This is important because, as P_H and P_L are the eigenstates of evolution, after we have tagged them at some time, they remain the same for all other times. Thus, if at time t_2 the meson in one side of the detector is found to be P_H, then at any later time, the meson in the opposite side of the detector will be P_H if the original state was Φ^+, or P_L if the original state was Φ^-.

This is all the more interesting in the CP-conserving case, when P_H and P_L are eigenstates of CP: $P_H = P_\pm$ and $P_L = P_\mp$. It then means that tagging the CP quantum number of the meson in the right side of the detector at some time automatically fixes the CP quantum number of the meson in the left side of the detector at any other time. We shall consider an application of this fact in § 9.10.4.

9.9 Uncorrelated mesons

We turn to the case in which the P^0–$\overline{P^0}$ pair is produced in an incoherent state. This occurs in $p\bar{p}$ collisions, Z decays, and on the e^+e^- continuum (for instance at LEP). In this case, there is an equal probability that at the initial instant the left-moving meson was a P^0 and the right-moving meson was a $\overline{P^0}$, or the opposite situation occurred; the two possibilities are incoherently superposed.

Considering as before that the left-moving meson decays at proper time t_1 into a final state f, while the right-moving meson decays at proper time t_2 into the final state g, the density of probability for this process is

$$
|\langle f, t_1; g, t_2|T|\Phi^u\rangle|^2
$$
$$
= \tfrac{1}{2}\left\{\Gamma[P^0(t_1) \to f]\Gamma[\overline{P^0}(t_2) \to g] + \Gamma[\overline{P^0}(t_1) \to f]\Gamma[P^0(t_2) \to g]\right\}. \quad (9.73)
$$

It may be checked that this is equal to

$$
\tfrac{1}{2}\left(\left|\langle f, t_1; g, t_2|T|\Phi^+\rangle\right|^2 + \left|\langle f, t_1; g, t_2|T|\Phi^-\rangle\right|^2\right). \quad (9.74)
$$

This must be the case since the uncorrelated state is equivalent to an incoherent superposition of even-parity and odd-parity correlated states. Equations (9.73) and (9.74) are two different algorithms to compute the decay rates of uncorrelated P^0–$\overline{P^0}$ pairs.

Thus, for instance, in order to obtain the probability that a decay into f occurs in the left half of the detector and a decay into g occurs in the right half of the detector, we must average eqn (9.66) for $\eta_c = +1$ and $\eta_c = -1$, obtaining

$$
|\langle f; g|T|\Phi^u\rangle|^2 = \frac{|A_f \bar{A}_g|^2}{4\Gamma^2}\left[\frac{\left(1 + |\lambda_f|^2\right)\left(1 + |\bar{\lambda}_g|^2\right) + 4y^2 \mathrm{Re}\lambda_f \mathrm{Re}\bar{\lambda}_g}{(1 - y^2)^2}\right.
$$
$$
+ \frac{\left(1 - |\lambda_f|^2\right)\left(1 - |\bar{\lambda}_g|^2\right) + 4x^2 \mathrm{Im}\lambda_f \mathrm{Im}\bar{\lambda}_g}{(1 + x^2)^2}\Bigg]
$$
$$
+ \frac{1}{2\Gamma^2}\left[-\frac{y\mathrm{Re}\left(a_+ b_+^*\right)}{(1 - y^2)^2} + \frac{x\mathrm{Im}\left(a_+ b_+^*\right)}{(1 + x^2)^2}\right]. \quad (9.75)
$$

This is equal to

$$
|\langle f; g|T|\Phi^u\rangle|^2 = \tfrac{1}{2}\left\{\Gamma[P^0 \to f]\Gamma[\overline{P^0} \to g] + \Gamma[\overline{P^0} \to f]\Gamma[P^0 \to g]\right\}, \quad (9.76)
$$

as is easily checked from eqns (9.10).

9.10 CP violation with neutral-meson pairs

Earlier in this chapter we have discussed some of the mixing and CP-violation measurements which are possible when one tags the initial flavour of one neutral meson. We now turn our attention to the case in which a P^0–$\overline{P^0}$ pair is produced, in either a coherent or incoherent state. The decay rates have been presented in the previous two sections. Again, we focus our attention on flavour-specific decays, and on decays into CP eigenstates.

9.10.1 Decays into a single flavour-specific final state

We have shown in § 9.8.1 that, for odd-parity correlated P^0–$\overline{P^0}$ pairs, the evolution of the meson in one side of the apparatus, after the meson on the opposite side has been tagged through its decay into a flavour-specific final state, is the same as the evolution of a tagged initial meson. In other words, for odd-parity correlated initial states, tagging one of the mesons effectively tags the other meson too, at that instant. Therefore, the analysis of mixing and CP is the same as the one in § 9.3 and 9.4. This is the reason why many authors disregard the correlated nature of the meson production and only discuss tagged, single-meson decay rates when discussing B-physics experiments at the $\Upsilon(4S)$. As explained in § 9.8.1, one must only be careful in this comparison when considering time-integrated decays.

The case of uncorrelated initial states is important, since it is the situation occuring at LEP, for both B_d^0–$\overline{B_d^0}$ and B_s^0–$\overline{B_s^0}$ pairs. Let us assume that the meson in the opposite side of the detector is found to be P^0 through its decay into o, at time t_2. If we integrate eqn (9.73) over t_2 we find

$$|\langle f, t_1; o|T|\Phi^u \rangle|^2 = \frac{|A_o|^2}{2} \left\{ \Gamma[P^0(t_1) \to f] \mathrm{TIP}[\overline{P^0} \to P^0] \right.$$
$$\left. + \Gamma[\overline{P^0}(t_1) \to f] \mathrm{TIP}[P^0 \to P^0] \right\}. \qquad (9.77)$$

This is what one would expect. We know the meson in the opposite side of the detector to be a P^0. In a number of times proportional to $\mathrm{TIP}[P^0 \to P^0]$, that meson has evolved from an initial P^0. In those cases, the meson in this side of the detector was originally a $\overline{P^0}$, and evolved as such, finally decaying into f at time t_1. But, the meson detected in the opposite side as a P^0 might have been at the initial instant a $\overline{P^0}$. This occurs a number of times proportional to $\mathrm{TIP}[\overline{P^0} \to P^0]$. In that case the meson in this side was originally a P^0. The two situations cannot be distinguished and must be incoherently averaged. In this context, experimentalists usually refer to $\mathrm{TIP}[P^0 \to P^0]$ and to $\mathrm{TIP}[\overline{P^0} \to P^0]$ as the 'right' and 'wrong' tags, respectively.

Analogously,

$$|\langle f, t_1; \bar{o}|T|\Phi^u \rangle|^2 = \frac{|\bar{A}_{\bar{o}}|^2}{2} \left\{ \Gamma[P^0(t_1) \to f] \mathrm{TIP}[\overline{P^0} \to \overline{P^0}] \right.$$
$$\left. + \Gamma[\overline{P^0}(t_1) \to f] \mathrm{TIP}[P^0 \to \overline{P^0}] \right\}. \qquad (9.78)$$

If enough statistics is made available, these decays may be used to look for mixing and CP violation. At LEP1, the statistics was such that the mixing variables x_d and x_s could be probed—see Chapter 10.

Another interesting use of the detection of a single flavour-specific decay occurs when we sum over all decay channels and integrate over all decay times in one side of the detector. The result is the same for either uncorrelated or correlated (with $\eta_c = +1$ or $\eta_c = -1$) initial states:

$$\int_0^\infty dt_2 \sum_g |\langle f, t_1; g, t_2|T|\Phi^u\rangle|^2 = \int_0^\infty dt_2 \sum_g |\langle f, t_1; g, t_2|T|\Phi^c\rangle|^2$$

$$= \tfrac{1}{2} \left\{ \Gamma[P^0(t_1) \to f] + \Gamma[\overline{P^0}(t_1) \to f] \right\}. \quad (9.79)$$

As for each event there was initially one P^0 and one $\overline{P^0}$, the decaying meson in the relevant side of the detector may have been initially either a P^0 or a $\overline{P^0}$, with equal probability. The result in eqn (9.79) still holds when CPT is violated (Yamamoto 1997a,b).

Since the final state f may be found in either side of the detector, the probability that one finds it gets an extra factor of 2,

$$\Gamma[\Phi(t_1) \to f] = \Gamma[P^0(t_1) \to f] + \Gamma[\overline{P^0}(t_1) \to f], \quad (9.80)$$

which is precisely the probability density $\Gamma_f(t_1)$ discussed in § 9.6. Then, the analysis of mixing and CP violation follows the one presented in that section. In particular, we may look for CP violation in the mixing through the single-tag asymmetry,

$$A_l(t) \equiv \frac{\Gamma[\Phi(t) \to o] - \Gamma[\Phi(t) \to \bar{o}]}{\Gamma[\Phi(t) \to o] + \Gamma[\Phi(t) \to \bar{o}]}. \quad (9.81)$$

The result is equal to the untagged asymmetry $A_M^U(t)$ in eqn (9.53). Integrating over time one finds, from eqn (9.54), the time-integrated single-tag asymmetry (Hagelin 1979)

$$A_l = A_M^U \equiv \frac{N[o] - N[\bar{o}]}{N[o] + N[\bar{o}]} \approx \delta \frac{x^2 + y^2}{1 + x^2}, \quad (9.82)$$

where we have assumed that δ is small, and have used the notation $N[f] \equiv \Gamma[\Phi \to f]$. We emphasize that this asymmetry does not depend on how the initial state is prepared: it is the same for correlated as well as for uncorrelated initial states.

For odd-parity initial states, there is another experimental procedure that yields a result proportional to $\Gamma_f(t_1)$. Using eqns (9.69) and (9.71), we find

$$\left| \langle f; o; t_-|T|\Phi^-\rangle \right|^2 + \left| \langle f; \bar{o}; t_-|T|\Phi^-\rangle \right|^2$$

$$= \frac{|A_o|^2}{4\Gamma} \left\{ \Gamma[P^0(t_-) \to f] + \Gamma[\overline{P^0}(t_-) \to f] \right\}, \quad (9.83)$$

for $t_- > 0$. We have assumed that there is no direct CP violation in the flavour-specific decays $(|A_o| = |\bar{A}_{\bar{o}}|)$.

9.10.2 *Decays into two flavour-specific final states*

We now consider cases in which both mesons decay into flavour-specific final states. If there were no mixing, the initial P^0–$\overline{P^0}$ pair would only be allowed to decay into $o\bar{o}$. Using eqn (9.66), we obtain the totally time-integrated rate

$$|\langle o; \bar{o}|T|\Phi^c\rangle|^2 = \frac{|A_o \bar{A}_{\bar{o}}|^2}{4\Gamma^2}\left[\frac{1+\eta_c y^2}{(1-y^2)^2} + \frac{1-\eta_c x^2}{(1+x^2)^2}\right]. \tag{9.84}$$

The fact that P^0 and $\overline{P^0}$ mix also allows for

$$|\langle o; o|T|\Phi^c\rangle|^2 = \frac{|A_o|^4}{4\Gamma^2}\left|\frac{p}{q}\right|^2\left[\frac{1+\eta_c y^2}{(1-y^2)^2} - \frac{1-\eta_c x^2}{(1+x^2)^2}\right],$$
$$|\langle \bar{o}; \bar{o}|T|\Phi^c\rangle|^2 = \frac{|\bar{A}_{\bar{o}}|^4}{4\Gamma^2}\left|\frac{q}{p}\right|^2\left[\frac{1+\eta_c y^2}{(1-y^2)^2} - \frac{1-\eta_c x^2}{(1+x^2)^2}\right]. \tag{9.85}$$

According to eqn (9.74), the numbers of events for uncorrelated initial pairs are obtained by averaging the above expressions for both signs of η_c. This is equivalent to setting η_c to zero.

Let us assume as before that $|A_o| = |\bar{A}_{\bar{o}}|$, and define $N[fg] \equiv |\langle f; g|T|\Phi\rangle|^2$, for both correlated and uncorrelated initial states. We introduce the mixing parameters

$$R \equiv \frac{N[\bar{o}\bar{o}] + N[oo]}{N[\bar{o}o] + N[o\bar{o}] + N[\bar{o}\bar{o}] + N[oo]}, \tag{9.86}$$

$$R' \equiv \frac{N[\bar{o}\bar{o}] + N[oo]}{N[\bar{o}o] + N[o\bar{o}]}, \tag{9.87}$$

which are related through

$$R = \frac{R'}{1+R'}. \tag{9.88}$$

For uncorrelated and for odd-parity initial states, R' is related to the Pais–Treiman parameters through (Fridman 1988)

$$R'_{\text{uncorr}} = \frac{r+\bar{r}}{1+r\bar{r}}, \tag{9.89}$$

$$R'_{\eta_c=-1} = \frac{r+\bar{r}}{2}, \tag{9.90}$$

where we have used eqns (9.84) and (9.85). In contrast, for an even-parity initial state,

$$R'_{\eta_c=+1} = \frac{1}{2}\left(\left|\frac{p}{q}\right|^2 + \left|\frac{q}{p}\right|^2\right)\frac{3x^2 + x^4 + y^2\left(3+x^4+x^2y^2-y^2\right)}{2+x^2+x^4+y^2\left(-1+4x^2+x^4-x^2y^2+y^2\right)}. \tag{9.91}$$

In general, this expression cannot be written in terms of r and $\bar{r}$.

We may probe CP violation in the mixing through the di-tag asymmetry

$$A_{ll} = \frac{N[oo] - N[\bar{o}\bar{o}]}{N[oo] + N[\bar{o}\bar{o}]} = \frac{|p/q|^2 - |q/p|^2}{|p/q|^2 + |q/p|^2} = \frac{2\delta}{1 + \delta^2}. \tag{9.92}$$

This result is independent of the way in which the initial state was prepared, and it coincides with the asymmetry A_M of eqn (9.27), which was defined for single tagged mesons.

Thus, for any initial state, we may use either the single-lepton asymmetry in eqn (9.82) or the dilepton asymmetry in eqn (9.92) to look for mixing CP violation. Which asymmetry will prove more accurate will depend on the values of x^2 and y^2; A_l is better for large mixing. Since the two methods involve data sets that are, to a large extent, statistically independent, they may be combined to improve the sensitivity (Yamamoto 1997a).

If there is direct CP violation in the decays tagging the neutral mesons, disentangling the different sources of CP violation should not be easy. In particular, for a given initial state, i.e., for some value of η_c, we cannot have a clean measurement of direct CP violation with dilepton time-integrated measurements alone. This is contrary to what happens when the flavour of a single meson is tagged through the associated production of a charged meson. In that case, we were able to define the asymmetry A_D measuring direct CP violation.

For uncorrelated initial states one may also consider a time-integrated tag on one side, and look for the time-dependence on the opposite side. Let us assume that $|A_o| = |\bar{A}_{\bar{o}}|$, and define

$$N[fg](t_1) \equiv |\langle f, t_1; g|T|\Phi^u\rangle|^2. \tag{9.93}$$

Using eqns (9.77) and (9.78), we find

$$\frac{N[\bar{o}\bar{o}](t_1) + N[oo](t_1)}{N[\bar{o}o](t_1) + N[o\bar{o}](t_1) + N[\bar{o}\bar{o}](t_1) + N[oo](t_1)} = \frac{1}{2}\left[1 - \frac{1 - y^2}{1 + x^2}\frac{\cos(\Delta m t_1)}{\cosh(\Delta \Gamma t_1/2)}\right]. \tag{9.94}$$

At LEP1 this method has been used to observe an oscillatory time dependence in the B_d^0–$\overline{B_d^0}$ system for the first time. Although the analyses of the LEP1 experiments have assumed $y = 0$, that assumption is not really needed. Indeed, y only appears in the prefactor $(1 - y^2)/(1 + x^2)$—which may be determined from time-integrated rates alone—and in the non-oscillatory denominator $\cosh(\Delta \Gamma t/2)$.

The prefactor $(1 - y^2)/(1 + x^2)$ is usually called 'dilution factor due to mixing' and denoted D_M by experimentalists. One might worry about the fact that the current bound $x_s > 9.5$ for the B_s^0–$\overline{B_s^0}$ system imposes a huge dilution. However, at LEP, HERA-B, or LHC, one produces all the combinations described in the introduction as cases A), B), C), and D) together. Therefore, the dilution factor in those cases should be obtained by averaging over all B species contributing to the tag signal. The average is dominated by the large values of D_M. This comment is also applicable to the tagging dilution found in the CP-violating asymmetries to be discussed in the next section.

9.10.3 *Decays into a flavour-specific state and a CP eigenstate*

Consider the case in which one neutral meson decays into a flavour-specific state, either o or $\bar{o}$, and the other meson decays into a CP eigenstate f. We shall assume that there is no direct CP violation in the tagging decays, $|A_o| = |\bar{A}_{\bar{o}}|$, and that there is no CP violation in the mixing, $|q/p| = 1$.

We first look at uncorrelated initial states Φ^u. We integrate over the time of the flavour-tagging decay, and follow the time dependence of the decay into the CP eigenstate. Using eqns (9.77) and (9.78), we find[23]

$$A^u_{CP}(t_1) \equiv \frac{N[\bar{o}f](t_1) - N[of](t_1)}{N[\bar{o}f](t_1) + N[of](t_1)}$$
$$= \frac{1 - y^2}{1 + x^2} A_{CP}(t_1). \tag{9.95}$$

Thus, for uncorrelated initial states, the mixing produces 'wrong tags' that dilute the asymmetry $A_{CP}(t)$ defined in eqn (9.35) for the case of tagged initial mesons.

Constructing the same asymmetry with time-integrated rates we find

$$A^u_{CP} \equiv \frac{N[\bar{o}f] - N[of]}{N[\bar{o}f] + N[of]}$$
$$= \frac{1 - y^2}{1 + x^2} A_{CP}. \tag{9.96}$$

Note that A_{CP}, defined in eqn (9.36) for tagged initial mesons, already has a factor $(1 - y^2)/(1 + x^2)$ in it.

We next consider decays from correlated states Φ^- or Φ^+. In this case it is more convenient to define

$$N^c[f; g; t_-] \equiv |\langle f; g; t_- | T | \Phi^c \rangle|^2 . \tag{9.97}$$

For Φ^- the analysis of this case is very simple, due to the quantum-mechanical effects discussed in § 9.8.1. One obtains

$$A^-_{CP}(t_-) \equiv \frac{N^-[\bar{o}; f; t_-] - N^-[o; f; t_-]}{N^-[\bar{o}; f; t_-] + N^-[o; f; t_-]}$$
$$= A_{CP}(t_-), \tag{9.98}$$

reproducing the asymmetry $A_{CP}(t_-)$ defined for tagged mesons—see eqn (9.35). However, we should stress that the time variable t_- lies in the interval $[-\infty, +\infty]$, while the time variable for the tagged decays of § 9.4.1 must be positive. This has a dramatic impact on the time-integrated rates. Indeed, if we separately

---

[23]The superscript u in A^u_{CP} refers to an uncorrelated initial state. It should not be confused with A^U_{CP}, which refers to untagged decays.

integrate each rate over t_- and construct the asymmetry analogous to the one in eqn (9.98), we obtain

$$A_{CP}^- = \frac{1-y^2}{1+x^2}\frac{1-|\lambda_f|^2}{1+|\lambda_f|^2},$$
(9.99)

which vanishes whenever $|\lambda_f| = 1$. This is responsible for the need to construct asymmetric B factories. With symmetric B factories of Φ^-, tagged decay rates can only be used to test direct CP violation (Deshpande and He 1996).

The time dependence for Φ^+ decays may be derived directly from eqn (9.64). The expressions are not as simple as in the previous case. Still, it is instructive to look at the dominant contributions to the asymmetry under the additional assumptions that $y = 0$ and $x \gg 1$. We find

$$A_{CP}^+(t_-) \equiv \frac{N^+[\bar o; f; t_-] - N^+[o; f; t_-]}{N^+[\bar o; f; t_-] + N^+[o; f; t_-]}$$

$$\approx \frac{x}{1+x^2}\left[\frac{|\lambda_f|^2 - 1}{|\lambda_f|^2 + 1}\sin(\Delta m|t_-|) + \frac{\mathrm{Im}\lambda_f}{|\lambda_f|^2 + 1}\cos(\Delta m|t_-|)\right].$$
(9.100)

This result, which is relevant for B_s^0–$\overline{B_s^0}$ pairs from the $\Upsilon(5S)$ resonance, exhibits an interesting feature: direct CP violation appears multiplied by $\sin(\Delta m|t_-|)$, while interference CP violation multiplies $\cos(\Delta m|t_-|)$. In particular, if there is no direct CP violation, we get

$$A_{CP}^+(t_-) = \frac{x}{2(1+x^2)}\mathrm{Im}\lambda_f \cos(\Delta m|t_-|),$$
(9.101)

rather than the usual dependence on $\sin(\Delta mt)$ found in eqn (9.35).

As for the asymmetry of time-integrated rates, we find the general expression

$$A_{CP}^+ = \left(\frac{1-y^2}{1+x^2}\right)^2 \frac{(1-x^2)(1-|\lambda_f|^2) + 4x\mathrm{Im}\lambda_f}{(1+y^2)(1+|\lambda_f|^2) - 4y\mathrm{Re}\lambda_f}.$$
(9.102)

For the particular case $y = 0$ and $x \gg 1$,

$$A_{CP}^+ \approx \frac{x^2\left(|\lambda_f|^2 - 1\right) + 4x\mathrm{Im}\lambda_f}{\left(|\lambda_f|^2 + 1\right)(1+x^2)^2}.$$
(9.103)

9.10.4 Decays into two CP eigenstates

We now turn our attention to the case in which both final states, f and g, are CP eigenstates (Bigi and Sanda 1987). Let us denote the CP-parities of f and g by η_f and η_g, respectively. We remind the reader that the initial state Φ^c has parity and C-parity equal to $\eta_c = (-1)^l$, where l is the spin of Φ^c. Hence, the CP-parity of Φ^c is $+1$. We want to prove

Theorem 9.1 *If, at any times t_1 and t_2, decays occur to eigenstates f and g of CP such that $\eta_c\eta_f\eta_g = -1$, then there is CP violation.*

Proof From eqn (7.28) we know that, if CP is conserved,

$$\lambda_g = \eta_f\eta_g\lambda_f. \tag{9.104}$$

Other necessary conditions for CP invariance are

$$|\lambda_f| = 1 \tag{9.105}$$

and

$$\text{Im}\lambda_f = 0. \tag{9.106}$$

Using eqns (9.104) and (9.105), the parameters a_c and b_c in eqns (9.60) become

$$a_c = A_f\bar{A}_g\left(\lambda_f + \eta_c\eta_f\eta_g\lambda_f^*\right),$$

$$b_c = A_f\bar{A}_g\left(1 + \eta_c\eta_f\eta_g\right).$$

Clearly, if $\eta_c\eta_f\eta_g = -1$ then $b_c = 0$. Also, $a_c \propto \text{Im}\lambda_f$ vanishes because of eqn (9.106). Thus, Φ^c is forbidden by CP invariance from decaying to f and g.
□

Recall eqns (9.72). If one starts with a Φ^-, whenever we have a P_H in one side of the detector, we must have a P_L in the opposite side, and vice versa. On the other hand, with a Φ^+, either we have P_L in both sides of the detector, or we have P_H in both of them.

Now, if CP is conserved, P_H and P_L coincide with the eigenstates of CP, and they have opposite CP-parities. Suppose that we start with a Φ^-, i.e., $\eta_c = -1$, and that at some time t_1 we observe the decay of one of the mesons to a CP eigenstate f with CP-parity η_f. We then learn that the meson in the opposite side of the detector must have the opposite CP-parity, and therefore it cannot decay, either at that time *or at any other time*, to a CP eigenstate g with CP-parity $\eta_g = \eta_f$.

On the other hand, if we start with Φ^+ and observe a CP eigenstate f in one side of the detector, we learn that the meson in the opposite side must have the same CP-parity η_f, and it cannot decay to a CP eigenstate g if $\eta_g = -\eta_f$.

Of course, the fact that there is CP violation whenever decays with $\eta_c\eta_f\eta_g = -1$ occur, does not inform us where exactly the CP violation lies. We must have some extra information, or some assumptions, if we want to explicitly determine a CP-violating quantity from the experimental data. Moreover, the branching ratios into exclusive CP eigenstates will in general be small, in the case of heavy mesons. It is then advisable to look for time-integrated rates.

As an example, let us assume that there is no CP violation in the mixing, $|q/p| = 1$, and that there is no direct CP violation in either of the decays. Then, $|\lambda_f| = |\lambda_g| = 1$. Let us moreover assume that eqn (9.104) also holds, i.e., $\lambda_g = \eta_f\eta_g\lambda_f$. (This occurs if the decays into f and g are dominated by the same diagram.) For instance, $\eta_f \eta_g = +1$ when $f = g$, and $\eta_f \eta_g = -1$ for $f = J/\Psi K_S$ and $g = J/\Psi K_L$.[24] With these simplifications and $\eta_c \eta_f \eta_g = -1$, eqns (9.60) yield

$$a_c = 2i A_f \bar{A}_g \mathrm{Im}\lambda_f, \tag{9.107}$$

$$b_c = 0, \tag{9.108}$$

from which eqn (9.66) becomes (Xing 1996)

$$|\langle f; g|T|\Phi^c\rangle|^2 = \frac{\left|A_f \bar{A}_g\right|^2}{\Gamma^2} \left[\frac{1 + \eta_c y^2}{(1 - y^2)^2} - \frac{1 - \eta_c x^2}{(1 + x^2)^2}\right] \mathrm{Im}^2 \lambda_f. \tag{9.109}$$

The observation of this time-integrated quantity determines $\mathrm{Im}\lambda_f$ up to its sign.

In particular, let us consider the decays of $\Upsilon(4S)$ into two identical CP eigenstates, such as $\Upsilon(4S) \to J/\Psi K_S \, J/\Psi K_S$ (Wolfenstein 1984; Gavela *et al.* 1985*b*; Bigi and Sanda 1987). If the decays are dominated by a single weak phase, one gets

$$|\langle f_{cp}; f_{cp}|T|\Phi^-\rangle|^2 = \frac{\left|A_f \bar{A}_f\right|^2}{\Gamma^2} \frac{x^2 + y^2}{(1 - y^2)(1 + x^2)} \mathrm{Im}^2 \lambda_f. \tag{9.110}$$

Thus, we may test for interference CP violation in time-integrated rates—and, therefore, even at symmetric B-factories. To do this, we must look for decays into two CP eigenstates rather than for decays into one CP eigenstate and one flavour-specific final state. Unfortunately, the corresponding rates are expected to be very small.

The case of uncorrelated initial states is less favorable, since the CP-forbidden transitions should be overwhelmed by CP-allowed transitions. In that case, signs of CP violation can only be sought through rate differences involving large cancellations. We shall not discuss them here.

---

[24]Strictly speaking, we should use the neutral-kaon CP eigenstates instead of K_S and K_L. The difference between the choices is equivalent to an effect of order 10^{-3}. We are implicitly assuming that this is negligible when compared to the intereference CP violation in the heavy-meson decays.

10

EXPERIMENTAL STATUS OF B^0–$\overline{B^0}$ MIXING

10.1 Introduction

In this chapter we discuss some experimental information on the B^0–$\overline{B^0}$ systems, where B^0 may be either B_d^0 or B_s^0. The experimental results that we shall discuss are important for two reasons. First, x_d (the mixing parameter x of the B_d^0–$\overline{B_d^0}$ system) provides a constraint on the standard model's CKM matrix, as we shall discuss in Chapter 18. Second, mixing measurements fix the values of the dilution factors that time-integrated experiments on CP violation will be faced with. For more details the reader may wish to consult the reviews by Fridman (1988) and by Schröder (1994), and the original articles of the various experimental collaborations.

10.2 Mixing variables in the B^0–$\overline{B^0}$ systems

Several sets of variables are commonly used to describe mixing in the B^0–$\overline{B^0}$ systems. The Pais–Treiman parameters refer to the evolution of a single, tagged, neutral meson and have been defined in eqns (9.20). Four other variables referring to the properties of a tagged neutral meson are introduced in this context:

$$\chi_0 \equiv \frac{x^2 + y^2}{2\left(1 + x^2\right)}, \tag{10.1}$$

$$\chi \equiv \frac{\text{TIP}[B^0 \to \overline{B^0}]}{\text{TIP}[B^0 \to B^0] + \text{TIP}[B^0 \to \overline{B^0}]} = \frac{|q/p|^2 \chi_0}{1 - \chi_0 + |q/p|^2 \chi_0}, \tag{10.2}$$

$$\bar{\chi} \equiv \frac{\text{TIP}[\overline{B^0} \to B^0]}{\text{TIP}[\overline{B^0} \to \overline{B^0}] + \text{TIP}[\overline{B^0} \to B^0]} = \frac{|p/q|^2 \chi_0}{1 - \chi_0 + |p/q|^2 \chi_0}, \tag{10.3}$$

$$\chi' \equiv \Gamma\,\text{TIP}[B^0 \to \overline{B^0}] = \left|\frac{q}{p}\right|^2 \frac{\chi_0}{1 - y^2}. \tag{10.4}$$

Notice that $0 \leq \chi_0 \leq 1/2$, because $0 \leq |y| \leq 1$. We want to emphasize once again that the time-integrated probabilities (TIP) that appear in eqns (10.2), (10.3), and (10.4) refer in practice to decays to flavour-tagging (semileptonic) modes, and not to real transitions. Also, they have dimensions of time; they are not probabilities. On the other hand, the parameter χ may be interpreted as a probability. Consider an initial tagged B^0 meson and its decays into flavour-specific final states o and $\bar{o}$. The parameter χ is the probability that the tagged B^0 will decay as $\overline{B^0}$; $1 - \chi$ is the probability that it will be observed as a B^0. A similar reasoning applies to $\bar{\chi}$.

The Pais–Treiman parameters may be written in terms of χ_0 as

$$r = \left|\frac{q}{p}\right|^2 \frac{\chi_0}{1 - \chi_0},$$

$$\bar{r} = \left|\frac{p}{q}\right|^2 \frac{\chi_0}{1 - \chi_0}. \tag{10.5}$$

Similarly, the dilution factor

$$D_M \equiv \frac{1 - y^2}{1 + x^2} = 1 - 2\chi_0. \tag{10.6}$$

The unitarity bound of eqn (6.54) becomes

$$\delta^2 \leq D_M. \tag{10.7}$$

In the case of production of (either correlated or uncorrelated) B^0-$\overline{B^0}$ pairs, one may measure the total numbers of flavour-specific final states, integrated over all decay times. From these one may form the parameters R and R', which have been defined in eqns (9.86) through (9.88), and are related to the Pais–Treiman parameters (or to x and y) by eqns (9.89)–(9.91).

One should be aware of the fact that, often, several unstated assumptions are made when experimental results are quoted. Typically, it is assumed that CP violation in the mixing is small ($|q| = |p|$ or $r = \bar{r}$), as predicted in the standard model for both the B^0_d-$\overline{B^0_d}$ and B^0_s-$\overline{B^0_s}$ systems. Henceforth we shall assume that $|q| = |p|$. Then,

$$\bar{\chi} = \chi = \chi_0,$$

$$\chi' = \frac{\chi}{1 - y^2}, \tag{10.8}$$

and $\chi \leq \chi'$. Since χ and $\bar{\chi}$ coincide in this limit, we may write symbolically

$$\mathrm{Prob}[B^0 \to \overline{B^0}] = \mathrm{Prob}[\overline{B^0} \to B^0] = \chi,$$

$$\mathrm{Prob}[B^0 \to B^0] = \mathrm{Prob}[\overline{B^0} \to \overline{B^0}] = 1 - \chi. \tag{10.9}$$

Also,

$$R_{\eta_c = -1} = \frac{r}{1 + r} = \chi,$$

$$R'_{\eta_c = -1} = r = \frac{\chi}{1 - \chi}, \tag{10.10}$$

and

$$R_{\mathrm{uncorr}} = \frac{2r}{(1 + r)^2} = 2\chi(1 - \chi),$$

$$R'_{\mathrm{uncorr}} = \frac{2r}{1 + r^2}. \tag{10.11}$$

That is, for C-odd initial states, $R_{\eta_c = -1} = \chi$ measures the probability that an initial B^0 decays as a $\overline{B^0}$. This result can be understood on physical grounds.

The original B_d^0-$\overline{B_d^0}$ pair from an $\Upsilon(4S)$ is in a p wave. Each neutral meson oscillates; if at a given instant one meson is detected as a $\overline{B_d^0}$, then the other meson must be B_d^0 at that instant. Therefore, detecting one meson as a $\overline{B_d^0}$ tags the other one as a B_d^0. The whole experiment can be reinterpreted[25] as if we had started with one tagged initial B_d^0. This is the reason why $R_{\eta_c = -1}$ must equal χ.

Of course, this does not hold when the initial mesons are uncorrelated. One then has $R_{\text{uncorr}} = 2\chi(1 - \chi)$. This result for uncorrelated B^0-$\overline{B^0}$ pairs is easy to understand, too. Using eqns (10.9), the probability to find mixed events is given by

$$\text{Prob}[B^0 \to \overline{B^0}]\text{Prob}[\overline{B^0} \to \overline{B^0}] + \text{Prob}[B^0 \to B^0]\text{Prob}[\overline{B^0} \to B^0] = 2\chi(1 - \chi). \tag{10.12}$$

Most often, results are quoted assuming also that $y = 0$, in addition to $|q/p| = 1$. In the SM, this is expected to hold to high precision in the B_d^0-$\overline{B_d^0}$ system. In the the B_s^0-$\overline{B_s^0}$ system, the theoretical estimate of Beneke *et al.* (1996) is

$$y_s = 0.08^{+0.06}_{-0.05}. \tag{10.13}$$

When $|q| = |p|$ and $y = 0$ one has

$$\chi = \bar{\chi} = \chi' = \chi_0 = \frac{x^2}{2(1 + x^2)}. \tag{10.14}$$

As a result, the four χs are often confused in the literature. We prefer to define them independently. The relation among them must be sought experimentally. In any case, if the measurements confirm that in the B_s^0-$\overline{B_s^0}$ system y is of the order of the upper limit in eqn (10.13), $|y| \sim 0.14$, then the correction to the equality $\chi' = \chi$ is very small: $\chi' = \chi/(1 - y_s^2) \approx 1.02\chi$.

10.3 Experiments at the $\Upsilon(4S)$

Experiments at the $\Upsilon(4S)$ provide a rather clean source of b quarks, since the branching ratio of $\Upsilon(4S)$ into non-$b\bar{b}$ states is measured to be less than 4%, at 95% confidence level (Particle Data Group 1996).

As mentioned before, in the SM, semileptonic decays respect the $\Delta B = \Delta Q$ selection rule. As a consequence, a B^0 meson (with a $\bar{b}$ antiquark) decaying semileptonicaly will yield a positively charged lepton l^+, while a $\overline{B^0}$ (with a b quark) will yield a negatively charged lepton l^-. The sign of the lepton's charge is the same as the sign of the heavy-quark's charge.

Unfortunately, in a typical event there are many sources of charged leptons. The leptons from the semileptonic decays of a b quark are selected through some specific high-momentum cut. There is a further source of confusion: in a given experimental situation one usually produces several types of hadrons containing b quarks. These cannot be separated, and one must estimate the

---

[25]See Kayser (1997) for a detailed, covariant explanation of this interpretation.

relative production rates in each experiment. In particular, the decays of the $\Upsilon(4S)$ yield both B_d^0-$\overline{B_d^0}$ pairs and B^+-B^- pairs. Roughly speaking, half of the $\Upsilon(4S)$ decay into B_d^0-$\overline{B_d^0}$, while the other half decay into B^+-B^- pairs. The 'mixed', same-sign-dilepton events may only come from a B_d^0-$\overline{B_d^0}$ pair. But, the 'unmixed', oposite-sign-dilepton events may come from either B_d^0-$\overline{B_d^0}$ or B^+-B^- pairs.

For this reason, we must be careful and distinguish between the R and R' of Chapter 9, and the quantities which are experimentally measured, which we shall denote R_{meas} and R'_{meas}. Experimentalists use for the quantities of Chapter 9 a slightly different notation:

$$R = \frac{N[\overline{B_d^0}\,\overline{B_d^0}] + N[B_d^0 B_d^0]}{N[\overline{B_d^0}B_d^0] + N[B_d^0\overline{B_d^0}] + N[\overline{B_d^0}\,\overline{B_d^0}] + N[B_d^0 B_d^0]}, \tag{10.15}$$

$$R' = \frac{N[\overline{B_d^0}\,\overline{B_d^0}] + N[B_d^0 B_d^0]}{N[\overline{B_d^0}B_d^0] + N[B_d^0\overline{B_d^0}]}. \tag{10.16}$$

This notation emphasizes the fact that we are interested in the numbers of flavour-tagging (semileptonic) decays originating in neutral B_d^0 and $\overline{B_d^0}$ mesons. On the other hand, the measured quantities involve semileptonic decays which may originate elsewhere.[26] We shall therefore use for those quantities the notation

$$R_{\text{meas}} = \frac{N[l^- l^-] + N[l^+ l^+]}{N[l^- l^+] + N[l^+ l^-] + N[l^- l^-] + N[l^+ l^+]}, \tag{10.17}$$

$$R'_{\text{meas}} = \frac{N[l^- l^-] + N[l^+ l^+]}{N[l^- l^+] + N[l^+ l^-]}, \tag{10.18}$$

where $R_{\text{meas}} = R'_{\text{meas}}/(1 + R'_{\text{meas}})$. We further introduce the experimentalists' parlance of 'mixed' and 'unmixed' events from B_d^0-$\overline{B_d^0}$ pairs:

$$\begin{aligned} N_0^m &\equiv N[\overline{B_d^0}\,\overline{B_d^0}] + N[B_d^0 B_d^0], \\ N_0^u &\equiv N[B_d^0\overline{B_d^0}] + N[B_d^0\overline{B_d^0}]. \end{aligned} \tag{10.19}$$

In their practical application these are the numbers of same-sign and opposite-sign dilepton events, respectively:

$$\begin{aligned} N_{\text{meas}}^m &\equiv N[l^- l^-] + N[l^+ l^+], \\ N_{\text{meas}}^u &\equiv N[l^+ l^-] + N[l^+ l^-]. \end{aligned} \tag{10.20}$$

Clearly,

---

[26] Actually, neutral B mesons may be identified by complete reconstruction of the event. However, this is extremely inefficient (see, for example, Schröder 1994), and one looks instead for semileptonic decays.

$$R = \frac{N_0^m}{N_0^m + N_0^u},$$
$$R' = \frac{N_0^m}{N_0^u}. \qquad (10.21)$$

Equations analogous to eqns (10.21) hold for the quantities having the subscript 'meas'.

Now,

$$N_{\text{meas}}^m = N_0^m,$$
$$N_{\text{meas}}^u = N_0^u + N_+^u. \qquad (10.22)$$

The unmixed events N_+^u from charged pairs are related to the events from neutral pairs through

$$\lambda \equiv \frac{N_+^u}{N_0^u + N_0^m} = \frac{f_+ \text{BR}_+^2}{f_0 \text{BR}_0^2}. \qquad (10.23)$$

Here, f_+ and f_0 are the probabilities, respectively, of charged and neutral pairs being produced at the $\Upsilon(4S)$. By definition, $f_0 + f_+ = 1$. The ratio of f_+ and f_0 has been measured to be (Particle Data Group 1996, p. 478)

$$\frac{f_+}{f_0} = 1.13 \pm 0.14 \pm 0.13 \pm 0.06, \qquad (10.24)$$

where the errors are, respectively, statistical, systematic, and due to the uncertainties in the ratio of B_d^0 and B^+ lifetimes. Often, experimental results are quoted assuming f_+/f_0 to be one, which is consistent with the fact that B_d^0 and B^+ have similar masses. The semileptonic branching ratios of charged- and neutral-mesons are denoted BR_+ and BR_0, respectively. Some authors also assume these branching ratios to be equal. Other authors assume that the semileptonic decay widths are the same for both, and then they trade BR_+/BR_0 by the ratio of lifetimes $\tau_{B^+}/\tau_{B_d^0}$, which has been measured to be 1.03 ± 0.06 (Particle Data Group 1996).

Therefore, the mixing parameters are related to the measured dilepton events through

$$R = (1 + \lambda) R_{\text{meas}},$$
$$R' = \frac{(1 + \lambda) R'_{\text{meas}}}{1 - \lambda R'_{\text{meas}}}. \qquad (10.25)$$

Also,

$$\left(\frac{N_{\text{meas}}^m}{N_{\text{meas}}^u} \right) \propto \left(\frac{f_0 \text{BR}_0^2 R}{f_0 \text{BR}_0^2 (1 - R) + f_+ \text{BR}_+^2} \right). \qquad (10.26)$$

The error in the production fractions is the largest source of uncertainty in these parameters. This dependence on λ may be reduced by reconstructing one of the B_d^0 mesons.

The combined results from the CLEO and ARGUS meaurements at the $\Upsilon(4S)$ are $\chi_d = 0.156 \pm 0.024$, where $|q| = |p|$ has been assumed. If one takes $y_d = 0$, this translates into a value for $x_d = \sqrt{2\chi_d/(1 - 2\chi_d)}$. One obtains $x_d = 0.67 \pm 0.08$, meaning that a complete oscillation period $2\pi/\Delta m_d$ takes about nine lifetimes.

CLEO has also looked for CP violation in the mixing through the dilepton asymmetry of eqn (9.92). They found (CLEO Collaboration 1993c)

$$|A_{ll}| = \frac{2|\delta_d|}{1 + \delta_d^2} = \left| \frac{N(l^+l^+) - N(l^-l^-)}{N(l^+l^+) + N(l^-l^-)} \right| < 0.18, \qquad (10.27)$$

leading to $|\delta_d| < 0.09$. This bound is still quite far from standard-model expectations, but is anyway much better than the one extracted from the unitarity bound $\delta_d^2 \leq 1 - 2\chi_d$. A better bound should be attained once these results are combined with the single-lepton asymmetry of eqn (9.82), $A_l = 2\delta\chi_0$, as proposed by Yamamoto (1997a). The idea behind this asymmetry is to look for the time-integrated rates of $\Upsilon(4S)$ into a single semileptonic final state, summing over all channels from the other decay.

Note that the B_d^0 and $\overline{B_d^0}$ move very slowly ($v/c \approx 0.06$) in the rest frame of the $\Upsilon(4S)$. Hence, for a $\Upsilon(4S)$ at rest in the laboratory frame, the B_d^0 mesons travel about 30 μm in one lifetime. Present technology precludes measurements of distances shorter than 50–100 μm. For this reason, with symmetric machines running on the $\Upsilon(4S)$ resonance, we can only perform time-integrated measurements. This is the reason why asymmetric colliders are built, so that the $\Upsilon(4S)$ moves in the laboratory frame. The difference between the beam energies makes the B_d^0 mesons travel around 200 μm in the laboratory frame. This allows the time difference between the decays, t_-, to be be measured.

10.4   Time-integrated experiments at high energy

In e^+e^- collisions at high energy, as at LEP, and in $p\bar{p}$ colliders, the situation is complicated by the fact that several $q\bar{q}$ pairs may arise from the colour field, allowing the $\bar{b}$ antiquark from the original $b\bar{b}$ pair to hadronize into $\bar{b}q$—with a probability f_q—or into Λ_b and other b-baryons—with a probability f_Λ. One usually assumes that $f_c = 0$, and, then, $f_d + f_u + f_s + f_\Lambda = 1$. Under several assumptions, in particular $f_d = f_u$, the Particle Data Group (1996) finds the values

$$f_d = f_u = 0.378 \pm 0.022, \qquad (10.28)$$
$$f_s = 0.112^{+0.018}_{-0.019}, \qquad (10.29)$$
$$f_\Lambda = 0.132 \pm 0.041. \qquad (10.30)$$

Assuming $|q| = |p|$, what is measured is a mixing parameter

$$\chi_B = \frac{\Gamma[b \to \overline{B^0} \to B^0 \to l^+X]}{\Gamma[b \to b\text{-hadron} \to l^\pm X]}, \qquad (10.31)$$

in which one cannot disentangle the oscillations of B_d^0 from those of B_s^0. Hence, the time-integrated mixing parameter measures

$$\chi_B = f_d \frac{BR_d}{\langle B \rangle} \chi_d + f_s \frac{BR_s}{\langle B \rangle} \chi_s,$$

(10.32)

where

$$\langle B \rangle \equiv f_d \, BR_d + f_u \, BR_u + f_s \, BR_s + f_\Lambda \, BR_\Lambda.$$

(10.33)

Here, BR_d, BR_u, BR_s, and BR_Λ are the branching ratios for B_d^0, $B^\pm$, B_s^0, and b-barions decaying into the observed mode, respectively. It is also possible to tag the flavour of each meson with decays other than the semileptonic ones. A common method uses the jet charge. The combined results of LEP1, $p\bar{p}$ collisions at 630 GeV, and $p\bar{p}$ collisions at 1.8 TeV, is $\chi_B = 0.126 \pm 0.008$ (Particle Data Group 1996).

These experiments are much more involved than those performed at the $\Upsilon(4S)$. This is in part due to the existence of more states with b-quark content, such as B_s^0 and b-baryons. Moreover, identifying the flavour of one jet does not tag the flavour of the other b at that instant. The tagging possibility exists for B_d^0–$\overline{B_d^0}$ pairs produced at the $\Upsilon(4S)$ due to the antisymmetry of the original wave function. At high energy, because of uncorrelated initial states, this possibility does not exist any more.

10.5 Time-dependent experiments at LEP1

At LEP one can measure the distance d between the primary vertex and the point where the B^0 meson decays. This is related to the meson's proper time t by $t = dm_B/p_B$; the momentum p_B of the B^0 meson is estimated from the momentum of its decay products. As we have seen before, the probabilities that a particle initially tagged as B^0 is still B^0, or is $\overline{B^0}$, at proper time t, are

$$\mathrm{Prob}[B^0(t) = B^0] = \mathrm{Prob}[\overline{B^0}(t) = \overline{B^0}] = \frac{e^{-\Gamma t}}{2} \left[\cosh \frac{\Delta\Gamma t}{2} + \cos(\Delta m t) \right],$$

$$\mathrm{Prob}[B^0(t) = \overline{B^0}] = \mathrm{Prob}[\overline{B^0}(t) = B^0] = \frac{e^{-\Gamma t}}{2} \left[\cosh \frac{\Delta\Gamma t}{2} - \cos(\Delta m t) \right],$$

(10.34)

respectively. (In the second equation we have assumed that there is no CP violation in the mixing, as is always done in the experimental analysis.) Furthermore, it is customary to assume that $\Delta\Gamma \ll \Gamma$. Since both B_d^0 and B_s^0 mesons are produced, the measurements are sensitive to both Δm_d and Δm_s. Extracting both parameters involves searching for two separate frequency components in the decay-time distributions.

These measurements are very complicated, and we shall only point out a few difficulties. Any given experiment measures the time evolution of only one meson. Its initial flavour may be tagged by the charged lepton in the opposite side of the detector.[27] This tagging has a probability χ of being incorrect, as

---

[27]At LEP, other tagging strategies were also used, such as the opposite jet charge, the same side jet charge, and the kaon tag. They are usually combined into a complicated multivariable analysis. The overall effect is to replace χ by the ratio of the wrong tags to the total number of events tagged (Moser, private communication).

we have seen before. Even if there were no complications due to the possible presence of several b hadrons, this would yield a ratio of dileptonic events given by (assuming $\Delta\Gamma = 0$ as usual)

$$\frac{N^m_{\text{meas}}}{N^m_{\text{meas}} + N^u_{\text{meas}}} = (1 - \chi)\frac{1 - \cos(\Delta mt)}{2} + \chi\frac{1 + \cos(\Delta mt)}{2}$$

$$= \frac{1 - (1 - 2\chi)\cos(\Delta mt)}{2}. \tag{10.35}$$

The same reasoning leads to the expression for $y \neq 0$,

$$\frac{N^m_{\text{meas}}}{N^m_{\text{meas}} + N^u_{\text{meas}}} = \frac{1}{2}\left[1 - (1 - 2\chi)\frac{\cos(\Delta mt)}{\cosh(\Delta\Gamma t/2)}\right], \tag{10.36}$$

which has been derived in eqn (9.94). The oscillatory term is not changed. Indeed, $\Delta\Gamma$ only shows up in $\cosh(\Delta\Gamma t/2)$ and implicitly in $\chi = (x^2 + y^2)/(2 + 2x^2)$. Since the experiments do observe an oscillation, removing the constraint $y_d = 0$ should not have a large impact on the central value and error bars of the x_d found in this way.[28] Hence, one can place a limit on y_d by combining these time-dependent determinations of Δm_d with the time-integrated bounds on χ_d.

In addition to the complication introduced by the mixing (i.e., by $\chi \neq 0$), one must deal with all the b hadrons, mixing mistags and backgrounds, some of which also depend on the amount of mixing (such as events with one lepton from a direct decay $b \to l$, and another from a chain $b \to c \to l$). When all this is taken into account, the time-dependent dileptonic fractions show a clear frequency component, consistent with the estimate of x_d from time-integrated measurements. The Particle Data Group (1996) finds $\Delta m_d = (0.50 \pm 0.04)\,\text{ps}^{-1}$. One may combine this with the information on the B^0_d lifetime to find x_d. Assuming $y_d = 0$ and combining all the time-independent and time-dependent determinations of x_d, the Particle Data Group (1996) finds $x_d = 0.73 \pm 0.05$.

In addition, the Particle Data Group (1996) gives $\Delta m_s > 5.6\,\text{ps}^{-1}$ (which corresponds to $x_s > 9.5$), where the assumptions $f_s = 11.2\%$ and $y_s = 0$ have been used in the fit. This results in $\chi_s > 0.49$, which is close to the upper bound $\chi < 0.5$. In this case there is also a bound on y_s coming from the comparison of the average B^0_d and B^0_s lifetimes. Assuming that $y_d = 0$, $\Gamma_{B_d} = \Gamma_{B_s}$, and that there is no bias in the experiments, the combined LEP and CDF data yields $y_s < 0.3$ (Moser, personal communication).

---

[28] Recent Monte Carlo experiments confirm this expectation (Moser, personal communication). Still, it would be nice to see it confirmed by the data.

Part II

CP violation in the standard model

GAUGE STRUCTURE OF THE STANDARD MODEL

11.1 Introduction

In this chapter we make an overview of the gauge structure of the standard model (SM), with the purpose of establishing notation and writing down the Feynman rules for all vertices and propagators in a general 't Hooft gauge. The part of the SM Lagrangian involving the fermions will be studied in the next chapter.

This chapter does not intend to be either pedagogical or self-contained. For a thorough derivation of the SM Lagrangian, the reader is advised to consult one of the texts existent on the subject—for instance Abers and Lee (1973); Fritzsch and Minkowski (1981); Leader and Predazzi (1982); Cheng and Li (1994). Some readers may prefer to skip this chapter altogether.

11.2 SU(2)

The Glashow–Weinberg–Salam (Glashow 1961; Weinberg 1967; Salam 1968) model for the electroweak interactions, termed the standard model, has gauge group SU(2)⊗U(1). The generators of SU(2) are denoted T_1, T_2, and T_3. They are Hermitian and obey the commutation relations

$$[T_k, T_l] = i\epsilon_{klm} T_m. \tag{11.1}$$

One defines

$$T_\pm \equiv \frac{T_1 \pm iT_2}{\sqrt{2}}. \tag{11.2}$$

Then, $T_- = T_+^\dagger$ and the commutation relations are

$$[T_+, T_-] = T_3,$$
$$[T_3, T_\pm] = \pm T_\pm. \tag{11.3}$$

In the doublet representation, the T_k are represented by the Pauli matrices τ_k divided by 2, i.e.,

$$T_1^{(2)} = \tfrac{1}{2}\begin{pmatrix} 0 & 1 \\ 1 & 0 \end{pmatrix}, \quad T_2^{(2)} = \tfrac{1}{2}\begin{pmatrix} 0 & -i \\ i & 0 \end{pmatrix}, \quad T_3^{(2)} = \tfrac{1}{2}\begin{pmatrix} 1 & 0 \\ 0 & -1 \end{pmatrix}. \tag{11.4}$$

Then,

$$T_+^{(2)} = \tfrac{1}{\sqrt{2}}\begin{pmatrix} 0 & 1 \\ 0 & 0 \end{pmatrix}, \quad T_-^{(2)} = \tfrac{1}{\sqrt{2}}\begin{pmatrix} 0 & 0 \\ 1 & 0 \end{pmatrix}. \tag{11.5}$$

In the triplet representation,

$$T_1^{(3)} = \frac{1}{\sqrt{2}} \begin{pmatrix} 0 & 1 & 0 \\ 1 & 0 & 1 \\ 0 & 1 & 0 \end{pmatrix}, \quad T_2^{(3)} = \frac{1}{\sqrt{2}} \begin{pmatrix} 0 & -i & 0 \\ i & 0 & -i \\ 0 & i & 0 \end{pmatrix}, \quad T_3^{(3)} = \begin{pmatrix} 1 & 0 & 0 \\ 0 & 0 & 0 \\ 0 & 0 & -1 \end{pmatrix}. \tag{11.6}$$

Then,

$$T_+^{(3)} = \begin{pmatrix} 0 & 1 & 0 \\ 0 & 0 & 1 \\ 0 & 0 & 0 \end{pmatrix}, \quad T_-^{(3)} = \begin{pmatrix} 0 & 0 & 0 \\ 1 & 0 & 0 \\ 0 & 1 & 0 \end{pmatrix}. \tag{11.7}$$

11.3 Covariant derivative

The covariant derivative is

$$\partial^\mu - ig \left(W_1^\mu T_1 + W_2^\mu T_2 + W_3^\mu T_3 \right) - ig' B^\mu Y. \tag{11.8}$$

Here, g denotes the SU(2) coupling constant and g' is the U(1) coupling constant. The U(1) charge Y is termed (weak) hypercharge; it is, for each irreducible representation of SU(2)⊗U(1), a real multiple of the unit matrix. In our normalization, the electric charge Q is given by $Q = T_3 + Y$. The three W_k^μ are the gauge fields of SU(2), while B^μ is the U(1) gauge field. We shall often omit the Lorentz indices on the gauge fields and on the derivatives, in order not to overload the notation.

One defines

$$W^\pm \equiv \frac{W_1 \mp iW_2}{\sqrt{2}}. \tag{11.9}$$

Instead of g and g' it is useful to introduce e (the electric-charge unit) and the angle θ_w defined through

$$\begin{aligned} g &= \frac{e}{s_w}, \\ g' &= -\frac{e}{c_w}, \end{aligned} \tag{11.10}$$

where $s_w \equiv \sin \theta_w$ and $c_w \equiv \cos \theta_w$. We also define the gauge fields A and Z to be the result of an orthogonal rotation of B and W_3:

$$\begin{pmatrix} B \\ W_3 \end{pmatrix} = \begin{pmatrix} c_w & s_w \\ -s_w & c_w \end{pmatrix} \begin{pmatrix} A \\ Z \end{pmatrix}. \tag{11.11}$$

The covariant derivative in eqn (11.8) may then be written

$$\partial + ieAQ - ig \left(W^+ T_+ + W^- T_- \right) - i\frac{g}{c_w} Z \left(T_3 - Qs_w^2 \right). \tag{11.12}$$

11.4 Self-interactions of the gauge bosons

Interactions among the gauge bosons are typical of a non-abelian gauge theory. In the SM, they arise because of the presence of the non-abelian SU(2) in the gauge group. If we denote

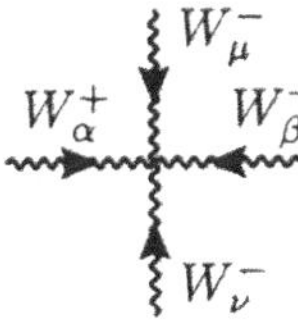

$$W_\alpha^+ \quad W_\beta^+ \quad = ig^2 \left(2g_{\alpha\beta}g_{\mu\nu} - g_{\alpha\mu}g_{\beta\nu} - g_{\alpha\nu}g_{\beta\mu}\right)$$

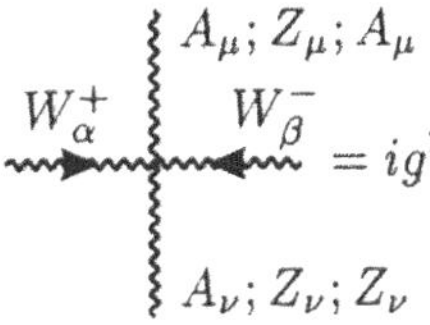

$$W_\alpha^+ \quad W_\beta^- \quad = ig^2 \left(g_{\alpha\mu}g_{\beta\nu} + g_{\alpha\nu}g_{\beta\mu} - 2g_{\alpha\beta}g_{\mu\nu}\right) \left(s_w^2; c_w^2; -s_w c_w\right)$$

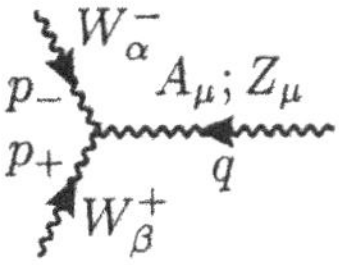

$$= ig \left[g_{\alpha\beta}\left(p_+ - p_-\right)_\mu + g_{\mu\beta}\left(q - p_+\right)_\alpha + g_{\mu\alpha}\left(p_- - q\right)_\beta\right] \times \left(s_w; -c_w\right)$$

FIG. 11.1. Self-interactions of the gauge bosons.

$$
\begin{aligned}
F_1^{\mu\nu} &\equiv \partial^\mu W_1^\nu - \partial^\nu W_1^\mu + g\left(W_2^\mu W_3^\nu - W_3^\mu W_2^\nu\right), \\
F_2^{\mu\nu} &\equiv \partial^\mu W_2^\nu - \partial^\nu W_2^\mu + g\left(W_3^\mu W_1^\nu - W_1^\mu W_3^\nu\right), \\
F_3^{\mu\nu} &\equiv \partial^\mu W_3^\nu - \partial^\nu W_3^\mu + g\left(W_1^\mu W_2^\nu - W_2^\mu W_1^\nu\right), \\
F_Y^{\mu\nu} &\equiv \partial^\mu B^\nu - \partial^\nu B^\mu,
\end{aligned}
\tag{11.13}
$$

then, the gauge-kinetic Lagrangian can be written

$$
\begin{aligned}
&-\tfrac{1}{4}\left(F_1^{\mu\nu}F_{1\mu\nu} + F_2^{\mu\nu}F_{2\mu\nu} + F_3^{\mu\nu}F_{3\mu\nu} + F_Y^{\mu\nu}F_{Y\mu\nu}\right) \\
= &- \left(\partial_\mu W_\nu^+\right)\left(\partial^\mu W^{\nu-}\right) + \left(\partial_\mu W_\nu^+\right)\left(\partial^\nu W^{\mu-}\right) \\
&- \tfrac{1}{2}\left(\partial_\mu A_\nu\right)\left(\partial^\mu A^\nu\right) + \tfrac{1}{2}\left(\partial_\mu A_\nu\right)\left(\partial^\nu A^\mu\right) \\
&- \tfrac{1}{2}\left(\partial_\mu Z_\nu\right)\left(\partial^\mu Z^\nu\right) + \tfrac{1}{2}\left(\partial_\mu Z_\nu\right)\left(\partial^\nu Z^\mu\right) \\
&+\text{non-quadratic terms.}
\end{aligned}
\tag{11.14}
$$

The non-quadratic terms in the last line of eqn (11.14) yield the vertices in Fig. 11.1.

11.5   Gauge interactions of the scalars

The scalar sector of the SM consists of only one doublet, ϕ, which has $Y = 1/2$. If one chooses a negative sign for the coefficient μ in the quadratic term $\mu\phi^\dagger\phi$ of the Higgs potential—see eqn (11.30)—then the SU(2)⊗U(1) gauge symmetry is broken into U(1), which is identified with the gauge group of the electromagnetic interaction. Without loss of generality, one can make an SU(2) rotation so that it is the lower component of ϕ which acquires a vacuum expectation value (VEV) v, which is a c-number, constant over the whole of Minkowski space. We write

$$\phi = \begin{pmatrix} \varphi^+ \\ \varphi^0 \end{pmatrix} = \begin{pmatrix} \varphi^+ \\ v + (H + i\chi)/\sqrt{2} \end{pmatrix}. \tag{11.15}$$

Here, H and χ are Hermitian Klein–Gordon fields, $\varphi^\pm$ are the Goldstone bosons to be absorbed in the longitudinal components of $W^\pm$, χ is the Goldstone boson to be absorbed in the longitudinal component of Z, and the physical Higgs particle is H. One introduces the conjugate SU(2) doublet

$$\tilde{\phi} \equiv i\tau_2 \phi^{\dagger T} = \begin{pmatrix} \varphi^{0\dagger} \\ -\varphi^- \end{pmatrix} = \begin{pmatrix} v + (H - i\chi)/\sqrt{2} \\ -\varphi^- \end{pmatrix}, \tag{11.16}$$

which has $Y = -1/2$.

From eqns (11.12), (11.4), and (11.5) it follows that the gauge-kinetic Lagrangian of ϕ is

$$\left[\partial\varphi^- - ieA\varphi^- + i\frac{g}{\sqrt{2}}W^-\varphi^{0\dagger} + i\frac{g}{2c_w}Z\left(c_w^2 - s_w^2\right)\varphi^- \right]$$
$$\times \left[\partial\varphi^+ + ieA\varphi^+ - i\frac{g}{\sqrt{2}}W^+\varphi^0 - i\frac{g}{2c_w}Z\left(c_w^2 - s_w^2\right)\varphi^+ \right]$$
$$+ \left(\partial\varphi^{0\dagger} + i\frac{g}{\sqrt{2}}W^+\varphi^- - i\frac{g}{2c_w}Z\varphi^{0\dagger} \right)$$
$$\times \left(\partial\varphi^0 - i\frac{g}{\sqrt{2}}W^-\varphi^+ + i\frac{g}{2c_w}Z\varphi^0 \right). \tag{11.17}$$

We use the relationship between v and the masses of the W and Z,

$$v = \frac{\sqrt{2}}{g}m_W = \frac{\sqrt{2}c_w}{g}m_Z, \tag{11.18}$$

and find that the expression in eqn (11.17) is

$$\left(\partial\varphi^-\right)\left(\partial\varphi^+\right) + \tfrac{1}{2}\left[(\partial H)^2 + (\partial\chi)^2\right] + m_W^2 W^- W^+ + \tfrac{1}{2}m_Z^2 Z^2$$
$$+ im_W\left(W^-\partial\varphi^+ - W^+\partial\varphi^-\right) + m_Z Z\partial\chi$$
$$+\text{non-quadratic terms}. \tag{11.19}$$

The terms in the second line of eqn (11.19) are cancelled out by the gauge-fixing Lagrangian, as we shall see next. The non-quadratic terms in the third line of eqn (11.19) give rise to the vertices in Fig. 11.2.

11.6   Gauge-fixing Lagrangian

A gauge fixing is required in order to define the propagators of the gauge bosons and of the Goldstone bosons. Here we shall consider only 't Hooft gauges. Let us first study the case of the gauge boson Z and the Goldstone boson χ. The

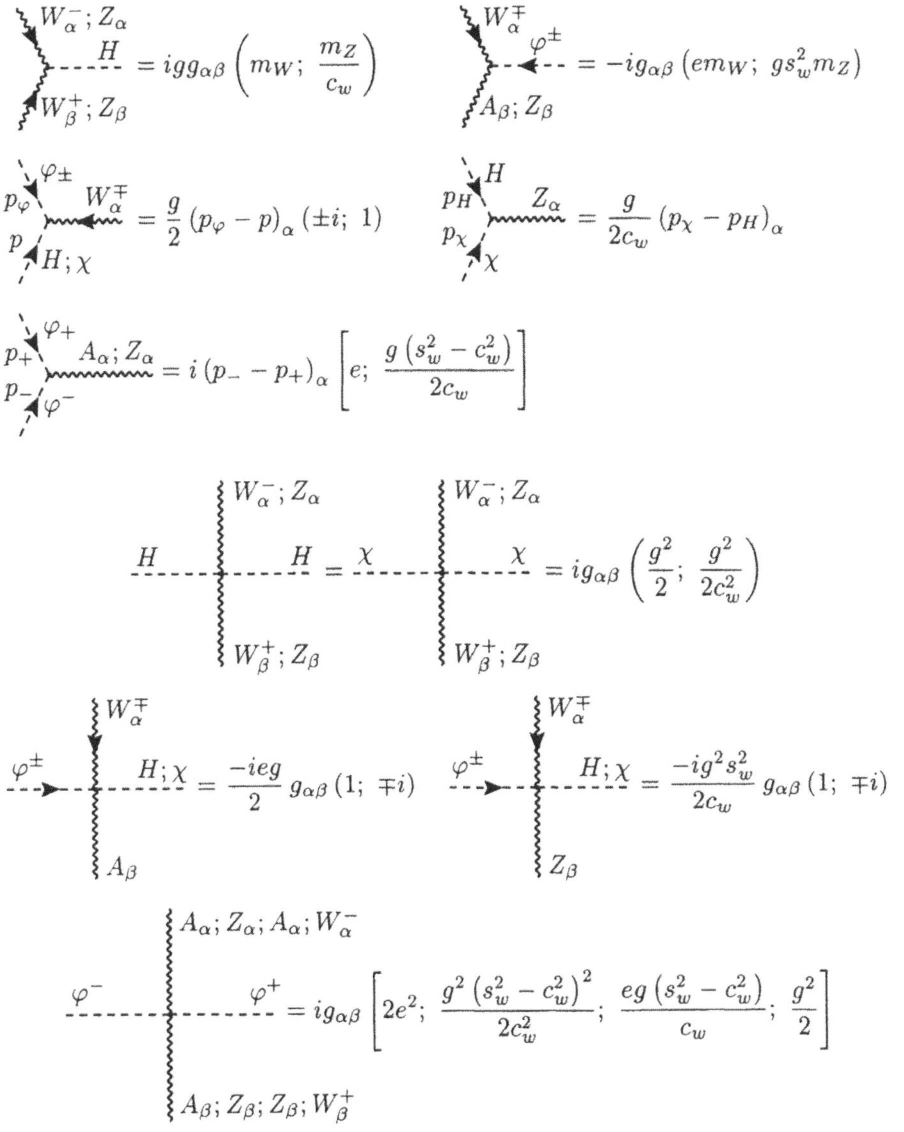

$$\begin{aligned}
& W_\alpha^-; Z_\alpha \\
& \qquad H \\
& W_\beta^+; Z_\beta
\end{aligned} = i g g_{\alpha\beta} \left(m_W; \ \frac{m_Z}{c_w} \right)$$

$$\begin{aligned}
& W_\alpha^\mp \\
& \quad \varphi^\pm \\
& A_\beta; Z_\beta
\end{aligned} = -i g_{\alpha\beta} \left(e m_W; \ g s_w^2 m_Z \right)$$

$$\begin{aligned}
& p_\varphi \quad \varphi^\pm \quad W_\alpha^\mp \\
& p \quad H; \chi
\end{aligned} = \frac{g}{2} \left(p_\varphi - p \right)_\alpha (\pm i; \ 1)$$

$$\begin{aligned}
& p_H \quad H \quad Z_\alpha \\
& p_\chi \quad \chi
\end{aligned} = \frac{g}{2 c_w} \left(p_\chi - p_H \right)_\alpha$$

$$\begin{aligned}
& p_+ \quad \varphi^+ \quad A_\alpha; Z_\alpha \\
& p_- \quad \varphi^-
\end{aligned} = i \left(p_- - p_+ \right)_\alpha \left[e; \ \frac{g \left(s_w^2 - c_w^2 \right)}{2 c_w} \right]$$

$$\begin{aligned}
& W_\alpha^-; Z_\alpha \\
H \cdots \Big| \cdots H &= \chi \cdots \Big| \cdots \chi = i g_{\alpha\beta} \left(\frac{g^2}{2}; \ \frac{g^2}{2 c_w^2} \right) \\
& W_\beta^+; Z_\beta
\end{aligned}$$

$$\begin{aligned}
& W_\alpha^\mp \\
\varphi^\pm \cdots \Big| \cdots H; \chi &= \frac{-i e g}{2} g_{\alpha\beta} (1; \ \mp i) \\
& A_\beta
\end{aligned}$$

$$\begin{aligned}
& W_\alpha^\mp \\
\varphi^\pm \cdots \Big| \cdots H; \chi &= \frac{-i g^2 s_w^2}{2 c_w} g_{\alpha\beta} (1; \ \mp i) \\
& Z_\beta
\end{aligned}$$

$$\begin{aligned}
& A_\alpha; Z_\alpha; A_\alpha; W_\alpha^- \\
\varphi^- \cdots \Big| \cdots \varphi^+ &= i g_{\alpha\beta} \left[2 e^2; \ \frac{g^2 \left(s_w^2 - c_w^2 \right)^2}{2 c_w^2}; \ \frac{e g \left(s_w^2 - c_w^2 \right)}{c_w}; \ \frac{g^2}{2} \right] \\
& A_\beta; Z_\beta; Z_\beta; W_\beta^+
\end{aligned}$$

FIG. 11.2. Gauge interactions of the scalars.

quadratic terms in the Lagrangian containing Z and χ are—see eqns (11.19) and (11.14)—

$$-\tfrac{1}{2} \left(\partial_\mu Z_\nu \right) \left(\partial^\mu Z^\nu \right) + \tfrac{1}{2} \left(\partial_\mu Z_\nu \right) \left(\partial^\nu Z^\mu \right) + \tfrac{1}{2} m_Z^2 Z_\mu Z^\mu + \tfrac{1}{2} \left(\partial_\mu \chi \right) \left(\partial^\mu \chi \right) + m_Z Z_\mu \partial^\mu \chi.$$
$$(11.20)$$

One uses the gauge-fixing Lagrangian

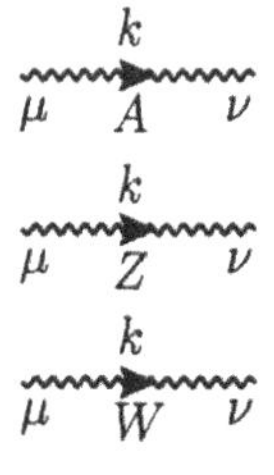

$$\underset{\mu \quad A \quad \nu}{k} = \frac{-ig_{\mu\nu}}{k^2} + (1 - \xi_A) \frac{ik_\mu k_\nu}{k^4}$$

$$\underset{\mu \quad Z \quad \nu}{k} = \frac{-ig_{\mu\nu}}{k^2 - m_Z^2} + \frac{ik_\mu k_\nu}{m_Z^2} \left(\frac{1}{k^2 - m_Z^2} - \frac{1}{k^2 - \xi_Z m_Z^2} \right)$$

$$\underset{\mu \quad W \quad \nu}{k} = \frac{-ig_{\mu\nu}}{k^2 - m_W^2} + \frac{ik_\mu k_\nu}{m_W^2} \left(\frac{1}{k^2 - m_W^2} - \frac{1}{k^2 - \xi_W m_W^2} \right)$$

FIG. 11.3. Propagators of the gauge bosons.

$$-\frac{1}{2\xi_Z} (\partial_\mu Z^\mu - \xi_Z m_Z \chi)^2 = -\frac{1}{2\xi_Z} (\partial_\mu Z^\mu)(\partial_\nu Z^\nu) + m_Z \chi \partial_\mu Z^\mu - \frac{\xi_Z}{2} m_Z^2 \chi^2,$$

$$(11.21)$$

where ξ_Z is an arbitrary real non-negative number, whose value is $\xi_Z = 0$ in the Landau gauge, $\xi_Z = 1$ in the Feynman gauge, and $\xi_Z = \infty$ in the unitary gauge. Physically meaningful quantities are independent of ξ_Z. Adding eqns (11.20) and (11.21) we find that χ has the usual propagator for a scalar boson, with squared mass $\xi_Z m_Z^2$. Also, after an integration by parts, the Z–χ mixing terms $m_Z Z_\mu \partial^\mu \chi$ and $m_Z \chi \partial_\mu Z^\mu$ cancel out. The remaining terms yield, after some integration by parts, the Z propagator. This propagator has two parts—see Fig. 11.3. The first part has its pole on the physical squared mass m_Z^2. The second part has its pole on the unphysical squared mass $\xi_Z m_Z^2$. The effects of this second part of the Z propagator must cancel out with the effects of the propagation of χ—and of the Z ghost, as we shall see in the next section—which have the same fictitious squared mass $\xi_Z m_Z^2$.

The gauge-fixing term for the $W^\pm$–$\varphi^\pm$ sector reads

$$-\frac{1}{\xi_W} \left(\partial^\mu W_\mu^+ - i\xi_W m_W \varphi^+ \right) \left(\partial^\nu W_\nu^- + i\xi_W m_W \varphi^- \right)$$

$$= -\frac{1}{\xi_W} \left(\partial^\mu W_\mu^+ \right) \left(\partial^\nu W_\nu^- \right) + i m_W \left(\varphi^+ \partial^\mu W_\mu^- - \varphi^- \partial^\mu W_\mu^+ \right)$$

$$- \xi_W m_W^2 \varphi^+ \varphi^-.$$

$$(11.22)$$

From eqns (11.19) and (11.22) we see that $\varphi^\pm$ has the usual propagator for a charged scalar, with unphysical squared mass $\xi_W m_W^2$. The $W^\pm$–$\varphi^\pm$ mixing terms cancel out. The gauge boson $W^\pm$ has a propagator with two parts, one with pole on the physical squared mass m_W^2, the other one with pole on the unphysical squared mass $\xi_W m_W^2$. Any physical quantity must be independent of the real non-negative number ξ_W. Note that ξ_W does not need to be equal to ξ_Z; we may choose different 't Hooft gauges for the Z–χ sector and for the $W^\pm$–$\varphi^\pm$ sector.

The gauge-fixing term for the photon is

$$-\frac{1}{2\xi_A} (\partial^\mu A_\mu)(\partial^\nu A_\nu).$$

$$(11.23)$$

$$\cdots\!\!\xrightarrow[H]{\;\;k\;\;}\!\!\cdots = \frac{i}{k^2 - m_H^2} \qquad \cdots\!\!\xrightarrow[\chi]{\;\;k\;\;}\!\!\cdots = \frac{i}{k^2 - \xi_Z m_Z^2} \qquad \cdots\!\!\xrightarrow[\varphi^\pm]{\;\;k\;\;}\!\!\cdots = \frac{i}{k^2 - \xi_W m_W^2}$$

FIG. 11.4. Propagators of the scalars.

Once again, the unphysical gauge parameter ξ_A does not have to be equal to either ξ_W or ξ_Z. From eqns (11.14) and (11.23) it follows that the photon propagator is as given in Fig. 11.3. It has a $k_\mu k_\nu$ part with gauge-dependent coefficient $1 - \xi_A$; that part of the propagator must give a vanishing contribution to any physical quantity.

The propagators of the scalars are collected in Fig. 11.4. The mass m_H of H originates in the scalar potential, to be treated in § 11.8.

11.7   Ghosts

The ghost Lagrangian depends on the gauge-fixing conditions and on the gauge transformations. An infinitesimal gauge transformation of the scalar doublet reads

$$
\begin{aligned}
\delta\phi &= i\left(g\alpha_k T_k^{(2)} + g'\frac{\beta}{2}\right)\phi, \\
\delta\tilde\phi &= i\left(g\alpha_k T_k^{(2)} - g'\frac{\beta}{2}\right)\tilde\phi,
\end{aligned}
\tag{11.24}
$$

where the α_k are the three infinitesimal parameters of the SU(2) gauge transformation, and β is the infinitesimal parameter of the U(1) gauge transformation. Writing

$$
\alpha^\pm = \frac{\alpha_1 \mp i\alpha_2}{\sqrt{2}},
$$

$$
\begin{pmatrix} \beta \\ \alpha_3 \end{pmatrix} = \begin{pmatrix} c_w & s_w \\ -s_w & c_w \end{pmatrix} \begin{pmatrix} \alpha_A \\ \alpha_Z \end{pmatrix}
\tag{11.25}
$$

—cf. eqns (11.9) and (11.11)—we find

$$
\begin{aligned}
\delta\varphi^+ &= im_W\alpha^+ + i\left[\left(-e\alpha_A + g\frac{c_w^2 - s_w^2}{2c_w}\alpha_Z\right)\varphi^+ + \frac{g}{2}\left(H + i\chi\right)\alpha^+\right], \\
\delta\varphi^- &= -im_W\alpha^- - i\left[\left(-e\alpha_A + g\frac{c_w^2 - s_w^2}{2c_w}\alpha_Z\right)\varphi^- + \frac{g}{2}\left(H - i\chi\right)\alpha^-\right], \\
\delta\chi &= -m_Z\alpha_Z - \frac{g}{2c_w}H\alpha_Z + \frac{g}{2}\left(\varphi^-\alpha^+ + \varphi^+\alpha^-\right).
\end{aligned}
\tag{11.26}
$$

The same infinitesimal transformation of the gauge fields reads

$$
\begin{aligned}
\delta A_\mu &= \partial_\mu\alpha_A + ie\left(\alpha^- W_\mu^+ - \alpha^+ W_\mu^-\right), \\
\delta Z_\mu &= \partial_\mu\alpha_Z - igc_w\left(\alpha^- W_\mu^+ - \alpha^+ W_\mu^-\right), \\
\delta W_\mu^\pm &= \partial_\mu\alpha^\pm \pm igW_\mu^\pm\left(-s_w\alpha_A + c_w\alpha_Z\right) \mp ig\left(-s_w A_\mu + c_w Z_\mu\right)\alpha^\pm.
\end{aligned}
\tag{11.27}
$$

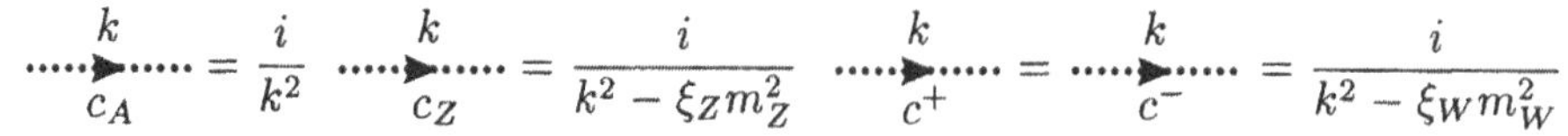

$$\cdots\!\!\cdots\!\!\!\!\underset{c_A}{\overset{k}{\longrightarrow}}\!\!\!\!\cdots\!\!\cdots = \frac{i}{k^2} \quad \cdots\!\!\cdots\!\!\!\!\underset{c_Z}{\overset{k}{\longrightarrow}}\!\!\!\!\cdots\!\!\cdots = \frac{i}{k^2 - \xi_Z m_Z^2} \quad \cdots\!\!\cdots\!\!\!\!\underset{c^+}{\overset{k}{\longrightarrow}}\!\!\!\!\cdots\!\!\cdots = \cdots\!\!\cdots\!\!\!\!\underset{c^-}{\overset{k}{\longrightarrow}}\!\!\!\!\cdots\!\!\cdots = \frac{i}{k^2 - \xi_W m_W^2}$$

FIG. 11.5. Propagators of the ghosts.

The Fadeev–Popov ghost Lagrangian for a general 't Hooft gauge reads, from eqns (11.21), (11.22), and (11.23),

$$\mathcal{L}_{\text{FP}} = -\sum_i \left[\bar{c}_Z \frac{\delta \left(\partial_\mu Z^\mu - \xi_Z m_Z \chi \right)}{\delta \alpha_i} c_i + \bar{c}_A \frac{\delta \left(\partial_\mu A^\mu \right)}{\delta \alpha_i} c_i \right.$$
$$\left. + \bar{c}^+ \frac{\delta \left(\partial^\mu W_\mu^+ - i\xi_W m_W \varphi^+ \right)}{\delta \alpha_i} c_i + \bar{c}^- \frac{\delta \left(\partial^\mu W_\mu^- + i\xi_W m_W \varphi^- \right)}{\delta \alpha_i} c_i \right] \quad (11.28)$$

The notation of the sum over i is the following: c_i denotes c_Z, c_A, c^+, and c^-, and the corresponding α_i are α_Z, α_A, α^+, and α^-, respectively. Notice that there are two distinct charged ghosts, c^+ and c^-, together with their distinct antighost fields, $\bar{c}^+$ and $\bar{c}^-$, respectively. We easily find

$$\mathcal{L}_{\text{FP}} = -\bar{c}_Z \left(\partial_\mu \partial^\mu + \xi_Z m_Z^2 \right) c_Z - \bar{c}_A \partial_\mu \partial^\mu c_A - \bar{c}^+ \left(\partial_\mu \partial^\mu + \xi_W m_W^2 \right) c^+$$
$$- \bar{c}^- \left(\partial_\mu \partial^\mu + \xi_W m_W^2 \right) c^- + \text{non-quadratic terms.} \quad (11.29)$$

From eqn (11.29) we find the ghost propagators in Fig. 11.5. The non-quadratic terms yield the vertices in Fig. 11.6. Note that some of those vertices are proportional to the gauge parameters ξ_W and ξ_Z.

11.8   Self-interactions of the scalars

The self-interactions of the scalars originate in the scalar potential—which appears in the Lagrangian density $\mathcal{L}$ with an overall minus sign, $\mathcal{L} = \cdots - V$:

$$V = \mu \phi^\dagger \phi + \lambda \left(\phi^\dagger \phi \right)^2, \quad (11.30)$$

where

$$\phi^\dagger \phi = v^2 + \sqrt{2} v H + \frac{H^2 + \chi^2}{2} + \varphi^- \varphi^+. \quad (11.31)$$

The vacuum stability condition is equivalent to the condition that the terms linear in H vanish: $\mu = -2\lambda v^2$. We trade λ by the Higgs mass m_H through $4\lambda v^2 = m_H^2$, and use the W mass m_W instead of v by having recourse to eqn (11.18). We obtain

$$V = -\frac{m_W^2 m_H^2}{2g^2} + \frac{m_H^2}{2} H^2 + \frac{g m_H^2}{2 m_W} \left(\frac{H^2 + \chi^2}{2} + \varphi^- \varphi^+ \right) H$$
$$+ \frac{g^2 m_H^2}{8 m_W^2} \left(\frac{H^2 + \chi^2}{2} + \varphi^- \varphi^+ \right)^2. \quad (11.32)$$

We find that, besides the mass term for H and the vacuum energy density $-m_W^2 m_H^2/(2g^2)$, the potential V contains cubic and quartic terms which yield the vertices in Fig. 11.7.

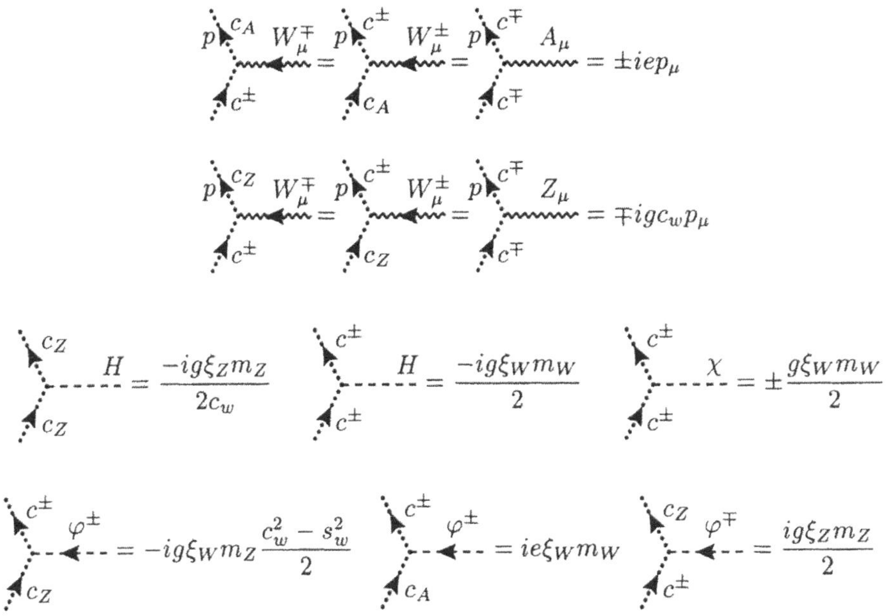

$$p \quad c_A \quad W_\mu^\mp \quad p \quad c^\pm \quad W_\mu^\pm \quad p \quad c^\mp \quad A_\mu$$
$$\Big|_{c^\pm} = \Big|_{c_A} = \Big|_{c^\mp} = \pm iep_\mu$$

$$p \quad c_Z \quad W_\mu^\mp \quad p \quad c^\pm \quad W_\mu^\pm \quad p \quad c^\mp \quad Z_\mu$$
$$\Big|_{c^\pm} = \Big|_{c_Z} = \Big|_{c^\mp} = \mp igc_w p_\mu$$

$$c_Z \quad H = \frac{-ig\xi_Z m_Z}{2c_w} \qquad c^\pm \quad H = \frac{-ig\xi_w m_W}{2} \qquad c^\pm \quad \chi = \pm\frac{g\xi_w m_W}{2}$$

$$c^\pm \quad \varphi^\pm = -ig\xi_w m_Z \frac{c_w^2 - s_w^2}{2} \qquad c^\pm \quad \varphi^\pm = ie\xi_w m_W \qquad c_Z \quad \varphi^\mp = \frac{ig\xi_Z m_Z}{2}$$

FIG. 11.6. Ghost vertices.

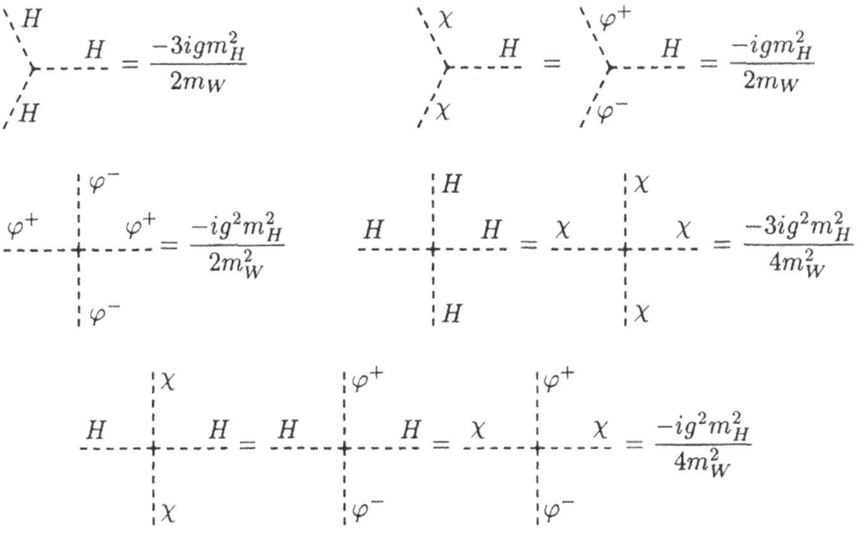

$$H \quad H = \frac{-3igm_H^2}{2m_W} \qquad \chi \quad H = \varphi^+ \quad H = \frac{-igm_H^2}{2m_W}$$

$$\varphi^+ \quad \varphi^+ = \frac{-ig^2 m_H^2}{2m_W^2} \qquad H \quad H = \chi \quad \chi = \frac{-3ig^2 m_H^2}{4m_W^2}$$

$$H \quad H = H \quad H = \chi \quad \chi = \frac{-ig^2 m_H^2}{4m_W^2}$$

FIG. 11.7. Self-interactions of the scalars.

12

THE FERMIONS IN THE STANDARD MODEL

12.1 Introduction

We proceed in our overview of the SM, now considering the part of the Lagrangian involving fermions. In the first two sections of this chapter we assume the existence of only one generation of fermions, and in the following sections we study the modifications introduced by the presence of $n_g > 1$ generations.

In this part of the book we shall mainly discuss quarks and their interactions, only occasionally commenting on leptons. This is because, in the SM, the leptonic sector is a simplified version of the quark sector: the absence of right-handed neutrinos and of Majorana masses leads to the absence of lepton mixing. In particular, the phenomenon that interests us most, CP violation, does not occur in the leptonic sector of the SM.

12.2 Gauge interactions of the fermions

The quarks of chirality -1 form SU(2) doublets with hypercharge 1/6:

$$Q_L = \begin{pmatrix} p_L \\ n_L \end{pmatrix}. \tag{12.1}$$

The quarks with chirality $+1$ are the singlets p_R, with $Q = Y = 2/3$, and n_R, with $Q = Y = -1/3$.

The charged- and neutral-current interactions of the quarks are derived from the general gauge-kinetic Lagrangian for a fermion multiplet f,

$$\bar{f}\gamma^\mu \left[i\partial_\mu - eA_\mu Q + g \left(W_\mu^+ T_+ + W_\mu^- T_- \right) + \frac{g}{c_w} Z_\mu \left(T_3 - Q s_w^2 \right) \right] f. \tag{12.2}$$

This yields the electromagnetic-interaction terms

$$\mathcal{L}_A^{(q)} = -eA_\mu J_{em}^\mu, \tag{12.3}$$

where

$$J_{em}^\mu \equiv \tfrac{2}{3} \left(\overline{p_L}\gamma^\mu p_L + \overline{p_R}\gamma^\mu p_R \right) - \tfrac{1}{3} \left(\overline{n_L}\gamma^\mu n_L + \overline{n_R}\gamma^\mu n_R \right). \tag{12.4}$$

The neutral-current interaction with the Z is given by

$$\mathcal{L}_Z^{(q)} = \frac{g}{2c_w} Z_\mu \left(\overline{p_L}\gamma^\mu p_L - \overline{n_L}\gamma^\mu n_L - 2s_w^2 J_{em}^\mu \right), \tag{12.5}$$

and the charged-current interaction with the $W^\pm$ is

$$\mathcal{L}_W^{(q)} = \frac{g}{\sqrt{2}} \left(W_\mu^+ \overline{p_L} \gamma^\mu n_L + W_\mu^- \overline{n_L} \gamma^\mu p_L \right). \tag{12.6}$$

For the leptons we have similar arrangements. A typical doublet of left-handed[29] leptons is

$$L_L = \begin{pmatrix} v_L \\ l_L \end{pmatrix}. \tag{12.7}$$

It has hypercharge $Y = -1/2$. The right-handed charged lepton l_R is an SU(2) singlet with $Q = Y = -1$. There are no right-handed neutrinos in the SM.

The charged-current interaction of the leptons is given by

$$\mathcal{L}_W^{(1)} = \frac{g}{\sqrt{2}} \left(W_\mu^+ \overline{v_L} \gamma^\mu l_L + W_\mu^- \overline{l_L} \gamma^\mu v_L \right), \tag{12.8}$$

and the neutral-current interactions, mediated by the Z and by the photon, are

$$\mathcal{L}_Z^{(1)} + \mathcal{L}_A^{(1)} = \frac{g}{2c_w} Z_\mu \left(\overline{v_L} \gamma^\mu v_L - \overline{l_L} \gamma^\mu l_L \right) + \left(\frac{g s_w^2}{c_w} Z_\mu + e A_\mu \right) \left(\overline{l_L} \gamma^\mu l_L + \overline{l_R} \gamma^\mu l_R \right). \tag{12.9}$$

12.3   The Yukawa Lagrangian

There are SU(2)⊗U(1)-invariant Yukawa couplings involving the left-handed doublets of fermions, the right-handed singlets, and the Higgs doublet:

$$\mathcal{L}_Y = - \left(\overline{Q_L} \Gamma \phi n_R + \overline{Q_L} \Delta \tilde{\phi} p_R + \overline{L_L} \Pi \phi l_R \right) + \text{H.c.} \tag{12.10}$$

$$= - \left[\left(\overline{p_L}\ \overline{n_L} \right) \Gamma \begin{pmatrix} \varphi^+ \\ \varphi^0 \end{pmatrix} n_R + \left(\overline{p_L}\ \overline{n_L} \right) \Delta \begin{pmatrix} \varphi^{0\dagger} \\ -\varphi^- \end{pmatrix} p_R \right.$$

$$\left. + \left(\overline{v_L}\ \overline{l_L} \right) \Pi \begin{pmatrix} \varphi^+ \\ \varphi^0 \end{pmatrix} l_R \right] + \text{H.c.}, \tag{12.11}$$

where Γ, Δ, and Π are arbitrary complex numbers. The notation '+H.c.' means 'plus the Hermitian conjugate'.

If we substitute φ^0 by its vacuum expectation value v, we obtain the mass terms

$$\mathcal{L}_{\text{mass}} = -\overline{n_L} M_n n_R - \overline{p_L} M_p p_R - \overline{l_L} M_l l_R + \text{H.c.}, \tag{12.12}$$

with $M_n = v\Gamma$, $M_p = v\Delta$, and $M_l = v\Pi$. The remaining terms in $\mathcal{L}_Y$ are the Yukawa interactions:

$$\mathcal{L}_Y - \mathcal{L}_{\text{mass}} = -\overline{n_L} \frac{M_n}{\sqrt{2}v} \left(H + i\chi \right) n_R - \overline{p_L} \frac{M_p}{\sqrt{2}v} \left(H - i\chi \right) p_R$$

$$- \overline{l_L} \frac{M_l}{\sqrt{2}v} \left(H + i\chi \right) l_R - \overline{p_L} \frac{M_n}{v} \varphi^+ n_R$$

$$+ \overline{n_L} \frac{M_p}{v} \varphi^- p_R - \overline{v_L} \frac{M_l}{v} \varphi^+ l_R + \text{H.c.}. \tag{12.13}$$

---

[29]From now on we adopt the standard use of terming fields with chirality -1 'left-handed' and fields with chirality $+1$ 'right-handed'.

12.4   Generations

One of the striking features of the spectrum of elementary fermions is the fact that there is in Nature a replication of the fermion multiplets. In the SM, the number n_g of fermion 'families' or 'generations' is not fixed by any symmetry principle. Experimentally, there is strong evidence that there are only three generations: $n_g = 3$. The couplings Γ, Δ, and Π in eqn (12.11) then become $n_g \times n_g$ matrices in generation space. Gauge invariance does not constrain the flavour structure of the Yukawa interactions and, as a result, Γ, Δ, and Π are completely arbitrary. This arbitrariness is actually responsible for most of the free parameters in the SM. This is the so-called flavour problem, one of the fundamental open questions in particle physics.

The Yukawa-coupling matrices are not necessarily Hermitian. They may be diagonalized by bi-unitary transformations

$$
\begin{aligned}
p_L &= U_L^p u_L, \\
p_R &= U_R^p u_R, \\
n_L &= U_L^n d_L, \\
n_R &= U_R^n d_R,
\end{aligned}
\tag{12.14}
$$

where $u_{L,R}$ and $d_{L,R}$ denote the $n_g \times 1$ column matrices with the chiral components of the quark mass eigenstates—of the physical quarks. The $n_g \times n_g$ unitary matrices U_L^p and U_R^p are chosen such as to bi-diagonalize M_p (or, equivalently, Δ), while U_L^n and U_R^n bi-diagonalize M_n (or, equivalently, Γ):

$$
\begin{aligned}
U_L^{p\dagger} M_p U_R^p &= M_u = \mathrm{diag}\,(m_u, m_c, m_t, \ldots), \\
U_L^{n\dagger} M_n U_R^n &= M_d = \mathrm{diag}\,(m_d, m_s, m_b, \ldots).
\end{aligned}
\tag{12.15}
$$

The matrices M_u and M_d are, by definition, diagonal; their diagonal matrix elements are real and non-negative.

If we define the Hermitian matrices

$$
\begin{aligned}
H_p &\equiv M_p M_p^\dagger, \\
H_n &\equiv M_n M_n^\dagger,
\end{aligned}
\tag{12.16}
$$

then we realize that the unitary matrices U_L^p and U_L^n diagonalize H_p and H_n:

$$
\begin{aligned}
U_L^{p\dagger} H_p U_L^p &= M_u^2, \\
U_L^{n\dagger} H_n U_L^n &= M_d^2.
\end{aligned}
\tag{12.17}
$$

The charged-current interaction in eqn (12.6), written in terms of the quark mass eigenstates, is

$$
\mathcal{L}_W^{(q)} = \frac{g}{\sqrt{2}} \left(W_\mu^+ \overline{u_L} \gamma^\mu V d_L + W_\mu^- \overline{d_L} \gamma^\mu V^\dagger u_L \right),
\tag{12.18}
$$

where

$$V = U_L^{p\dagger} U_L^n \qquad (12.19)$$

is the Cabibbo–Kobayashi–Maskawa (CKM) matrix (Cabibbo 1963; Kobayashi and Maskawa 1973). It is written

$$V = \begin{pmatrix} V_{ud} & V_{us} & V_{ub} & \cdots \\ V_{cd} & V_{cs} & V_{cb} & \cdots \\ V_{td} & V_{ts} & V_{tb} & \cdots \\ \vdots & \vdots & \vdots & \ddots \end{pmatrix}. \qquad (12.20)$$

The appearance of a non-trivial CKM matrix in the charged current reflects the fact that the Hermitian matrices H_p and H_n are in general diagonalized by different unitary matrices.

In general, we will designate up-type ($Q = 2/3$) quarks by Greek letters $\alpha, \beta, \ldots$, which may assume the values u, c, t, and so on (if there are more than three generations). Down-type ($Q = -1/3$) quarks will be designated by Latin letters $i, j, \ldots$, which represent d, s, b, and so on. A general matrix element of V will be denoted $V_{\alpha i}$.

The neutral-current Lagrangian preserves the form in eqns (12.3)–(12.5), with the weak eigenstates $p_{L,R}$ and $n_{L,R}$ substituted by the mass eigenstates $u_{L,R}$ and $d_{L,R}$, respectively. This means that no mixing matrix analogous to V arises in the current which couples to the Z. In the SM there are no flavour-changing neutral currents (FCNC) at tree level. This is a consequence of the fact that all fermions of a given charge and helicity have the same value of T_3, the third component of weak isospin (Glashow and Weinberg 1977; Paschos 1977).

In the leptonic sector, we bi-diagonalize M_l by performing unitary transformations of the fields, analogously to what is done in the quark sector. However, as the neutrinos are massless in the SM, we are free to transform them in such a way that a mixing matrix does not arise in the leptonic sector:

$$\begin{aligned} v_L &= U_L^l \nu_L, \\ l_L &= U_L^l e_L, \\ l_R &= U_R^l e_R, \end{aligned} \qquad (12.21)$$

where e and ν denote the mass eigenstates of the leptons. Notice that the same matrix U_L^l is used to transform both l_L and v_L. The unitary matrices U_L^l and U_R^l are chosen such that

$$U_L^{l\dagger} M_l U_R^l = M_e = \mathrm{diag}\,(m_e, m_\mu, m_\tau, \ldots). \qquad (12.22)$$

In terms of the lepton mass eigenstates, both the charged-current interaction in eqn (12.8) and the neutral-current interactions in eqn (12.9) preserve their form, with the weak eigenstates $l_{L,R}$ and v_L substituted by the mass eigenstates $e_{L,R}$ and ν_L, respectively. In the SM there is no mixing matrix in the leptonic sector due to the fact that there is no mass matrix for the neutrinos. The absence of

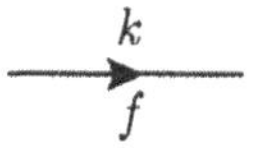

$$\xrightarrow[f]{k} = i \frac{\not{k} + m_f}{k^2 - m_f^2}$$

FIG. 12.1. Propagator of a fermion f with mass m_f.

such a mass matrix allows one to transform v_L into ν_L by using the same unitary matrix U_L^l which is used to transform l_L into e_L. In the SM, neutrinos have zero Dirac mass due to the fact that no right-handed neutrinos are introduced. Since neutrinos have zero electric charge, one could in principle have Majorana mass terms (see Chapter 25) of the type $\nu_{Li}^T C^{-1} \nu_{Lj}$. In the SM, tree-level Majorana mass terms do not arise due to the absence of Higgs triplets. Majorana mass terms might be induced by higher-order diagrams or by non-perturbative effects. However, in the SM the quantum number $B - L$ is exactly conserved—B is the baryon number and L is the lepton number. As Majorana mass terms violate $B - L$, we are sure that they do not arise.

The Feynman rules for the fermion propagators are displayed in Fig. 12.1. The rules for the fermion gauge vertices are in Fig. 12.2.

12.5 Yukawa interactions

Let us return to the Yukawa interactions in eqn (12.13) and rewrite them in terms of the physical-fermion fields. One obtains

$$\mathcal{L}_Y - \mathcal{L}_{\text{mass}} = -\overline{d_L} \frac{M_d}{\sqrt{2}v} (H + i\chi) d_R - \overline{u_L} \frac{M_u}{\sqrt{2}v} (H - i\chi) u_R$$

$$-\overline{e_L} \frac{M_e}{\sqrt{2}v} (H + i\chi) e_R - \overline{u_L} V \frac{M_d}{v} \varphi^+ d_R$$

$$+\overline{d_L} V^\dagger \frac{M_u}{v} \varphi^- u_R - \overline{\nu_L} \frac{M_e}{v} \varphi^+ e_R + \text{H.c.}$$

$$= -\frac{H}{\sqrt{2}v} \left(\overline{d} M_d d + \overline{u} M_u u + \overline{e} M_e e\right) \tag{12.23}$$

$$-\frac{i\chi}{\sqrt{2}v} \left(\overline{d} M_d \gamma_5 d - \overline{u} M_u \gamma_5 u + \overline{e} M_e \gamma_5 e\right) \tag{12.24}$$

$$+\frac{\varphi^+}{v} \left[\overline{u} \left(M_u V \gamma_L - V M_d \gamma_R\right) d - \overline{\nu} M_e \gamma_R e\right] \tag{12.25}$$

$$+\frac{\varphi^-}{v} \left[\overline{d} \left(V^\dagger M_u \gamma_R - M_d V^\dagger \gamma_L\right) u - \overline{e} M_e \gamma_L \nu\right]. \tag{12.26}$$

Here we have introduced the quark fields $d \equiv d_L + d_R$ and $u \equiv u_L + u_R$, and similarly for the leptons, $e \equiv e_L + e_R$ and $\nu \equiv \nu_L$. Equations (12.23)–(12.26) yield the Feynman rules for the Yukawa interactions of the fermions with the physical and unphysical scalars, shown in Fig. 12.3.

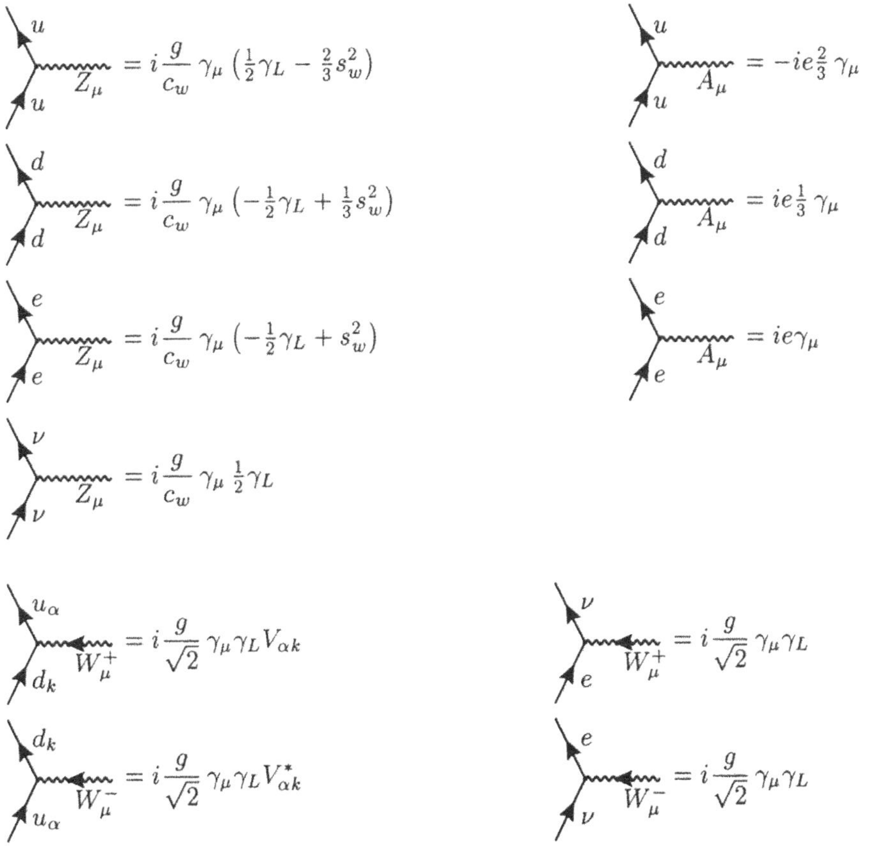

$$\bar{u}\,u\,Z_\mu = i\frac{g}{c_w}\,\gamma_\mu\left(\tfrac{1}{2}\gamma_L - \tfrac{2}{3}s_w^2\right)$$

$$\bar{d}\,d\,Z_\mu = i\frac{g}{c_w}\,\gamma_\mu\left(-\tfrac{1}{2}\gamma_L + \tfrac{1}{3}s_w^2\right)$$

$$\bar{e}\,e\,Z_\mu = i\frac{g}{c_w}\,\gamma_\mu\left(-\tfrac{1}{2}\gamma_L + s_w^2\right)$$

$$\bar{\nu}\,\nu\,Z_\mu = i\frac{g}{c_w}\,\gamma_\mu\,\tfrac{1}{2}\gamma_L$$

$$\bar{u}\,u\,A_\mu = -ie\tfrac{2}{3}\,\gamma_\mu$$

$$\bar{d}\,d\,A_\mu = ie\tfrac{1}{3}\,\gamma_\mu$$

$$\bar{e}\,e\,A_\mu = ie\gamma_\mu$$

$$\bar{u}_\alpha\,d_k\,W_\mu^+ = i\frac{g}{\sqrt{2}}\,\gamma_\mu\gamma_L V_{\alpha k}$$

$$\bar{d}_k\,u_\alpha\,W_\mu^- = i\frac{g}{\sqrt{2}}\,\gamma_\mu\gamma_L V_{\alpha k}^*$$

$$\bar{\nu}\,e\,W_\mu^+ = i\frac{g}{\sqrt{2}}\,\gamma_\mu\gamma_L$$

$$\bar{e}\,\nu\,W_\mu^- = i\frac{g}{\sqrt{2}}\,\gamma_\mu\gamma_L$$

FIG. 12.2. Gauge interactions of the fermions.

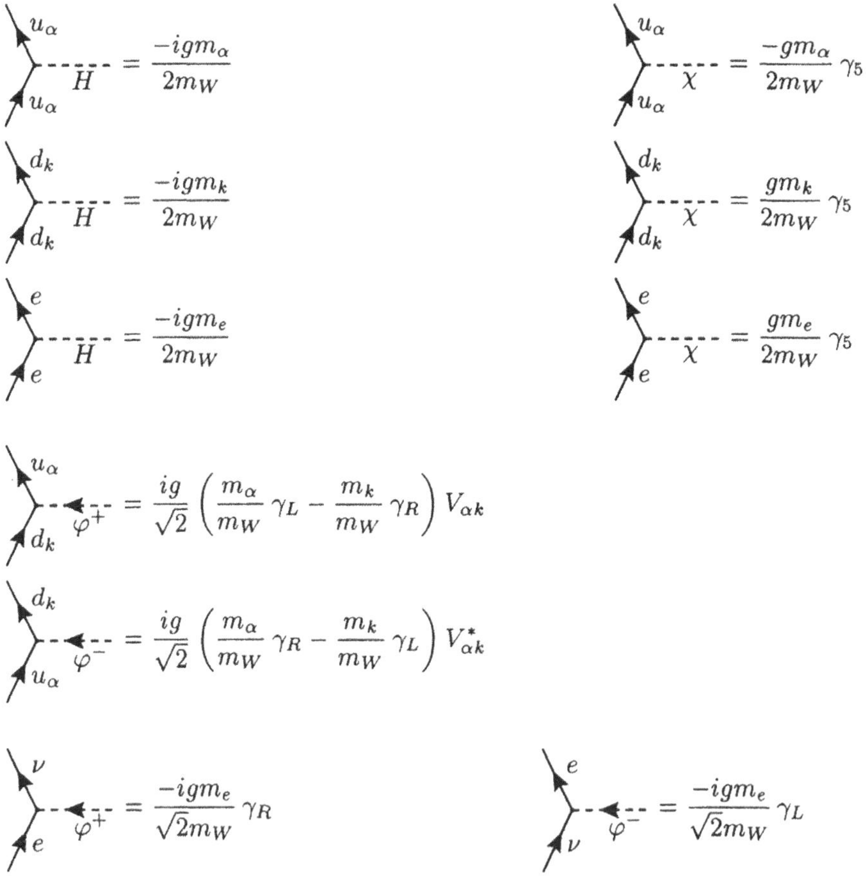

$$\begin{array}{c} u_\alpha \\ \diagup \\ \diagdown \quad H \\ u_\alpha \end{array} = \frac{-igm_\alpha}{2m_W}$$

$$\begin{array}{c} u_\alpha \\ \diagup \\ \diagdown \quad \chi \\ u_\alpha \end{array} = \frac{-gm_\alpha}{2m_W}\,\gamma_5$$

$$\begin{array}{c} d_k \\ \diagup \\ \diagdown \quad H \\ d_k \end{array} = \frac{-igm_k}{2m_W}$$

$$\begin{array}{c} d_k \\ \diagup \\ \diagdown \quad \chi \\ d_k \end{array} = \frac{gm_k}{2m_W}\,\gamma_5$$

$$\begin{array}{c} e \\ \diagup \\ \diagdown \quad H \\ e \end{array} = \frac{-igm_e}{2m_W}$$

$$\begin{array}{c} e \\ \diagup \\ \diagdown \quad \chi \\ e \end{array} = \frac{gm_e}{2m_W}\,\gamma_5$$

$$\begin{array}{c} u_\alpha \\ \diagup \\ \diagdown \quad \varphi^+ \\ d_k \end{array} = \frac{ig}{\sqrt{2}}\left(\frac{m_\alpha}{m_W}\,\gamma_L - \frac{m_k}{m_W}\,\gamma_R\right)V_{\alpha k}$$

$$\begin{array}{c} d_k \\ \diagup \\ \diagdown \quad \varphi^- \\ u_\alpha \end{array} = \frac{ig}{\sqrt{2}}\left(\frac{m_\alpha}{m_W}\,\gamma_R - \frac{m_k}{m_W}\,\gamma_L\right)V_{\alpha k}^*$$

$$\begin{array}{c} \nu \\ \diagup \\ \diagdown \quad \varphi^+ \\ e \end{array} = \frac{-igm_e}{\sqrt{2}m_W}\,\gamma_R$$

$$\begin{array}{c} e \\ \diagup \\ \diagdown \quad \varphi^- \\ \nu \end{array} = \frac{-igm_e}{\sqrt{2}m_W}\,\gamma_L$$

FIG. 12.3. Yukawa interactions.

13

FUNDAMENTAL PROPERTIES OF THE CKM MATRIX

13.1 Rephasing-invariance

The CKM matrix V is complex, but some of the phases in it do not have physical meaning. Indeed, one has the freedom to rephase the quark fields,

$$
\begin{aligned}
u_\alpha &= e^{i\psi_\alpha} u'_\alpha, \\
d_k &= e^{i\psi_k} d'_k,
\end{aligned}
\tag{13.1}
$$

with n_g arbitrary phases ψ_α and n_g arbitrary phases ψ_k. Under eqn (13.1), V transforms as

$$
V'_{\alpha k} = e^{i(\psi_k - \psi_\alpha)} V_{\alpha k}.
\tag{13.2}
$$

One may thus arbitrarily change, and in particular altogether eliminate, the phases of $2n_g - 1$ matrix elements of V. (One cannot eliminate $2n_g$ phases because, if all ψ_α and all ψ_k are equal, no $V_{\alpha k}$ changes—a common rephasing of all quark fields has no effect on V.) In particular, we may choose an arbitrary row and an arbitrary column of V to be real non-negative; this is true for any number of families.

Physically meaningful quantities must be invariant under a rephasing of the fields; only functions of V which are rephasing-invariant may be measurable. The simplest invariants are the moduli of the matrix elements. We denote

$$
U_{\alpha i} \equiv |V_{\alpha i}|^2 .
\tag{13.3}
$$

The next-simplest invariants are the 'quartets'

$$
Q_{\alpha i \beta j} \equiv V_{\alpha i} V_{\beta j} V^*_{\alpha j} V^*_{\beta i}.
\tag{13.4}
$$

We require $\alpha \neq \beta$ and $i \neq j$, lest the quartet reduces to the product of two squared moduli. Clearly, $Q_{\alpha i \beta j} = Q_{\beta j \alpha i} = Q^*_{\alpha j \beta i} = Q^*_{\beta i \alpha j}$. We denote

$$
\omega_{\alpha i \beta j} \equiv \arg Q_{\alpha i \beta j}.
\tag{13.5}
$$

The phases $\omega_{\alpha i \beta j}$ are invariant under eqn (13.2).

Invariants of higher order may in general be written as functions of the quartets and of the moduli. For instance,

$$
V_{\alpha i} V_{\beta j} V_{\gamma k} V^*_{\alpha j} V^*_{\beta k} V^*_{\gamma i} = \frac{Q_{\alpha i \beta j} Q_{\beta i \gamma k}}{U_{\beta i}}.
\tag{13.6}
$$

The procedure may fail in singular cases in which some CKM matrix elements are zero. We shall not consider such singular cases here.

13.2   CP violation

A pure gauge Lagrangian is necessarily CP-invariant (Grimus and Rebelo 1997);
as an instance of this, we have shown in § 3.7 that the Lagrangian of QCD is
CP-invariant.[30] The scalar potential of the SM, in which only one Higgs doublet
exists, automatically conserves CP. As a result, CP violation can only arise from
the simultaneous presence of Yukawa interactions and gauge interactions.

We have seen in Chapter 3 that, under CP, $\partial^\mu \to \partial_\mu$ and

$$(\mathcal{CP})\, A^\mu\,(t,\vec{r})\,(\mathcal{CP})^\dagger = -A_\mu\,(t,-\vec{r})\,. \tag{13.7}$$

Also, from eqns (3.12) and (3.20),

$$\begin{aligned}
(\mathcal{CP})\, \varphi^+\,(t,\vec{r})\,(\mathcal{CP})^\dagger &= e^{i\xi_W}\,\varphi^-\,(t,-\vec{r})\,, \\
(\mathcal{CP})\, \varphi^-\,(t,\vec{r})\,(\mathcal{CP})^\dagger &= e^{-i\xi_W}\,\varphi^+\,(t,-\vec{r})\,,
\end{aligned} \tag{13.8}$$

where ξ_W is an arbitrary phase. Analysing the interactions in eqn (11.17) we
find that, in order to obtain CP invariance of that part of the Lagrangian, one
must postulate

$$\begin{aligned}
(\mathcal{CP})\, W^{+\mu}\,(t,\vec{r})\,(\mathcal{CP})^\dagger &= -e^{i\xi_W}\, W_\mu^-\,(t,-\vec{r})\,, \\
(\mathcal{CP})\, W^{-\mu}\,(t,\vec{r})\,(\mathcal{CP})^\dagger &= -e^{-i\xi_W}\, W_\mu^+\,(t,-\vec{r})\,,
\end{aligned} \tag{13.9}$$

together with

$$(\mathcal{CP})\, Z^\mu\,(t,\vec{r})\,(\mathcal{CP})^\dagger = -Z_\mu\,(t,-\vec{r})\,, \tag{13.10}$$

$$(\mathcal{CP})\, H\,(t,\vec{r})\,(\mathcal{CP})^\dagger = H\,(t,-\vec{r})\,, \tag{13.11}$$

$$(\mathcal{CP})\, \chi\,(t,\vec{r})\,(\mathcal{CP})^\dagger = -\chi\,(t,-\vec{r})\,. \tag{13.12}$$

One finds that all interactions in Chapter 11 are invariant under this CP transformation, which must be extended to the ghosts:

$$\begin{aligned}
(\mathcal{CP})\, c^+\,(t,\vec{r})\,(\mathcal{CP})^\dagger &= \pm e^{i\xi_W}\, c^-\,(t,-\vec{r})\,, \\
(\mathcal{CP})\, c^-\,(t,\vec{r})\,(\mathcal{CP})^\dagger &= \pm e^{-i\xi_W}\, c^+\,(t,-\vec{r})\,, \\
(\mathcal{CP})\, \bar{c}^+\,(t,\vec{r})\,(\mathcal{CP})^\dagger &= \pm e^{-i\xi_W}\, \bar{c}^-\,(t,-\vec{r})\,, \\
(\mathcal{CP})\, \bar{c}^-\,(t,\vec{r})\,(\mathcal{CP})^\dagger &= \pm e^{i\xi_W}\, \bar{c}^+\,(t,-\vec{r})\,, \\
(\mathcal{CP})\, c_Z\,(t,\vec{r})\,(\mathcal{CP})^\dagger &= \pm c_Z\,(t,-\vec{r})\,, \\
(\mathcal{CP})\, \bar{c}_Z\,(t,\vec{r})\,(\mathcal{CP})^\dagger &= \pm \bar{c}_Z\,(t,-\vec{r})\,, \\
(\mathcal{CP})\, c_A\,(t,\vec{r})\,(\mathcal{CP})^\dagger &= \pm c_A\,(t,-\vec{r})\,, \\
(\mathcal{CP})\, \bar{c}_A\,(t,\vec{r})\,(\mathcal{CP})^\dagger &= \pm \bar{c}_A\,(t,-\vec{r})\,.
\end{aligned} \tag{13.13}$$

---

[30] We forget for the moment the possibility of strong CP violation, which will be studied in
Chapter 27.

From Chapter 3 we know that

$$
\begin{aligned}
(\mathcal{CP})\, u_\alpha\, (t, \vec{r})\, (\mathcal{CP})^\dagger &= e^{i\xi_\alpha}\gamma^0 C \overline{u_\alpha}^T (t, -\vec{r})\,, \\
(\mathcal{CP})\, \overline{u_\alpha}\, (t, \vec{r})\, (\mathcal{CP})^\dagger &= -e^{-i\xi_\alpha} u_\alpha^T (t, -\vec{r})\, C^{-1}\gamma^0\,, \\
(\mathcal{CP})\, d_k\, (t, \vec{r})\, (\mathcal{CP})^\dagger &= e^{i\xi_k}\gamma^0 C \overline{d_k}^T (t, -\vec{r})\,, \\
(\mathcal{CP})\, \overline{d_k}\, (t, \vec{r})\, (\mathcal{CP})^\dagger &= -e^{-i\xi_k} d_k^T (t, -\vec{r})\, C^{-1}\gamma^0\,.
\end{aligned}
\tag{13.14}
$$

Now consider the interactions in eqn (12.25). We derive

$$
(\mathcal{CP})\, \varphi^+ \overline{u_\alpha} m_\alpha V_{\alpha k} \gamma_L d_k\, (\mathcal{CP})^\dagger = e^{i(\xi_W + \xi_k - \xi_\alpha)} \varphi^- \overline{d_k} m_\alpha V_{\alpha k} \gamma_R u_\alpha\,, \tag{13.15}
$$

and compare this result with the term $\varphi^- \overline{d_k} V_{\alpha k}^* m_\alpha \gamma_R u_\alpha$ in eqn (12.26). We conclude that, in order for the sum of eqns (12.25) and (12.26) to be CP-invariant, one must have

$$
V_{\alpha k}^* = e^{i(\xi_W + \xi_k - \xi_\alpha)} V_{\alpha k}\,. \tag{13.16}
$$

We would have arrived at this same condition if, instead of imposing CP invariance on the interactions in eqns (12.25) and (12.26), we had imposed CP invariance on the interactions in eqn (12.18).

Equation (13.16) can always be made to hold if one considers a single matrix element of V, because the CP-transformation phases ξ_W, ξ_α, and ξ_k are arbitrary. However, if one simultaneously considers many matrix elements of V, one realizes that eqn (13.16) forces the quartets, and all other rephasing-invariant functions of V, to be real. In general, *there is CP violation in the SM if and only if any of the rephasing-invariant functions of the CKM matrix is not real.*

Because of the absence of a mixing matrix, the interactions of the leptons in the SM are always CP-invariant.

13.3 Parameter counting

In the SM with n_g generations the CKM matrix is $n_g \times n_g$ unitary. It would therefore, in general, be parametrized by n_g^2 parameters. However, $2n_g - 1$ phases may be absorbed or changed at will by rephasing the quark fields. Therefore, the number of physical parameters in V is

$$
N_{\text{param}} = n_g^2 - (2n_g - 1) = (n_g - 1)^2\,. \tag{13.17}
$$

An $n_g \times n_g$ orthogonal matrix is parametrized by $n_g (n_g - 1)/2$ rotation angles, which are sometimes called Euler angles. An unitary matrix is a complex extension of an orthogonal matrix. Therefore, out of the N_{param} parameters of V,

$$
N_{\text{angle}} = \tfrac{1}{2} n_g (n_g - 1) \tag{13.18}
$$

should be identified with rotation angles. The remaining

$$
N_{\text{phase}} = N_{\text{param}} - N_{\text{angle}} = \tfrac{1}{2}(n_g - 1)(n_g - 2) \tag{13.19}
$$

parameters of V are physical phases.

Thus, in the two-generation SM there is no physical phase—no phase which cannot be eliminated by a quark-field rephasing. CP violation is therefore absent when $n_g = 2$. One may confirm this in another way. When $n_g = 2$ there is only one potentially complex quartet, Q_{udcs}. However, the orthogonality relation

$$V_{ud}V_{cd}^* + V_{us}V_{cs}^* = 0 \qquad (13.20)$$

implies, when we multiply it by $V_{us}^*V_{cs}$, that $Q_{udcs} = -U_{us}U_{cs}$, which is real. The quartet being real, there is no CP violation.

According to eqn (13.19), there is one physical phase in V when $n_g = 3$. That phase generates CP violation, as was first pointed out by Kobayashi and Maskawa (1973). Let us confirm in a different way the uniqueness of the origin of CP violation when $n_g = 3$. The orthogonality relation for the first two rows of V reads

$$V_{ud}V_{cd}^* + V_{us}V_{cs}^* + V_{ub}V_{cb}^* = 0. \qquad (13.21)$$

Multiplying it by $V_{us}^*V_{cs}$ and taking the imaginary part, one obtains

$$\operatorname{Im} Q_{udcs} = -\operatorname{Im} Q_{ubcs}. \qquad (13.22)$$

Thus, the two quartets Q_{udcs} and Q_{ubcs} have symmetrical imaginary parts. Proceeding in the same way, one easily shows that, because of the orthogonality of any pair of different rows or columns of V, the imaginary parts of all quartets are equal up to their sign (Jarlskog 1985a,b; Dunietz et al. 1985). One may therefore define

$$J \equiv \operatorname{Im} Q_{uscb} = \operatorname{Im}\left(V_{us}V_{cb}V_{ub}^*V_{cs}^*\right). \qquad (13.23)$$

The imaginary parts of all quartets are equal to J up to their sign. This is a counterpart of $N_{\text{phase}} = 1$.

13.4 Unitarity conditions on the moduli

The SM prediction that the CKM matrix is unitary must be experimentally tested. Non-fulfilment of the unitarity constraints would signal physics beyond the SM. (It must however be kept in mind that the unitarity conditions depend on the number of families. Therefore, it may happen that the physics beyond the SM simply consists of extra generations, without any change in the structure of the model.)

The rows and columns of V are normalized:

$$\sum_{\alpha=1}^{n_g} U_{\alpha i} = \sum_{i=1}^{n_g} U_{\alpha i} = 1, \qquad (13.24)$$

and therefore only $(n_g - 1)^2$ moduli are independent. Thus, the number of parameters N_{param} is equal to the number of independent moduli. This is a special feature of the SM which does not hold in some other models. It suggests the possibility of parametrizing the CKM matrix by the moduli of its matrix elements (Branco and Lavoura 1988a; Lavoura 1989a,b).

For $n_g = 3$, let us take U_{us}, U_{ub}, U_{cs}, and U_{cb} as the independent squared moduli. The other $U_{\alpha i}$ are easily found from these:

$$
\begin{aligned}
U_{ud} &= 1 - U_{us} - U_{ub}, \\
U_{cd} &= 1 - U_{cs} - U_{cb}, \\
U_{ts} &= 1 - U_{us} - U_{cs}, \\
U_{tb} &= 1 - U_{ub} - U_{cb}, \\
U_{td} &= U_{us} + U_{ub} + U_{cs} + U_{cb} - 1.
\end{aligned}
\tag{13.25}
$$

The four independent squared moduli must of course be non-negative. They must also be such that their linear combinations in the right-hand-sides of eqns (13.25) are non-negative.

Now consider the orthogonality condition in eqn (13.21). From it we derive

$$
2\mathrm{Re}\, Q_{uscb} = U_{ud}U_{cd} - U_{us}U_{cs} - U_{ub}U_{cb}.
\tag{13.26}
$$

Using eqns (13.25) for U_{ud} and for U_{cd}, one obtains

$$
2\mathrm{Re}\, Q_{uscb} = 1 - U_{us} - U_{ub} - U_{cs} - U_{cb} + U_{us}U_{cb} + U_{ub}U_{cs}.
\tag{13.27}
$$

Therefore,

$$
4U_{us}U_{cb}U_{ub}U_{cs} \geq (1 - U_{us} - U_{ub} - U_{cs} - U_{cb} + U_{us}U_{cb} + U_{ub}U_{cs})^2.
\tag{13.28}
$$

Thus, if we take the squared moduli U_{us}, U_{ub}, U_{cs}, and U_{cb} as independent, we must check not only that they are non-negative and that their linear combinations in eqns (13.25) are non-negative, but also that eqn (13.28) holds. As a matter of fact, it may be shown (Branco and Lavoura 1988a) that, if the four independent moduli are non-negative and if eqn (13.28) holds, then the linear combinations in eqns (13.25) automatically are non-negative.

One may write $|J|$ as a function of the moduli:

$$
\begin{aligned}
4J^2 &= 4 \left(\mathrm{Im}\, Q_{uscb}\right)^2 \\
&= 4U_{us}U_{cb}U_{ub}U_{cs} - 4 \left(\mathrm{Re}\, Q_{uscb}\right)^2 \\
&= 4U_{us}U_{cb}U_{ub}U_{cs} - (1 - U_{us} - U_{ub} - U_{cs} - U_{cb} + U_{us}U_{cb} + U_{ub}U_{cs})^2.
\end{aligned}
\tag{13.29}
$$

One sees that J^2 is a quadratic function of the $U_{\alpha i}$. For instance, if one keeps U_{us}, U_{ub}, and U_{cb} fixed, and computes J^2 as a function of U_{cs} by using eqn (13.29), one finds the following. There are two values of U_{cs} for which $J^2 = 0$ vanishes; call them $U_{cs}^{\min}$ and $U_{cs}^{\max}$. For $U_{cs} < U_{cs}^{\min}$ and for $U_{cs} > U_{cs}^{\max}$ the inequality (13.28) is violated, and there is no unitary matrix corresponding to the assumed values of the $U_{\alpha i}$. For $U_{cs} = U_{cs}^{\min}$ and for $U_{cs} = U_{cs}^{\max}$, V is an orthogonal matrix, with $J = 0$. For $U_{cs}^{\min} < U_{cs} < U_{cs}^{\max}$, J^2 first increases and then decreases, forming a parabola as a function of U_{cs}, and the matrix V is unitary non-real.

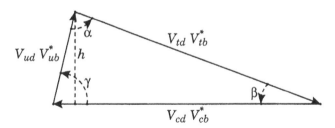

FIG. 13.1. Representation of eqn (13.30) as a triangle in the complex plane.

13.5   The unitarity triangle

Consider the orthogonality condition between the first and third columns of V:

$$V_{ud}V_{ub}^* + V_{cd}V_{cb}^* + V_{td}V_{tb}^* = 0. \tag{13.30}$$

This equation may be interpreted as representing a triangle in the complex plane. That triangle is depicted in Fig. 13.1. It is rotated as a whole when the CKM matrix is rephased as in eqn (13.2):

$$V'_{\alpha d}V'^*_{\alpha b} = e^{i(\psi_d - \psi_b)} V_{\alpha d}V_{\alpha b}.$$

However, the shape of the triangle remains unchanged, because both its inner angles and the length of its sides are rephasing-invariant. In drawing Fig. 13.1 we have adopted a phase convention in which $V_{cd}V_{cb}^*$ is real and negative.

The inner angles of the triangle, α, β, and γ, are defined by

$$\alpha \equiv \arg\left(-\frac{V_{td}V_{tb}^*}{V_{ud}V_{ub}^*}\right) = \arg\left(-Q_{ubtd}\right),$$

$$\beta \equiv \arg\left(-\frac{V_{cd}V_{cb}^*}{V_{td}V_{tb}^*}\right) = \arg\left(-Q_{tbcd}\right), \tag{13.31}$$

$$\gamma \equiv \arg\left(-\frac{V_{ud}V_{ub}^*}{V_{cd}V_{cb}^*}\right) = \arg\left(-Q_{cbud}\right).$$

They satisfy, by definition,

$$\alpha + \beta + \gamma = \arg(-1) = \pi \bmod 2\pi. \tag{13.32}$$

It should be emphasized that eqn (13.32) holds even if eqn (13.30) is not valid— i.e., even if there are deviations from 3×3 unitarity—provided that one sticks to the definitions in eqns (13.31).

It is useful to rescale the triangle by defining

$$R_t \equiv \left|\frac{V_{td}V_{tb}}{V_{cd}V_{cb}}\right|,$$

$$R_b \equiv \left|\frac{V_{ud}V_{ub}}{V_{cd}V_{cb}}\right|. \tag{13.33}$$

One then obtains a triangle with sides of length 1, R_t, and R_b, which is depicted in Fig. 13.2. In that figure we have marked the complex coordinates of

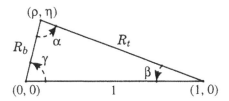

FIG. 13.2. The triangle of Fig. 13.1 after division of its sides by $|V_{cd}V_{cb}|$.

the vertices of the triangle as $(0,0)$, $(1,0)$, and (ρ,η), after we have made the phase convention that the side with length 1 is real. Clearly, $R_b = \sqrt{\rho^2 + \eta^2}$ and $R_t = \sqrt{(1-\rho)^2 + \eta^2}$. This phase convention and parametrization will be explained in Chapter 16.

Trigonometry yields

$$\cos\alpha = \frac{R_t^2 + R_b^2 - 1}{2R_t R_b},$$

$$\cos\beta = \frac{1 + R_t^2 - R_b^2}{2R_t}, \tag{13.34}$$

$$\cos\gamma = \frac{1 - R_t^2 + R_b^2}{2R_b},$$

and

$$\sin\alpha = \frac{\sqrt{\Sigma}}{2R_t R_b},$$

$$\sin\beta = \frac{\sqrt{\Sigma}}{2R_t}, \tag{13.35}$$

$$\sin\gamma = \frac{\sqrt{\Sigma}}{2R_b},$$

where

$$\Sigma \equiv -1 - R_t^4 - R_b^4 + 2\left(R_t^2 + R_b^2 + R_t^2 R_b^2\right) \tag{13.36}$$

must be non-negative. The condition $\Sigma \geq 0$ is equivalent to eqn (13.28). This is an instance of the theorem which states that it is possible to build a triangle with sides of length a, b, and c if and only if those three numbers are non-negative and satisfy

$$\lambda(a,b,c) \equiv -a^4 - b^4 - c^4 + 2\left(a^2 b^2 + a^2 c^2 + b^2 c^2\right) \geq 0. \tag{13.37}$$

The law of sines reads

$$\sin\alpha : \sin\beta : \sin\gamma = 1 : R_b : R_t. \tag{13.38}$$

Other useful relations are

$$R_t = \frac{\sin\gamma}{\sin(\gamma+\beta)},$$

$$R_b = \frac{\sin\beta}{\sin(\gamma+\beta)}, \qquad (13.39)$$

$$R_t^2 - R_b^2 = \frac{\sin(\gamma-\beta)}{\sin(\gamma+\beta)}.$$

13.6   Geometrical interpretation of J

Consider again the triangle in Fig. 13.1. Its height is $h = |V_{ud}V_{ub}\sin\gamma|$ and the area is $|V_{cd}V_{cb}|\,h/2$. Therefore, the area of the triangle is $(1/2)\,|Q_{udcb}\sin\gamma|$, i.e.,

$$\text{Area} = \frac{|\text{Im}\,Q_{udcb}|}{2} = \frac{|J|}{2}. \qquad (13.40)$$

Thus, $|J|$ may be geometrically interpreted as twice the area of the unitarity triangle.

Similar considerations apply to the other orthogonality relations and to the corresponding unitarity triangles. The six orthogonality relations for the three-generation CKM matrix,

$$V_{ud}V_{us}^* + V_{cd}V_{cs}^* + V_{td}V_{ts}^* = 0, \qquad (13.41)$$

$$V_{ud}V_{ub}^* + V_{cd}V_{cb}^* + V_{td}V_{tb}^* = 0, \qquad (13.42)$$

$$V_{us}V_{ub}^* + V_{cs}V_{cb}^* + V_{ts}V_{tb}^* = 0, \qquad (13.43)$$

$$V_{ud}V_{cd}^* + V_{us}V_{cs}^* + V_{ub}V_{cb}^* = 0, \qquad (13.44)$$

$$V_{ud}V_{td}^* + V_{us}V_{ts}^* + V_{ub}V_{tb}^* = 0, \qquad (13.45)$$

$$V_{cd}V_{td}^* + V_{cs}V_{ts}^* + V_{cb}V_{tb}^* = 0, \qquad (13.46)$$

may be represented by six triangles in the complex plane. All these triangles have the same area $|J|/2$.

At this stage, one should note that, as we shall see in Chapter 15, the orders of magnitude of the moduli $|V_{\alpha i}|$ may be given as powers of a small parameter $\lambda = 0.22$ in the following way:

$$|V| \sim \begin{pmatrix} 1 & \lambda & \lambda^3 \\ \lambda & 1 & \lambda^2 \\ \lambda^3 & \lambda^2 & 1 \end{pmatrix}. \qquad (13.47)$$

If we do this, we see that the sides of the triangles corresponding to eqns (13.42) and (13.45) are all of comparable size λ^3. Those are 'normal' triangles. On the other hand, the other four unitarity triangles are all very flat, in each case one of their sides being much smaller than the other two—the triangles in eqns (13.43) and (13.46) have two sides $\sim \lambda^2$ and one side $\sim \lambda^4$; the triangles in eqns (13.41) and (13.44) have two sides $\sim \lambda$ and one side $\sim \lambda^5$. In spite of the very different shapes, all six triangles end up having the same area. This follows from the unique character of $|J|$ as the imaginary part of all quartets in the three-generation SM.

All imaginary parts of rephasing-invariant products of CKM matrix elements are proportional to $|J|$. Hence, $|J|$ is a measure of the strength of CP violation. Its order of magnitude is

$$|J| = |V_{ud}V_{ub}V_{cd}V_{cb} \sin \gamma| \sim \lambda^6 |\sin \gamma| \lesssim 10^{-4}. \qquad (13.48)$$

It is clear from eqn (13.48) that, in the framework of the three-generation SM, the smallness of $|J|$ is due to the smallness of the moduli of the off-diagonal matrix elements of V, and not to the fact that any phase is particularly small. It is also interesting to note that, if we did not have any experimental knowledge about the moduli of the CKM matrix elements, the maximum possible value of $|J|$ would be $1/\left(6\sqrt{3}\right) \approx 0.096$, which is attained for

$$V = \tfrac{1}{\sqrt{3}} \begin{pmatrix} 1 & 1 & 1 \\ 1 & \exp(2i\pi/3) & \exp(-2i\pi/3) \\ 1 & \exp(-2i\pi/3) & \exp(2i\pi/3) \end{pmatrix}. \qquad (13.49)$$

13.7 The parameters λ_α

In the standard-model discussion of the neutral-meson systems, certain combinations of CKM matrix elements will be useful. Those combinations are different for the K^0–$\overline{K^0}$, B_d^0–$\overline{B_d^0}$, and B_s^0–$\overline{B_s^0}$ systems; however, it is customary to use the same notation λ_α in all three cases. We define

$$\lambda_\alpha \equiv V_{\alpha s}^* V_{\alpha d} \quad \text{for the } K^0\text{–}\overline{K^0} \text{ system}, \qquad (13.50)$$

$$\lambda_\alpha \equiv V_{\alpha b}^* V_{\alpha d} \quad \text{for the } B_d^0\text{–}\overline{B_d^0} \text{ system}, \qquad (13.51)$$

$$\lambda_\alpha \equiv V_{\alpha b}^* V_{\alpha s} \quad \text{for the } B_s^0\text{–}\overline{B_s^0} \text{ system}. \qquad (13.52)$$

In all three cases, α may be either u, c, or t. Unitarity of the 3×3 CKM matrix implies

$$\lambda_u + \lambda_c + \lambda_t = 0. \qquad (13.53)$$

The basic CP-violating quantity J is

$$J = \mathrm{Im}\,(\lambda_t \lambda_u^*) = \mathrm{Im}\,(\lambda_u \lambda_c^*) \qquad (13.54)$$

in the K^0–$\overline{K^0}$ and in the B_s^0–$\overline{B_s^0}$ cases; in the B_d^0–$\overline{B_d^0}$ case, we have to invert the sign of J given by eqn (13.54).

13.8 Main conclusions

- As the quark fields may be freely rephased, there are phases in the CKM matrix which have no physical significance.
- The simplest non-real rephasing-invariant functions of the CKM matrix are the quartets $V_{\alpha i}V_{\beta j}V_{\alpha j}^*V_{\beta i}^*$.
- CP is conserved in the SM if and only if all rephasing-invariant functions of the CKM matrix are real.

- Unitarity of the CKM matrix implies that there can be no CP violation in the SM for $n_g = 2$.
- Unitarity of the CKM matrix implies that there is only one basic CP-violating quantity for $n_g = 3$. That quantity is called J.
- For $n_g = 3$, the orthogonality relation between any pair of rows or columns of the CKM matrix may be represented as a triangle in the complex plane. All triangles have the same area $2|J|$, in spite of their very different shapes.
- The triangle with sides $|V_{ud}V_{ub}^*|$, $|V_{cd}V_{cb}^*|$, and $|V_{td}V_{tb}^*|$ is usually called 'the unitarity triangle' and is particularly important. Its three sides have the same order of magnitude. Its inner angles are called α, β, and γ.
- The physical CKM matrix can be parametrized by $\frac{1}{2}n_g(n_g - 1)$ Euler angles and $\frac{1}{2}(n_g - 1)(n_g - 2)$ physical phases.

14

WEAK-BASIS INVARIANTS AND CP VIOLATION

14.1 Introduction

In the previous two chapters we have taken the usual path to CP violation
in the SM. We have bi-diagonalized the quark mass matrices and noticed the
appearance of a mixing matrix V in the charged-current interaction. We saw
that, for three or more fermion generations, some of the phases in V cannot
be removed by rephasing the quark fields; in general, some rephasing-invariant
combinations of CKM matrix elements are not real. We have considered the
CP transformations of the physical quark fields and have concluded that all
rephasing-invariant combinations of CKM matrix elements must be real in order
for CP symmetry to hold.

In this chapter, we shall describe what we believe to be the most natural path
to the study of the CP properties of any gauge theory. We have seen in § 2.2
that the CP-transformation properties of the fields in a given Lagrangian are
defined by the part of that Lagrangian which conserves CP. Thus, the complete
Lagrangian is written as

$$\mathcal{L} = \mathcal{L}_{\text{CP}} + \mathcal{L}_{\text{remaining}}, \tag{14.1}$$

where $\mathcal{L}_{\text{CP}}$ is CP-conserving. Typically, it is the part of the Lagrangian which
involves the gauge interactions of the various fields—in particular, their electromagnetic interaction. In order to analyse whether the whole Lagrangian violates
CP, one has to check whether the CP transformation under which $\mathcal{L}_{\text{CP}}$ is invariant implies non-trivial restrictions—i.e., restrictions which are not automatically
satisfied—on the rest of the Lagrangian, $\mathcal{L}_{\text{remaining}}$. The Lagrangian leads to CP
violation if and only if such restrictions exist and are not satisfied.

We have also seen that one should allow for the most general CP transfor-
mation. Typically, $\mathcal{L}_{\text{CP}}$ leaves a large freedom of choice in the definition of the
CP transformation. This is due to the existence of internal symmetries. CP is
violated only when there is no possible choice of CP transformation which leaves
$\mathcal{L}_{\text{remaining}}$ invariant.

We find this approach (Bernabéu et al. 1986a) aesthetically more appealing
than the previous one, since it gives new insight into the nature of CP violation
in the SM, and allows for interesting generalizations to other models.

14.2 Conditions for CP invariance

Let us consider the Lagrangian of the SM with one Higgs doublet and an arbitrary
number n_g of fermion generations. The original SU(2)⊗U(1) gauge symmetry

has been spontaneously broken into the U(1) of electromagnetism. We know how the gauge fields, the scalar fields, and the ghost fields transform under CP—see eqns (13.7)–(13.13). The gauge interactions of the quarks are given by eqns (12.3)–(12.6).

One starts by choosing that part of the Lagrangian which will be used to define CP. This will be the part of the Lagrangian involving the gauge interactions. One should consider the CP transformations of the eigenstates of the electroweak interaction, i.e., the fields p and n. The most general CP transformation of the quarks which leaves the gauge interactions invariant is

$$
\begin{aligned}
(\mathcal{CP})\, p_L\, (t, \vec{r})\, (\mathcal{CP})^\dagger &= e^{i\xi w}\, K_L \gamma^0 C \overline{p_L}^T\, (t, -\vec{r})\,, \\
(\mathcal{CP})\, n_L\, (t, \vec{r})\, (\mathcal{CP})^\dagger &= K_L \gamma^0 C \overline{n_L}^T\, (t, -\vec{r})\,, \\
(\mathcal{CP})\, p_R\, (t, \vec{r})\, (\mathcal{CP})^\dagger &= K_R^p \gamma^0 C \overline{p_R}^T\, (t, -\vec{r})\,, \\
(\mathcal{CP})\, n_R\, (t, \vec{r})\, (\mathcal{CP})^\dagger &= K_R^n \gamma^0 C \overline{n_R}^T\, (t, -\vec{r})\,,
\end{aligned}
\tag{14.2}
$$

where K_L, K_R^p, and K_R^n are $n_g \times n_g$ unitary matrices acting in family space. It follows that

$$
\begin{aligned}
(\mathcal{CP})\, \overline{p_L}\, (t, \vec{r})\, (\mathcal{CP})^\dagger &= -e^{-i\xi w}\, p_L^T\, (t, -\vec{r})\, C^{-1} \gamma^0 K_L^\dagger\,, \\
(\mathcal{CP})\, \overline{n_L}\, (t, \vec{r})\, (\mathcal{CP})^\dagger &= -n_L^T\, (t, -\vec{r})\, C^{-1} \gamma^0 K_L^\dagger\,, \\
(\mathcal{CP})\, \overline{p_R}\, (t, \vec{r})\, (\mathcal{CP})^\dagger &= -p_R^T\, (t, -\vec{r})\, C^{-1} \gamma^0 K_R^{p\,\dagger}\,, \\
(\mathcal{CP})\, \overline{n_R}\, (t, \vec{r})\, (\mathcal{CP})^\dagger &= -n_R^T\, (t, -\vec{r})\, C^{-1} \gamma^0 K_R^{n\,\dagger}\,.
\end{aligned}
\tag{14.3}
$$

The appearance in the CP transformation of unitary matrices mixing the families should not surprise us—remember § 2.2.2. Before one introduces the Yukawa couplings, all fermion generations have identical weak interaction and the flavours are indistinguishable. As a result, the generations may in general mix under CP. This is the reason for the appearance of matrices K_L, K_R^p, and K_R^n in the CP transformation. Those matrices are unitary because we must preserve the normalization of the Lagrangian, in particular of the fermion kinetic terms. Furthermore, the presence of the left-handed charged current $\overline{p_L}\gamma^\mu n_L$ in $\mathcal{L}_W^{(q)}$ constrains p_L and n_L to transform with the same unitary matrix K_L. On the other hand, since there is no right-handed charged current, p_R and n_R may transform differently.

We next consider the Yukawa Lagrangian of eqn (12.11) and test it for invariance under the CP transformation in eqns (14.2) and (14.3). One readily finds that, in order to obtain CP invariance, the Yukawa-coupling matrices Γ and Δ should satisfy

$$
\begin{aligned}
K_L^\dagger \Delta K_R^p &= \Delta^*\,, \\
K_L^\dagger \Gamma K_R^n &= \Gamma^*\,.
\end{aligned}
\tag{14.4}
$$

It is convenient to work with the mass matrices $M_p = v\Delta$ and $M_n = v\Gamma$ instead of working with Δ and Γ:

$$K_L^\dagger M_p K_R^p = M_p^*,$$
$$K_L^\dagger M_n K_R^n = M_n^*. \tag{14.5}$$

One may thus state

Theorem 14.1 *The Lagrangian of the SM is CP-invariant if and only if the quark mass matrices M_p and M_n are such that unitary matrices K_L, K_R^p, and K_R^n exist, which satisfy eqns (14.5).*

The non-trivial point is the simultaneous appearance of the matrix K_L in both eqns (14.5). This is a consequence of the existence of a charged current connecting the left-handed up-type and down-type quarks. We shall see that, for three or more generations, in general one cannot find an unitary matrix K_L satisfying both eqns (14.5), and therefore CP can be violated. In contrast, the lepton sector is CP-invariant because unitary matrices K_L^l and K_R^l can always be found such that

$$K_L^l{}^\dagger M_l K_R^l = M_l^*. \tag{14.6}$$

Indeed, from eqn (12.22) one has

$$M_l = U_L^l M_e U_R^l{}^\dagger, \tag{14.7}$$

where M_e is a real diagonal matrix; therefore, the matrices $K_L^l = U_L^l U_L^l{}^T$ and $K_R^l = U_R^l U_R^l{}^T$ satisfy eqn (14.6).

Remembering the Hermitian matrices H_p and H_n defined in eqns (12.16), one sees that eqns (14.5) imply

$$K_L^\dagger H_p K_L = H_p^*,$$
$$K_L^\dagger H_n K_L = H_n^*. \tag{14.8}$$

Conversely, one may prove that eqns (14.8) imply eqns (14.5). If eqns (14.8) hold, then the matrices

$$K_R^p = M_p^{-1} K_L M_p^*,$$
$$K_R^n = M_n^{-1} K_L M_n^* \tag{14.9}$$

are unitary and satisfy eqns (14.5). One has thus proved

Theorem 14.2 *The Lagrangian of the SM is CP-invariant if and only if the matrices $H_p \equiv M_p M_p^\dagger$ and $H_n \equiv M_n M_n^\dagger$ are such that a unitary matrix K_L exists which satisfies eqns (14.8).*

14.3   Weak-basis transformations

Since M_p and M_n are complex $n_g \times n_g$ matrices, they contain a total of $4n_g^2$ real numbers. However, there is a lot of unphysical information in those matrices, due to the freedom that one has to make weak-basis transformations (WBT).

When a theory has several fields with the same quantum numbers—e.g. the n_g multiplets Q_L—one is free to rewrite the Lagrangian in terms of new fields, obtained from the original ones by means of a unitary transformation which mixes them. Such changes of weak basis (WB) are an extension of the freedom that one generally has to rephase a field. A WBT is a transformation of the fermion fields which leaves invariant the kinetic-energy terms as well as the gauge interactions. The WBT depend on the gauge theory that one is considering because, if there are more gauge interactions then, in principle, there will be less freedom to make WBT. In the SM a WBT is defined by

$$
\begin{aligned}
Q_L &= W_L Q'_L, \\
p_R &= W_R^p p'_R, \\
n_R &= W_R^n n'_R,
\end{aligned}
\tag{14.10}
$$

where W_L, W_R^p, and W_R^n are $n_g \times n_g$ unitary matrices acting in family space. Under this WBT, the matrices M_p and M_n in eqn (12.10) transform as

$$
\begin{aligned}
M'_p &= W_L^\dagger M_p W_R^p, \\
M'_n &= W_L^\dagger M_n W_R^n.
\end{aligned}
\tag{14.11}
$$

The transformed mass matrices M'_n and M'_p have the same physical content as the original matrices M_n and M_p. One may use the freedom of making WBT to put M_n and/or M_p in a special form. For example, in the SM, through a WBT one can make M_n Hermitian and, simultaneously, M_p diagonal with real non-negative diagonal elements. In particular, there is a WB in which

$$
\begin{aligned}
H_p &= \mathrm{diag}\left(m_u^2, m_c^2, m_t^2, \ldots\right), \\
H_n \equiv H'_n &= V \mathrm{diag}\left(m_d^2, m_s^2, m_b^2, \ldots\right) V^\dagger.
\end{aligned}
\tag{14.12}
$$

Alternatively, one may consider another WB, in which

$$
\begin{aligned}
H_p \equiv H'_p &= V^\dagger \mathrm{diag}\left(m_u^2, m_c^2, m_t^2, \ldots\right) V, \\
H_n &= \mathrm{diag}\left(m_d^2, m_s^2, m_b^2, \ldots\right).
\end{aligned}
\tag{14.13}
$$

In a WBT one just rewrites the Lagrangian in terms of new fields. Physics must be invariant under such a change. The result of any physical process can only depend on WB-invariant quantities. Those quantities are especially useful in the analysis of CP violation. Indeed, the possibility of rephasing each field means that there are spurious phases in the Lagrangian. Those phases must be carefully distinguished from the physical phases that are the hallmark of CP violation.

Suppose that, in a particular WB, matrices K_L, K_R^p, and K_R^n exist which satisfy eqns (14.5). In another WB the Yukawa-coupling matrices are given by eqns (14.11). Then the unitary matrices

$$
\begin{aligned}
K'_L &\equiv W_L K_L W_L^T, \\
K_R^{p\,'} &\equiv W_R^p K_R^p W_R^{pT}, \\
K_R^{n\,'} &\equiv W_R^n K_R^n W_R^{nT},
\end{aligned}
\tag{14.14}
$$

satisfy

$$K_L'^\dagger M_p' K_R^{p\prime} = M_p'^*,$$
$$K_L'^\dagger M_n' K_R^{n\prime} = M_n'^*,$$

(14.15)

analogously to eqns (14.5). Thus, Theorem 14.1 is invariant under a change of WB, as it should.

14.4   Weak-basis invariants

Physically meaningful quantities must be invariant under the transformation in eqns (14.11). In order to find such quantities one must get rid of the arbitrary unitary matrices W_L, W_R^p, and W_R^n. We do this by first considering the matrices H_p and H_n, which allow us to eliminate the unitary matrices W_R^p and W_R^n. Indeed,

$$H_p' = W_L^\dagger H_p W_L,$$
$$H_n' = W_L^\dagger H_n W_L.$$

(14.16)

Then, traces of arbitrary polynomials of H_p and H_n are WB-invariant. For instance, $\operatorname{tr} H_p$, $\operatorname{tr}\left(H_n^2 H_p\right)$, and $\operatorname{tr}\left(H_n H_p H_n^2 H_p^2\right)$ are invariant under a WBT and, therefore, they are in principle physically meaningful quantities. Considering any of the weak bases in eqns (14.12) and (14.13), one sees that

$$\operatorname{tr} H_p^a = m_u^{2a} + m_c^{2a} + m_t^{2a} + \cdots,$$
$$\operatorname{tr} H_n^a = m_d^{2a} + m_s^{2a} + m_b^{2a} + \cdots$$

(14.17)

are physically meaningful quantities. In the same way,

$$\operatorname{tr}\left(H_p^r H_n^s\right) = \sum_{\alpha=u,c,t,\ldots} \sum_{i=d,s,b,\ldots} m_\alpha^{2r} m_i^{2s} U_{\alpha i}$$

(14.18)

is measurable too. Thus, WB-invariant quantities in general are physical, i.e., measurable.

14.5   Weak-basis-invariant conditions for CP invariance

We next want to derive necessary conditions for CP invariance, expressed in terms of weak-basis invariants. As the matrices K_L, K_R^p, and K_R^n are WB-dependent, see eqns (14.14), this is in practice equivalent to eliminating those matrices from the CP-invariance conditions.

We start from eqns (14.8), in which K_R^p and K_R^n are already absent. As $H_p^* = H_p^T$ and $H_n^* = H_n^T$, we obtain

$$K_L^\dagger [H_p, H_n] K_L = \left[H_p^T, H_n^T\right]$$

$$= -[H_p, H_n]^T . \tag{14.19}$$

We multiply eqn (14.19) by itself an odd number of times to obtain

$$K_L^\dagger [H_p, H_n]^r K_L = -[H_p, H_n]^{rT} \text{ for } r \text{ odd.} \tag{14.20}$$

Taking the trace, one has

$$\text{tr}\,[H_p, H_n]^r = 0 \text{ for } r \text{ odd.} \tag{14.21}$$

Equation (14.21) is non-trivial if $r \geq 3$. For $r = 1$ it is trivial, since the trace of a commutator is identically zero.

Equation (14.21) is a necessary condition for CP invariance, valid for an arbitrary number of generations. Indeed, in its derivation we have not specified n_g.

One easily finds

$$\text{tr}\,[H_p, H_n]^3 = 6i\,\text{Im tr}\,\left(H_p^2 H_n^2 H_p H_n\right). \tag{14.22}$$

$$= 6i \sum_{\alpha,\beta=u,c,t,\dots} \sum_{i,j=d,s,b,\dots} m_\alpha^4 m_\beta^2 m_i^4 m_j^2 \text{Im}\,Q_{\alpha i \beta j}. \tag{14.23}$$

14.6 Two and three generations

14.6.1 $n_g = 2$

The conditions in eqn (14.21) should be automatically satisfied in the two-generation case, because there can be no CP violation in the SM when $n_g = 2$, as seen in § 13.3. Indeed, for arbitrary 2×2 Hermitian matrices H_1 and H_2, $[H_1, H_2]^2$ is a multiple of the unit matrix. As a consequence, eqn (14.21) is trivially satisfied when $n_g = 2$.

14.6.2 $n_g = 3$

Starting from eqn (14.23) and taking into account that, when $n_g = 3$, all quartets have identical imaginary parts, up to the sign, one finds

$$\text{tr}\,[H_p, H_n]^3 = 6i\,\left(m_t^2 - m_c^2\right)\left(m_t^2 - m_u^2\right)\left(m_c^2 - m_u^2\right)$$
$$\left(m_b^2 - m_s^2\right)\left(m_b^2 - m_d^2\right)\left(m_s^2 - m_d^2\right) J. \tag{14.24}$$

One concludes that, in order for there to be CP violation, all three up-type quarks must be non-degenerate, all three down-type quarks must be non-degenerate, and J cannot vanish.

The fact that the quark masses should not be degenerate has to do with the definition of J. Let us suppose, for instance, that m_b was equal to m_s. Then, we would be free to mix the quarks s and b by means of a 2×2 unitary matrix U_2. This would correspond to a change of the CKM matrix,

$$V \rightarrow V \begin{pmatrix} 1 & 0 \\ 0 & U_2 \end{pmatrix}.$$

It is easily seen that, by manipulating U_2, one may arbitrarily change the value of J, and in particular set $J = 0$. Thus, J becomes arbitrary when any two quarks of equal charge are mass-degenerate. That is the reason why CP is conserved in that limit.

14.6.2.1 *Sufficient condition for CP conservation*   We have seen that the condition

$$\operatorname{tr}[H_p, H_n]^3 = 0 \tag{14.25}$$

is necessary for CP invariance in the SM, for an arbitrary number of generations. We shall next show that, for $n_g = 3$, eqn (14.25) is not only a necessary condition for CP invariance, it is also a sufficient condition.

For that purpose, it is advantageous to consider the WB of eqns (14.12). One has

$$\operatorname{tr}[H_p, H_n']^3 \propto (m_t^2 - m_c^2)(m_t^2 - m_u^2)(m_c^2 - m_u^2)\operatorname{Im}(H'_{n12}H'_{n23}H'_{n31}). \tag{14.26}$$

Let us consider first the case where the up-type quarks are not mass-degenerate. In that case eqn (14.25) implies

$$\operatorname{Im}(H'_{n12}H'_{n23}H'_{n31}) = 0, \tag{14.27}$$

from which it follows that there exist three phases α_1, α_2, and α_3, such that

$$\arg H'_{njk} = \alpha_j - \alpha_k \bmod \pi. \tag{14.28}$$

In this case, a unitary matrix K_L satisfying eqns (14.8) can easily be found. Remembering that H_p is diagonal, we have

$$K_{Ljk} = \delta_{jk}\exp(2i\alpha_j). \tag{14.29}$$

Let us now consider the case in which two up-type-quark masses, say m_α and m_β, are equal. We may then use a WBT to obtain $H'_{n\alpha\beta} = 0$. The condition of eqn (14.27) is then trivially satisfied, and the unitary matrix K_L of eqn (14.29) will once again satisfy eqns (14.8).

We have thus shown that, in the SM with three families, eqn (14.25) implies eqns (14.8), and therefore CP conservation.[31]

---

[31] For $n_g = 3$, as $[H_p, H_n]$ is a 3×3 traceless matrix,

$$\operatorname{tr}[H_p, H_n]^3 = 3\det[H_p, H_n], \tag{14.30}$$

and therefore the necessary and sufficient condition for CP invaraince may be written

$$\det[H_p, H_n] = 0. \tag{14.31}$$

As a matter of fact, the condition in eqn (14.31) (Jarlskog 1985a,b) anticipated historically the one in eqn (14.25) (Bernabéu et al. 1986a). However, it must be stressed that eqn (14.25) is a necessary condition for CP invariance for an arbitrary number of families, while the condition in eqn (14.31) is only valid for n_g odd—as one easily sees from eqn (14.19). Thus, *eqn (14.25) is more general than eqn (14.31)*.

14.7   More than three generations

14.7.1   *CP restrictions in a special weak basis*

It is useful to analyse the form of the CP restrictions, for an arbitrary number of generations, in a special WB, namely the WB where one of the quark mass matrices is diagonal and real. For definiteness, let us consider the WB in eqns (14.12). From our analysis of the CP-violating phases in the CKM matrix, one expects to obtain $(n_g - 1)(n_g - 2)/2$ independent CP restrictions. The first eqn (14.8), in that WB, constrains the matrix K_L to be of the form

$$K_L = \text{diag}\left(e^{i\delta_1}, e^{i\delta_2}, \ldots, e^{i\delta_{n_g}}\right). \tag{14.32}$$

In deriving eqn (14.32) we have assumed that there is no mass-degeneracy in the up-quark sector. The phases δ_i are arbitrary. Substituting eqn (14.32) in the second eqn (14.8) one obtains the constraint

$$\arg H'_{nij} = \frac{\delta_i - \delta_j}{2} \text{ mod } \pi. \tag{14.33}$$

The meaning of eqn (14.33) is clear: in the SM and for an arbitrary number of generations, a necessary and sufficient condition for CP invariance is that H_n has cyclic phases in a WB where H_p is diagonal.[32] Since H'_n is an $n_g \times n_g$ Hermitian matrix, it has in general $n_g (n_g - 1)/2$ independent phases. Therefore, eqn (14.33) implies the existence of $(n_g - 1)(n_g - 2)/2$ independent CP restrictions, as expected.

14.7.2   *Sufficient conditions for CP invariance*

The question of finding a minimal set of WB-invariant conditions which are necessary and sufficient for CP invariance becomes quite involved for $n_g > 3$. For $n_g = 4$, this problem has been solved by Gronau *et al.* (1986).

One can also derive WB-invariant conditions relevant to CP violation using the following argument (Roldán, unpublished thesis). Comparing eqns (14.5) and (14.11), one observes that CP transformations do exactly the same thing as weak-basis transformations, except for the fact that they transform c-numbers into their complex conjugates. One concludes that WB invariants are transformed into their complex conjugates by a CP transformation.[33] It follows that a non-zero imaginary part of any WB invariant signals CP violation. However, many invariants are real independently of CP being violated or not. For instance, $\text{tr} \, H_p^r$ is real because the diagonal elements of an Hermitian matrix are real; and $\text{tr} \left(H_p^r H_n^s\right)$ is real because it is the trace of the product of two Hermitian matrices. The simplest invariant which may have a non-zero imaginary part is $\text{tr} \left(H_p^2 H_n^2 H_p H_n\right) = -\text{tr} \left(H_n^2 H_p^2 H_n H_p\right)$, cf. eqn (14.23).

---

[32] Conversely, H_p should have cyclic phases in a WB in which H_n is diagonal.

[33] One excludes the case in which phases are introduced by hand into the weak-basis invariant quantities. For instance, the imaginary part of $(1 + i) \, \text{tr} \left(H_p H_n\right)$ is non-vanishing and invariant under a WBT, yet it does not signal CP violation.

It is important to stress that *it is only when a quantity is WB invariant that its imaginary part signals CP violation*. Non-invariant quantities may in general be rephased at will, and their phases have no significance whatsoever.

14.8   Conclusions

CP violation arises when the kinetic-gauge part of the Lagrangian implies CP-transformation properties of the fields which are not satisfied by some other part of the Lagrangian. In the SM, this happens because of a clash between the Yukawa couplings and the CP-transformation properties required by the charged-current weak interaction.

It is often said in the literature that, in the SM, CP violation arises from complex Yukawa couplings. This is a misleading statement, since complex Yukawa couplings by themselves do not lead to CP breaking. It is the simultaneous presence of the charged weak interaction and of complex Yukawa couplings which leads to CP violation in the SM. Moreover, complex Yukawa couplings do not necessarily lead to CP violation; rather, it is the weak-basis-invariant quantities which must be non-real in order for CP violation to arise.

The main conclusions of this chapter are the following:

- CP is conserved if and only if there is a unitary matrix K_L such that eqns (14.8) hold.
- For three generations, CP is conserved if and only if the Hermitian mass matrices satisfy $\mathrm{tr}\,[H_p, H_n]^3 = 0$. This happens either when there are two mass-degenerate quarks in the same charge sector, or when $J = 0$.

The WB-invariant conditions that we have derived may be useful in model-building. In the SM, the Yukawa couplings are completely arbitrary, and there is no relationship between the quark masses and the elements of the CKM mixing matrix. However, additional 'horizontal' symmetries may be introduced, either in the SM or in extensions thereof, with the aim of restricting the number of free parameters and obtaining relations among the quark masses and CKM matrix elements. The presence of these extra horizontal symmetries can have an impact on CP violation. For example, it may be interesting to find the strength of CP violation in the presence of horizontal symmetries. The CP invariants that we have derived enable one to do this without having to diagonalize the Yukawa-coupling matrices.

15

MODULI OF THE CKM MATRIX ELEMENTS

15.1 Introduction

Experimentally, the rephasing-invariant functions of the CKM matrix to which one has most direct access are the moduli of its matrix elements. The relevant processes are semileptonic production and decay rates which are, under very broad assumptions about possible physics beyond the SM, dominated by tree-level amplitudes with a $W^\pm$ as intermediate state. New particles are more likely to contribute to loop processes. The information on the CKM matrix that we may gather from the analysis of the mass difference and of CP violation in neutral-meson systems—as we shall be doing in Chapter 18—is much more sensitive to the possible presence of new heavy particles in the theory.

In this chapter we make an overview of the methods used to determine the $U_{\alpha i}$. We follow the reviews by Rosner (1994) and by the Particle Data Group (1996). Readers may want to skip this chapter.

15.2 $|V_{ud}|$

The matrix element V_{ud} only involves the first-generation quarks, and is for this reason the one which can be determined with best precision. Three different methods have been used in its determination.

The first method is the most exact one to date. It makes use of superallowed Fermi transitions, which are beta decays connecting two $J^P = 0^+$ nuclides in the same isospin multiplet. In these transitions only the vector current is involved. Denoting the lifetime of the decaying nucleus by t, the Coulomb correction factor by f, the nuclear-dependent ('outer') radiative corrections by Δ_{outer}, and the universal ('inner') radiative corrections by Δ_{inner}, one has

$$|V_{ud}|^2 = \frac{\pi^3 \ln 2}{G_F^2 m_e^5 ft\,(1 + \Delta_{\text{outer}})\,(1 + \Delta_{\text{inner}})} = \frac{2984.4 \pm 0.1\,\text{s}}{ft\,(1 + \Delta_{\text{outer}})\,(1 + \Delta_{\text{inner}})}.$$
(15.1)

The value of G_F used in eqn (15.1) is computed from muon decay (one must be careful to take into account the radiative corrections to that process). The universal radiative corrections have been computed (Marciano 1991; Marciano and Sirlin 1986) to be $\Delta_{\text{inner}} = 0.0234 \pm 0.0012$. The values of $Ft \equiv ft\,(1 + \Delta_{\text{outer}})$ differ among the various nuclides undergoing superallowed Fermi transitions: ^{10}C, ^{14}O, ^{26m}Al, ^{34}Cl, ^{38m}K, ^{42}Sc, ^{46}V, ^{50}Mn, and ^{54}Co. New and more comprehensive analyses of the radiative corrections regularly appear in the literature (Marciano and Sirlin 1986; Sirlin and Zucchini 1986; Jaus and Rasche 1987; Sirlin 1987; Brown and Ormand 1989; Hardy et al. 1990; Barker 1992; Towner 1992;

Barker 1994; Towner and Hardy 1998). Taking into account the results of a recent experiment at Chalk River Laboratory (Hagberg *et al.* 1997), Towner and Hardy (1998) have obtained

$$|V_{ud}| = 0.9740 \pm 0.0005. \tag{15.2}$$

The value given by the Particle Data Group (1996) is

$$|V_{ud}| = 0.9736 \pm 0.0010. \tag{15.3}$$

The second method relates $|V_{ud}|$ to the free-neutron lifetime τ_n. On the one hand, the beta decay of the free neutron has the advantage that one does not have to deal with nuclear-structure-dependent corrections; on the other hand, there is the disadvantage that neutron decay also receives a contribution from the axial-vector current. Moreover, the experimental input needed is more challenging to obtain. One equates the vector coupling in neutron decay with the one in muon decay. Denoting by g_A the ratio of the axial-vector and vector effective couplings, one has (Wilkinson 1982; Marciano 1991)

$$|V_{ud}|^2 = \frac{4904.0 \pm 5.0\,\mathrm{s}}{\tau_n \left(1 + 3g_A^2\right)}. \tag{15.4}$$

Using (Mampe *et al.* 1989; Alfimenkov *et al.* 1990; Byrne *et al.* 1990) $\tau_n = 888.9 \pm 1.9\,\mathrm{s}$ and, from the decay asymmetries (Bopp *et al.* 1986; Klemt *et al.* 1988; Erozolimskii *et al.* 1991), $g_A = 1.257 \pm 0.003$, one finds $|V_{ud}| = 0.9804 \pm 0.0005 \pm 0.0010 \pm 0.0020$. The first source of uncertainty is radiative corrections; the second one is τ_n; the third one is g_A. Notice that the error bar on the result from free-neutron decay is some four times larger than the error bar on the result from superallowed nuclear decays.

The value of $|V_{ud}|$ obtained from neutron decay is higher than the one obtained from superallowed Fermi transitions. This is of interest, because the former value yields the correct normalization for the first row of the CKM matrix,

$$|V_{ud}|^2 + |V_{us}|^2 + |V_{ub}|^2 = 1 \tag{15.5}$$

while the latter one does not. The value of $|V_{ub}|$ is so low that it does not contribute effectively to eqn (15.5).

The third method relies on an analysis of the rate of the decay $\pi^+ \to \pi^0 e^+ \nu_e$. This is a purely vector transition between two hadronic states with $J^P = 0^-$. The major disadvantage is the extremely small branching ratio ($\sim 10^{-8}$) for this particular decay of the charged pion. One obtains (McFarlane *et al.* 1985; Marciano and Parsa 1986; Towner and Hardy 1998) $|V_{ud}| = 0.9670 \pm 0.0161$. The uncertainty is substantially larger than in the other two methods, and is overwhelmingly dominated by the experimental error in the branching ratio.

15.3 $|V_{us}|$

This matrix element is obtained from semileptonic decays of strange particles. Consider first K_{e3} decays. The hadronic matrix element is a function of two form factors. For instance,

$$\langle \pi^+(p')|J_\mu|\overline{K^0}(p)\rangle = f_+(q^2)(p+p')_\mu + f_-(q^2)q_\mu, \qquad (15.6)$$

where $q_\mu = p_\mu - p'_\mu$. The form factor f_- gives a contribution to the decay rate proportional to the square of the mass of the electron, which may be neglected. One has

$$\frac{d\Gamma(\overline{K^0} \to \pi^+ e^- \bar{\nu}_e)}{dx_\pi} = \frac{G_F^2 m_K^5}{192\pi^3}|V_{us}|^2 \left|f_+(q^2)\right|^2 \left(x_\pi^2 - 4\frac{m_\pi^2}{m_K^2}\right)^{3/2}, \qquad (15.7)$$

where $x_\pi \equiv 2E_\pi/m_K$. One makes a fit to the Dalitz-plot distribution using a model for $f_+(q^2)$ based on a K^*-pole approximation,

$$f_+(q^2) = \frac{f_+(0)}{1 - q^2/m_{K^*}^2}, \qquad (15.8)$$

with $f_+(0)$ calculated from the quark model; this calculation is the main source of theoretical error. From both charged- and neutral-K_{l3} decays one obtains (Shrock and Wang 1978; Leutwyler and Roos 1984; Barker 1992)

$$|V_{us}| = 0.2196 \pm 0.0023. \qquad (15.9)$$

A different starting point is furnished by semileptonic hyperon decays, which include $\Lambda \to pX_e$, $\Sigma^- \to nX_e$, $\Xi^- \to \Lambda X_e$, and $\Xi^- \to \Sigma^0 X_e$, where $X_e \equiv e^- \bar{\nu}_e$. In order to extract information on $|V_{us}|$ one compares those decays, using flavour-SU(3) symmetry, with the strangeness-conserving processes $n \to pX_e$ and $\Sigma^- \to \Lambda X_e$. While only the vector current contributes to K_{l3} transitions, for which therefore SU(3) violation only arises at second order (Ademollo and Gatto 1964), that violation is first order and much larger in hyperon decays, which receive contributions from both vector and axial-vector currents. The analysis (Donoghue et al. 1987) of the data of the WA2 Collaboration (1983) gave

$$|V_{us}| = 0.220 \pm 0.001 \pm 0.003, \qquad (15.10)$$

in good agreement with eqn (15.9).[34]

The Particle Data Group (1996) recommends the value

$$|V_{us}| = 0.2205 \pm 0.0018. \qquad (15.11)$$

---

[34]García et al. (1992) have used different SU(3)-breaking corrections and obtained higher $|V_{us}|$.

15.4 $|V_{cd}|$ and $|V_{cs}|$

The direct determination of $|V_{cd}|$ and $|V_{cs}|$ is both poor and fraught with theo-
retical uncertainties. The first source of information on these matrix elements is
deep inelastic neutrino excitation of charm, in reactions such as $\nu_\mu d \to \mu^- c$ for
$|V_{cd}|$ and $\nu_\mu s \to \mu^- c$ for $|V_{cs}|$ (the latter process involves strange quarks in the
parton sea, the modelling of which is somewhat uncertain). The experimental
signature is a dimuon event, because the semileptonic decay of the charm quark
yields a second muon. The method has been followed by the CDHS Collaboration (1982) and later by the CCFR Collaboration (1993, 1995a). A reanalysis of
their results (Particle Data Group 1996) lead to

$$|V_{cd}| = 0.224 \pm 0.016,$$
$$|V_{cs}| \geq 0.59. \tag{15.12}$$

The second source of information are semileptonic decays of charmed parti-
cles. These include $D^0 \to \pi^- \bar{X}_e$ for $|V_{cd}|$, and $D^0 \to K^- \bar{X}_e$ and $D^+ \to \overline{K^0} \bar{X}_e$
for $|V_{cs}|$ (where $\bar{X}_e \equiv e^+ \nu_e$). The theoretical framework is the same as in the
previous section; one uses models for the form factors with poles on the masses
of the vector mesons—D^* and D_s^*, respectively—and one computes the values
of $f_+^\pi(0)$ and $f_+^K(0)$. For instance, from the experimental data by the Mark-III
Collaboration (1989, 1991) and CLEO Collaboration (1993b) one obtains

$$\left| f_+^K(0) V_{cs} \right|^2 = 0.495 \pm 0.036; \tag{15.13}$$

using (Aliev et al. 1984; Bauer et al. 1985; Grinstein et al. 1986, 1989) $f_+^K(0) = 0.7 \pm 0.1$, one finds

$$|V_{cs}| = 1.01 \pm 0.18. \tag{15.14}$$

Notice that the central value $|V_{cs}| = 1.01$ is above what unitarity of the CKM
matrix allows.

15.5 $|V_{cb}|$

The bottom quark decays predominantly to the charm quark. The CKM-matrix
element $|V_{cb}|$ may be obtained by using a spectator approximation for B-meson
decays:

$$\Gamma(B \to X_c l \nu_l) = \frac{\mathrm{BR}(B \to X_c l \nu_l)}{\tau_B} \approx \Gamma\left(b \to cl^- \bar{\nu}_l\right), \tag{15.15}$$

where B is any bottom-flavoured meson and X_c is any charmed set of particles.
One only has to compute $\Gamma(b \to cl^- \bar{\nu}_l)$.

The decay $b \to u_\alpha l^- \bar{\nu}_l$ (where u_α may be either the up or the charm quark)
has, if one takes $m_l = 0$, a differential decay rate

$$\frac{d\Gamma(b \to u_\alpha l^- \bar{\nu}_l)}{dx} = \frac{G_F^2 m_b^5}{192\pi^3} |V_{\alpha b}|^2 \left[2x^2 \left(\frac{1-x-\zeta}{1-x} \right)^2 \left(3 - 2x + \zeta + \frac{2\zeta}{1-x} \right) \right], \tag{15.16}$$

where $\zeta = m_\alpha^2/m_b^2$ and $x \equiv 2E_l/m_b$, with E_l the lepton's energy in the b rest frame. The maximum allowed value for x is $1 - \zeta$. Integrating over x one gets

$$\Gamma = \frac{G_F^2 m_b^5}{192\pi^3} |V_{ab}|^2 f(\zeta) = \left(7.14 \times 10^{-11} \text{ GeV}\right) \left(\frac{m_b}{5 \text{ GeV}}\right)^5 |V_{ab}|^2 f(\zeta), \quad (15.17)$$

where

$$f(\zeta) = 1 - 8\zeta + 8\zeta^3 - \zeta^4 - 12\zeta^2 \ln \zeta \qquad (15.18)$$

is related to the reduction in phase space due to m_α not being zero.

Equation (15.17) displays a dependence of the decay rate on the fifth power of the bottom mass. This makes the determination of $|V_{cb}|$ very sensitive to the value that one uses for that mass. An analysis (Ball *et al.* 1995) of the inclusive decays yields

$$|V_{cb}| \sqrt{\frac{\tau_b}{1.5 \text{ ps}}} = 0.041 \pm 0.002. \qquad (15.19)$$

For the average lifetime one may use (Patterson 1995) $\tau_b = 1.55 \pm 0.06$ ps.

Another way to extract $|V_{cb}|$ is from exclusive decays $B \to Dl\nu_l$ (ARGUS Collaboration 1993; CCFR Collaboration 1995b; Scott *et al.* 1995), using corrections based on heavy-quark effective theory. One obtains (Neubert 1995)

$$|V_{cb}| = 0.041 \pm 0.003 \pm 0.002. \qquad (15.20)$$

The Particle Data Group (1996) recommends

$$|V_{cb}| = 0.041 \pm 0.003. \qquad (15.21)$$

Continuous progress is being made on the determination of $|V_{cb}|$, and the uncertainty bar can be expected to become smaller in the next few years.

15.6 $|V_{ub}|/|V_{cb}|$

From eqn (15.17) it follows that

$$\frac{\Gamma\left(b \to ul^-\bar{\nu}_l\right)}{\Gamma\left(b \to cl^-\bar{\nu}_l\right)} = \frac{f\left(m_u^2/m_b^2\right)}{f\left(m_c^2/m_b^2\right)} \left|\frac{V_{ub}}{V_{cb}}\right|^2 = \frac{1}{0.41\text{--}0.54} \left|\frac{V_{ub}}{V_{cb}}\right|^2. \qquad (15.22)$$

From the observation of the endpoint of the lepton-energy spectrum in semileptonic B decays one may conclude that $|V_{ub}| \neq 0$. That endpoint is given, as seen in the previous section, by $E_l^{\max} = (m_b/2)(1 - m_\alpha^2/m_b^2)$. As m_u is smaller than m_c, the leptons from $b \to ul^-\bar{\nu}_l$ may attain a higher energy than those from $b \to cl^-\bar{\nu}_l$. Studying the electron spectrum the ARGUS Collaboration (1990, 1991) and the CLEO Collaboration (1993a) have observed electrons with higher energies than allowed for $b \to ce^-\bar{\nu}_e$, and concluded that the quantity in eqn (15.22) is 0.01–0.02.

Still, the determination of $|V_{ub}|/|V_{cb}|$ is plagued with theoretical uncertainties. The highest energy attainable by the leptons is determined by the low-mass states formed from the final u quark and the spectator quark $\bar{q}$ in $B = b\bar{q}$. The decay $B \to \pi l \nu_l$ should be the dominant source of high-energy leptons, yet the various models (Altarelli *et al.* 1982; Bauer *et al.* 1985; Grinstein *et al.* 1986, 1989) differ in their predictions for this decay, and in general in the modelling of the lepton-energy spectrum. The Particle Data Group (1996) suggests using

$$\left| \frac{V_{ub}}{V_{cb}} \right| = 0.08 \pm 0.02, \tag{15.23}$$

but the theoretical uncertainty quoted is probably somewhat optimistic (Pène, personal communication).

The theoretical models agree better on the exclusive modes $B \to \rho l \nu_l$ and $B \to \omega l \nu_l$. The CLEO Collaboration (1996) has measured BR $\left(B^0 \to \pi^- l^+ \nu_l \right)$ and BR $\left(B^0 \to \rho^- l^+ \nu_l \right)$, which are at the level of 10^{-4}. The experimental uncertainties are still relatively large, but this has already allowed the derivation of

$$|V_{ub}| = 0.0033 \pm 0.0008, \tag{15.24}$$

which agrees with eqns (15.23) and (15.21).

Another method to measure $|V_{ub}|$ would be using the decay $B^+ \to \mu^+ \nu_\mu$, which proceeds with rate

$$\Gamma \left(B^+ \to \mu^+ \nu_\mu \right) = \frac{G_F^2 f_B^2 m_B m_\mu^2}{8\pi} \left(1 - \frac{m_\mu^2}{m_B^2} \right)^2 |V_{ub}|^2, \tag{15.25}$$

cf. eqn (C.41). However, this would require a good estimate of f_B, which is at present unavailable.

15.7 Consequences of unitarity

The assumption of 3×3 unitarity of the CKM matrix allows one to further constrain the $U_{\alpha i}$, starting from the results in the previous sections of this chapter. Let us use the 90%-confidence-level values

$$0.219 \leq |V_{us}| \leq 0.224,$$
$$0.036 \leq |V_{cb}| \leq 0.046, \tag{15.26}$$
$$0.002 \leq |V_{ub}| \leq 0.005.$$

We want to investigate the consequences of the assumption that V is 3×3 unitary for the moduli of the other matrix elements. As the direct determination of $|V_{cd}|$ and $|V_{cs}|$ is poor, we expect unitarity to produce better constraints on them than the experimental ones. We also want to have a prediction for the matrix elements of the third row of V— for the couplings of the top quark—to which direct experimental access is not yet possible (see however Swain and Taylor 1997, 1998). We use

$$|V_{ud}| = \sqrt{1 - |V_{us}|^2 - |V_{ub}|^2},$$
$$|V_{tb}| = \sqrt{1 - |V_{cb}|^2 - |V_{ub}|^2}, \tag{15.27}$$

and eqns (15.26) to obtain

$$0.9745 \leq |V_{ud}| \leq 0.9757,$$
$$0.9989 \leq |V_{tb}| \leq 0.9993. \tag{15.28}$$

From eqns (15.28) we infer the following. Firstly, $|V_{tb}|$ is the best known of all nine moduli, in spite of not being directly measured. Secondly, in order to have the first row of V duly normalized, $|V_{ud}|$ should be at least one standard deviation higher than the central value in eqn (15.3). The simplest alternative is allowing for the existence of extra generations of quarks, in which case $|V_{ud}|^2 + |V_{us}|^2 + |V_{ub}|^2 \leq 1$. This has been suggested by Marciano and Sirlin (1986).

In order to constrain the remaining four $U_{\alpha i}$ one uses the unitarity condition in eqn (13.28), which yields a lower and an upper limit on $|V_{cs}|$. From this it is easy to obtain bounds on $|V_{cd}|$, $|V_{td}|$, and $|V_{ts}|$, by using the normalization of the rows and columns of V. The result is summarized in the following matrix of moduli:

$$|V| = \begin{pmatrix} 0.9745\text{--}0.9757 & 0.219\text{--}0.224 & 0.002\text{--}0.005 \\ 0.218\text{--}0.224 & 0.9736\text{--}0.9751 & 0.036\text{--}0.046 \\ 0.003\text{--}0.015 & 0.034\text{--}0.046 & 0.9989\text{--}0.9993 \end{pmatrix}. \tag{15.29}$$

This is the consequence of the assumption of eqns (15.26) and of 3×3 unitarity. Let us make a few comments:

- The ranges for the $|V_{\alpha i}|$ in eqn (15.29) are correlated among themselves. One is in general not allowed to choose any value for the modulus of any one matrix element independently of the values chosen for the moduli of other matrix elements.
- Equation (15.29) displays the approximate pattern of moduli anticipated in eqn (13.47). As a matter of fact, $|V_{ub}|$ is smaller than λ^3, and $|V_{ub}| \sim \lambda^4$ might be closer to reality. The same is true for $|V_{td}|$.
- In the limit $V_{ub} = 0$ unitarity implies $|V_{td}| = |V_{us}||V_{cb}| \sim \lambda^3$. Similarly, if V_{td} happened to vanish one would have $|V_{ub}| = |V_{cd}||V_{ts}| \sim \lambda^3$.
- Normalization of the rows and columns of V implies

$$U_{us} - U_{cd} = U_{cb} - U_{ts} = U_{td} - U_{ub}. \tag{15.30}$$

It is likely that the difference of squared moduli in eqn (15.30) is positive.

16

PARAMETRIZATIONS OF THE CKM MATRIX

16.1 Introduction

The CKM matrix is usually parametrized in some specific way. The purpose of the parametrizations is to incorporate the constraints of 3×3 unitarity. Some parametrizations also incorporate experimental information, in particular the pattern of moduli in eqn (13.47); this is the case of the Wolfenstein parametrization and of related parametrizations, which in practice are the ones most often used.

Rephasing-invariance is the possibility of changing the overall phase of any row, or of any column, of the CKM matrix, without changing the physics contained in that matrix. We may use this freedom to constrain five matrix elements to be real, or else to fix their phase in any other desirable way. It follows that the 3×3 unitary CKM matrix should in principle be parametrized by three rotation angles and one phase.[35]

Even if five matrix elements may have their phases fixed, it is important to notice that those five matrix elements cannot be chosen at will. This is because the quartets are rephasing-invariant. One must be careful not to implicitly fix the phase of any quartet when choosing a phase convention for the CKM matrix.

All phase conventions used in practice have one thing in common: the matrix elements V_{ud} and V_{us} are chosen real and positive. The reason for this choice is the central role played by $\lambda_u \equiv V_{ud} V_{us}^*$ in the physics of the neutral-kaon system. If λ_u is real and spurious phases are neglected, then $\Gamma_{12} \approx A_0^* \bar{A}_0$ is real at tree level, as will be shown in Chapter 17. This is an advantageous simplification. Still, this phase convention is in no way necessary, because physics is rephasing-invariant.[36]

16.2 Parametrizations with Euler angles

It seems natural to parametrize V by means of three Euler angles—the angles of three successive rotations about different axes—and one phase. The phase must be introduced in such a way that it cannot be eliminated by means of a rephasing of the quark fields.

---

[35]There are parametrizations in which none of the four parameters can be interpreted as a rotation angle.

[36]It is important to keep in mind that many of the formulae for the neutral-kaon system found in the literature are not rephasing-invariant. Those formulae should not be used together with a phase convention in which λ_u is not real.

16.2.1 *Kobayashi–Maskawa parametrization*

The first parametrization of the CKM matrix was put forward by Kobayashi and Maskawa (1973). They wrote

$$
V = \begin{pmatrix} 1 & 0 & 0 \\ 0 & c_2 & -s_2 \\ 0 & s_2 & c_2 \end{pmatrix} \begin{pmatrix} c_1 & -s_1 & 0 \\ s_1 & c_1 & 0 \\ 0 & 0 & e^{i\delta} \end{pmatrix} \begin{pmatrix} 1 & 0 & 0 \\ 0 & c_3 & s_3 \\ 0 & s_3 & -c_3 \end{pmatrix}
$$

$$
= \begin{pmatrix} c_1 & -s_1 c_3 & -s_1 s_3 \\ s_1 c_2 & c_1 c_2 c_3 - s_2 s_3 e^{i\delta} & c_1 c_2 s_3 + s_2 c_3 e^{i\delta} \\ s_1 s_2 & c_1 s_2 c_3 + c_2 s_3 e^{i\delta} & c_1 s_2 s_3 - c_2 c_3 e^{i\delta} \end{pmatrix}. \tag{16.1}
$$

Here, c_i and s_i are shorthands for $\cos\theta_i$ and $\sin\theta_i$, respectively, where θ_1, θ_2, and θ_3 are Euler angles. One of the rotations is on the xy plane, and the other two rotations are on the yz plane. The phase δ appears as a rephasing of the third generation; as the rephasing occurs in between two rotations involving that generation, it is impossible to identify δ with a rephasing of the quark fields.

The first row and the first column of V have implicitly been chosen to be real, by use of the rephasing freedom of the CKM matrix. Without loss of generality θ_1, θ_2, and θ_3 may be constrained to lie in the first quadrant, provided one allows δ to be free, $0 \le \delta < 2\pi$. Indeed, putting one of the Euler angles in any other quadrant is equivalent to a physically meaningless rephasing of V, sometimes coupled with the transformation $\delta \to \delta + \pi$. For instance, if θ_1 was chosen to lie in the second quadrant, then we might bring it into the first quadrant by means of the transformation $c_1 \to -c_1$. This transformation is equivalent to $e^{i\delta} \to -e^{i\delta}$ together with

$$
V \to \mathrm{diag}\,(-1, 1, 1)\, V\, \mathrm{diag}\,(1, -1, -1),
$$

which is a change of the sign of the fields u, s, and b.

In the Kobayashi–Maskawa parametrization

$$
J = c_1 s_1^2 c_2 s_2 c_3 s_3 \sin\delta. \tag{16.2}
$$

From this it is easy to derive that the maximum possible value of J is $1/\left(6\sqrt{3}\right)$, which is obtained when $\delta = \pi/2$, $\theta_2 = \theta_3 = \pi/4$, and $c_1 = 1/\sqrt{3}$, i.e., when all matrix elements of V have modulus $3^{-1/2}$, cf. eqn (13.49).

16.2.2 *Chau–Keung parametrization*

Chau and Keung (1984) have introduced a different parametrization, the use of which has been advocated by the Particle Data Group (1996):

$$
V = \begin{pmatrix} 1 & 0 & 0 \\ 0 & c_{23} & s_{23} \\ 0 & -s_{23} & c_{23} \end{pmatrix} \begin{pmatrix} c_{13} & 0 & s_{13} e^{-i\delta_{13}} \\ 0 & 1 & 0 \\ -s_{13} e^{i\delta_{13}} & 0 & c_{13} \end{pmatrix} \begin{pmatrix} c_{12} & s_{12} & 0 \\ -s_{12} & c_{12} & 0 \\ 0 & 0 & 1 \end{pmatrix}
$$

$$
= \begin{pmatrix} c_{12} c_{13} & s_{12} c_{13} & s_{13} e^{-i\delta_{13}} \\ -s_{12} c_{23} - c_{12} s_{23} s_{13} e^{i\delta_{13}} & c_{12} c_{23} - s_{12} s_{23} s_{13} e^{i\delta_{13}} & s_{23} c_{13} \\ s_{12} s_{23} - c_{12} c_{23} s_{13} e^{i\delta_{13}} & -c_{12} s_{23} - s_{12} c_{23} s_{13} e^{i\delta_{13}} & c_{23} c_{13} \end{pmatrix}. \tag{16.3}
$$

Here, c_{ij} and s_{ij} are shorthands for $\cos\theta_{ij}$ and $\sin\theta_{ij}$, respectively. The three rotation angles θ_{12}, θ_{13}, and θ_{23} may be restricted to lie in the first quadrant provided one allows the phase δ_{13} to be free. Only four matrix elements are chosen to be real; still, only one physical phase appears in the parametrization.

The s_{ij} are simply related to directly measurable quantities:

$$s_{13} = |V_{ub}|,$$

$$s_{12} = \frac{|V_{us}|}{\sqrt{1 - U_{ub}}} \approx |V_{us}|, \tag{16.4}$$

$$s_{23} = \frac{|V_{cb}|}{\sqrt{1 - U_{ub}}} \approx |V_{cb}|,$$

because experimentally $|V_{ub}|$ is very small. On the other hand,

$$J = c_{12}s_{12}c_{13}^2 s_{13}c_{23}s_{23}\sin\delta_{13}, \tag{16.5}$$

and therefore

$$\sin\delta_{13} = \frac{(1 - U_{ub})\,J}{|V_{ud}V_{us}V_{ub}V_{cb}V_{tb}|} \tag{16.6}$$

is related in a complicated way to rephasing-invariant quantities.

16.3   Rephasing-invariant parametrizations

We call a parametrization 'rephasing-invariant' when its parameters are defined to be rephasing-invariant quantities, for instance the moduli of some matrix elements, or the phases of some quartets. In contrast, the rotation angles in the previous section can be related to measurable quantities—see for instance eqns (16.4)—but they are not directly defined to be rephasing-invariant quantities.

We shall next present three rephasing-invariant parametrizations. As they are not used very often, some readers may prefer to skip this section.

16.3.1   *Branco–Lavoura parametrization*

Branco and Lavoura (1988a) have suggested parametrizing the CKM matrix by means of four linearly independent $U_{\alpha i}$. This is a convenient choice, because the moduli constitute the most reliable information on the CKM matrix. A convenient set of four squared moduli is

$$U_{us}, \ U_{ub}, \ U_{cb}, \ \text{and} \ U_{td}. \tag{16.7}$$

Indeed, $|V_{us}|$, $|V_{ub}|$, and $|V_{cb}|$ are small and relatively well measured, while $|V_{td}|$ is crucial in B_d^0–$\overline{B}_d^0$ mixing. In order to reconstruct the full CKM matrix from the parameters in eqn (16.7), one first uses the normalization of the rows and columns of V, which allows one to compute the values of the remaining five moduli. One then uses the relation

$$2\mathrm{Re}\,Q_{\alpha i\beta j} = 1 - U_{\alpha i} - U_{\beta j} - U_{\alpha j} - U_{\beta i} + U_{\alpha i}U_{\beta j} + U_{\alpha j}U_{\beta i}, \qquad (16.8)$$

which is derived in an analogous way as for eqn (13.27). One can thus find the real part of each quartet. The imaginary part of the quartets, J, is given, as in eqn (13.29), by

$$4J^2 = 4U_{\alpha i}U_{\beta j}U_{\alpha j}U_{\beta i} - \left(2\mathrm{Re}\,Q_{\alpha i\beta j}\right)^2, \qquad (16.9)$$

together with eqn (16.8). The sign of J cannot be found from the moduli alone. This is because the transformation $V \to V^*$ leaves the moduli invariant but changes the sign of J. As a consequence, the parametrization of Branco and Lavoura requires that sign J be given together with the parameters in eqn (16.7).

16.3.2 Bjorken–Dunietz parametrization

Bjorken and Dunietz (1987) were the first authors to put forward a rephasing-invariant parametrization. They chose the following phase convention:

$$V_{ud}, \; V_{us}, \; V_{cs}, \; V_{cb}, \text{ and } V_{tb} \text{ real and positive.} \qquad (16.10)$$

They used as parameters the rephasing-invariant quantities

$$U_{us}, \; U_{ub}, \; U_{cb}, \text{ and } \phi \equiv \omega_{uscb}. \qquad (16.11)$$

It follows from the definition of ϕ and from the phase convention in eqn (16.10) that $V_{ub} = \sqrt{U_{ub}}\exp\left(-i\phi\right)$, while $V_{us} = \sqrt{U_{us}}$ and $V_{cb} = \sqrt{U_{cb}}$.

The full CKM matrix may be reconstructed in the following way. Firstly, as V_{ud} and V_{tb} are real and positive by convention,

$$\begin{aligned} V_{ud} &= \sqrt{1 - U_{us} - U_{ub}}, \\ V_{tb} &= \sqrt{1 - U_{cb} - U_{ub}}. \end{aligned} \qquad (16.12)$$

Then,

$$\begin{aligned} 2\mathrm{Re}\,Q_{uscb} &= 2\sqrt{U_{us}U_{cb}U_{ub}U_{cs}}\cos\phi \\ &= 1 - U_{us} - U_{cb} - U_{ub} - U_{cs} + U_{us}U_{cb} + U_{ub}U_{cs} \end{aligned} \qquad (16.13)$$

constitutes a quadratic equation for $|V_{cs}|$, which gives (remember eqn 16.10)

$$V_{cs} = \frac{1}{1 - U_{ub}}\left[-\left(U_{us}U_{cb}U_{ub}\right)^{1/2}\cos\phi + \left(1 - U_{us} - U_{cb} + U_{us}U_{cb} - 2U_{ub} \right.\right.$$
$$\left.\left. + U_{us}U_{ub} + U_{cb}U_{ub} + U_{ub}^2 - U_{us}U_{cb}U_{ub}\sin^2\phi\right)^{1/2}\right]. \qquad (16.14)$$

The orthogonality conditions yield the remaining three matrix elements of V:

$$V_{cd} = -\frac{V_{cs}V_{us}^* + V_{cb}V_{ub}^*}{V_{ud}^*},$$

$$V_{ts} = -\frac{V_{us}V_{ub}^* + V_{cs}V_{cb}^*}{V_{tb}^*}, \qquad (16.15)$$

$$V_{td} = -\frac{V_{ts}V_{us}^* + V_{tb}V_{ub}^*}{V_{ud}^*}.$$

16.3.3   Aleksan–Kayser–London parametrization

In the SM with n_g generations, one may eliminate $2n_g - 1$ phases from the initial n_g^2 phases of the matrix elements of V through a rephasing of the quark fields. The number of rephasing-invariant phases is thus

$$n_{\text{phases}} = n_g^2 - (2n_g - 1) = (n_g - 1)^2. \qquad (16.16)$$

At this stage we are not yet imposing unitarity. It is remarkable that n_{phases} equals the number of parameters necessary to parametrize the $n_g \times n_g$ unitary matrix V: $n_{\text{phases}} = N_{\text{param}}$.

The idea of Aleksan *et al.* (1994) was to parametrize V by four $\omega_{\alpha i \beta j}$. We already know that $\omega_{\alpha i \beta j} = \omega_{\beta j \alpha i} = -\omega_{\alpha j \beta i} = -\omega_{\beta i \alpha j}$. Therefore, for three generations one needs to consider only nine phases: ω_{tbud}, ω_{tbcd}, ω_{cbud}, ω_{tbus}, ω_{tbcs}, ω_{cbus}, ω_{tsud}, ω_{tscd}, and ω_{csud}. From these nine phases only four are linearly independent. We may choose as parameters

$$\omega_{tbcd}, \ \omega_{cbud}, \ \omega_{tbcs}, \ \text{and} \ \omega_{csud}. \qquad (16.17)$$

The first two phases are related to β and γ in eqns (13.31) by

$$\omega_{tbcd} = \beta - (\text{sign}\,\beta)\,\pi,$$
$$\omega_{cbud} = \gamma - (\text{sign}\,\gamma)\,\pi, \qquad (16.18)$$

where we have taken the argument of a complex number to lie between $-\pi$ and $+\pi$. Similarly, ω_{tbcs} and ω_{csud} can be related to the two phases

$$\epsilon \equiv \arg\left(-\frac{V_{cb}V_{cs}^*}{V_{tb}V_{ts}^*}\right) \qquad (16.19)$$

$$\epsilon' \equiv \arg\left(-\frac{V_{us}V_{ud}^*}{V_{cs}V_{cd}^*}\right), \qquad (16.20)$$

through[37]

$$\omega_{tbcs} = -\epsilon + (\text{sign}\,\epsilon)\,\pi,$$
$$\omega_{csud} = -\epsilon' + (\text{sign}\,\epsilon')\,\pi. \qquad (16.21)$$

In the SM, these four angles obey a strong hierarchy (Aleksan *et al.* 1994): although β and γ may be large, ϵ and ϵ' must be small: $\epsilon \lesssim 0.05$ and $\epsilon' \lesssim 0.0025$.

---

[37] Aleksan *et al.* (1994) used α, β, ϵ, and ϵ' as parameters. We prefer to use γ instead of α, for reasons that will become apparent in § 16.4.2.

The four phases β, γ, ϵ, and ϵ'—or, equivalently, ω_{tbcd}, ω_{cbud}, ω_{tbcs}, and ω_{csud}—can be used to parametrize the 3×3 unitary CKM matrix. The other five $\omega_{\alpha i \beta j}$ can be readily obtained from

$$\omega_{tbud} = \omega_{tbcd} + \omega_{cbud},$$
$$\omega_{tbus} = \omega_{cbud} + \omega_{tbcs} - \omega_{csud},$$
$$\omega_{cbus} = \omega_{cbud} - \omega_{csud}, \qquad (16.22)$$
$$\omega_{tsud} = \omega_{tbcd} - \omega_{tbcs} + \omega_{csud},$$
$$\omega_{tscd} = \omega_{tbcd} - \omega_{tbcs}.$$

Equations (16.22) follow from the algebra of complex numbers. Unitarity is only needed in order to compute the moduli of the matrix elements from the phases of the quartets. It follows from the normalization of the ith column of V that

$$U_{ui} = \frac{1}{1 + (U_{ci}/U_{ui}) + (U_{ti}/U_{ui})},$$
$$U_{ci} = \frac{(U_{ci}/U_{ui})}{1 + (U_{ci}/U_{ui}) + (U_{ti}/U_{ui})}, \qquad (16.23)$$
$$U_{ti} = \frac{(U_{ti}/U_{ui})}{1 + (U_{ci}/U_{ui}) + (U_{ti}/U_{ui})}.$$

We therefore need to know the ratios U_{ti}/U_{ui} and U_{ci}/U_{ui}. They are found by applying the law of sines to the unitarity triangles. Let (i,j,k) be a permutation of the indices (d,s,b) and consider the unitarity triangles arising from the orthogonality of the columns of the CKM matrix. Then,

$$\frac{U_{ci}}{U_{ui}} = \left| \frac{V_{ci}V_{cj}^*}{V_{ui}V_{uj}^*} \right| \cdot \left| \frac{V_{uj}V_{uk}^*}{V_{cj}V_{ck}^*} \right| \cdot \left| \frac{V_{ck}V_{ci}^*}{V_{uk}V_{ui}^*} \right|$$
$$= \left| \frac{\sin \omega_{tiuj}}{\sin \omega_{ticj}} \right| \cdot \left| \frac{\sin \omega_{tjck}}{\sin \omega_{tjuk}} \right| \cdot \left| \frac{\sin \omega_{tkui}}{\sin \omega_{tkci}} \right|,$$
$$\frac{U_{ti}}{U_{ui}} = \left| \frac{V_{ti}V_{tj}^*}{V_{ui}V_{uj}^*} \right| \cdot \left| \frac{V_{uj}V_{uk}^*}{V_{tj}V_{tk}^*} \right| \cdot \left| \frac{V_{tk}V_{ti}^*}{V_{uk}V_{ui}^*} \right| \qquad (16.24)$$
$$= \left| \frac{\sin \omega_{ciuj}}{\sin \omega_{ticj}} \right| \cdot \left| \frac{\sin \omega_{tjck}}{\sin \omega_{cjuk}} \right| \cdot \left| \frac{\sin \omega_{ckui}}{\sin \omega_{tkci}} \right|.$$

By using eqns (16.23) and (16.24) one obtains the moduli of all matrix elements as functions of the sines of linear combinations of the parameters in eqn (16.17). Of course, since the elements of the CKM matrix are not rephasing-invariant, one must choose a specific phase convention before the matrix elements themselves can be written in terms of the manifestly rephasing-invariant moduli and $\omega_{\alpha i \beta j}$. Clearly, once one knows the moduli of all matrix elements and the phases of all quartets, we are in possession of all the physical information in the CKM matrix.

16.4 Wolfenstein parametrization

In 1983 it was realized that the bottom quark decays predominantly to the charm quark: $|V_{cb}| \gg |V_{ub}|$. Wolfenstein (1983) then noticed that $|V_{cb}| \sim |V_{us}|^2$ and introduced an approximate parametrization of V—a parametrization in which unitarity only holds approximately—which has since become very popular. He wrote

$$V = \begin{pmatrix} 1 - \lambda^2/2 & \lambda & A\lambda^3 (\rho - i\eta) \\ -\lambda & 1 - \lambda^2/2 & A\lambda^2 \\ A\lambda^3 (1 - \rho - i\eta) & -A\lambda^2 & 1 \end{pmatrix} + O(\lambda^4). \qquad (16.25)$$

The parameter $\lambda \approx 0.22$ is small and serves as an expansion parameter. On the other hand, $A \sim 1$ because $|V_{cb}| \sim |V_{us}|^2$. Finally, $|V_{ub}| / |V_{cb}| \sim \lambda/2$ and therefore ρ and η should be smaller than one. Thus, one may estimate the order of magnitude of any function of the matrix elements of V by considering the leading term of its expansion in λ.[38]

One easily checks that the unitarity relations—normalization of each row and column of V, and orthogonality of each pair of different rows or columns—are satisfied up to order λ^3 by the matrix in eqn (16.25). An expansion of V up to a higher power of λ must be made if one wants to obtain a better approximation to unitarity.

The Wolfenstein parametrization is original for two main reasons. Firstly, it incorporates as ingredients not only unitarity, but also experimental information: $|V_{us}| \ll 1$, $|V_{cb}| \sim |V_{us}|^2$, and $|V_{ub}| \ll |V_{cb}|$. Secondly, it is only approximately unitary, with the approximation to exact unitarity being achieved in a series expansion.

In the Wolfenstein parametrization, to leading order,

$$\frac{V_{ud}V_{ub}^*}{|V_{cd}V_{cb}|} = \rho + i\eta,$$
$$\frac{V_{cd}V_{cb}^*}{|V_{cd}V_{cb}|} = -1, \qquad (16.26)$$
$$\frac{V_{td}V_{tb}^*}{|V_{cd}V_{cb}|} = 1 - \rho - i\eta.$$

This is the justification for the coordinates of the vertices of the unitarity triangle in Fig. 13.2.

While $\lambda = 0.2205 \pm 0.0018$ and $A = 0.824 \pm 0.075$ are relatively well known, the parameters ρ and η—or, equivalently, the angles α, β, and γ—are much more uncertain. The main goal of CP-violation experiments is to over-constrain these parameters and, possibly, to find inconsistencies suggesting the existence of physics beyond the SM.

---

[38] One should keep in mind the possibility of additional suppressions because ρ and/or η may be very small.

16.4.1   *Exact version of the parametrization*

Sometimes the expansion up to order λ^3 in eqn (16.25) is not sufficient and one may want to use terms of higher order in λ. One knows for instance that the imaginary parts of all quartets should be equal in absolute value. This is however not true when using eqn (16.25): Q_{udcs} and Q_{cstb} are real, while $\text{Im}\,Q_{tdcb} = A^2\lambda^6\eta$ and $\text{Im}\,Q_{uscb} = A^2\lambda^6\eta\left(1 - \lambda^2/2\right)$. Such imprecisions may become misleading and/or constitute a source of error when using eqn (16.25).

Expanding the Wolfenstein parametrization to a higher order in λ is easier and more systematic when one is guided by an exact parametrization, i.e., by an exactly unitary matrix V. Indeed, one needs a definition of the way in which the series expansion in λ is to be carried out to higher orders. A way to do this has been suggested by Branco and Lavoura (1988b). They have used as a guide the Bjorken–Dunietz parametrization. They have *defined* the parameters by means of the equations

$$
\begin{aligned}
V_{us} &= \lambda, \\
V_{cb} &= A\lambda^2, \\
V_{ub} &= A\mu\lambda^3 e^{-i\phi},
\end{aligned}
\tag{16.27}
$$

together with the phase convention in eqn (16.10). In this way, $\lambda = |V_{us}|$, $A = |V_{cb}/V_{us}^2|$, $\mu = |V_{ub}/(V_{us}V_{cb})|$, and $\phi = \omega_{uscb}$ are directly related to measurable quantities. It is important to stress that eqns (16.27) are *exact* by definition: the expressions for V_{us}, V_{cb}, and V_{ub} are not corrected by terms of higher order in λ.

We may reconstruct the full CKM matrix just as was done in the Bjorken–Dunietz parametrization. Thus,

$$
\begin{aligned}
V_{ud} &= \sqrt{1 - \lambda^2 - A^2\mu^2\lambda^6}, \\
V_{tb} &= \sqrt{1 - A^2\lambda^4 - A^2\mu^2\lambda^6}, \\
V_{cs} &= \Big\{-A^2\mu\lambda^6\cos\phi + \big[1 - \lambda^2 - A^2\lambda^4 + A^2\left(1 - 2\mu^2\right)\lambda^6 + A^2\mu^2\lambda^8 \\
&\quad + A^4\mu^2\lambda^{10} + A^4\mu^2\left(\mu^2 - \sin^2\phi\right)\lambda^{12}\big]^{1/2}\Big\} / \left(1 - A^2\mu^2\lambda^6\right).
\end{aligned}
\tag{16.28}
$$

Together with eqns (16.15) this fixes the CKM matrix. We may now perform the expansion as a series in λ up to any desired order.[39] We present here the result of the expansion up to order λ^5. For ease of comparison with eqn (16.25), we substitute μ and ϕ by $\rho \equiv \mu\cos\phi$ and $\eta \equiv \mu\sin\phi$. We obtain

---

[39]It should be noted that, for each individual matrix element, the expansion parameter is not really λ but rather $\lambda^2 \sim 1/20$. The series expansion is thus, as a matter of fact, much more precise when one considers individual functions of the matrix elements of V.

$$V_{ud} = 1 - \tfrac{1}{2}\lambda^2 - \tfrac{1}{8}\lambda^4 + O\left(\lambda^6\right),$$

$$V_{cd} = -\lambda + A^2 \left(\tfrac{1}{2} - \rho - i\eta\right)\lambda^5 + O\left(\lambda^7\right),$$

$$V_{cs} = 1 - \tfrac{1}{2}\lambda^2 - \tfrac{1}{8}\left(1 + 4A^2\right)\lambda^4 + O\left(\lambda^6\right),$$

$$V_{td} = A\left(1 - \rho - i\eta\right)\lambda^3 + \tfrac{1}{2}A\left(\rho + i\eta\right)\lambda^5 + O\left(\lambda^7\right),$$

$$V_{ts} = -A\lambda^2 + A\left(\tfrac{1}{2} - \rho - i\eta\right)\lambda^4 + O\left(\lambda^6\right),$$

$$V_{tb} = 1 - \tfrac{1}{2}A^2\lambda^4 + O\left(\lambda^6\right).$$

(16.29)

Equations (16.27) and (16.29) coincide with eqn (16.25) up to order λ^3.

Buras *et al.* (1994) have used the Chau–Keung parametrization as the basis for a different exact version of the Wolfenstein parametrization. They defined the parameters by means of the equations

$$s_{12} = \lambda,$$

$$s_{23} = A\lambda^2,$$

$$s_{13}e^{-i\delta_{13}} = A\lambda^3\left(\rho - i\eta\right).$$

(16.30)

Then,

$$c_{12} = \sqrt{1 - \lambda^2},$$

$$c_{23} = \sqrt{1 - A^2\lambda^4},$$

$$c_{13} = \sqrt{1 - A^2\lambda^6\left(\rho^2 + \eta^2\right)}.$$

(16.31)

Substituting these expressions in eqn (16.3) one obtains an exact parametrization of the CKM matrix, which one may then proceed to expand as a power series in λ. In practice, the differences between the parametrizations of Buras *et al.* (1994) and of Branco and Lavoura (1988b) first arise only at order λ^6: eqns (16.29) are valid in both parametrizations.

In the parametrization of Branco and Lavoura (1988b)

$$J = A^2\lambda^6\eta\, V_{cs} \approx A^2\lambda^6\eta\left(1 - \tfrac{1}{2}\lambda^2\right).$$

(16.32)

The parameters λ_α for the K^0–$\overline{K^0}$ system, defined in eqn (13.50), are

$$\lambda_u = \lambda\left(1 - \tfrac{1}{2}\lambda^2\right) + O\left(\lambda^5\right),$$

$$\lambda_c = -\lambda\left(1 - \tfrac{1}{2}\lambda^2\right) + O\left(\lambda^5\right),$$

$$\lambda_t = -A^2\lambda^5\left[1 - \rho - i\eta - \tfrac{1}{2}\lambda^2 + 2\lambda^2\rho - \lambda^2\left(\rho^2 + \eta^2\right)\right] + O\left(\lambda^9\right).$$

(16.33)

For the B_d^0–$\overline{B_d^0}$ system,

$$\lambda_u = A\lambda^3\left(\bar{\rho} + i\bar{\eta}\right) + O\left(\lambda^7\right),$$

$$\lambda_c = -A\lambda^3 + O\left(\lambda^7\right),$$

$$\lambda_t = A\lambda^3\left(1 - \bar{\rho} - i\bar{\eta}\right) + O\left(\lambda^7\right),$$

(16.34)

where

$$\bar{\rho} \equiv \rho \left(1 - \tfrac{1}{2}\lambda^2\right),$$

$$\bar{\eta} \equiv \eta \left(1 - \tfrac{1}{2}\lambda^2\right). \qquad (16.35)$$

For the $B_s^0\text{-}\overline{B_s^0}$ system,

$$\lambda_u = A\lambda^4 \left(\rho + i\eta\right),$$

$$\lambda_c = A\lambda^2 \left(1 - \tfrac{1}{2}\lambda^2\right) + O\left(\lambda^6\right), \qquad (16.36)$$

$$\lambda_t = A\lambda^2 \left[-1 + \lambda^2 \left(\tfrac{1}{2} - \rho - i\eta\right)\right] + O\left(\lambda^6\right).$$

16.4.2 Parametrization with R_t and R_b

It is useful to introduce a Wolfenstein-type parametrization of the CKM matrix with the following four parameters: $\lambda \equiv |V_{us}|$, $A \equiv |V_{cb}| / |V_{us}|^2$, $R_t \equiv |V_{td}V_{tb}| / |V_{cd}V_{cb}|$, and $R_b \equiv |V_{ud}V_{ub}| / |V_{cd}V_{cb}|$. As usual, we make the phase convention that V_{ud} and V_{us} are real and positive; we also choose V_{cd} negative and V_{cb} positive, so that the product $V_{cd}V_{cb}^*$ is real and negative as in the unitarity triangle in Fig. 13.1. Finally, we choose V_{tb} positive. In this phase convention, the phase of V_{ub} is $-\gamma$ and the phase of V_{td} is $-\beta$ (remember eqns 13.31).

Working out this parametrization, and making the usual series expansion in λ, one obtains

$$
\begin{aligned}
V_{ud} &= 1 - \tfrac{1}{2}\lambda^2 - \tfrac{1}{8}\lambda^4 + O\left(\lambda^6\right), \\
V_{ub} &= \left[AR_b\lambda^3 + \tfrac{1}{2}AR_b\lambda^5 + O\left(\lambda^7\right)\right] e^{-i\gamma}, \\
V_{cd} &= -\lambda + \tfrac{1}{2}A^2 \left(R_t^2 - R_b^2\right)\lambda^5 + O\left(\lambda^7\right), \\
V_{cs} &= 1 - \tfrac{1}{2}\lambda^2 - \tfrac{1}{8}\left(1 + 4A^2 + 4iA^2\sqrt{\Sigma}\right)\lambda^4 + O\left(\lambda^6\right), \\
V_{td} &= \left[AR_t\lambda^3 + O\left(\lambda^7\right)\right] e^{-i\beta}, \\
V_{ts} &= -A\lambda^2 + \tfrac{1}{2}A\left(R_t^2 - R_b^2 - i\sqrt{\Sigma}\right)\lambda^4 + O\left(\lambda^6\right), \\
V_{tb} &= 1 - \tfrac{1}{2}A^2\lambda^4 + O\left(\lambda^6\right).
\end{aligned}
\qquad (16.37)
$$

If one uses for $\exp\left(i\beta\right)$ and for $\exp\left(i\gamma\right)$ the expressions in eqns (13.34)–(13.36), one has a parametrization of the CKM matrix in terms of λ, A, R_t, and R_b. All matrix elements have been given up to order λ^5.

Using this parametrization only up to order λ^3, one has the simple result (Buras and Fleischer 1998)

$$
V = \begin{pmatrix}
1 - \lambda^2/2 & \lambda & AR_b\lambda^3 e^{-i\gamma} \\
-\lambda & 1 - \lambda^2/2 & A\lambda^2 \\
AR_t\lambda^3 e^{-i\beta} & -A\lambda^2 & 1
\end{pmatrix} + O\left(\lambda^4\right), \qquad (16.38)
$$

which will be extensively used in Part IV.

16.5   Main results

- One may parametrize the 3×3 unitary CKM matrix by means of three rotation angles and one phase. Examples are the Kobayashi–Maskawa parametrization in eqn (16.1) and the Chau–Keung parametrization in eqn (16.3).

- The most commonly used parametrization nowadays is the Wolfenstein parametrization in eqn (16.25). This is a series expansion in a parameter $\lambda \approx 0.22$, and takes into account the experimental data. The parameter A is of order unity, while ρ and η are probably smaller than 0.5.

- Sometimes one may need to use a version of the Wolfenstein parametrization in which the expansion in λ is taken to higher order than λ^3. One possibility is given in eqns (16.27) and (16.29).

17

$$\epsilon$$

17.1 Introduction

We study in this chapter the standard-model computation of the CP-violating parameters ϵ and δ. These are the only non-zero CP-violating parameters measured to date, but they are equivalent, as seen in eqn (8.92). Their fit constitutes a crucial test for any model of CP violation.

In the Kobayashi–Maskawa model ϵ provides a constraint on the CKM matrix. That constraint depends on the assumption of the inexistence of sources of CP violation beyond the complexity of the 3×3 CKM matrix.

In the SM all the CP-violating quantities are proportional to J. Their sign is fixed by the sign of J. The experimentally measured sign of ϵ fixes the sign of J, which in turn fixes the sign of any other CP-violating quantity. Thus, the prediction of the signs of the CP-violating asymmetries in the B_d^0–$\overline{B_d^0}$ system hinges on the fit of ϵ.

17.2 Γ_{12} and q_K/p_K

We first consider the decay amplitudes of K^0 and $\overline{K^0}$ to $2\pi, I = 0$. These amplitudes, after the final-state-interaction phase δ_0 has been factored out, have been denoted A_0 and $\bar{A}_0$, respectively, in eqns (8.73). They are given, at tree level, by the diagrams in Fig. 17.1. If we define the parameters $\lambda_\alpha \equiv V_{\alpha s}^* V_{\alpha d}$ as in eqn (13.50), and the Dirac-matrix combination

$$\Gamma^\mu \equiv \gamma^\mu \gamma_L, \tag{17.1}$$

then it follows from Fig. 17.1 that

$$\frac{A_0}{\bar{A}_0} = \frac{\lambda_u \langle 2\pi, I = 0 | \left(\bar{s}\Gamma^\mu u\right) \left(\bar{u}\Gamma_\mu d\right) | K^0 \rangle}{\lambda_u^* \langle 2\pi, I = 0 | \left(\bar{u}\Gamma_\mu s\right) \left(\bar{d}\Gamma^\mu u\right) | \overline{K^0} \rangle}. \tag{17.2}$$

The matrix elements in eqn (17.2) are determined by the hadronization mechanism of the strong interactions, and they are difficult to compute. They may be

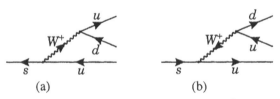

(a)   (b)

FIG. 17.1. Diagrams responsible for the decays (a) $K^0 \to 2\pi, I = 0$, and (b) $\overline{K^0} \to 2\pi, I = 0$, at tree level.

related to each other by means of CP, which is a good symmetry of the strong interactions. As we know,

$$
\begin{aligned}
\mathcal{CP}|K^0\rangle &= e^{i\xi_K}|\overline{K^0}\rangle, \\
\mathcal{CP}|2\pi, I = 0\rangle &= |2\pi, I = 0\rangle
\end{aligned}
\tag{17.3}
$$

(the two-pion state is CP-even); also,

$$
(\mathcal{CP})\, q\, (\mathcal{CP})^{\dagger} = e^{i\xi_q}\gamma^0 C\bar{q}^T,
\tag{17.4}
$$

for any quark field q. Most authors assume either $\xi_K = 0$ or $\xi_K = \pi$, while the CP-transformation phases ξ_q of the quark fields are usually neglected. We display all these phases explicitly so that independence of the final results from them becomes evident. CP-invariance of the strong interactions thus implies

$$
\langle 2\pi, I = 0|\,(\bar{s}\Gamma^{\mu}u)\,(\bar{u}\Gamma_{\mu}d)\,|K^0\rangle = e^{i(\xi_K+\xi_d-\xi_s)}\langle 2\pi, I = 0|\,(-\bar{u}\Gamma_{\mu}s)\,(-\bar{d}\Gamma^{\mu}u)\,|\overline{K^0}\rangle.
$$

Hence,

$$
\frac{A_0}{\bar{A}_0} = \frac{\lambda_u}{\lambda_u^*}e^{i(\xi_K+\xi_d-\xi_s)}.
\tag{17.5}
$$

It then follows from eqn (8.90) that

$$
\arg\Gamma_{12}^* = \xi_K + \xi_d - \xi_s + 2\arg\lambda_u.
\tag{17.6}
$$

Remembering eqn (6.70), one obtains

$$
\begin{aligned}
\frac{q_K}{p_K} &= -\frac{\lambda_u}{\lambda_u^*}e^{i(\xi_K+\xi_d-\xi_s)}\sqrt{\frac{1-\delta}{1+\delta}}\,\frac{u-i\delta}{\sqrt{u^2+\delta^2}} \\
&\approx -\frac{\lambda_u}{\lambda_u^*}e^{i(\xi_K+\xi_d-\xi_s)},
\end{aligned}
\tag{17.7}
$$

because $\delta \approx 3.3 \times 10^{-3}$ is very small, while $u \equiv -\Delta\Gamma/(2\Delta m) \approx 1$.

17.3 Master formula for ϵ

Using eqns (8.98) and (17.5),

$$
3.224 \times 10^{-3} \approx -\frac{\text{Im}\left[M_{12}e^{i(\xi_K+\xi_d-\xi_s)}\lambda_u^2\right]}{\Delta m\,|\lambda_u|^2}.
\tag{17.8}
$$

The computation of ϵ thus reduces to the computation of M_{12}.

Equation (17.8) holds whenever A_0 and $\bar{A}_0$ are dominated by the diagrams in Fig. 17.1. This is true in most extensions of the standard model. A more precise formula for ϵ should include the loop contributions (notably from penguin diagrams) to A_0 and $\bar{A}_0$. Some of those contributions are proportional to λ_t, which has a phase different from that of λ_u. One should then return to eqn (8.98) and compute the ensuing corrections to ϵ. However, that exercise is at present

ϵ

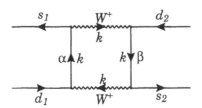

FIG. 17.2. Box diagram effecting the transition $K^0 \to \overline{K^0}$.

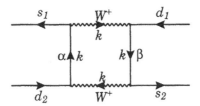

FIG. 17.3. Another box diagram effecting the transition $K^0 \to \overline{K^0}$.

rather futile, because we know from the experimental value of ϵ'/ϵ that the loop contributions to the phases of A_0 and $\bar{A}_0$ are very small. Besides, the theoretical uncertainties in the computation of M_{12} are sufficiently large that it does not really make sense to be worrying about a small deviation of the phase of $A_0 \bar{A}_0^*$ from its tree-level value.

17.4 The box diagram

We want to compute M_{12} in the SM. One first performs the weak-interaction, perturbative part of the computation, which is done in terms of quarks. The matrix element M_{12} corresponds to the transition $\overline{K^0} \to K^0$. One interprets $\overline{K^0}$ as $s\bar{d}$ and K^0 and $\bar{s}d$. Then, M_{12} arises from the box diagrams in Figs. 17.2 and 17.3. Those diagrams are computed in Appendix B. They are gauge-independent and translate into an effective Hamiltonian

$$\mathcal{H}_{\text{eff}} = \frac{G_F^2 m_W^2}{4\pi^2} (\bar{s}\Gamma^\mu d)(\bar{s}\Gamma_\mu d)\,\mathcal{F}_0 + \text{H.c.}, \qquad (17.9)$$

where

$$\mathcal{F}_0 = \lambda_c^2 S_0(x_c) + \lambda_t^2 S_0(x_t) + 2\lambda_c \lambda_t S_0(x_c, x_t). \qquad (17.10)$$

Equation (13.53) has been used to eliminate λ_u in favour of λ_c and λ_t; moreover, we have made the approximation of taking the up-quark mass to be zero. The functions $S_0(x, y)$ and $S_0(x) = \lim_{y \to x} S_0(x, y)$ are given in eqns (B.15) and (B.16), respectively. They are functions of the up-type-quark masses through $x_\alpha \equiv m_\alpha^2/m_W^2$.

Then, M_{12} is given by

$$M_{12} = \frac{G_F^2 m_W^2}{4\pi^2} \langle K^0| \left(\bar{d}\Gamma^\mu s\right) \left(\bar{d}\Gamma_\mu s\right) |\overline{K^0}\rangle \mathcal{F}_0^*. \tag{17.11}$$

The computation of the hadronic matrix element $\langle K^0| \left(\bar{d}\Gamma^\mu s\right) \left(\bar{d}\Gamma_\mu s\right) |\overline{K^0}\rangle$ involves the strong interaction and the corresponding hadronization process. This is one of the awkward steps in the computation of M_{12}: the matrix element cannot be reliably computed.

Still, an order-of-magnitude estimate may be obtained by using the vacuum-insertion approximation (VIA), which consists in the insertion of the vacuum state in all possible ways in between bilinear quark operators. With the VIA one reduces the problem of calculating the matrix element of a quartic operator into the problem of calculating the matrix elements of two bilinear operators. These may be calculated in some model or, sometimes, directly determined from experiment. One uses the fact that the strong interactions enjoy P, CP, and isospin symmetries, in order to simplify and relate among themselves various matrix elements. This is illustrated in Appendix C. One obtains there the result

$$\langle K^0| \left(\bar{d}\Gamma^\mu s\right) \left(\bar{d}\Gamma_\mu s\right) |\overline{K^0}\rangle_{\text{VIA}} = -\tfrac{1}{3} e^{i(\xi_s - \xi_d - \xi_K)} f_K^2 m_K. \tag{17.12}$$

The true matrix element is usually parametrized as the product of its VIA estimate and a corrective factor B_K,

$$\langle K^0| \left(\bar{d}\Gamma^\mu s\right) \left(\bar{d}\Gamma_\mu s\right) |\overline{K^0}\rangle = -\tfrac{1}{3} e^{i(\xi_s - \xi_d - \xi_K)} f_K^2 m_K B_K, \tag{17.13}$$

in the hope that B_K does not differ too much from unity. Of course, computing B_K is the same as computing the original matrix element. The 'bag parameter' B_K must be real because of CP conservation by the strong interactions, but it may be either positive or negative. If it is negative, then the VIA has failed badly in approximating the matrix element. Most authors now agree that B_K is positive. This is good, because the sign of ϵ and the sign predicted for all CP asymmetries in the B^0-$\overline{B^0}$ systems hinge on sign B_K.

From eqns (17.11) and (17.13) we obtain

$$M_{12} = -\frac{G_F^2 m_W^2}{12\pi^2} f_K^2 m_K B_K e^{i(\xi_s - \xi_d - \xi_K)} \mathcal{F}_0^*. \tag{17.14}$$

It is important to note the proportionality of M_{12} to $\exp\left(-i\xi_K\right)$. Comparing the first eqn (6.11) and the first eqn (6.20), one sees that they agree in the way that M_{12} transforms under a rephasing of $|K^0\rangle$ and $|\overline{K^0}\rangle$. Notice that M_{12} is also proportional to the difference of the CP-transformation phases of the s and d quarks.

17.5   QCD corrections to the $|\Delta S| = 2$ effective Hamiltonian

When QCD corrections are taken into account, the $|\Delta S| = 2$ effective Hamiltonian in eqn (17.9) becomes

$$\mathcal{H}_{\text{eff}} = \frac{G_F^2 m_W^2}{4\pi^2} \left(\bar{s}\Gamma^\mu d\right) \left(\bar{s}\Gamma_\mu d\right) \mathcal{F} \left[\alpha_s^{(3)}(\mu)\right]^{-2/9} \left[1 + \frac{\alpha_s^{(3)}(\mu)}{4\pi} J_{(3)}\right] + \text{H.c.}, \quad (17.15)$$

where

$$\mathcal{F} = \eta_1 \lambda_c^2 S_0\left(x_c\right) + \eta_2 \lambda_t^2 S_0\left(x_t\right) + 2\eta_3 \lambda_c \lambda_t S_0\left(x_c, x_t\right). \quad (17.16)$$

Several features are worth notice:

- There are three QCD correction factors η_1, η_2, and η_3 in $\mathcal{F}$, which were absent in $\mathcal{F}_0$—cf. eqns (17.10) and (17.16).
- The gluonic corrections do not lead to the appearance of any operators beyond the one that was already present, $(\bar{s}\Gamma^\mu d)(\bar{s}\Gamma_\mu d)$.[40]
- The effective Hamiltonian depends on a renormalization scale μ. The scale dependence is brought in by $\alpha_s(\mu)$. The scale μ should be taken below the charm-quark threshold, $\mu < m_c$; there, only the three light quarks u, d, and s are dynamic degrees of freedom—thence the notation $\alpha_s^{(3)}(\mu)$.
- There is also a quantity $J_{(3)}$ which depends, through the beta function and the anomalous dimension of the operator, on the renormalization scheme.

Since physical amplitudes cannot depend on an arbitrary renormalization scale, the μ-dependence must be cancelled out by a μ-dependence of the hadronic matrix element of the operator. Thus, the coefficient B_K, describing the deviation of the true value of the matrix element from its VIA value, is now a scale-dependent quantity:

$$\langle K^0| \left(\bar{d}\Gamma^\mu s\right) \left(\bar{d}\Gamma_\mu s\right) |\overline{K^0}\rangle = -\tfrac{1}{3} e^{i(\xi_s - \xi_d - \xi_K)} f_K^2 m_K B_K(\mu). \quad (17.17)$$

The renormalization-scheme-dependence must also be cancelled out by that of $B_K(\mu)$. One may hide the μ-dependence and the renormalization-scheme-dependence of the matrix element by defining a μ-independent parameter B_K:

$$B_K = B_K(\mu) \left[\alpha_s^{(3)}(\mu)\right]^{-2/9} \left[1 + \frac{\alpha_s^{(3)}(\mu)}{4\pi} J_{(3)}\right]. \quad (17.18)$$

Its value can be found from lattice calculations, $1/N$ expansion, and a number of other methods. A conservative estimate is $1/3 \leq B_K \leq 1$, but we shall use the value $B_K = 0.75 \pm 0.15$ suggested by Buras and Fleischer (1998), which is warranted by recent lattice computations (Gupta 1998).

As a result,

$$M_{12} = -\frac{G_F^2 m_W^2}{12\pi^2} f_K^2 m_K B_K e^{i(\xi_s - \xi_d - \xi_K)} \mathcal{F}^*. \quad (17.19)$$

The interpretation of B_K is refined, but the final formula is the same as in eqn (17.14), only with the QCD-corrected function $\mathcal{F}$ instead of $\mathcal{F}_0$.

---

[40]This nice feature is somewhat illusory, since the computation of the coefficients η_i involves the $|\Delta S| = 1$ effective Hamiltonian, which includes many operators.

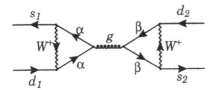

F IG. 17.4.   Double-penguin diagram for K^0–$\overline{K^0}$ mixing.

The computation of the coefficients η in eqn (17.16) is presented in the review article by Buchalla *et al.* (1996), who quote

$$\eta_1 = 1.38 \pm 0.20,$$
$$\eta_2 = 0.57 \pm 0.01, \qquad\qquad (17.20)$$
$$\eta_3 = 0.47 \pm 0.04.$$

17.6   Other contributions to M_{12}

In the standard model there are other important contributions to M_{12}, beyond the one given by the box diagrams.

First, there is another type of diagram which may contribute to M_{12}, the so-called double-penguin diagram (see Fig. 17.4), which is a two-loop diagram with a gluon connecting the two weak-interaction pieces. They are estimated to be negligible in the case of neutral-kaon mixing (Donoghue *et al.* 1986*d*; Eeg and Picek 1987, 1988), but they may give sizeable contributions to M_{12} in other neutral-meson systems, in particular in the D^0–$\overline{D^0}$ system (Petrov 1997).

Second, there are long-distance contributions to M_{12}, in which the intermediate states in the transition $\overline{K^0} \to K^0$ are mesons instead of up-type quarks and $W^\pm$ gauge bosons. The intermediate states may for instance be off-shell two-pion or three-pion states, or a π^0, η, η', and so on. The corresponding contribution to M_{12} has been estimated using flavour-SU(3) symmetry by Donoghue *et al.* (1986*a*). In any case, the dominant intermediate meson states are analogous, in quark terms, to a state $u\bar{u}$; hence, the long-distance contribution to M_{12} should have a phase given by $\arg \lambda_u^{*\,2} + \xi_s - \xi_d - \xi_K$,[41] and then it does not modify ϵ. However, $M_{12}^{\text{long distance}}$ is expected to contribute significantly to $|M_{12}|$ or, equivalently, to Δm. This is the reason why we are unable to compute Δm in the neutral-kaon system; as a consequence, in the denominator of eqn (17.8) one uses the experimental value of Δm instead of $2|M_{12}|$ computed from the box diagram.

---

[41]It is sometimes stated that there are long-distance contributions to Re M_{12}, but not to Im M_{12}. This is not a rephasing-invariant statement. The phase conventions involved are the following: firstly, λ_u is taken to be real, as in the usual parametrizations of the CKM matrix; secondly, the phases ξ_s and ξ_d are assumed to be equal; thirdly, ξ_K is chosen to be either 0 or π. It is within these assumptions that the long-distance contribution to M_{12} is real.

17.7 Fit of ϵ

From eqns (17.8) and (17.19),

$$3.224 \times 10^{-3} = \frac{G_F^2 m_W^2}{12\pi^2} f_K^2 m_K B_K \frac{\mathrm{Im}\left(\mathcal{F}^* \lambda_u^2\right)}{\Delta m \left|\lambda_u\right|^2}. \tag{17.21}$$

The phase $\xi_K + \xi_d - \xi_s$ has cancelled out, marking the invariance of the computation under a change of the spurious CP-transformation phases.

Using eqns (17.16) and (13.54), one gets

$$\mathrm{Re}\left(\lambda_c^* \lambda_u\right)\left[\eta_1 S_0\left(x_c\right) - \eta_3 S_0\left(x_c, x_t\right)\right] + \mathrm{Re}\left(\lambda_t^* \lambda_u\right)\left[\eta_3 S_0\left(x_c, x_t\right) - \eta_2 S_0\left(x_t\right)\right]$$

$$= 3.224 \times 10^{-3} \frac{6\pi^2 \Delta m \left|\lambda_u\right|^2}{G_F^2 m_W^2 f_K^2 m_K B_K J}$$

$$= 5.949 \times 10^{-8} \frac{\left|\lambda_u\right|^2}{B_K J}. \tag{17.22}$$

We shall treat in detail the uncertainty introduced by the parameter $B_K = 0.75 \pm 0.15$, and only illustrate the main other sources of uncertainty. We have used the following values: $\Delta m = 3.491 \times 10^{-15}$ GeV, $G_F = 1.16639 \times 10^{-5}$ GeV^{-2}, $m_W = 80.4$ GeV, $f_K = 0.160$ GeV, and $m_K = 0.497672$ GeV.

We use the Wolfenstein parametrization. From eqns (16.32) and (16.33), together with $\lambda = \left|V_{us}\right|$ and $A\lambda^2 = \left|V_{cb}\right|$, we obtain $J \approx \left|V_{us}\right|^2 \left|V_{cb}\right|^2 \eta$, while $\left|\lambda_u\right|^2 \approx -\mathrm{Re}\left(\lambda_c^* \lambda_u\right) \approx \left|V_{us}\right|^2$ and $\mathrm{Re}\left(\lambda_t^* \lambda_u\right) \approx -\left|V_{us}\right|^2 \left|V_{cb}\right|^2 (1 - \rho)$. Thus,

$$B_K \eta \left|V_{us}\right|^2 \left\{\left|V_{cb}\right|^2 \left[\eta_3 S_0\left(x_c, x_t\right) - \eta_1 S_0\left(x_c\right)\right] \right.$$

$$\left. + \left|V_{cb}\right|^4 (1 - \rho) \left[\eta_2 S_0\left(x_t\right) - \eta_3 S_0\left(x_c, x_t\right)\right]\right\} = 5.949 \times 10^{-8}. \tag{17.23}$$

Up to now, there is no substantial source of uncertainty in the computation; all the values used are relatively well known, with small associated errors.

With $m_c = 1.25 \pm 0.25$ GeV and $m_t = 175.5 \pm 5.5$ GeV,

$$S_0\left(x_c\right) = 2.42^{+1.06}_{-0.87} \times 10^{-4},$$

$$S_0\left(x_t\right) = 2.59^{+0.12}_{-0.13}, \tag{17.24}$$

$$S_0\left(x_c, x_t\right) = 2.17^{+0.83}_{-0.72} \times 10^{-3}.$$

One should remember that $S_0\left(x_c\right) \approx x_c$ and $S_0\left(x_t\right) = x_t f\left(x_t\right)$, where f is a slowly varying function of order unity. The main source of uncertainty in $S_0\left(x_c, x_t\right)$ is m_c; the relatively small uncertainty from the top-quark mass is almost immaterial. Together with the values in eqns (17.20), we obtain

$$\eta_3 S_0\left(x_c, x_t\right) - \eta_1 S_0\left(x_c\right) = 6.86^{+4.33}_{-3.07} \times 10^{-4},$$

$$\eta_2 S_0\left(x_t\right) - \eta_3 S_0\left(x_c, x_t\right) = 1.48^{+0.09}_{-0.10}. \tag{17.25}$$

The first quantity is very uncertain, the second one only has a small error bar, which one may neglect. Using $\left|V_{us}\right| = 0.2205$ in eqn (17.23), one obtains

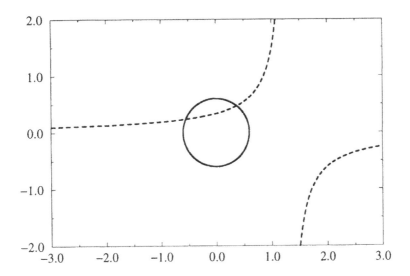

FIG. 17.5. The curves in eqns. (17.29) (dashed line) and (17.30) (full line). The parameter ρ is on the horizontal axis, η on the vertical axis.

$$B_K \eta \left[6.86^{+4.33}_{-3.07} \times 10^{-4} \, |V_{cb}|^2 + 1.48 \, |V_{cb}|^4 \, (1 - \rho) \right] = 1.223 \times 10^{-6}. \qquad (17.26)$$

The value of $|V_{cb}|$ is a relevant source of uncertainty, which however should diminish with an increasingly better theoretical understanding of the decays of b-flavoured mesons. Using $|V_{cb}| = 0.040$ and neglecting the uncertainty arising from the charm-quark mass, one obtains

$$\eta \, (1.29 - \rho) = \frac{0.32}{B_K}. \qquad (17.27)$$

We see that the Kobayashi–Maskawa mechanism of CP violation gives the right order of magnitude for ϵ. This fact in itself should be considered a success of the model.[42]

Equation (17.27) is of the form

$$\eta \, (A - \rho) = B, \qquad (17.28)$$

which may be depicted in the ρ–η plane as a hyperbola with focus $(\rho, \eta) = (A, 0)$. We depict in Fig. 17.5 the curve

$$\eta \, (1.29 - \rho) = \frac{0.32}{0.75}, \qquad (17.29)$$

together with the circle which follows from the constraint in eqn (15.23),

---

[42] A quick estimate of the order of magnitude of ϵ in the Kobayashi–Maskawa model might have been produced in the following way: $\epsilon = \langle 0|T|K_L \rangle \langle 0|T|K_S \rangle^* / \, |\langle 0|T|K_S \rangle|^2$; the numerator is proportional to $J \sim \lambda^6$; the denominator is proportional to λ^2; therefore, $\epsilon \propto \lambda^4 \approx 2.5 \times 10^{-3}$.

$$\rho^2 + \eta^2 = 0.36. \tag{17.30}$$

One sees that η should be positive. Knowledge of the sign of η enables one to predict the sign of any other calculable CP asymmetry in the Kobayashi–Maskawa model. Remember however that, if the sign of B_K were negative, the sign of η would turn out negative too, and the predictions for all CP asymmetries would see their signs inverted.

17.8   Main conclusions

- The spurious phase $\xi_K + \xi_d - \xi_s$ arises in both $A_0 \bar{A}_0^*$ and M_{12}; the physical parameter ϵ, on the other hand, is independent of that phase.
- The computation of ϵ boils down to the computation of M_{12}. Long-distance contributions to this parameter may be sizeable, but they have a phase such that they do not modify ϵ. The relevant contributions are those given by the box diagrams.
- In the ρ-η plane, the fit of ϵ leads to hyperbolae of the form in eqn (17.28). Together with the experimental value of $|V_{ub}/V_{cb}|$, this implies that η and J are both positive.
- A very uncertain matrix element, parametrized by the so-called 'bag parameter' B_K, is involved in the exact fit of ϵ, but it is usually agreed that B_K is positive.

18

MIXING IN THE B_q^0–$\overline{B_q^0}$ SYSTEMS

18.1 M_{12}

Let us consider the B_q^0–$\overline{B_q^0}$ systems, where q may be either d or s. The meson B_q^0 is made up of a heavy antiquark $\bar{b}$ and a light quark q, while $\overline{B_q^0} = b\bar{q}$. We want to compute the mixing-matrix element M_{12} for these neutral-meson systems. It corresponds to the transition $\overline{B_q^0} \to B_q^0$ or, in quark language, $b\bar{q} \to \bar{b}q$. In the SM, the main short-distance contribution to this amplitude arises from box diagrams like those in Fig. 17.2. The computation of those diagrams and of the corresponding hadronic matrix element is analogous to the one presented in Chapter 17 and in Appendices B and C. The result is, just as in eqn (17.19),

$$M_{12} = -\frac{G_F^2 m_W^2}{12\pi^2} f_{B_q}^2 m_{B_q} B_{B_q} e^{i(\xi_b - \xi_q - \xi_{B_q})} \mathcal{F}^*, \tag{18.1}$$

where

$$\mathcal{F} = \eta_1 \lambda_c^2 S_0\left(x_c\right) + \eta_2 \lambda_t^2 S_0\left(x_t\right) + 2\eta_3 \lambda_c \lambda_t S_0\left(x_c, x_t\right). \tag{18.2}$$

The combinations λ_c and λ_t of CKM-matrix elements now are

$$\lambda_\alpha = V_{\alpha b}^* V_{\alpha q}. \tag{18.3}$$

Their values in the Wolfenstein parametrization are given in eqns (16.34)–(16.36). We see that, when $q = d$, λ_c and λ_t both have order of magnitude λ^3; when $q = s$, they both have order of magnitude λ^2.[43] Explicit values of the function S_0 are given in eqns (17.24). The fact that

$$S_0\left(x_t\right) \gg S_0\left(x_c, x_t\right) > S_0\left(x_c\right) \tag{18.4}$$

is crucial. Indeed, as λ_c and λ_t have the same order of magnitude, we may approximate eqn (18.2) by

$$\mathcal{F} \approx \eta_{B_q} \lambda_t^2 S_0\left(x_t\right), \tag{18.5}$$

where we now designate the QCD-correction coefficient η_2 by η_{B_q}. We shall use the value (Buchalla *et al.* 1996; see also Buras *et al.* 1990)

$$\eta_{B_q} = 0.55, \tag{18.6}$$

---

[43]This situation should be contrasted with what happens in the K^0–$\overline{K^0}$ system, where $\lambda_c \sim \lambda$ is much larger than $\lambda_t \sim \lambda^5$.

which is approximately the same for both B_q^0–$\overline{B_q^0}$ systems. One thus has

$$M_{12} \approx -\frac{G_F^2 m_W^2}{12\pi^2} f_{B_q}^2 m_{B_q} B_{B_q} \eta_{B_q} S_0\left(x_t\right) \left(V_{tb} V_{tq}^*\right)^2 e^{i\left(\xi_b - \xi_q - \xi_{B_q}\right)}, \qquad (18.7)$$

which is an approximation valid up to order $S_0(x_c, x_t)/S_0(x_t) \sim 10^{-3}$.

As before, there is a dependence on a renormalization scale μ, coming from factors $\alpha_s(\mu)$ in the QCD corrections to the box diagrams. That dependence of the effective Hamiltonian on μ should be cancelled out by the μ-dependence of the hadronic matrix element

$$\langle B_q^0| \left(\bar q \Gamma^\mu b\right)\left(\bar q \Gamma_\mu b\right) |\overline{B_q^0}\rangle \left(\mu\right) = -\tfrac{1}{3} e^{i\left(\xi_b - \xi_q - \xi_{B_q}\right)} f_{B_q}^2 m_{B_q} B_{B_q}\left(\mu\right). \qquad (18.8)$$

The matrix element has been normalized by its VIA value. In order to hide the renormalization-scale- and renormalization-scheme-dependence, one defines

$$B_{B_q} \equiv B_{B_q}\left(\mu\right) \left[\alpha_s^{(5)}\left(\mu\right)\right]^{-6/23} \left[1 + \frac{\alpha_s^{(5)}\left(\mu\right)}{4\pi} J_{(5)}\right], \qquad (18.9)$$

just as in eqn (17.18). Notice that μ should now be taken to be of order $m_{B_q} \sim$ 5 GeV, where five quark flavours are still dynamical degrees of freedom; for that reason we now denote the strong coupling constant by $\alpha_s^{(5)}$.

Contrary to what happens in the kaon system, the long-distance contributions to M_{12} are estimated to be negligible in the B_q^0–$\overline{B_q^0}$ systems. This is because the relevant mass scale m_b is much larger than the mass scale $\Lambda_{\rm QCD}$ below which quarks cease to provide a reasonable picture of hadronic physics.

The non-diagonal mass term M_{12}, which is of second order in the weak interactions, can compete with the diagonal mass terms, which are dominated by the bottom-quark mass and therefore are much larger, only because the latter are degenerate as a consequence of CPT symmetry. If the diagonal mass terms of B_q^0 and $\overline{B_q^0}$ were substantially different, then the box diagram connecting B_q^0 to $\overline{B_q^0}$ would be irrelevant. As an illustration of this fact, one may point out that there are box diagrams connecting B_d^0 to B_s^0, but one never takes them into account. There is a mass matrix connecting B_d^0, $\overline{B_d^0}$, B_s^0, and $\overline{B_s^0}$ (and also other states, like K^0, $\overline{K^0}$, and so on); however, the difference between the masses of B_d^0 and B_s^0 is sufficiently large that the boxes connecting B_d^0–$\overline{B_d^0}$ to B_s^0–$\overline{B_s^0}$ are irrelevant. As $m_{B_d} \neq m_{B_s}$ the 4×4 mass matrix effectively breaks down into two 2×2 submatrices; one of them describes B_d^0–$\overline{B_d^0}$ mixing, the other one describes B_s^0–$\overline{B_s^0}$ mixing.

18.2   A note on CP invariance

From eqn (18.7),

$$\frac{M_{12}^*}{M_{12}} = e^{2i(\xi_{B_q} + \xi_q - \xi_b)} \frac{\left(V_{tb}^* V_{tq}\right)^2}{\left(V_{tb} V_{tq}^*\right)^2}. \qquad (18.10)$$

We know from Table 6.1 that CP conservation in the mixing implies

$$\frac{M_{12}^*}{M_{12}} = e^{2i\xi_{B_q}} . \tag{18.11}$$

Comparing eqns (18.10) and (18.11) we find that CP conservation requires

$$(V_{tb}V_{tq}^*)^2 = e^{2i(\xi_q - \xi_b)}(V_{tb}^*V_{tq})^2 . \tag{18.12}$$

If we had not neglected the contribution to M_{12}^* of the terms proportional to λ_c^2 and to $\lambda_c\lambda_t$, and since the quark masses are in principle arbitrary, we would have concluded that CP invariance also requires

$$(V_{cb}V_{cq}^*)^2 = e^{2i(\xi_q - \xi_b)}(V_{cb}^*V_{cq})^2 . \tag{18.13}$$

Equations (18.12) and (18.13) are precisely those that follow from the general condition for CP invariance of the SM in eqn (13.16). This is a good check of the consistency of the whole scheme.

18.3 Γ_{12}

Now consider Γ_{12}, which is given by

$$\Gamma_{12} = \sum_f \langle f|T|B_q^0\rangle^* \langle f|T|\overline{B_q^0}\rangle, \tag{18.14}$$

where f are the physical states to which both B_q^0 and $\overline{B_q^0}$ decay. This relation may be interpreted, in quark terms, as the absorptive part of the box diagrams with intermediate c or u quarks. As the mass of the top quark is much larger than m_{B_q}, B_q^0 and $\overline{B_q^0}$ cannot decay to any top-flavoured hadron; therefore, the box diagrams with intermediate top quarks have vanishing absorptive part. The value of the absorptive part of the box diagram, or indeed of any other diagram contributing to Γ_{12}, must be dominated by the mass available in the decays of B_q^0 and $\overline{B_q^0}$, i.e., by $m_{B_q} \approx m_b$.[44] As $M_{12} \propto S_0(x_t) \propto x_t \propto m_t^2$, one arrives at the prediction

$$\left|\frac{\Gamma_{12}}{M_{12}}\right| \sim \frac{m_b^2}{m_t^2} \sim 10^{-3} . \tag{18.15}$$

Thus, while M_{12} incorporates a GIM-enhancement and increases with the mass of the heaviest up-type quark, Γ_{12} is bound to remain $\sim m_b$.

From eqns (18.15), (6.61), and (6.62), one then finds

$$\Delta m = 2|M_{12}|, \tag{18.16}$$

$$\Delta\Gamma = \frac{2\mathrm{Re}\,(M_{12}^*\Gamma_{12})}{|M_{12}|}, \tag{18.17}$$

with $|\Delta\Gamma| \ll \Delta m$.

---

[44]The computation of the absorptive part requires that the masses and momenta of the external quarks not be neglected, contrary to what was done in Appendix B.

If the masses of the up and charm quarks were equal, then the absorptive part of the box diagram would be proportional to the CKM factor $(\lambda_u + \lambda_c)^2 = \lambda_t^2$, and the angle between Γ_{12} and M_{12} would vanish. Therefore, we expect

$$\arg\left(M_{12}^*\Gamma_{12}\right) \sim \frac{m_c^2 - m_u^2}{m_b^2} \sim 10^{-1}, \tag{18.18}$$

leading to

$$\delta \sim \frac{m_c^2}{m_t^2} \sim 10^{-4}. \tag{18.19}$$

CP violation in the mixing is then very small (Hagelin 1981; Bigi *et al.* 1989; Soares, unpublished thesis; see, on the other hand, Altomari *et al.* 1988.), and eqn (18.17) simplifies into

$$\Delta\Gamma = 2\left|\Gamma_{12}\right| \operatorname{sign} \cos \arg\left(M_{12}^*\Gamma_{12}\right). \tag{18.20}$$

We may find the sign of $\Delta\Gamma$ once we know whether $\arg\left(M_{12}^*\Gamma_{12}\right)$ is close to 0 or close to π.

The above results have a simple qualitative interpretation (Kayser 1997). Let us choose the phase convention for the CP transformation $\xi_{B_q} + \xi_q - \xi_b = 0$. We know that R_{12} is computed from the same box diagram as R_{21}, with all quarks exchanged by their respective antiquarks. Hence the CKM phases get complex-conjugated. Since the mass of the B^0 mesons is much smaller than the $t\bar{t}$ production threshold, there is no absorptive part in the t-boxes. These facts imply that $R_{12} = R_{21}^*$ when the c- and u-boxes are neglected. Then, $\Gamma_{12} = 0$ and $\delta = 0$. Also, $\Delta\mu = 2\sqrt{R_{12}R_{21}}$ is real, and therefore $\Delta\Gamma = 0$. Hence, if it were not for the absorptive parts of the boxes with up and charm quarks, there would be no CP violation in the mixing, and the widths of B_H and B_L would be equal. One therefore expects both $\Delta\Gamma$ and δ to be small.

The smallness of mixing CP violation—cf. eqn (18.18)—has a deeper justification in the B_s^0–$\overline{B_s^0}$ system. There, M_{12} arises from the two-top-quarks box diagram, while Γ_{12} is dominated by the decays $b \to c\bar{c}s$. Both M_{12} and Γ_{12} involve mainly the last two families, hence they cannot exhibit CP violation. This is clear in the Wolfenstein parametrization of the CKM matrix: there are no phases to lowest order in λ between $\lambda_c \approx A\lambda^2$ and $\lambda_t \approx -A\lambda^2$. In order to get a phase one must introduce the first family, via the contributions to Γ_{12} of the suppressed decays $b \to \bar{u}cs, \bar{c}us, \bar{u}us$.

18.4   The mass difference in the B_d^0–$\overline{B_d^0}$ system

The mass difference between the eigenstates of mixing in the B_d^0–$\overline{B_d^0}$ system has been measured to be $\Delta m_{B_d} = (3.12 \pm 0.20) \times 10^{-13}\,\text{GeV}$.[45] Fitting this

---

[45] The mass difference is thirteen orders of magnitude lower than the mass of the B_d^0, $m_{B_d} = (5.2792 \pm 0.0018)\,\text{GeV}$.

mass difference in the SM provides an important constraint on the CKM matrix. From eqns (18.16) and (18.7) we obtain

$$\Delta m_{B_d} = \frac{G_F^2 m_W^2}{6\pi^2} \eta_{B_d} m_{B_d} f_{B_d}^2 B_{B_d} S_0\left(x_t\right) \left|V_{tb} V_{td}\right|^2 . \tag{18.21}$$

The matrix element $|V_{tb}|$ must be very close to 1, as seen in § 15.7. Therefore, the measurement of Δm_{B_d} can be used to constrain $|V_{td}|$. The uncertainty is due mainly to the poorly known f_{B_d} and B_{B_d}. For simplicity we shall take only these quantities to be uncertain. Using $\Delta m_{B_d} = 3.12 \times 10^{-13}$ GeV, $G_F = 1.16639 \times 10^{-5}$ GeV^{-2}, $m_W = 80.4$ GeV, $\eta_{B_d} = 0.55$, $m_{B_d} = 5.2792$ GeV, $S_0\left(x_t\right) = 2.59$, and

$$\left|V_{tb} V_{td}\right|^2 \approx A^2 \lambda^6 \left[(1-\rho)^2 + \eta^2\right] \approx |V_{us}|^2 |V_{cb}|^2 \left[(1-\rho)^2 + \eta^2\right], \tag{18.22}$$

with $|V_{us}| = 0.2205$ and $|V_{cb}| = 0.040$, one obtains

$$f_{B_d}^2 B_{B_d} \left[(1-\rho)^2 + \eta^2\right] = 0.036 \text{ GeV}^2. \tag{18.23}$$

We need the values of the meson decay constant f_{B_d}, which has not yet been measured, and of the matrix element in eqn (18.8). Various theoretical computations are available; we shall use (Flynn 1997)

$$\begin{aligned} f_{B_d} &= (175 \pm 25) \text{ MeV}, \\ B_{B_d} &= 1.31 \pm 0.03; \end{aligned} \tag{18.24}$$

these values may be combined (Buras and Fleischer 1998) into

$$f_{B_d} \sqrt{B_{B_d}} = (200 \pm 40) \text{ MeV}. \tag{18.25}$$

One then has

$$(1-\rho)^2 + \eta^2 = 0.90^{+0.51}_{-0.28}. \tag{18.26}$$

18.5   The ρ-η plane and the unitarity triangle

In the Wolfenstein parametrization there are four parameters: λ, A, ρ, and η. The first two are rather well determined: $\lambda = |V_{us}| = 0.2205 \pm 0.0018$ and $A = |V_{cb}| / |V_{us}|^2 = 0.824 \pm 0.075$. On the other hand, ρ and η are poorly determined. It is convenient to picture the constraints on these two parameters in the ρ-η plane. Those constraints are:

- The value of $|V_{ub}| / |V_{cb}|$ in eqn (15.23) implies

$$0.27 < \sqrt{\rho^2 + \eta^2} < 0.45. \tag{18.27}$$

In the ρ-η plane this is the area in between two circumferences with centre $(\rho, \eta) = (0, 0)$.

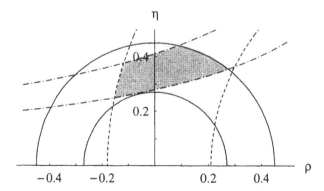

FIG. 18.1. The area in between the full lines is the one determined by eqn (18.27); the area in between the dashed lines is the one determined by eqn (18.28); the area in between the dashed–dotted lines is the one determined by eqn (18.29). The intersection of all three areas, which is shown shadowed, is the allowed domain for ρ and η.

- The constraint in eqn (18.26) from the measured value of Δm_B may be rewritten

$$0.79 < \sqrt{(1-\rho)^2 + \eta^2} < 1.18. \tag{18.28}$$

In the ρ–η plane this is the area in betweeen two circumferences with centre $(\rho, \eta) = (1, 0)$.

- The constraint in eqn (17.27) from the measured value of ϵ may be rewritten, as $0.6 < B_K < 0.9$,

$$0.36 < \eta\,(1.29 - \rho) < 0.53. \tag{18.29}$$

In the ρ–η plane this is the area in between two hyperbolae with focus $(\rho, \eta) = (1.29, 0)$.

We have depicted these constraints in Fig. 18.1. The precise boundaries of the shadowed area should not be taken too seriously; on the one hand, because they are dominated by the estimated values of theoretical errors, like the uncertainties in the values of B_K, of B_{B_d}, and of $|V_{ub}| / |V_{cb}|$; on the other hand, because some experimental errors are correlated—a fact that we have not taken into account when intersecting the three domains in eqns (18.27)–(18.29). In any case, we gather important information from Fig. 18.1. The parameter η is found to be 0.33 ± 0.10, but ρ is not that well determined: $\rho = 0.04 \pm 0.22$.

It is important to translate this information into expected values for the angles of the unitarity triangle in Fig. 18.2. One sees that the angle β is rather well determined: it should be between $10°$ and $30°$, approximately, corresponding to

$$0.4 < \sin 2\beta < 0.9. \tag{18.30}$$

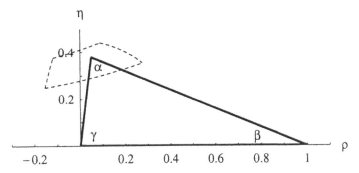

FIG. 18.2. The unitarity triangle of Fig. 13.2 and the allowed area of Fig. 18.1.

The angle γ should lie between $50°$ and $125°$, which constrains $\sin^2 \gamma > 0.58$.[46]
Finally, α is expected to be between $35°$ and $120°$, which only constrains $\sin 2\alpha$
to be larger than -0.9. These limits, though, should not be taken too literally.

It is interesting to notice that, if one accepts only the bounds shown in
eqns (18.27) and in (18.28), then one can already conclude that η must be non
zero, even if we discard the bound from CP violation in the $K^0 - \overline{K^0}$ system. This
means that, if we trust the Standard Model, then we can assert that there is CP
violation by looking exclusively at CP-conserving observables.

18.6 The mass difference in the $B_s^0 - \overline{B_s^0}$ system

Clearly,

$$\frac{\Delta m_{B_s}}{\Delta m_{B_d}} \approx \frac{f_{B_s}^2}{f_{B_d}^2} \frac{B_{B_s}}{B_{B_d}} \left| \frac{V_{ts}}{V_{td}} \right|^2 . \tag{18.31}$$

The ratio of hadronic parameters should be one in the flavour-SU(3) limit, and
is much less uncertain than the numerator and denominator individually (Flynn
1997):

$$\frac{f_{B_s} \sqrt{B_{B_s}}}{f_{B_d} \sqrt{B_{B_d}}} = 1.15 \pm 0.05. \tag{18.32}$$

There is also a quark-model bound $f_{B_s}/f_{B_d} < 1.25$ (Rosner 1990; Amundson
et al. 1993). Therefore, once Δm_{B_s} is known, one will get a much better constraint on the CKM matrix from $\Delta m_{B_s}/\Delta m_{B_d}$. At present, there is only a bound
$\Delta m_{B_s}/\Delta m_{B_d} > 21.2$ from LEP, which allows one to find $|1 - \rho - i\eta| < 1.2$. This
is already encroaching on the allowed region in the $\rho - \eta$ plane, cf. eqn (18.28).

18.7 Main conclusions

- The ratio $\Gamma_{12}/M_{12} \sim 10^{-3}$ is very small in the $B_q^0 - \overline{B_q^0}$ systems.

---

[46]This result is especially sensitive to the error bars that one assumes for the various input
parameters and, thus, to the numerical values in eqns (18.27)–(18.29). The numerical values
used and the error analysis vary significantly from one article to the next. As a consequence,
our results should be taken as merely illustrative. In Part IV we shall use the conservative
bound $\sin^2 \gamma > 0.33$ given by Ali and London (1997).

- Mixing CP violation, $\delta \sim 10^{-4}$, is very small too.
- The mass difference $\Delta m_{B_d} \propto |1 - \rho - i\eta|^2$ provides an important constraint on ρ and η.
- The measured values of Δm_{B_d}, of ϵ, and of $|V_{ub}/V_{cb}|$, allow us to determine a closed domain in which ρ and η should lie.
- In the future, the experimental value of $\Delta m_{B_s}/\Delta m_{B_d}$ may also turn out to provide a strong constraint on ρ and η.

19

$K_L \to \pi^0 \nu \bar{\nu}$

19.1 Introduction

The rare kaon decay $K_L \to \pi^0 \nu \bar{\nu}$ may become very important in the study of CP violation. In the context of the Kobayashi–Maskawa model, it gives direct access to the parameter η of the CKM matrix, with little associated theoretical uncertainty—unknown hadronic matrix elements or other poorly known parameters. On the other hand, this decay is extremely challenging for experimentalists, due to the presence of two neutrinos and no charged particle in the final state. The process $K_L \to \pi^0 \nu \bar{\nu}$ has not yet been observed; the 90%-confidence-level experimental bound is (E799 Collaboration 1994)

$$\mathrm{BR}\left(K_L \to \pi^0 \nu \bar{\nu}\right) < 5.8 \times 10^{-5}. \tag{19.1}$$

This is six orders of magnitude above the SM prediction (Buchalla 1997; Buras 1997)

$$\mathrm{BR}\left(K_L \to \pi^0 \nu \bar{\nu}\right) = (2.8 \pm 1.7) \times 10^{-11}, \tag{19.2}$$

and therefore much progress on the experimental side is still needed before we are able to vindicate theory. Inverting the argument, there is ample room for the discovery of new physics by observation of a branching ratio larger than the SM prediction.

In the standard model $K_L \to \pi^0 \nu \bar{\nu}$ violates CP. The argument leading to this conclusion still holds in many extensions of the SM (Grossman and Nir 1997). The first assumption is conservation of the individual lepton numbers. Then, the neutrino and the antineutrino in $K_L \to \pi^0 \nu \bar{\nu}$ have the same flavour, i.e., they are the antiparticle of each other. This is a necessary condition for the final state to be an eigenstate of CP.

If CP is conserved, K_L does not decay into two pions and has CP-parity -1. Then, the CP-parity of $\pi^0 \nu \bar{\nu}$ should be -1 too. In the rest frame of K_L

$$\mathrm{CP}\left(\pi^0 \nu \bar{\nu}\right) = \mathrm{CP}\left(\pi^0\right) \mathrm{CP}\left(\nu \bar{\nu}\right) (-1)^L, \tag{19.3}$$

where L is the relative angular momentum of π^0 and the $\nu \bar{\nu}$ pair. As both K_L and π^0 are spinless, L is equal to J, the total angular momentum of the $\nu \bar{\nu}$ pair. We know—see § 4.2 and 4.4—that π^0 has CP-parity -1. Therefore,

$$\mathrm{CP}\left(\pi^0 \nu \bar{\nu}\right) = - \mathrm{CP}\left(\nu \bar{\nu}\right) (-1)^J. \tag{19.4}$$

One now assumes that the neutrino is left-handed and the antineutrino is right-handed. Then, in the rest frame of the $\nu \bar{\nu}$ pair, the projection of angular momentum on the direction of flight of those massless particles is 1. Moreover, the dominant operator creating the $\nu\bar{\nu}$ pair out of the vacuum is $\bar{\nu}\gamma^\mu\gamma_L\nu$.[47] Let us separate this operator into its time and space components: $\bar{\nu}\gamma^0\gamma_L\nu$ and $\bar{\nu}\vec{\gamma}\gamma_L\nu$ (Kayser 1997). The time component creates a $\nu\bar{\nu}$ pair with $J = 0$, and therefore does not contribute to $K_L \to \pi^0\nu\bar{\nu}$, as one easily sees in the rest frame of the $\nu\bar{\nu}$ pair. The space component has CP$=+1$ (see eqn 3.80) and creates a $\nu\bar{\nu}$ pair with $J = 1$. The product CP $(\nu\bar{\nu})(-1)^J$ is -1. We thus conclude that

$$\text{CP}\left(\pi^0\nu\bar{\nu}\right) = +1. \tag{19.5}$$

As the CP-parities of K_L and of $\pi^0\nu\bar{\nu}$ are different, CP is violated in $K_L \to \pi^0\nu\bar{\nu}$. Notice that eqn (19.5) depends on ν and $\bar{\nu}$ being each other's antiparticle, and on the operator which creates them being $\bar{\nu}\gamma^\mu\gamma_L\nu$; in general, the three-particle state $\pi^0\nu\bar{\nu}$ would not have a well-defined CP-parity.

19.2 $\lambda_{\pi\nu\bar{\nu}}$

In the rest of this chapter we study the SM prediction of BR $\left(K_L \to \pi^0\nu\bar{\nu}\right)$. Some readers may want to skip this.

The transition $K_L \to \pi^0\nu\bar{\nu}$ corresponds to either $K^0 \to \pi^0\nu\bar{\nu}$ or $\overline{K^0} \to \pi^0\nu\bar{\nu}$. In terms of quarks, the first decay is $\bar{s} \to \bar{d}\nu\bar{\nu}$, and the second decay is $s \to d\nu\bar{\nu}$. The standard-model effective Hamiltonian for these transitions has been computed by Inami and Lim (1981)—see Appendix D. It is

$$\mathcal{H}_{\text{eff}} = \frac{g^4}{16\pi^2 m_W^2} \left[\lambda_c X\left(x_c\right) + \lambda_t X\left(x_t\right)\right] \left(\bar{s}\gamma^\mu\gamma_L d\right)\left(\bar{\nu}\gamma_\mu\gamma_L\nu\right) + \text{H.c.}, \tag{19.6}$$

where $\lambda_\alpha \equiv V_{\alpha s}^* V_{\alpha d}$, $x_\alpha \equiv m_\alpha^2/m_W^2$, and

$$X\left(x\right) \equiv \frac{x}{8\left(x-1\right)}\left(x+2+\frac{3x-6}{x-1}\ln x\right). \tag{19.7}$$

With $m_c = 1.25 \pm 0.25$ GeV and $m_t = 175.5 \pm 5.5$ GeV, one has

$$\begin{aligned} X\left(x_c\right) &= (1.44 \pm 0.45) \times 10^{-3}, \\ X\left(x_t\right) &= 1.615 \pm 0.058. \end{aligned} \tag{19.8}$$

We shall also need the Hamiltonian for the tree-level decay $K^+ \to \pi^0 e^+\nu$—in terms of quarks, $\bar{s} \to \bar{u}e^+\nu$—, which is

$$\mathcal{H}'_{\text{eff}} = \frac{g^2}{2m_W^2} V_{us}^* \left(\bar{s}\gamma^\mu\gamma_L u\right)\left(\bar{\nu}\gamma_\mu\gamma_L e\right) + \text{H.c.}. \tag{19.9}$$

---

[47] Other operators, which create the pair in such a state that CP is conserved, may be present. For instance, in the SM the box diagram in Fig. D.1 (Appendix D) yields the operator $\bar{\nu}\gamma^\alpha\gamma_L\partial^\beta\nu - \left(\partial^\beta\bar{\nu}\right)\gamma^\alpha\gamma_L\nu$ when the four-momenta of the external particles are not neglected. However, the coefficients of those operators are suppressed by factors $m_K^2/m_W^2 \sim 10^{-4}$, and therefore these CP-conserving contributions to $K_L \to \pi^0\nu\bar{\nu}$ may safely be neglected. For a detailed account, see Buchalla and Isidori (1998).

The crucial parameter to be computed is

$$\lambda_{\pi\nu\bar\nu} \equiv \frac{q_K}{p_K} \frac{A(\overline{K^0} \to \pi^0\nu\bar\nu)}{A(K^0 \to \pi^0\nu\bar\nu)}. \tag{19.10}$$

Equation (19.6) tells us that

$$\frac{A(\overline{K^0} \to \pi^0\nu\bar\nu)}{A(K^0 \to \pi^0\nu\bar\nu)} = \frac{[\lambda_c^* X(x_c) + \lambda_t^* X(x_t)] \langle \pi^0\nu\bar\nu| (\bar{d}\gamma^\mu\gamma_L s)(\bar\nu\gamma_\mu\gamma_L\nu) |\overline{K^0}\rangle}{[\lambda_c X(x_c) + \lambda_t X(x_t)] \langle \pi^0\nu\bar\nu| (\bar{s}\gamma_\mu\gamma_L d)(\bar\nu\gamma^\mu\gamma_L\nu) |K^0\rangle}$$

$$= \frac{\lambda_c^* X(x_c) + \lambda_t^* X(x_t)}{\lambda_c X(x_c) + \lambda_t X(x_t)} e^{i(\xi_s - \xi_d - \xi_K)}. \tag{19.11}$$

We have used CP symmetry together with eqn (19.5) to evaluate the ratio of matrix elements.

We know the value of q_K/p_K from eqn (17.7). Thus,

$$\lambda_{\pi\nu\bar\nu} = -\frac{\lambda_u}{\lambda_u^*} \sqrt{\frac{1-\delta}{1+\delta}} \frac{u - i\delta}{\sqrt{u^2 + \delta^2}} \frac{\lambda_c^* X(x_c) + \lambda_t^* X(x_t)}{\lambda_c X(x_c) + \lambda_t X(x_t)}$$

$$\approx -\frac{\lambda_u}{\lambda_u^*} \frac{\lambda_c^* X(x_c) + \lambda_t^* X(x_t)}{\lambda_c X(x_c) + \lambda_t X(x_t)}. \tag{19.12}$$

The parameter $\lambda_{\pi\nu\bar\nu}$ is independent of the spurious phases ξ, as it should be.

From eqn (7.25) we know that $\lambda_{\pi\nu\bar\nu} \neq \pm 1$ implies CP violation. This may happen because of indirect CP violation ($|q/p| \neq 1$), direct CP violation ($|\bar{A}/A| \neq 1$), or interference CP violation ($\sin \arg \lambda_{\pi\nu\bar\nu} \neq 0$). In the case at hand there is no direct CP violation; strong final-state-interaction phases are absent, because there is no state scattering strongly to $\pi^0\nu\bar\nu$; absorptive parts of Feynman diagrams could in principle be present, but only when the intermediate quark is the up quark, and the GIM suppression makes it that diagrams with intermediate up quarks hardly contribute to the decay amplitude at all (see Appendix D). There is indirect CP violation ($\delta \approx 3.3 \times 10^{-3}$ does not vanish), but it is very small. The main reason for $\lambda_{\pi\nu\bar\nu} \neq \pm 1$ is interference between mixing and decay: the phases of q/p and of $\bar{A}/A$ do not match, as $\lambda_u\lambda_t^*$ and $\lambda_u\lambda_c^*$ are not real. This is different from CP violation in the two-pion decays of the neutral kaons; there, the phases of the mixing and decay amplitudes are practically equal and CP violation arises almost exclusively from δ. CP violation in $K_L \to \pi^0\nu\bar\nu$ may be much larger precisely because it is mainly interference CP violation; the measured value of δ would lead by itself alone to a branching ratio much smaller than the prediction in eqn (19.2). Thus, if that prediction is experimentally vindicated the superweak theory of Wolfenstein (1964) will be disproved.[48]

---

[48]Some authors would call this 'direct CP violation', because they interpret this expression as meaning any form of CP violation which goes beyond the superweak model. In our terminology, direct CP violation is something different. From our point of view, $K_L \to \pi^0\nu\bar\nu$ originates mainly in interference CP violation, not direct CP violation.

At this point it is convenient to introduce

$$R \equiv X(x_t) \operatorname{Re}(\lambda_t \lambda_u^*) + X(x_c) \operatorname{Re}(\lambda_c \lambda_u^*),$$
$$I \equiv \operatorname{Im}(\lambda_t \lambda_u^*) [X(x_t) - X(x_c)]. \tag{19.13}$$

Then, from eqn (19.12),

$$\lambda_{\pi \nu \bar{\nu}} \approx -\frac{R - iI}{R + iI}. \tag{19.14}$$

Using the Wolfenstein parametrization and the fact that $X(x_t) \gg X(x_c)$, one has

$$R \approx -A^2 \lambda^6 (1 - \rho) X(x_t) - \lambda^2 X(x_c),$$
$$I \approx A^2 \lambda^6 \eta X(x_t). \tag{19.15}$$

19.3   Prediction of the branching ratio

The relevant decay amplitude is

$$A(K_L \to \pi^0 \nu \bar{\nu}) = p_K A(K^0 \to \pi^0 \nu \bar{\nu}) + q_K A(\overline{K^0} \to \pi^0 \nu \bar{\nu})$$
$$= p_K (1 + \lambda_{\pi \nu \bar{\nu}}) A(K^0 \to \pi^0 \nu \bar{\nu}). \tag{19.16}$$

We remind that $|p_K|^2 = (1 + \delta)/2$. Therefore,

$$\begin{aligned}
|A(K_L \to \pi^0 \nu \bar{\nu})|^2 &= \left(\frac{g^4}{16\pi^2 m_W^2}\right)^2 |\langle \pi^0 \nu \bar{\nu}| (\bar{s} \gamma_\mu \gamma_L d)(\bar{\nu} \gamma^\mu \gamma_L \nu) |K^0\rangle|^2 \\
&\quad \times \frac{1 + \delta}{2} |1 + \lambda_{\pi \nu \bar{\nu}}|^2 |\lambda_c X(x_c) + \lambda_t X(x_t)|^2 \\
&\approx \left(\frac{g^4}{16\pi^2 m_W^2}\right)^2 |\langle \pi^0 \nu \bar{\nu}| (\bar{s} \gamma_\mu \gamma_L d)(\bar{\nu} \gamma^\mu \gamma_L \nu) |K^0\rangle|^2 \\
&\quad \times \frac{|\lambda_u^* [\lambda_c X(x_c) + \lambda_t X(x_t)] - \lambda_u [\lambda_c^* X(x_c) + \lambda_t^* X(x_t)]|^2}{2 |\lambda_u|^2} \\
&= \left(\frac{g^4}{16\pi^2 m_W^2}\right)^2 |\langle \pi^0 \nu \bar{\nu}| (\bar{s} \gamma_\mu \gamma_L d)(\bar{\nu} \gamma^\mu \gamma_L \nu) |K^0\rangle|^2 \frac{2I^2}{|\lambda_u|^2}.
\end{aligned} \tag{19.17}$$

We cannot compute the matrix element, but we may equate it, using isospin symmetry, to the matrix element for the tree-level decay $K^+ \to \pi^0 e^+ \nu$. From eqn (19.9),

$$|A(K^+ \to \pi^0 e^+ \nu)|^2 = \left(\frac{g^2}{2m_W^2}\right)^2 |V_{us}|^2 |\langle \pi^0 e^+ \nu| (\bar{s} \gamma^\mu \gamma_L u)(\bar{\nu} \gamma_\mu \gamma_L e) |K^+\rangle|^2. \tag{19.18}$$

Thus,

$$\left| \frac{A\left(K_L \to \pi^0 \nu \bar{\nu}\right)}{A\left(K^+ \to \pi^0 e^+ \nu\right)} \right|^2 = \left(\frac{g^2}{8\pi^2}\right)^2 \frac{2I^2}{|\lambda_u V_{us}|^2}. \tag{19.19}$$

On the other hand,

$$\left| \frac{A\left(K_S \to \pi^0 \nu \bar{\nu}\right)}{A\left(K^+ \to \pi^0 e^+ \nu\right)} \right|^2 = \left(\frac{g^2}{8\pi^2}\right)^2 \frac{2R^2}{|\lambda_u V_{us}|^2}. \tag{19.20}$$

Remembering $g^2 = 4\pi\alpha/s_w^2$, one has

$$\begin{aligned}
\frac{\Gamma\left(K_L \to \pi^0 \nu \bar{\nu}\right)}{\Gamma\left(K^+ \to \pi^0 e^+ \nu\right)} &= 3\frac{\alpha^2}{4\pi^2 s_w^4} \frac{2I^2}{|V_{ud} V_{us}^2|^2}, \\
\frac{\Gamma\left(K_S \to \pi^0 \nu \bar{\nu}\right)}{\Gamma\left(K^+ \to \pi^0 e^+ \nu\right)} &= 3\frac{\alpha^2}{4\pi^2 s_w^4} \frac{2R^2}{|V_{ud} V_{us}^2|^2}.
\end{aligned} \tag{19.21}$$

The factor 3 is because there are three neutrino species. With $|V_{ud}| \approx 1$, $|V_{us}| = \lambda$, and $|V_{cb}| = A\lambda^2$, one may write, because of the second eqn (19.15),

$$\frac{\Gamma\left(K_L \to \pi^0 \nu \bar{\nu}\right)}{\Gamma\left(K^+ \to \pi^0 e^+ \nu\right)} = \frac{3\alpha^2}{2\pi^2 s_w^4} |V_{cb}|^4 \eta^2 \left[X\left(x_t\right)\right]^2. \tag{19.22}$$

19.4 $K^+ \to \pi^+ \nu \bar{\nu}$

The rare decay $K^+ \to \pi^+ \nu \bar{\nu}$ also originates in the effective Hamiltonian in eqn (19.6). Once again, one must compare it to the dominant decay in order to get rid of the unknown matrix element:

$$\begin{aligned}
\frac{\mathrm{BR}\left(K^+ \to \pi^+ \nu \bar{\nu}\right)}{\mathrm{BR}\left(K^+ \to \pi^0 e^+ \nu\right)} &= 3\left(\frac{g^2}{8\pi^2}\right)^2 \left| \frac{\lambda_c X\left(x_c\right) + \lambda_t X\left(x_t\right)}{V_{us}} \right|^2 \\
&\quad \times \left| \frac{\langle \pi^+ \nu \bar{\nu}| \left(\bar{s}\gamma^\mu \gamma_L d\right) \left(\bar{\nu}\gamma_\mu \gamma_L \nu\right) |K^+\rangle}{\langle \pi^0 e^+ \nu| \left(\bar{s}\gamma^\mu \gamma_L u\right) \left(\bar{\nu}\gamma_\mu \gamma_L e\right) |K^+\rangle} \right|^2 \\
&= \frac{3\alpha^2}{2\pi^2 s_w^4} \left| \frac{\lambda_c X\left(x_c\right) + \lambda_t X\left(x_t\right)}{V_{us}} \right|^2, \tag{19.23}
\end{aligned}$$

where we have taken into account that, as $\pi^0 \sim \left(\bar{u}u - \bar{d}d\right)/\sqrt{2}$ while $\pi^+ \sim \bar{d}u$, the ratio of matrix elements is equal to $\sqrt{2}$ when isospin symmetry is exact.

19.5 Explicit values

Before proceeding to explicit numerical predictions, one must take into account various corrections.

Firstly, $X\left(x_t\right)$ receives a QCD correction

$$X\left(x_t\right) \to X\left(x_t\right) + \frac{\alpha_s}{4\pi} X_1\left(x_t\right), \tag{19.24}$$

where the function X_1 has been computed by Buchalla and Buras (1993a,b). In practice, eqn (19.24) amounts to making $X\left(x_t\right) \to 0.985\, X\left(x_t\right)$.

Secondly, there are QCD corrections to $X(x_c)$ too. These are much more uncertain, because for $\mu \sim m_c$ the strong interactions are largely non-perturbative. Besides, the effective Hamiltonian for $K_L \to \pi^0 \nu\bar{\nu}$ depends on the neutrino flavour, because the box diagram has an internal charged-lepton propagator. In Appendix D we have assumed the charged leptons to be massless, but this is not a good approximation for the τ lepton. In practice, the ensuing dependence on m_τ proves to be very small in the case of $X(x_t)$, but is important in $X(x_c)$. Instead of $X(x_c)$ it is better to use $(2/3)X_{\rm NL}^e + (1/3)X_{\rm NL}^\tau$ with the function $X_{\rm NL}^l$ computed by Buchalla and Buras (1994) for any mass m_l of the charged lepton l.[49] In practice,

$$X(x_c) \to \tfrac{2}{3}X_{\rm NL}^e + \tfrac{1}{3}X_{\rm NL}^\tau = (9.5 \pm 1.4) \times 10^{-4}, \qquad (19.25)$$

which is somewhat smaller than $X(x_c)$ in eqn (19.8).

Thirdly, the isospin symmetry used to equate matrix elements is not exact. It is convenient to separate the isospin-breaking corrections (Marciano and Parsa 1996) into three factors. The first factor originates in the different phase space for different decays; the second factor comes from isospin violation in the $K \to \pi$ form factors; the third factor stems from electromagnetic radiative corrections. The latter factor is equal to 0.979 for both rare kaon decays considered; the phase-space correction is 1.0522 for the decay of K_L and 0.9614 for the decay of K^+; the form-factor correction is 0.9166 for the former and 0.9574 for the latter. Thus, $\Gamma(K_L \to \pi^0 \nu\bar{\nu})$ is reduced by $1.0522 \times 0.9166 \times 0.979 = 0.944$ relative to the original computation, while $\Gamma(K^+ \to \pi^+ \nu\bar{\nu})$ is reduced by $0.9614 \times 0.9574 \times 0.979 = 0.901$.

After introducing these corrections we may proceed to the numerical computations. We use BR $(K^+ \to \pi^0 e^+ \nu) = 0.0482$, $\tau(K_L) = 5.17 \times 10^{-8}$ s, $\tau(K^+) = 1.2386 \times 10^{-8}$ s, $\alpha = 1/127.9$, and $s_w^2 = 0.2315$. We get

$$\text{BR}(K_L \to \pi^0 \nu\bar{\nu}) = 3.4878 \times 10^{-5} |V_{cb}|^4 \eta^2 [X(x_t)]^2$$
$$= 2.8 \times 10^{-11}, \qquad (19.26)$$

in which we have used $|V_{cb}| = 0.04$, $\eta = 0.35$, and $X(x_t) = 0.985 \times 1.615$. We thus reproduce the prediction in eqn (19.2). Notice that BR $(K_L \to \pi^0 \nu\bar{\nu})$ depends strongly on $|V_{cb}|$, and is proportional to η^2.

---

[49]This substitution is conceptually wrong. The right computation would involve a sum over the three neutrino flavours of the decay rates, which would be of the form

$$2|A_1 + B|^2 + |A_2 + B|^2,$$

with B the amplitude from the diagrams with intermediate top quarks, and A_1 or A_2 the amplitude from the diagrams with intermediate charm quarks. Instead, we are performing the computation as

$$3\left|\tfrac{2}{3}A_1 + \tfrac{1}{3}A_2 + B\right|^2.$$

However, the error involved is $2/3|A_1 - A_2|^2$, which is negligible in practice.

Now consider the case of the superweak model. In that model the ratios of the decay widths of K_L and of K_S to CP eigenstates with CP-parity $+1$ are all equal—see § 7.3.4. Thus,

$$\frac{\Gamma\left(K_L \to \pi^0 \nu \bar{\nu}\right)}{\Gamma\left(K_S \to \pi^0 \nu \bar{\nu}\right)} = \frac{\Gamma\left(K_L \to 2\pi, I = 0\right)}{\Gamma\left(K_S \to 2\pi, I = 0\right)} = |\epsilon|^2 . \tag{19.27}$$

Comparing eqns (19.21) and (19.27), we see that

$$\mathrm{BR}\left(K_L \to \pi^0 \nu \bar{\nu}\right)_{\mathrm{superweak}} = 3.4878 \times 10^{-5} \frac{R^2}{|V_{ud} V_{us}^2|^2} |\epsilon|^2 . \tag{19.28}$$

We take $|\epsilon|^2 \approx 10^{-5}/2$ from the two-pion decays and, from eqn (19.15),

$$\frac{-R}{|V_{ud} V_{us}^2|} \approx A^2 \lambda^4 \left(1 - \rho\right) X\left(x_t\right) + X\left(x_c\right)$$

$$\approx 3.5 \times 10^{-3}. \tag{19.29}$$

We obtain

$$\mathrm{BR}\left(K_L \to \pi^0 \nu \bar{\nu}\right)_{\mathrm{superweak}} \approx 2 \times 10^{-15}. \tag{19.30}$$

This means that experimental vindication of eqn (19.2) would disprove the superweak theory.

For the charged-kaon decay $K^+ \to \pi^+ \nu \bar{\nu}$ one has, from eqn (19.23), and using the Wolfenstein parametrization,

$$\mathrm{BR}\left(K^+ \to \pi^+ \nu \bar{\nu}\right) = 7.53 \times 10^{-6} \left| \frac{\lambda_c X\left(x_c\right) + \lambda_t X\left(x_t\right)}{V_{us}} \right|^2$$

$$= 7.53 \times 10^{-6} |V_{cb}|^4 \left[X\left(x_t\right)\right]^2$$

$$\times \left\{ \eta^2 + \left[1 - \rho + \frac{X\left(x_c\right)}{|V_{cb}|^2 X\left(x_t\right)}\right]^2 \right\}$$

$$= 9.8 \times 10^{-11}, \tag{19.31}$$

where we have used the same values as above, together with $\rho = 0$ and $X\left(x_c\right) = 9.5 \times 10^{-4}$. A careful analysis yields (Buchalla 1997)

$$\mathrm{BR}\left(K^+ \to \pi^+ \nu \bar{\nu}\right) = (9 \pm 3) \times 10^{-11}. \tag{19.32}$$

The computation of $\mathrm{BR}\left(K^+ \to \pi^+ \nu \bar{\nu}\right)$ has much larger theoretical uncertainties than that of $\mathrm{BR}\left(K_L \to \pi^0 \nu \bar{\nu}\right)$. This is because the value of $X\left(x_c\right)$ is relevant in $K^+ \to \pi^+ \nu \bar{\nu}$ while it is mostly immaterial in $K_L \to \pi^0 \nu \bar{\nu}$. This is a consequence of the CP-violating character of the latter transition; as $K_L \to \pi^0 \nu \bar{\nu}$ violates CP, the top and charm quarks must contribute to it with opposite signs, and the relevant quantity is $X\left(x_t\right) - X\left(x_c\right) \approx X\left(x_t\right)$.

EFFECTIVE HAMILTONIANS

20.1 Current–current operators

We study in this chapter the effective low-energy Hamiltonian which drives the nonleptonic decays of the kaons, as a preliminary for the computation of ϵ'/ϵ in the next chapter. Readers less interested in detailed theoretical computations may want to skip both chapters.

We base our analysis on the review articles by Buchalla *et al.* (1996) and by Buras and Fleischer (1998). The reader should also consult the books by Donoghue *et al.* (1992) and by Weinberg (1995), which contain useful introductory chapters on the operator product expansion, the renormalization group, and the determination of Wilson coefficients. Finally, the TASI93 lectures by Cohen (1994) constitute a very nice and pedagogical introduction to the use of effective Hamiltonians.

Consider the decay $K^0 \to 2\pi$. At tree level, it occurs through the W-exchange diagram in Fig. 20.1 (a). When the masses of the external quarks are neglected, the corresponding amplitude is

$$\text{Fig. 20.1 (a)} = \frac{ig^2}{2(k^2 - m_W^2)} V_{us}^* V_{ud} \, (\bar{s}\gamma^\mu \gamma_L u)(\bar{u}\gamma_\mu \gamma_L d)$$

$$= -i \frac{G_F}{\sqrt{2}} \lambda_u Q_2 + O\left(k^2/m_W^2\right), \tag{20.1}$$

where $\lambda_\alpha \equiv V_{\alpha s}^* V_{\alpha d}$. We have defined

$$Q_2 \equiv (\bar{s}u)_{V-A} (\bar{u}d)_{V-A} . \tag{20.2}$$

Here, V denotes the vector coupling and A denotes the axial-vector coupling, with the notation

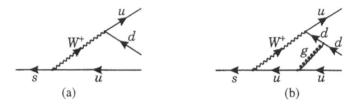

FIG. 20.1. (a) Tree-level quark diagram for $K^0 \to 2\pi$, and (b) a typical QCD-correction diagram.

$$(\bar{q}q')_{V \pm A} \equiv \bar{q}\gamma^{\mu} (1 \pm \gamma_5) \, q'. \tag{20.3}$$

The Lorentz index μ is summed over in the product of two quark bilinears.

Since the momentum k of the internal W-boson line is small compared with m_W, we may neglect the higher-order terms in the k^2/m_W^2 expansion. The resulting amplitude, $-i \left(G_F/\sqrt{2}\right) \lambda_u Q_2$, may be obtained from the effective Hamiltonian

$$\mathcal{H}_{\text{eff}} = \frac{G_F}{\sqrt{2}} \lambda_u Q_2 + \text{H.c.}. \tag{20.4}$$

The higher-order terms in the k^2/m_W^2 expansion may be taken into account by including extra operators in the effective Hamiltonian. Those operators generally involve derivatives of the fields.

One has thus removed the W-boson degree of freedom from the theory.[50] The result is a set of local operators multiplied by effective coupling constants, called Wilson coefficients. The operators do not involve the heavy degrees of freedom—in our elementary example, the W field; information about them is hidden in the Wilson coefficients—in our case, in $G_F/\sqrt{2} = g^2/\left(8m_W^2\right)$.

At this stage, the effective Hamiltonian is just a convenient way of parametrizing the low-energy effects of the full theory. One may compute the relevant processes using either the full theory or the effective Hamiltonian. However, it may be more convenient to use an effective Hamiltonian since only a finite number of operators appear up to a given order. Once the Wilson coefficients are known, the same effective Hamiltonian may be used for a variety of low-energy processes, as has been done, for instance, in the previous chapter.

Next consider the QCD corrections to $K^0 \to 2\pi$. The simplest diagram is depicted in Fig. 20.1 (b). It is computed setting to zero the external masses and momenta, analogously to what was done in Appendices B and D. Due to the zero mass of the gluon, an infrared divergence arises, which must be regulated. The operator

$$\sum_{a=1}^{8} \left[\bar{s}_w \gamma^{\mu} (1 - \gamma_5) \, \lambda^a_{wz} u_z\right] \left[\bar{u}_y \gamma_\mu (1 - \gamma_5) \, \lambda^a_{yx} d_x\right] \tag{20.5}$$

is generated, where λ^a are the Gell-Mann matrices, and x, y, w, and z are colour indices. Using the Fierz transformation in eqn (C.10), we find that the operator in eqn (20.5) is equal to $-(2/3)Q_2 + 2Q_1$, with

$$Q_1 \equiv (\bar{s}_x u_y)_{V-A} \, (\bar{u}_y d_x)_{V-A} \,. \tag{20.6}$$

Thus, in order to take QCD effects into account one needs at least two operators in the effective Hamiltonian, Q_1 and Q_2, with Wilson coefficients C_1 and C_2, respectively:

$$\mathcal{H}_{\text{eff}} = \frac{G_F}{\sqrt{2}} \lambda_u \left(C_1 Q_1 + C_2 Q_2\right) + \text{H.c.}. \tag{20.7}$$

---

[50]This is sometimes referred to as 'integrating out' the degree of freedom. The expression originates in the formal path-integral derivation of the procedure.

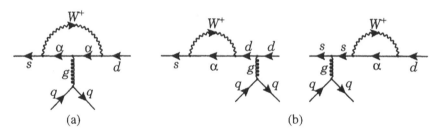

FIG. 20.2. (a) Gluonic penguin for $K^0 \to 2\pi$, and (b) self-energy diagrams which must be added to it in order to obtain a finite, gauge-invariant result.

For obvious reasons, Q_1 and Q_2 are known as current–current operators. At tree level, i.e., without QCD, the Wilson coefficients are $C_1 = 0$ and $C_2 = 1$.

20.2   Penguin operators

Besides the diagram in Fig. 20.1 (b), there are other diagrams involving a gluon which contribute to $K^0 \to 2\pi$. One example is shown in Fig. 20.2 (a). This diagram is known as 'gluonic penguin'. Strictly speaking, one must also include in its computation the self-energy corrections to the external lines, depicted in Fig. 20.2 (b), and only then does one obtain finite effective vertices. In the limit in which all external masses and momenta are set to zero, the gluonic penguin brings into play four new operators:

$$Q_3 \equiv (\bar{s}d)_{V-A} \sum_q (\bar{q}q)_{V-A}, \qquad (20.8)$$

$$Q_4 \equiv (\bar{s}_x d_y)_{V-A} \sum_q (\bar{q}_y q_x)_{V-A}, \qquad (20.9)$$

$$Q_5 \equiv (\bar{s}d)_{V-A} \sum_q (\bar{q}q)_{V+A}, \qquad (20.10)$$

$$Q_6 \equiv (\bar{s}_x d_y)_{V-A} \sum_q (\bar{q}_y q_x)_{V+A}. \qquad (20.11)$$

There are two differences between the gluonic-penguin operators and the current–current operators. The first difference is the appearance of the Dirac structure $(V - A) \times (V + A)$ in some operators. This happens because the gluon couples vectorially, and that coupling may be split into a right-handed and a left-handed part. The second difference is the sum over quark flavours. In particular, we have operators generating the transition $\bar{s} \to d\bar{d}d$ instead of $\bar{s} \to u\bar{u}d$.

Depending on the virtual up-type quark α in the loop in Fig. 20.2, the CKM factor may now be any of the three λ_α, instead of only λ_u. Using eqn (13.53), this may be re-expressed in terms of only two CKM factors, which may conveniently be chosen to be λ_u and λ_t.

The Wilson coefficients depend on the mass m_α of the quark α. (In the computation of the gluonic penguin one may ignore m_α-independent terms, because

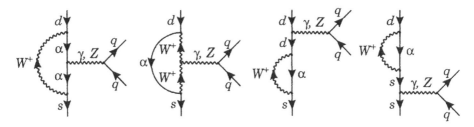

FIG. 20.3. Electroweak-penguin diagrams for $K^0 \to 2\pi$.

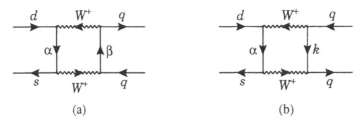

(a)   (b)

FIG. 20.4. Box diagrams which must be added to Fig. 20.3 in order to obtain a gauge-invariant result. The diagram (a) holds in case q is a down-type quark; the diagram (b) is for q an up-type quark.

eqn (13.53) ensures that they give a vanishing contribution. If all up-type-quark masses were equal the gluonic penguin would vanish.) Generically, the coefficients grow with m_α. Explicit calculation of the gluonic penguin with an intermediate top quark yields a dominant logarithmic dependence.

There are other penguin diagrams, in which the gluon is replaced by either the photon or the Z, as shown in Fig. 20.3. They must be computed together with the box diagrams in Fig. 20.4 if we want to obtain a gauge-independent result. (Also, when calculating any diagram involving either W or Z bosons the contributions of the pseudo-Goldstone bosons must be included.) They bring in four new operators,

$$Q_7 \equiv \tfrac{3}{2}\,(\bar{s}d)_{V-A} \sum_q e_q\,(\bar{q}q)_{V+A}\,, \tag{20.12}$$

$$Q_8 \equiv \tfrac{3}{2}\,(\bar{s}_x d_y)_{V-A} \sum_q e_q\,(\bar{q}_y q_x)_{V+A}\,, \tag{20.13}$$

$$Q_9 \equiv \tfrac{3}{2}\,(\bar{s}d)_{V-A} \sum_q e_q\,(\bar{q}q)_{V-A}\,, \tag{20.14}$$

$$Q_{10} \equiv \tfrac{3}{2}\,(\bar{s}_x d_y)_{V-A} \sum_q e_q\,(\bar{q}_y q_x)_{V-A}\,. \tag{20.15}$$

Electroweak penguins with different final-state $q\bar{q}$ pairs acquire different factors e_q, the electric charge of q. Thus, $e_q = 2/3$ for $q = u$, while $e_q = -1/3$ for $q = d$

and for $q = s$.

The electroweak penguins are second-order electroweak effects, and one might expect them to be suppressed with respect to the gluonic penguin. However, some of their Wilson coefficients are considerably enhanced due to the large top-quark mass, when $\alpha = t$. There is an m_t^2 enhancement, like the well-known one found in the process $Z \to b\bar{b}$ (Akundov et al. 1986; Beenakker and Hollik 1988; Bernabéu et al. 1988).

20.3   The effective Hamiltonian

After QCD and QED corrections have been taken into account, the effective Hamiltonian takes the form

$$\mathcal{H}_{\text{eff}} = \frac{G_F}{\sqrt{2}} \left[\lambda_u \sum_{n=1}^{10} z_n(\mu) Q_n - \lambda_t \sum_{n=3}^{10} y_n(\mu) Q_n \right] + \text{H.c.}. \tag{20.16}$$

Equation (20.16) gives the general structure of the effective Hamiltonian. It is the sum of pieces of the type $\left(G_F/\sqrt{2} \right) V_{\text{CKM}} C_n(\mu) Q_n$, each with a definite CKM factor V_{CKM}, which may be either λ_u or λ_t. We have used eqn (13.53) to remove terms proportional to λ_c. Notice that the operators Q_1 and Q_2 always have CKM factor λ_u; no contribution proportional to λ_t exists for those operators.

We have indicated explicitly that there is a dependence of the Wilson coefficients C_n on the renormalization scale μ. The Wilson coefficients are computed by matching the standard model and the effective theory at a scale $\mu \sim m_W$. One thus obtains $C_n(m_W)$. The perturbative evolution of $C_n(\mu)$ is then computed, with the help of renormalization-group techniques, down to scales $\mu \sim 1\,\text{GeV}$. At those scales α_s becomes so large that the perturbative renormalization group breaks down and non-perturbative QCD effects, especially hadronization, must be taken into account.

When computing the evolution of the Wilson coefficients from $\mu \sim m_W$ down to $\mu \sim 1\,\text{GeV}$, one has to face three important technical difficulties: the need to sum large logarithms, the presence of operator mixing, and the existence of quark thresholds.

- Large logarithms arise due to the presence of two very different scales. In fact, there are terms in the perturbative expansion which are proportional to α_s, but there are also terms proportional to $\alpha_s \ln \left(m_W^2/\mu^2 \right)$. With μ sufficiently low, $\alpha_s \ln \left(m_W^2/\mu^2 \right)$ may be large even when α_s is small. The large logarithms must be summed to all orders in α_s. This can be done efficiently using the renormalization group, and results in a 'renormalization-group-improved' perturbative expansion for the Wilson coefficients. At leading order, terms of the type $\alpha_s^n \ln^n \left(m_W^2/\mu^2 \right)$ are summed to all orders; at next-to-leading order, the terms $\alpha_s^n \ln^{n-1} \left(m_W^2/\mu^2 \right)$ are summed to all orders. These approximations are known as leading logarithmic approximation (LLA) and next-to-leading logarithmic approximation (NLLA), respectively.

- The operators mix under renormalization due to the matrix of anomalous dimensions not being diagonal. As a consequence, the relation between the coefficients at different scales is a matrix relation:

$$C_n\left(\mu\right) = \sum_m U\left(\mu, m_W\right)_{nm} C_m\left(m_W\right). \qquad (20.17)$$

Each coefficient at the scale μ depends on many, sometimes on all, coefficients at the scale m_W.

- When evolving the coefficients from $\mu \sim m_W$ down to $\mu \sim 1\,\mathrm{GeV}$ one crosses several quark thresholds. This is accommodated by matching the effective theories above the threshold—with the relevant quark as a degree of freedom—and below the threshold—with that quark integrated out. This is done at a scale μ of the order of the mass of the quark. Below that scale, the operators do not involve that quark any more; thus, the sum over q in operators Q_3–Q_{10} runs over different ranges according to the scale μ at which one is writing down $\mathcal{H}_{\mathrm{eff}}$.

The Wilson coefficients $y_n\left(\mu\right)$ and $z_n\left(\mu\right)$, calculated in the NLLA, may be found in the review by Buchalla *et al.* (1996).

20.4 Calculating amplitudes with the effective Hamiltonian

With the effective Hamiltonian in eqn (20.16) a physical amplitude reads

$$A\left(i \to f\right) = \frac{G_F}{\sqrt{2}} \left[\lambda_u \sum_{n=1}^{10} z_n\left(\mu\right) \langle f|Q_n|i\rangle\left(\mu\right) - \lambda_t \sum_{n=3}^{10} y_n\left(\mu\right) \langle f|Q_n|i\rangle\left(\mu\right) \right].$$
$$(20.18)$$

The μ-dependence of the Wilson coefficients must be offset by the μ-dependence of the matrix elements of the operators. Similarly, any renormalization-scheme dependence of the Wilson coefficients[51] must drop out in the amplitudes. These cancellations may in principle be quite complicated, involving several Wilson coefficients and matrix elements simultaneously.

The crucial feature of the operator product expansion (OPE) is that it allows a separation of two regimes: the hard-gluon contributions are included in the Wilson coefficients; the non-perturbative, soft-gluon effects are included in the hadronic matrix elements of the operators. This is achieved by choosing for the renormalization scale at which the matching of the two regimes is done a value $\mu \sim 1\,\mathrm{GeV}$.[52]

---

[51] This dependence arises because of the different prescriptions on how to deal with the matrix γ_5 in dimensional regularization.

[52] When we shall be dealing with the effective Hamiltonian for bottom-meson decays, in Chapter 32, we will choose $\mu \sim 5\,\mathrm{GeV}$. This is because, there, the mass scale for the hadronization effects is dominated by the bottom-quark mass, and not by the scale at which QCD becomes non-perturbative.

Taking into account eqn (20.17), one may write the product of Wilson coefficients and matrix elements of the corresponding operators as

$$\sum_n C_n\left(\mu\right)\langle f|Q_n|i\rangle\left(\mu\right) = \sum_{n,m}\langle f|Q_n|i\rangle\left(\mu\right) U\left(\mu, m_W\right)_{nm} C_m\left(m_W\right). \qquad (20.19)$$

The transfiguration of the Wilson coefficients due to mixing and matching hides their dependence on the quark masses, notably on the mass of the top quark. Buchalla et $al.$ (1991) have proposed a 'Penguin Box Expansion' (PBE) in order to eliminate this inconvenience and highlight the dependence on m_t. The PBE couples together the first two terms in the right-hand-side of eqn (20.19), and rewrites the $C_m\left(m_W\right)$ in terms of a set of simple process-independent functions.[53]

20.5   Hadronic matrix elements

It only remains to compute the hadronic matrix elements $\langle f|Q_n|i\rangle\left(\mu\right)$. This is the awkward part of the calculation, as would be expected, since the OPE has swept all the hadronization effects into those matrix elements. Eventually they may be reliably calculated in the lattice; at the moment, only rough estimates are possible. So, although the calculations of the Wilson coefficients have become very developed, one still faces huge hadronic uncertainties in the evaluation of decay amplitudes.

The simplest procedure consists in breaking the matrix elements of four-fermion operators into the product of matrix elements of two quark bilinears by inserting the vacuum in all possible ways. This is the vacuum-insertion approximation (VIA), and assumes that that factorization is possible.[54] The resulting matrix elements of quark bilinears are parametrized with form factors, one for each momentum structure consistent with Lorentz invariance and parity. When possible, the form factors are determined directly from experiment.[55]

In the next chapter the matrix elements of the operators Q_1–Q_{10} for the decays of K^0 to $2\pi, I = 0$ and $2\pi, I = 2$ will be needed. As the gluonic-penguin operators Q_3–Q_6 do not differentiate between a final state $u\bar{u}$ and a final state $d\bar{d}$, they only contribute to final states of zero isospin. Thus, $\langle 2|Q_n|K^0\rangle = 0$ for $n = 3$, 4, 5, and 6. The other matrix elements may be computed in the VIA. In an adequate phase convention for $|K^0\rangle$, one obtains (Bertolini et $al.$ 1998b)

---

[53] A similar procedure was used in the calculation of ϵ. The μ-dependence of the coefficients was cancelled by the μ-dependence of the matrix element when defining a scale-independent parameter B_K in eqn (17.18). The same was done in the definition of the parameter B_{B_q} in eqn (18.9).

[54] Gluons can be exchanged between quarks in different bilinears, invalidating factorization. The reader should keep the possible existence of nonfactorizable terms in mind.

[55] An example of this can be found in Appendix C, where $\langle 0|\bar{s}\Gamma^\mu d|K^0\rangle$ is related to $\langle 0|\bar{s}\Gamma^\mu u|K^+\rangle$ by isospin symmetry; the latter matrix element is then parametrized by a parameter f_K, which is determined from experiment.

$$\langle 2|Q_1|K^0\rangle_{\text{VIA}} = \langle 2|Q_2|K^0\rangle_{\text{VIA}} = \tfrac{4\sqrt{6}}{9}X,$$
$$\langle 2|Q_7|K^0\rangle_{\text{VIA}} = -\tfrac{\sqrt{6}}{2}X + \tfrac{\sqrt{6}}{3}Z,$$
$$\langle 2|Q_8|K^0\rangle_{\text{VIA}} = -\tfrac{\sqrt{6}}{6}X + \sqrt{6}Z, \tag{20.20}$$
$$\langle 2|Q_9|K^0\rangle_{\text{VIA}} = \langle 2|Q_{10}|K^0\rangle_{\text{VIA}} = \tfrac{2\sqrt{6}}{3}X,$$

and

$$\langle 0|Q_1|K^0\rangle_{\text{VIA}} = -\tfrac{\sqrt{3}}{9}X,$$
$$\langle 0|Q_2|K^0\rangle_{\text{VIA}} = \tfrac{5\sqrt{3}}{9}X,$$
$$\langle 0|Q_3|K^0\rangle_{\text{VIA}} = \tfrac{\sqrt{3}}{3}X,$$
$$\langle 0|Q_4|K^0\rangle_{\text{VIA}} = \sqrt{3}X,$$
$$\langle 0|Q_5|K^0\rangle_{\text{VIA}} = -\tfrac{4\sqrt{3}}{3}Y,$$
$$\langle 0|Q_6|K^0\rangle_{\text{VIA}} = -4\sqrt{3}Y, \tag{20.21}$$
$$\langle 0|Q_7|K^0\rangle_{\text{VIA}} = \tfrac{\sqrt{3}}{2}X + \tfrac{2\sqrt{3}}{3}Y + \tfrac{2\sqrt{3}}{3}Z,$$
$$\langle 0|Q_8|K^0\rangle_{\text{VIA}} = \tfrac{\sqrt{3}}{6}X + 2\sqrt{3}Y + 2\sqrt{3}Z,$$
$$\langle 0|Q_9|K^0\rangle_{\text{VIA}} = -\tfrac{\sqrt{3}}{3}X,$$
$$\langle 0|Q_{10}|K^0\rangle_{\text{VIA}} = \tfrac{\sqrt{3}}{3}X,$$

where

$$X \equiv f_\pi \left(m_K^2 - m_\pi^2\right),$$
$$Y \equiv \frac{(f_K - f_\pi)\, m_K^4}{(m_s + m_d)^2}, \tag{20.22}$$
$$Z \equiv \frac{f_\pi m_K^4}{(m_s + m_d)^2}.$$

We remind the reader that in our normalization $f_\pi \approx 131\,\text{MeV}$ and $f_K \approx 160\,\text{MeV}$ (Particle Data Group 1996, p. 319).

One may normalize the true matrix elements by their VIA value:

$$\langle 0|Q_n|K^0\rangle = B_n^{(1/2)}\langle 0|Q_n|K^0\rangle_{\text{VIA}},$$
$$\langle 2|Q_n|K^0\rangle = B_n^{(3/2)}\langle 2|Q_n|K^0\rangle_{\text{VIA}}. \tag{20.23}$$

The computation of the matrix elements thus translates into the computation of the values of the bag parameters B_n. By definition, those parameters are 1 in the VIA. The B_n depend both on the renormalization scheme and on the renormalization scale μ.

21

$$\epsilon'/\epsilon$$

21.1 Introduction

In this chapter we review the standard-model computation of

$$\epsilon' \equiv \frac{\langle 2|T|K_L\rangle\langle 0|T|K_S\rangle - \langle 0|T|K_L\rangle\langle 2|T|K_S\rangle}{\sqrt{2}\langle 0|T|K_S\rangle^2}$$

$$\approx \frac{i}{\sqrt{2}}e^{i(\delta_2-\delta_0)}\frac{\text{Im}\,(A_2 A_0^*)}{|A_0|^2}. \tag{21.1}$$

Experimentally, the ratio

$$\frac{\epsilon'}{\epsilon} \approx \frac{\eta_{+-} - \eta_{00}}{2\eta_{+-} + \eta_{00}} \tag{21.2}$$

is more useful. Indeed, η_{+-} and η_{00} are approximately equal:

$$|\eta_{+-}| \approx |\eta_{00}| \approx |\epsilon| = (2.280 \pm 0.013) \times 10^{-3},$$
$$\phi_{+-} \approx \phi_{00} \approx \phi_{\text{sw}} = 43.49°. \tag{21.3}$$

As a consequence, ϵ'/ϵ is small:

$$\frac{\epsilon'}{\epsilon} \approx \tfrac{1}{3}\left(1 - \frac{\eta_{00}}{\eta_{+-}}\right). \tag{21.4}$$

Experimentally, one measures $|\eta_{00}/\eta_{+-}|^2$. In this way only $\text{Re}\,\epsilon'/\epsilon$ is obtained:

$$\text{Re}\,\frac{\epsilon'}{\epsilon} \approx \tfrac{1}{6}\left(1 - \left|\frac{\eta_{00}}{\eta_{+-}}\right|^2\right). \tag{21.5}$$

For many years the experimental situation was unclear—the NA31 Collaboration (1993) and the E731 Collaboration (1997) had produced results that were difficult to conciliate:

$$\text{Re}\,\epsilon'/\epsilon = (23 \pm 3.6 \pm 5.4) \times 10^{-4} \quad (\text{NA31}),$$
$$\text{Re}\,\epsilon'/\epsilon = (7.4 \pm 5.2 \pm 2.9) \times 10^{-4} \quad (\text{E731}). \tag{21.6}$$

This gave no clear evidence for a non-vanishing $\text{Re}\,\epsilon'/\epsilon$. A recent result by the KTeV Collaboration has changed this situation; they have obtained

$$\text{Re}\,\epsilon'/\epsilon = (28 \pm 4.1) \times 10^{-4}. \tag{21.7}$$

It is expected that error bars $\sim 1\text{-}2 \times 10^{-4}$ in $\text{Re}\,\epsilon'/\epsilon$ will be attained by the present generation of experiments (Iconomidou-Fayard 1997).

As emphasized in eqn (8.89), ϵ' originates in the difference between the parameters λ_2 and λ_0. That difference is CP-violating—remember eqn (7.28). In the superweak theory all parameters λ are equal and ϵ' vanishes. Now, it is clear that CP violation in the standard model is not superweak in nature, and ϵ' is in principle non-zero; yet, because of an *accidental cancellation* among the various contributions, ϵ' may turn out to be extremely small. This explains why the superweak theory was difficult to disprove using the measurement of the quantity in eqn (21.5).

The theoretical analysis of ϵ'/ϵ has been subject to increasing refinement for more than twenty years (for a recent review see Bertolini *et al.* 1998*b*). At tree level, both A_0 and A_2 originate in the diagram in Fig. 20.1 (a); they then have the same phase, and $\epsilon' \propto \text{Im}(A_2 A_0^*) = 0$. One must compute the amplitudes at loop level in order to obtain a non-vanishing ϵ'. In the pioneering works of Vainshtein *et al.* (1975, 1977) and of Gilman and Wise (1979) it was recognized that the gluonic penguin plays a central role in generating a *positive* ϵ'/ϵ. The gluonic penguin contributes to A_0 but not to A_2. As it has a component proportional to λ_t, it has a different phase from the tree-level diagram. This contribution to the phase of A_0 generates a non-vanishing ϵ'; it also changes $\epsilon \propto \text{Im}(M_{12} A_0 \bar{A}_0^*)$—$\epsilon$ gets a small correction, proportional to ϵ', when the phases of A_0 and $\bar{A}_0$ change.

Later, due to the realization that the top-quark is very heavy, attention was called to the fact that the electroweak penguins increase with m_t (Bijnens and Wise 1984; Donoghue *et al.* 1986*b*; Buras and Gérard 1987; Lusignoli 1989); when the top quark is sufficiently heavy the electroweak penguins become important. Those diagrams break isospin and therefore they contribute to A_2. It was found that they tend to counter the effect of the gluonic penguin, cancelling the change of the phase of A_0 by a change of the phase of A_2 in the same direction. In this way, ϵ' becomes smaller when the top quark is heavier. This observation became particularly interesting when it was speculated that the top quark might be so heavy ($\sim 200\,\text{GeV}$) that it would lead to a vanishing ϵ'/ϵ, because of the almost complete cancellation between the effects of the electroweak penguins on A_2 and of the gluonic penguin on A_0 (Flynn and Randall 1989; Buchalla *et al.* 1990; Paschos and Wu 1991; Lusignoli *et al.* 1992).

More recently, the next-to-leading order computation of the Wilson coefficients (Buras *et al.* 1992, 1993*a,b,c*; Ciuchini *et al.* 1993, 1994), the experimental determination of the top-quark mass, and increasing efforts to estimate the relevant matrix elements, ushered in a more mature phase in the computation of ϵ'/ϵ. Yet, because of the strong cancellation among the various contributions, and because of the uncertainties in the hadronic matrix elements, an accurate prediction of ϵ'/ϵ is not yet possible (Buras and Fleischer 1998).

It is important to stress that ϵ'/ϵ, being the ratio of two CP-violating parameters, is not in itself CP-violating. (When CP is conserved, both ϵ and ϵ' vanish, and their ratio becomes indeterminate.) Thus, in the standard model ϵ'/ϵ *is not proportional to* J, and it neither tends to zero when J tends to zero, nor does it increase when J increases. The usual theoretical computation of ϵ'/ϵ regrettably distorts this fact, in that what is really computed is ϵ', which is of course proportional to J, while the value of ϵ is just inputted from experiment. If one takes into account the fact that the phases of ϵ and ϵ' should be approximately equal—see Chapter 8—one derives from eqn (21.1) that

$$\frac{\epsilon'}{\epsilon} \approx \frac{\operatorname{Im}(A_2 A_0^*)}{\sqrt{2}\,|\epsilon A_0^2|}. \tag{21.8}$$

As $\operatorname{Im}(A_2 A_0^*) \propto J$, some authors state that $\epsilon'/\epsilon \propto J$ too. This statement is not correct.

On the other hand, the value of J is found, in practice, from the fit of ϵ. As $\epsilon \propto B_K J$, when fitting ϵ one obtains a value of J which is inversely proportional to the inputed value of B_K. One may then correctly state that $\epsilon'/\epsilon \propto 1/B_K$.

21.2   Master formula for ϵ'/ϵ

Let us start from the simple observation that

$$|\omega| = \frac{\operatorname{Re}(A_2 A_0^*)}{|A_0|^2}$$
$$= \frac{\operatorname{Re}(A_2\lambda_u^*)\operatorname{Re}(A_0\lambda_u^*) + \operatorname{Im}(A_2\lambda_u^*)\operatorname{Im}(A_0\lambda_u^*)}{[\operatorname{Re}(A_0\lambda_u^*)]^2 + [\operatorname{Im}(A_0\lambda_u^*)]^2}$$
$$= 0.045, \tag{21.9}$$
$$\epsilon' \propto \frac{\operatorname{Im}(A_2 A_0^*)}{|A_0|^2}$$
$$= \frac{\operatorname{Im}(A_2\lambda_u^*)\operatorname{Re}(A_0\lambda_u^*) - \operatorname{Re}(A_2\lambda_u^*)\operatorname{Im}(A_0\lambda_u^*)}{[\operatorname{Re}(A_0\lambda_u^*)]^2 + [\operatorname{Im}(A_0\lambda_u^*)]^2}. \tag{21.10}$$

The dominant contributions to the amplitudes are proportional to λ_u; one thus expects $|\operatorname{Re}(A_I\lambda_u^*)| \gg |\operatorname{Im}(A_I\lambda_u^*)|$. Then,

$$|\omega| \approx \frac{\operatorname{Re}(A_2\lambda_u^*)}{\operatorname{Re}(A_0\lambda_u^*)} \tag{21.11}$$

and

$$\frac{\epsilon'}{\epsilon} \approx \frac{\operatorname{Im}(A_2\lambda_u^*) - |\omega|\operatorname{Im}(A_0\lambda_u^*)}{\sqrt{2}\,|\epsilon|\operatorname{Re}(A_0\lambda_u^*)}$$
$$\approx \frac{\operatorname{Im}(A_2\lambda_u^*) - |\omega|\operatorname{Im}(A_0\lambda_u^*)}{\sqrt{2}\,|\epsilon A_0\lambda_u|}. \tag{21.12}$$

Following eqn (20.18),

$$A_I e^{i\delta_I} = \frac{G_F}{\sqrt{2}} \sum_{n=1}^{10} [\lambda_u z_n(\mu) - \lambda_t y_n(\mu)]\,\langle I|Q_n|K^0\rangle(\mu), \tag{21.13}$$

where $y_1(\mu) = y_2(\mu) \equiv 0$. The final-state strong-interaction phases δ_I arise from complex matrix elements $\langle I|Q_n|K^0\rangle$. Thus,

$$\text{Im}\,(A_I\lambda_u^*)\cos\delta_I = -\frac{G_F}{\sqrt{2}}\text{Im}\,(\lambda_t\lambda_u^*)\sum_{n=3}^{10} y_n\,(\mu)\,\text{Re}\,\langle I|Q_n|K^0\rangle\,(\mu). \qquad (21.14)$$

Then, from eqn (21.12),

$$\frac{\epsilon'}{\epsilon} = \frac{G_F J}{2\,|\epsilon A_0\lambda_u|}\sum_{n=3}^{10} y_n\,(\mu)\left[\frac{|\omega|\,\text{Re}\,\langle 0|Q_n|K^0\rangle\,(\mu)}{\cos\delta_0} - \frac{\text{Re}\,\langle 2|Q_n|K^0\rangle\,(\mu)}{\cos\delta_2}\right]. \qquad (21.15)$$

Using the Wolfenstein parametrization and the values of $|\omega|$ in eqn (21.9), of $|\epsilon|$ in eqn (21.3), and of $|A_0|$ in eqn (8.51), one obtains

$$\frac{\epsilon'}{\epsilon} \approx \frac{A^2\lambda^5\eta}{5.43\times10^{-6}\,\text{MeV}^3}\sum_{n=3}^{10} y_n\,(\mu)\left[\frac{0.045\,\text{Re}\,\langle 0|Q_n|K^0\rangle\,(\mu)}{\cos\delta_0}\right.$$
$$\left. - \frac{\text{Re}\,\langle 2|Q_n|K^0\rangle\,(\mu)}{\cos\delta_2}\right]. \qquad (21.16)$$

At present, it is not possible to perform a reliable theoretical computation of $|A_0|$ and $|A_2|$. In particular, the smallness of $|\omega|$ reflects the $|\Delta I| = 1/2$ rule, which one is unable to explain fully and consistently, presumably because its origin lies in non-perturbative hadronic physics. This is the reason why we have inputted the experimental values of $|\omega|$ and of $|A_0|$ in eqn (21.16). On the other hand, we trust the computation of $\text{Im}\,(A_I\lambda_u^*)$ in eqn (21.14), which is equivalent to saying that we believe that the unknown contributions to A_0 and to A_2 should be proportional to λ_u. This is reasonable, because only λ_u should intervene in the low-energy, long-distance part of the strong interactions, which presumably is responsible for our inability to obtain the moduli of the decay amplitudes.

In eqn (21.16) notice that one only needs the coefficients y_n for the computation of ϵ'/ϵ. The coefficients z_n are unnecessary. The matrix elements of Q_1 and Q_2 are not necessary either.

In eqn (21.16) we have taken into account the fact that the matrix elements $\langle I|Q_n|K^0\rangle$ will in general have an absorptive part, i.e., be complex. Their complex values build up the final-state strong-interaction phases δ_I. Correspondingly, we have introduced the cosines of those phases as denominators. We thus follow, for the sake of generality, the Trieste group (Bertolini *et al.* 1996, 1998b). Other groups omit this detail, implicitly setting $\cos\delta_0 = \cos\delta_2 = 1$. A fit of the experimental data (Basdevant *et al.* 1974, 1975; Froggatt and Petersen 1977) yields $\delta_0 = 37° \pm 3°$ and $\delta_2 = -7° \pm 1°$. Thus, $\cos\delta_2 \approx 1$ but $\cos\delta_0 \approx 0.8$. This effect tends to enhance $\langle 0|Q_n|K^0\rangle$ relative to $\langle 2|Q_n|K^0\rangle$. This is one reason why the Trieste group obtains a higher value for ϵ'/ϵ than other groups.

One still needs to introduce a further refinement in eqn (21.16). The different masses of the up and down quarks lead to an isospin-breaking mixing between the mesons π^0, η, and η'. This mixing generates a contribution to A_2 proportional to A_0. This may be parametrized by writing, instead of eqn (21.16),

$$\frac{\epsilon'}{\epsilon} = \frac{A^2\lambda^5\eta}{5.43\times10^{-6}\,\text{MeV}^3}\sum_{n=3}^{10} y_n\,(\mu)\left[\frac{0.045\,\text{Re}\,\langle 0|Q_n|K^0\rangle\,(\mu)}{\cos\delta_0}\,(1-\Omega_{\eta+\eta'})\right.$$

$$-\frac{\text{Re}\,\langle 2|Q_n|K^0\rangle\,(\mu)}{\cos\delta_2}\Bigg],\qquad\qquad (21.17)$$

where (Bertolini *et al.* 1998*b*)

$$\Omega_{\eta+\eta'} = 0.25 \pm 0.10. \qquad\qquad (21.18)$$

Equation (21.17) is the master formula for ϵ'/ϵ.

21.3 Matrix elements

Using the values in eqns (15.11) and (15.21), and $\eta \approx 0.35$ as found in § 18.5, one obtains

$$A^2\lambda^5\eta = \left|V_{us}V_{cb}^2\right|\eta \approx 1.3 \times 10^{-4}. \qquad\qquad (21.19)$$

Extensive tables for the coefficients $y_n\,(\mu)$ can be found in Buchalla *et al.* (1996). As for the final-state phases, one may use $\cos\delta_0 = 0.8$ and $\cos\delta_2 = 1$, as advocated by the Trieste group (Bertolini *et al.* 1998*b*), or one may neglect the issue altogether and use $\cos\delta_0 = \cos\delta_2 = 1$, as other experts usually do. It remains to discuss the matrix elements. This is the crucial issue.

There are many approaches to the problem. Buras *et al.* (1993*c*) have tried to determine as many matrix elements as possible from the experimental data on the CP-conserving decays $K \to 2\pi$. However, the matrix elements of the $(V - A) \times (V + A)$ operators Q_5–Q_8 cannot be constrained in this way, and one must rely on theoretical methods to evaluate them. This is regrettable, because the operators Q_6 and Q_8 have large Wilson coefficients and their matrix elements are fundamental in the computation of ϵ'/ϵ.

The strength of the lattice approach is precisely the direct computation of the crucial parameters $B_6^{(1/2)}$ and $B_8^{(3/2)}$; from Bernard and Soni (1989), Franco *et al.* (1989), Kilcup (1991), and Sharpe (1991) one gathers that those parameters are both equal to 1.0 ± 0.2. However, this value for $B_6^{(1/2)}$ has been questioned and may suffer large corrections, even in the quenched approximation (Gupta 1998). In the chiral quark model (Bertolini *et al.* 1996, 1998*a*) $B_6^{(1/2)} = 1.0 \pm 0.4$ and $B_8^{(3/2)} = 0.92 \pm 0.02$. In general that model predicts $B_8^{(3/2)} < B_6^{(1/2)}$, as advocated also by Heinrich *et al.* (1992).

Still, various problems remain. The values quoted for the matrix elements are taken at different scales μ. The methods used to evaluate the matrix elements are at present unable to give their μ-dependence and renormalization-scheme-dependence. Because of these problems, the different approaches cannot be directly compared. It should also be pointed out that most approaches do not predict the imaginary parts of the matrix elements.

The parameters $B_{5,6}^{(1/2)}$ and $B_{7,8}^{(3/2)}$ depend only very weakly on μ (Buras *et al.* 1993*c*). Therefore, the μ-dependence of $\langle 0|Q_{5,6}|K^0\rangle\,(\mu)$ and of $\langle 2|Q_{7,8}|K^0\rangle\,(\mu)$

is dominated by the μ-dependence of the strange-quark mass, see eqns (20.21)–(20.23). At present, the value of $m_s(\mu)$ constitutes an important source of uncertainty in the calculation of the matrix elements, though future lattice computations might eliminate this issue of m_s. For the moment, we have to take that uncertainty into account. In particular, one may use a 'high' m_s, say,

$$m_s = (150 \pm 20)\ \text{MeV}, \qquad (21.20)$$

as found by Allton et al. (1994), Jamin and Münz (1995), Chetyrkin et al. (1995), and Narison (1995); or one may prefer a 'low' m_s,

$$m_s = (100 \pm 20)\ \text{MeV}, \qquad (21.21)$$

as advocated by Onogi et al. (1997) and by Gupta and Bhattacharya (1997). The choice between eqns (21.20) and (21.21) has disturbing implications for the theoretical prediction of ϵ'/ϵ. Of course, this effect should disappear when the matrix elements are computed directly from the lattice.

21.4   Final result

The final predictions for ϵ'/ϵ are the following. The Munich group (Buras et al. 1996) gives

$$\begin{aligned} -1.2 \times 10^{-4} &\leq \epsilon'/\epsilon \leq 16.0 \times 10^{-4} \quad \text{for } m_s = (150 \pm 20)\ \text{MeV}, \\ 0 \times 10^{-4} &\leq \epsilon'/\epsilon \leq 43.0 \times 10^{-4} \quad \text{for } m_s = (100 \pm 20)\ \text{MeV}. \end{aligned} \qquad (21.22)$$

The Rome group has used lattice methods to evaluate the matrix elements and has given (Ciuchini 1997)

$$\epsilon'/\epsilon = (4.6 \pm 3.0 \pm 0.4) \times 10^{-4}. \qquad (21.23)$$

The Trieste group used the chiral quark model to compute the matrix elements and found (Bertolini et al. 1998a)

$$\epsilon'/\epsilon = \left(1.7^{+1.4}_{-1.0}\right) \times 10^{-3}. \qquad (21.24)$$

It is clear that, although all three groups agree that ϵ'/ϵ is most probably positive and not larger than 3×10^{-3}, there is still a long way to go before a reasonably precise prediction may be claimed. Thus, it is not to be expected that the forthcoming high-precision experimental results on $\text{Re}\,\epsilon'/\epsilon$ will allow any constraint on the CKM matrix to be derived.

For more details, the reader may consult the reviews by Buras and Fleischer (1998) and by Bertolini et al. (1998b), which we have extensively used in writing this account.

Part III

CP violation beyond the standard model

MULTI-HIGGS-DOUBLET MODELS

22.1 Introduction

The scalar sector of the standard model (SM) consists of only one doublet with weak hypercharge $Y = 1/2$. Most extensions of the SM include an enlargement of the Higgs sector. There are many theoretical motivations to enlarge the scalar sector of the standard electroweak theory, even if one only considers extensions of that theory based on the standard $SU(3)_c \otimes SU(2) \otimes U(1)$ gauge group. Among the specially important theoretical motivations, one may include:

Supersymmetry—in a supersymmetric extension of the SM (for a review, see e.g. Nilles 1984) a minimum of two Higgs doublets, with weak hypercharges $Y = 1/2$ and $Y = -1/2$, are necessary.[56] This is done, on the one hand because of the need to give masses to both the up-type and the down-type quarks, on the other hand in order to eliminate the gauge anomalies generated by the fermionic supersymmetric partners of the scalars.

Spontaneous CP violation—if one wishes to have CP as a good symmetry of the Lagrangian, only broken by the vacuum, then an extension of the Higgs sector is required. This will be explained in detail in the next chapter, where specific examples are presented.

Strong CP problem—most of the proposed solutions for this problem (see Chapter 27), and in particular the Peccei–Quinn solution in any of its variations, require an enlargement of the Higgs sector.

Baryogenesis—one of the exciting features of the electroweak gauge theories is the fact that they have all the necessary ingredients (Sakharov 1967)— namely baryon-number violation, C and CP violation, and departure from thermal equilibrium—to generate a net baryon asymmetry in the early Universe. However, it is by now clear that the SM cannot provide the observed baryon asymmetry, for various reasons which include

1. the fact that the electroweak phase transition is not strongly first order (Anderson and Hall 1992; Buchmüller *et al.* 1994; Kajantie *et al.* 1996), and as a result any baryon asymmetry generated during the transition would be subsequently washed out by unsuppressed B-violating processes in the broken phase;

---

[56]In a non-supersymmetric theory, a scalar doublet with $Y = 1/2$ is equivalent to a scalar doublet with $Y = -1/2$, cf. eqns (11.15) and (11.16). In a supersymmetric theory this is not true any more, because each scalar multiplet belongs to a chiral supermultiplet which also includes a fermion multiplet of definite chirality. Indeed, the C-conjugate of a left-handed fermion is a right-handed antifermion, and not a left-handed antifermion.

2. CP-violating effects arising through the Kobayashi–Maskawa mechanism in the three-generation SM are too small (Gavela *et al.* 1994; Huet and Sather 1995).

Therefore, the need of having extra sources of CP violation which could lead to a successful baryogenesis is an important motivation to consider physics beyond the SM. As will be seen in this chapter and in the following one, the enlargement of the Higgs sector is one of the simplest ways of having new sources of CP violation beyond the Kobayashi–Maskawa mechanism.

No fundamental scalars have yet been experimentally observed, and therefore at present one only has experimental bounds on the masses and coupling constants of the scalar sector. The experimental search—see e.g. Gunion *et al.* (1990)—for Higgs particles is one of the most important tasks of particle physics.

An important constraint on the enlargement of the Higgs sector arises from the experimentally well-established relationship $m_W = c_W m_Z$. This equality holds at classical level if the scalar fields which get a vacuum expectation value (VEV) are either singlets of SU(2)⊗U(1)—whose VEVs contribute neither to m_W nor to m_Z—or the neutral components of doublets of SU(2). Almost any other neutral scalar getting a VEV will make $m_W \neq c_W m_Z$ at tree level.[57] Hence their VEVs must be sufficiently small. Thus, from all types of scalar multiplets that we may think of adding to the SM, two are outstanding: SU(2) doublets with $Y = \pm 1/2$, and SU(2) singlets with $Y = 0$. Both types of multiplets have the advantage of having relatively few components; in the case of doublets, there is the added advantage that they may have Yukawa couplings to the usual fermions, allowing some interesting effects to arise.

In this chapter we dwell on multi-Higgs-doublet models (MHDMs). These models have gauge group SU(2)⊗U(1) and the usual fermion content: n_g families of left-handed doublets Q_L and L_L and of right-handed singlets p_R, n_R, and l_R. The scalar sector of the model consists of $n_d > 1$ doublets ϕ_a ($a = 1, 2, \ldots, n_d$) with $Y = 1/2$. Thus,

$$\phi_a = \begin{pmatrix} \varphi_a^+ \\ \varphi_a^0 \end{pmatrix}. \tag{22.1}$$

Then,

$$\tilde{\phi}_a \equiv i\tau_2 \phi_a^{\dagger T} = \begin{pmatrix} \varphi_a^{0\dagger} \\ -\varphi_a^- \end{pmatrix} \tag{22.2}$$

are doublets of SU(2) with $Y = -1/2$.

In the next four sections we study the general features of the MHDMs. We give special attention to the two-Higgs-doublet model (THDM), which is the object of § 22.3 and 22.5. We start the study of CP violation in multi-Higgs-doublet models in § 22.6.

---

[57]In general (Tsao 1980), Higgs multiplets with weak isospin T and weak hypercharge Y lead to the relation $m_W = c_W m_Z$ provided $T(T+1) = 3Y^2$. Solutions to this equation apart from $T = Y = 0$ and $T = Y = 1/2$ are usually not considered, because they correspond to large scalar multiplets which cannot have Yukawa couplings to the known fermions.

22.2   General multi-Higgs-doublet model

The Yukawa Lagrangian reads

$$\mathcal{L}_Y = -\overline{Q_L}\left(\Gamma_a \phi_a n_R + \Delta_a \tilde{\phi}_a p_R\right) - \overline{L_L}\Pi_a \phi_a l_R + \text{H.c.}. \tag{22.3}$$

The matrices Γ_a, Δ_a, and Π_a have dimension $n_g \times n_g$. A sum over a from 1 to n_d is implicit in eqn (22.3). The scalar potential is

$$V = Y_{ab}\phi_a^\dagger \phi_b + Z_{abcd}\left(\phi_a^\dagger \phi_b\right)\left(\phi_c^\dagger \phi_d\right). \tag{22.4}$$

Once again, sums from 1 to n_d over the indices a, b, c, and d are implicit. The coefficients Y_{ab} have dimensions of mass squared; the coefficients Z_{abcd} are dimensionless. We assume

$$Z_{abcd} = Z_{cdab} \tag{22.5}$$

without loss of generality. Hermiticity of V implies

$$Y_{ab} = Y_{ba}^*, \\ Z_{abcd} = Z_{badc}^*. \tag{22.6}$$

Hence, there are n_d^2 independent real parameters in the quadratic couplings Y_{ab}, while the quartic couplings Z_{abcd} are parametrized by $n_d^2\left(n_d^2+1\right)/2$ real quantities.

We *assume* that the vacuum preserves a U(1) gauge symmetry corresponding to electromagnetism. Thus,

$$\langle 0|\phi_a|0\rangle = \begin{pmatrix} 0 \\ v_a e^{i\theta_a} \end{pmatrix}, \tag{22.7}$$

with the v_a real and non-negative. Without loss of generality, we may use a U(1) gauge transformation to make the VEV of φ_1^0 real and positive, just as in the SM. We shall assume this from now on. Thus, $\theta_1 = 0$.

We write

$$\phi_a = e^{i\theta_a}\begin{pmatrix} \varphi_a^+ \\ v_a + (\rho_a + i\eta_a)/\sqrt{2} \end{pmatrix}, \\ \tilde{\phi}_a = e^{-i\theta_a}\begin{pmatrix} v_a + (\rho_a - i\eta_a)/\sqrt{2} \\ -\varphi_a^- \end{pmatrix}. \tag{22.8}$$

We define the $n_d \times n_d$ Hermitian matrix V_{ab} as

$$V_{ab} \equiv v_a v_b e^{i(\theta_a - \theta_b)}. \tag{22.9}$$
$$V_{ab} = V_{ba}^*. \tag{22.10}$$

Let us define

$$v = \sqrt{v_1^2 + v_2^2 + \cdots + v_{n_d}^2} > 0. \tag{22.11}$$

As m_W^2 and m_Z^2 receive additive contributions from $v_1^2, v_2^2, \ldots, v_{n_d}^2$, we find that v is related to m_W and m_Z by the same eqns (11.18) as in the SM.

22.3   The two-Higgs-doublet model

The simplest example of a MHDM is the THDM, in which only two scalar doublets, ϕ_1 and ϕ_2, are introduced. The most general renormalizable scalar potential invariant under SU(2)⊗U(1) then is

$$
\begin{aligned}
V = {} & m_1\phi_1^\dagger\phi_1 + m_2\phi_2^\dagger\phi_2 + m_3\left(e^{i\delta_3}\phi_1^\dagger\phi_2 + e^{-i\delta_3}\phi_2^\dagger\phi_1\right) \\
& + a_1\left(\phi_1^\dagger\phi_1\right)^2 + a_2\left(\phi_2^\dagger\phi_2\right)^2 + a_3\left(\phi_1^\dagger\phi_1\right)\left(\phi_2^\dagger\phi_2\right) + a_4\left(\phi_1^\dagger\phi_2\right)\left(\phi_2^\dagger\phi_1\right) \\
& + a_5\left[e^{i\delta_5}\left(\phi_1^\dagger\phi_2\right)^2 + e^{-i\delta_5}\left(\phi_2^\dagger\phi_1\right)^2\right] + a_6\left(\phi_1^\dagger\phi_1\right)\left(e^{i\delta_6}\phi_1^\dagger\phi_2 + e^{-i\delta_6}\phi_2^\dagger\phi_1\right) \\
& + a_7\left(\phi_2^\dagger\phi_2\right)\left(e^{i\delta_7}\phi_1^\dagger\phi_2 + e^{-i\delta_7}\phi_2^\dagger\phi_1\right).
\end{aligned}
\tag{22.12}
$$

The coupling constants m_i (with i from 1 to 3) and a_j (with j from 1 to 7) are real; all phases have been explicitly displayed and, as a matter of fact, m_3, a_5, a_6, and a_7 may be taken to be non-negative without loss of generality.

In the language of the previous section, the tensors Y_{ab} and Z_{abcd} are given by

$$
\begin{aligned}
Y_{11} &= m_1, & Y_{12} &= m_3 e^{i\delta_3}, \\
Y_{21} &= m_3 e^{-i\delta_3}, & Y_{22} &= m_2;
\end{aligned}
\tag{22.13}
$$

$$
\begin{aligned}
Z_{1111} &= a_1, & Z_{2222} &= a_2, \\
Z_{1122} = Z_{2211} &= a_3/2, & Z_{1221} = Z_{2112} &= a_4/2, \\
Z_{1212} &= a_5 e^{i\delta_5}, & Z_{2121} &= a_5 e^{-i\delta_5}, \\
Z_{1112} = Z_{1211} &= a_6 e^{i\delta_6}/2, & Z_{1121} = Z_{2111} &= a_6 e^{-i\delta_6}/2, \\
Z_{2212} = Z_{1222} &= a_7 e^{i\delta_7}/2, & Z_{2221} = Z_{2122} &= a_7 e^{-i\delta_7}/2.
\end{aligned}
\tag{22.14}
$$

The VEVs are given by

$$
\langle 0|\phi_1|0\rangle = \begin{pmatrix} 0 \\ v_1 \end{pmatrix}, \quad \langle 0|\phi_2|0\rangle = \begin{pmatrix} 0 \\ v_2 e^{i\theta} \end{pmatrix},
\tag{22.15}
$$

with v_1 and v_2 real and positive by definition. The expectation value of the potential in the vacuum is

$$
\begin{aligned}
V_0 \equiv \langle 0|V|0\rangle = {} & m_1 v_1^2 + m_2 v_2^2 + 2m_3 v_1 v_2 \cos\left(\delta_3 + \theta\right) \\
& + a_1 v_1^4 + a_2 v_2^4 + (a_3 + a_4) v_1^2 v_2^2 + 2a_5 v_1^2 v_2^2 \cos\left(\delta_5 + 2\theta\right) \\
& + 2a_6 v_1^3 v_2 \cos\left(\delta_6 + \theta\right) + 2a_7 v_1 v_2^3 \cos\left(\delta_7 + \theta\right).
\end{aligned}
\tag{22.16}
$$

The stability of the vacuum requires that

$$
\begin{aligned}
0 &= \frac{-1}{2v_1 v_2}\frac{\partial V_0}{\partial\theta} \\
&= m_3 \sin\left(\delta_3 + \theta\right) + 2a_5 v_1 v_2 \sin\left(\delta_5 + 2\theta\right)
\end{aligned}
$$

$$+a_6 v_1^2 \sin\left(\delta_6 + \theta\right) + a_7 v_2^2 \sin\left(\delta_7 + \theta\right). \qquad (22.17)$$

The quark Yukawa Lagrangian is

$$\mathcal{L}_Y^{(q)} = -\overline{Q_L}\left[\left(\Gamma_1 \phi_1 + \Gamma_2 \phi_2\right) n_R + \left(\Delta_1 \tilde{\phi}_1 + \Delta_2 \tilde{\phi}_2\right) p_R\right] + \text{H.c..} \qquad (22.18)$$

The quark mass matrices are

$$\begin{aligned} M_p &= v_1 \Delta_1 + v_2 \Delta_2 e^{-i\theta}, \\ M_n &= v_1 \Gamma_1 + v_2 \Gamma_2 e^{i\theta}. \end{aligned} \qquad (22.19)$$

They are bi-diagonalized in the usual way, see eqns (12.14) and (12.15), and the CKM matrix is in eqn (12.19).

22.4   The Higgs basis

In any MHDM there is an advantageous basis for the scalar doublets, which we shall refer to as 'the Higgs basis'. We use for the doublets in the Higgs basis the notation $H_1, H_2, \ldots, H_{n_d}$. The Higgs basis is defined in the following way: the doublet H_1 has VEV v; all other doublets have zero VEV. The defining property of the Higgs basis is that only H_1 has a VEV, which is real and positive.

The Higgs basis is obtained by means of a unitary transformation—a weak-basis transformation—of the original scalar doublets $\phi_1, \phi_2, \ldots, \phi_{n_d}$, which mixes them without altering the gauge-kinetic Lagrangian. However, the Higgs basis is not completely well defined, because one has the freedom to redefine the doublets with vanishing VEV by means of an $(n_d - 1) \times (n_d - 1)$ unitary transformation.

The Higgs basis is useful because, when using it, the Goldstone bosons are isolated as components of H_1. Thus,

$$\begin{aligned} H_1 &= \begin{pmatrix} \varphi^+ \\ v + (H + i\chi)/\sqrt{2} \end{pmatrix}, \\ \tilde{H}_1 &= \begin{pmatrix} v + (H - i\chi)/\sqrt{2} \\ -\varphi^- \end{pmatrix}, \end{aligned} \qquad (22.20)$$

just as in eqns (11.15) and (11.16). The fields $\varphi^\pm$ and χ are the Goldstone bosons. Contrary to what happens in the SM, the Hermitian neutral field H is not, in general, an eigenstate of mass, rather it mixes with the neutral components of $H_2, H_3, \ldots, H_{n_d}$.

22.5   The Higgs basis in the THDM

In the THDM, with the VEVs given by eqn (22.15), one reaches the Higgs basis by performing the following unitary transformation of the scalar multiplets:

$$\begin{pmatrix} H_1 \\ H_2 \end{pmatrix} = \frac{1}{v} \begin{pmatrix} v_1 & v_2 \\ v_2 & -v_1 \end{pmatrix} \begin{pmatrix} \phi_1 \\ e^{-i\theta}\phi_2 \end{pmatrix}. \qquad (22.21)$$

We write

$$H_2 = \begin{pmatrix} C^+ \\ (N + iA)/\sqrt{2} \end{pmatrix},$$

$$\tilde{H}_2 = \begin{pmatrix} (N - iA)/\sqrt{2} \\ -C^- \end{pmatrix}. \tag{22.22}$$

The Hermitian fields N and A mix with H to form the three physical neutral scalars S_1, S_2, and S_3, as we shall soon see.

Thus,

$$\begin{pmatrix} \varphi^+ \\ C^+ \end{pmatrix} = \frac{1}{v} \begin{pmatrix} v_1 & v_2 \\ v_2 & -v_1 \end{pmatrix} \begin{pmatrix} \varphi_1^+ \\ \varphi_2^+ \end{pmatrix},$$

$$\begin{pmatrix} H \\ N \end{pmatrix} = \frac{1}{v} \begin{pmatrix} v_1 & v_2 \\ v_2 & -v_1 \end{pmatrix} \begin{pmatrix} \rho_1 \\ \rho_2 \end{pmatrix}, \tag{22.23}$$

$$\begin{pmatrix} \chi \\ A \end{pmatrix} = \frac{1}{v} \begin{pmatrix} v_1 & v_2 \\ v_2 & -v_1 \end{pmatrix} \begin{pmatrix} \eta_1 \\ \eta_2 \end{pmatrix}.$$

22.5.1   The potential

The scalar potential in the Higgs basis is

$$V = \mu_1 H_1^\dagger H_1 + \mu_2 H_2^\dagger H_2 + \left(\mu_3 H_1^\dagger H_2 + \text{H.c.} \right) + \lambda_1 \left(H_1^\dagger H_1 \right)^2 + \lambda_2 \left(H_2^\dagger H_2 \right)^2$$

$$+ \lambda_3 \left(H_1^\dagger H_1 \right) \left(H_2^\dagger H_2 \right) + \lambda_4 \left(H_1^\dagger H_2 \right) \left(H_2^\dagger H_1 \right)$$

$$+ \left[\left(\lambda_5 H_1^\dagger H_2 + \lambda_6 H_1^\dagger H_1 + \lambda_7 H_2^\dagger H_2 \right) \left(H_1^\dagger H_2 \right) + \text{H.c.} \right]. \tag{22.24}$$

The μ_i (with i from 1 to 3) have dimensions of mass squared, while the λ_j (with j from 1 to 7) are dimensionless. All coupling constants are real but for μ_3, λ_5, λ_6, and λ_7, which are not in general real.

The vacuum state is assumed to be a stability point of the potential; therefore, the terms of V linear in the fields H, N, and A vanish. This yields

$$\mu_1 = -2\lambda_1 v^2,$$

$$\mu_3 = -\lambda_6 v^2. \tag{22.25}$$

We shall use eqns (22.25) to trade μ_1 and μ_3 by λ_1 and λ_6, respectively.

One expands the potential V in terms of the fields, after making the substitutions in eqns (22.25). The terms quadratic in the fields are the mass terms of the scalars:

$$V = -\lambda_1 v^4 + m_C^2 C^+ C^- + \tfrac{1}{2} \left(H \ N \ A \right) \mathcal{M} \begin{pmatrix} H \\ N \\ A \end{pmatrix} + \text{cubic and quartic terms,}$$

$$\tag{22.26}$$

where $m_C^2 = \mu_2 + v^2 \lambda_3$ is the squared mass of the charged scalars $C^\pm$, and

$$\mathcal{M} = \begin{pmatrix} 4v^2 \lambda_1 & 2v^2 \text{Re} \lambda_6 & -2v^2 \text{Im} \lambda_6 \\ 2v^2 \text{Re} \lambda_6 & m_C^2 + (\lambda_4 + 2\text{Re} \lambda_5) v^2 & -2v^2 \text{Im} \lambda_5 \\ -2v^2 \text{Im} \lambda_6 & -2v^2 \text{Im} \lambda_5 & m_C^2 + (\lambda_4 - 2\text{Re} \lambda_5) v^2 \end{pmatrix}. \tag{22.27}$$

The symmetric matrix $\mathcal{M}$ is diagonalized by an orthogonal matrix T. The diagonalization yields the masses m_1, m_2, and m_3 of the physical neutral scalars of the THDM. Thus,

$$\mathcal{M} = T \mathrm{diag}\left(m_1^2, m_2^2, m_3^2\right) T^T \tag{22.28}$$

and

$$\begin{pmatrix} H \\ N \\ A \end{pmatrix} = T \begin{pmatrix} S_1 \\ S_2 \\ S_3 \end{pmatrix}. \tag{22.29}$$

Without loss of generality, we may choose the determinant of T to be 1.

The cubic and quartic terms in eqn (22.26) give rise to scalar self-interactions. For instance, there is a cubic interaction between the neutral and the charged scalars:

$$V = \cdots + (c_1 H + c_2 N + c_3 A) C^+ C^- = \cdots + \sum_{k=1}^{3} f_k S_k C^+ C^-, \tag{22.30}$$

where $f_k \equiv T_{1k} c_1 + T_{2k} c_2 + T_{3k} c_3$ for $k = 1, 2$, and 3, and

$$\begin{pmatrix} c_1 & c_2 & c_3 \end{pmatrix} = \sqrt{2} v \left(\lambda_3 \ \mathrm{Re}\,\lambda_7 \ -\mathrm{Im}\,\lambda_7\right). \tag{22.31}$$

22.5.2   The Yukawa interactions

The Yukawa interactions with the quarks read

$$\mathcal{L}_Y^{(q)} = -\frac{\overline{Q_L}}{v} \left[(M_n H_1 + Y_n H_2)\, n_R + \left(M_p \tilde{H}_1 + Y_p \tilde{H}_2\right) p_R \right] + \mathrm{H.c.}, \tag{22.32}$$

where the matrices Y_n and Y_p are in principle arbitrary and unrelated to the mass matrices M_n and M_p. Namely,

$$\begin{aligned} Y_p &\equiv v_2 \Delta_1 - v_1 \Delta_2 e^{-i\theta}, \\ Y_n &\equiv v_2 \Gamma_1 - v_1 \Gamma_2 e^{i\theta}. \end{aligned} \tag{22.33}$$

We define

$$\begin{aligned} Y_u &\equiv U_L^{p\dagger} Y_p U_R^p, \\ Y_d &\equiv U_L^{n\dagger} Y_n U_R^n. \end{aligned} \tag{22.34}$$

While M_u and M_d are diagonal, real and positive by definition, Y_u and Y_d in general are arbitrary $n_g \times n_g$ matrices. Then,

$$\begin{aligned} \mathcal{L}_Y^{(q)} = &-\left(1 + \frac{H}{\sqrt{2}v}\right) \left(\bar{d} M_d d + \bar{u} M_u u\right) + \frac{i\chi}{\sqrt{2}v} \left(\bar{u} M_u \gamma_5 u - \bar{d} M_d \gamma_5 d\right) \\ &+ \frac{\varphi^+}{v} \left(\overline{u_R} M_u V d_L - \overline{u_L} V M_d d_R\right) + \frac{\varphi^-}{v} \left(\overline{d_L} V^\dagger M_u u_R - \overline{d_R} M_d V^\dagger u_L\right) \\ &-\frac{N}{\sqrt{2}v} \left[\bar{d}\left(Y_d \gamma_R + Y_d^\dagger \gamma_L\right) d + \bar{u}\left(Y_u \gamma_R + Y_u^\dagger \gamma_L\right) u\right] \end{aligned}$$

$$-\frac{iA}{\sqrt{2}v}\left[\bar{d}\left(Y_d\gamma_R-Y_d^\dagger\gamma_L\right)d-\bar{u}\left(Y_u\gamma_R-Y_u^\dagger\gamma_L\right)u\right]$$
$$+\frac{C^+}{v}\left(\overline{u_R}Y_u^\dagger Vd_L-\overline{u_L}VY_dd_R\right)+\frac{C^-}{v}\left(\overline{d_L}V^\dagger Y_u u_R-\overline{d_R}Y_d^\dagger V^\dagger u_L\right)\quad(22.35)$$

The first two lines of eqn (22.35) display the same interactions as in the SM, cf. eqns (12.23)–(12.26). The last three lines of eqn (22.35) include the Yukawa interactions of N, A, and $C^\pm$, which depend on the non-diagonal, arbitrary matrices Y_u and Y_d. This is one of the most important features of multi-Higgs-doublet models: in general, there are flavour-changing neutral Yukawa interactions (FCNYI), mediated by neutral scalars. We shall come back to this important question in § 22.10.

22.6   CP transformation

In this chapter, our starting point when defining the CP transformation is the usual one: we require the gauge-kinetic terms of the Lagrangian to be CP-invariant, and this requirement fixes the most general CP transformation allowed. In particular, the pattern of spontaneous symmetry breaking—of the VEVs—influences the gauge interactions of the various fields, notably, those involving the scalar fields, and therefore the explicit values of the VEVs must be taken into account in the definition of the CP transformation—see eqn (22.37) below.[58]

Thus, from the requirement of CP invariance of the gauge interactions of the fermions, one finds that they transform as in eqns (14.2).[59] We write $\varphi_a^0 = v_a e^{i\theta_a}+H_a^0$, where the H_a^0 are quantum fields, while the VEVs are c-numbers constant over space–time. From the requirement of CP invariance of the gauge interactions of the scalars, one finds that they transform as

$$(\mathcal{CP})\,\varphi_a^+(t,\vec{r})\,(\mathcal{CP})^\dagger=U_{ab}^{CP}\varphi_b^-(t,-\vec{r}),$$
$$(\mathcal{CP})\,H_a^0(t,\vec{r})\,(\mathcal{CP})^\dagger=U_{ab}^{CP}H_b^{0\dagger}(t,-\vec{r}),\quad(22.36)$$

where the $n_d\times n_d$ unitary matrix U^{CP} must be chosen such that

$$v_a e^{i\theta_a}=U_{ab}^{CP}v_b e^{-i\theta_b}.\quad(22.37)$$

Equations (22.36) and (22.37) may be put together as

$$(\mathcal{CP})\,\phi_a(t,\vec{r})\,(\mathcal{CP})^\dagger=U_{ab}^{CP}\phi_b^{\dagger^T}(t,-\vec{r}).\quad(22.38)$$

In the Higgs basis, eqn (22.37) constrains U^{CP} to be of the form

---

[58]In the next chapter, which will be dedicated to spontaneous CP violation, the starting point will be different, in that we shall *postulate* the invariance of the Lagrangian, before spontaneous symmetry breaking, under a certain CP transformation, fixed *a priori*, and require that, after spontaneous symmetry breaking, there is no CP transformation under which the Lagrangian is invariant.

[59]We shall set the phase ξ_W to zero. This leads to a simplification in some equations, without thereby losing generality.

$$U^{CP} = \begin{pmatrix} 1 & 0_{1 \times (n_d - 1)} \\ 0_{(n_d - 1) \times 1} & K^{CP} \end{pmatrix}, \tag{22.39}$$

with K^{CP} an *arbitrary* $(n_d - 1) \times (n_d - 1)$ unitary matrix. Thus, the fields $\varphi^{\pm}$, H, and χ transform under CP in exactly the same way as in the SM—see eqns (13.8), (13.11), and (13.12), respectively.

22.7 CP violation in the scalar potential: simple examples

22.7.1 *THDM with a discrete symmetry*

Consider the THDM with a discrete symmetry under which $\phi_2 \to -\phi_2$. The scalar potential in eqn (22.12) then has

$$m_3 = a_6 = a_7 = 0. \tag{22.40}$$

Then, there is only one θ-dependent term in the vacuum potential of eqn (22.16). As a_5 is positive by definition, the minimum is attained when

$$\cos{(\delta_5 + 2\theta)} = -1. \tag{22.41}$$

The matrix U^{CP} is fixed by eqn (22.37): $U^{CP} = \text{diag}\left(1, e^{2i\theta}\right)$. Therefore, from eqn (22.38),

$$(\mathcal{CP}) \left(\phi_1^{\dagger} \phi_2\right)^2 (\mathcal{CP})^{\dagger} = e^{4i\theta} \left(\phi_2^{\dagger} \phi_1\right)^2. \tag{22.42}$$

CP-invariance of the a_5-term of the potential then requires

$$e^{i(\delta_5 + 4\theta)} = e^{-i\delta_5}. \tag{22.43}$$

But, eqn (22.41) implies eqn (22.43). One concludes that CP is conserved in this simple model—as long as no Yukawa couplings are introduced, at least.

Notice that this happens in spite of the potential not being real and in spite of the existence of a complex phase between the VEVs of the two doublets. This implies that neither condition by itself alone, or even both of them together, leads to CP violation. The crucial point is that the vacuum phase θ is determined by *only one term* in the potential: there is only one θ-dependent term in V_0. Such a situation, in which there is only one term in V_0 for each (relative) phase in the vacuum, usually leads to CP invariance.

22.7.2 *Softly broken discrete symmetry*

The situation changes when one allows the discrete symmetry $\phi_2 \to -\phi_2$ to be softly broken. A symmetry is said to be broken softly when all terms which break it have dimension lower than four. In this specific case, allowing for soft breaking of the symmetry will lead to the presence of only one extra quadratic term in the potential, the one with coefficient m_3 in eqn (22.12). CP-invariance would now require

$$e^{2i(\delta_5 + 2\theta)} = 1,$$
$$e^{2i(\delta_3 + \theta)} = 1. \tag{22.44}$$

However, θ is now determined by the stability condition in eqn (22.17), with $a_6 = a_7 = 0$, which will in general yield θ satisfying neither of the two eqns (22.44).

Therefore, CP is violated. The point is that V_0 now contains two clashing terms depending on the sole vacuum phase θ; under such conditions, one may in general expect CP violation to occur.

This two-Higgs-doublet model, in which the reflection symmetry $\phi_2 \rightarrow -\phi_2$ is softly broken by the quadratic terms proportional to m_3, has been used as a toy model for CP violation in the scalar sector (Branco and Rebelo 1985; Weinberg 1990).

22.7.3 *Weinberg model*

In the three-Higgs-doublet model of Weinberg (1976) there are two distinct symmetries: the first one transforms $\phi_2 \rightarrow -\phi_2$ and leaves all other fields unchanged; the second one transforms $\phi_3 \rightarrow -\phi_3$ and leaves all other fields unchanged. These two symmetries are assumed not to be softly broken—though they end up being spontaneously broken by the non-vanishing VEVs of ϕ_2 and ϕ_3. The scalar potential is

$$
\begin{aligned}
V = {} & m_1 \phi_1^\dagger \phi_1 + m_2 \phi_2^\dagger \phi_2 + m_3 \phi_3^\dagger \phi_3 + a_1 \left(\phi_1^\dagger \phi_1 \right)^2 + a_2 \left(\phi_2^\dagger \phi_2 \right)^2 + a_3 \left(\phi_3^\dagger \phi_3 \right)^2 \\
& + b_1 \left(\phi_2^\dagger \phi_2 \right) \left(\phi_3^\dagger \phi_3 \right) + b_2 \left(\phi_3^\dagger \phi_3 \right) \left(\phi_1^\dagger \phi_1 \right) + b_3 \left(\phi_1^\dagger \phi_1 \right) \left(\phi_2^\dagger \phi_2 \right) \\
& + c_1 \left(\phi_2^\dagger \phi_3 \right) \left(\phi_3^\dagger \phi_2 \right) + c_2 \left(\phi_3^\dagger \phi_1 \right) \left(\phi_1^\dagger \phi_3 \right) + c_3 \left(\phi_1^\dagger \phi_2 \right) \left(\phi_2^\dagger \phi_1 \right) \\
& + d_1 \left[e^{i\varepsilon_1} \left(\phi_2^\dagger \phi_3 \right)^2 + e^{-i\varepsilon_1} \left(\phi_3^\dagger \phi_2 \right)^2 \right] + d_2 \left[e^{i\varepsilon_2} \left(\phi_3^\dagger \phi_1 \right)^2 + e^{-i\varepsilon_2} \left(\phi_1^\dagger \phi_3 \right)^2 \right] \\
& + d_3 \left[e^{i\varepsilon_3} \left(\phi_1^\dagger \phi_2 \right)^2 + e^{-i\varepsilon_3} \left(\phi_2^\dagger \phi_1 \right)^2 \right],
\end{aligned}
\tag{22.45}
$$

with real and positive d_1, d_2, and d_3. The vacuum potential is

$$
\begin{aligned}
V_0 = {} & m_1 v_1^2 + m_2 v_2^2 + m_3 v_3^2 + a_1 v_1^4 + a_2 v_2^4 + a_3 v_3^4 \\
& + (b_1 + c_1)\, v_2^2 v_3^2 + (b_2 + c_2)\, v_3^2 v_1^2 + (b_3 + c_3)\, v_1^2 v_2^2 \\
& + 2 d_1 v_2^2 v_3^2 \cos\left(2\theta_2 - 2\theta_3 - \varepsilon_1 \right) + 2 d_2 v_3^2 v_1^2 \cos\left(2\theta_3 - \varepsilon_2 \right) \\
& + 2 d_3 v_1^2 v_2^2 \cos\left(2\theta_2 + \varepsilon_3 \right).
\end{aligned}
\tag{22.46}
$$

In the vacuum there are two gauge-invariant relative phases, θ_2 and θ_3, but in V_0 there are three terms which depend on them. The fact that there are less phases than phase-dependent terms in V_0 leads, once again, to CP violation in the self-interactions of the scalars.

22.8 General treatment of CP violation

CP violation is associated with the presence of irremovable phases in the Lagrangian of the theory. However, a weak-basis transformation of the fields—which includes the rephasing of the fields as a particular case—can bring new phases in and out of the Lagrangian. The spurious phases thus generated or eliminated have no bearing on CP violation. Therefore, it is important to find quantities which characterize CP violation in a given theory and which do not depend on the weak basis chosen to write the Lagrangian. We have encountered this problem in Chapter 14, where we have derived the weak-basis (WB) invariant $\text{tr}\,[H_p, H_n]^3$ for the three-generation SM using a general method (Bernabéu et al. 1986a) to construct CP-violating WB invariants. This method has been applied to some extensions of the SM, such as models with vector-like quarks (Branco and Lavoura 1986), models with Majorana neutrinos (Branco et al. 1986), and left-right-symmetric models (Branco and Rebelo 1985). In all these applications, the CP-violating WB invariants were related to clashes between the CP-transformation properties required by the gauge interactions, on the one hand, and the fermion mass terms, on the other hand. Botella and Silva (1995) have extended the method to the Higgs sector. The method involves the construction of tensors of increasing complexity, whose indices lie in the various family spaces—the families of identical multiplets which may be mixed by weak-basis transformations (WBT). By taking traces over all those indices one obtains weak-basis-invariant quantities, thus removing all the spurious phases created by WBT. Any remaining imaginary part constitutes a hallmark of CP violation. This method for constructing weak-basis-invariant quantities is quite general: it works for any gauge group, and for any (basic or effective) Lagrangian. The method can also be extended to provide weak-basis-invariant measures for the breaking of other discrete symmetries, like R-parity in supersymmetric theories (Davidson and Ellis 1997).

22.8.1 *Weak-basis transformations*

Weak-basis transformations of the fermion fields are identical to the ones in the SM, see eqns (14.10). In an MHDM there are several identical scalar multiplets; therefore, we may perform a WBT of the scalar fields too:

$$\phi'_a = U_{ab}\phi_b, \tag{22.47}$$

where U is an $n_d \times n_d$ unitary matrix, so that the gauge-kinetic Lagrangian of the scalars does not get changed. The VEVs transform in the same way as the doublets, therefore

$$V'_{cd} = U_{ca}U^*_{db}V_{ab}, \tag{22.48}$$

where the matrix V_{ab} was defined in eqn (22.9). The couplings in the scalar potential transform as

$$\begin{aligned}
Y'_{cd} &= U_{ca}U^*_{db}Y_{ab}, \\
Z'_{efgh} &= U_{ea}U^*_{fb}U_{gc}U^*_{hd}Z_{abcd}.
\end{aligned} \tag{22.49}$$

In a simultaneous WBT of the scalar doublets and of the fermion multiplets—eqns (22.47) and (14.10)—the Yukawa-coupling matrices transform as

$$\begin{aligned}
\Gamma'_a &= U^*_{ab}W^\dagger_L\Gamma_b W^n_R, \\
\Delta'_a &= U_{ab}W^\dagger_L\Delta_b W^p_R.
\end{aligned} \tag{22.50}$$

22.8.2   *Weak-basis invariants*

Meaningful physical quantities must be invariant under a WBT. In order to construct such quantities, one first considers the matrices

$$
\begin{aligned}
H_{\Gamma ab} &\equiv \Gamma_b \Gamma_a^\dagger, \\
H_{\Delta ab} &\equiv \Delta_a \Delta_b^\dagger.
\end{aligned}
\tag{22.51}
$$

From eqns (22.50), one has

$$
\begin{aligned}
H'_{\Gamma cd} &= U_{ca} U_{db}^* \left(W_L^\dagger H_{\Gamma ab} W_L \right), \\
H'_{\Delta cd} &= U_{ca} U_{db}^* \left(W_L^\dagger H_{\Delta ab} W_L \right).
\end{aligned}
\tag{22.52}
$$

Taking traces over the fermionic indices one obtains quantities which are invariant under a WBT of the fermion fields, but are tensors under a WBT of the scalar fields. Taking traces over the indices $a, b, c, \ldots$ too, one finally obtains the weak-basis invariants (Botella and Silva 1995). Simple examples are $V_{ab} Y_{ba}$, $V_{ab} \mathrm{tr}\, H_{\Gamma ba}$, and $V_{ab} Y_{bc} \mathrm{tr}\, H_{\Delta ca}$. For instance, since the mass matrix of the down-type quarks is $M_n = v_a \Gamma_a e^{i\theta_a}$,

$$
V_{ab} \mathrm{tr}\, H_{\Gamma ba} = \mathrm{tr}\left(M_n M_n^\dagger \right) = \mathrm{tr}\, M_d^2 = m_d^2 + m_s^2 + m_b^2 + \cdots .
\tag{22.53}
$$

22.8.3   *CP violation*

The CP-invariance conditions for the Yukawa-coupling matrices Γ_a and Δ_a and for the couplings of the scalar potential Y_{ab} and Z_{abcd} contain several CP-transformation matrix elements and are not very transparent. However, when considering weak-basis-invariant combinations of the couplings and of the VEVs, one obtains the simple result that *all weak-basis-invariant quantities must be real in order for CP symmetry to hold.*[60]

If, for definiteness, one wants to study the CP-invariance conditions for the scalar potential only, under the given pattern of spontaneous symmetry breaking, one must take into account the tensors V_{ab}, Y_{ab}, and Z_{abcd}. With these three tensors one may construct various weak-basis-invariant quantities. Some of these are real, whether CP is conserved or not, because of eqns (22.5), (22.6), and (22.10); others are not necessarily real. The simplest non-real invariants are (Lavoura and Silva 1994; Botella and Silva 1995)

$$
\begin{aligned}
I_1 &\equiv Y_{ab} Z_{bccd} V_{da}, \\
I_2 &\equiv V_{ab} Y_{bc} V_{de} Y_{ef} Z_{cafd}.
\end{aligned}
\tag{22.54}
$$

If either I_1 or I_2 is not real, then the scalar potential together with the vacuum structure violates CP.

---

[60] We are excluding the artificial procedure in which one would be introducing by hand some phases in the definition of otherwise real WB invariants.

If one also considers the Yukawa interactions, extra tensors come into play, for instance $\operatorname{tr} H_{\Gamma ab}$. If one only considers this extra tensor, one may construct two more weak-basis invariants which may, in principle, be non-real:

$$\begin{aligned}
I_3 &\equiv V_{ab} Y_{bc} \operatorname{tr} H_{\Gamma ca}, \\
I_4 &\equiv V_{ab} \operatorname{tr} H_{\Gamma bc} V_{de} \operatorname{tr} H_{\Gamma ef} Z_{cafd}.
\end{aligned} \tag{22.55}$$

If either I_3 or I_4 is not real, there is CP violation in the clash between the Yukawa interactions and the scalar sector of the model.

22.9 CP violation in the two-Higgs-doublet model

In this section we apply the general methods of the previous section to the study of CP violation in the THDM. Our aim is to identify sources of CP violation in the THDM which are not present in the SM.

The tensor V_{ab} has a very simple form in the Higgs basis: $V_{11} = v^2$ and all other $V_{ab} = 0$. This simplifies considerably the computation of the weak-basis-invariants and, therefore, the analysis of CP violation becomes much simpler in the Higgs basis. In this section we shall use the Higgs basis throughout.

22.9.1 I_1 and I_2

Computing the invariants I_1 and I_2 one finds

$$\begin{aligned}
I_1 &= v^2 Y_{1b} Z_{bcc1} \\
&= -v^4 \left(2\lambda_1 Z_{1cc1} + \lambda_6 Z_{2cc1}\right) \\
&= -v^4 \left(2\lambda_1^2 + \lambda_1\lambda_4 + |\lambda_6|^2/2 + \lambda_6\lambda_7^*/2\right),
\end{aligned} \tag{22.56}$$

$$\begin{aligned}
I_2 &= v^4 Y_{1c} Y_{1f} Z_{c1f1} \\
&= -v^6 Y_{1f} \left(2\lambda_1 Z_{11f1} + \lambda_6 Z_{21f1}\right) \\
&= v^8 \left[2\lambda_1 \left(2\lambda_1 Z_{1111} + \lambda_6 Z_{2111}\right) + \lambda_6 \left(2\lambda_1 Z_{1121} + \lambda_6 Z_{2121}\right)\right] \\
&= v^8 \left(4\lambda_1^3 + 2\lambda_1 |\lambda_6|^2 + \lambda_6^2\lambda_5^*\right).
\end{aligned} \tag{22.57}$$

Thus, $\operatorname{Im}(\lambda_6\lambda_7^*) \propto \operatorname{Im} I_1$ and $\operatorname{Im}(\lambda_6^2\lambda_5^*) \propto \operatorname{Im} I_2$ are CP-violating quantities.

Using eqns (22.27) and (22.31), one may reproduce these quantities. One easily finds

$$\mathcal{M}_{12}c_3 - \mathcal{M}_{13}c_2 = 2\sqrt{2}v^3 \operatorname{Im}(\lambda_6\lambda_7^*),$$
$$\mathcal{M}_{12}\mathcal{M}_{13}(\mathcal{M}_{22} - \mathcal{M}_{33}) + \mathcal{M}_{23}(\mathcal{M}_{13}^2 - \mathcal{M}_{12}^2) = -8v^6 \operatorname{Im}(\lambda_6^2\lambda_5^*). \tag{22.58}$$

Using eqn (22.28) and the choice $\det T = 1$, one obtains

$$\begin{aligned}
\mathcal{M}_{12}c_3 - \mathcal{M}_{13}c_2 &= f_1 T_{12}T_{13}\left(m_3^2 - m_2^2\right) + f_2 T_{11}T_{13}\left(m_1^2 - m_3^2\right) \\
&\quad + f_3 T_{11}T_{12}\left(m_2^2 - m_1^2\right),
\end{aligned} \tag{22.59}$$

$$\begin{aligned}
\mathcal{M}_{12}\mathcal{M}_{13}(\mathcal{M}_{22} - \mathcal{M}_{33}) & \\
+ \mathcal{M}_{23}(\mathcal{M}_{13}^2 - \mathcal{M}_{12}^2) &= \left(m_1^2 - m_2^2\right)\left(m_1^2 - m_3^2\right)\left(m_2^2 - m_3^2\right) T_{11}T_{12}T_{13}.
\end{aligned} \tag{22.60}$$

The fact that the quantity in eqn (22.60) violates CP was first noticed by Méndez and Pomarol (1991); the proof of that fact was later given by Lavoura and Silva

(1994), who also discovered the quantity in eqn (22.59). Equation (22.60) means that the mixing of the neutral scalars violates CP if the masses of the three physical scalars are all different and if, moreover, all three matrix elements of the first row of T are different from zero. Notice the similarity of eqn (22.60) with the expression for the CP-violating invariant of the SM, cf. eqn (14.24).

22.9.2 I_3 and I_4

Comparing the Yukawa Lagrangians in eqns (22.3) and (22.32), one concludes that, in the Higgs basis,

$$
\begin{aligned}
v^2 \mathrm{tr}\, H_{\Gamma 11} &= \mathrm{tr}\,(M_d^2), \\
v^2 \mathrm{tr}\, H_{\Gamma 22} &= \mathrm{tr}\,(Y_d Y_d^\dagger), \\
v^2 \mathrm{tr}\, H_{\Gamma 12} &= \mathrm{tr}\,(M_d Y_d), \\
v^2 \mathrm{tr}\, H_{\Gamma 21} &= \mathrm{tr}\,(M_d Y_d^\dagger).
\end{aligned}
\tag{22.61}
$$

One then finds, after a tedious yet straightforward computation,

$$
\mathrm{Im}\, I_3 = \sum_{i=d,s,b,\ldots} \frac{m_i}{2} \sum_{k=1}^{3} m_k^2 T_{1k} \left[T_{2k}\mathrm{Im}\,(Y_d)_{ii} + T_{3k}\mathrm{Re}\,(Y_d)_{ii} \right].
\tag{22.62}
$$

Similarly,

$$
\begin{aligned}
\mathrm{Im}\, I_4 - \frac{\mathrm{tr}\,(M_d^2)}{v^2}\mathrm{Im}\, I_3 = \sum_{i,j=d,s,b,\ldots} \frac{m_i m_j}{2v^2} \sum_{k=1}^{3} m_k^2 & \left[T_{2k}\mathrm{Im}\,(Y_d)_{ii} + T_{3k}\mathrm{Re}\,(Y_d)_{ii} \right] \\
& \times \left[T_{2k}\mathrm{Re}\,(Y_d)_{jj} - T_{3k}\mathrm{Im}\,(Y_d)_{jj} \right].
\end{aligned}
\tag{22.63}
$$

22.9.3   *Feynman rules and CP violation: I_1 and I_2*

We call a scalar field $S(t, \vec{r})$ 'CP-even' when

$$
(\mathcal{CP})\, S(t, \vec{r})\, (\mathcal{CP})^\dagger = S(t, -\vec{r});
\tag{22.64}
$$

on the other hand, S is 'CP-odd' when

$$
(\mathcal{CP})\, S(t, \vec{r})\, (\mathcal{CP})^\dagger = -S(t, -\vec{r}).
\tag{22.65}
$$

From eqn (13.11) one knows that H is CP-even. The CP transformation of N and A is determined by eqns (22.38) and (22.39):

$$
(\mathcal{CP})\, [N(t, \vec{r}) + iA(t, \vec{r})]\, (\mathcal{CP})^\dagger = e^{i\vartheta}\, [N(t, -\vec{r}) - iA(t, -\vec{r})].
\tag{22.66}
$$

Therefore, $N\cos(\vartheta/2) + A\sin(\vartheta/2)$ is CP-even, while $N\sin(\vartheta/2) - A\cos(\vartheta/2)$ is CP-odd. (The phase ϑ is arbitrary, and therefore the exact determination of which linear combination of N and A is CP-odd is meaningless. However, the reasoning is not affected by this.) If CP were conserved, the CP-even scalars

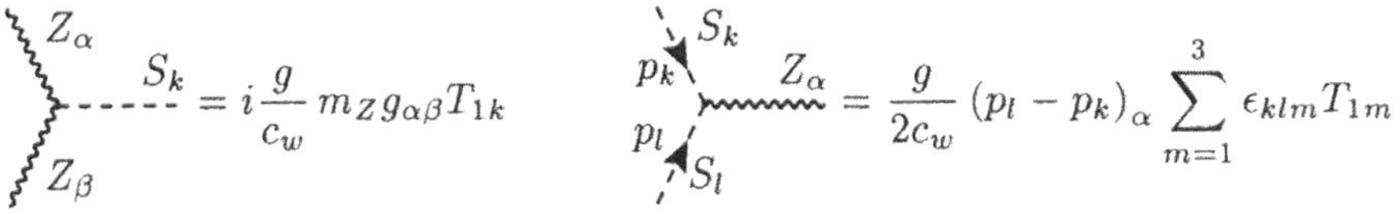

$$\genfrac{}{}{0pt}{}{Z_\alpha}{Z_\beta} \!\!\!- - - - S_k = i\,\frac{g}{c_w}\,m_Z\,g_{\alpha\beta}T_{1k}$$

$$\genfrac{}{}{0pt}{}{p_k\,S_k}{p_l\,S_l}\!\!\! Z_\alpha = \frac{g}{2c_w}\,(p_l - p_k)_\alpha \sum_{m=1}^{3}\epsilon_{klm}T_{1m}$$

FIG. 22.1. Feynman rules for the $Z_\alpha Z_\beta S_k$ and $Z_\alpha S_k S_l$ vertices in the THDM. (In the latter case, k must be different from l.) Notice the similarity with the vertices $Z_\alpha Z_\beta H$ and $Z_\alpha \chi H$, respectively, of the SM, in Fig. 11.2.

would not mix with the CP-odd scalar in the mass matrix $\mathcal{M}$. Then, out of the three physical neutral scalars, two would be CP-even and one would be CP-odd.

The Feynman rules for the vertices of one neutral scalar with two Z bosons, and of two *distinct* neutral scalars with one Z boson, are given in Fig. 22.1. One sees that the vertex $S_k Z_\alpha Z_\beta$ is proportional to $T_{1k}\,g_{\alpha\beta}$. Under CP, $Z_\alpha \to -Z^\alpha$. As $g_{\alpha\beta} = g^{\alpha\beta}$, it follows that this vertex only exists if S_k is CP-even. If T_{11}, T_{12}, and T_{13} are all non-zero, all three neutral scalars S_1, S_2, and S_3 couple to $Z_\alpha Z_\beta$ in this way, and therefore all of them are CP-even. However, we have seen in the previous paragraph that, when CP is conserved, one out of the three neutral scalars must be CP-odd. We thus conclude that $T_{11}T_{12}T_{13} \neq 0$ implies CP violation. This coincides with what we deduced from eqn (22.60).

The same may be seen in yet another way. Suppose that both S_1 and S_2 couple to $Z_\alpha Z_\beta$, with vertices proportional to $g_{\alpha\beta}$. Then, both S_1 and S_2 must be CP-even. Now suppose that there also is a vertex $S_1 S_2 Z_\alpha$ which, according to Fig. 22.1, is proportional to $T_{13}\,(p_2 - p_1)_\alpha$. Under CP, $Z_\alpha \to -Z^\alpha$, but $(p_2 - p_1)_\alpha \to (p_2 - p_1)^\alpha$. It follows that either S_1 is CP-even and S_2 is CP-odd, or vice versa. This contradicts the above conclusion that both S_1 and S_2 are CP-even. Thus, if $T_{11}T_{12}T_{13} \neq 0$ all three above-mentioned Feynman vertices exist and there is CP violation.

We may thus construct three simple cases in which the co-existence of three different Feynman vertices displays CP violation:

$$\left.\begin{aligned} S_1 S_2 Z_\alpha &\propto (p_1 - p_2)_\alpha \\ S_1 S_3 Z_\alpha &\propto (p_1 - p_3)_\alpha \\ S_2 S_3 Z_\alpha &\propto (p_2 - p_3)_\alpha \end{aligned}\right\} \Rightarrow \text{CP violation}, \qquad (22.67)$$

$$\left.\begin{aligned} S_1 S_2 Z_\alpha &\propto (p_1 - p_2)_\alpha \\ S_1 Z_\alpha Z_\beta &\propto g_{\alpha\beta} \\ S_2 Z_\alpha Z_\beta &\propto g_{\alpha\beta} \end{aligned}\right\} \Rightarrow \text{CP violation}, \qquad (22.68)$$

$$\left.\begin{aligned} S_1 Z_\alpha Z_\beta &\propto g_{\alpha\beta} \\ S_2 Z_\alpha Z_\beta &\propto g_{\alpha\beta} \\ S_3 Z_\alpha Z_\beta &\propto g_{\alpha\beta} \end{aligned}\right\} \Rightarrow \text{CP violation}. \qquad (22.69)$$

One should notice however that, while eqns (22.67) and (22.68) are valid *in any model*, eqn (22.69) holds only in the THDM, because in the THDM there are only three neutral scalars and one of them must be CP-odd. In the context of, say, a three-Higgs-doublet model, the existence of three scalars with gauge interactions as in eqn (22.69) would not imply CP violation.

One may interpret eqn (22.59) along similar lines. The interaction $f_k S_k C^+ C^-$ in eqn (22.30) implies that S_k is CP-even. Similarly, the interaction $T_{1k} S_k Z^\mu Z_\mu$ implies that S_k is CP-even. Hence, if for instance the product $f_1 T_{12} T_{13} \neq 0$, this means that S_1, S_2, and S_3 all are CP-even, which is impossible in the THDM. For this reason, the quantity in eqn (22.59) signals CP violation in the THDM.

22.9.4   *Feynman rules and CP violation:* I_3 *and* I_4

In order to understand the results in eqns (22.62) and (22.63), let us study in more detail the Yukawa interactions of the neutral scalars with the down-type quarks. From eqn (22.35), they read

$$\mathcal{L}_Y^{(q)} = \cdots - \sum_{k=1}^{3} \frac{S_k}{\sqrt{2}v} \overline{d} \left[T_{1k} M_d + (T_{2k} + iT_{3k}) Y_d \gamma_R + (T_{2k} - iT_{3k}) Y_d^\dagger \gamma_L \right] d.$$
(22.70)

Let us consider only the diagonal interactions, in which the incoming and outgoing quarks have the same flavour. They are

$$\mathcal{L}_Y^{(q)} = \cdots - \sum_{k=1}^{3} \frac{S_k}{\sqrt{2}v} \sum_i \overline{d_i} \left(a_{ki} + i\gamma_5 b_{ki} \right) d_i,$$
(22.71)

where

$$\begin{aligned} a_{ki} &\equiv T_{1k} m_i + T_{2k} \mathrm{Re}\,(Y_d)_{ii} - T_{3k} \mathrm{Im}\,(Y_d)_{ii}, \\ b_{ki} &\equiv T_{2k} \mathrm{Im}\,(Y_d)_{ii} + T_{3k} \mathrm{Re}\,(Y_d)_{ii}. \end{aligned}$$
(22.72)

Now, a glance at eqns (3.74) and (3.75) tells us that, in order for the interactions in eqn (22.71) to be CP-invariant, S_k must be CP-even when $a_{ki} \neq 0$, but it must be CP-odd when $b_{ki} \neq 0$. Clearly, if a_{ki} and b_{kj} are simultaneously non-zero for any two down-type quarks d_i and d_j—even when $d_i \neq d_j$—then S_k does not have a definite CP-parity, and CP is violated. This is precisely what is reflected in eqns (22.62) and (22.63).

CP violation in the Yukawa couplings of the neutral scalars leads to effects like the generation of electric-dipole moments—in the coupling of a fermion with the photon— or weak electric-dipole moments—in the coupling of a fermion with the Z—at one-loop level. Weak electric-dipole moments have a particularly rich variety of contributing diagrams; some examples are presented in Fig. 22.2.

22.10   Flavour-changing neutral Yukawa interactions

This and the next section, on the problem of flavour-changing neutral currents, may be skipped by the reader.

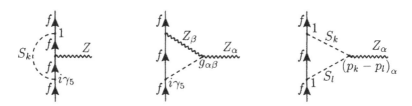

FIG. 22.2. Some diagrams which may contribute to the weak electric-dipole moment in the coupling of a fermion f to the Z boson. In some vertices we have explicitly written down the form that that vertex might assume in order for the diagram to violate CP.

We have seen in § 22.5.2 that the matrices Y_u and Y_d defined by eqns (22.34) are not in general diagonal, and therefore the neutral scalars have flavour-changing neutral Yukawa interactions (FCNYI) with quarks of identical electric charge. Let us introduce the Hermitian matrices X_k and X'_k, defined by[61]

$$X_k \equiv \frac{(T_{2k} + iT_{3k})\, Y_d + (T_{2k} - iT_{3k})\, Y_d^\dagger}{2\sqrt{2}v},$$

$$X'_k \equiv \frac{-i\,(T_{2k} + iT_{3k})\, Y_d + i\,(T_{2k} - iT_{3k})\, Y_d^\dagger}{2\sqrt{2}v}.$$

(22.73)

We may then rewrite eqn (22.70) in the form

$$\mathcal{L}_Y^{(q)} = \cdots - \sum_{k=1}^{3} S_k \bar{d} \left(T_{1k} \frac{M_d}{\sqrt{2}v} + X_k + iX'_k \gamma_5 \right) d.$$

(22.74)

There is for instance a Yukawa interaction connecting the s quark with the d quark,

$$\mathcal{L}_Y^{(q)} = \cdots - \sum_{k=1}^{3} S_k \left\{ \bar{s}\,[(X_k)_{21} + i\,(X'_k)_{21}\,\gamma_5]\,d + \bar{d}\,[(X_k)_{21}^* + i\,(X'_k)_{21}^*\,\gamma_5]\,s \right\}.$$

(22.75)

This interaction leads to a contribution M_{21}^S to the off-diagonal matrix element M_{21} in the neutral-kaon system. That contribution reads—see Appendix B—

$$M_{21}^S = e^{i(\xi_K + \xi_d - \xi_s)} \sum_{k=1}^{3} \frac{f_K^2 m_K}{24 m_k^2} \left\{ [(X'_k)_{21}]^2 \left[1 - \frac{11 m_K^2}{(m_s + m_d)^2} \right] \right.$$

$$\left. + [(X_k)_{21}]^2 \left[1 - \frac{m_K^2}{(m_s + m_d)^2} \right] \right\},$$

(22.76)

where $m_K \approx 498\,\text{MeV}$, $f_K \approx 160\,\text{MeV}$, and $m_s + m_d \approx 180\,\text{MeV}$.

[61]The Yukawa interactions of the scalars with the up-type quarks in general also display FCNYI. However, the strongest experimental bounds on this type of interaction arise in the down-type-quark sector.

We know that the mass difference between K_L and K_S is $2|M_{21}| \approx 3.49 \times 10^{-12}\,\mathrm{MeV}$. This suggests that the masses m_k should be rather high. Indeed, let us assume that M_{21}^S is the largest contribution to M_{21}, so that $|M_{21}| \approx |M_{21}^S|$. We would then have

$$\sum_{k=1}^{3} \frac{1}{m_k^2} \left\{ 83.20\,[(X_k')_{21}]^2 + 6.65\,[(X_k)_{21}]^2 \right\} = 3.29 \times 10^{-6}\,\mathrm{TeV}^{-2}. \qquad (22.77)$$

The complex numbers $(X_k)_{21}$ and $(X_k')_{21}$ are in principle arbitrary. However, we may reasonably guess that $|(Y_d)_{21}|$ and $|(Y_d)_{12}|$ should be of order $\sqrt{m_s m_d} \approx 40\,\mathrm{MeV}$. If this is so, and as $v = 174\,\mathrm{GeV}$, we have $|(X_k')_{21}|, |(X_k)_{21}| \sim 10^{-4}$.

Let us further assume that in the sum in the left-hand side of eqn (22.77) there are no large cancellations. It is then reasonable to estimate that, for each value of k, $6 \times 10^{-7}/m_k^2 \sim 3 \times 10^{-6}\,\mathrm{TeV}^{-2}$, i.e.,

$$m_k \gtrsim 0.5\,\mathrm{TeV}. \qquad (22.78)$$

This is a rather high value for the masses of the Higgs scalars. Of course, the derivation of eqn (22.78) involved various *ad hoc* assumptions Still, it is clear that we are confronted with a potential problem for the THDM: unless the neutral scalars have masses of order $1\,\mathrm{TeV}$, their contribution to the mass difference of the neutral kaons may be too large.

The Yukawa interactions of the neutral scalars are a potential source of CP violation, too. The contribution M_{21}^S to M_{21} might generate, not only a large mass difference Δm, but also a large CP-violating parameter ϵ. In principle, the constraints on the masses of the scalars from consideration of the contribution of M_{21}^S to ϵ will be stronger than the ones from the contribution of M_{21}^S to Δm. However, there are natural ways of suppressing the imaginary parts in the FCNYI (Branco and Rebelo 1985), and thus their contribution to ϵ. We therefore stick to the bound in eqn (22.78), which is somewhat more difficult to avoid.

The FCNYI are a general problem of multi-Higgs-doublet models. Those models in general have neutral scalar particles whose Yukawa couplings are not flavour-diagonal. Then, in order to satisfy experimental constraints arising from K^0–$\overline{K^0}$, B^0–$\overline{B^0}$, and D^0–$\overline{D^0}$ mixing, as well as from some rare decays, either one has to find a natural mechanism to suppress the non-diagonal couplings, or the masses of the neutral scalars have to be rather high, in the TeV range.

22.11   Mechanisms for natural suppression of the FCNYI

22.11.1   *Natural flavour conservation*

In order to solve the problem of FCNYI, the concept of natural flavour conservation (NFC) was developed. With NFC, one avoids the FCNYI by imposing some extra symmetry on the Lagrangian of the MHDM; the extra symmetry should constrain the Yukawa interactions of the neutral scalars in such a way that they turn out diagonal. Glashow and Weinberg (1977) and Paschos (1977) have shown that the only way to achieve NFC is to ensure that only one Higgs doublet has Yukawa interactions with quarks of a given charge—see also Ecker *et al.* (1988).

Consider the Yukawa interactions in eqn (22.3). FCNYI arise because not all Γ_a can be simultaneously bi-diagonalized, i.e., diagonalized by the same two unitary matrices. When one bi-diagonalizes the particular linear combination of the Γ_a which constitutes M_n, the down-type-quark mass matrix, one is not bi-diagonalizing other linear combinations of the Γ_a, orthogonal to M_n. A simple solution to this problem is the following: all Γ_a, except one of them, should be identically zero; the same thing happening in the up-quark sector, where all Δ_a except one should vanish.

This leads to two possibilities: either the matrices Γ_a and Δ_b which do not vanish correspond to different Higgs doublets ($a \neq b$), or they correspond to the same Higgs doublet ($a = b$). The first situation—which in the context of two-Higgs-doublet models is sometimes called 'model 1', while the second choice is called 'model 2'—is more interesting from the theoretical point of view, in particular because it automatically arises in a supersymmetric theory. Let us study it in more detail.

We start from eqn (22.3). We assume that

$$\begin{aligned} \Gamma_2 &= \Delta_1 = 0, \\ \Gamma_a &= \Delta_a = 0 \text{ for } a > 2. \end{aligned} \tag{22.79}$$

(We leave open the possibility that there are more than two doublets; we just assume that the extra doublets, if they exist, do not have Yukawa couplings to the quarks.) Equation (22.79) may be enforced by two discrete symmetries:

$$\begin{aligned} D_1 &: \quad \phi_a \rightarrow -\phi_a \text{ for } a > 2, \\ D_2 &: \quad \phi_2 \rightarrow -\phi_2 \text{ and } p_R \rightarrow -p_R, \end{aligned} \tag{22.80}$$

cf. § 22.7.3. The Yukawa Lagrangian in eqn (22.18), under the condition in eqn. (22.79), is

$$\begin{aligned} \mathcal{L}_Y^{(q)} = &-\left(1 + \frac{\rho_1}{\sqrt{2}v_1}\right)\bar{d}M_d d - \left(1 + \frac{\rho_2}{\sqrt{2}v_2}\right)\bar{u}M_u u \\ &-\frac{i\eta_1}{\sqrt{2}v_1}\bar{d}M_d\gamma_5 d + \frac{i\eta_2}{\sqrt{2}v_2}\bar{u}M_u\gamma_5 u \\ &+\left(\frac{\varphi_2^+}{v_2}\bar{u}_R M_u V d_L - \frac{\varphi_1^+}{v_1}\bar{u}_L V M_d d_R + \text{H.c.}\right), \end{aligned} \tag{22.81}$$

cf. eqns (12.23)–(12.26). All neutral Yukawa interactions are flavour-diagonal and proportional to the mass of the quark. FCNYI do not arise because only one Yukawa-coupling matrix must be bi-diagonalized in each quark sector.

22.11.2 *Non-vanishing but naturally small FCNYI*

In this section we shall consider the possibility of having non-vanishing but naturally suppressed FCNYI. By natural suppression we mean that whatever mechanism is responsible for the suppression, it should result from either an exact or a softly broken symmetry of the Lagrangian. This naturalness requirement is essential in order to guarantee that the suppression mechanism is stable under radiative corrections.

A possible suppression mechanism could arise if the flavour-changing couplings of the neutral scalars were entirely fixed by quark masses and elements of the CKM matrix V. Since some of these matrix elements are experimentally known to be very small, one could then have a suppression. For definiteness, let us consider the flavour-changing neutral coupling vertex connecting two down-type quarks d_i and d_j with a scalar. Let us assume that the corresponding Yukawa coupling Y_{ij} depends only on quark masses and on matrix elements of V. Of course, Y_{ij} will have to be invariant under rephasing of the fields d_i and d_j. This restricts the functional dependence of Y_{ij} on the CKM-matrix elements. The simplest dependence which conforms to the constraint of rephasing invariance is $V_{\alpha i}^* V_{\alpha j}$, where u_α denotes any of the up-type quarks. If one considers the specific case $d_i = d$ and $d_j = s$, and if u_α turns out to be the top quark, then one has a very strong suppression factor (Joshipura and Rindani 1991) $|V_{ts} V_{td}|^2 \propto \lambda^{10} \sim 10^{-8}$ in the neutral-scalar contribution to the $|\Delta S| = 2$ effective Hamiltonian.

The important question is whether it is possible to have such a functional dependence as a result of an exact or softly broken symmetry of the full Lagrangian. It has been shown (Branco *et al.* 1996) that this is indeed possible through the introduction of a symmetry Z_n. There is some freedom in the choice of n, and the quark u_α may be any of the up-type quarks, depending on the specific transformation properties of the quark fields under Z_n. Within this class of models the Higgs bosons may often be relatively light, with masses $\sim 100\,\text{GeV}$. The important point that we want to emphasize is that the constraint of NFC may be too restrictive and other interesting scenarios are possible, with non-vanishing but naturally suppressed FCNYI. The interest in this possibility has been revived by the suggestion (Antaramian *et al.* 1992; Hall and Weinberg 1993) that some suppression factors could result from approximate family symmetries.

22.12 Main conclusions

We next collect the main conclusions of this chapter on what has to do with CP violation.

- CP violation in the self-interactions of the scalars arises when the number of gauge-invariant phases between the vacuum expectation values is smaller than the number of terms in the scalar potential which feel those phases.

- In the two-Higgs-doublet model, there is CP violation in the mixing of the neutral scalar fields if and only if

  1. All neutral scalar fields S_k have an interaction with two Z bosons of the form $S_k Z_\mu Z^\mu$;
  2. All pairs of neutral scalar fields have an interaction with a Z boson of the form $Z_\mu \left(S_k \partial^\mu S_l - S_l \partial^\mu S_k \right)$.

3. For any pair of neutral scalar fields S_k and S_l, the three interactions $Z_\mu \left(S_k \partial^\mu S_l - S_l \partial^\mu S_k \right)$, $Z_\mu Z^\mu S_k$, and $Z_\mu Z^\mu S_l$, are simultaneously present.

4. All neutral scalar fields S_k have an interaction with a charged scalar $C^\pm$ of the form $S_k C^+ C^-$.

• CP is violated if any neutral scalar S has an interaction with a quark q of the form $S \bar{q} \left(a + i b \gamma_5 \right) q$, with the real numbers a and b simultaneously non-zero.

23

SPONTANEOUS CP VIOLATION

23.1 Introduction

Spontaneous CP violation (SCPV), also termed spontaneous CP breaking, occurs when CP is a symmetry of the original Lagrangian but, after spontaneous symmetry breaking, no CP symmetry remains. This means the following:

- There is a transformation that may be physically interpreted as CP under which the Lagrangian is invariant.

- There is no transformation that may be physically interpreted as CP under which both the Lagrangian and the vacuum are invariant.

The idea of spontaneous CP breaking was put forward by Lee (1973) in the same year in which Kobayashi and Maskawa (1973) suggested the possible existence of a third generation and of the corresponding CP-violating phase in the charged gauge interaction. At that time, only two—incomplete—generations were known, and it was impossible to explain CP violation in the context of the standard model (SM).

In the framework of relativistic quantum field theory, the CPT theorem tells us that if CP is a good symmetry of a Lagrangian, then T will also be an invariance of that Lagrangian. Also, if a given vacuum breaks CP, it will necessarily break T too. As emphasized by Lee (1973), the idea of having spontaneous CP and T violation is especially attractive from a theoretical point of view. On the one hand, if time reversal and space inversion (identified as CP, not as P) are good symmetries of the Lagrangian, then the full Poincaré group of space–time transformations, including the discrete ones, is a group of symmetries of the Lagrangian. On the other hand, renormalizable gauge theories are based on the spontaneous-symmetry-breaking mechanism, and it is natural to have the spontaneous breaking of CP and T as an integral part of that mechanism.

Only scalar fields may have non-zero vacuum expectation values (VEVs), lest Lorentz invariance be broken. Therefore, the scalar sector and the scalar potential are crucial in any model of spontaneous CP breaking. We shall first prove that

Theorem 23.1 *SCPV is not possible in the SU(2)⊗U(1) gauge theory with only one Higgs doublet ϕ.*

Proof In order to preserve Poincaré invariance, the VEV of ϕ must be constant over space–time, i.e., independent of both t and $\vec{r}$. By means of an SU(2) gauge rotation, one may bring that VEV to the form

$$\langle 0|\phi|0\rangle = \begin{pmatrix} 0 \\ ve^{i\psi} \end{pmatrix}. \tag{23.1}$$

The most general CP transformation of ϕ is (see eqn 22.38)

$$(CP)\,\phi\,(t,\vec{r})\,(CP)^{\dagger} = e^{i\vartheta}\phi^{\dagger T}(t,-\vec{r})\,. \tag{23.2}$$

The phase ϑ is arbitrary. Choosing $\vartheta = 2\psi$, the VEV in eqn (23.1) is invariant under the CP transformation in eqn (23.2). $\qquad\qquad\qquad\qquad\qquad\square$

Notice that this argument is independent of the form of the scalar potential.

In this chapter we shall study the two basic models of SCPV in the framework of the SU(2)⊗U(1) gauge theory: the two-Higgs-doublet model (THDM) of Lee and the three-Higgs-doublet model of Branco. We leave to later chapters models in which SCPV appears in the context of a theory either with an extended gauge sector, or with an extended spectrum of fermions.

23.2   CP invariance and reality of the coupling constants

One must first specify the CP transformation under which the model is assumed to be invariant before spontaneous symmetry breaking. In this chapter, we always assume that the Lagrangian is invariant under a trivial CP transformation, which mixes neither the various fermion generations nor the various scalar doublets among themselves, i.e.,

$$\begin{aligned}
(CP)\,p_{Li}\,(t,\vec{r})\,(CP)^{\dagger} &= \gamma^{0}C\overline{p_{Li}}^{T}\,(t,-\vec{r})\,, \\
(CP)\,n_{Li}\,(t,\vec{r})\,(CP)^{\dagger} &= \gamma^{0}C\overline{n_{Li}}^{T}\,(t,-\vec{r})\,, \\
(CP)\,p_{Ri}\,(t,\vec{r})\,(CP)^{\dagger} &= \gamma^{0}C\overline{p_{Ri}}^{T}\,(t,-\vec{r})\,, \\
(CP)\,n_{Ri}\,(t,\vec{r})\,(CP)^{\dagger} &= \gamma^{0}C\overline{n_{Ri}}^{T}\,(t,-\vec{r})\,,
\end{aligned} \tag{23.3}$$

cf. eqns (14.2). The index i is a generation index, running from 1 to n_g. Notice that we have assumed the $n_g \times n_g$ matrices K_L, K_R^p, and K_R^n to be equal to the unit matrix. Without loss of generality, we have omitted the phase ξ_W. Similarly, instead of the general eqn (22.38), we assume

$$(CP)\,\phi_a\,(t,\vec{r})\,(CP)^{\dagger} = \phi_a^{\dagger T}(t,-\vec{r})\,. \tag{23.4}$$

As we are considering the Lagrangian *before spontaneous symmetry breaking*, we do not have to restrict the CP transformation by means of eqn (22.37).

The assumption of CP invariance of the Lagrangian forces the matrices of Yukawa couplings to be real. For instance,

$$\begin{aligned}
(CP)\,\overline{Q_{Li}}\,(\Gamma_a)_{ij}\,\phi_a n_{Rj}\,(CP)^{\dagger} &= (\Gamma_a)_{ij}\left(-Q_{Li}^T C^{-1}\gamma^0\right)\phi_a^{\dagger T}\gamma^0 C\overline{n_{Rj}}^T \\
&= -(\Gamma_a)_{ij}\,Q_{Li}^T\phi_a^{\dagger T}\overline{n_{Rj}}^T \\
&= (\Gamma_a)_{ij}\,\overline{n_{Rj}}\phi_a^{\dagger}Q_{Li}
\end{aligned}$$

$$= \left[\overline{Q_{Li}} \left(\Gamma_a \right)_{ij}^* \phi_a n_{Rj} \right]^\dagger.$$

Thus, CP symmetry together with the Hermiticity of the Lagrangian require $(\Gamma_a)_{ij}^* = (\Gamma_a)_{ij}$. In the same way, the scalar potential must have real coupling constants:

$$(\mathcal{CP}) \, Y_{ab} \phi_a^\dagger \phi_b \, (\mathcal{CP})^\dagger = Y_{ab} \phi_a^T \phi_b^{\dagger\,T} = Y_{ab} \phi_b^\dagger \phi_a = \left(Y_{ab}^* \phi_a^\dagger \phi_b \right)^\dagger.$$

Thus, $Y_{ab}^* = Y_{ab}$ if the Lagrangian is to be simultaneously Hermitian and CP-invariant. Analogously, $Z_{abcd} = Z_{abcd}^*$.

Thus, the assumption of CP invariance of the Lagrangian under the trivial CP transformation will in practice correspond to setting all coupling constants real, both in the scalar potential and in the Yukawa interactions.

23.3 Lee model

23.3.1 The Higgs potential

The model of Lee (1973) is a THDM. Since the model is built to achieve SCPV, the Lagrangian is assumed to be CP- and T-invariant. No other discrete or continuous symmetry—except of course for the gauge symmetry—is assumed. The scalar potential is therefore the one in eqn (22.12), with

$$e^{2i\delta_3} = e^{2i\delta_5} = e^{2i\delta_6} = e^{2i\delta_7} = 1, \tag{23.5}$$

so that all coupling constants are real.

One may perform a weak-basis transformation of ϕ_1 and ϕ_2, rotating them by means of an orthogonal transformation, so that the form of the CP transformation in eqn (23.4) does not change:

$$\begin{pmatrix} \phi_1' \\ \phi_2' \end{pmatrix} = \begin{pmatrix} \cos\sigma & \sin\sigma \\ -\sin\sigma & \cos\sigma \end{pmatrix} \begin{pmatrix} \phi_1 \\ \phi_2 \end{pmatrix}. \tag{23.6}$$

The angle σ may be chosen in such a way as to simplify the potential, making for instance $m_3 = 0$. In his original paper Lee adopted this simplification.

For any values of ϕ_1 and ϕ_2, a gauge transformation may be performed which makes, at every space–time point, the upper component of ϕ_2 equal to zero and the two components of ϕ_1 real and positive. This can be done in particular for the vacuum state, which has ϕ_1 and ϕ_2 constant over space–time, in order to preserve Poincaré invariance. One may therefore choose a gauge in which

$$\langle 0 | \phi_1 (t, \vec{r}) | 0 \rangle = \begin{pmatrix} u \\ v_1 \end{pmatrix},$$

$$\langle 0 | \phi_2 (t, \vec{r}) | 0 \rangle = \begin{pmatrix} 0 \\ v_2 e^{i\theta} \end{pmatrix}, \tag{23.7}$$

with u, v_1, and v_2 real and positive. Then,

$$V_0 \equiv \langle 0 | V | 0 \rangle = m_1 \left(u^2 + v_1^2 \right) + m_2 v_2^2 + a_1 \left(u^2 + v_1^2 \right)^2 + a_2 v_2^4 + a_3 \left(u^2 + v_1^2 \right) v_2^2$$

$$+(a_4 - 2a_5)v_1^2 v_2^2 + 4a_5 v_1^2 v_2^2 (\cos\theta - 2\Delta)\cos\theta, \qquad (23.8)$$

where

$$\Delta \equiv -\frac{m_3 + a_6 \left(u^2 + v_1^2\right) + a_7 v_2^2}{4 a_5 v_1 v_2}. \qquad (23.9)$$

Contrary to what happens in the SM, the preservation by the vacuum of the unbroken U(1) gauge symmetry of electromagnetism is not automatic in a THDM. If u, v_1, and v_2 all are non-zero, no U(1) remains unbroken and all four gauge bosons turn out massive. On the other hand, if $v_1 = 0$ there is a gauge transformation which eliminates the phase θ, and no spontaneous CP breaking occurs, because the vacuum state is invariant under eqn (23.4). We therefore want to find a range of parameters such that $u = 0$ but $v_1 \neq 0$ and $v_2 \neq 0$ at the minimum.

Theorem 23.2 *There is a range of parameters of the Lee-model potential for which its minimum both preserves a U(1) gauge invariance and is not invariant under the CP transformation in eqn (23.4).*

Proof It follows from eqn (23.8) that

$$\frac{\partial V_0}{\partial \theta} = 8 a_5 v_1^2 v_2^2 (\Delta - \cos\theta)\sin\theta, \qquad (23.10)$$

$$\frac{\partial V_0}{\partial u^2} - \frac{\partial V_0}{\partial v_1^2} = (2a_5 - a_4)v_2^2 + 4a_5 v_2^2 (\Delta - \cos\theta)\cos\theta. \qquad (23.11)$$

The minimum of V_0 must be at a point where $\partial V_0/\partial\theta = 0$. There, from eqn (23.10), either $\sin\theta = 0$ or $\cos\theta = \Delta$. If $\sin\theta = 0$ the CP symmetry in eqn (23.4) remains unbroken. We would therefore like to have the minimum at $\cos\theta = \Delta$. We see in eqn (23.8) that this occurs whenever $a_5 > 0$.

As u is non-negative, the minimum of V_0 must either be at $u = 0$ or at a point where $\partial V_0/\partial u^2 = 0$. Similarly, the minimum of V_0 must have either $v_1 = 0$ or $\partial V_0/\partial v_1^2 = 0$. If $v_1 = 0$ the phase θ can be gauged away, and SCPV is not achieved. We therefore assume that $\partial V_0/\partial v_1^2 = 0$. Then, from eqn (23.11) with $\cos\theta = \Delta$ we find that, if we also assume that $2a_5 - a_4$ is positive, then $\partial V_0/\partial u^2 > 0$ and the minimum is at $u = 0$. $\qquad \square$

Thus, for $a_5 > 0$ and $2a_5 > a_4$ the upper components of $\langle 0|\phi_1|0\rangle$ and $\langle 0|\phi_2|0\rangle$ are simultaneously zero, and a U(1) gauge symmetry—electromagnetism—is preserved. Moreover, $\cos\theta = \Delta$ implies the non-preservation of any CP symmetry after spontaneous symmetry breaking.

It remains to solve the equations $\partial V_0/\partial v_1^2 = \partial V_0/\partial v_2^2 = 0$ and show that there is a range of parameters of the potential for which the minimum has non-zero v_1 and v_2; moreover, those values of v_1 and v_2 must be such that $|\Delta| < 1$, because $\cos\theta = \Delta$ at the minimum. That task is not very instructive and we shall not pursue it here.

23.3.2   *The Yukawa interactions*

The quark Yukawa Lagrangian of the Lee model is in eqn (22.18). The matrices Γ_a and Δ_a (for $a = 1$ and $a = 2$) are real so that $\mathcal{L}_Y^{(q)}$ is CP-invariant.

As the matrices Γ_a and Δ_a are real and arbitrary, the matrices M_p and M_n in eqns (22.19) are complex and arbitrary, even though all their phases arise from the single phase θ in the VEVs. Hence, the CKM matrix is complex, and for three or more fermion generations the charged gauge interaction violates CP.

For specific choices of the matrices Γ_a and Δ_a one may verify whether a physical phase in the CKM matrix is generated or not. From eqns (22.19) and from the fact that the matrices Γ_a and Δ_a are real, it follows that

$$
\begin{aligned}
H_p &= v_1^2 \Delta_1 \Delta_1^T + v_2^2 \Delta_2 \Delta_2^T + v_1 v_2 \left(\Delta_1 \Delta_2^T + \Delta_2 \Delta_1^T \right) \cos \theta \\
&\quad + i v_1 v_2 \left(\Delta_1 \Delta_2^T - \Delta_2 \Delta_1^T \right) \sin \theta, \\
H_n &= v_1^2 \Gamma_1 \Gamma_1^T + v_2^2 \Gamma_2 \Gamma_2^T + v_1 v_2 \left(\Gamma_2 \Gamma_1^T + \Gamma_1 \Gamma_2^T \right) \cos \theta \\
&\quad + i v_1 v_2 \left(\Gamma_2 \Gamma_1^T - \Gamma_1 \Gamma_2^T \right) \sin \theta.
\end{aligned}
\tag{23.12}
$$

One may then evaluate $\operatorname{tr} [H_p, H_n]^3$; if that trace is non-zero, the charged gauge interaction violates CP. (In the Lee model there are of course extra sources of CP violation, notably the Yukawa interactions of the charged and neutral Higgs scalars.)

23.3.3   *Effective superweak models*

From eqns (23.12) one gets a clue on how to construct an effective superweak model, within the framework of gauge theories, in which the CKM matrix is real and the charged gauge interaction does not violate CP. One may achieve this through the introduction of additional symmetries leading to $\Gamma_2 \Gamma_1^T = \Gamma_1 \Gamma_2^T$ and $\Delta_2 \Delta_1^T = \Delta_1 \Delta_2^T$, thus obtaining H_p and H_n real and thereby guaranteeing the vanishing of $\operatorname{tr} [H_p, H_n]^3$. In order to illustrate the main features of effective superweak models, we describe below an example which has been suggested by Lavoura (1994), who proposed a THDM in which CP violation arises only from the interactions of neutral Higgs bosons, through the mechanism of scalar-pseudoscalar mixing. As we shall see, in Lavoura's model there is no CP violation in the exchange of either $W^\pm$ or charged Higgs bosons. Lavoura's model is a THDM where one introduces a symmetry under which

$$
\begin{aligned}
\phi_2 &\to -\phi_2, \\
p_{R3} &\to -p_{R3}, \tag{23.13} \\
n_{R3} &\to -n_{R3}. \tag{23.14}
\end{aligned}
$$

As a result of this symmetry, the Yukawa-coupling matrices have the following structure:

$$
\Delta_1, \Gamma_1 \sim \begin{pmatrix} \times & \times & 0 \\ \times & \times & 0 \\ \times & \times & 0 \end{pmatrix} ; \quad \Delta_2, \Gamma_2 \sim \begin{pmatrix} 0 & 0 & \times \\ 0 & 0 & \times \\ 0 & 0 & \times \end{pmatrix},
\tag{23.15}
$$

where the crosses denote non-vanishing matrix elements. From eqn (23.15) one obtains $\Delta_2 \Delta_1^T = \Delta_1 \Delta_2^T = \Gamma_2 \Gamma_1^T = \Gamma_1 \Gamma_2^T = 0$. Therefore, in Lavoura's model $\text{tr}\,[H_p, H_n]^3$ vanishes and the CKM matrix is real, implying a null-area unitarity triangle.

We now remind the reader of the matrices Y_u and Y_d in eqns (22.34), which control the flavour-changing neutral Yukawa interactions (FCNYI) mediated by N and A (see eqn 22.35). Taking into account the special structure of the Γ_a in Lavoura's model, one obtains

$$Y_d = \begin{pmatrix} m_d \left(\frac{v_2}{v_1} - K x^2 \right) & -m_d K x y & -m_d K x z \\ -m_s K x y & m_s \left(\frac{v_2}{v_1} - K y^2 \right) & -m_s K y z \\ -m_b K x z & -m_b K y z & m_b \left(\frac{v_2}{v_1} - K z^2 \right) \end{pmatrix}, \qquad (23.16)$$

where $K \equiv v^2 / (v_1 v_2)$ and $x^2 + y^2 + z^2 = 1$. The matrix Y_u has similar form.

The strength of the FCNYI contributing to Δm_K, Δm_{B_d}, and Δm_{B_s} depend on the off-diagonal matrix elements of Y_d. In the literature one often finds some *ad hoc* assumptions about how the strength of the FCNYI scales from one neutral-meson system to the next. In the explicit example of Lavoura's model one sees that the strength of the superweak interactions does scale with the quark masses, but this scaling may be irrelevant, since the additional parameters x, y and z are arbitrary. This is actually a generic feature of superweak-like models: the relative strength of the FCNYI contributing to $|\Delta S| = 2$ and to $|\Delta B| = 2$ processes is to a large extent arbitrary. Thus, one should be cautious when attempting to constrain or to rule out superweak-like models by using the experimental data, in particular the data on the mass differences in neutral-meson systems. Often, those analyses use *ad hoc* assumptions about the relative strength of the FCNYI; the example of Lavoura's model shows that those assumptions may not be valid in general.

23.4   Natural flavour conservation and SCPV

23.4.1   *Mechanism for generating ϵ*

We now consider the multi-Higgs-doublet model (MHDM) with natural flavour conservation (NFC) of § 22.11.1, and examine the consequences of assuming that it is CP-conserving at Lagrangian level. The Yukawa interactions are given by eqns (22.3) and (22.79). The mass matrices then are

$$\begin{aligned} M_p &= v_2 \Delta_2 e^{-i\theta_2}, \\ M_n &= v_1 \Gamma_1. \end{aligned} \qquad (23.17)$$

We assume that Δ_2 and Γ_1 are real. Then, they may be bi-diagonalized by orthogonal matrices:

$$\begin{aligned} O_L^{p\,T} \Delta_2 O_R^p &= \frac{M_u}{v_2}, \\ O_L^{n\,T} \Gamma_1 O_R^n &= \frac{M_d}{v_1}. \end{aligned} \qquad (23.18)$$

Therefore, the mass matrices may be bi-diagonalized while keeping U_L^p and U_L^n real:

$$
\begin{aligned}
U_L^{p\dagger} &= O_L^{p\,T}, \\
U_R^p &= e^{i\theta_2} O_R^p, \\
U_L^{n\dagger} &= O_L^{n\,T}, \\
U_R^n &= O_R^n,
\end{aligned}
\tag{23.19}
$$

cf. eqns (12.15). The CKM matrix $V = U_L^{p\dagger} U_L^n = O_L^{p\,T} O_L^n$ is real (orthogonal). Thus, NFC, together with the assumption of CP-invariance of the Lagrangian before spontaneous symmetry breaking, lead to a real CKM matrix, thereby eliminating the standard mechanism of CP violation (Branco 1980a).

Taking into account eqn (17.8), it is then obvious that, if one wants to fit ϵ in the MHDM with NFC, one must have extra contributions to M_{12}. Diagrams with the exchange of neutral scalar fields cannot help, because—by definition of NFC—the neutral scalars have diagonal Yukawa couplings with the quarks, i.e., they cannot change the quark flavour and therefore they cannot contribute to M_{12}.

Still, in a MHDM there are contributions to K^0–$\overline{K^0}$ mixing beyond the box diagrams of Figs. 17.2 and 17.3. Namely, one may substitute one or both $W^\pm$ in those diagrams by charged scalars. The extra diagrams may in principle lead to $\epsilon \neq 0$, even when the CKM matrix V is real, because the Yukawa couplings of the charged scalars are not necessarily real.

The Yukawa couplings of the charged scalars are given in the last line of eqn (22.81). They are determined by the product of V and the diagonal quark mass matrices and, apparently, they are real. However, as a complex phase may be present in the mixing of $\varphi_1^\pm$ and $\varphi_2^\pm$, the Yukawa couplings of the physical charged scalars may be complex.

Unfortunately, in a *two*-Higgs-doublet model with NFC this mechanism for generating ϵ does not exist. Indeed, in a THDM the physical charged scalars are given by the first eqn (22.23):

$$
C^\pm = \frac{v_2 \varphi_1^\pm - v_1 \varphi_2^\pm}{v}.
\tag{23.20}
$$

This is a *real* linear combination of $\varphi_1^\pm$ and of $\varphi_2^\pm$. Therefore, the Yukawa couplings of $C^\pm$ are real, just as those of $\varphi_1^\pm$ and of $\varphi_2^\pm$, and no non-spurious phase in M_{12} can be obtained in this way.

We conclude that in a multi-Higgs-doublet model with NFC the CP violation which explains the observed ϵ of the kaon system may only originate from box diagrams with charged scalars, when the mixing of $\varphi_1^\pm$ and $\varphi_2^\pm$ is complex. However, this can only happen with more than two Higgs doublets, because in a THDM the mixing of $\varphi_1^\pm$ and $\varphi_2^\pm$ to form the physical charged scalars $C^\pm$ is real. Thus, we must go to three scalar doublets in order to implement this mechanism of CP violation. This is what was done in Branco model.

23.4.2 *Symmetries of the scalar potential and SCPV*

In the previous subsection we have shown that a THDM with NFC and SCPV cannot generate ϵ, because both the charged gauge interaction and the charged Yukawa interactions are determined by real mixing matrices, while the neutral Yukawa interactions are flavour-diagonal and cannot contribute to M_{12}.

A different argument shows that, as a matter of fact, a THDM with NFC and SCPV cannot exist, if the scalar potential is renormalizable. Indeed, eqns (22.79) must be enforced by some symmetry, else they will be unstable under renormalization. The symmetry must treat differently ϕ_1 and ϕ_2. The simplest possibility is a reflection symmetry like the one mentioned in eqns (22.80). But then, one may prove that

Theorem 23.3 *Once a reflection symmetry is imposed on the scalar potential of the two-Higgs-doublet model, SCPV cannot be obtained.*

Proof A reflection symmetry forces m_3, a_6, and a_7 to be zero, and then Δ in eqn (23.9) vanishes. Hence, the candidate vacuum for SCPV, identified in the previous section, has $\cos\theta = 0$, i.e.,

$$\phi_1 = \begin{pmatrix} 0 \\ v_1 \end{pmatrix}, \quad \phi_2 = \begin{pmatrix} 0 \\ \pm i v_2 \end{pmatrix}, \tag{23.21}$$

with v_1 and v_2 real. The corresponding CP transformation is

$$(CP)\,\phi_1\,(t,\vec{r})\,(CP)^\dagger = \phi_1^{\dagger\,T}(t,-\vec{r}),$$
$$(CP)\,\phi_2\,(t,\vec{r})\,(CP)^\dagger = -\phi_2^{\dagger\,T}(t,-\vec{r}). \tag{23.22}$$

It so happens that the potential with $m_3 = a_6 = a_7 = 0$ is symmetric under this CP transformation too, and not only under the one in eqn (23.4). Moreover, we may extend this CP transformation to the fermion sector by assuming that

$$(CP)\,p_{Li}\,(t,\vec{r})\,(CP)^\dagger = \gamma^0 C\overline{p_{Li}}^{\,T}(t,-\vec{r}),$$
$$(CP)\,n_{Li}\,(t,\vec{r})\,(CP)^\dagger = \gamma^0 C\overline{n_{Li}}^{\,T}(t,-\vec{r}),$$
$$(CP)\,p_{Ri}\,(t,\vec{r})\,(CP)^\dagger = -\gamma^0 C\overline{p_{Ri}}^{\,T}(t,-\vec{r}),$$
$$(CP)\,n_{Ri}\,(t,\vec{r})\,(CP)^\dagger = \gamma^0 C\overline{n_{Ri}}^{\,T}(t,-\vec{r}) \tag{23.23}$$

(notice the minus sign in the third equation), instead of eqns (23.3), and then the Yukawa interactions are CP-invariant too. $\qquad\Box$

The crucial point is that the potential with $m_3 = a_6 = a_7 = 0$ and $\exp(2i\delta_5) = 1$, together with the Yukawa interactions of eqn (22.79), allow for two choices of CP symmetry, instead of only one choice: CP may either be given by eqns (23.3) and (23.4), or by eqns (23.22) and (23.23). As a consequence, the vacuum state must satisfy extra requirements if it is to break CP. This illustrates the fact that all possibilities must be tried for the CP transformation before one can be sure that there is CP violation.

In general, whenever one imposes on a Lagrangian a discrete symmetry D together with the CP symmetry in eqn (23.4), one must keep in mind that, automatically, the product of D and the CP transformation in eqn (23.4) constitutes another CP transformation under which the Lagrangian is invariant too. Therefore, in order for SCPV to occur, the vacuum state must violate not only the original CP symmetry in eqn (23.4), but also its product with D, which is also a CP symmetry. This is usually difficult, both because a more constrained potential has a lesser capacity to generate a vacuum with non-trivial phases, and because those phases have to satisfy extra requirements in order to correspond to SCPV. Examples of this general problem have been given by Branco *et al.* (1984).

23.5   Branco model

We saw in § 23.4.1 that at least three doublets are needed in order to achieve a model with NFC and SCPV in which ϵ is generated by box diagrams with charged scalars. Such a model was put forward by Branco (1980b). It is identical to the Weinberg model of § 22.7.3, except for the assumption that CP violation is spontaneous.

The Yukawa couplings are given by eqns (22.3) and (22.79). They are protected by the discrete symmetries in eqns (22.80). Those symmetries also constrain the scalar potential to be the one in eqn (22.45). As CP symmetry is imposed at the Lagrangian level, we assume

$$\varepsilon_1 = \varepsilon_2 = \varepsilon_3 = 0, \tag{23.24}$$

while d_1, d_2, and d_3 may be either positive or negative. The vacuum value of the potential is the one in eqn (22.46) with the proviso in eqns (23.24). The stationarity equations are

$$
\begin{aligned}
\frac{-1}{4v_2^2}\frac{\partial V_0}{\partial \theta_2} &= 0 = d_3 v_1^2 \sin 2\theta_2 + d_1 v_3^2 \sin\left(2\theta_2 - 2\theta_3\right), \\
\frac{1}{4v_3^2}\frac{\partial V_0}{\partial \theta_3} &= 0 = d_1 v_2^2 \sin\left(2\theta_2 - 2\theta_3\right) - d_2 v_1^2 \sin 2\theta_3,
\end{aligned}
\tag{23.25}
$$

and

$$
\begin{aligned}
\frac{\partial V_0}{\partial v_1^2} &= 0 = m_1 + 2a_1 v_1^2 + \left(b_2 + c_2\right) v_3^2 + \left(b_3 + c_3\right) v_2^2 \\
&\quad + 2d_2 v_3^2 \cos 2\theta_3 + 2d_3 v_2^2 \cos 2\theta_2, \\
\frac{\partial V_0}{\partial v_2^2} &= 0 = m_2 + 2a_2 v_2^2 + \left(b_3 + c_3\right) v_1^2 + \left(b_1 + c_1\right) v_3^2 \\
&\quad + 2d_3 v_1^2 \cos 2\theta_2 + 2d_1 v_3^2 \cos\left(2\theta_2 - 2\theta_3\right), \\
\frac{\partial V_0}{\partial v_3^2} &= 0 = m_3 + 2a_3 v_3^2 + \left(b_1 + c_1\right) v_2^2 + \left(b_2 + c_2\right) v_1^2 \\
&\quad + 2d_1 v_2^2 \cos\left(2\theta_2 - 2\theta_3\right) + 2d_2 v_1^2 \cos 2\theta_3.
\end{aligned}
\tag{23.26}
$$

Equations (23.26) will be used to trade the coefficients of the potential m_1, m_2, and m_3 by the VEVs v_1^2, v_2^2, and v_3^2.

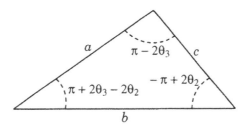

FIG. 23.1. Geometrical representation of eqns (23.27).

23.5.1 *CP-breaking vacuum*

Equations (23.25) admit four trivial solutions:

1. $2\theta_2 = 0 \pmod{2\pi}$ and $2\theta_3 = 0 \pmod{2\pi}$;
2. $2\theta_2 = \pi \pmod{2\pi}$ and $2\theta_3 = \pi \pmod{2\pi}$;
3. $2\theta_2 = 0 \pmod{2\pi}$ and $2\theta_3 = \pi \pmod{2\pi}$;
4. $2\theta_2 = \pi \pmod{2\pi}$ and $2\theta_3 = 0 \pmod{2\pi}$.

These solutions are CP-conserving. This is because the Lagrangian enjoys not only the trivial CP symmetry in eqns (23.3) and (23.4), but also three other CP symmetries, which are the products of the trivial CP transformation with D_1 and/or D_2 transformations. Fortunately, for a range of the parameters of the potential there is another solution to the system of eqns (23.25), and that solution is the absolute minimum of V_0.

In order to see this, note that eqns (23.25) may be written in the form

$$a \sin\left(\pi + 2\theta_3 - 2\theta_2\right) = c \sin\left(-\pi + 2\theta_2\right),$$
$$b \sin\left(\pi + 2\theta_3 - 2\theta_2\right) = c \sin\left(\pi - 2\theta_3\right),$$
(23.27)

where $a \equiv d_1 d_2 v_3^2$, $b \equiv d_1 d_3 v_2^2$, and $c \equiv d_2 d_3 v_1^2$. If a, b, c, and $\lambda\left(a, b, c\right)$—remember eqn (13.37)—are positive, then eqns (23.27) may be geometrically interpreted as meaning that a triangle with sides a, b, and c and with internal angles $\pi + 2\theta_3 - 2\theta_2$, $-\pi + 2\theta_2$, and $\pi - 2\theta_3$ exists, as depicted in Fig. 23.1. This geometrical interpretation yields

$$\cos 2\theta_2 = \frac{a^2 - b^2 - c^2}{2bc}, \qquad \sin 2\theta_2 = \frac{\mp\sqrt{\lambda\left(a, b, c\right)}}{2bc},$$

$$\cos 2\theta_3 = \frac{b^2 - c^2 - a^2}{2ca}, \qquad \sin 2\theta_3 = \frac{\pm\sqrt{\lambda\left(a, b, c\right)}}{2ca}, \qquad (23.28)$$

$$\cos\left(2\theta_2 - 2\theta_3\right) = \frac{c^2 - a^2 - b^2}{2ab}, \quad \sin\left(2\theta_2 - 2\theta_3\right) = \frac{\pm\sqrt{\lambda\left(a, b, c\right)}}{2ab}.$$

If $\lambda\left(a, b, c\right) > 0$ this solution of eqns (23.25) has non-trivial vacuum angles (the triangle has non-zero area), and CP is spontaneously broken, as desired.

It remains to see whether this solution is the minimum of V_0, and not just a stability point. It is easily found that when d_1, d_2, and d_3 are all positive this solution is the absolute minimum of V_0, because V_0 is smaller here than at any other stability point.

23.5.2 Scalar mass matrices

Let us compute the mass matrices of the scalars. One develops the potential in eqn (22.45), using the form of the doublets in eqn (22.8). The terms linear in the fields ρ_a and η_a vanish because of the stability eqns (23.25) and (23.26). The terms bilinear in the fields may be simplified by use of the same equations. One obtains

$$
\begin{aligned}
V = {} & -a_1 v_1^4 - a_2 v_2^4 - a_3 v_3^4 - (b_1 + c_1)\, v_2^2 v_3^2 - (b_2 + c_2)\, v_3^2 v_1^2 - (b_3 + c_3)\, v_1^2 v_2^2 \\
& + \left(d_1^2 d_2^2 v_3^4 + d_3^2 d_1^2 v_2^4 + d_2^2 d_3^2 v_1^4 \right) / (d_1 d_2 d_3) \\[2mm]
& - \begin{pmatrix} \dfrac{\varphi_1^-}{v_1} & \dfrac{\varphi_2^-}{v_2} & \dfrac{\varphi_3^-}{v_3} \end{pmatrix}
\begin{pmatrix}
X_{12} + X_{31} & -X_{12} + iY & -X_{31} - iY \\
-X_{12} - iY & X_{12} + X_{23} & -X_{23} + iY \\
-X_{31} + iY & -X_{23} - iY & X_{31} + X_{23}
\end{pmatrix}
\begin{pmatrix} \varphi_1^+ / v_1 \\ \varphi_2^+ / v_2 \\ \varphi_3^+ / v_3 \end{pmatrix} \\[2mm]
& - \frac{1}{2} \begin{pmatrix} \dfrac{\eta_1}{v_1} & \dfrac{\eta_2}{v_2} & \dfrac{\eta_3}{v_3} \end{pmatrix}
\begin{pmatrix}
D_{12} + D_{31} & -D_{12} & -D_{31} \\
-D_{12} & D_{12} + D_{23} & -D_{23} \\
-D_{31} & -D_{23} & D_{31} + D_{23}
\end{pmatrix}
\begin{pmatrix} \eta_1 / v_1 \\ \eta_2 / v_2 \\ \eta_3 / v_3 \end{pmatrix} \\[2mm]
& - \frac{1}{2} \begin{pmatrix} \dfrac{\rho_1}{v_1} & \dfrac{\rho_2}{v_2} & \dfrac{\rho_3}{v_3} \end{pmatrix} M_\rho
\begin{pmatrix} \rho_1 / v_1 \\ \rho_2 / v_2 \\ \rho_3 / v_3 \end{pmatrix} \\[2mm]
& - 2Y \begin{pmatrix} \dfrac{\eta_1}{v_1} & \dfrac{\eta_2}{v_2} & \dfrac{\eta_3}{v_3} \end{pmatrix}
\begin{pmatrix}
0 & 1 & -1 \\
-1 & 0 & 1 \\
1 & -1 & 0
\end{pmatrix}
\begin{pmatrix} \rho_1 / v_1 \\ \rho_2 / v_2 \\ \rho_3 / v_3 \end{pmatrix}
\end{aligned}
$$

+cubic and quartic terms.   (23.29)

Here,

$$
\begin{aligned}
D_{23} &\equiv 4 d_1 v_2^2 v_3^2 \cos(2\theta_2 - 2\theta_3), & X_{23} &\equiv c_1 v_2^2 v_3^3 + D_{23}/2, \\
D_{31} &\equiv 4 d_2 v_3^2 v_1^2 \cos 2\theta_3, & X_{31} &\equiv c_2 v_3^2 v_1^3 + D_{31}/2, \\
D_{12} &\equiv 4 d_3 v_1^2 v_2^2 \cos 2\theta_2, & X_{12} &\equiv c_3 v_1^2 v_2^3 + D_{12}/2;
\end{aligned}
\tag{23.30}
$$

the matrix M_ρ is 3×3 real and symmetric; finally,

$$
Y \equiv \frac{\sqrt{\lambda}}{d_1 d_2 d_3}
\tag{23.31}
$$

is a CP-violating quantity proportional to the area of the triangle in Fig. 23.1. The Goldstone bosons are

$$
\begin{aligned}
\varphi^\pm &= \frac{v_1 \varphi_1^\pm + v_2 \varphi_2^\pm + v_3 \varphi_3^\pm}{v}, \\
\chi &= \frac{v_1 \eta_1 + v_2 \eta_2 + v_3 \eta_3}{v}.
\end{aligned}
\tag{23.32}
$$

The interesting point in eqn (23.29) is the role played by Y. We see that that quantity, on the one hand, causes mixing of the ρ_a with the η_a, and on the other hand, it generates a complex mass matrix for the $\varphi_a^{\pm}$. Both effects are CP-violating.

In order to see that the mixing of the ρ_a and η_a violates CP, consider for instance the Yukawa interactions of ρ_1 and η_1 with the strange quark, which are given by (see eqn 22.81)

$$-\frac{m_s}{\sqrt{2}v_1}\bar{s}\left(\rho_1 + i\eta_1\gamma_5\right)s. \tag{23.33}$$

As $\bar{s}s \overset{\text{CP}}{\to} \bar{s}s$ and $\bar{s}\gamma_5 s \overset{\text{CP}}{\to} -\bar{s}\gamma_5 s$, we conclude that ρ_1 is CP-even while η_1 is CP-odd. When $Y \neq 0$ each physical neutral scalar contains simultaneously ρ_1 and η_1 components, hence it does not have a definite CP-parity, and CP is violated. A consequence of CP violation in this case would be the generation of an electric dipole moment for the strange quark.

In the charged-scalar sector, the eigenstates of mass are $\varphi^{\pm}$, which are Goldstone bosons; $X_1^{\pm}$, with mass m_1; and $X_2^{\pm}$, with mass m_2. They are related to the $\varphi_a^{\pm}$ by an unitary matrix. We may write this relation as

$$\begin{pmatrix} \varphi_1^+ \\ \varphi_2^+ \\ \varphi_3^+ \end{pmatrix} = \begin{pmatrix} v_1/v & \left(-cv_2 e^{i\zeta} - s\frac{v_1 v_3}{v}\right)/v_{12} & \left(sv_2 e^{i\zeta} - c\frac{v_1 v_3}{v}\right)/v_{12} \\ v_2/v & \left(cv_1 e^{i\zeta} - s\frac{v_2 v_3}{v}\right)/v_{12} & \left(-sv_1 e^{i\zeta} - c\frac{v_2 v_3}{v}\right)/v_{12} \\ v_3/v & s\,v_{12}/v & c\,v_{12}/v \end{pmatrix} \begin{pmatrix} \varphi^+ \\ X_1^+ \\ X_2^+ \end{pmatrix}. \tag{23.34}$$

Here, $v_{12} \equiv \sqrt{v_1^2 + v_2^2}$; also, $c \equiv \cos\alpha$ and $s \equiv \sin\alpha$, where α is a mixing angle; finally, ζ is a phase which arises in the mixing once $Y \neq 0$. The Yukawa interaction of the charged scalars are given by—see eqn (22.81)—

$$\bar{u}\left(\frac{\varphi_2^+}{v_2}M_u V\gamma_L - \frac{\varphi_1^+}{v_1}V M_d\gamma_R\right)d + \text{H.c.}$$

$$= \frac{X_1^+}{v_{12}}\bar{u}\left[\left(c\frac{v_1}{v_2}e^{i\zeta} - s\frac{v_3}{v}\right)M_u V\gamma_L + \left(c\frac{v_2}{v_1}e^{i\zeta} + s\frac{v_3}{v}\right)V M_d\gamma_R\right]d$$

$$- \frac{X_2^+}{v_{12}}\bar{u}\left[\left(s\frac{v_1}{v_2}e^{i\zeta} + c\frac{v_3}{v}\right)M_u V\gamma_L + \left(s\frac{v_2}{v_1}e^{i\zeta} - c\frac{v_3}{v}\right)V M_d\gamma_R\right]d$$

$$+ \frac{\varphi^+}{v}\bar{u}\left(M_u V\gamma_L - V M_d\gamma_R\right)d + \text{H.c.}. \tag{23.35}$$

One sees that the couplings of the massive charged scalars carry different phases in the parts proportional to $M_u V\gamma_L$ and to $V M_d\gamma_R$. These different phases lead to CP violation if the two charged scalars have different masses, i.e., if $m_1 \neq m_2$. (If $m_1 = m_2$ one may redefine $X_1^{\pm}$ and $X_2^{\pm}$ in such a way as to eliminate the phase ζ and the mixing angle α, which then lose their meaning.)

23.6 Model with one doublet and one singlet

It is worthwhile making a small detour and considering the case of a scalar sector even simpler than the one of the Lee model. The simple extension of the SM that we shall now study is instructive in itself, and will be useful in Chapters 24 and 27.

Let us add a non-Hermitian $SU(2) \otimes U(1)$-singlet scalar S to the SM. The singlet S does not have any Yukawa interactions, because all left-handed fields in the SM are in doublets. Therefore, no FCNYI arise. Let us assume that

$$(\mathcal{CP}) \, \phi \, (t, \vec{r}) \, (\mathcal{CP})^{\dagger} = \phi^{\dagger T} (t, -\vec{r}) ,$$
$$(\mathcal{CP}) \, S \, (t, \vec{r}) \, (\mathcal{CP})^{\dagger} = S^{\dagger} \, (t, -\vec{r}) . \tag{23.36}$$

We also assume the existence of a discrete symmetry $S \to -S$. This symmetry is superfluous at the moment, but will be necessary in Chapter 27. The most general renormalizable scalar potential invariant under this discrete symmetry and under the CP symmetry in eqns (23.36) is

$$V = \nu_1 \phi^{\dagger} \phi + l_1 \left(\phi^{\dagger} \phi \right)^2 + \nu_2 |S|^2 + l_2 |S|^4 + l_3 \left(\phi^{\dagger} \phi \right) |S|^2$$
$$+ \left[\nu_3 + l_4 |S|^2 + l_5 \left(\phi^{\dagger} \phi \right) \right] \left(S^2 + S^{\dagger 2} \right) + l_6 \left(S^4 + S^{\dagger 4} \right) . \tag{23.37}$$

The VEVs of ϕ and S may be written

$$\langle 0 | \phi | 0 \rangle = \begin{pmatrix} 0 \\ v \end{pmatrix} ,$$
$$\langle 0 | S | 0 \rangle = V e^{i \alpha} . \tag{23.38}$$

Then,

$$\langle 0 | V | 0 \rangle = \nu_1 v^2 + l_1 v^4 + \nu_2 V^2 + l_2 V^4 + l_3 v^2 V^2$$
$$+ 2 \left(\nu_3 + l_4 V^2 + l_5 v^2 \right) V^2 \cos 2\alpha + 2 l_6 V^4 \cos 4\alpha . \tag{23.39}$$

When $l_6 > 0$, the minimum has

$$\cos 2\alpha = - \frac{\nu_3 + l_4 V^2 + l_5 v^2}{4 l_6 V^2} , \tag{23.40}$$

provided the modulus of the right-hand-side of this equation does not exceed 1.

The vacuum is invariant under the CP symmetry in eqns (23.36) provided $\exp (2 i \alpha) = 1$. In general, α given by eqn (23.40) does not satisfy this condition, and we might believe that this simple extension of the SM exhibits SCPV. However, this is not so. It is crucial to remember that all symmetries of the Lagrangian which may be physically interpreted as CP must be broken by the vacuum before a CP-violating effect arises. As S is a singlet of $SU(2) \otimes U(1)$ it does not have any gauge interactions. Hence, there is a large degree of arbitrariness in the definition of the action of C on S. In particular, the transformation

$$(\mathcal{CP}) \, \phi \, (t, \vec{r}) \, (\mathcal{CP})^{\dagger} = \phi^{\dagger T} (t, -\vec{r}) ,$$
$$(\mathcal{CP}) \, S \, (t, \vec{r}) \, (\mathcal{CP})^{\dagger} = S \, (t, -\vec{r}) , \tag{23.41}$$

in which both Hermitian components of S are CP-even, is just as good a representation of CP as eqns (23.36). Now, eqns (23.41) constitute a symmetry of the

Lagrangian, which cannot be broken by the VEVs in eqns (23.38), whatever the value of α turns out to be. Therefore, this model is CP-conserving.

This illustrates the pitfalls that gauge singlets may lead to when discussing CP violation. In Chapter 24 we shall see that, once one adds left-handed singlet fermions to this model, the vacuum in eqns (23.38) will already imply SCPV; the Lagrangian then having extra Yukawa interactions, it will no longer be possible to interpret the transformation in eqns (23.41) as CP.

23.7 Summary

In this chapter and in the preceding one we have considered extensions of the SM where the gauge and fermion sectors are standard but the scalar sector is enlarged. We have examined models with two and three Higgs doublets, paying special attention to the identification of new sources of CP violation beyond the Kobayashi–Maskawa (KM) mechanism. Our main conclusions have been the following:

- In a two-Higgs-doublet model with no extra symmetry imposed on the Lagrangian, CP can be violated by the scalar sector, at the Lagrangian level.

- With the introduction of an exact reflection symmetry which can guarantee NFC, the Higgs sector is no longer able to generate CP violation by itself.

- CP can be violated provided that one allows the reflection symmetry to be softly broken. This is important, since one can thus have NFC, while at the same time obtaining new sources of CP violation.

- In two-Higgs-doublet model where CP is imposed at the Lagrangian level, CP can be spontaneously broken provided that no other discrete symmetry is imposed on the Lagrangian. This is Lee's model. Although the Yukawa couplings are real at the Lagrangian level, and the only physical phase is the relative phase between the VEVs of the two scalar doublets, that phase enters the quark mass matrices in such a way that it generates a complex CKM mixing matrix. Lee's model also has other sources of CP violation, such as scalar-pseudoscalar mixing.

- Three-Higgs-doublet models with NFC implemented through an exact reflection symmetry may have CP violation either at the Lagrangian level or spontaneously generated. In the latter case the CKM matrix is real and CP violation arises exclusively from scalar-particle exchange. This is the Branco model.

- In an extension of the Higgs sector of the SM, with only one scalar doublet together with a non-Hermitian scalar field S, singlet under SU(2)⊗U(1), although the minimum of the scalar potential may have $\alpha = \arg\langle 0|S|0\rangle$ different from zero, this fact does not correspond to CP violation, when only standard fermions are introduced.

One may put the main findings of this and the preceding chapter in the form of a short summary of models:

TWO HIGGS DOUBLETS

- **No NFC**: It is possible to obtain spontaneous CP violation. In general, one has two sources of CP violation: neutral-scalar exchange, and the Kobayashi–Maskawa mechanism in the charged gauge interaction. However, superweak-like models with a real CKM matrix can be constructed through the introduction of appropriate extra symmetries.
- **Exact NFC**: It is not possible to achieve spontaneous CP violation. There are no new sources of CP violation from the scalar sector. In order to have CP violation, one has to allow for complex Yukawa couplings, leading to the Kobayashi–Maskawa mechanism.
- **Softly broken NFC**: It is possible to achieve spontaneous CP violation, but neutral scalars do not contribute to K^0–$\overline{K^0}$ transitions. If CP is only spontaneously broken, the CKM matrix is real. If CP is not imposed at the Lagrangian level, the Kobayashi–Maskawa mechanism takes place, and furthermore there are new sources of CP violation in the Higgs sector which may be relevant for baryogenesis.

THREE HIGGS DOUBLETS

- **NFC and explicit CP violation**: There are three sources of CP violation: the Kobayashi–Maskawa mechanism, charged-scalar exchange, and neutral-scalar exchange.
- **NFC and spontaneous CP violation**: The CKM matrix is real. CP is violated through charged-scalar and neutral-scalar exchange.

23.7.1 *Phenomenological consequences*

The phenomenological implications of multi-Higgs models of CP violation may vary widely. As far as the masses of the scalars are concerned, their allowed range strongly depends on the mechanism—if any—which is adopted to suppress FC-NYI. If no suppression mechanism is introduced, then the scalars are constrained to have masses in the TeV range. However, much smaller values, in the 100 GeV range, are allowed if one adopts a suppression mechanism for the FCNYI, as described in § 22.11.2.

The generic situation in multi-Higgs-doublet models is having three sources of CP violation:

- the usual KM mechanism,
- CP violation in the exchange of charged scalar fields,
- CP violation in the exchange of neutral scalar fields.

There are contributions to neutral-meson mixing from the standard box diagram with $W^\pm$ exchange, as well as from box diagrams with one or two charged scalars, and also from tree-level transitions mediated by neutral scalars. These new contributions to the mixing may lead to significant departures from the SM predictions for CP asymmetries in the decays of B_d^0 and B_s^0 and their antiparticles. In a generic MHDM one has a unitarity triangle which is not flat, i.e., $J \neq 0$, but the SM expressions giving the CP asymmetries in neutral-B decays in terms of the angles of the unitarity triangle no longer hold. However, in some variants of the MHDM the phenomenological situation may be quite different; for example, in a superweak-like model the unitarity triangle will be flat, i.e., J will vanish, but there will still be CP asymmetries in neutral-B decays, with predictions which differ from those of the SM.

MODELS WITH VECTOR-LIKE QUARKS

24.1 Motivation

There are many extensions of the standard model in which fermions with non-standard $SU(2) \otimes U(1)$ quantum numbers—'exotic fermions'—naturally occur.[62] For example, in the grand unified theory based on the Lie algebra E_6, each family of left-handed fermions is in the $\underline{27}$ representation of that algebra. If we consider the $SU(3)_c \otimes SU(2) \otimes U(1)$ subgroup of E_6, that representation has the following branching rule:

$$\underline{27} = (3,2)_{1/6} + (\bar{3},1)_{-2/3} + (\bar{3},1)_{1/3} + (1,2)_{-1/2} + (1,1)_1$$
$$+ (3,1)_{-1/3} + (\bar{3},1)_{1/3} + (1,2)_{1/2} + (1,2)_{-1/2} + (1,1)_0 + (1,1)_0. \quad (24.1)$$

The $\underline{27}$ contains, apart from the fifteen usual chiral fields in the first line of eqn (24.1)—the doublet of quarks, the up antiquark, the down antiquark, the doublet of leptons, and the positron—, a vector-like isosinglet quark of charge $-1/3$, a vector-like isodoublet of leptons, and two isosinglet neutrinos. Here, we denote by 'vector-like fermions' those fermions whose left-handed and right-handed components transform in the same way under $SU(3)_c \otimes SU(2) \otimes U(1)$.

In this chapter we concentrate our attention on vector-like isosinglet quarks. Our interest in extensions of the SM with those exotic fermions is justified for the following reasons:

1. They provide a framework for having a naturally small violation of the 3×3 unitarity of the CKM matrix. This leads to non-vanishing, but naturally suppressed, flavour-changing neutral currents (FCNC). The presence of FCNC opens up many interesting possibilities for rare K and B decays, as well as for CP asymmetries in neutral-B decays.

2. Adding isosinglet quarks to the SM leads to new sources of CP violation. In particular, isosinglet quarks enable one to achieve spontaneous CP violation (SCPV) using a very simple set of scalar fields, with only one non-Hermitian $SU(2) \otimes U(1)$-singlet scalar added to the usual doublet—see § 24.7.

3. Extensions of the SM with isosinglet quarks may solve the strong CP problem through an implementation of the mechanism of Barr (1984) and Nelson (1984)—see § 27.7.1.

---

[62] Generic properties of models with exotic quarks have been studied, for example, by Langacker and London (1988), Joshipura (1992), Lavoura and Silva (1993a,b), and Nardi et al. (1995).

Models with vector-like quarks have been considered extensively in the literature. Early references include the papers by Aguila and Cortés (1985), Fishbane *et al.* (1985, 1986), Barger *et al.* (1986), Branco and Lavoura (1986), and Enqvist *et al.* (1986). A revived interest in these models was generated by the study of the impact of heavy isosinglet quarks on the CP-violating asymmetries in neutral-B decays, both in the limit where Z-exchange gives the dominant contribution to B_d^0-$\overline{B_d^0}$ mixing (Nir and Silverman 1990) and in the general case where both Z-exchange and the standard box-diagram contributions are taken into account (Branco *et al.* 1993). Recent analyses of the experimental constraints on these models may be found in the papers by Silverman (1996) and by Barenboim and Botella (1998).

24.2 The model

24.2.1 *Quark spectrum*

Our extension of the SM has n_g doublets of left-handed quarks,

$$\begin{pmatrix} p_{Lo} \\ n_{Lo} \end{pmatrix}, \text{ with } o = 1, 2, \ldots, n_g, \tag{24.2}$$

together with the following singlet quark fields: n_p charge 2/3 left-handed P_L, n_n charge $-1/3$ left-handed N_L, $n_g + n_p$ charge 2/3 right-handed P_R, and $n_g + n_n$ charge $-1/3$ right-handed N_R:

$$\begin{aligned}
&P_{Lr}, \text{ with } r = 1, 2, \ldots, n_p, \\
&N_{Ls}, \text{ with } s = 1, 2, \ldots, n_n, \\
&P_{R\alpha}, \text{ with } \alpha = 1, 2, \ldots, n_g + n_p, \\
&N_{Rk}, \text{ with } k = 1, 2, \ldots, n_g + n_n.
\end{aligned} \tag{24.3}$$

Therefore, the electromagnetic current is

$$\begin{aligned}
J_{\text{em}}^\mu \equiv{}& \tfrac{2}{3} \left(\overline{p_{Lo}} \gamma^\mu p_{Lo} + \overline{P_{Lr}} \gamma^\mu P_{Lr} + \overline{P_{R\alpha}} \gamma^\mu P_{R\alpha} \right) \\
&- \tfrac{1}{3} \left(\overline{n_{Lo}} \gamma^\mu n_{Lo} + \overline{N_{Ls}} \gamma^\mu N_{Ls} + \overline{N_{Rk}} \gamma^\mu N_{Rk} \right),
\end{aligned} \tag{24.4}$$

and the electroweak gauge interactions are

$$\begin{aligned}
\mathcal{L}_{\text{gauge}} ={}& -e A_\mu J_{\text{em}}^\mu + \frac{g}{\sqrt{2}} \left(W_\mu^+ \overline{p_{Lo}} \gamma^\mu n_{Lo} + W_\mu^- \overline{n_{Lo}} \gamma^\mu p_{Lo} \right) \\
&+ \frac{g}{2c_w} Z_\mu \left(\overline{p_{Lo}} \gamma^\mu p_{Lo} - \overline{n_{Lo}} \gamma^\mu n_{Lo} - 2 s_w^2 J_{\text{em}}^\mu \right).
\end{aligned} \tag{24.5}$$

The quark mass terms may in general be written

$$\begin{aligned}
\mathcal{L}_{\text{M}} ={}& -\overline{p_{Lo}} \left(m_p \right)_{o\alpha} P_{R\alpha} - \overline{n_{Lo}} \left(m_n \right)_{ok} N_{Rk} \\
&- \overline{P_{Lr}} \left(M_p \right)_{r\alpha} P_{R\alpha} - \overline{N_{Ls}} \left(M_n \right)_{sk} N_{Rk} + \text{H.c.}.
\end{aligned} \tag{24.6}$$

We shall use the letter x to denote either p or n in equations which hold for both $x = p$ and $x = n$. The m_x are $n_g \times (n_g + n_x)$ matrices. They contain the usual $|\Delta T| = 1/2$ mass terms (T is the weak isospin), arising from the VEV of one or more Higgs doublets. Their mass scale should be $m \sim v$. The M_x are $n_x \times (n_g + n_x)$ matrices. They contain $|\Delta T| = 0$ mass terms. Since these are SU(2)⊗U(1)-invariant, they may be present in the Lagrangian prior to the spontaneous breaking of the gauge symmetry. Not being protected by that symmetry, their mass scale M may be significantly larger than v.

We may denote $\mathcal{M}_p$ and $\mathcal{M}_n$ the full mass matrices for the up-type and down-type quarks, respectively:

$$\mathcal{M}_x = \begin{pmatrix} m_x \\ M_x \end{pmatrix}. \tag{24.7}$$

These are $(n_g + n_x) \times (n_g + n_x)$ matrices.

24.2.2 Mixing matrices

We denote the quark mass eigenstates by u_α and d_k, where α and k run over the ranges in eqns (24.3). The unitary matrices W_p and W_n relate the left-handed-quark weak and mass eigenstates:

$$\begin{aligned} \begin{pmatrix} p_L \\ P_L \end{pmatrix} &= W_p u_L, \\ \begin{pmatrix} n_L \\ N_L \end{pmatrix} &= W_n d_L. \end{aligned} \tag{24.8}$$

The W_x are $(n_g + n_x) \times (n_g + n_x)$ matrices. It is useful to write them as

$$W_x = \begin{pmatrix} A_x \\ B_x \end{pmatrix}, \tag{24.9}$$

where the A_x are rectangular matrices consisting of the first n_g rows of W_x, while the B_x consist of the last n_x rows of W_x. The unitarity of W_x implies, on the one hand,

$$A_x^\dagger A_x + B_x^\dagger B_x = 1_{n_g+n_x}, \tag{24.10}$$

and, on the other hand,

$$\begin{aligned} A_x A_x^\dagger &= 1_{n_g}, \\ B_x B_x^\dagger &= 1_{n_x}, \\ A_x B_x^\dagger &= 0_{n_g \times n_x}. \end{aligned} \tag{24.11}$$

The W_x must be chosen such as to diagonalize the Hermitian mass matrices $\mathcal{M}_x \mathcal{M}_x^\dagger$:

$$\mathcal{M}_x \mathcal{M}_x^\dagger W_x = W_x D_x^2. \tag{24.12}$$

The real diagonal matrices D_p and D_n contain the masses of the up-type and down-type quarks, respectively.

The electromagnetic current in eqn (24.4) may be written in terms of the mass eigenstates as

$$J^\mu_{em} = \tfrac{2}{3}\overline{u_\alpha}\gamma^\mu u_\alpha - \tfrac{1}{3}\overline{d_k}\gamma^\mu d_k. \tag{24.13}$$

The gauge interactions in eqn (24.5) are

$$\mathcal{L}_{gauge} = -eA_\mu J^\mu_{em} + \frac{g}{\sqrt{2}}\left(W^+_\mu \overline{u_{L\alpha}}\gamma^\mu V_{\alpha k} d_{Lk} + W^-_\mu \overline{d_{Lk}}\gamma^\mu V^*_{\alpha k} u_{L\alpha}\right)$$

$$+\frac{g}{2c_w}Z_\mu\left[\overline{u_{L\alpha}}\gamma^\mu \left(Z_p\right)_{\alpha\beta} u_{L\beta} - \overline{d_{Lk}}\gamma^\mu \left(Z_n\right)_{kj} d_{Lj} - 2s^2_w J^\mu_{em}\right], \tag{24.14}$$

where the mixing matrices are

$$\begin{aligned} V &= A^\dagger_p A_n \\ Z_x &= A^\dagger_x A_x. \end{aligned} \tag{24.15}$$

The (generalized) CKM matrix V is an $(n_g + n_p) \times (n_g + n_n)$ rectangular matrix. The CKM matrix in this model is not unitary as in the SM. The mixing matrices for the neutral currents Z_x are Hermitian $(n_g + n_x) \times (n_g + n_x)$ matrices. In general, they are not diagonal, i.e., flavour-changing neutral currents are present.

24.2.3   Non-unitarity of V and its relation to the FCNC

The generalized CKM matrix is not, in general, unitary. Indeed, if $n_p \neq n_n$ it is not even a square matrix. However, from the unitarity of W_p and W_n it follows (Branco et al. 1993) that there is an $(n_g + n_p + n_n) \times (n_g + n_p + n_n)$ matrix

$$Y \equiv \begin{pmatrix} V & B^\dagger_p \\ B_n & 0 \end{pmatrix} = \begin{pmatrix} A^\dagger_p A_n & B^\dagger_p \\ B_n & 0 \end{pmatrix} \tag{24.16}$$

which is unitary. This may easily be checked using eqns (24.10) and (24.11).

The deviations of V from unitarity are closely related to the presence of FCNC. Indeed, because of the first eqn (24.11),

$$\begin{aligned} Z_p &= A^\dagger_p A_p = A^\dagger_p A_n A^\dagger_n A_p = VV^\dagger, \\ Z_n &= A^\dagger_n A_n = A^\dagger_n A_p A^\dagger_p A_n = V^\dagger V. \end{aligned} \tag{24.17}$$

Thus, if V was unitary then Z_p and Z_n would be the unit matrix, and FCNC would be absent. From the first eqn (24.11) it also follows that

$$Z^2_x = Z_x. \tag{24.18}$$

Equations (24.17) may be elegantly derived without ever referring to the diagonalization of the quark mass matrices (Bamert et al. 1996). From eqns (11.12) and (11.3), we know that the gauge interactions in the $SU(2) \otimes U(1)$ gauge theory are given by

$$-eAQ + g\left(W^+ T_+ + W^- T_-\right) + \frac{g}{c_w}Z\left([T_+, T_-] - Qs^2_w\right). \tag{24.19}$$

In the vector space spanned by the $u_{L\alpha}$ and the d_{Lk}, the matrices T_+ and T_- are given by

$$T_+ = \tfrac{1}{\sqrt{2}} \begin{pmatrix} 0 & V \\ 0 & 0 \end{pmatrix}, \quad T_- = \tfrac{1}{\sqrt{2}} \begin{pmatrix} 0 & 0 \\ V^\dagger & 0 \end{pmatrix}, \tag{24.20}$$

cf. eqn (24.14). Indeed, eqns (24.20) may be looked upon as the *definition* of the CKM matrix. From eqn (24.19), the matrix which determines that part of the neutral current which is not proportional to J_{em}^μ is

$$\tfrac{1}{2} \begin{pmatrix} Z_p & 0 \\ 0 & -Z_n \end{pmatrix} \equiv [T_+, T_-] = \tfrac{1}{2} \begin{pmatrix} VV^\dagger & 0 \\ 0 & -V^\dagger V \end{pmatrix}. \tag{24.21}$$

This result is very general: it holds in the SU(2)⊗U(1) gauge theory for any standard or exotic $Q = 2/3$ and $Q = -1/3$ quarks. It is valid if one adds to the SM not only vector-like isosinglet quarks, but also vector-like isodoublet quarks and mirror quarks (Lavoura and Silva 1993b).

24.3 Natural suppression of the FCNC

It is known experimentally that the FCNC are very much suppressed, although the detailed bounds depend on the flavours involved—see for example Lavoura and Silva (1993a) and Silverman (1996). Therefore, in order for the class of models considered here to be plausible, they must have a mechanism for natural suppression of the FCNC. In order to show that such a mechanism exists, we have to explicitly diagonalize the quark mass matrices.

In the discussion that follows, it will be convenient to have $\mathcal{M}_p$ and $\mathcal{M}_n$ in a special form, which we can obtain by using the freedom that one has to make weak-basis transformations (WBT). These are transformations of the quark fields which leave $\mathcal{L}_{\text{gauge}}$ in eqn (24.5) invariant. It is easy to convince oneself that, by making a WBT, one can bring both $\mathcal{M}_p$ and $\mathcal{M}_n$ to the form

$$\mathcal{M}_x = \begin{pmatrix} G_x & J_x \\ 0 & \hat{M}_x \end{pmatrix}, \tag{24.22}$$

where the $\hat{M}_x$ are $n_x \times n_x$ matrices, diagonal and real by definition. The matrices G_x and J_x are complex and have dimensions $n_g \times n_g$ and $n_g \times n_x$, respectively. There is still some freedom left to make a WBT such that either G_p or G_n is made diagonal and real; we shall take G_p to be diagonal and real.

Let us write

$$W_x = \begin{pmatrix} A_x \\ B_x \end{pmatrix} = \begin{pmatrix} K_x & R_x \\ S_x & T_x \end{pmatrix}. \tag{24.23}$$

The matrices K_x and T_x have dimensions $n_g \times n_g$ and $n_x \times n_x$, respectively. The matrix R_x is $n_g \times n_x$, while S_x is $n_x \times n_g$. Equation (24.10) now reads

$$\begin{aligned} K_x^\dagger K_x + S_x^\dagger S_x &= 1_{n_g}, \\ R_x^\dagger R_x + T_x^\dagger T_x &= 1_{n_x}, \\ K_x^\dagger R_x + S_x^\dagger T_x &= 0_{n_g \times n_x}. \end{aligned} \tag{24.24}$$

We shall also write the matrices D_p and D_n as

$$D_x = \begin{pmatrix} \bar{m}_x & 0 \\ 0 & \hat{M}_x \end{pmatrix}. \qquad (24.25)$$

Equation (24.12) reads

$$
\begin{aligned}
\left(G_x G_x^\dagger + J_x J_x^\dagger\right) K_x + J_x \hat{M}_x S_x &= K_x \bar{m}_x^2, \\
\left(G_x G_x^\dagger + J_x J_x^\dagger\right) R_x + J_x \hat{M}_x T_x &= R_x \bar{M}_x^2, \\
\hat{M}_x J_x^\dagger K_x + \hat{M}_x^2 S_x &= S_x \bar{m}_x^2, \\
\hat{M}_x J_x^\dagger R_x + \hat{M}_x^2 T_x &= T_x \bar{M}_x^2.
\end{aligned}
\qquad (24.26)
$$

All the above equations are exact, no approximations have been done yet. We now *assume* that G_x and J_x are $\sim m$, while $\hat{M}_x \sim M$, with $M \gg m$. One obtains the following solution of eqns (24.24) and (24.26) to leading order in m/M:

$$
\begin{aligned}
T_x &= 1_{n_x}, \\
R_x &= J_x \hat{M}_x^{-1}, \\
S_x &= -\hat{M}_x^{-1} J_x^\dagger K_x,
\end{aligned}
\qquad (24.27)
$$

together with $\bar{M}_x^2 = \hat{M}_x^2$; while K_x is the unitary matrix which diagonalizes $G_x G_x^\dagger$:

$$K_x^\dagger G_x G_x^\dagger K_x = \bar{m}_x^2. \qquad (24.28)$$

In particular, since we have chosen to work in the weak basis where G_p is diagonal, $K_p = 1_{n_g}$ to leading order.

The quark mixing matrix entering the charged current may now be computed and one obtains

$$V = \begin{pmatrix} K_n & J_n \hat{M}_n^{-1} \\ \hat{M}_p^{-1} J_p^\dagger K_n & \hat{M}_p^{-1} J_p^\dagger J_n \hat{M}_n^{-1} \end{pmatrix}. \qquad (24.29)$$

It is clear from eqn (24.29) that the mixing of standard quarks with isosinglet quarks is suppressed by m/M. Also, the matrix K_n, which gives the interactions among the usual quarks, is unitary up to terms $\sim m^2/M^2$.

The mixing matrices for the weak neutral current are

$$
\begin{aligned}
Z_n &= \begin{pmatrix} 1 - K_n^\dagger J_n \hat{M}_n^{-2} J_n^\dagger K_n & K_n^\dagger J_n \hat{M}_n^{-1} \\ \hat{M}_n^{-1} J_n^\dagger K_n & \hat{M}_n^{-1} J_n^\dagger J_n \hat{M}_n^{-1} \end{pmatrix}, \\
Z_p &= \begin{pmatrix} 1 - J_p \hat{M}_p^{-2} J_p^\dagger & J_p \hat{M}_p^{-1} \\ \hat{M}_p^{-1} J_p^\dagger & \hat{M}_p^{-1} J_p^\dagger J_p \hat{M}_p^{-1} \end{pmatrix}.
\end{aligned}
\qquad (24.30)
$$

The $n_g \times n_g$ upper-left submatrix, which governs the neutral-current interaction among the usual quarks, is the unit matrix up to a correction of the form $K_x^\dagger J_x \hat{M}_x^{-2} J_x^\dagger K_x$. Thus, flavour-changing neutral currents do arise, but they are naturally suppressed by a factor m^2/M^2, because $J_x \sim m$ while $\hat{M}_x \sim M$.

24.4   CP-violating phases

When the SM is extended through the addition of vector-like quarks, the CKM matrix is no longer unitary. Counting the independent CP-violating phases in V becomes complicated, especially when there are both $Q = -1/3$ and $Q = 2/3$ vector-like quarks. We shall return to this question later; at this stage, we only want to count the number of phases in V when that matrix is evaluated in the approximation leading to eqn (24.29). Since in this approximation K_n is unitary, the number of physical phases in it is just the same as in the SM, i.e., $(n_g - 1)(n_g - 2)/2$. In the upper-right block of V, we have the matrix J_n which, being an $n_g \times n_n$ complex matrix, would in general contain $n_g n_n$ phases. However, n_n phases may be eliminated by rephasing the last n_n fields d_L. Analogously, J_p, in the lower-left block of V, has $(n_g - 1) n_p$ phases. Therefore, one has altogether

$$\frac{(n_g - 1)(n_g - 2)}{2} + (n_g - 1)(n_p + n_n) \tag{24.31}$$

physical phases in V. We shall later show that this is indeed the correct number of CP-violating phases.

24.5   The invariant approach

24.5.1   CP-invariance conditions

The most general CP transformation of the quark fields leaving eqn (24.5) invariant is[63]

$$\begin{aligned}
(\mathcal{CP})\, p_L \,(\mathcal{CP})^\dagger &= U_L \gamma^0 C \overline{p_L}^T, \\
(\mathcal{CP})\, n_L \,(\mathcal{CP})^\dagger &= U_L \gamma^0 C \overline{n_L}^T, \\
(\mathcal{CP})\, P_R \,(\mathcal{CP})^\dagger &= W_R^p \gamma^0 C \overline{P_R}^T, \\
(\mathcal{CP})\, N_R \,(\mathcal{CP})^\dagger &= W_R^n \gamma^0 C \overline{N_R}^T, \\
(\mathcal{CP})\, P_L \,(\mathcal{CP})^\dagger &= W_L^p \gamma^0 C \overline{P_L}^T, \\
(\mathcal{CP})\, N_L \,(\mathcal{CP})^\dagger &= W_L^n \gamma^0 C \overline{N_L}^T.
\end{aligned} \tag{24.32}$$

We have suppressed the space–time variables for the sake of clarity. The matrix U_L is $n_g \times n_g$ unitary, and the matrices W_R^z and W_L^z are $(n_g + n_z) \times (n_g + n_z)$ and $n_z \times n_z$ unitary, respectively.

Requiring CP invariance of the mass terms in eqn (24.6) leads to the following conditions:

$$\begin{aligned}
U_L^\dagger m_p W_R^p &= m_p^*, \\
U_L^\dagger m_n W_R^n &= m_n^*, \\
W_L^{p\dagger} M_p W_R^p &= M_p^*, \\
W_L^{n\dagger} M_n W_R^n &= M_n^*.
\end{aligned} \tag{24.33}$$

The existence of unitary matrices U_L, W_R^p, W_R^n, W_L^p, and W_L^n which satisfy eqns (24.33) is necessary and sufficient for CP invariance of $\mathcal{L}_{\text{gauge}} + \mathcal{L}_{\text{M}}$. The

---

[63]We have eliminated the inconsequential phase ξ_W from the CP transformation of $W^\pm$ in eqn (13.9).

fulfilment of this condition in any particular weak basis is equivalent to its fulfilment in any other weak basis, as one easily checks.

24.5.2 Analysis in a specific weak basis

In order to find the number of independent CP restrictions, we shall again use
the weak basis in which the mass matrices have the form in eqn (24.22). We
remind the reader that $\hat{M}_p$, $\hat{M}_n$, and G_p are diagonal and real by definition.
In order to eliminate exceptional cases, of null measure in parameter space, we
assume that the diagonal matrix elements of each of these three matrices are
all different—the matrices are non-degenerate. Under these conditions, we easily
find that eqns (24.33) imply

$$
\begin{aligned}
U_L &= \mathrm{diag}\left(e^{i\alpha_o}\right), \\
W_L^p &= \mathrm{diag}\left(e^{i\beta_r}\right), \\
W_L^n &= \mathrm{diag}\left(e^{i\gamma_s}\right),
\end{aligned}
\tag{24.34}
$$

while the matrices W_R^p and W_R^n must take the block form

$$
\begin{aligned}
W_R^p &= \begin{pmatrix} U_L & 0 \\ 0 & W_L^p \end{pmatrix}, \\
W_R^n &= \begin{pmatrix} X_L & 0 \\ 0 & W_L^n \end{pmatrix},
\end{aligned}
\tag{24.35}
$$

with X_L an arbitrary $n_g \times n_g$ unitary matrix. The CP-invariance conditions of
eqns (24.33) get reduced to

$$
\begin{aligned}
\mathrm{diag}\left(e^{-i\alpha_o}\right) G_n X_L &= G_n^*, \\
\mathrm{diag}\left(e^{-i\alpha_o}\right) J_n \, \mathrm{diag}\left(e^{i\gamma_s}\right) &= J_n^*, \\
\mathrm{diag}\left(e^{-i\alpha_o}\right) J_p \, \mathrm{diag}\left(e^{i\beta_r}\right) &= J_p^*.
\end{aligned}
\tag{24.36}
$$

As X_L may be chosen at will, the first eqn (24.36) is in fact equivalent to the
simpler condition (see the argument in § 14.2)

$$
\mathrm{diag}\left(e^{-i\alpha_o}\right) \left(G_n G_n^\dagger\right) \mathrm{diag}\left(e^{i\alpha_o}\right) = \left(G_n G_n^\dagger\right)^*.
\tag{24.37}
$$

Thus, CP invariance imposes restrictions on the complex phases in the matrices
J_p, J_n, and $G_n G_n^\dagger$. Remember that the J_x are complex matrices of dimension
$n_g \times n_x$, while $G_n G_n^\dagger$ is an $n_g \times n_g$ Hermitian matrix. The number of independent
phases in these three matrices is

$$
n_g \left(n_p + n_n\right) + \frac{n_g \left(n_g - 1\right)}{2}.
\tag{24.38}
$$

CP invariance constrains these phases to be equal to differences of n_g phases
α_o, n_p phases β_r, and n_n phases γ_s. The number of independent CP restrictions
therefore is

$$n_g \left(n_p + n_n\right) + \frac{n_g \left(n_g - 1\right)}{2} - \left(n_g + n_p + n_n - 1\right) = \left(n_g - 1\right) \left(\frac{n_g}{2} - 1 + n_p + n_n\right).$$

(24.39)

This coincides with the number of CP-violating phases in the CKM matrix, given by eqn (24.31).

24.5.3 Invariant CP restrictions

From the first two eqns (24.33) one readily derives the following necessary conditions for CP invariance:

$$\operatorname{tr} \left[h_p^a, h_n^b\right]^f = 0,$$

(24.40)

where a and b are integers, f is an odd integer, and $h_x \equiv m_x m_x^\dagger$. The conditions in eqn (24.40) are entirely analogous to those obtained in the SM, cf. eqn (14.21). In models with vector-like quarks, there are necessary conditions for CP invariance of a different type, for example

$$\operatorname{Im} \operatorname{tr} \left(h_p h_n H_n\right) = 0,$$
$$\operatorname{Im} \operatorname{tr} \left(H_p h_p h_n\right) = 0,$$

(24.41)

where $H_x \equiv m_x M_x^\dagger M_x m_x^\dagger$.

The question of finding a set of necessary and sufficient conditions for CP invariance, expressed in terms of weak-basis invariants, is in general very complicated. For the case of a minimal extension in which only one vector-like isosinglet down-type quark is added to the three-generation SM—i.e., $n_g = 3$, $n_n = 1$, and $n_p = 0$—it has been shown (Aguila et al. 1998) that

$$\operatorname{Im} \operatorname{tr} \left(h_p^2 h_n h_p h_n^2\right) = 0,$$
$$\operatorname{Im} \operatorname{tr} \left(h_p h_n H_n\right) = 0,$$
$$\operatorname{Im} \operatorname{tr} \left(h_p^2 h_n H_n\right) = 0,$$
$$\operatorname{Im} \operatorname{tr} \left(h_p h_n^2 H_n\right) = 0,$$
$$\operatorname{Im} \operatorname{tr} \left(h_p^2 h_n^2 H_n\right) = 0,$$
$$\operatorname{Im} \operatorname{tr} \left[h_p^2 \left(h_p h_n - h_n h_p\right) H_n\right] = 0,$$
$$\operatorname{Im} \operatorname{tr} \left[h_p^2 \left(h_p h_n^2 - h_n^2 h_p\right) H_n\right] = 0,$$

(24.42)

are necessary and sufficient conditions for CP conservation.

These invariant conditions are especially useful for analysing the limit where some of the quark masses are effectively degenerate. This is the case in very-high-energy collisions, where the natural asymptotic states no longer are hadronic states but rather quark jets. At high energies, it should be very difficult, if not impossible, to identify the flavour of the quark jets. In the extreme chiral limit $m_u = m_c = m_d = m_s = 0$ there is no CP violation in the SM, because there are identical-charge quarks which are degenerate. However, in the model with one down-type vector-like quark, there is CP violation even in that limit. Then, the strength of CP violation is controlled by the second invariant in eqns (24.42), all other invariants being proportional to that one (Aguila et al. 1998).

24.6   Parametrization of the CKM matrix

With the addition of vector-like quarks to the SM, parametrizing the CKM matrix V becomes rather involved, essentially due to the fact that it no longer is a unitary matrix. Still, in some cases one may use the fact that V is a submatrix of the larger unitary matrix Y in eqn (24.16). Various approaches to this problem have been proposed, including:

1. parametrization through Euler angles and phases (Branco and Lavoura 1986);

2. parametrization through the moduli of matrix elements and the arguments of quartets (Branco et al. 1993);

3. Wolfenstein-type parametrization (Lavoura and Silva 1993a).

Here we shall only describe the first of these approaches, which is interesting because it yields a different way of counting the number of CP-violating phases in V.

24.6.1   Parametrization with Euler angles and phases

This type of parametrization is possible when there are vector-like quarks of only one electric charge, either $Q = -1/3$ or $Q = 2/3$. In this case V consists of the first n_g rows of a unitary matrix which, in principle, does not have any zero matrix elements. This is no longer true when both n_p and n_n are non-zero; then, the unitary matrix Y defined in eqn (24.16) has a zero submatrix, and its parametrization through Euler angles and phases is awkward.

For definiteness, let us assume that there are n_n down-type vector-like quarks, but no up-type vector-like quarks, i.e., that $n_p = 0$. In this case, one may choose, without loss of generality, a weak basis in which the up-type-quark mass matrix is diagonal and real non-negative. Then, V consists of the first n_g rows of the $(n_g + n_n) \times (n_g + n_n)$ unitary matrix W_n which diagonalizes $\mathcal{M}_n \mathcal{M}_n^\dagger$. The task then consists of finding a parametrization of W_n with only $(n_g - 1)(n_g/2 - 1 + n_n)$ phases in its first n_g rows, according to eqns (24.31) and (24.39).

In order to construct such a parametrization, let us introduce the orthogonal matrices

$$
O_{ij} \equiv \begin{pmatrix}
1 & & & & & \\
& \ddots & & & & \\
& & \cos\theta_{ij} & \cdots & -\sin\theta_{ij} & \\
& & \vdots & & \vdots & \\
& & \sin\theta_{ij} & \cdots & \cos\theta_{ij} & \\
& & & & & \ddots \\
& & & & & & 1
\end{pmatrix}
\begin{matrix}
\\ \\ \leftarrow i \\ \\ \leftarrow j \\ \\ \\
\end{matrix}
\tag{24.43}
$$

and the rephasing matrices

$$I_j\left(\delta_k\right) \equiv \begin{pmatrix} 1 & & & & & & \\ & \ddots & & & & & \\ & & 1 & & & & \\ & & & e^{i\delta_k} & & & \\ & & & & 1 & & \\ & & & & & \ddots & \\ & & & & & & 1 \end{pmatrix} \leftarrow j\,. \qquad (24.44)$$

We may write W_n as a product of 'complex rotations' Ω_{ij}:

$$W_n = \prod_{i=\mathcal{N}-1}^{1} \prod_{j=i+1}^{\mathcal{N}} \Omega_{ij}, \qquad (24.45)$$

where $\mathcal{N} = n_g + n_n$, and the unitary matrices Ω_{ij} mix the ith and j^{th} rows and columns, and contain one rotation angle and three phases:

$$\Omega_{ij} = I_i\left(\delta_a\right) I_j\left(\delta_b\right) O_{ij} I_i\left(\delta_c\right). \qquad (24.46)$$

Most of the rephasings $I_j\left(\delta_k\right)$ in the product in eqn (24.45) may be exchanged in position in such a way that they are brought out of the product of rotations O_{ij}, and then they become equivalent to rephasings of the quark fields. If we omit writing down those trivial rephasings, it can be shown (Anselm *et al.* 1985) that the matrix W_n may be written as

$$\begin{aligned} W_n = \ & O_{(\mathcal{N}-1)\mathcal{N}} I_{\mathcal{N}} O_{(\mathcal{N}-2)(\mathcal{N}-1)} O_{(\mathcal{N}-2)\mathcal{N}} I_{\mathcal{N}-1} I_{\mathcal{N}} \\ & \times O_{(\mathcal{N}-3)(\mathcal{N}-2)} O_{(\mathcal{N}-3)(\mathcal{N}-1)} O_{(\mathcal{N}-3)\mathcal{N}} I_{\mathcal{N}-2} I_{\mathcal{N}-1} I_{\mathcal{N}} \cdots O_{23} \cdots O_{2\mathcal{N}} \\ & \times I_3 \cdots I_{\mathcal{N}} O_{12} \cdots O_{1\mathcal{N}}. \end{aligned} \qquad (24.47)$$

Explicit computation of the $\mathcal{N} \times \mathcal{N}$ unitary matrix in eqn (24.47) yields that it has $(n_g - 1)\left(\mathcal{N} - n_g/2 - 1\right)$ phases in its upper n_g rows. This coincides with eqn (24.31), as $n_n = \mathcal{N} - n_g$.

Let us consider the example of E_6 with three families in the **27** representation. In each **27** there is one standard quark doublet together with one vector-like isosinglet down-type quark; therefore, $n_g = n_n = 3$ and $n_p = 0$. The quark mixing matrix V consists of the first three rows of the 6×6 unitary matrix W_n, in the weak basis in which the up-type-quark mass matrix is diagonal. We parametrize

$$\begin{aligned} W_n = \ & O_{56} I_6\left(\delta_{10}\right) O_{45} O_{46} I_5\left(\delta_9\right) I_6\left(\delta_8\right) O_{34} O_{35} O_{36} I_4\left(\delta_7\right) I_5\left(\delta_6\right) I_6\left(\delta_5\right) \\ & \times O_{23} O_{24} O_{25} O_{26} I_3\left(\delta_4\right) I_4\left(\delta_3\right) I_5\left(\delta_2\right) I_6\left(\delta_1\right) O_{12} O_{13} O_{14} O_{15} O_{16}. \end{aligned} \qquad (24.48)$$

The distribution of phases among the rows of this matrix turns out to be

$$\begin{pmatrix} \text{no phases} \\ \delta_1, \delta_2, \delta_3, \delta_4 \\ \delta_1, \delta_2, \ldots, \delta_7 \\ \delta_1, \delta_2, \ldots, \delta_9 \\ \delta_1, \delta_2, \ldots, \delta_{10} \\ \delta_1, \delta_2, \ldots, \delta_{10} \end{pmatrix}, \qquad (24.49)$$

i.e., the first row of V is real, the second row depends on four phases, the third row depends on seven phases, and so on. This coincides with the formula $(n_g - 1)(\mathcal{N} - n_g/2 - 1)$ for $\mathcal{N} = 6$.

24.7   Spontaneous CP violation: a simple model

Models with vector-like quarks provide one of the simplest scenarios for spontaneous CP violation. We shall illustrate this feature by considering a minimal model proposed by Bento *et al.* (1991). This consists of the SM supplemented with only one charge $-1/3$ vector-like quark and one non-Hermitian scalar singlet S. For later use, we shall also introduce a discrete symmetry under which S, N_L, and N_{R4} change sign. This discrete symmmetry will not, however, play any role in the discussion of SCPV; it will only be important in the discussion of the strong CP problem in § 27.7.1.

The scalar potential and the pattern of vacuum expectation values (VEVs) for this model have been discussed in § 23.6. One obtains the VEVs for ϕ and S in eqns (23.38), with a non-trivial phase α. However, we have emphasized in § 23.6 that, in the context of the SM with standard quarks only, the non-trivial vacuum phase does not lead to spontaneous CP breaking. In the extension of the SM with vector-like quarks the situation is different, because the presence of extra interactions means that there is less freedom in the choice of a CP transformation and, in particular, eqns (23.41) are no longer a valid CP transformation. This can readily be seen by examining the most general quark Yukawa couplings and the mass term consistent with the SU(2)⊗U(1) gauge symmetry and with the discrete symmetry:

$$\mathcal{L}_Y = -\overline{Q_L}\Gamma\phi n_R - \overline{Q_L}\Delta\tilde{\phi}p_R - \mu\overline{N_L}N_R - \overline{N_L}\left(FS + F'S^\dagger\right)n_R + \text{H.c..} \quad (24.50)$$

Here, we have reserved the notation N_R only for N_{R4}, the singlet quark which changes sign under the discrete symmetry, while for the other P_R and N_R fields we have returned to the SM notation and denoted them p_R and n_R, respectively. We assume the CP transformation in eqn (23.36), together with

$$\begin{aligned}
(\mathcal{CP})\, p_{Ro}\, (\mathcal{CP})^\dagger &= \gamma^0 C \overline{p_{Ro}}^T, \\
(\mathcal{CP})\, n_{Ro}\, (\mathcal{CP})^\dagger &= \gamma^0 C \overline{n_{Ro}}^T, \\
(\mathcal{CP})\, N_R\, (\mathcal{CP})^\dagger &= \gamma^0 C \overline{N_R}^T, \\
(\mathcal{CP})\, \overline{p_{Lo}}\, (\mathcal{CP})^\dagger &= -p_{Lo}^T C^{-1} \gamma^0, \\
(\mathcal{CP})\, \overline{n_{Lo}}\, (\mathcal{CP})^\dagger &= -n_{Lo}^T C^{-1} \gamma^0, \\
(\mathcal{CP})\, \overline{N_L}\, (\mathcal{CP})^\dagger &= -N_L^T C^{-1} \gamma^0.
\end{aligned} \qquad (24.51)$$

Hence, the 1×3 matrices of Yukawa couplings F and F' are real, just as the 3×3 matrices Γ and Δ, and the $\Delta T = 0$ mass μ. On the other hand, the transformation in eqn (23.41) would be a symmetry of the Lagrangian only if F' were equal to F^*. We thus see that the presence of extra Yukawa interactions ensures that exactly the same vacuum structure now leads to CP violation, while in § 23.6 it did not.

Since the model has no charge 2/3 vector-like quarks, we may choose a weak basis in which the up-type-quark mass matrix is diagonal and real. The down-type-quark mass matrix is

$$\mathcal{M}_n = \begin{pmatrix} v\Gamma & 0 \\ V\mathcal{F} & \mu \end{pmatrix}, \tag{24.52}$$

where

$$\mathcal{F} \equiv e^{i\alpha} F + e^{-i\alpha} F'. \tag{24.53}$$

(Notice that $\mathcal{F}$ would be real if F' were equal to F^*.) The matrix in eqn (24.52) is not in the standard form of eqn (24.22), but we can bring it to that form by means of a unitary transformation of the right-handed quark fields. Let us define

$$M \equiv \sqrt{\mu^2 + V^2 \mathcal{F}\mathcal{F}^\dagger}. \tag{24.54}$$

We may then construct the 4×4 unitary matrix

$$U = \begin{pmatrix} X & V\mathcal{F}^\dagger/M \\ Y & \mu/M \end{pmatrix}, \tag{24.55}$$

where X is a 3×3 matrix and Y is a 1×3 matrix. The unitarity of U implies, in particular,

$$XX^\dagger = 1_3 - \frac{V^2}{M^2}\mathcal{F}^\dagger\mathcal{F} \tag{24.56}$$

and

$$V\mathcal{F}X + \mu Y = 0. \tag{24.57}$$

We then effect the weak-basis transformation

$$\mathcal{M}_n \to \mathcal{M}_n U = \begin{pmatrix} v\Gamma X & (vV/M)\Gamma\mathcal{F}^\dagger \\ 0 & M \end{pmatrix}, \tag{24.58}$$

where we have used eqns (24.54) and (24.57). We thus obtain a matrix of the form in eqn (24.22). One may now use the results of § 24.3, provided that one assumes that $M \gg v$ and $M^2 \gg vV$.[64] The CKM matrix is a 3×4 matrix;

---

[64]The assumption is indeed that the vacuum expectation value V and the $\Delta T = 0$ mass μ are of the same order of magnitude, which is much larger than v. Then, both X and Y have matrix elements ~ 1.

its block connecting the standard quarks is a 3×3 matrix K which, in a first approximation, is the unitary matrix which diagonalizes

$$(v\Gamma X)(v\Gamma X)^\dagger = v^2 \Gamma\Gamma^T - \frac{v^2 V^2}{M^2}\Gamma\mathcal{F}^\dagger\mathcal{F}\Gamma^T, \tag{24.59}$$

where we have used eqn (24.56). Because of the presence of the complex matrix $\mathcal{F}^\dagger\mathcal{F}$ in eqn (24.59), it is clear that, if V and M are of the same order of magnitude—or, equivalently, if V and μ are of the same order of magnitude—then the matrix K will in general contain a CP-violating phase and, moreover, this phase will not be suppressed by small mass ratios (Bento et al. 1991).

It is worth recalling the main features of the model that we have described. The scalar sector consists of the standard doublet and a non-Hermitian singlet S. In the fermion sector, only one charge $-1/3$ isosinglet quark N is added. One imposes CP invariance of the Lagrangian. CP is spontaneously broken through the phase of $\langle 0|S|0\rangle$. The crucial role played by the Yukawa couplings connecting N with the standard quarks should be emphasized. On the one hand, those couplings render the phase of $\langle 0|S|0\rangle$ genuinely CP-violating, by restricting the allowed CP transformations of the Lagrangian; on the other hand, it is through these couplings that that phase leads to the complexity of the 3×3 block of the CKM matrix connecting the standard quarks.

It is important to stress that this model may also be important because, if μ and V are made higher than the cosmological-inflation scale, then domain walls, which are a problem which typically plagues models of SCPV, will be absent. Spontaneous CP breaking can occur at such a high energy scale that domain walls disappear during inflation, while CP violation in the CKM matrix remains unsuppressed.

24.8 Phenomenological implications

As we have emphasized, one of the salient features of models with vector-like quarks is the fact that the 3×3 CKM matrix connecting the standard quarks is no longer unitary and, as a result, there are FCNC coupling to the Z.

24.8.1 *Experimental bounds on Z_{kj}*

We shall first present some experimental constraints on FCNC in the down-type-quark sector. The relevant Lagrangian is given by

$$\mathcal{L}_{\text{gauge}} = \cdots - \frac{g}{2c_w} Z_\mu \overline{d_{Lk}}\gamma^\mu Z_{kj} d_{Lj}. \tag{24.60}$$

The flavour-changing parameters Z_{kj} are closely related to the deviation of the CKM matrix V from unitarity, through the relations

$$\begin{aligned}
V_{ud}^* V_{us} + V_{cd}^* V_{cs} + V_{td}^* V_{ts} &= Z_{ds}, \\
V_{ub}^* V_{ud} + V_{cb}^* V_{cd} + V_{tb}^* V_{td} &= Z_{bd}, \\
V_{ub}^* V_{us} + V_{cb}^* V_{cs} + V_{tb}^* V_{ts} &= Z_{bs}.
\end{aligned} \tag{24.61}$$

The strongest constraint on Z_{ds} stems from the experimental upper bound

$$\text{BR}\left(K^+ \to \pi^+ \nu \bar{\nu}\right) < 5.2 \times 10^{-9}. \tag{24.62}$$

As in Appendix D, we compare the process $K^+ \to \pi^+ \nu \bar{\nu}$ with the charged-current process $K^+ \to \pi^0 e^+ \nu_e$, which has

$$\text{BR}\left(K^+ \to \pi^0 e^+ \nu_e\right) = 0.482 \pm 0.006, \tag{24.63}$$

and using flavour-SU(3) symmetry, one obtains (Lavoura and Silva 1993b)

$$\frac{\text{BR}\left(K^+ \to \pi^+ \nu \bar{\nu}\right)}{\text{BR}\left(K^+ \to \pi^0 e^+ \nu_e\right)} = \frac{3 \left|Z_{ds}\right|^2}{2 \left|V_{us}\right|^2}, \tag{24.64}$$

where the factor 3 corresponds to the sum over the three neutrino flavours. From eqns (24.62)–(24.64) one obtains

$$\left|Z_{ds}\right| < 5.9 \times 10^{-5}. \tag{24.65}$$

There are other limits, on $\left|\text{Re}\left(Z_{ds}^2\right)\right|$ and on $\left|\text{Im}\left(Z_{ds}^2\right)\right|$, arising from Δm_K and from ϵ_K, but eqn (24.65) is the stringest bound on Z_{ds}.

The best limit on $|Z_{db}|$ and on $|Z_{sb}|$ is derived from the experimental bound (Particle Data Group 1996)

$$\frac{\text{BR}\left(B \to \mu^+ \mu^- X\right)}{\text{BR}\left(B \to \mu \nu_\mu X\right)} < \frac{5 \times 10^{-5}}{(10.3 \pm 0.5) \times 10^{-2}}. \tag{24.66}$$

The FCNC contributes to the decay $B \to \mu^+ \mu^- X$, where X is an arbitrary set of particles. A straightforward calculation gives (Parada 1996)

$$\frac{\text{BR}\left(B \to \mu^+ \mu^- X\right)}{\text{BR}\left(B \to \mu \nu_\mu X\right)} = \frac{\left[\left(\frac{1}{2} - \sin^2 \theta_w\right)^2 + \sin^4 \theta_w\right]\left(\left|Z_{db}\right|^2 + \left|Z_{sb}\right|^2\right)}{\left|V_{ub}\right|^2 + \left|V_{cb}\right|^2 f\left(m_c^2/m_b^2\right)}, \tag{24.67}$$

where the function f has been given in eqn (15.18). From eqns (24.66) and (24.67) one obtains

$$\left|Z_{db}\right|^2 + \left|Z_{sb}\right|^2 < 5 \times 10^{-6}. \tag{24.68}$$

24.8.2 *Implications for CP asymmetries*

The new contribution to B^0–$\overline{B^0}$ mixing arising from the Z-mediated $|\Delta B| = 2$ tree-level diagram can have a significant impact on CP asymmetries in B^0 decays. Let us write the complete matrix element M_{12} of B_d^0–$\overline{B_d^0}$ mixing as

$$M_{12} = M_{12}^{\text{SM}} \left(r e^{-i\theta}\right)^2, \tag{24.69}$$

where M_{12}^{SM} is the standard-model box-diagram contribution, while (Barenboim *et al.* 1998)

$$\left(re^{i\theta}\right)^2 = 1 + \frac{4\pi s_w^2}{\alpha S_0\left(x_t\right)}\left(\frac{Z_{bd}}{V_{td}V_{tb}^*}\right)^2 - \frac{R_0\left(x_t\right)}{S_0\left(x_t\right)}\frac{Z_{bd}}{V_{td}V_{tb}^*}, \tag{24.70}$$

where $x_t = m_t^2/m_W^2$, the Inami–Lim (1981) function S_0 has been defined in eqn (B.16), and

$$R_0\left(x\right) = \frac{x}{1-x}\left(4 - x + \frac{3x\ln x}{1-x}\right) \tag{24.71}$$

is another Inami–Lim function. The first term in eqn (24.70) corresponds to the box diagram; the second term results from the Z-exchange tree-level diagram; the third term arises from a one-loop Z-vertex correction, which is relevant for small values of Z_{bd} (Barenboim and Botella 1998). Analogous expressions hold in the case of B_s^0–$\overline{B}_s^0$ mixing. From eqn (24.68) it can be readily verified (Branco et al. 1993) that in the case of B_d^0–$\overline{B}_d^0$ mixing the Z contribution can be the dominant one, while in the case of B_s^0–$\overline{B}_s^0$ mixing it can at most compete with the usual box diagram.

The study of CP asymmetries should provide a very sensitive probe of FCNC. The measurement of those asymmetries with the expected experimental uncertainties may detect FCNC effects (Branco et al. 1993; Barenboim et al. 1998) even for rather small values of Z_{bd}, at the level

$$\left|\frac{Z_{bd}}{V_{td}V_{tb}^*}\right| \sim 10^{-2}. \tag{24.72}$$

24.8.3 Baryogenesis

It has been shown (Branco et al. 1998) that, in a model with an extra singlet complex scalar, the first-order electroweak phase transition can be strong enough to avoid the baryon-asymmetry washout by sphalerons. The crucial point is the fact that in this model there is CP violation even in the extreme chiral limit where $m_d = m_s = m_u = m_c = 0$. As a result, and contrary to what happens in the SM, there is no strong suppression of the baryon asymmetry by the ratio of light-quark masses over the critical temperature T_c. The dominant contribution to the baryon asymmetry was estimated to be (Branco et al. 1998)

$$\frac{n_B}{s} \sim -10^{-3}v_w\frac{\operatorname{Im}\operatorname{tr}\left(h_p h_n H_n\right)}{T_c^8}, \tag{24.73}$$

where v_w is the bubble-wall velocity. The appearance of the weak-basis (WB) invariant in the numerator of eqn (24.73) was to be expected, since it is the invariant which controls the strength of CP violation in the extreme chiral limit. That WB invariant can be expressed in terms of quark masses and mixing angles, and one obtains in leading order

$$\frac{n_B}{s} \sim -10^{-3}v_w\frac{m_t^2 m_b^2 m_D^4}{T_c^8}\operatorname{Im}\left(V_{tb}V_{4D}V_{tD}^*V_{4b}^*\right), \tag{24.74}$$

when there is only one vector-like $Q = -1/3$ quark D, with mass m_D. If $m_D = 200\,\text{GeV}$ and $|\operatorname{Im}\left(V_{tb}V_{4D}V_{tD}^*V_{4b}^*\right)| = 10^{-4}$ one obtains the right order of magnitude for n_B/s.

24.9   Main conclusions

- Models with vector-like quarks have new sources of CP violation and provide a well-defined framework to study deviations from unitarity of the 3×3 CKM matrix, as well as flavour-changing neutral currents mediated by the Z. Those models have a rich phenomenology, which may be tested through the search for rare K and B decays, and through the study of CP asymmetries at B factories.

- With the addition of at least one vector-like quark to the SM, spontaneous CP violation can be achieved with a very simple Higgs system, consisting of only one non-Hermitian scalar, singlet under SU(2)⊗U(1). CP violation originates in the phase of the VEV of the singlet scalar. Through the mixing of the vector-like quarks with standard quarks, this phase generates an unsuppressed CP-violating phase in the 3×3 block of the CKM matrix connecting the usual quarks.

We are now in a position to add another item to the summary of models in § 23.7:

ONE HIGGS DOUBLET AND ONE SCALAR SINGLET
(CP SYMMETRY IMPOSED ON THE LAGRANGIAN)

- **No extra fermions**: The vacuum contains one irremovable phase, but that phase does not lead to spontaneous CP breaking. The scalar sector does not lead to any new sources of CP violation.

- **Isosinglet quarks**: The physical phase in the VEV of the singlet non-Hermitian scalar leads to spontaneous CP violation—it generates a complex CKM matrix. This is a minimal realization of the Nelson–Barr mechanism.

MASSIVE NEUTRINOS AND CP VIOLATION IN THE LEPTONIC SECTOR

25.1 Introduction

Neutrinos are fascinating particles which have played an important role in the understanding of the structure of both the charged- and neutral-weak-current interactions. They may also play an important role in astrophysics and cosmology. A complete treatment of all aspects of neutrino physics is beyond the scope of this book; there are excellent books and review articles on the subject (Bilenky and Petcov 1987; Kayser *et al.* 1989; Mohapatra and Pal 1991; Boehm and Vogel 1992; Kim and Pevsner 1993). Our emphasis here will be on those aspects of neutrino physics having to do with CP violation. First we shall describe some of the motivations for considering massive neutrinos.

25.1.1 *Theoretical motivations*

In the standard model (SM), the masslessness of neutrinos is essentially due to the fact that no $SU(2) \otimes U(1)$-singlet leptons, i.e., no right-handed neutrinos, are introduced. This is an unnatural feature of the SM and breaks quark–lepton symmetry. If right-handed neutrinos, i.e., leptons which are singlets under the gauge group, are introduced, then quark–lepton symmetry is re-established, and the most general renormalizable and $SU(2) \otimes U(1)$-invariant Lagrangian includes Yukawa interactions which generate neutrino masses upon spontaneous gauge-symmetry breaking.

Neutrino masses naturally arise in most extensions of the SM. For example, if one requires a grand-unified theory (GUT) of the electroweak and strong interactions in which each fermion family is contained in a single representation of the gauge group, then SO(10) is the minimal GUT. In SO(10) each fermion family is unified in a 16-dimensional spinor representation where, together with the fifteen chiral fermions of the SM, a right-handed neutrino is included. Neutrino masses are then unavoidable.

25.1.2 *Phenomenological motivations*

The solar-neutrino data obtained by several different experiments—for a review see Cleveland *et al.* (1995)—indicate a deficit in the observed neutrinos in comparison to the predictions for the neutrino fluxes of the standard solar model (Bahcall and Pinsonneault 1992, 1995; Turck-Chièse *et al.* 1993; Dar and Shaviv 1996; Bahcall *et al.* 1998). This solar-neutrino deficit may be explained as resulting from oscillations of the electron neutrino into some other neutrino species. A particularly plausible solution to the solar-neutrino problem is provided by the

MSW mechanism (Wolfenstein 1978, 1979; Mikheev and Smirnov 1985), where the oscillations are enhanced by the material medium of the solar interior (Bethe 1986).

Various experiments—see for instance the Super-Kamiokande Collaboration (1998), and references therein—have measured the ratio of the number of muon neutrinos to the number of electron neutrinos produced in the atmosphere from the decays of pions and kaons originating in cosmic rays. The measured ratio is much smaller than the predicted one. The double ratio

$$R \equiv \frac{\left(n_{\nu_\mu}\right)_{\text{observed}}}{\left(n_{\nu_e}\right)_{\text{observed}}} \frac{\left(n_{\nu_e}\right)_{\text{predicted}}}{\left(n_{\nu_\mu}\right)_{\text{predicted}}} \tag{25.1}$$

is on average 60% and, most importantly, depends on the direction from which neutrinos come. This anomaly can be explained by oscillations of the muon neutrinos, on their way through the Earth, into another type of neutrino (see for instance Fogli et al. 1999).

Neutrino oscillations are most probably due to neutrinos being massive. There is another possibility, though: they might arise from flavour-changing Yukawa interactions among the various neutrino flavours even in the case of massless neutrinos.

Another motivation for considering massive neutrinos is the fact that neutrinos may constitute the hot dark matter in the Universe, provided neutrino masses are of a few eV. See, for example, the excellent book by Bahcall (1989).

25.2 Dirac and Majorana masses

25.2.1 Dirac masses

If right-handed neutrinos are introduced in the SM, Dirac mass terms can be constructed for the neutrinos, entirely analogous to those for the quarks and for the charged leptons. In general, one may write

$$\mathcal{L}_{\text{Dirac}} = -\overline{\upsilon_L} M_D \upsilon_R + \text{H.c.}, \tag{25.2}$$

where

$$\upsilon_{L,R} = \begin{pmatrix} \nu_e \\ \nu_\mu \\ \nu_\tau \\ \vdots \end{pmatrix}_{L,R}. \tag{25.3}$$

Here, $\nu_e, \nu_\mu, \nu_\tau, \ldots$ are the eigenstates of the weak interaction, which get mixed through the mass terms. One can find the neutrino mass eigenstates through bi-diagonalization of M_D:

$$V_L^\dagger M_D V_R = \text{diag}\,(m_1, m_2, m_3, \ldots), \tag{25.4}$$

where V_L and V_R are the unitary matrices relating the weak eigenstates to the mass eigenstates:

$$\begin{pmatrix} \nu_e \\ \nu_\mu \\ \nu_\tau \\ \vdots \end{pmatrix}_{L,R} = V_{L,R} \begin{pmatrix} \nu_1 \\ \nu_2 \\ \nu_3 \\ \vdots \end{pmatrix}_{L,R}. \tag{25.5}$$

In the SM, due to the absence of neutrino masses, the lepton-flavour numbers $L_e, L_\mu, L_\tau, \ldots$ are conserved and, of course, the total lepton number $L = L_e + L_\mu + L_\tau + \cdots$ is conserved too.[65] In the presence of Dirac mass terms, each individual lepton-flavour number will be violated, but the total lepton number will remain conserved. Indeed, $\mathcal{L}_{\text{Dirac}}$ is invariant under the global U(1) transformation

$$\nu_{L,R} \to e^{i\alpha} \nu_{L,R}. \tag{25.6}$$

This situation is entirely analogous to the one in the quark sector. Individual quark flavours are violated by the charged-current weak interaction, while baryon number is conserved at the perturbative level.

25.2.2 Majorana masses

Neutrinos are the only known fermions with no electric charge. As a result, one may have different mass terms involving neutrinos—the Majorana mass terms. Let ψ be the field associated with a neutral spin-1/2 particle and let $\psi^c = C\overline{\psi}^T$ be the charge-conjugate field. Then, from eqns (3.45) and (3.54) it follows that $\overline{\psi}\psi$, $\overline{\psi^c}\psi$, $\overline{\psi}\psi^c$, and $\overline{\psi^c}\psi^c$ are Lorentz-invariant quantities. As ψ and $\overline{\psi}$ anticommute, $\overline{\psi^c}\psi^c = -\psi^T\overline{\psi}^T = \overline{\psi}\psi$. Also, $\overline{\psi^c}\psi = -\psi^T C^{-1}\psi$ and $\overline{\psi}\psi^c = \overline{\psi}C\overline{\psi}^T$ do not vanish, in spite of the matrix C being antisymmetric, because of the anticommuting nature of the spinors. The Majorana mass terms are

$$\begin{aligned}
\mathcal{L}_{\text{Majorana}} &= -\frac{m_L}{2}\left[\overline{(\psi_L)^c}\psi_L + \overline{\psi_L}(\psi_L)^c\right] - \frac{m_R}{2}\left[\overline{(\psi_R)^c}\psi_R + \overline{\psi_R}(\psi_R)^c\right] \\
&= \frac{m_L}{2}\left(\psi_L^T C^{-1}\psi_L - \overline{\psi_L}C\overline{\psi_L}^T\right) + \frac{m_R}{2}\left(\psi_R^T C^{-1}\psi_R - \overline{\psi_R}C\overline{\psi_R}^T\right) \tag{25.7}
\end{aligned}$$

The factors 1/2 are inserted in order that, in the equation of motion, m_L can be interpreted as the mass of ψ_L, while m_R is the mass of ψ_R (see below). Note that a Majorana mass term may be introduced even when ψ is a Weyl spinor. The Majorana mass term is more economical than the Dirac mass term, because it does not require the existence of both ψ_L and ψ_R. It is clear that Majorana mass terms are not invariant under the transformation in eqn (25.6):

$$\psi \to e^{i\alpha}\psi \Rightarrow \begin{cases} \overline{\psi^c}\psi \to e^{2i\alpha}\overline{\psi^c}\psi, \\ \overline{\psi}\psi^c \to e^{-2i\alpha}\overline{\psi}\psi^c. \end{cases} \tag{25.8}$$

---

[65]Strictly speaking, L is violated by global electroweak anomalies, and it is only $B - L$ which is conserved (B is the baryon number). However, at temperatures much lower than v—the electroweak-phase-transition scale—the violation of L is negligible.

Therefore, a Majorana mass term can only be used for particles which do not carry any conserved quantum number like electric charge. If one introduces Majorana mass terms for the neutrinos, then the lepton number L is no longer conserved.

Let us consider that there are n_g generations of left-handed neutrinos. The Majorana mass terms may be written

$$\mathcal{L}_{\text{Majorana}} = \tfrac{1}{2} v_L^T C^{-1} M_L v_L + \text{H.c.}, \tag{25.9}$$

where v_L denotes as before the column matrix with the neutrino weak eigenstates, and M_L is an $n_g \times n_g$ matrix. Without loss of generality, M_L is symmetric. (One takes into account that C is antisymmetric and that fermion fields anticommute.) In general, a symmetric matrix may be diagonalized by the following transformation:

$$V_L^T M_L V_L = \text{diag}\,(m_1, m_2, m_3, \ldots), \tag{25.10}$$

where V_L is the unitary matrix which relates the weak and mass eigenstates:

$$v_L = V_L \nu_L. \tag{25.11}$$

The masses $m_1, m_2, \ldots$ are real and non-negative. The mass terms may then be written

$$\mathcal{L}_{\text{Majorana}} = -\frac{m_i}{2} \left[\overline{(\nu_{Li})^c} \nu_{Li} + \overline{\nu_{Li}} \left(\nu_{Li}\right)^c \right]. \tag{25.12}$$

The index i runs from 1 to n_g.

Let us now derive the Dirac equation for a spin-1/2 fermion with Majorana mass. In the absence of interactions, the Lagrangian is

$$\mathcal{L} = \frac{i}{2}\, \overline{\psi}\gamma_\mu \left(\partial^\mu \psi\right) - \frac{i}{2} \left(\partial^\mu \overline{\psi}\right) \gamma_\mu \psi - \frac{m}{2}\, e^{i\varsigma} \overline{\psi} C \overline{\psi}^T + \frac{m}{2}\, e^{-i\varsigma} \psi^T C^{-1} \psi. \tag{25.13}$$

The equation of motion is derived in the standard fashion:

$$0 = \partial^\mu \frac{\delta \mathcal{L}}{\delta \left(\partial^\mu \overline{\psi}\right)} - \frac{\delta \mathcal{L}}{\delta \overline{\psi}}. \tag{25.14}$$

One must be careful to remember that all derivatives must be taken from the same side (say, they must be left-derivatives), because one is dealing with anticommuting fields. The result is

$$0 = \partial^\mu \left(-\frac{i}{2}\, \gamma_\mu \psi \right) - \frac{i}{2}\, \gamma_\mu \left(\partial^\mu \psi\right) + m e^{i\varsigma} C \overline{\psi}^T$$
$$= -i\, \gamma_\mu \partial^\mu \psi + m e^{i\varsigma} C \overline{\psi}^T. \tag{25.15}$$

One obtains the usual Dirac equation, but for a field which is identical with its charge conjugate:

$$\psi = e^{i\varsigma} C \overline{\psi}^T. \tag{25.16}$$

Thus, the Majorana mass terms naturally lead to Majorana fermions (see Chapter 3).

Suppose that a field ψ has (real) Majorana mass m. Then, $\psi = \psi^c \equiv C\overline{\psi}^T$. Now, it can easily be shown that

$$(\mathcal{CP})\,\psi\,(\mathcal{CP})^\dagger = e^{i\theta}\psi^c \Leftrightarrow (\mathcal{CP})\,\psi^c\,(\mathcal{CP})^\dagger = -e^{-i\theta}\psi. \qquad (25.17)$$

As $\psi = \psi^c$, one must have $\exp(i\theta) = -\exp(-i\theta) = \pm i$. One has thus arrived at the conclusion that *the CP-parity of a Majorana field must be either i or $-i$*.

25.3 The seesaw mechanism

The inclusion of right-handed neutrinos may lead to difficulties in understanding the smallness of neutrino masses. Fortunately, a natural explanation for the smallness of those masses has been found through the seesaw mechanism (Gell-Mann *et al.* 1979; Yanagida 1979; Mohapatra and Senjanović 1980), which we shall now describe.

Let us first consider that there is only one generation—one left-handed neutrino v_L and one right-handed neutrino v_R. The most general Dirac and Majorana mass terms may be written

$$\mathcal{L}_{\text{mass}} = -\tfrac{1}{2}\left(\overline{(v_L)^c}\ \overline{v_R}\right)\begin{pmatrix} m_L & m_D \\ m_D & m_R \end{pmatrix}\begin{pmatrix} v_L \\ (v_R)^c \end{pmatrix} + \text{H.c..} \qquad (25.18)$$

Assuming for definiteness that the 2×2 neutrino mass matrix above is real, one may diagonalize it through the transformation

$$\begin{pmatrix} v_L \\ (v_R)^c \end{pmatrix} = \begin{pmatrix} -i\cos\theta & \sin\theta \\ i\sin\theta & \cos\theta \end{pmatrix}\begin{pmatrix} \nu_{1L} \\ \nu_{2L} \end{pmatrix}, \qquad (25.19)$$

with the angle θ given by

$$\sin^2\theta = \frac{\sqrt{(m_L - m_R)^2 + 4m_D^2} + m_L - m_R}{2\sqrt{(m_L - m_R)^2 + 4m_D^2}},$$

$$\cos^2\theta = \frac{\sqrt{(m_L - m_R)^2 + 4m_D^2} + m_R - m_L}{2\sqrt{(m_L - m_R)^2 + 4m_D^2}}, \qquad (25.20)$$

$$\sin\theta\cos\theta = \frac{m_D}{\sqrt{(m_L - m_R)^2 + 4m_D^2}}.$$

Then,

$$\begin{pmatrix} -i\cos\theta & i\sin\theta \\ \sin\theta & \cos\theta \end{pmatrix}\begin{pmatrix} m_L & m_D \\ m_D & m_R \end{pmatrix}\begin{pmatrix} -i\cos\theta & \sin\theta \\ i\sin\theta & \cos\theta \end{pmatrix} = \text{diag}\,(m_-, m_+), \qquad (25.21)$$

with

$$m_\pm = \pm\tfrac{1}{2}\left[m_L + m_R \pm \sqrt{(m_L - m_R)^2 + 4m_D^2}\right]. \qquad (25.22)$$

So far, no approximation has been done. Let us now consider the physically interesting limiting case where $m_L = 0$ and $m_R \gg m_D$. Within gauge theories this is the natural situation. Then,

$$\sqrt{(m_L - m_R)^2 + 4m_D^2} \approx m_R + \frac{2m_D^2}{m_R}, \qquad (25.23)$$

and the masses are given by

$$m_- \approx \frac{m_D^2}{m_R},$$
$$m_+ \approx m_R + \frac{m_D^2}{m_R} \approx m_R. \qquad (25.24)$$

This result illustrates the seesaw mechanism. Assuming m_D to have the same mass scale as the charged leptons and quarks, one sees that the neutrino masses are suppressed by a factor m_D/m_R. Moreover, from eqns (25.20) and (25.23) it follows that the mixing between v_L and $(v_R)^c$ is suppressed by the same factor.

These results may be generalized to an arbitrary number of generations. Let us suppose that n_g left-handed neutrinos v_L and n_g right-handed neutrinos v_R have the following general Dirac and Majorana mass terms:

$$\mathcal{L}_{\text{mass}} = -\tfrac{1}{2} \left(\overline{(v_L)^c} \; \overline{v_R} \right) \begin{pmatrix} M_L & M_D \\ M_D^T & M_R \end{pmatrix} \begin{pmatrix} v_L \\ (v_R)^c \end{pmatrix} + \text{H.c.}. \qquad (25.25)$$

The $2n_g \times 2n_g$ symmetric mass matrix

$$\mathcal{M} = \begin{pmatrix} M_L & M_D \\ M_D^T & M_R \end{pmatrix} \qquad (25.26)$$

may be diagonalized by the transformation

$$\mathcal{U}^\dagger \mathcal{M} \mathcal{U}^* = \begin{pmatrix} D_L & 0 \\ 0 & D_R \end{pmatrix}, \qquad (25.27)$$

where D_L and D_R are diagonal, real, and non-negative $n_g \times n_g$ matrices. We write

$$\mathcal{U} = \begin{pmatrix} V_L & V_{LR} \\ V_{RL} & V_R \end{pmatrix}, \qquad (25.28)$$

and assume that $M_L = 0$, that M_R is a non-singular matrix (Branco *et al.* 1989), and that $M_D \ll M_R$. Under this approximation, we obtain

$$V_R^\dagger M_R V_R^* = D_R,$$
$$V_L^\dagger \left(-M_D M_R^{-1} M_D^T \right) V_L^* = D_L. \qquad (25.29)$$

As seen in eqns (25.29), the masses of the heavy neutrinos are $\sim M_R$, while the masses of the light neutrinos are $\sim M_D^2/M_R$, instead of being $\sim M_D$. The matrices V_R and V_L are approximately unitary, while V_{LR} and V_{RL} are $\sim M_D/M_R$.

The light neutrinos are, to a good approximation, mixtures of the ν_L only. These are the two relevant accomplishments of the seesaw mechanism: a suppression of the masses of the light neutrinos relative to the masses of the other fermions, and a suppression of the mixing of the doublet neutrinos—of the ν_L—with the singlet neutrinos—with the $(\nu_R)^c$.

25.4 Neutrino masses in SU(2)⊗U(1) gauge theories

In the SM neutrinos are strictly massless. They do not have Dirac masses because no SU(2)⊗U(1)-singlet fermions are introduced. No Majorana neutrino masses arise at tree level because such masses would have $|\Delta T_3| = 1$, and scalar triplets are absent in the SM. No Majorana neutrino masses are generated either perturbatively or non-perturbatively, due to the presence of an exact $B - L$ symmetry.

We consider an arbitrary extension of the SM in which the gauge structure and the interactions of the leptons with the $W^\pm$, photon, and Z, are identical to the SM case—see eqns (12.8) and (12.9):

$$
\mathcal{L}_{\text{gauge}} = \frac{g}{\sqrt{2}} \left(W_\mu^+ \overline{\nu_L} \gamma^\mu l_L + W_\mu^- \overline{l_L} \gamma^\mu \nu_L \right) + eA_\mu \left(\overline{l_L} \gamma^\mu l_L + \overline{l_R} \gamma^\mu l_R \right)
$$
$$
+ \frac{g}{c_w} Z_\mu \left[\tfrac{1}{2} \overline{\nu_L} \gamma^\mu \nu_L + \left(s_w^2 - \tfrac{1}{2} \right) \overline{l_L} \gamma^\mu l_L + s_w^2 \overline{l_R} \gamma^\mu l_R \right]. \quad (25.30)
$$

We introduce n' right-handed neutrinos ν_R, which are singlets of SU(2)⊗U(1). The most general Dirac and Majorana mass terms for neutrinos and charged leptons are

$$
\mathcal{L}_{\text{mass}} = -\overline{l_L} M_l l_R - \overline{\nu_L} M_\nu \nu_R + \tfrac{1}{2} \nu_L^T C^{-1} M_L \nu_L + \tfrac{1}{2} \nu_R^T C^{-1} M_R \nu_R + \text{H.c.}, \quad (25.31)
$$

where M_l is the $n_g \times n_g$ charged-lepton mass matrix, M_ν is the $n_g \times n'$ neutrino Dirac mass matrix, M_L is the $n_g \times n_g$ matrix of Majorana masses for the ν_L, and M_R is the $n' \times n'$ matrix of Majorana masses for the ν_R. Both M_L and M_R are symmetric without loss of generality—their antisymmetric parts give vanishing contributions to eqn (25.31). The Dirac masses M_l and M_ν are generated in the usual way, through the Yukawa couplings of the standard Higgs doublet. The right-handed mass terms in M_R are SU(2)⊗U(1)-invariant, and therefore they should in general be included independently of the assumed spectrum of scalar fields. At tree level, M_L can arise through the Yukawa couplings of a Higgs triplet. However, even in a minimal extension of the SM, where ν_R are introduced but the Higgs structure is standard—i.e., no scalar triplets—the Majorana masses for the left-handed neutrinos in M_L will be perturbatively generated, since L is no longer a symmetry of the Lagrangian, being explicitly broken by the right-handed-neutrino Majorana masses in M_R (provided $M_\nu \neq 0$ too).

It is often useful to rewrite the neutrino mass terms in eqn (25.31) in terms of left-handed fields only, or of right-handed fields only. This may be done by defining

$$v'_R \equiv (v_L)^c = C\overline{v_L}^T,$$
$$v''_L \equiv (v_R)^c = C\overline{v_R}^T. \tag{25.32}$$

One can then write the neutrino mass terms in terms of right-handed fields only:

$$\mathcal{L}_{\text{mass}} = -\overline{l_L} M_l l_R + \tfrac{1}{2} \left(v'^T_R \ v^T_R \right) C^{-1} \begin{pmatrix} M_L^\dagger & M_v \\ M_v^T & M_R \end{pmatrix} \begin{pmatrix} v'_R \\ v_R \end{pmatrix} + \text{H.c.}; \tag{25.33}$$

or, alternatively, in terms of left-handed fields only:

$$\mathcal{L}_{\text{mass}} = -\overline{l_L} M_l l_R + \tfrac{1}{2} \left(v^T_L \ v''^T_L \right) C^{-1} \begin{pmatrix} M_L & M_v^* \\ M_v^\dagger & M_R^\dagger \end{pmatrix} \begin{pmatrix} v_L \\ v''_L \end{pmatrix} + \text{H.c.}. \tag{25.34}$$

Without loss of generality, we may choose to work in a weak basis where the charged-lepton mass matrix M_l is diagonal and real. The $(n_g + n') \times (n_g + n')$ Dirac–Majorana mass matrix can be diagonalized with the help of a unitary matrix; the lepton mixing matrix appearing in the charged current will consist of the first n_g columns of that unitary matrix. The charged-current interaction of the leptons can be written

$$\mathcal{L}_W^{(l)} = \frac{g}{\sqrt{2}} W_\mu^+ \left(\overline{v_1} \ \overline{v_2} \cdots \overline{v_{n_g+n'}} \right) K \gamma^\mu \gamma_L \begin{pmatrix} e \\ \mu \\ \tau \\ \vdots \end{pmatrix} + \text{H.c.}. \tag{25.35}$$

The lepton mixing matrix K is an $(n_g + n') \times n_g$ matrix. One may ask how many CP-violating phases appear in it. That number may be evaluated with the help of the method in Chapter 24, leading to

$$n_{\text{phases}} = \tfrac{1}{2} (n_g - 1)(n_g - 2 + 2n') + n_g + n' - 1 \tag{25.36}$$
$$= n'n_g + \tfrac{1}{2} n_g (n_g - 1). \tag{25.37}$$

Indeed, right-handed neutrinos in the leptonic sector are analogous to vector-like isosinglets in the quark sector—they correspond to left-handed fields which are singlets of SU(2). Thus, in eqn (25.36), $\tfrac{1}{2} (n_g - 1)(n_g - 2 + 2n')$ is the number of CP-violating phases in a model with n_g standard quark generations together with n' isosinglet vector-like quarks of like charge. The extra $n_g + n' - 1$ phases in eqn (25.36) result from the Majorana character of the neutrino fields. Indeed, the physical neutrinos having Majorana masses, one is no longer free to rephase them, since that would introduce phases in the masses. There are thus $n_g + n' - 1$ phases in the mixing matrix that we are unable to absorb through redefinitions of the neutrino fields; those phases have physical consequences, viz., they are CP-violating.

In the special case of no right-handed neutrinos, i.e., when $n' = 0$, one has $n_{\text{phases}} = \tfrac{1}{2} n_g (n_g - 1)$. In particular, a model with two neutrinos with Majorana masses may violate CP. It is worth giving a simple explanation why this is so, while three generations are needed in order to have CP violation in the quark sector. The crucial difference lies in the Majorana masses. If only Dirac masses exist, CP violation in the quark and lepton sectors arises under the same conditions. If, however, neutrinos have Majorana masses, there are extra observable phases in the lepton mixing matrix K. Indeed, the Majorana mass terms for the physical neutrinos,

$$\frac{m_1}{2}\left(\nu_1^T C^{-1}\nu_1 - \overline{\nu_1}C\overline{\nu_1}^T\right) + \frac{m_2}{2}\left(\nu_2^T C^{-1}\nu_2 - \overline{\nu_2}C\overline{\nu_2}^T\right), \qquad (25.38)$$

do not allow the neutrino fields to be rephased, lest m_1 and m_2 become complex. The charged-lepton fields may be rephased, and we may thus eliminate two phases from the lepton mixing matrix; for instance, we may set K_{11} and K_{12} to be real by rephasing the e and μ fields, respectively. One physical phase still remains in K, and that phase is the source of CP violation.

25.5 Conditions for CP invariance

The absence of neutrino masses in the SM leads to CP conservation in the leptonic sector of that theory. We want to consider the case of non-vanishing neutrino masses, and find out the conditions for CP invariance of $\mathcal{L} = \mathcal{L}_{\text{gauge}} + \mathcal{L}_{\text{mass}}$. For definiteness, we shall deal only with the CP violation arising from the clash between $\mathcal{L}_{\text{gauge}}$ and $\mathcal{L}_{\text{mass}}$, which manifests itself as physical CP-violating phases in the charged-current and neutral-current weak interactions. The interactions of the scalar fields, which are needed to break the gauge symmetry but which we have not introduced explicitly, may constitute extra sources of CP violation in the complete theory (beyond $\mathcal{L}_{\text{gauge}}$ and $\mathcal{L}_{\text{mass}}$).[66]

We start as usual with $\mathcal{L}_{\text{gauge}}$. The CP transformation of the gauge bosons was written down in eqns (13.7), (13.9), and (13.10). It must be accompanied by the following CP transformation of the lepton fields:

$$\begin{aligned} (\mathcal{CP})\, l_L\, (\mathcal{CP})^\dagger &= ie^{-i\xi_W} U_L \gamma^0 C \overline{l_L}^T, \\ (\mathcal{CP})\, \overline{l_L}\, (\mathcal{CP})^\dagger &= ie^{i\xi_W} l_L^T C^{-1} \gamma^0 U_L^\dagger, \end{aligned} \qquad (25.39)$$

$$\begin{aligned} (\mathcal{CP})\, \nu_L\, (\mathcal{CP})^\dagger &= iU_L \gamma^0 C \overline{\nu_L}^T, \\ (\mathcal{CP})\, \overline{\nu_L}\, (\mathcal{CP})^\dagger &= i\nu_L^T C^{-1} \gamma^0 U_L^\dagger, \end{aligned} \qquad (25.40)$$

$$\begin{aligned} (\mathcal{CP})\, l_R\, (\mathcal{CP})^\dagger &= ie^{-i\xi_W} U_R^l \gamma^0 C \overline{l_R}^T, \\ (\mathcal{CP})\, \overline{l_R}\, (\mathcal{CP})^\dagger &= ie^{i\xi_W} l_R^T C^{-1} \gamma^0 U_R^l{}^\dagger. \end{aligned} \qquad (25.41)$$

We have omitted the space–time coordinates, and introduced arbitrary $n_g \times n_g$ unitary matrices U_L and U_R^l. For later convenience, we inserted explicit phase

---

[66]An analysis of the sources of CP violation in the Yukawa interactions of the Gelmini–Roncadelli model (Gelmini and Roncadelli 1981) was given by Bernabéu *et al.* (1986b). They concluded that no extra sources of CP violation beyond those already present in $\mathcal{L}_{\text{gauge}} + \mathcal{L}_{\text{mass}}$ occur in that specific model. As those authors have emphasized, that result cannot be generalized to other models for Majorana masses.

factors i in the definition of those unitary matrices. We have also inserted phase factors $\exp(\pm i\xi_W)$ which cancel out similar phase factors in the CP transformation of $W^\pm$. Notice that, just as in Chapter 14, the same matrix U_L appears in the CP transformation of both l_L and v_L, because those fields are connected by the charged-current interaction.

We require that $\mathcal{L}_{\text{mass}}$ be invariant under CP. We first face the problem that the CP transformation of v_R is not defined, because that field does not enter $\mathcal{L}_{\text{gauge}}$. But, invariance of the Dirac mass term $\overline{v_L}M_v v_R + \overline{v_R}M_v^\dagger v_L$, together with eqns (25.40), implies

$$
\begin{aligned}
(\mathcal{CP})\, v_R\, (\mathcal{CP})^\dagger &= iU_R^v \gamma^0 C \overline{v_R}^T, \\
(\mathcal{CP})\, \overline{v_R}\, (\mathcal{CP})^\dagger &= iv_R^T C^{-1}\gamma^0 U_R^{v\dagger},
\end{aligned}
\tag{25.42}
$$

with

$$
U_L^\dagger M_v U_R^v = M_v^*.
\tag{25.43}
$$

Analogously, we have

$$
U_L^\dagger M_l U_R^l = M_l^*.
\tag{25.44}
$$

For the Majorana mass terms the CP-invariance conditions are the following:

$$
U_L^T M_L U_L = M_L^*,
\tag{25.45}
$$

$$
U_R^{v\,T} M_R U_R^v = M_R^*.
\tag{25.46}
$$

The existence of matrices U_L, U_R^l, and U_R^v, such that eqns (25.43)–(25.46) hold, is necessary and sufficient for CP invariance of $\mathcal{L}_{\text{gauge}} + \mathcal{L}_{\text{mass}}$.

The unitary matrix U_R^l may be eliminated from eqn (25.44) by considering the Hermitian matrix $H_l \equiv M_l M_l^\dagger$. The CP-conservation conditions then simplify to the following: an $n_g \times n_g$ unitary matrix U_L and an $n' \times n'$ unitary matrix U_R^v should exist, such that

$$
U_L^\dagger H_l U_L = H_l^*,
\tag{25.47}
$$

$$
U_L^T M_L U_L = M_L^*,
\tag{25.48}
$$

$$
U_L^\dagger M_v U_R^v = M_v^*,
\tag{25.49}
$$

$$
U_R^{v\,T} M_R U_R^v = M_R^*.
\tag{25.50}
$$

These CP-invariance conditions are invariant under a change of weak basis. If we transform

$$
\begin{aligned}
l_L &= Z_L \tilde{l}_L, \\
v_L &= Z_L \tilde{v}_L, \\
v_R &= Z_R^v \tilde{v}_R,
\end{aligned}
\tag{25.51}
$$

with an $n_g \times n_g$ unitary matrix Z_L and an $n' \times n'$ unitary matrix Z_R^v, we obtain

$$
\begin{aligned}
\tilde{H}_l &= Z_L^\dagger M_l Z_L, \\
\tilde{M}_L &= Z_L^T M_L Z_L, \\
\tilde{M}_v &= Z_L^\dagger M_v Z_R^v, \\
\tilde{M}_R &= Z_R^{v\,T} M_R Z_R^v.
\end{aligned}
\tag{25.52}
$$

Now, if the conditions in eqns (25.43)–(25.46) hold, analogous conditions with tilded matrices also hold, provided that we define

$$\tilde{U}_L \equiv Z_L^{\dagger} U_L Z_L^*,$$
$$\tilde{U}_R^v \equiv Z_R^{v\dagger} U_R^v Z_R^{v*}. \tag{25.53}$$

Since eqns (25.47)–(25.50) are weak-basis independent, one may analyse the restrictions on the lepton mass matrices that they imply in a conveniently chosen weak basis (WB). Let us consider the WB where M_L and M_R are diagonal, with diagonal elements real and non-negative. In general, those diagonal matrix elements will be non-zero and non-degenerate. In that case, eqns (25.48) and (25.50) constrain U_L and U_R^v to be of the form

$$U_L = \text{diag}\left(e^{i\pi p_1}, e^{i\pi p_2}, \ldots, e^{i\pi p_{n_g}}\right),$$
$$U_R^v = \text{diag}\left(e^{i\pi q_1}, e^{i\pi q_2}, \ldots, e^{i\pi q_{n'}}\right), \tag{25.54}$$

where the numbers p_i and q_j are integers. The index j runs from 1 to n'. From eqns (25.47) and (25.49) one concludes that CP invariance constrains H_l and M_v to have the following phase structure:

$$\arg\left(H_l\right)_{ii'} = (p_i - p_{i'})\,\pi/2,$$
$$\arg\left(M_v\right)_{ij} = (p_i - q_j)\,\pi/2. \tag{25.55}$$

Since H_l is an $n_g \times n_g$ Hermitian matrix, while M_v is an arbitrary $n_g \times n'$ matrix, the number of independent restrictions implied by eqns (25.55) is precisely equal to n_{phases} in eqn (25.37).

25.6   The case with no right-handed neutrinos

We now concentrate on the simple case in which there are no right-handed neutrinos v_R. CP invariance of $\mathcal{L}_{\text{gauge}} + \mathcal{L}_{\text{mass}}$ is then equivalent to the existence of an $n_g \times n_g$ unitary matrix U such that

$$U^{\dagger} H U = H^*, \tag{25.56}$$
$$U^T M U = M^*, \tag{25.57}$$

where we have unloaded the notation by writing H instead of H_l and M instead of M_L.

From eqns (25.56) and (25.57) one may derive, through elimination of the matrix U, necessary conditions for CP invariance. For instance, if r is an odd integer, and $a, \ldots, e$ are arbitrary integers, then

$$\text{tr}\left[(M^*M)^a, H^b\right]^r = 0, \tag{25.58}$$

$$\text{Im}\,\text{tr}\left[H^c\,(M^*M)^d\,(M^*H^*M)^e\right] = 0. \tag{25.59}$$

A particular case of eqn (25.59) is

$$\text{Im}\,\text{tr}\left(HM^*MM^*H^*M\right) = 0. \tag{25.60}$$

Equation (25.60) is a necessary condition for CP invariance.

The charged-current-interaction Lagrangian is as in eqn (25.35), only with $n' = 0$. The matrix K is the lepton mixing matrix. Thus,

$$\mathcal{L}_W^{(l)} = \frac{g}{\sqrt{2}} W_\mu^+ \left(\overline{\nu_e} \ \overline{\nu_\mu} \ \cdots \right) \gamma^\mu \gamma_L \begin{pmatrix} e \\ \mu \\ \vdots \end{pmatrix} + \text{H.c.} \tag{25.61}$$

$$= \frac{g}{\sqrt{2}} W_\mu^+ \left(\overline{\nu_1} \ \overline{\nu_2} \ \cdots \right) K \gamma^\mu \gamma_L \begin{pmatrix} e \\ \mu \\ \vdots \end{pmatrix} + \text{H.c..} \tag{25.62}$$

We see that $\left(\overline{\nu_e} \ \overline{\nu_\mu} \ \cdots \right) = \left(\overline{\nu_1} \ \overline{\nu_2} \ \cdots \right) K$. Therefore, in the WB in which the charged-lepton mass matrix is diagonal, one has

$$\begin{aligned} H &= \text{diag} \left(m_e^2, m_\mu^2, \ldots \right), \\ M &= K^T \text{diag} \left(m_1, m_2, \ldots \right) K, \end{aligned} \tag{25.63}$$

where the physical-neutrino Majorana masses m_i are real and non-negative by definition. Alternatively, it may be useful to consider the WB in which M is diagonal:

$$\begin{aligned} M &= \text{diag} \left(m_1, m_2, \ldots \right), \\ H &= \bar{H} = K \text{diag} \left(m_e^2, m_\mu^2, \ldots \right) K^\dagger. \end{aligned} \tag{25.64}$$

We denote $\bar{H}$ the special form that H has in the WB in which M is diagonal.

For $n_g = 3$, the matrix K may be parametrized using, for instance, the Kobayashi–Maskawa parametrization in eqn (16.1). One must however remember that, since the Majorana masses do not allow the neutrino fields to be rephased, there are two extra physical phases. Thus,

$$K = \text{diag} \left(1, e^{i\phi_1}, e^{i\phi_2} \right) \times \begin{pmatrix} c_1 & -s_1 c_3 & -s_1 s_3 \\ s_1 c_2 & c_1 c_2 c_3 - s_2 s_3 e^{i\delta} & c_1 c_2 s_3 + s_2 c_3 e^{i\delta} \\ s_1 s_2 & c_1 s_2 c_3 + c_2 s_3 e^{i\delta} & c_1 s_2 s_3 - c_2 c_3 e^{i\delta} \end{pmatrix}. \tag{25.65}$$

For $n_g = 2$, one may use

$$K = \text{diag} \left(1, e^{i\phi} \right) \times \begin{pmatrix} \cos\theta & -\sin\theta \\ \sin\theta & \cos\theta \end{pmatrix}. \tag{25.66}$$

25.6.1 The CP-parities of the neutrinos

Let us analyse the case in which CP is conserved. For definiteness, we work out the three-generation case; the extension of our results to $n_g \neq 3$ will be obvious. If CP is conserved, then the phase δ in the mixing matrix K of eqn (25.65) must either vanish or be equal to π, i.e., $\exp(i\delta) = \pm 1$. One then has, in the WB of eqn (25.64),

$$\begin{aligned} M &= \text{diag} \left(m_1, m_2, m_3 \right), \\ H &= \text{diag} \left(1, e^{i\phi_1}, e^{i\phi_2} \right) O \, \text{diag} \left(m_e^2, m_\mu^2, m_\tau^2 \right) O^T \text{diag} \left(1, e^{-i\phi_1}, e^{-i\phi_2} \right), \end{aligned} \tag{25.67}$$

where O is an orthogonal (real) matrix. Assuming that m_1, m_2, and m_3 are all different, eqn (25.57) tells us that

$$U = \pm \text{diag} \left[1, (-1)^a, (-1)^b\right], \qquad (25.68)$$

a and b being integers. Remembering eqns (25.40), one thus sees that the CP-parities of the physical neutrinos are $\pm i$,

$$
\begin{aligned}
(\mathcal{CP})\, \nu_1 \,(\mathcal{CP})^\dagger &= \pm i \gamma^0 C \overline{\nu_1}^T, \\
(\mathcal{CP})\, \nu_2 \,(\mathcal{CP})^\dagger &= \pm (-1)^a i \gamma^0 C \overline{\nu_2}^T, \\
(\mathcal{CP})\, \nu_3 \,(\mathcal{CP})^\dagger &= \pm (-1)^b i \gamma^0 C \overline{\nu_3}^T,
\end{aligned}
\qquad (25.69)
$$

which agrees with the conclusion in § 25.2.2. Now, the absolute CP-parity of each neutrino—whether it is $+i$ or $-i$—is a matter of convention, but the relative CP-parities of any two neutrinos—the numbers $(-1)^a$ and $(-1)^b$—are physically relevant. Indeed, the CP-invariance condition in eqn (25.56), when applied to the matrix H of the second eqn (25.67), tells us that

$$
\begin{aligned}
(-1)^a &= e^{2i\phi_1}, \\
(-1)^b &= e^{2i\phi_2}.
\end{aligned}
\qquad (25.70)
$$

Thus, CP is conserved if $\exp(i\phi_1)$ and $\exp(i\phi_2)$ are either ± 1 or $\pm i$. If one of those exponentials is $\pm i$, this does not represent any CP violation, it just means that there are neutrinos with opposite CP-parities.

One may look at the situation from a different perspective if one uses the WB of eqns (25.63). There, one has

$$
\begin{aligned}
H &= \text{diag} \left(m_e^2, m_\mu^2, m_\tau^2\right), \\
M &= O^T \text{diag} \left(m_1, m_2 e^{2i\phi_1}, m_3 e^{2i\phi_2}\right) O.
\end{aligned}
\qquad (25.71)
$$

In this WB one realizes that the physical phases are not really $\exp(i\phi_1)$ and $\exp(i\phi_2)$, rather they are $\exp(2i\phi_1)$ and $\exp(2i\phi_2)$. One sees moreover that the presence of neutrinos with opposite CP-parities is equivalent to the Majorana mass matrix M having eigenvalues with different signs. Indeed, if M has a negative eigenvalue, then one must multiply the corresponding neutrino field by $\pm i$ in order to make its Majorana mass become positive; that phase transformation of $\pm i$ does *not* correspond to CP violation (Wolfenstein 1981).

25.6.2   Two generations

Let us now consider the case $n_g = 2$. Equation (25.60) reads, in the weak basis of eqns (25.64),

$$
\begin{aligned}
0 &= \text{Im} \, \text{tr} \left(H M^* M M^* H^* M\right) \\
&= m_1 m_2 \left(m_2^2 - m_1^2\right) \text{Im} \, \bar{H}_{12}^2 \\
&= m_1 m_2 \left(m_2^2 - m_1^2\right) \text{Im} \left[\left(m_e^2 K_{11} K_{21}^* + m_\mu^2 K_{12} K_{22}^*\right)^2\right] \\
&= m_1 m_2 \left(m_2^2 - m_1^2\right) \left(m_e^2 - m_\mu^2\right)^2 \text{Im} \left(K_{11}^2 K_{21}^{*2}\right),
\end{aligned}
\qquad (25.72)
$$

in which use was made of $K_{12} K_{22}^* = -K_{11} K_{21}^*$.

Theorem 25.1 *Equation (25.72) is a necessary and sufficient condition for CP conservation.*

Proof We already know that eqn (25.72) is necessary for CP conservation, therefore we only have to show that it is sufficient too. We have to demonstrate that, if eqn (25.72) holds, then a unitary matrix U exists such that eqns (25.56) and (25.57) are satisfied. If eqn (25.72) holds then either $m_1 = 0$, or $m_2 = 0$, or $m_1 = m_2$, or $\bar{H}_{12}$ is real, or $\bar{H}_{12}$ is purely imaginary. If $\bar{H}_{12}$ is real we choose U to be the unit matrix; if $\bar{H}_{12}$ is purely imaginary we choose $U = \mathrm{diag}\,(1, -1)$; if $m_1 = 0$ we choose $U = \mathrm{diag}\left(e^{2i\,\mathrm{arg}\,\bar{H}_{12}}, 1\right)$; if $m_2 = 0$ we choose $U = \mathrm{diag}\left(1, e^{-2i\,\mathrm{arg}\,\bar{H}_{12}}\right)$. Finally, if $m_1 = m_2$ we choose

$$U = \begin{pmatrix} \cos\zeta & \sin\zeta \\ \sin\zeta & -\cos\zeta \end{pmatrix}, \tag{25.73}$$

with the angle ζ given by

$$\tan\zeta = \frac{2\,\mathrm{Re}\,\bar{H}_{12}}{\bar{H}_{11} - \bar{H}_{22}}. \tag{25.74}$$

$\square$

Thus, the invariant $\mathrm{Im}\,\mathrm{tr}\,(HM^*MM^*H^*M)$ occupies, in the case of two generations with Majorana masses for the left-handed neutrinos, the same central role in CP violation that $\mathrm{Im}\,\mathrm{tr}\,[H_p, H_n]^3$ has in the three-generation SM. If we parametrize the lepton mixing matrix K as in eqn (25.66), one obtains, from eqn (25.72),

$$\mathrm{Im}\,\mathrm{tr}\,(HM^*MM^*H^*M) = -\tfrac{1}{4}m_1 m_2 \left(m_2^2 - m_1^2\right)\left(m_e^2 - m_\mu^2\right)^2 \sin^2 2\theta \sin 2\phi. \tag{25.75}$$

One sees once again that $\exp(i\phi) = \pm i$ does not lead to CP violation.

25.6.3 *Neutrinoless double beta decay*

Let us proceed with the explicit computation of an effect in which the Majorana character of the neutrinos is essential: neutrinoless double beta decay. This is a process in which two neutrons in a nucleus decay into two protons by emitting two electrons, with no neutrino emission. This can only occur if L is violated, which typically happens because neutrinos are Majorana particles. The Feynman diagram for neutrinoless double beta decay is presented in Fig. 25.1 (a). It is convenient to redraw it as in Fig. 25.1 (b), with all leptonic arrows pointing in the same direction. The computation can then be easily performed. First we note that

$$\bar{e}\gamma_\mu \gamma_L \nu = -\nu^T C^{-1} C \gamma_L^T \gamma_\mu^T C^{-1} C \bar{e}^T = -\overline{\nu^c}\gamma_L \gamma_\mu e^c = -\overline{\nu^c}\gamma_\mu \gamma_R e^c. \tag{25.76}$$

Thus, as $\nu = \nu^c$, one may write the leptonic part of the diagram in Fig. 25.1 (b) as

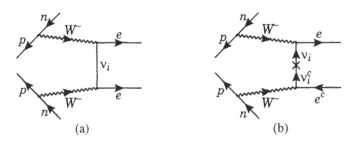

FIG. 25.1. Feynman diagram for neutrinoless double beta decay.

$$-\left(iK_{ei}^{\dagger}\frac{g}{\sqrt{2}}\right)^2\gamma_\mu\gamma_L\left(i\frac{\not{k}+m_i}{k^2-m_i^2}\right)\gamma_\nu\gamma_R = i\frac{g^2}{2}K_{ie}^{2*}\gamma_\mu\gamma_\nu\gamma_R\frac{m_i}{k^2-m_i^2}, \qquad (25.77)$$

where k^α is the momentum flowing along the neutrino line. Taking the simple case $n_g = 2$ for the sake of illustration, one has

$$K_{ie}^{2*}\frac{m_i}{k^2-m_i^2} = \frac{m_1\cos^2\theta}{k^2-m_1^2} + \frac{m_2\sin^2\theta}{k^2-m_2^2}e^{-2i\phi}. \qquad (25.78)$$

This simple computation illustrates two points. Firstly, the effects of the Majorana nature of the neutrinos are proportional to the neutrino masses in a theory with left-handed currents only. Those effects vanish in the limit of vanishing neutrino masses.[67] Secondly, the masses arise multiplied by phase factors $\exp(2i\phi)$, just as in eqns (25.71). CP violation arises when $\sin 2\phi \neq 0$, but not necessarily when $\sin\phi \neq 0$.

It is clear from eqn (25.78) that one may have destructive interference between the contributions of two given neutrinos even if there is CP invariance, provided the two neutrinos have opposite CP-parities (Wolfenstein 1981).

25.6.4 Mass-degenerate neutrinos

We now return to the general case $n_g > 2$, and prove the following theorem:

Theorem 25.2 *When all n_g neutrinos have the same mass m, CP conservation is not automatic if $n_g > 2$.*

Proof We work in the WB of eqns (25.64), where now $M = m1_{n_g}$. We may make an extra change of WB, $H \to Z^{\dagger}HZ$ and $M \to Z^T M Z$, cf. eqns (25.52), which preserves the form of M if Z is orthogonal. We choose Z such as to diagonalize the symmetric matrix $\mathrm{Re}\,H$. Thus, there is a WB in which M is proportional to the unit matrix and $\mathrm{Re}\,H$ is diagonal. In that WB, the matrix U should, because of eqn (25.57), be orthogonal. Equation (25.56) then reads

$$U^T(\mathrm{Re}\,H)U = \mathrm{Re}\,H, \qquad (25.79)$$

---

[67] Kayser (1982) has called this 'the practical Dirac–Majorana confusion theorem': in the limit of vanishing neutrino masses, and in a theory with no right-handed currents, it is irrelevant whether one considers the neutrinos to be Majorana particles or not. See also Kayser and Shrock (1982) and Nieves (1982).

$$U^T (\operatorname{Im} H) U = -\operatorname{Im} H. \tag{25.80}$$

In general, $\operatorname{Re} H$ is non-degenerate. Equation (25.79) then implies

$$U = \operatorname{diag} (\epsilon_1, \epsilon_2, \dots, \epsilon_{n_g}), \tag{25.81}$$

where each ϵ_i may be either $+1$ or -1. In general, no off-diagonal matrix element of $\operatorname{Im} H$ vanishes. Equation (25.80) then implies that $\epsilon_i \epsilon_{i'} = -1$ for all pairs $i \neq i'$. Clearly, this is impossible for $n_g > 2$, although for $n_g = 2$ it is possible (with $\epsilon_1 = +1$ and $\epsilon_2 = -1$). Thus, there is in general no matrix U satisfying both eqns (25.79) and (25.80), and therefore CP is violated. □

The interest in the limit of exact mass degeneracy is more than just academic. The data on the atmospheric-neutrino anomaly and on the solar-neutrino deficit, together with the assumption that relic neutrinos constitute the hot dark matter in the Universe, suggest, in a framework with three left-handed neutrinos and no other neutral fermions, that neutrinos are almost mass-degenerate. Therefore, it is worth analysing in more detail the features of mixing and CP violation for three Majorana neutrinos in the case of exact mass degeneracy. Let us now work in the WB of eqn (25.63), where

$$M = m K^T K. \tag{25.82}$$

The Majorana mass terms only allow us to redefine the mass eigenstates through an *orthogonal* matrix, and not through an arbitrary unitary matrix. Since K is in general not orthogonal, the leptonic mixing cannot be rotated away. From eqn (25.82) it follows that M is proportional to a unitary symmetric matrix. Such a matrix may be parametrized by two angles and one phase. Starting from eqn (25.65), one makes $c_3 = c_2$, $s_3 = s_2$, and $\exp(i\phi_1) = \exp(i\phi_2) = -1$, and one obtains

$$\frac{M}{m} = \begin{pmatrix} c_1 & -s_1 c_2 & -s_1 s_2 \\ -s_1 c_2 & -c_1 c_2^2 + s_2^2 e^{i\delta} & -c_1 c_2 s_2 - c_2 s_2 e^{i\delta} \\ -s_1 s_2 & -c_1 c_2 s_2 - c_2 s_2 e^{i\delta} & -c_1 s_2^2 + c_2^2 e^{i\delta} \end{pmatrix}. \tag{25.83}$$

The phase δ is CP-violating.

It may be shown (Branco *et al.* 1999) that a necessary and sufficient condition for CP invariance in the case of three generations with degenerate Majorana masses is

$$0 = \operatorname{tr} [MHM^*, H^*]^3 \tag{25.84}$$

$$= m^6 \operatorname{tr} [K^T K H K^\dagger K^*, H^*]^3$$

$$= 6 i m^6 (m_\tau^2 - m_\mu^2)^2 (m_\tau^2 - m_e^2)^2 (m_\mu^2 - m_e^2)^2$$

$$\times \operatorname{Im} [(K^T K)_{11} (K^T K)_{22} (K^T K)_{12}^* (K^T K)_{21}^*]. \tag{25.85}$$

This is a weak-basis-invariant condition.

25.7 Main conclusions

- As neutrinos do not carry any conserved quantum number, they may have Majorana mass terms. Those mass terms cause the physical neutrinos to be Majorana particles, i.e., to be identical with their antiparticles.

- The presence of Majorana masses does not allow the physical-neutrino fields to be rephased at will. This causes the appearance of $n - 1$ extra physical phases ϕ in the theory, if n is the total number of Majorana neutrinos.

- The phases ϕ violate CP when $\sin 2\phi \neq 1$. The case $\exp(i\phi) = \pm i$ does not correspond to CP violation.

- In the SM with Majorana masses for the neutrinos, CP violation in the leptonic sector may arise already for $n_g = 2$, and there may be CP violation even when the neutrinos are mass-degenerate for $n_g \geq 3$.

THE LEFT–RIGHT-SYMMETRIC MODEL

26.1 Overview of the model

26.1.1 *Introduction*

The main motivation for considering a left–right-symmetric model (LRSM) (Pati and Salam 1974; Mohapatra and Pati 1975; Senjanović and Mohapatra 1975) is having an extension of the standard model in which parity is a spontaneously broken symmetry. This means that the Lagrangian is symmetric under a parity transformation and this symmetry is only broken by the vacuum. In the SM the left-handed fermions are in doublets and the right-handed fermions are singlets of an SU(2) gauge group. This arrangement is not parity-symmetric, because parity must interchange left-handed and right-handed fermions. The simplest way of having a left–right-symmetric extension of the SM is through the introduction of a second SU(2) gauge group which transforms the right-handed fermions as doublets and the left-handed fermions as singlets. The gauge group of the LRSM is thus $SU(2)_L \otimes SU(2)_R \otimes U(1)$. When one considers grand unified theories, the gauge group of the LRSM can be elegantly interpreted as a subgroup of $SO(10)$. The lepton fields are in doublets

$$L_L = \begin{pmatrix} \upsilon_L \\ l_L \end{pmatrix}; \; L_R = \begin{pmatrix} \upsilon_R \\ l_R \end{pmatrix} \tag{26.1}$$

of $SU(2)_L$ and $SU(2)_R$, respectively. The quark fields are in doublets

$$Q_L = \begin{pmatrix} p_L \\ n_L \end{pmatrix}; \; Q_R = \begin{pmatrix} p_R \\ n_R \end{pmatrix} \tag{26.2}$$

of $SU(2)_L$ and $SU(2)_R$, respectively. Parity interchanges fermionic and bosonic multiplets of $SU(2)_L$ with analogous multiplets of $SU(2)_R$; for each multiplet of $SU(2)_L$ there is a similar multiplet of $SU(2)_R$. Therefore, under parity the fermion fields transform as

$$L_L \leftrightarrow L_R,$$
$$Q_L \leftrightarrow Q_R, \tag{26.3}$$

while the gauge bosons W_{Lk} and W_{Rk} (k from 1 to 3) associated with the gauge groups $SU(2)_L$ and $SU(2)_R$, respectively, transform as

$$W_{Lk} \leftrightarrow W_{Rk}. \tag{26.4}$$

Parity invariance of the Lagrangian constrains the gauge coupling constants g_L of $SU(2)_L$ and g_R of $SU(2)_R$ to be equal. So one has $g_L = g_R = g$. The covariant derivative is

$$\partial - ig \sum_{k=1}^{3} (W_{Lk} T_{Lk} + W_{Rk} T_{Rk}) - ig' BY, \tag{26.5}$$

where g' is the U(1) coupling constant and Y denotes the (weak) hypercharge, which in the LRSM takes values different from the SM ones. The T_{Lk} and T_{Rk} are the generators of SU(2)$_L$ and of SU(2)$_R$, respectively.

The formula for the electric charge should be a left–right-symmetric extension of the SM expression $Q = T_3 + Y$. An attractive feature of the LRSM is that, if one makes the obvious extension $Q = T_{L3} + T_{R3} + Y$, then the hypercharge Y acquires a simple physical meaning. As the lepton doublets in eqn (26.1) have hypercharge $-1/2$ while the quark doublets in eqn (26.2) have hypercharge $1/6$, one concludes that a general formula for the hypercharge is $Y = (B - L)/2$, where B and L denote the baryon number and the lepton number, respectively. Therefore in the LRSM one has

$$Q = T_{L3} + T_{R3} + \frac{B - L}{2}. \tag{26.6}$$

In view of eqn (26.6), the U(1) factor of the LRSM gauge group is often denoted U(1)$_{B-L}$.

26.1.2 Gauge couplings

We proceed in a fashion similar to what was done in the treatment of the SM in Chapter 11. We define

$$T_{L\pm} \equiv \frac{T_{L1} \pm iT_{L2}}{\sqrt{2}}, \quad T_{R\pm} \equiv \frac{T_{R1} \pm iT_{R2}}{\sqrt{2}},$$
$$W_L^{\pm} \equiv \frac{W_{L1} \mp iW_{L2}}{\sqrt{2}}, \quad W_R^{\pm} \equiv \frac{W_{R1} \mp iW_{R2}}{\sqrt{2}}. \tag{26.7}$$

Then,

$$\sum_{k=1}^{2} (W_{Lk} T_{Lk} + W_{Rk} T_{Rk}) = W_L^+ T_{L+} + W_L^- T_{L-} + W_R^+ T_{R+} + W_R^- T_{R-}. \tag{26.8}$$

The gauge bosons $W_L^{\pm}$ and $W_R^{\pm}$ in general mix, i.e., they are not eigenstates of mass.

In the neutral sector, instead of g and g' it is convenient to use the angle θ_w and the electric-charge unit e, defined by

$$s_w \equiv \sin \theta_w = -\frac{g'}{\sqrt{g^2 + 2g'^2}},$$
$$c_w \equiv \cos \theta_w = \sqrt{\frac{g^2 + g'^2}{g^2 + 2g'^2}}, \tag{26.9}$$
$$e = gs_w.$$

It is useful to introduce the neutral gauge bosons A, X_1, and X_2, defined by the orthogonal transformation

$$\begin{pmatrix} W_{L3} \\ W_{R3} \\ B \end{pmatrix} = \begin{pmatrix} -s_w & c_w & 0 \\ -s_w & -s_w^2/c_w & \sqrt{c_w^2 - s_w^2}/c_w \\ \sqrt{c_w^2 - s_w^2} & s_w\sqrt{c_w^2 - s_w^2}/c_w & s_w/c_w \end{pmatrix} \begin{pmatrix} A \\ X_1 \\ X_2 \end{pmatrix}, \quad (26.10)$$

where A is to be identified with the photon, and we choose the sign of $\sqrt{c_w^2 - s_w^2}$ such that it equals $g/\sqrt{g^2 + 2g'^2}$. The neutral gauge couplings are given by

$$g\left(W_{L3}T_{L3} + W_{R3}T_{R3}\right) + g'BY = -eAQ + \frac{g}{c_w}X_1\left(T_{L3} - Qs_w^2\right)$$

$$+ \frac{g\sqrt{c_w^2 - s_w^2}}{c_w}X_2\left(T_{R3} - Y\frac{s_w^2}{c_w^2 - s_w^2}\right).(26.11)$$

With the above definitions, X_1 interacts like the Z of the SM. However, X_1 is not in general an eigenstate of mass, it mixes with X_2.

26.1.3   Scalar multiplets

The Higgs sector has several important functions to perform:

1. It should lead to an appropriate spontaneous breaking of the SU(2)$_L\otimes$ SU(2)$_R\otimes$U(1) gauge group. In view of the left-handed character of the observed charged-current interaction, the breaking of the LRSM gauge group should occur in two steps (Senjanović 1979): at a first stage—at high energy—the breaking should be to SU(2)$_L\otimes$U(1)$_Y$, where Y denotes the hypercharge of the SM. Parity invariance is broken at this stage. The non-observation of right-handed charged currents at low energies requires that the mass m_{W_R} of $W_R^\pm$ be substantially larger than m_{W_L}.[68] At a second stage—at a lower energy—the SM gauge group should be broken to the U(1) of electromagnetism.

2. It should give quarks and charged leptons a mass, while at the same time giving either zero or naturally small masses to the neutrinos.

Let us first consider the requirement of fermion masses. A general SU(2)$_L$ transformation is represented in the doublet representation by the 2×2 unitary matrix U_L. Similarly, an SU(2)$_R$ transformation is represented in the doublet representation by the 2×2 unitary matrix U_R. The quark doublets transform as $Q_L \to U_L Q_L$ and $Q_R \to U_R Q_R$. Therefore, a scalar multiplet ϕ which gives mass to the quarks via a Yukawa coupling $\overline{Q_L}\phi Q_R$ must be a 2×2 matrix of fields transforming as $\phi \to U_L\phi U_R^\dagger$. Moreover, as Q_L and Q_R have the same hypercharge, the hypercharge of ϕ must be zero. Then,

---

[68]Explicit lower bounds on m_{W_R} vary between 240 GeV and 1600 GeV, depending on the experimental technique and on the specific variant of the LRSM that one is considering. The Particle Data Group (1996, p. 231) recommends $m_{W_R} > 406$ GeV.

$$\phi = \begin{pmatrix} \varphi_1^{0\dagger} & \varphi_2^+ \\ -\varphi_1^- & \varphi_2^0 \end{pmatrix}, \tag{26.12}$$

where we have already displayed the electric charge of each component field.

The multiplet

$$\tilde{\phi} \equiv \tau_2 \phi^{\dagger^T} \tau_2 = \begin{pmatrix} \varphi_2^{0\dagger} & \varphi_1^+ \\ -\varphi_2^- & \varphi_1^0 \end{pmatrix} \tag{26.13}$$

transforms under a gauge transformation in the same way as ϕ. Notice that

$$\begin{pmatrix} \varphi_1^+ \\ \varphi_1^0 \end{pmatrix} \quad \text{and} \quad \begin{pmatrix} \varphi_2^+ \\ \varphi_2^0 \end{pmatrix} \tag{26.14}$$

are doublets of $SU(2)_L$; from this point of view, the LRSM is like a two-Higgs-doublet model with an extra $SU(2)$ symmetry. Indeed,

$$\begin{pmatrix} \varphi_1^+ \\ -\varphi_2^{0\dagger} \end{pmatrix} \quad \text{and} \quad \begin{pmatrix} \varphi_2^+ \\ -\varphi_1^{0\dagger} \end{pmatrix} \tag{26.15}$$

are doublets of $SU(2)_R$.

The VEV of φ_1^0 is k_1 and the VEV of φ_2^0 is k_2. Both k_1 and k_2 are in general complex. We assume that the scalar potential is such that the other components of ϕ do not acquire a VEV. The Yukawa couplings of ϕ and $\tilde{\phi}$ will generate Dirac masses for all fermions, including neutrinos. However, ϕ is not sufficient, other Higgs multiplets have to be introduced. Both φ_1^0 and φ_2^0 have $T_{L3} = -1/2$, $T_{R3} = 1/2$, and $Y = 0$. Therefore, when they acquire a VEV they keep unbroken the two $U(1)$ groups generated by $T_{L3} + T_{R3}$ and by Y. Thus, k_1 and k_2 keep two neutral gauge bosons massless, instead of giving mass to every gauge boson but the photon. Furthermore, k_1 and k_2 cannot perform the spontaneous breaking of parity symmetry. This is because both φ_1^0 and φ_2^0 are components of doublets of $SU(2)_L$ and $SU(2)_R$, and they cannot distinguish between the two gauge groups. We must introduce some extra Higgs multiplets which distinguish between the two $SU(2)$ groups. In the first versions of the LRSM, this was achieved by introducing, apart from ϕ, multiplets χ_L and χ_R transforming as doublets of $SU(2)_L$ and of $SU(2)_R$, respectively, while being singlets of the other $SU(2)$ gauge group, and having $B - L = 1$. At present, a more attractive choice is introducing a triplet of $SU(2)_L$ which is a singlet of $SU(2)_R$, together with a triplet of $SU(2)_R$ which is a singlet of $SU(2)_L$. Both triplets are chosen to have $B - L = 2$. The advantage of using these triplets is that their Yukawa couplings can generate $|\Delta L| = 2$ Majorana masses, thus leading to naturally small neutrino masses through the seesaw mechanism (see § 25.3). We may write a triplet Δ' of $SU(2)$ as

$$\Delta' = \begin{pmatrix} \Delta_1 \\ \Delta_2 \\ \Delta_3 \end{pmatrix}. \tag{26.16}$$

An infinitesimal SU(2) transformation of Δ' reads

$$\Delta' \rightarrow \left[1 + i\left(\theta_+ T_+^{(3)} + \theta_+^* T_-^{(3)} + \theta_3 T_3^{(3)}\right)\right]\Delta'. \tag{26.17}$$

The infinitesimal parameter θ_3 is real but the infinitesimal parameter θ_+ is complex. It is convenient to write the triplet in the form of a 2×2 traceless matrix

$$\Delta \equiv \sqrt{2}\left(\Delta_3 T_-^{(2)} + \Delta_2 T_3^{(2)} - \Delta_1 T_+^{(2)}\right) = \begin{pmatrix} \Delta_2/\sqrt{2} & -\Delta_1 \\ \Delta_3 & -\Delta_2/\sqrt{2} \end{pmatrix}. \tag{26.18}$$

By using the commutation algebra of the SU(2) generators, we find that the SU(2) transformation in eqn (26.17) may be written

$$\Delta \rightarrow \left[1 + i\left(\theta_+ T_+^{(2)} + \theta_+^* T_-^{(2)} + \theta_3 T_3^{(2)}\right)\right]\Delta\left[1 - i\left(\theta_+ T_+^{(2)} + \theta_+^* T_-^{(2)} + \theta_3 T_3^{(2)}\right)\right]. \tag{26.19}$$

Generalizing to a non-infinitesimal SU(2) transformation, represented in the doublet representation by the unitary matrix U, this means that $\Delta \rightarrow U\Delta U^\dagger$.

Thus, the triplets are 2×2 traceless matrices

$$\Delta_L = \begin{pmatrix} \Delta_L^+/\sqrt{2} & -\Delta_L^{++} \\ \Delta_L^0 & -\Delta_L^+/\sqrt{2} \end{pmatrix} \text{ and } \Delta_R = \begin{pmatrix} \Delta_R^+/\sqrt{2} & -\Delta_R^{++} \\ \Delta_R^0 & -\Delta_R^+/\sqrt{2} \end{pmatrix}, \tag{26.20}$$

which transform as $\Delta_L \rightarrow U_L \Delta_L U_L^\dagger$ and $\Delta_R \rightarrow U_R \Delta_L U_R^\dagger$, and have $B - L = 2$. This form of writing triplets is particularly convenient since eventually we want to build gauge singlets out of tensor products of triplets and doublets of SU(2)$_L$ and SU(2)$_R$. The VEV of Δ_L^0 is F_L and the VEV of Δ_R^0 is F_R. We assume that all other Higgs fields have vanishing VEV, so that the U(1) of electromagnetism remains unbroken.

26.1.4   Gauge-boson masses

The covariant derivative of φ_1^+ is

$$\partial\varphi_1^+ - i\left[g\left(\frac{W_L^+}{\sqrt{2}}\varphi_1^0 - \frac{W_R^+}{\sqrt{2}}\varphi_2^{0\dagger}\right) - eA\varphi_1^+\right.$$
$$\left. + \frac{g\left(c_w^2 - s_w^2\right)}{2c_w}X_1\varphi_1^+ + \frac{g\sqrt{c_w^2 - s_w^2}}{2c_w}X_2\varphi_1^+\right]. \tag{26.21}$$

Substituting the fields by their VEVs and taking the squared modulus, we obtain

$$\left|\frac{g}{\sqrt{2}}\left(W_L^+ k_1 - W_R^+ k_2^*\right)\right|^2. \tag{26.22}$$

Analogously, from the covariant derivatives of φ_2^+, φ_1^0, and φ_2^0, we obtain

$$\left|\frac{g}{\sqrt{2}}\left(W_L^+ k_2 - W_R^+ k_1^*\right)\right|^2,$$

$$\left|\frac{gk_1}{2c_w}\left(-X_1 + \sqrt{c_w^2 - s_w^2}\,X_2\right)\right|^2, \qquad (26.23)$$

$$\left|\frac{gk_2}{2c_w}\left(-X_1 + \sqrt{c_w^2 - s_w^2}\,X_2\right)\right|^2,$$

respectively. The covariant derivatives of Δ_L^{++} and of Δ_R^{++} do not yield any mass term for the gauge bosons, but the covariant derivatives of Δ_L^+, Δ_R^+, Δ_L^0, and Δ_R^0, yield

$$\left|gW_L^+ F_L\right|^2,$$

$$\left|gW_R^+ F_R\right|^2,$$

$$\left|\frac{g}{c_w}\left(X_1 + \frac{s_w^2}{\sqrt{c_w^2 - s_w^2}}X_2\right)F_L\right|^2, \qquad (26.24)$$

$$\left|\frac{gc_w}{\sqrt{c_w^2 - s_w^2}}X_2 F_R\right|^2,$$

respectively. Putting everything together, we obtain the mass terms for the charged gauge bosons,

$$g^2\left(W_L^-\ W_R^-\right)\begin{pmatrix}\dfrac{|k_1|^2 + |k_2|^2}{2} + |F_L|^2 & -k_1^* k_2^* \\[2mm] -k_1 k_2 & \dfrac{|k_1|^2 + |k_2|^2}{2} + |F_R|^2\end{pmatrix}\begin{pmatrix}W_L^+ \\ W_R^+\end{pmatrix}, \quad (26.25)$$

and for the neutral gauge bosons,

$$\frac{g^2}{c_w^2}\left(X_1\ \ \sqrt{c_w^2 - s_w^2}\,X_2\right)\mathcal{M}\begin{pmatrix}X_1 \\ \sqrt{c_w^2 - s_w^2}\,X_2\end{pmatrix}, \qquad (26.26)$$

where

$$\mathcal{M} = \begin{pmatrix}\dfrac{|k_1|^2 + |k_2|^2}{4} + |F_L|^2 & -\dfrac{|k_1|^2 + |k_2|^2}{4} + \dfrac{s_w^2}{c_w^2 - s_w^2}|F_L|^2 \\[3mm] -\dfrac{|k_1|^2 + |k_2|^2}{4} + \dfrac{s_w^2}{c_w^2 - s_w^2}|F_L|^2 & \dfrac{|k_1|^2 + |k_2|^2}{4} + \dfrac{s_w^4|F_L|^2 + c_w^4|F_R|^2}{(c_w^2 - s_w^2)^2}\end{pmatrix}.$$

$$(26.27)$$

In eqn (26.25) we see that $W_L^{\pm}$ and $W_R^{\pm}$ mix if k_1 and k_2 are simultaneously non-zero.[69] The product $k_1 k_2$ has $T_{L3} = -1$ and $T_{R3} = +1$. We may therefore without loss of generality choose a gauge in which $k_1 k_2$ is real, rendering the

[69]It is difficult to have $k_2 = 0$ when $k_1 \neq 0$ (or vice versa), because the scalar potential contains terms linear in k_2 when k_1, F_L, and F_R do not vanish. Those terms draw k_2 away from $k_2 = 0$. See eqn (26.36) below.

$W_L^\pm$–$W_R^\pm$ mixing real. The physical charged gauge bosons are the eigenstates of the mass matrix in eqn (26.25) and can be written

$$\begin{pmatrix} W_1 \\ W_2 \end{pmatrix} = \begin{pmatrix} \cos\zeta & -\sin\zeta \\ \sin\zeta & \cos\zeta \end{pmatrix} \begin{pmatrix} W_L \\ W_R \end{pmatrix}. \tag{26.28}$$

It will be shown in § 26.2 that there is a region of parameters of the Higgs potential which leads to a minimum with $|F_R| \neq |F_L|$. In this case, parity is spontaneously broken. The crucial assumption of the LRSM is that $|F_R|$ is much larger than $|k_1|$, $|k_2|$, and $|F_L|$. In this case, the mixing angle ζ and the masses of W_1 and W_2 are approximately given by

$$\zeta \approx \frac{k_1 k_2}{|F_R|^2},$$

$$m_{W_1}^2 \approx g^2 \left(|F_L|^2 + \frac{|k_1|^2 + |k_2|^2}{2} \right), \tag{26.29}$$

$$m_{W_2}^2 \approx g^2 \left(|F_R|^2 + \frac{|k_1|^2 + |k_2|^2}{2} \right).$$

Since $\zeta \ll 1$, W_1 and W_2 coincide, to a good approximation, with W_L and W_R, respectively. Similarly, X_1 and X_2 are approximate eigenstates of mass, with squared masses

$$m_{X_1}^2 \approx \frac{g^2}{c_w^2} \left(\frac{|k_1|^2 + |k_2|^2}{2} + 2|F_L|^2 \right),$$

$$m_{X_2}^2 \approx \frac{g^2}{c_w^2} \left[\frac{(|k_1|^2 + |k_2|^2)(c_w^2 - s_w^2)}{2} + 2\frac{s_w^4 |F_L|^2 + c_w^4 |F_R|^2}{c_w^2 - s_w^2} \right]. \tag{26.30}$$

Both $W_R^\pm$ and X_2 are much heavier than $W_L^\pm$ and X_1, and therefore the interactions mediated by the former gauge bosons are suppressed when compared to the ones mediated by the latter. In particular, the charged gauge interactions of the right-handed fermions are much weaker than those among the left-handed fermions. The gauge boson $W_L^\pm$ is identified with the $W^\pm$ of the SM, and X_1 is identified with the Z of the SM.

We would like the mass of X_1 to be approximately equal to the mass of $W_L^\pm$ divided by c_w, because this is an experimental fact. From eqns (26.29) and (26.30) we see that, in order to obtain this, we must assume $|F_L|^2$ to be much smaller than $|k_1|^2 + |k_2|^2$. Indeed, the SM relationship $m_W = c_w m_Z$ is a consequence of the fact that the breaking of SU(2)$_L$ is effected by doublets. The VEV of a triplet of SU(2)$_L$ must be very small compared to the VEV of at least one of the doublets.

26.2   Spontaneous symmetry breaking

This section contains an analysis of the scalar potential of the LRSM, and of the conditions under which spontaneous P and CP breaking may be obtained. Some readers may prefer to skip all but the first subsection.

26.2.1 *The scalar potential*

Under a gauge transformation,

$$
\phi \to U_L \phi U_R^\dagger, \qquad \tilde{\phi} \to U_L \tilde{\phi} U_R^\dagger,
$$
$$
\phi^\dagger \to U_R \phi^\dagger U_L^\dagger, \qquad \tilde{\phi}^\dagger \to U_R \tilde{\phi}^\dagger U_L^\dagger. \tag{26.31}
$$

Parity interchanges U_L and U_R. Therefore, ϕ should transform under parity into some unitary combination of $\phi^\dagger$ and $\tilde{\phi}^\dagger$. We shall assume that ϕ is transformed into $\phi^\dagger$. It can be shown (Ecker *et al.* 1981*a,b*) that this is the only choice leading to realistic quark masses and mixings. Thus, we assume that the Higgs multiplets transform under parity in the following way:

$$
\phi \overset{\text{P}}{\to} \phi^\dagger, \quad \Delta_L \overset{\text{P}}{\leftrightarrow} \Delta_R. \tag{26.32}
$$

After elimination of all redundant terms, the scalar potential may then be written

$$
\begin{aligned}
V = {}& \mu_1 \text{tr}\left(\phi^\dagger \phi\right) + \mu_2 \text{tr}\left(\tilde{\phi}^\dagger \phi + \phi^\dagger \tilde{\phi}\right) \\
& + \lambda_1 \left[\text{tr}\left(\phi^\dagger \phi\right)\right]^2 + \lambda_2 \text{tr}\left(\tilde{\phi}^\dagger \phi \tilde{\phi}^\dagger \phi + \phi^\dagger \tilde{\phi} \phi^\dagger \tilde{\phi}\right) \\
& + \lambda_3 \text{tr}\left(\phi^\dagger \phi\right) \text{tr}\left(\tilde{\phi}^\dagger \phi + \phi^\dagger \tilde{\phi}\right) + \lambda_4 \text{tr}\left(\tilde{\phi}^\dagger \phi\right) \text{tr}\left(\phi^\dagger \tilde{\phi}\right) \\
& + \mu_3 \text{tr}\left(\Delta_L^\dagger \Delta_L + \Delta_R^\dagger \Delta_R\right) + \lambda_5 \text{tr}\left(\Delta_L^\dagger \Delta_L \Delta_L^\dagger \Delta_L + \Delta_R^\dagger \Delta_R \Delta_R^\dagger \Delta_R\right) \\
& + \lambda_6 \text{tr}\left(\Delta_L^\dagger \Delta_L^\dagger \Delta_L \Delta_L + \Delta_R^\dagger \Delta_R^\dagger \Delta_R \Delta_R\right) + \lambda_7 \text{tr}\left(\Delta_L^\dagger \Delta_L\right) \text{tr}\left(\Delta_R^\dagger \Delta_R\right) \\
& + \lambda_8 \left[\text{tr}\left(\Delta_L \Delta_L\right) \text{tr}\left(\Delta_R^\dagger \Delta_R^\dagger\right) + \text{tr}\left(\Delta_R \Delta_R\right) \text{tr}\left(\Delta_L^\dagger \Delta_L^\dagger\right)\right] \\
& + \lambda_9 \text{tr}\left(\Delta_L^\dagger \Delta_L \phi \phi^\dagger + \Delta_R^\dagger \Delta_R \phi^\dagger \phi\right) + \lambda_{10} \text{tr}\left(\Delta_L^\dagger \Delta_L \tilde{\phi} \tilde{\phi}^\dagger + \Delta_R^\dagger \Delta_R \tilde{\phi}^\dagger \tilde{\phi}\right) \\
& + \lambda_{11} \text{tr}\left(\Delta_L^\dagger \Delta_L \phi \tilde{\phi}^\dagger + \Delta_R^\dagger \Delta_R \phi^\dagger \tilde{\phi}\right) + \lambda_{11}^* \text{tr}\left(\Delta_L^\dagger \Delta_L \tilde{\phi} \phi^\dagger + \Delta_R^\dagger \Delta_R \tilde{\phi}^\dagger \phi\right) \\
& + \lambda_{12} \text{tr}\left(\Delta_L^\dagger \phi \Delta_R \phi^\dagger + \Delta_R^\dagger \phi^\dagger \Delta_L \phi\right) + \lambda_{13} \text{tr}\left(\Delta_L^\dagger \tilde{\phi} \Delta_R \phi^\dagger + \Delta_R^\dagger \tilde{\phi}^\dagger \Delta_L \phi\right) \\
& + \lambda_{14} \text{tr}\left(\Delta_L^\dagger \phi \Delta_R \tilde{\phi}^\dagger + \Delta_R^\dagger \phi^\dagger \Delta_L \tilde{\phi}\right).
\end{aligned} \tag{26.33}
$$

We have used the letter μ for couplings with dimension of mass squared, and the letter λ for dimensionless couplings. All couplings except maybe λ_{11} are real because of Hermiticity together with parity.

Let us consider the vacuum expectation value of the potential. The terms in the potential with coefficients λ_6 and λ_8 have zero VEV. We introduce the notation

$$
\begin{aligned}
|k_1| &= K \cos s, \\
|k_2| &= K \sin s, \\
|F_R| &= F \cos r, \\
|F_L| &= F \sin r,
\end{aligned} \tag{26.34}
$$

where K and F are two positive quantities with mass dimension, while s and r are two angles of the first quadrant. Also, we denote the two gauge-invariant vacuum phases

$$\delta \equiv \arg(k_1 k_2 F_L^* F_R),$$
$$\alpha \equiv \arg(k_1 k_2^*). \tag{26.35}$$

One then obtains

$$\langle 0|V|0\rangle = \mu_1 K^2 + \lambda_1 K^4 + \mu_3 F^2 + \lambda_5 F^4 + \frac{\lambda_9 + \lambda_{10}}{2} F^2 K^2$$
$$+ 2\mu_2 K^2 \sin 2s \cos\alpha + \lambda_2 K^4 \sin^2 2s \cos 2\alpha$$
$$+ 2\lambda_3 K^4 \sin 2s \cos\alpha + \lambda_4 K^4 \sin^2 2s$$
$$+ \frac{\lambda_7 - 2\lambda_5}{4} F^4 \sin^2 2r + \frac{\lambda_9 - \lambda_{10}}{2} F^2 K^2 \cos 2s$$
$$+ F^2 K^2 \sin 2s \, (\mathrm{Re}\,\lambda_{11} \cos\alpha - \mathrm{Im}\,\lambda_{11} \cos 2r \sin\alpha)$$
$$+ F^2 K^2 \sin 2r \, [\lambda_{12} \sin s \cos s \cos\delta$$
$$+ \lambda_{13} \cos^2 s \cos(\delta + \alpha) + \lambda_{14} \sin^2 s \cos(\delta - \alpha)]. \tag{26.36}$$

26.2.2 Spontaneous breaking of P

In order to analyse whether parity can be spontaneously broken, one has to examine in detail the r-dependence of the Higgs potential. We shall assume that the parameters of the scalar potential are chosen so that F is much larger than K. In order to obtain this, one clearly has to make some fine-tuning of the couplings in the potential. We shall assume, though, that no other fine-tuning of parameters beyond this one is done, i.e., that no coupling, and no combination of couplings, is assumed to be $\sim K^2/F^2$ in order to obtain spontaneous breaking of either parity or CP.

As $F \gg K$, the r-dependent part of the potential is dominated by the term $(\lambda_7 - 2\lambda_5) F^4 \sin^2 r \cos^2 r$. If $\lambda_7 - 2\lambda_5 < 0$, the minimum of this term occurs for

$$\sin r = \cos r = \tfrac{1}{\sqrt{2}} \Leftrightarrow |F_R| = |F_L|. \tag{26.37}$$

For $\lambda_7 - 2\lambda_5 > 0$, one has two possible minima:

$$\sin r = 0 \Leftrightarrow F_L = 0,$$
$$\sin r = 1 \Leftrightarrow F_R = 0. \tag{26.38}$$

The extremum of eqn (26.37) is parity-conserving, while those of eqns (26.38) lead to spontaneous parity violation. Obviously, we are interested in the minimum corresponding to $F_L = 0$, at this level of approximation.

We next consider the subleading r-dependent terms in the potential. There are two such terms, one proportional to $F^2 K^2 \cos 2r$ and another proportional to $F^2 K^2 \sin 2r$. These subleading terms pull the minimum of the potential to $\sin 2r \sim K^2/F^2$, thus leading to

$$F_L \sim \frac{K^2}{F} \approx \frac{K^2}{F_R}. \tag{26.39}$$

One has thus a minimum characterized by $F_R^2 \gg |k_1|^2 + |k_2|^2 \gg F_L^2$, ensuring that $m_W \approx c_w m_Z$, as desired. The vacuum state is related by parity symmetry to another possible vacuum with the same energy density, characterized by $|F_R/F_L| \sim K^2/F^2$. We assume that it is the first vacuum which is realized in nature, or at least in that part of the Universe in which we live. Then, the lowest-energy charged-current weak interaction is among the left-handed fermions and not among the right-handed ones, as experimentally observed.

26.2.3 *Spontaneous CP breaking*

We next investigate whether CP can be spontaneously broken in the minimal LRSM. In the LRSM it is natural to assume CP to be a spontaneously broken symmetry, since parity is spontaneously broken too. We impose CP at the Lagrangian level, assuming the trivial CP transformation:

$$\phi \overset{\text{CP}}{\to} \phi^{\dagger\,T}, \quad \Delta_L \overset{\text{CP}}{\to} \Delta_L^{\dagger\,T}, \quad \Delta_R \overset{\text{CP}}{\to} \Delta_R^{\dagger\,T}. \tag{26.40}$$

CP invariance then constrains λ_{11} to be real. It is convenient to write the VEV of the potential as

$$
\begin{aligned}
\langle 0|V|0 \rangle = {}& \mu_1 K^2 + \lambda_1 K^4 + \mu_3 F^2 + \lambda_5 F^4 + \frac{\lambda_9 + \lambda_{10}}{2} F^2 K^2 \\
& + \lambda_4 K^4 \sin^2 2s + \frac{\lambda_9 - \lambda_{10}}{2} F^2 K^2 \cos 2s \\
& + (m_3 K^2 + l_6 K^4 + l_7 F^2 K^2) \sin 2s \cos\alpha + l_8 K^4 \sin^2 2s \cos 2\alpha \\
& + V_r,
\end{aligned}
\tag{26.41}
$$

with

$$
\begin{aligned}
V_r &= l_9 F^4 \sin^2 2r + F^2 K^2 V_\delta \sin 2r, \\
V_\delta &= l_{10} \sin 2s \cos\delta + l_{11} \sin^2 s \cos(\delta - \alpha) + l_{12} \cos^2 s \cos(\delta + \alpha).
\end{aligned}
\tag{26.42}
$$

We have changed the notation in the following way: $2\mu_2 \to m_3$, $2\lambda_3 \to l_6$, $\operatorname{Re}\lambda_{11} \to l_7$, $\lambda_2 \to l_8$, $\lambda_7 - 2\lambda_5 \to 4l_9$, $\lambda_{12} \to 2l_{10}$, $\lambda_{14} \to l_{11}$, and $\lambda_{13} \to l_{12}$. All the δ-dependent terms are in V_δ, and all the r-dependent terms are in V_r.

Consider the minimization of V_r as a function of r. We choose $l_9 > 0$ as in the previous subsection. We may always obtain $V_\delta < 0$, if necessary by transforming $\delta \to \pi + \delta$. With $V_\delta < 0$, the minimum occurs at the extremum

$$\sin 2r = -\frac{V_\delta K^2}{2l_9 F^2} \ll 1. \tag{26.43}$$

This gives $V_r = -V_\delta^2 K^4/(4l_9)$.

We proceed with the minimization relative to δ. Defining

$$\Delta \equiv \left[l_{10} \sin 2s + \left(l_{11} \sin^2 s + l_{12} \cos^2 s \right) \cos\alpha \right]^2 + \left(l_{11} \sin^2 s - l_{12} \cos^2 s \right)^2 \sin^2 \alpha, \tag{26.44}$$

the minimum is given by

$$\sin \delta = \frac{\left(l_{11} \sin^2 s - l_{12} \cos^2 s\right) \sin \alpha}{\sqrt{\Delta}},$$

$$\cos \delta = \frac{l_{10} \sin 2s + \left(l_{11} \sin^2 s + l_{12} \cos^2 s\right) \cos \alpha}{\sqrt{\Delta}}, \tag{26.45}$$

which leads to $V_\delta = \sqrt{\Delta}$ at the minimum. As we want $V_\delta < 0$, we choose the negative sign for the square root.

Under these conditions, the extremum condition for α is

$$
\begin{aligned}
0 = -\frac{\partial \langle 0|V|0 \rangle}{\partial \alpha} &= K^2 \sin 2s \sin \alpha \left[-\frac{l_{10}}{2l_9} \left(l_{11} \sin^2 s + l_{12} \cos^2 s\right) K^2 \right. \\
&\quad + m_3 + l_6 K^2 + l_7 F^2 \\
&\quad \left. + \left(4 l_8 - \frac{l_{11} l_{12}}{2 l_9}\right) K^2 \sin 2s \cos \alpha \right].
\end{aligned} \tag{26.46}
$$

The trivial solution of eqn (26.46) is $\sin \alpha = 0$, which implies $\sin \delta = 0$, see eqns (26.45), and is therefore a CP-conserving solution. If we want to have spontaneous CP breaking we must choose the non-trivial solution of eqn (26.46),

$$\cos \alpha = \frac{2 l_9 \left(m_3 + l_6 K^2 + l_7 F^2\right) - l_{10} \left(l_{11} \sin^2 s + l_{12} \cos^2 s\right) K^2}{\left(l_{11} l_{12} - 8 l_8 l_9\right) K^2 \sin 2s}. \tag{26.47}$$

However, because $|\cos \alpha|$ cannot exceed 1, this solution only exists if $l_7 \sim K^2/F^2$, or else if some cancellation occurs in the numerator such that it ends up being of order K^2 instead of being $\sim F^2$, as one should *a priori* expect. We conclude that having spontaneous CP violation in the minimal LRSM requires fine-tuning.

Our derivation was based on the assumption that under CP the scalar fields transform as in eqn (26.40). We might try and define CP symmetry in a different way. However, it can be shown that no spontaneous-CP-breaking solution ever exists in the minimal LRSM unless some unnatural fine-tuning is assumed, in the sense that some quantity which should be of order F^2 is assumed to be of order K^2 instead. Thus, spontaneous CP violation is unnatural in the minimal LRSM (Branco and Lavoura 1985).

There are however extensions of the LRSM, with an enlarged scalar sector, in which spontaneous CP breaking is possible without any contrived fine-tuning. The simplest extensions of the minimal LRSM which can lead to natural spontaneous CP breaking are:

1. Add a real scalar singlet η which transforms under P and under CP in the following way (Chang *et al.* 1984):

$$\eta \overset{\text{P}}{\to} -\eta, \quad \eta \overset{\text{CP}}{\to} -\eta, \tag{26.48}$$

while the fields ϕ, Δ_L, and Δ_R still have the transformation properties in eqns (26.32) and (26.40). In this case there are new terms in the Higgs potential, in particular $i\eta \, \text{tr} \left(\tilde{\phi}^\dagger \phi - \phi^\dagger \tilde{\phi} \right)$ and $\eta^2 \, \text{tr} \left(\tilde{\phi}^\dagger \phi + \phi^\dagger \tilde{\phi} \right)$, which make it possible to have spontaneous CP violation without fine-tuning of couplings.

2. Add another scalar multiplet ϕ' transforming under the gauge group in the same way as ϕ. In this case, there are many more phase-dependent terms in the Higgs potential and, again, a CP-breaking vacuum can be obtained without fine-tuning.

26.3   Quark masses and mixing matrices

26.3.1   *Mass matrices*

The Yukawa couplings of the quarks are given by

$$\mathcal{L}_Y^{(q)} = \overline{Q_L}\left(\phi\Gamma_1 + \tilde{\phi}\Gamma_2\right)Q_R + \overline{Q_R}\left(\phi^\dagger\Gamma_1^\dagger + \tilde{\phi}^\dagger\Gamma_2^\dagger\right)Q_L, \tag{26.49}$$

where Γ_1 and Γ_2 are $n_g \times n_g$ matrices in generation space. We shall assume that ϕ transforms under parity as in eqn (26.32). As we have mentioned, this is the only choice leading to a realistic quark spectrum. Taking into account that parity interchanges Q_{La} and Q_{Ra}—where $a = 1, 2, \ldots, n_g$ is a generation index—it is clear that parity invariance constrains the Yukawa-coupling matrices to be Hermitian, i.e., $\Gamma_1 = \Gamma_1^\dagger$ and $\Gamma_2 = \Gamma_2^\dagger$.

The mass matrices for the up-type and down-type quarks, defined by the mass terms $\overline{p_L}M_p p_R$ and $\overline{n_L}M_n n_R$, are, respectively,

$$\begin{aligned} M_p &= k_1^* \Gamma_1 + k_2^* \Gamma_2 \\ M_n &= k_2 \Gamma_1 + k_1 \Gamma_2. \end{aligned} \tag{26.50}$$

These are not the most general complex matrices, still they do not lead to any constraints on observable quantities like the quark masses.

26.3.2   *Mixing matrices and CP-violating phases*

The quark mass matrices are bi-diagonalized in the usual way, see eqns (12.14) and (12.15). The charged-current Lagrangian can be written in terms of the quark mass eigenstates as

$$\frac{g}{\sqrt{2}}\overline{u}\gamma^\mu\left(W_{L\mu}^+ V_L \frac{1-\gamma_5}{2} + W_{R\mu}^+ V_R \frac{1+\gamma_5}{2}\right)d + \text{H.c.}, \tag{26.51}$$

where

$$\begin{aligned} V_L &\equiv U_L^{p\dagger} U_L^n, \\ V_R &\equiv U_R^{p\dagger} U_R^n \end{aligned} \tag{26.52}$$

are the charged-current mixing matrices. They are $n_g \times n_g$ unitary matrices. Just as in the SM, the neutral-current Lagrangian does not change its form when expressed in terms of the quark mass eigenstates u_α and d_k. This is due to the fact that all quark fields of a given charge and helicity have the same T_{L3}, T_{R3}, and Y. Hence, the neutral gauge interaction is not a source of CP violation in the LRSM.

In the charged-current Lagrangian we should take into account the fact that $W_L^\pm$ and $W_R^\pm$ are not the eigenstates of propagation, because they mix. However, as we have stressed before, the $W_L^\pm$–$W_R^\pm$ mixing may be made real by a gauge choice, and therefore it is not a source of CP violation.

In general, the two mixing matrices V_L and V_R contain a total of $n_g(n_g + 1)$ phases. However, we may rephase the quark fields,

$$
\begin{aligned}
u'_{L\alpha} &= \exp\left(i\theta_{L\alpha}\right) u_{L\alpha}, \\
u'_{R\alpha} &= \exp\left(i\theta_{R\alpha}\right) u_{R\alpha}, \\
d'_{Lk} &= \exp\left(i\chi_{Lk}\right) d_{Lk}, \\
d'_{Rk} &= \exp\left(i\chi_{Rk}\right) d_{Rk}.
\end{aligned}
\tag{26.53}
$$

Then,

$$
\begin{aligned}
(V'_L)_{\alpha k} &= \exp\left[i\left(\theta_{L\alpha} - \chi_{Lk}\right)\right] (V_L)_{\alpha k}, \\
(V'_R)_{\alpha k} &= \exp\left[i\left(\theta_{R\alpha} - \chi_{Rk}\right)\right] (V_R)_{\alpha k}, \\
m'_\alpha &= \exp\left[i\left(\theta_{L\alpha} - \theta_{R\alpha}\right)\right] m_\alpha, \\
m'_k &= \exp\left[i\left(\chi_{Lk} - \chi_{Rk}\right)\right] m_k.
\end{aligned}
\tag{26.54}
$$

We see that, if we require the mass m_α of the up-type quark α to remain real and positive, we must impose $\theta_{L\alpha} = \theta_{R\alpha}$. Similarly, we must set $\chi_{Lk} = \chi_{Rk}$ in order that the mass m_k of the down-type quark k remains real and positive. This means that the rephasings of the left-handed and right-handed quark fields must be identical, and as a consequence we can only eliminate $2n_g - 1$ phases from V_L and/or from V_R by means of rephasings. The total number of meaningful phases is thus

$$
N_{\text{phase}}^{\text{LR}} = n_g(n_g + 1) - (2n_g - 1) = n_g^2 - n_g + 1.
\tag{26.55}
$$

Therefore, even in the one-generation case there is one CP-violating phase remaining. We shall soon see the origin of this phase.

In general, the total number of CP-violating phases in the LRSM is given in eqn (26.55). However, the distribution of these phases by the mixing matrices V_L and V_R has some arbitrariness, since it depends on which phases we choose to eliminate. A convenient choice consists in using the rephasing freedom to eliminate the maximum number of phases from V_L. With this choice the number of CP-violating phases in V_L and in V_R will be given by, respectively,

$$
\begin{aligned}
N_L &= \frac{(n_g - 1)(n_g - 2)}{2}, \\
N_R &= \frac{n_g(n_g + 1)}{2}.
\end{aligned}
\tag{26.56}
$$

For two generations one has $N_L = 0$ and $N_R = 3$, and in this case all CP violation arises from the right-handed charged currents.[70]

---

[70]The fact that even for two generations one can obtain CP violation in the LRSM provided some of the original motivation to introduce this model. One should keep in mind that at the

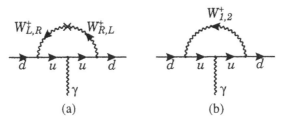

FIG. 26.1. Diagram generating an electric dipole moment for the down quark in the LRSM with one generation of quarks. (a) The diagram is drawn in the basis of the unphysical gauge bosons $W_L^\pm$ and $W_R^\pm$. (b) The diagram is drawn in the basis of the physical gauge bosons $W_1^\pm$ and $W_2^\pm$.

The origin of the extra CP-violating phases in the LRSM lies in the fact that we now have two mixing matrices, V_L and V_R, with the same rephasing properties. This means that $\arg\left[(V_L)_{\alpha k} (V_R)_{\alpha k}^*\right]$, for any up-type quark α and for any down-type quark k, is a meaningful phase; the sine of any of these phases is CP-violating. In practical calculations these phases always appear multiplied by the masses of the corresponding quarks. Indeed, from eqns (26.54) we see that the rephasing-invariant quantities are $m_\alpha^* m_k (V_L)_{\alpha k} (V_R)_{\alpha k}^*$. Of course, in practice one takes m_α and m_k to be real and positive.

We now see the reason why in the LRSM there is one CP-violating phase even in the case $n_g = 1$. Indeed, if there were only the up and the down quarks, then the quantity $(V_L)_{ud} (V_R)_{ud}^*$ might be complex and cause CP violation. In a practical calculation such a quantity might arise in the interference of two diagrams, one of them with a $V - A$ interaction and the other one with a $V + A$ interaction; or in a single diagram with both interactions, like for instance a diagram for the electric dipole moment of a quark mediated by a $W_L^\pm$ mixing with a $W_R^\pm$ (Fig. 26.1).

26.3.3 *Manifest and pseudo-manifest left–right symmetry*

In the above analysis, we have assumed that the quark mass matrices are arbitrary, i.e., that they are neither Hermitian nor symmetric. This leads to V_L and V_R being completely independent, with different mixing angles and phases in the left-handed and right-handed charged currents. This case is often referred to as having 'non-manifest left–right symmetry' and it can be realized if Γ_1 and Γ_2 are complex and $\alpha \neq 0$. Next we shall consider two particular scenarios:

---

time when the model was proposed, only two incomplete generations were known, since the charm quark had not yet been discovered.

When there are only two generations, eqns (26.56) imply that we may set V_L real. All CP violation then arises from the right-handed gauge interactions and, as such, it is expected to be suppressed by $\left(m_{W_L}/m_{W_R}\right)^2$. This suppression of CP-violating effects (Chang 1983a,b) was another attractive feature of the LRSM; however, it does not hold for three or more generations (Branco *et al.* 1983; Mohapatra 1985).

1. *CP is explicitly broken, but $\alpha = 0$.* This means that we may choose k_1 and k_2 simultaneously real which, in view of the fact that the Yukawa-coupling matrices are Hermitian, leads to $M_p = M_p^\dagger$ and $M_n = M_n^\dagger$. This case is called 'manifest left–right symmetry'.

2. *Spontaneously broken CP ($\alpha \neq 0$).* Let us assume that CP is a good symmetry of the Lagrangian, with the scalars transforming as in eqn (26.40). CP invariance then constrains Γ_1 and Γ_2 to be real, while P invariance enforces Hermiticity. The mass matrices M_p and M_n are then symmetric. However, they will in general be complex, since $\alpha \neq 0$. This case is often referred to as displaying 'pseudo-manifest left–right symmetry'.

One should emphasize that the scenario of manifest left–right symmetry is quite contrived. Indeed, once CP is not assumed to be a symmetry of the Lagrangian, there are in general terms in the vacuum potential dependent on $\sin\alpha$ and other terms dependent on $\cos\alpha$ (see eqn 26.36), and it is difficult to see how α can turn out to vanish.

In the case of manifest left–right symmetry M_p and M_n are Hermitian. They are diagonalized by unitary transformations, i.e., $U_L^p = U_R^p X^p$ and $U_L^n = U_R^n X^n$, where X^p and X^n are diagonal orthogonal matrices, needed to render the quark masses non-negative. As a result $V_L = X^p V_R X^n$, i.e., the mixing matrices appearing in the left-handed and right-handed charged currents are essentially the same. In this case, the number of CP-violating phases coincides with the corresponding number in the SM.

In the case of pseudo-manifest left–right symmetry the quark mass matrices are symmetric. From the first eqn (12.15) one obtains

$$U_L^{p\dagger} M_p M_p^\dagger U_L^p = U_R^{p\dagger} M_p^\dagger M_p U_R^p = M_u^2. \tag{26.57}$$

Taking into account that M_p is symmetric, one derives from eqn (26.57)

$$U_L^{pT} M_p^* M_p U_L^{p*} = U_R^{p\dagger} M_p^* M_p U_R^p = M_u^2. \tag{26.58}$$

From eqn (26.58) one concludes that $U_L^{p*} = U_R^p K^p$, where K^p is a diagonal unitary matrix. Similarly, for the down sector one has $U_L^{n*} = U_R^n K^n$. This leads to

$$V_R = K^p V_L K^{n*}. \tag{26.59}$$

Taking into account that K^p and K^n contain $2n_g - 1$ meaningful phases, one concludes that the total number of phases in this case is $n_g(n_g + 1)/2$. If $n_g = 1$ or $n_g = 2$ this is the same as the general case in eqn (26.55): one phase for one generation, three phases for two generations. For three generations, however, the general case admits seven CP-violating phases, while the case of pseudo-manifest left–right symmetry only has six phases.

Another way to look at this case consists in using the rephasing freedom of the quark fields to transform

$$\begin{aligned} V_R' &= (K^p)^{-1/2} V_R (K^n)^{1/2}, \\ V_L' &= (K^p)^{-1/2} V_L (K^n)^{1/2}. \end{aligned} \tag{26.60}$$

One then obtains $V'_R = V'^*_L$.

Thus, in the case of manifest left–right symmetry the two mixing matrices may be chosen to be equal; in the case of pseudo-manifest left–right symmetry, they may be chosen to be the complex-conjugate of each other.

26.4   Weak-basis invariants and CP violation

In this section we study the CP properties of the LRSM through the method of weak-basis invariants, which was derived and applied to the SM in Chapter 14.

26.4.1   *Conditions for CP invariance*

Let us consider the Lagrangian of the LRSM, assuming that the original gauge symmetry has already been broken into the U(1) of electromagnetism. The most general CP transformation of the quark fields which leaves invariant the gauge interactions is

$$
\begin{aligned}
(\mathcal{CP})\, p_L\, (t, \vec{r})\, (\mathcal{CP})^\dagger &= K_L \gamma^0 C \overline{p_L}^T\, (t, -\vec{r})\,, \\
(\mathcal{CP})\, n_L\, (t, \vec{r})\, (\mathcal{CP})^\dagger &= K_L \gamma^0 C \overline{n_L}^T\, (t, -\vec{r})\,, \\
(\mathcal{CP})\, p_R\, (t, \vec{r})\, (\mathcal{CP})^\dagger &= K_R \gamma^0 C \overline{p_R}^T\, (t, -\vec{r})\,, \\
(\mathcal{CP})\, n_R\, (t, \vec{r})\, (\mathcal{CP})^\dagger &= K_R \gamma^0 C \overline{n_R}^T\, (t, -\vec{r})\,,
\end{aligned}
\tag{26.61}
$$

where K_L and K_R are $n_g \times n_g$ unitary matrices acting in family space. Note that, due to the presence of the right-handed charged current, p_R and n_R must transform in the same way under CP; this is the crucial difference between the CP properties of the LRSM and of the SM.

One readily concludes that, in order for CP invariance to hold, the quark mass matrices must satisfy

$$
\begin{aligned}
K_L^\dagger M_p K_R &= M_p^*\,, \\
K_L^\dagger M_n K_R &= M_n^*.
\end{aligned}
\tag{26.62}
$$

One may thus state

Theorem 26.1 *The gauge interactions of the LRSM, together with the quark mass terms, are CP-invariant if and only if the quark mass matrices M_p and M_n are such that unitary matrices K_L and K_R exist which satisfy eqns (26.62).*

26.4.2   *Weak-basis transformations*

In the LRSM, a weak-basis transformation is more restricted than in the SM:

$$
\begin{aligned}
Q_L &= W_L Q'_L\,, \\
Q_R &= W_R Q'_R.
\end{aligned}
\tag{26.63}
$$

The right-handed quarks p_R and n_R now transform in the same way, because of the presence of the right-handed charged current. The quark mass matrices transform as

$$M'_p = W_L^\dagger M_p W_R,$$
$$M'_n = W_L^\dagger M_n W_R. \tag{26.64}$$

Following a line analogous to the one in § 14.3, it is easy to show that Theorem 26.1 is weak-basis independent.

26.4.3 CP restrictions in a special weak basis

In order to understand the restrictions that eqns (26.62) imply on M_p and M_n, it is useful to work in the weak basis in which $M_p = M_u$ is diagonal and real. Such a weak basis always exists in the LRSM, as was also the case in the SM. Assuming the up-type-quark masses to be non-degenerate, one then finds

$$K_L = K_R = \mathrm{diag}\,[\exp\,(i\delta_j)], \tag{26.65}$$

and

$$\arg\,(M_n)_{jk} = \frac{\delta_j - \delta_k}{2}\ (\mathrm{mod}\ \pi). \tag{26.66}$$

Thus, CP invariance constrains M_n to have cyclic phases in the weak basis where M_p is diagonal and real.

The matrix M_n is in general complex, having n_g^2 independent phases. The requirement that it has cyclic phases corresponds to $n_g^2 - (n_g - 1)$ independent restrictions. As expected, this number coincides with the number $N_{\mathrm{phase}}^{\mathrm{LR}}$ of independent CP-violating phases appearing in the left-handed and right-handed charged currents, in the mass-eigenstate basis.

It is useful to compare the LRSM with the SM. In the SM, CP invariance requires the *Hermitian* matrix $M_n M_n^\dagger$ to have cyclic phases in the weak basis in which $M_p M_p^\dagger$ is diagonal. In the LRSM, CP constrains an *arbitrary* complex matrix M_n to have cyclic phases in the weak basis in which M_p is diagonal. Thus, CP invariance is a much stronger requirement on the LRSM than on the SM. One understands in a simple way why in the LRSM it is possible to have CP violation for one or two generations, while in the SM this is not possible. Indeed, 1×1 and 2×2 Hermitian matrices automatically have cyclic phases, while a 1×1 or a 2×2 general matrix does not necessarily have phases with that property.

26.4.4 Weak-basis invariants

In this subsection we shall construct necessary conditions for CP invariance, expressed in terms of weak-basis invariants (Branco and Rebelo 1985). From eqns (26.62) one obtains

$$K_L^\dagger H_p K_L = H_p^T,$$
$$K_L^\dagger H_n K_L = H_n^T, \tag{26.67}$$

where $H_p \equiv M_p M_p^\dagger$ and $H_n \equiv M_n M_n^\dagger$. It follows that

$$\mathrm{tr}\,[H_p, H_n]^r = 0 \ \text{for } r \text{ odd}. \tag{26.68}$$

These are the weak-basis-invariant conditions for CP invariance which were already encountered in the SM. The LRSM has, however, other WB invariants which must also vanish in order for CP to be conserved. Indeed, one also obtains from eqns (26.62) that

$$K_R^\dagger H_p' K_R = H_p'^{\,T},$$
$$K_R^\dagger H_n' K_R = H_n'^{\,T},$$

(26.69)

with $H_p' \equiv M_p^\dagger M_p$ and $H_n' \equiv M_n^\dagger M_n$. It follows that

$$\text{tr}\left[H_p', H_n'\right]^r = 0 \text{ for } r \text{ odd.}$$

(26.70)

The matrices H_p' and H_n' are diagonalized by unitary transformations of the right-handed quark fields. Therefore, the conditions of eqn (26.70), when written in terms of quark masses and mixing angles, have the same form as those of eqn (26.68), with the mixing angles of V_L substituted by those of V_R.

Obviously, the lowest value of r for which the conditions of eqn (26.68) and eqn (26.70) are non-trivial is 3. Therefore, those invariants have mass dimensions of at least twelve. In the LRSM there are other WB invariants, relevant for the study of CP violation, which have no counterpart in the SM and have a much lower mass dimension. Indeed, from eqns (26.62) one obtains

$$K_L^\dagger M_p M_n^\dagger K_L = M_p^* M_n^T.$$

(26.71)

Taking the trace, one obtains

$$\text{Im} \, \text{tr} \left(M_p M_n^\dagger\right) = 0.$$

(26.72)

This is a necessary condition for CP invariance of the LRSM for any number of generations. The remarkable feature of eqn (26.72) is that it is non-trivial even for one generation. Indeed, eqn (26.72) reads

$$\sum_\alpha \sum_k m_\alpha m_k \text{Im} \left[(V_R)_{\alpha k}(V_L)_{\alpha k}^*\right] = 0.$$

(26.73)

In particular, for one quark generation, one has

$$\text{Im} \left[(V_R)_{ud}(V_L)_{ud}^*\right] = 0.$$

(26.74)

This result agrees with our considerations in § 26.3.2.

26.5   Phenomenological implications

The phenomenological implications of left–right-symmetric models depend to a large extent on whether the left–right symmetry is manifest or pseudo-manifest. Considering the latter case, with spontaneously broken CP, a detailed phenomenological analysis was carried out by Ecker and Grimus (1985), who extended earlier work by Chang (1983a,b) and by Branco et al. (1983). Imposing the requirement that the absolute value of the new contribution to Δm in the K^0–$\overline{K^0}$ system should not exceed the experimental value, they obtained the bound $m_{W_R} > 2.5\,\mathrm{TeV}$, which improved the earlier bound derived by Beall *et al.* (1982). It has also been pointed out by Ecker and Grimus (1985) that the ratio ϵ'/ϵ in the LRSM can have either sign, and is correlated with the value of the neutron electric dipole moment. The possibility of having a small value of ϵ'/ϵ was, at the time of their calculation, very interesting, since the prediction of the SM was thought to be $\epsilon'/\epsilon \geq 2 \times 10^{-3}$. At present, and taking into account the high value of the top-quark mass, values $\sim 10^{-4}$ for $|\epsilon'/\epsilon|$ can be obtained within the SM, and therefore that ratio is no longer a very useful parameter to distinguish between the SM and the LRSM.

26.6   Main conclusions

- Due to the presence of a charged current with the right-handed quarks, there are new sources of CP violation in the left–right-symmetric model. These arise from the possibility that the corresponding matrix elements for the left-handed-quarks mixing matrix and for the right-handed-quarks mixing matrix might have different phases.
- In the LRSM, CP violation may be present even in the case of only one family.
- In the LRSM, there are weak-basis invariants relevant for CP violation of much lower mass dimension than in the SM.

It should also be pointed out that the scalar sector of the LRSM is quite complicated, and may be the source of various effects, including CP violation.

THE STRONG CP PROBLEM

27.1 Introduction

In this chapter we make a short introduction to an intriguing problem of the standard model of the strong and electroweak interactions: the 'strong CP problem'. This problem might as well be called the 'strong P problem', because its crux is the possible presence in the strong-interaction Lagrangian of a P- and CP-violating term called the θ-term. We shall focus our attention on possible implications of the strong CP problem to model-building. Although the strong CP problem is, to a great extent, still an open question in particle physics, it is possible to construct models for the electroweak interaction which keep the strong CP problem at bay. The hope is that the strong CP problem may provide a clue to the 'true' model of CP violation.

We shall consider only briefly the origin of the strong CP problem. Thus, in §§ 27.2 and 27.3 we make a rather perfunctory introduction to the $U(1)_A$ problem and to instantons; indeed, it is the $U(1)_A$ problem which forces the inclusion into $\mathcal{L}_{QCD}$ of the θ-term, which is justified by the existence of special solutions to the equations of motion of the Yang–Mills theory in the Euclidean metric, called instantons. Readers who do not want to delve into these side issues may prefer to skip both sections. We proceed to explain, in § 27.4, the strong CP problem itself, and in the following sections we review some possible ways of solving it.

27.2 The U(1)$_A$ problem

The strong CP problem is closely related to the $U(1)_A$ problem of quantum chromodynamics (QCD) and to its solution as proposed by 't Hooft (1976). Therefore, we shall start by briefly recalling the main features of the $U(1)_A$ problem. For a detailed analysis of that problem and its solution the reader may consult the reviews on the subject (Coleman 1978; Crewther 1978; Christos 1984; 't Hooft 1986).

Let us consider the QCD Lagrangian for two *massless* flavours u and d:

$$\mathcal{L}_{QCD} = -\tfrac{1}{4}F^a_{\mu\nu}F^{a\mu\nu} + \sum_{q=u,d}\left(\overline{q_x}i\gamma^\mu\partial_\mu q_x + g_s\overline{q_x}\gamma^\mu G^a_\mu\frac{\lambda^a_{xy}}{2}q_y\right), \qquad (27.1)$$

cf. eqn (3.90). We have omitted the quark mass terms. In this limit, $\mathcal{L}_{QCD}$ has a global symmetry $SU(2)_R\otimes SU(2)_L\otimes U(1)_R\otimes U(1)_L$. The vectorial part of that symmetry is $SU(2)_V\equiv SU(2)_{R+L}$ and $U(1)_V\equiv U(1)_{R+L}$. This corresponds to the transformations

$$\text{SU(2)}_\text{V} : \begin{pmatrix} u \\ d \end{pmatrix} \rightarrow \exp\left(i\alpha^a \frac{\tau^a}{2}\right) \begin{pmatrix} u \\ d \end{pmatrix},$$

$$\text{U(1)}_\text{V} : \begin{pmatrix} u \\ d \end{pmatrix} \rightarrow \exp\left(i\beta\right) \begin{pmatrix} u \\ d \end{pmatrix},$$

$$(27.2)$$

where τ^a $(a = 1, 2, 3)$ are the Pauli matrices. The symmetries in eqns (27.2) are manifestly realized in nature: SU(2)_V is isospin, and U(1)_V corresponds to baryon-number conservation.

The symmetry $\text{SU(2)}_\text{A} \equiv \text{SU(2)}_\text{R-L}$, on the other hand, is spontaneously broken by quark condensates

$$\langle 0|\bar{u}u|0\rangle \approx \langle 0|\bar{d}d|0\rangle, \tag{27.3}$$

leading to three Goldstone bosons, which are identified as the isotriplet of pions. The pions are massless in the chiral (massless quarks) limit; once non-vanishing but small quark masses m_u and m_d are included, the pions acquire a mass which is small when compared to the strong-interaction mass scale Λ_QCD.

We now turn our attention to the symmetry $\text{U(1)}_\text{A} \equiv \text{U(1)}_\text{R-L}$, which corresponds to the transformation

$$\begin{pmatrix} u \\ d \end{pmatrix} \rightarrow \exp\left(i\gamma_5\sigma\right) \begin{pmatrix} u \\ d \end{pmatrix}. \tag{27.4}$$

If U(1)_A stayed unbroken, this would imply a parity doubling of the hadron spectrum, which is not observed in Nature. On the other hand, if U(1)_A were spontaneously broken, one would expect a fourth Goldstone boson, with mass comparable to the one of the pions. Indeed, taking into account the quark masses m_u and m_d, Weinberg (1975) has shown that the mass m_0 of the Goldstone boson associated to U(1)_A should satisfy the relation

$$m_0 \leq \sqrt{3}m_\pi. \tag{27.5}$$

However, there is no experimentally observed particle which can be identified with this Goldstone boson. The meson η has the right quantum numbers, since it is a zero-isospin pseudoscalar, but it is too heavy: $m_\eta = 547\,\text{MeV}$ while $m_\pi = 138\,\text{MeV}$, grossly violating eqn (27.5).

If one considers QCD with three flavours instead of two, the spontaneous breakdown of the chiral symmetry SU(3)_A leads to eight Goldstone bosons, which acquire mass once the light-quark masses m_u, m_d, and m_s are introduced. There is again a symmetry U(1)_A corresponding to the transformation of eqn (27.4), where now $s \rightarrow \exp\left(i\gamma_5\sigma\right) s$ too. The spontaneous breakdown of this symmetry should lead to a ninth Goldstone boson, which is not observed. This is the U(1)_A problem: the spontaneous breakdowm of the symmetry U(1)_A predicts the existence of a ninth light pseudoscalar, beyond the eight light pseudoscalars resulting from the spontaneous breakdown of SU(3)_A. Experimentally one only observes eight light pseudoscalar mesons: the three pions, the four kaons, and the η. The ninth light pseudoscalar is not found in Nature.

27.3   Instantons

The solution of the $U(1)_A$ problem is related to the complex structure of the vacuum of QCD. An important role is played by topologically non-trivial solutions of QCD named instantons. Instantons (for a review see Rajaraman 1982) are finite-action solutions of the classical field equations for non-Abelian gauge theories *in Euclidean space.*

For simplicity, let us consider an SU(2) Yang–Mills theory (Yang and Mills 1954), whose structure constants are given by the completely antisymmetric tensor ϵ^{abc}:

$$\left[\frac{\tau^a}{2}, \frac{\tau^b}{2} \right] = i\epsilon^{abc} \frac{\tau^c}{2}, \tag{27.6}$$

where τ^a denote the Pauli matrices. There are three gauge fields, G_μ^a. The field-strength tensor is

$$F_{\mu\nu}^a = \partial_\mu G_\nu^a - \partial_\nu G_\mu^a + g\epsilon^{abc} G_\mu^b G_\nu^c, \tag{27.7}$$

cf. eqn (3.92). It is convenient to write both the gauge fields and the field-strength tensor in the form of 2×2 matrices:

$$G_\mu \equiv -ig G_\mu^a \frac{\tau^a}{2},$$
$$F_{\mu\nu} \equiv -ig F_{\mu\nu}^a \frac{\tau^a}{2}. \tag{27.8}$$

Then, eqn (27.7) may be rewritten as

$$F_{\mu\nu} = \partial_\mu G_\nu - \partial_\nu G_\mu + [G_\mu, G_\nu]. \tag{27.9}$$

The gauge Lagrangian is

$$-\tfrac{1}{4} F_{\mu\nu}^a F^{a\mu\nu} = \frac{1}{2g^2} \operatorname{tr} F_{\mu\nu} F^{\mu\nu}. \tag{27.10}$$

Under a gauge transformation given by the matrix of SU(2)

$$U = \exp\left(ig\alpha^a \frac{\tau^a}{2} \right) \tag{27.11}$$

one has, if ϕ is a doublet of SU(2), $\phi \to U\phi$. The covariant derivative $\partial_\mu \phi + G_\mu \phi$ should transform in the same way as ϕ itself; therefore,

$$G_\mu \to U G_\mu U^{-1} - (\partial_\mu U) U^{-1}. \tag{27.12}$$

If the parameters α^a are infinitesimal, this translates into

$$G_\mu^a \to G_\mu^a + g\epsilon^{abc} G_\mu^b \alpha^c + \partial_\mu \alpha^a, \tag{27.13}$$

cf. eqns (11.27).

One is looking for solutions of the classical Yang–Mills theory for which the Euclidean action is finite, i.e.,

$$\left| \int d^4x \operatorname{tr} F_{\mu\nu} F^{\mu\nu} \right| < \infty. \tag{27.14}$$

In order to have this condition satisfied, $F_{\mu\nu}$ should approach zero when $|x|$ goes to infinity. This implies that G_μ is a gauge transform of zero at infinity, i.e.,

$$G_\mu(x) = -\left(\partial_\mu U\right) U^{-1} \tag{27.15}$$

for x at infinity in the four-dimensional Euclidean space.[71] Those points x form a three-dimensional sphere, S^3. Therefore, we must have a mapping of S^3 (the points x at infinity) into SU(2) (the matrix U). Now, SU(2) is topologically equivalent to S^3. This is because a 2×2 unitary matrix with unit determinant can always be written as $z^0 + iz^a\tau^a$, with $\left(z^0\right)^2 + \left(z^1\right)^2 + \left(z^2\right)^2 + \left(z^3\right)^2 = 1$. Therefore, one has to consider mappings of S^3 into S^3. Those mappings can be classified in homotopic classes. Two mappings belong to the same homotopic class if one of them can be continuously deformed into the other one. Each homotopic class is characterized by a winding number. In order to understand this concept, it is useful to examine the simple case of a mapping $S^1 \to S^1$. Take the function

$$f(\varphi) = \exp\left[i\left(n\varphi + \theta\right)\right], \tag{27.16}$$

where n is an integer. Consider two such functions, $f_1(\varphi)$ and $f_2(\varphi)$, having the same value of n but two different values θ_1 and θ_2 of θ. Then, f_1 and f_2 belong to the same homotopic class, since f_1 can be continuously deformed into f_2 by letting θ_1 approach θ_2. On the other hand, two functions with different values of n belong to different homotopic classes. The function f corresponds to a mapping of a circle onto another circle, and n tells us how many times one winds around the second circle when one goes around the first circle once. In this case, one may write the winding number as

$$n = \frac{1}{2\pi i} \int_0^{2\pi} d\varphi \, \frac{1}{f} \frac{df}{d\varphi}. \tag{27.17}$$

Analogously, in the case of mappings $S^3 \to S^3$, a winding number exists which may be expressed as (Coleman 1978; Rajaraman 1982)

$$n = \frac{1}{16\pi^2} \int d^4x \operatorname{tr} F_{\mu\nu} \tilde{F}^{\mu\nu} = -\frac{g^2}{32\pi^2} \int d^4x \, F_{\mu\nu}^a \tilde{F}^{a\mu\nu}, \tag{27.18}$$

where $\tilde{F}^{\mu\nu}$ is the dual of $F^{\mu\nu}$:

$$\tilde{F}^{\mu\nu} \equiv \tfrac{1}{2} \epsilon^{\mu\nu\rho\sigma} F_{\rho\sigma}. \tag{27.19}$$

Here, $\epsilon^{\mu\nu\rho\sigma}$ is the totally antisymmetric tensor with four indices.

---

[71] Notice that G_μ must have the form in eqn (27.15) only for $|x| \to \infty$. In particular, the gauge function U need not be well defined over the whole four-dimensional Euclidean space. It is the possible existence of singularities in U that leads to the introduction of topological indices.

In order to gain some familiarity with the above expression, it is useful to show that it is a total divergence. Indeed,

$$
\begin{aligned}
\operatorname{tr} F_{\mu\nu}\tilde{F}^{\mu\nu} &= \tfrac{1}{2}\epsilon^{\mu\nu\rho\sigma}\operatorname{tr} F_{\mu\nu}F_{\rho\sigma} \\
&= \tfrac{1}{2}\epsilon^{\mu\nu\rho\sigma}\operatorname{tr}\left\{(\partial_\mu G_\nu - \partial_\nu G_\mu + [G_\mu, G_\nu])(\partial_\rho G_\sigma - \partial_\sigma G_\rho + [G_\rho, G_\sigma])\right\} \\
&= 2\epsilon^{\mu\nu\rho\sigma}\operatorname{tr}\left\{(\partial_\mu G_\nu)(\partial_\rho G_\sigma) + (\partial_\mu G_\nu)[G_\rho, G_\sigma] + \tfrac{1}{4}[G_\mu, G_\nu][G_\rho, G_\sigma]\right\} \\
&= 2\epsilon^{\mu\nu\rho\sigma}\operatorname{tr}\left\{\partial_\mu[G_\nu(\partial_\rho G_\sigma)] + \tfrac{1}{3}\partial_\mu(G_\nu[G_\rho, G_\sigma])\right\} \\
&= 2\partial_\mu \epsilon^{\mu\nu\rho\sigma}\operatorname{tr}\left\{G_\nu(\partial_\rho G_\sigma + \tfrac{1}{3}[G_\rho, G_\sigma])\right\}.
\end{aligned}
\tag{27.20}
$$

We have used the cyclic property of the trace and the fact that, due to

$$
[G_\nu,[G_\rho,G_\sigma]] + [G_\sigma,[G_\nu,G_\rho]] + [G_\rho,[G_\sigma,G_\nu]] = 0,
\tag{27.21}
$$

one has

$$
\epsilon^{\mu\nu\rho\sigma}\operatorname{tr}([G_\mu,G_\nu][G_\rho,G_\sigma]) = 0.
\tag{27.22}
$$

Thus, we have found that the integrand in eqn (27.18) is a total divergence:

$$
\operatorname{tr} F_{\mu\nu}\tilde{F}^{\mu\nu} = \partial_\mu J^\mu,
\tag{27.23}
$$

with

$$
\begin{aligned}
J^\mu &\equiv 2\epsilon^{\mu\nu\rho\sigma}\operatorname{tr}\left\{G_\nu(\partial_\rho G_\sigma + \tfrac{1}{3}[G_\rho, G_\sigma])\right\} \\
&= \epsilon^{\mu\nu\rho\sigma}\operatorname{tr}\left[G_\nu(F_{\rho\sigma} - \tfrac{2}{3}G_\rho G_\sigma)\right].
\end{aligned}
\tag{27.24}
$$

Out of all finite-action gauge fields, i.e., fields which satisfy eqn (27.14), there are some which solve the Yang–Mills equations of motion. In particular, Belavin *et al.* (1975) found the following $n = 1$ solution:

$$
G_\mu(x) = -\frac{r^2}{r^2 + \lambda^2}(\partial_\mu U)U^{-1},
\tag{27.25}
$$

where $r^2 = (x^0)^2 + (x^1)^2 + (x^2)^2 + (x^3)^2$ and U is an SU(2) matrix:

$$
U = \frac{x^0 + ix^a\tau^a}{r}.
\tag{27.26}
$$

The solution of Belavin *et al.* (1975) also involves an arbitrary positive number λ^2. It is clear that, for points x at infinity, i.e., for $r \to \infty$, the field in eqn (27.25) approaches a pure gauge.

It has been shown by Atiyah and Ward (1977) that there are no other instantons with $n = 1$ in the SU(2) gauge theory.

Although Belavin *et al.* (1975) have only considered the case of the SU(2) gauge theory, their results apply to any non-Abelian simple group G, due to Bott's theorem, which states that any continuous mapping of S^3 into G can be continuously deformed into a mapping of S^3 into an SU(2) subgroup of G.

27.4 The strong CP problem

Within the framework of the path-integral approach to the quantization of non-Abelian gauge theories, instantons correspond to paths connecting initial and final vacuum states with different winding number, thus leading to tunnelling between topologically distinct vacuum configurations ('t Hooft 1976; Callan *et al.* 1976; Jackiw and Rebbi 1976). It was shown by 't Hooft (1976) that the inclusion of instantons in the path integral leads to the breaking of the U(1)$_A$ symmetry without producing a Goldstone boson, thus solving the U(1)$_A$ problem. An essential feature of this solution to the U(1)$_A$ problem is that the QCD Lagrangian must include a term

$$\mathcal{L}_\theta = \theta_{\text{QCD}} \frac{g_s^2}{32\pi^2} F^{a\mu\nu} \tilde{F}^a_{\mu\nu}, \tag{27.27}$$

where θ_{QCD} is a free parameter. The inclusion of the θ-term is essential for the solution of the U(1)$_A$ problem, yet it creates another problem. The point is that $\mathcal{L}_\theta$ violates P and T. Indeed, it follows from eqns (3.96) that

$$\mathcal{P} F^a_{\mu\nu} F^a_{\rho\sigma} \mathcal{P}^\dagger = F^{a\mu\nu} F^{a\rho\sigma},$$
$$\mathcal{T} F^a_{\mu\nu} F^a_{\rho\sigma} \mathcal{T}^{-1} = F^{a\mu\nu} F^{a\rho\sigma}, \tag{27.28}$$
$$\mathcal{C} F^a_{\mu\nu} F^a_{\rho\sigma} \mathcal{C}^\dagger = F^a_{\mu\nu} F^a_{\rho\sigma}.$$

As $\epsilon^{\mu\nu\rho\sigma} = -\epsilon_{\mu\nu\rho\sigma}$, P and T are violated by $\mathcal{L}_\theta$, while C is conserved. This is puzzling, since prior to these developments it was considered that one of the nice features of QCD is the fact that it automatically enjoys the symmetries C, P, and T, in agreement with all our experimental knowledge of the strong interaction.

One might think that the inclusion of $\mathcal{L}_\theta$ in the QCD Lagrangian would not affect physics, since the θ-term is a total divergence (see eqn 27.23). However, the complex nature of the QCD vacuum, with gauge fields which do not fall off to zero at infinity, both requires the presence of the θ-term and justifies its importance to physics.

One could try and avoid the strong CP problem by postulating the P or CP invariance of the Lagrangian, thereby setting $\theta_{\text{QCD}} = 0$ by hand. However, the situation is more involved, for the following reason. In QCD, although the axial vector current J^5_μ is conserved at classical level, at loop level one has, due to the chiral anomaly (Adler 1969; Bardeen 1969; Bell and Jackiw 1969),

$$\partial^\mu J^5_\mu = \frac{n_f g_s^2}{16\pi^2} F^{a\mu\nu} \tilde{F}^a_{\mu\nu}, \tag{27.29}$$

where n_f is the number of flavours. It follows from eqn (27.29) that a U(1)$_A$ transformation induces in the Lagrangian a term proportional to $F^{a\mu\nu} \tilde{F}^a_{\mu\nu}$, i.e., a θ-term. Now, in the standard model (SM) the quark mass matrices M_p and M_n, generated upon the spontaneous breaking of SU(2)$\otimes$U(1), are in general arbitrary complex matrices, which are diagonalized by the bi-unitary transformations of eqns (12.15). These transformations include in particular the chiral transformation necessary to make M_u and M_d real. This induces a contribution to θ_{QCD},

$$\theta_{\mathrm{QCD}} \to \bar{\theta} \equiv \theta_{\mathrm{QCD}} + \theta_{\mathrm{QFD}}, \tag{27.30}$$

where

$$\theta_{\mathrm{QFD}} = \arg \det (M_p M_n). \tag{27.31}$$

Under a chiral transformation of the quark fields both θ_{QCD} and θ_{QFD} change, but their sum $\bar{\theta}$ remains invariant. Indeed, $\bar{\theta}$ is the physical parameter which measures the strength of the P- and CP-violating effects in QCD.

In the SM, in order to have CP violation in the charged weak interaction one has to allow for complex Yukawa couplings, and thus M_p and M_n are in general complex matrices. Therefore, there is no reason for θ_{QFD} to vanish. In fact, within the context of the SM, even if one puts $\theta_{\mathrm{QFD}} = 0$ at tree level, this condition will not be stable under renormalization, since θ_{QFD} receives divergent contributions at higher orders of perturbation theory (Ellis and Gaillard 1979).

The most stringent constraint on the θ-term arises from its contribution to the electric dipole moment of the neutron, D_n, which is P- and T-violating. Indeed, using a bag model Baluni (1979) obtained

$$D_n = \left(2.7 \times 10^{-16}\,\bar{\theta}\right) \mathrm{e\,cm}. \tag{27.32}$$

Crewther et al. (1979) evaluated D_n in chiral perturbation theory and obtained

$$D_n = \left(3.6 \times 10^{-16}\,\bar{\theta}\right) \mathrm{e\,cm}. \tag{27.33}$$

Using the experimental limit (Altarev et al. 1992) $D_n < 1.1 \times 10^{-25}\,\mathrm{e\,cm}$, one thus finds

$$\bar{\theta} < 3 \times 10^{-10}. \tag{27.34}$$

Why should a dimensionless free parameter have such a small value? The constraint in eqn (27.34) implies a serious fine-tuning problem for the SM. It is unnatural, in the sense of 't Hooft (1980), to choose an extremely small value for $\bar{\theta}$, since the SM does not acquire a larger symmetry when $\bar{\theta} = 0$. This is the essence of the strong CP problem (for reviews see Kim 1987; Cheng 1988; Peccei 1989).

27.5 $m_u = 0$ solution

Various solutions to the strong CP problem have been proposed. Most of them belong to one of the following three categories:

- $m_u = 0$ solution;
- Peccei–Quinn solution;
- solution with calculable and naturally small $\bar{\theta}$.

The first category is the simplest one: we just assume that one of the quarks (the best candidate is the up quark) has vanishing mass. If the up-quark mass m_u is zero, then $\bar{\theta}$ may be eliminated through a chiral transformation of the up-quark field. One may view this in another way: if one of the quarks has zero mass, then θ_{QFD} is arbitrary because $\det{(M_p M_n)} = 0$. This means that we can set $\bar{\theta} = 0$ with no loss of generality, and there is no strong CP violation. This is the simplest solution to the strong CP problem.

However, a careful analysis by Gasser and Leutwyler (1982) and by Leutwyler (1996) indicates that none of the quark masses vanishes. Although this analysis is generally considered to be the most reliable one, it should be mentioned that a different point of view has been put forward by Kaplan and Manohar (1986) and by Choi *et al.* (1988).

27.6   Peccei–Quinn solution

An elegant solution of the strong CP problem was suggested by Peccei and Quinn (PQ), who have pointed out that one could have a global chiral symmetry $U(1)_{\text{PQ}}$ under which both the quarks and the Higgs multiplets transform non-trivially. The parameter $\bar{\theta}$ then becomes a dynamical variable and it was shown (Peccei and Quinn 1977) in specific examples that $\bar{\theta}$ is dynamically set to zero.

Soon after the PQ proposal, Weinberg (1978) and Wilczek (1978) pointed out that, since $U(1)_{\text{PQ}}$ is a global continuous symmetry spontaneously broken by the vacuum, there is a Goldstone boson, named the axion, whose bare mass is zero, but which acquires a small mass through instanton effects. For definiteness, let us consider the SM with two Higgs doublets ϕ_1 and ϕ_2, and let us introduce the following global continuous symmetry:

$$\begin{aligned}
\phi_1 &\rightarrow e^{i\gamma}\phi_1, \\
\phi_2 &\rightarrow e^{-i\gamma}\phi_2, \\
p_R &\rightarrow e^{i\gamma}p_R, \\
n_R &\rightarrow e^{i\gamma}n_R.
\end{aligned} \tag{27.35}$$

The most general Yukawa couplings of the quarks may then be written

$$\mathcal{L}_Y^{(q)} = -\overline{Q_L}\left(\Gamma_2\phi_2 n_R + \Delta_1\tilde{\phi}_1 p_R\right) + \text{H.c.,} \tag{27.36}$$

cf. eqn (22.18). After spontaneous symmetry breaking the neutral components of the doublets acquire vacuum expectation values (VEVs) $\langle 0|\phi_a^0|0\rangle = v_a$. Both v_1 and v_2 are real and positive, without loss of generality, because there is no term in the scalar potential which feels the relative phase of $\langle 0|\phi_1^0|0\rangle$ and $\langle 0|\phi_2^0|0\rangle$. We expand the neutral components (see Chapter 22) as

$$\phi_k^0 = v_k + \frac{\rho_k + i\eta_k}{\sqrt{2}}. \tag{27.37}$$

Since for non-vanishing v_1 and v_2 the rephasing symmetry in eqns (27.35) is broken by the vacuum, just as the gauge symmetry, one is led to two Goldstone bosons. One of them is

$$\chi = \frac{v_1\eta_1 + v_2\eta_2}{v}, \tag{27.38}$$

with $v = \sqrt{v_1^2 + v_2^2}$, and is absorbed by the longitudinal component of the Z; the other one is

$$A = \frac{v_2 \eta_1 - v_1 \eta_2}{v}, \tag{27.39}$$

which is a physical particle, the axion. Using current algebra, Weinberg (1978), Wilczek (1978), Bardeen and Tye (1978), and Kandaswamy et al. (1978) have obtained the following estimate for the axion mass:

$$\begin{aligned} m_A &\approx \frac{n_f m_\pi f_\pi v^2}{4 v_1 v_2} \sqrt{\frac{\sqrt{2} G_F m_u m_d m_s}{(m_u m_d + m_u m_s + m_d m_s)(m_u + m_d)}} \\ &\approx \frac{n_f m_\pi f_\pi v^2}{4 v_1 v_2} \frac{\sqrt{\sqrt{2} G_F m_u m_d}}{m_u + m_d}. \end{aligned} \tag{27.40}$$

For $n_f = 6$, and taking $m_s/m_d = 20$ and $m_d/m_u = 1.8$, one obtains

$$m_A \approx 51 \frac{v^2}{v_1 v_2} \text{ keV}. \tag{27.41}$$

This initial version of the PQ model and the corresponding axion have been ruled out by experiment (see Kim 1987; Raffelt 1996).

A simple way of avoiding the experimental bounds is by noting that both the axion mass and its Yukawa couplings are inversely proportional to the scale v_{PQ} at which the PQ symmetry is broken. In the initial version of the PQ model v_{PQ} was identified with the weak scale, $v_{PQ} \sim v \sim G_F^{-1/2}$. If one considers instead that v_{PQ} is much larger than the weak scale, then one has very light and very weakly coupled axions. Models with $v_{PQ} \gg G_F^{-1/2}$ are often designated 'invisible-axion models'. Two important instances are the model of Kim (1979) and of Shifman et al. (1980) (KSVZ model) and the model of Zhitnitsky (1980) and of Dine et al. (1981) (DFSZ model). We shall describe the latter, which is especially interesting since only standard quarks are introduced, and which furthermore can be easily incorporated into grand-unified theories.

27.6.1   The DFSZ model

We add to the two-Higgs-doublet model a complex scalar singlet of SU(2)⊗U(1), denoted S. The symmetry U(1)$_{PQ}$ in eqns (27.35) is extended by

$$S \to e^{i\gamma} S. \tag{27.42}$$

The most general scalar potential consistent with U(1)$_{PQ}$ and with the gauge symmetry includes only one term depending on the relative phases of ϕ_1, ϕ_2, and S. That term is

$$e^{i\alpha} \left(\phi_1^\dagger \phi_2 \right) S^2 + e^{-i\alpha} \left(\phi_2^\dagger \phi_1 \right) S^{\dagger 2}. \tag{27.43}$$

Then, $\langle 0|S|0 \rangle = V e^{-i\alpha/2}$ and we may expand S as

$$S = e^{-i\alpha/2}\left(V + \frac{R_0 + iI_0}{\sqrt{2}}\right). \tag{27.44}$$

The Goldstone boson which is absorbed by the longitudinal component of the Z is χ in eqn (27.38); the axion is

$$A_0 = \frac{2v_1v_2A + vVI_0}{\sqrt{4v_1^2v_2^2 + v^2V^2}}, \tag{27.45}$$

and there is a massive scalar $N_0 = \left(4v_1^2v_2^2 + v^2V^2\right)^{-1/2}\left(vVA - 2v_1v_2I_0\right)$. The couplings of the axion to the quarks are given by

$$\mathcal{L}_Y^{(q)} = \cdots - 2iA_0\sqrt{\frac{\sqrt{2}G_F}{4v_1^2v_2^2 + v^2V^2}}\left(v_2^2\bar{u}M_u\gamma_5u + v_1^2\bar{d}M_d\gamma_5d\right). \tag{27.46}$$

The mass of the axion is

$$m_{A_0} \approx \frac{n_f m_\pi f_\pi v^2}{2\sqrt{4v_1^2v_2^2 + v^2V^2}}\sqrt{\frac{\sqrt{2}G_F m_u m_d m_s}{(m_u m_d + m_u m_s + m_d m_s)(m_u + m_d)}}. \tag{27.47}$$

Since S is a singlet, it is natural to have $V \gg v$. It then follows from eqns (27.46) and (27.47) that the axion is both very light and very weakly coupled to the quarks.

In spite of their weak coupling to matter, there has been a serious effort in searching for 'invisible' axions—see the contributions by Murayama, Raffelt, Hagmann et al. to the Particle Data Group (1998, pp. 264–271). Axions could be produced in hot plasmas, leading to energy loss by stars. A limit on the interactions of axions with photons, electrons, and nucleons can then be derived from the requirement that axions do not affect the observed stellar evolution. Axions are also a good candidate to constitute the hot dark matter of the Universe, and various microwave-cavity experiments, as well as telescope searches, have been performed under that assumption. In spite of all these experimental efforts, no axion has ever been found, leading to the bound $m_{\text{axion}} \leq 10^{-2}\,\text{eV}$.

27.7   Solutions with calculable and naturally small $\bar{\theta}$

In this class of solutions of the strong CP problem one introduces a symmetry through which $\bar{\theta}$ no longer is a free parameter, rather it is a calculable and naturally small quantity. In order to achieve this, it is necessary that $\bar{\theta}$ vanishes at tree level. Since $F^a_{\mu\nu}\tilde{F}^{a\mu\nu}$ is both P- and CP-violating, the simplest way to have $\theta_{\text{QCD}} = 0$ is to impose either P or CP invariance at the Lagrangian level. P or CP are then spontaneously broken. When this happens, one has to guarantee, by means of some other symmetry, that θ_{QFD} also vanishes at tree level.

27.7.1   A model with spontaneous CP breaking

The first models of this type which were suggested (Bég and Tsao 1978; Mohapatra and Senjanović 1978; Barr and Langacker 1979; Bég et al. 1979; Segré and Weldon 1979; Georgi 1981) had CP as a good symmetry of the Lagrangian, broken at the electroweak scale. Some of those models were based on low-energy left-right symmetry while others were based on SU(2)⊗U(1), but introduced two or more Higgs doublets, in order to obtain spontaneous CP breaking.

A different and very interesting subclass of models was suggested by Nelson (1984) and generalized by Barr (1984). Among the essential features of this subclass of models are the existence of heavy exotic fermions, and the fact that CP is spontaneously broken at a high-energy scale. The Nelson–Barr (NB) criteria for models of this type are the following:

1. The fermions are divided into two sets belonging to representations F and R of the gauge group. The representation F is complex and contains fermions with the same SU(3)⊗SU(2)⊗U(1) quantum numbers as the SM fermions. On the other hand, R is a real representation with respect to SU(3)⊗SU(2)⊗U(1).

2. The VEVs which break SU(2)⊗U(1) only contribute to fermion mass terms of the type F–F, and not to F–R, R–F, or R–R mass terms.

3. Scalars having complex VEVs only contribute to F–R and R–F mass terms, but not to F–F and R–R mass terms.

The initial model proposed by Nelson (1984) was based on SU(5), with an additional SO(3) flavour symmetry. An interesting realization based on SO(10) was given by Barr (1984). In order to show how the NB mechanism works, we shall describe a minimal extension of the SM in which the mechanism is implemented (Bento *et al.* 1991).

Let us consider the SM with three generations, extended through the addition of an isosinglet vector-like quark with charge $-1/3$, and of a complex electroweak-singlet scalar S. This model was previously considered in § 23.6 and 24.7, where it was described as an example of a minimal extension of the SM in which spontaneous CP violation can be achieved. So we shall impose CP symmetry at the Lagrangian level, but the Higgs potential and the Yukawa couplings will be such that the vacuum spontaneously breaks CP through a complex value $\langle 0|S|0 \rangle = V e^{i\alpha}$. The fact that CP is imposed as a symmetry of the Lagrangian leads to $\theta_{\mathrm{QCD}} = 0$. In order to make the model fulfil the NB criteria, we introduce a Z_2 symmetry under which all the SM fields are even, while the new isosinglet quark N and the complex scalar S change sign.

The scalar potential has been given in eqn (23.37). The most general Yukawa couplings invariant under SU(2)⊗U(1)⊗Z_2 were given in eqn (24.50). Upon spontaneous symmetry breaking by the VEVs of ϕ and S in eqns (23.38), masses are generated for the $Q = 2/3$ and $Q = -1/3$ quarks. The 4×4 mass matrix for the $Q = -1/3$ quarks is given by eqn (24.52):

$$\mathcal{M}_n = \begin{pmatrix} v\Gamma & 0 \\ V\mathcal{F} & \mu \end{pmatrix}, \tag{27.48}$$

where

$$\mathcal{F} \equiv e^{i\alpha} F + e^{-i\alpha} F'. \tag{27.49}$$

Note that, although there is only one CP-violating phase α, $\mathcal{F}$ is an arbitrary 1×3 complex matrix.

Let us describe how the NB criteria are realized in this simple extension of the SM. The fermions F are the SM ones; the fermions R are N_L and N_R. The scalar S, which has a complex VEV, contributes to the F–R mass term $V\mathcal{F}$, but it contributes neither to F–F nor to R–R mass terms. The VEV of ϕ, which breaks SU(2)⊗U(1), only contributes to the F–F mass terms $v\Gamma$. The discrete symmetry Z_2 prevents the appearance of couplings of the type $\overline{Q_L}\phi N_R$, thus leading to the zero column submatrix in eqn (27.48).

Because of the CP invariance imposed at the Lagrangian level, the Yukawa-coupling matrix Δ is real; hence, the up-type-quark mass matrix is real. Furthermore, since both the 3×3 matrix $v\Gamma$ and the 1×1 matrix μ are real, det $\mathcal{M}_n$ is real. Therefore,

$$(\theta_{\text{QFD}})_{\text{tree}} = 0. \tag{27.50}$$

Since the vanishing of $\bar{\theta}$ at tree level results from a symmetry of the Lagrangian, and not from the particular values assumed by the coupling constants, higher-order contributions to $\bar{\theta}$ are finite and calculable.

It is worth recalling that, in the limit $V, \mu \gg v$, this model is equivalent to the SM in what respects CP violation. This is, in fact, a general feature of models which satisfy the NB criteria.

27.7.1.1 Evaluation of $\bar{\theta}$

At loop level, the quark mass matrix M becomes $M + \Sigma$, where Σ originates in the quark self-energies:

$$\mathcal{L}^{(q)}_{\text{mass}} = \overline{q_L} \left(M + \Sigma \right) q_R + \overline{q_R} \left(M + \Sigma \right)^\dagger q_L. \tag{27.51}$$

The parameter θ_{QFD} then becomes

$$
\begin{aligned}
\theta_{\text{QFD}} &= \arg\left[\det\left(M + \Sigma\right)\right] \\
&\approx \arg\left\{ \left(\det M\right) \left[1 + \text{tr}\left(M^{-1}\Sigma\right)\right] \right\} \\
&= \arg\det M + \arg\left[1 + \text{tr}\left(M^{-1}\Sigma\right)\right] \\
&\approx \text{Im tr}\left(M^{-1}\Sigma\right).
\end{aligned}
\tag{27.52}
$$

where we have used the fact that Σ is small compared to M, and that det M is real.

The self-energies of the quarks receive, at one-loop level, two kinds of contributions: tadpoles, and diagrams in which the quark emits and then reabsorbs a boson. Tadpole diagrams are equivalent to a change in the VEVs of the scalar fields; however, the fact that the determinant of the mass matrix is real does not depend on the specific values of the VEVs. Hence, tadpole diagrams do not contribute to θ_{QFD}. Diagrams in which the quark emits and reabsorbs a $W^\pm$ do not contribute to the quark mass matrix, since the couplings of that boson are purely left-handed. Diagrams in which the quark emits and reabsorbs a photon yield $\Sigma \propto M$, because the couplings of the photon are flavour-conserving; they therefore do not contribute to $\theta_{\text{QFD}} \approx \text{Im tr}\left(M^{-1}\Sigma\right)$. The couplings of the Z have two components: one of them is flavour-conserving, and the other one is purely left-handed; as a consequence, diagrams in which quarks emit and reabsorb a Z do not contribute to θ_{QFD} either.

The only diagrams that can lead to a non-vanishing θ_{QFD} are those with a scalar in the loop. Bento et al. (1991) have evaluated those diagrams and concluded that

$$\bar{\theta} \sim \frac{FF^T - F'F'^T}{16\pi^2} \frac{v^2}{V^2} \sin 2\alpha \qquad (27.53)$$

is suppressed by the small ratio v/V.

27.7.2   A model with spontaneous P breaking

The strong CP problem is also a strong P problem, because $\mathcal{L}_\theta$ violates not only CP but also P. Thus, if we want to have $\theta_{\text{QCD}} = 0$, we may impose P symmetry at the Lagrangian level, instead of imposing CP symmetry. This possibility has been advocated by Babu and Mohapatra (1990) and by Barr et al. (1991); here we shall present a simple example suggested by Lavoura (1997).

The model has gauge group $SU(2)_L \otimes SU(2)_R \otimes U(1)$ (see Chapter 26). The electric-charge formula is $Q = T_{L3} + T_{R3} + Y$. Besides the quark doublets in eqns (26.2), in this model one also has n_g quarks N_L and n_g quarks N_R which are singlets of both $SU(2)_L$ and $SU(2)_R$ and have $Q = Y = -1/3$. Parity symmetry interchanges $Q_L \leftrightarrow Q_R$ and $N_L \leftrightarrow N_R$.

The scalar sector of the model consists of a bi-doublet ϕ as in eqn (26.12), together with a doublet χ_L of $SU(2)_L$ and a doublet χ_R of $SU(2)_R$; these doublets have $Y = -1/2$ so that their lower components are neutral. Under parity $\phi \to \phi^\dagger$ and $\chi_L \leftrightarrow \chi_R$.

One imposes a discrete symmetry Z_4, under which

$$
\begin{aligned}
\phi &\to i\phi, \\
\chi_L &\to -\chi_L, \\
\chi_R &\to i\chi_R, \\
Q_R &\to iQ_R, \\
N_R &\to -N_R,
\end{aligned}
\qquad (27.54)
$$

while Q_L and N_L remain invariant. Because of this discrete symmetry, and of parity, the Yukawa interactions of the quarks are given by

$$\mathcal{L}_Y^{(q)} = -\overline{Q_L}\tilde{\phi}\Delta Q_R - \overline{Q_R}\tilde{\phi}^\dagger \Delta Q_L - \left(\overline{Q_L}\chi_L G N_R + \overline{Q_R}\chi_R G N_L + \text{H.c.}\right). \quad (27.55)$$

The $n_g \times n_g$ matrix of Yukawa couplings Δ is Hermitian, while G is an unconstrained $n_g \times n_g$ matrix.

The most general scalar potential consistent with parity and with the discrete symmetry in eqns (27.54) has the property that it leads to the following VEVs:

$$\langle 0|\phi|0\rangle = \begin{pmatrix} 0 & 0 \\ 0 & k_2 \end{pmatrix},$$

$$\langle 0|\chi_L|0\rangle = \begin{pmatrix} 0 \\ v_L \end{pmatrix}, \tag{27.56}$$

$$\langle 0|\chi_L|0\rangle = \begin{pmatrix} 0 \\ v_R \end{pmatrix},$$

with k_2, v_L, and v_R simultaneously *real*. The VEV of the neutral field φ_1^0 (see eqn 26.12) is assumed to vanish. This is a self-consistent assumption, because there is no term in the vacuum potential linear in φ_1^0.

The mass terms of the quarks are

$$\mathcal{L}_{\text{mass}}^{(q)} = -\overline{p_L}k_2^*\Delta p_R - \overline{n_L}v_L G N_R - \overline{N_L}v_R^* G^\dagger n_R + \text{H.c.}. \tag{27.57}$$

As the VEVs are real and Δ is Hermitian, the determinant of the mass matrix is real, and $\theta_{\text{QFD}} = 0$ at tree level.

This model has the following interesting features (Lavoura 1997):

1. it does not have $W_L^\pm$–$W_R^\pm$ mixing;
2. it does not have flavour-changing neutral Yukawa interactions;
3. it does not have CP violation in the scalar sector: neutral scalars are either pure CP-even or pure CP-odd, they couple to fermions with a Dirac matrix which is either 1 or $i\gamma_5$;
4. the lepton sector may be arranged in such a way that neutrinos are exactly massless.

In this model, because of the absence of CP violation in the scalar sector, the neutral scalars do not generate a complex Σ. CP violation lies in the complex mixing matrix (CKM matrix) of the charged gauge interactions, just as in the SM. A non-zero θ_{QFD} is generated only at three-loop level, and it is expected to be no larger than $\sim 10^{-16}$ (Lavoura 1997), and thus the model provides a possible solution to the strong CP problem.

27.8 Appraisal

The strong CP problem is far from being fully understood, and it is often ignored by present-day research on CP violation. Yet, the strong CP problem may give us a clue to the origin of CP violation. At present, most of the proposed solutions to the strong CP problem belong to one of the following categories:

- A Peccei–Quinn symmetry, giving rise to a (visible or 'invisible') axion.
- CP invariance is imposed at the Lagrangian level (guaranteeing $\theta_{\text{QCD}} = 0$) together with an extra symmetry which leads to the natural vanishing of $\theta_{\text{QFD}} = \arg \det (M_p M_n)$, after spontaneous breaking of CP and of the gauge symmetry.
- The same as before, with P symmetry instead of CP symmetry.

The first kind of solution leads to $\bar\theta$ being dynamically driven to zero; the other solutions lead to a calculable $\bar\theta = \theta_{\text{QFD}} \approx \text{Im tr} (M^{-1}\Sigma)$, which may arise either at one-loop level or, preferably, at higher order in the perturbative expansion.

It seems clear that all solutions proposed so far are somewhat *ad hoc* and have shortcomings. The Peccei–Quinn solution has the virtue of leading to the prediction of a new particle, the axion. However, this virtue is also the source of difficulties, since no axions have been experimentally found up to now. The other solutions have the disadvantage of requiring the introduction of an additional symmetry which has no other rôle, apart from leading to the natural vanishing of $\bar{\theta}$ in tree approximation. One would like to have a solution to the strong CP problem which could shed some light into other open questions of the SM, such as the origin of CP violation, or the pattern of quark masses and mixings. No such solution exists at present.

Part IV

CP violation in B decays

INTRODUCTION

28.1   B decays as a testing ground for the SM

In this part of the book we apply the observables described in Part I to the study of CP violation in the B_d^0–$\bar{B}_d^0$ and B_s^0–$\bar{B}_s^0$ systems. These mesons, together with $B^+ = \bar{b}u$, can be found in Table 28.1. There is also experimental evidence for

Table 28.1 *Masses and lifetimes of the B mesons.*

Meson	Quark content	Mass (MeV)	Lifetime (10^{-12} s)
B^+	$\bar{b}u$	5278.9 ± 1.8	1.62 ± 0.06
B_d^0	$\bar{b}d$	5279.8 ± 1.6	1.56 ± 0.06
B_s^0	$\bar{b}s$	5369.6 ± 2.4	$1.61^{+0.10}_{-0.09}$

$B_c^+ = \bar{b}c$, but at present little is known about its properties. We shall sometimes use the notation B^0 to refer to either B_d^0 or B_s^0. When specifying the flavour of the light quark clarifies a particular expression, we shall use instead B_q^0, where q may be either d or s.

The resonance $\Upsilon(4S) = \bar{b}b$, with mass $10580.0 \pm 3.5\,\mathrm{MeV}$, width $21 \pm 4\,\mathrm{MeV}$, and parity and C-parity -1, is experimentally very important. It is copiously produced at the e^+e^- colliders called B factories, like those at SLAC and KEK. The branching ratios for the decays of $\Upsilon(4S)$ into B^+–B^- pairs and into B_d^0–$\bar{B}_d^0$ pairs are close to 50% each.

The most compelling reason to study the decays of B mesons is to learn more about the mechanism behind CP violation. In particular, we would like to test the Standard Model with three generations of quarks and leptons (SM). In the SM, CP violation is accommodated through a single irremovable complex phase present in the Cabibbo–Kobayashi–Maskawa (CKM) matrix V. This may be represented in a rephasing-invariant manner by $J \equiv \mathrm{Im}\,(V_{us}V_{cb}V_{ub}^*V_{cs}^*)$ (Jarl-skog 1985; Dunietz *et al.* 1985)—see eqn (13.23). The fact that there is only one independent CP-violating quantity in the SM means that this is a very predictive model. Thus far, the SM picture of CP violation has only been tested to the extent that it is consistent with the experimental value of the CP-violating parameter ϵ_K in the neutral-kaon system. Experiments with B decays will enable us to learn more about the CKM matrix.

Unfortunately, this is not a straightforward endeavour. First, there are experimental errors and uncertainties. Besides, the SM and other theoretical models are written in terms of *quarks*, while experiments are performed with *hadrons*. The relation between quarks and hadrons involves low-energy strong interactions and, as a consequence, the extraction of CKM parameters from experiment is generally plagued by theoretical uncertainties. Furthermore, $|J| \sim 3 \times 10^{-5}$ is very small. As a result, large CP-violating asymmetries should only be found in channels with small branching ratios; conversely, channels with large branching ratios are likely to display small CP-violating asymmetries. This fact drives the need for large statistics and, therefore, for experiments producing large numbers of B mesons—the B factories.

As a consequence of both the theoretical and the experimental uncertainties, one should try to determine the CKM matrix elements in various different ways. Our final aim should be to overconstrain the SM and, thus, test it.

In B physics in general, and in particular when discussing the B_d^0–$\overline{B_d^0}$ system, an important role is played by the unitarity triangle discussed in § 13.5. This triangle represents the unitarity equation

$$V_{ud}V_{ub}^* + V_{cd}V_{cb}^* + V_{td}V_{tb}^* = 0 \tag{28.1}$$

in the complex plane—see Fig. 13.1. The angles between the sides of the unitarity triangle are

$$
\begin{aligned}
\alpha &\equiv \arg\left(-V_{td}V_{ub}V_{tb}^*V_{ud}^*\right), \\
\beta &\equiv \arg\left(-V_{cd}V_{tb}V_{cb}^*V_{td}^*\right), \\
\gamma &\equiv \arg\left(-V_{ud}V_{cb}V_{ub}^*V_{cd}^*\right).
\end{aligned}
\tag{28.2}
$$

In § 18.5 we have summarized the constraints on the unitarity triangle. Those constraints follow from the available experimental information on the moduli of the CKM matrix elements, together with the SM fits of the mass difference in the B_d^0–$\overline{B_d^0}$ system and of ϵ_K in the K^0–$\overline{K^0}$ system. This results in[72]

$$
\begin{aligned}
-0.9 &< \sin 2\alpha, \\
0.4 &< \sin 2\beta < 0.9, \\
0.33 &< \sin^2 \gamma.
\end{aligned}
\tag{28.3}
$$

The main goal of experiments at B factories will be to measure the phases in eqns (28.2), thereby testing the consistency of the SM. Towards this end, a number of experiments will be performed looking for CP-violating asymmetries in the decays of neutral B mesons into CP eigenstates.

28.2   CP-violating asymmetries

28.2.1   The parameters λ_f

We have defined in Chapter 7 the parameters λ_f for the decays of two mixing neutral mesons P^0 and $\overline{P^0}$ into a final state f:

$$\lambda_f \equiv \frac{q}{p}\frac{\bar{A}_f}{A_f}, \tag{28.4}$$

---

[72]We recall that the bound on $\sin^2 \gamma$ is very sensitive to the input parameters and to the treatment of errors. We use the rather conservative bound derived by Ali and London (1997).

where $\bar{A}_f = A(\overline{P^0} \to f)$ and $A_f = A(P^0 \to f)$ are the decay amplitudes of $\overline{P^0}$ and P^0, respectively. The parameters q and p describe the transformation from the flavour basis into the basis of the eigenstates of evolution:

$$
\begin{aligned}
|P_H\rangle &= p|P^0\rangle + q|\overline{P^0}\rangle, \\
|P_L\rangle &= p|P^0\rangle - q|\overline{P^0}\rangle.
\end{aligned}
\tag{28.5}
$$

In B decays one generally assumes, based on both experimental and theoretical arguments, that there is no CP violation in the mixing: $|q/p| = 1$. There are then two possible forms of CP violation: direct CP violation, and CP violation in the interference between the mixing and the decays. For decays into a CP eigenstate f, these two forms of CP violation are measured by $|\bar{A}_f| - |A_f|$ and by $\mathrm{Im}\,\lambda_f$, respectively. Most of the study of CP violation in B decays hinges on measuring these quantities, and interpreting the meaning of the measured values, for various CP-eigenstate final states f_{cp}.

A good gauge to evaluate our progress in the study of CP violation is to see whether we are able to disprove the superweak theory of Wolfenstein (1964). As described in § 7.3, the superweak theory asserts that there is no CP violation in the decay amplitudes. This has two consequences. Firstly, there is no direct CP violation: $|\bar{A}_f| = |A_f|$ when f is a CP eigenstate. Secondly, all parameters λ_f are equal up to their sign. Thus, if f and g are two CP eigenstates with CP-parities $\eta_f = \pm 1$ and $\eta_g = \pm 1$, respectively, then the superweak theory predicts $\lambda_f = \eta_f \eta_g \lambda_g$.

28.2.2 *Definition of the CP asymmetries*

We have derived in Chapter 9 the basic formulae for the decays of tagged mesons P^0 and $\overline{P^0}$ into final states f which are common to both mesons. In this part of the book we shall usually work under the approximations $\Delta\Gamma = 0$ and $|q/p| = 1$.[73] Using these approximations we get

$$
\Gamma[B^0(t) \to f] = \frac{|A_f|^2 e^{-\Gamma t}}{2} \left[1 + |\lambda_f|^2 + \left(1 - |\lambda_f|^2\right) \cos \Delta mt + 2\mathrm{Im}\,\lambda_f \sin \Delta mt \right],
$$

$$
\Gamma[\overline{B^0}(t) \to f] = \frac{|A_f|^2 e^{-\Gamma t}}{2} \left[1 + |\lambda_f|^2 - \left(1 - |\lambda_f|^2\right) \cos \Delta mt - 2\mathrm{Im}\,\lambda_f \sin \Delta mt \right].
\tag{28.6}
$$

For the time-integrated decay rates one has

$$
\Gamma[B^0 \to f] = \frac{|A_f|^2}{2\Gamma} \left[1 + |\lambda_f|^2 + \frac{1 - |\lambda_f|^2 + 2x\mathrm{Im}\,\lambda_f}{1 + x^2} \right],
$$

$$
\Gamma[\overline{B^0} \to f] = \frac{|A_f|^2}{2\Gamma} \left[1 + |\lambda_f|^2 - \frac{1 - |\lambda_f|^2 + 2x\mathrm{Im}\,\lambda_f}{1 + x^2} \right].
\tag{28.7}
$$

We can see from these expressions that the parameters λ_f are observables. They are crucial in the study of CP violation in neutral-meson systems.

---

[73]The general formulae, with no approximations, are given in eqns (9.9) and (9.3), for the time-dependent decay rates, and in eqns (9.10) and (9.4), for the time-integrated decay rates.

Most articles in the literature refer to the CP-violating asymmetry built from tagged decays into CP eigenstates f. Using the decay rates in eqns (28.6) we find

$$A_{CP}(t) \equiv \frac{\Gamma[B^0(t) \to f] - \Gamma[\overline{B^0}(t) \to f]}{\Gamma[B^0(t) \to f] + \Gamma[\overline{B^0}(t) \to f]}$$
$$= a^{\text{dir}} \cos \Delta mt + a^{\text{int}} \sin \Delta mt, \qquad (28.8)$$

where we define

$$a^{\text{dir}} = \frac{1 - |\lambda_f|^2}{1 + |\lambda_f|^2}, \qquad (28.9)$$

$$a^{\text{int}} = \frac{2\text{Im}\,\lambda_f}{1 + |\lambda_f|^2}. \qquad (28.10)$$

Similarly, from eqns (28.7) one derives

$$A_{CP} \equiv \frac{\Gamma[B^0 \to f] - \Gamma[\overline{B^0} \to f]}{\Gamma[B^0 \to f] + \Gamma[\overline{B^0} \to f]}$$
$$= \frac{a^{\text{dir}} + x a^{\text{int}}}{1 + x^2}. \qquad (28.11)$$

The asymmetries $A_{CP}(t)$ and A_{CP} violate CP when the final state f is an eigenstate of CP, which we denote f_{cp}: indeed, $a^{\text{int}} \propto \text{Im}\,\lambda_f$ measures interference CP violation, and $a^{\text{dir}} \propto 1 - |\lambda_f|^2 \propto |A_f|^2 - |\bar{A}_f|^2$ (remember that we are assuming $|q/p| = 1$) measures direct CP violation.

28.3 Decays dominated by a single weak phase

Let us consider the decays of B^0 and $\overline{B^0}$ into a CP eigenstate f, and assume that they are dominated by a single weak phase ϕ_A:

$$\begin{aligned} A_f &= A e^{i(\phi_A + \delta)}, \\ \bar{A}_f &= \eta_f A e^{i(-\phi_A + \delta)}. \end{aligned} \qquad (28.12)$$

We have used the CP symmetry of the strong interactions to go from the first to the second equation; in this chapter we neglect spurious phases in the CP transformation, which will however be carefully taken into account in the following chapters. The weak phase ϕ_A can usually be determined directly from the Lagrangian. In the SM, it is the phase of a certain combination of CKM matrix elements. On the other hand, the computation of the modulus A and of the strong phase δ of the decay amplitudes is usually plagued by uncertain or unknown hadronic matrix elements.

Since $|q/p|=1$, we may write $q/p = -e^{2i\phi_M}$. As a result,

$$\lambda_f = -e^{2i\phi_M} \frac{\eta_f A e^{i(-\phi_A + \delta)}}{A e^{i(\phi_A + \delta)}}$$
$$= -\eta_f e^{-2i(\phi_A - \phi_M)}. \qquad (28.13)$$

We conclude that, under these conditions, λ_f is a pure phase. This occurs because we may relate the numerator $\bar{A}_f$ and the denominator A_f via the CP symmetry.

Due to the presence of a single weak phase ϕ_A, both the magnitude A and the strong phase δ of the decay amplitudes cancel out in the ratio. This is welcome, because these quantities suffer from hadronic uncertainties. In λ_f only the weak phase $2\,(\phi_A - \phi_M)$ remains. Notice that ϕ_A and ϕ_M are not separately rephasing-invariant; however, their difference $\phi_A - \phi_M$ is rephasing-invariant, and it can be measured, as shown by eqns (28.6) and (28.13).

When eqn (28.13) holds, there is no direct CP violation, because $|\lambda_f| = 1$ and then $a^{\mathrm{dir}} = 0$. Moreover, the interference CP violation is given by $\mathrm{Im}\,\lambda_f = \eta_f \sin 2\,(\phi_A - \phi_M)$. As a result, the tagged, time-dependent CP asymmetry in eqn (28.8) reduces to

$$A_{CP}(t) = \eta_f \sin 2\,(\phi_A - \phi_M) \sin \Delta m t. \qquad (28.14)$$

This is the Holy Grail of CP-violation measurements in B decays: when the decays into a CP eigenstate are dominated by a single weak phase, the CP asymmetry measures directly a weak phase in the Lagrangian. For certain decays in the SM, that phase is related to α, β, and γ.

It is important to notice the oscillatory dependence on the *sine* of $\Delta m t$ of the CP asymmetry in eqn (28.14). As seen in eqn (28.8), this happens because direct CP violation, represented by a^{dir}, vanishes when the decays are dominated by a single weak phase. In the presence of direct CP violation there is another oscillatory term, but now involving the *cosine* of $\Delta m t$.

It is also important to stress the advantages of working with final states which are CP eigenstates. If $\bar{f} \neq f$, we must compare the decays into f and into $\bar{f}$ in order to study CP violation, and the experimental task becomes more demanding.

28.4   Penguin pollution

Thus far, CP violation has only been detected in the quark sector, and we are unavoidably confronted with strong interactions. They bring with them several difficulties to be faced:

- Besides tree-level diagrams there are also gluonic penguins. These diagrams complicate the analysis because, typically, they carry a weak phase which differs from the weak phase of the tree-level diagrams. (Electroweak penguins also play a crucial role in certain decays in which the tree-level diagrams are very much suppressed.)

- Though we can compute reliably Feynman diagrams and effective Hamiltonians incorporating short-distance effects, the hadronic matrix elements are non-perturbative and they are not known to the desirable precision. This is responsible for the large errors in the determination of the sides of the unitarity triangle and, unless those matrix elements cancel out—as happens in eqn (28.13)—they also produce errors in the interpretation of CP asymmetries.

- Strong phases are induced by the final-state interactions (FSI). At least two such phases are needed if there is to be direct CP violation in a decay channel. On the other hand, the FSI phases obscure the interpretation of the interference CP violation measured in a^{int}.

We now discuss the impact that these effects have on the measurement of weak phases with CP asymmetries.

In § 28.3 we have shown that, if there is only one weak phase contributing to the decay amplitudes, then the CP asymmetry measures that weak phase. In general, the presence of another diagram with a different weak phase destroys that simple result. In most cases, this is due to the presence of penguin diagrams, in which case this effect is referred to as *penguin pollution*.

When there are two weak phases contributing to the decay amplitudes, the latter may be written as[74]

$$
\begin{aligned}
A_f &= A_1 e^{i(\phi_{A1}+\delta_1)} + A_2 e^{i(\phi_{A2}+\delta_2)}, \\
\bar{A}_f &= \eta_f \left[A_1 e^{i(-\phi_{A1}+\delta_1)} + A_2 e^{i(-\phi_{A2}+\delta_2)} \right].
\end{aligned}
\tag{28.15}
$$

The real numbers A_1 and A_2 are the moduli of the interfering amplitudes. The weak phases ϕ_{A1}, ϕ_{A2}, and ϕ_M are not rephasing-invariant. On the other hand, the differences $\phi_1 \equiv \phi_{A1} - \phi_M$, $\phi_2 \equiv \phi_{A2} - \phi_M$, and $\Delta \equiv \delta_2 - \delta_1$ can be measured.

Let us see how the observables in eqns (28.9) and (28.10) are related to the weak and strong phases. In the approximation where $r = A_2/A_1$ is small, we find

$$
\begin{aligned}
\lambda_f &= -\eta_f e^{-2i\phi_1} \frac{1 + r e^{i(\Delta-\phi_2+\phi_1)}}{1 + r e^{i(\Delta+\phi_2-\phi_1)}} \\
&\approx -\eta_f e^{-2i\phi_1} \left[1 + 2r \sin\Delta \sin(\phi_2 - \phi_1) - 2ir \cos\Delta \sin(\phi_2 - \phi_1) \right].
\end{aligned}
\tag{28.16}
$$

Then,

$$
a^{\text{dir}} \approx 2r \sin(\phi_1 - \phi_2) \sin\Delta,
\tag{28.17}
$$
$$
a^{\text{int}} \approx \eta_f \left[\sin 2\phi_1 - 2r \cos 2\phi_1 \sin(\phi_1 - \phi_2) \cos\Delta \right].
\tag{28.18}
$$

When the decays are dominated by a single weak phase, i.e., when either $r \approx 0$ or $\phi_2 \approx \phi_1$, there is no direct CP violation and a^{int} measures a single weak phase ϕ_1. This is the situation discussed in § 28.3. However, it is a very particular situation; in general, a^{int} does not measure a single weak phase.

This problem is not rooted in the presence of direct CP violation. Indeed, if we assume the FSI to be vanishingly small, we have $\Delta = 0$ and there is no direct CP violation—$a^{\text{dir}} = 0$. The parameter λ_f is then a pure phase but, still, a^{int} does not measure a single weak phase (Gronau 1993). Indeed, when $\Delta = 0$ we can write

$$
\lambda_f = -\eta_f e^{-2i\left(\phi_1 - \delta_{\phi_1}\right)},
\tag{28.19}
$$

---

[74]In the SM, the decays $b \to k$ (where k may be either d or s) may always be brought to the form in eqns (28.15), with only two interfering amplitudes. This occurs because there are three relevant combinations of CKM matrix elements—$V_{ub}V_{uk}^*$, $V_{cb}V_{ck}^*$, and $V_{tb}V_{tk}^*$—but their sum is zero because of the unitarity of the CKM matrix. As a consequence, there are only two independent weak phases.

where δ_{ϕ_1} is defined by

$$\tan \delta_{\phi_1} = \frac{r \sin (\phi_1 - \phi_2)}{1 + r \cos (\phi_1 - \phi_2)}. \tag{28.20}$$

Therefore,

$$a^{\text{int}} = \eta_f \sin 2 (\phi_1 - \delta_{\phi_1}). \tag{28.21}$$

The presence of a second amplitude with a different weak phase, $\phi_2 \neq \phi_1$, may spoil the measurement of $\sin 2\phi_1$, even if the second amplitude has the same strong phase as the first one. This occurs even for moderate values of r (Gronau 1993).

28.5 The four phases in the CKM matrix

We shall allow for new phases to be present in B^0–$\overline{B^0}$ mixing, see § 24.8.2 and 30.4.2. On the other hand, we shall in general *assume* that the decay amplitudes are given by SM diagrams. For decays that occur through unsuppressed SM tree-level diagrams, it is likely that no new physics can occur at a competing level, since that new physics should have been detected elsewhere. The situation is of course different for decays that are strongly suppressed in the SM.

We shall also allow for a possible non-unitarity of the CKM matrix. If the CKM matrix is not unitary, eqn (28.1) does not hold; the usual relations between the angles α, β, and γ and the sides of the unitarity triangle are destroyed, and the bounds in eqn (28.3) cease to be valid. Nevertheless, if the decay amplitudes are dominated by SM diagrams, as we assume, then the weak phases in those decay amplitudes will still be controlled by the phases in the CKM matrix.

The three phases introduced in eqn (28.2) are not independent; they satisfy by definition

$$\alpha + \beta + \gamma = \arg (-1) = \pi \pmod{2\pi}. \tag{28.22}$$

It is important to stress that using these three phases is redundant, because they are linearly dependent. In general, we shall take β and γ to be the fundamental phases, and we shall consider α as just a linear combination of β and γ.

In any model, as in the SM, in which eqn (28.1) holds, the phases α, β, and γ may be geometrically pictured as the angles between the sides of the unitarity triangle. Then, $\alpha + \beta + \gamma$ may be either π or 5π, but not 3π (Grossman *et al.* 1997a). Indeed, if the angles are interior to the triangle, then they lie in the range $[0, \pi]$ and add up to π; if the angles are exterior to the triangle then they lie in the range $[\pi, 2\pi]$ and add up to 5π. This is a restriction on eqn (28.22).

Aleksan *et al.* (1994) have introduced two further phases, ϵ and ϵ':[75]

$$\epsilon \equiv \arg (-V_{cb} V_{ts} V_{cs}^* V_{tb}^*),$$
$$\epsilon' \equiv \arg (-V_{us} V_{cd} V_{ud}^* V_{cs}^*). \tag{28.23}$$

They have shown—see § 16.3.3—that, in the SM, one can parametrize the 3×3 unitary CKM matrix V, moduli and phases alike, with only four phases β, γ, ϵ,

---

[75] In spite of the identical notation, these phases have nothing to do with the parameters ϵ_K and ϵ'_K measuring CP violation in the two-pion decays of the neutral kaons.

and ϵ' (Aleksan *et al.* 1994). These four phases are useful even in the presence of new physics. Indeed, it is easy to see that β, γ, ϵ, *and* ϵ' *parametrize all the phases in the usual* 3×3 *submatrix of the generalized CKM matrix.* This happens because the 3×3 submatrix of the generalized CKM matrix has nine phases. By rephasing the six quarks it is possible to redefine away five phases. Therefore, there are only four physical phases in the submatrix. We may choose a phase convention such that its phase structure is given by

$$\arg V \sim \begin{pmatrix} 0 & \epsilon' & -\gamma \\ \pi & 0 & 0 \\ -\beta & \pi + \epsilon & 0 \end{pmatrix}. \qquad (28.24)$$

The choice of phases in eqn (28.24) is useful in order to identify rapidly the phase of any rephasing-invariant combination of CKM matrix elements.

28.5.1 *SM values of* ϵ *and* ϵ'

We have seen in § 18.5 that, in the SM, the angle β is quite well determined: $10° < \beta < 30°$. The angle γ should lie between $35°$ and $135°$. Finally, α is expected to lie between $35°$ and $120°$. These values correspond to the bounds in eqns (28.3).

Thus, α, β, and γ are in principle large angles. On the other hand, as Aleksan *et al.* (1994) pointed out, within the three-generation SM ϵ and ϵ' must be very small: $\epsilon \lesssim \lambda^2 \sim 0.05$ and $\epsilon' \lesssim \lambda^4 \sim 0.0025$. This can easily be checked by considering the Wolfenstein parametrization of the CKM matrix, and in particular eqns (16.27) and (16.29). One sees that

$$\epsilon = \arg\left[1 - \lambda^2 \left(\tfrac{1}{2} - \rho - i\eta\right) + O\left(\lambda^4\right)\right]$$
$$\approx \lambda^2 \eta, \qquad (28.25)$$
$$\epsilon' = \arg\left[1 - A^2 \lambda^4 \left(\tfrac{1}{2} - \rho - i\eta\right) + O\left(\lambda^6\right)\right]$$
$$\approx A^2 \lambda^4 \eta. \qquad (28.26)$$

Notice that $\eta = 0.33 \pm 0.10$—see § 18.5—entails a further suppression of the values of ϵ and ϵ', beyond the one given by the powers of $\lambda = 0.22$.

We conclude that, in the SM, the CKM matrix contains only two independent large phases. Once β is measured, only one large phase—say, γ—remains to be determined. From our point of view, measuring α is just a different way of measuring γ.

28.5.2 *Smallness of* ϵ' *in models of new physics*

When there is new physics the bounds on α, β, and γ given in § 28.5.1 are relaxed. Also, ϵ can in principle be large. On the other hand, in most models of new physics the 3×3 CKM matrix is a submatrix of a larger $n \times n$ unitary matrix $\hat{V}$; this holds, in particular, in the SM with more than three generations, as well as in a model with vector-like isosinglet or isodoublet quarks, or with mirror quarks. In those models, one may use the fact that, experimentally, $|V_{ud}|$

and $|V_{cs}|$ are very close to one, together with the orthonormality of the first and second columns of $\hat{V}$, to prove that ϵ' will still be rather small. One derives

$$\left|\hat{V}_{11}\hat{V}_{12}^* + \hat{V}_{21}\hat{V}_{22}^*\right|^2 = \left|\sum_{\alpha=3}^{n}\hat{V}_{\alpha 1}\hat{V}_{\alpha 2}^*\right|^2$$

$$\leq \sum_{\alpha=3}^{n}\left|\hat{V}_{\alpha 1}\right|^2 \sum_{\beta=3}^{n}\left|\hat{V}_{\beta 2}\right|^2$$

$$= \left(1 - \left|\hat{V}_{11}\right|^2 - \left|\hat{V}_{21}\right|^2\right)\left(1 - \left|\hat{V}_{12}\right|^2 - \left|\hat{V}_{22}\right|^2\right). \quad (28.27)$$

As a consequence,

$$\cos\epsilon' \geq \frac{-1 + |V_{ud}|^2 + |V_{us}|^2 + |V_{cd}|^2 + |V_{cs}|^2 - |V_{ud}V_{cs}|^2 - |V_{us}V_{cd}|^2}{2|V_{ud}V_{us}V_{cd}V_{cs}|}$$

$$\geq 0.983; \quad (28.28)$$

in the last step we have used the lower bounds from Chapter 15,

$$\begin{aligned}|V_{ud}| &\geq 0.9726, \\ |V_{us}| &\geq 0.2187, \\ |V_{cd}| &\geq 0.208, \\ |V_{cs}| &\geq 0.83.\end{aligned} \quad (28.29)$$

Thus, $|\epsilon'| < 0.2$ (Kurimoto and Tomita 1997). This is a much poorer bound than in the SM, where $|\epsilon'|$ is two orders of magnitude smaller, but it holds in most models of new physics. Only rather contrived models, which change the normalization of the CKM matrix, might avoid this conclusion; it is difficult to imagine such models which pass all the experimental constraints. In particular, a large ϵ' may lead to problems with the CP-violating parameter ϵ_K of the neutral-kaon system: this is because $\Gamma_{21} \propto \lambda_u^2$, while M_{21} has pieces proportional to λ_u^2, $\lambda_u\lambda_c$, and λ_c^2, where $\lambda_\alpha \equiv V_{us}^*V_{ud}$. If ϵ' is large then the phase difference between λ_u and λ_c is large, and the phase of M_{21} may turn out too different from the one of Γ_{21}, destroying the fit of ϵ_K in the SM.

28.6   Outline of Part IV

Several important issues arise when one considers the precise conditions to be faced in a real experiment. Extracting information on λ_f is not as simple as the discussions in Chapter 9 might lead us to believe. In general, there are several meson configurations in the initial state; tagging the initial flavour of a decaying meson may not be trivial; and the ability to trace the time dependence of the decays is limited by the capabilities of vertex reconstruction. Some of these problems are discussed briefly in Chapter 29.

In order to compute the rephasing-invariant parameters λ_f one needs to compute the phase-convention-dependent quantities q/p and $\bar{A}_f/A_f$ separately. Each of these two quantities depends on the CP-transformation phases for the quark fields and for the hadron state vectors; in their product, however, those spurious phases cancel out. The computation of q/p for the neutral-meson systems K^0–$\overline{K^0}$, B_d^0–$\overline{B_d^0}$, and B_s^0–$\overline{B_s^0}$ is the subject of Chapter 30. We show that, in all these cases, $|q/p|$ is very close to unity.[76] This is done using only phenomenological arguments and the known experimental values. We then proceed to find out the phase of q/p in the SM; in other models, we parametrize the phase of q/p in the B_q^0–$\overline{B_q^0}$ system by means of an extra phase θ_q.

The decay amplitudes are the subject of Chapters 31 and 32. We discuss the decay amplitudes in the SM from a diagrammatic point of view in Chapter 31. A more formal presentation, using effective Hamiltonians, is left to Chapter 32, which however is inessential for most of the rest of the book. The elementary quark-decay-diagram analysis of Chapter 31 is most convenient in order to identify the weak phases present in each amplitude.

Throughout Part IV, we assume that *the strong-interaction rescatterings among the various final states of the weak decays are negligible* or, at least, they do not affect the result of a given calculation. This is a central assumption in most treatments of CP violation in the B^0–$\overline{B^0}$ systems; indeed, no one really knows how to treat the FSI, both because of their non-perturbative nature and because they may mix many different final states. Another restriction in this book is that we concentrate on decays into two-meson final states. These are the easiest final states to describe, and the ones most commonly treated in the literature. We shall concentrate on decays into final states that are either flavour-specific or CP eigenstates. We also consider some decays into non-CP eigenstates. We mention only briefly other possibilities like inclusive decays, semi-inclusive decays, and decays into states where an angular decomposition permits the distinction between CP-even and CP-odd components. These have been summarized by Dunietz (1994), where references to the original literature can be found.

We emphasize that, for the methods analysed in this book, we assume that the strong-interaction final-state rescatterings do not mix decays which occur through different quark processes. This allows us to estimate the relative contributions of the different quark-level diagrams, as is done by most authors. Although useful, such estimates should not be taken too seriously. Rescattering effects may alter them—see, for example, Blok *et al.* (1997), Ciuchini *et al.* (1997a), Gérard and Weyers (1997), and Neubert (1997).

Chapters 29–32 contain detailed discussions which may be skipped in a cursory reading. Readers eager to get a quick acquaintance with CP violation in B decays may want to proceeed to Chapter 33 immediately, and refer to earlier chapters as the need arises. Chapter 33 is the crucial one of Part IV. In it, we put together the results for q/p and $\bar{A}_f/A_f$ and compute the parameters

---

[76]In the neutral-kaon system the observed CP violation is essentially mixing CP violation, i.e., $1 - |q_K/p_K| \approx \delta_K \approx 3.3 \times 10^{-3}$. However, the CP asymmetries in the B^0–$\overline{B^0}$ systems are often expected to be of order 1; the effect on them of $\delta_K \sim 10^{-3}$ is small and may be neglected. For this reason, here we take $|q_K/p_K|$ to be 1.

λ_f. In particular, in § 33.1 we assume that only the tree-level SM amplitudes contribute to the decays, and show that the CP asymmetries in $B_d^0 \to \pi^+\pi^-$, $B_d^0 \to J/\psi K_S$, and $B_s^0 \to \rho^0 K_S$ measure $\sin 2\alpha$, $\sin 2\beta$, and $\sin 2\gamma$, respectively. Unfortunately, the penguin pollution spoils the theoretical interpretation of the first and third asymmetries as $\sin 2\alpha$ and $\sin 2\gamma$, respectively. On the other hand, the CP asymmetry in the decay $B_d^0 \to J/\psi K_S$ measures $\sin 2\beta$ almost without hadronic uncertainties.

In Chapter 34 we study decay chains that involve an intermediate neutral-meson system. As shown in Appendix E, the mixing in the D^0–$\overline{D^0}$ system is very small in the SM. Therefore, we assume that this mixing vanishes when we consider decay chains which include D^0 and/or $\overline{D^0}$ at an intermediate stage. This restriction will be lifted in § 34.6, where we discuss the most general cascade decay chain. That decay chain requires the introduction of new CP-violating parameters, beyond the ones used in most of this book.

Several methods have been devised to overcome the problem of penguin pollution. Decays of the type $B \to D \to f$ have been used by some authors to gain access to the phase γ. Other possibilities to determine weak phases involve the use of isospin or SU(3)-flavour symmetries in order to relate different decay channels. Methods using the decays of the B_d^0–$\overline{B_d^0}$ system are treated in Chapters 35 and 36; those using the decays of the B_s^0–$\overline{B_s^0}$ system are presented in Chapter 37.

All these methods give access to trigonometric functions of the weak phases, rather than to the phases themselves. The extraction of the values of phases from the values of trigonometric functions thereof suffers from discrete ambiguities. In Chapter 38 we discuss ways to resolve these ambiguities.

Chapters 35–38 include mostly relatively recent contributions by a variety of authors. Many proposals must still be tested in order to check their feasibility. As a result, the subjects covered in those chapters are likely to evolve more rapidly than those in the rest of the book.

29

SOME EXPERIMENTAL ISSUES

29.1 Introduction

Phenomenologically, the study of the mixing and CP-violation observables involves three steps:

1. preparation of the initial state;
2. time evolution;
3. detection of the final states.

In the subsequent chapters we shall focus almost entirely on questions related with the second step. However, it is important to appreciate the impact that the form of the initial state and the detection capabilities have on the CP-violation experiments. For example, one usually refers to 'tagging' as if that were a clear and straightforward procedure. But, as we have seen in Chapter 10, the presence of several b-hadrons in the initial state complicates the analysis considerably. Similarly, the need to follow the time dependence of the decays is subject to experimental constraints. Indeed, the technological limits on vertex reconstruction mean that we can only measure time intervals larger than a certain amount. This has a decisive impact on the need to build asymmetric B factories. In this chapter we make some elementary remarks about these experimental issues.

29.2 Preparation of the initial state

Since the strong and electromagnetic interactions preserve flavour, the experiments always produce a $\bar{b}$ antiquark in association with a b quark. Hence, there are always two b-hadrons of opposite beauty flavour. The various possibilities have been discussed in Chapter 9. They can be represented schematically as (Bigi and Sanda 1981, 1987)

A) $B^+ B^-$,
B) $B_q^0 B^-$, $\overline{B_q^0} B^+$, $B_q^0 \Lambda_b$, $\overline{B_q^0}\, \overline{\Lambda_b}$,
C) $B_q^0 \overline{B_q^0}$,
D) $B_d^0 \overline{B_s^0}$, $B_s^0 \overline{B_d^0}$,

where we have only shown the two b-hadrons involved. These will be produced in connection with other particles, in such a way as to preserve charge and flavour.

The problem is that these different states evolve differently: only the neutral mesons mix and, in case C), the two neutral mesons can even be correlated. The analysis is complicated whenever there are several of these initial states in a given experimental sample.

29.2.1   *Experiments at the $\Upsilon(4S)$*

The CLEO, Babar, and Belle experiments start with the production of $\Upsilon(4S)$ from e^+e^- collisions, which then decays into a mixture of A) and C).

The $\Upsilon(4S)$ decays about 50% of the times into B_d^0–$\overline{B_d^0}$ pairs. The resulting $B_d^0\overline{B_d^0}$ state has parity and C-parity -1, and may be written

$$|\Phi^-\rangle = \tfrac{1}{\sqrt{2}}\left[|B_d^0(\vec{k})\rangle \otimes |\overline{B_d^0}(-\vec{k})\rangle - |\overline{B_d^0}(\vec{k})\rangle \otimes |B_d^0(-\vec{k})\rangle\right], \qquad (29.1)$$

where $\vec{k}$ and $-\vec{k}$ are the three-momenta of the left-moving and right-moving meson, respectively. The meson with momentum $\vec{k}$ decays at time t_1 into the final state f, and the meson with momentum $-\vec{k}$ decays at time t_2 into the final state g. The amplitude for this process is given by $\langle f, t_1; g, t_2|T|\Phi^-\rangle$, as studied in § 9.7.

The branching ratio of the decay of $\Upsilon(4S)$ into B^+B^- is also close to 50%. In this case, there is no mixing and all we can test is direct CP violation. Still, the presence of B^+B^- in the experimental sample may render the tagging of B_d^0 and $\overline{B_d^0}$ difficult. As illustrated in § 10.3, this difficulty is present when one studies observables involving the semileptonic decay of both mesons. There, the relative production fraction of B^+–B^- and B_d^0–$\overline{B_d^0}$ pairs must be known before one can extract information about the mixing of neutral mesons from the experimental results.

29.2.2   *Other experiments*

Another important class of experiments takes place at $p\bar{p}$ collisions, Z^0 decays, and on the e^+e^- continuum. This includes studies made at LEP, LHC, Tevatron, as well as the HERA-B experiment. Generically, in these experiments one produces a combination of all the particle contents in A), B), C), and D). Tagging is then rather involved.

Here, the B^0–$\overline{B^0}$ pairs of case C) are no longer correlated. They are produced in an uncorrelated initial state, Φ^u. There is an equal probability that at the initial instant the left-moving meson was a B^0 and the right-moving meson was a $\overline{B^0}$, or vice versa, and the two possibilities are incoherently superposed—see § 9.9.

29.3   Tagged decays

29.3.1   *Tagging one meson with semileptonic decays*

Our aim is to study the observables involved in the tagged decays of eqns (28.6) and (28.7). In order to measure those observables one uses the fact that a b quark is always produced in association with a $\bar{b}$ antiquark. The flavour-specific decay of one meson is used as a tag, and one looks for the decay into a CP eigenstate of the other neutral meson.[77]

---

[77] In specific environments, other tagging strategies are possible, such as same-side jet charge, opposite-side jet charge, and kaon tag. In particular, the same-side jet charge tagging strategy,

One can use the semileptonic decay of the $b/\bar{b}$ to tag the flavour of one B meson. We denote by l^+ a state of the type $l^+\nu_l X$, and by l^- a state of the type $l^-\bar{\nu}_l X$. The leptons may be identified as coming from the decay of a $b/\bar{b}$ with the aid of some high-p_T cut. The sign of the charge of the charged lepton in the final state is equal to the sign of the charge of the decaying b or $\bar{b}$. In general, one assumes that there is no direct CP violation in these semileptonic decays, as predicted by the SM. As a result, one substitutes $|\langle X^- l^+ \nu_l |T| B^0 \rangle|$ and $|\langle X^+ l^- \bar{\nu}_l |T| \overline{B^0} \rangle|$ by a common value $|A_l|$.

We stress that detecting only the charged lepton does identify the flavour of the $b/\bar{b}$, but not the meson in which this quark is contained. For example, a positively charged electron may come from either B_d^0, B_s^0, or B^+, since they all contain $\bar{b}$. If there is no other information about the final state, then we must know the production fractions of the various initial states that might be present in a given experiment. In this respect, studies performed at the $\Upsilon(4S)$ in which one tags the flavour in one side of the detector and also reconstructs the final state f in the other side, are much cleaner than others. The reason is that the charge of f tells us whether we initially had a $B^+ B^-$ pair or a $B_d^0 \overline{B_d^0}$ pair.

29.3.2 *Tagging neutral mesons*

Even after we allow for the several initial meson configurations present in a given experiment, we still have to take into account the fact that, in most cases, detecting the flavour of the meson in one side of the detector does not identify the flavour of the neutral meson in the other side. Indeed, suppose that one tags the meson in one side through its semileptonic decay at time t_2; this identifies the flavour of that meson at that time. If the tagged meson is neutral, then we do not know its flavour at the time of the creation of the $b\bar{b}$ pair. This occurs because the tagged neutral meson might have oscillated in between the time of its creation and the time of its decay. Therefore, we also do not know the initial flavour of the meson in the opposite side of the detector—the one that we wished to tag in order to follow its tagged, time-dependent decay rate.

There are only two situations in which the semileptonic decay of one meson really identifies the flavour of the other meson. The first situation is when the neutral B meson is produced together with a charged B meson (or with a b-baryon)—case B)—, and we look for the semileptonic decay of the latter. Here the tagging is effective because the charged meson does not oscillate. Strictly speaking, eqns (28.6) and (28.7) only apply to this case. The second situation is when we produce two neutral B mesons in an antisymmetric wave, as at the $\Upsilon(4S)$. Here, the tagging is due to the fact that two identical bosons can never be in an overall antisymmetric wave; determining the flavour of one neutral meson through its semileptonic decay at time t_2 ensures that the other neutral meson has the opposite flavour at t_2, and evolves thereafter as a tagged meson, with time variable $t_- = t_1 - t_2$. Thus,

which has been applied successfully at LEP and CDF, does not make use of the opposite side in the associated production.

$$|\langle f; l^-; t_- |T|\Upsilon(4S)\rangle|^2 = \frac{|A_l|^2}{4\Gamma}\Gamma[B^0(t_-) \to f],$$

$$|\langle f; l^+; t_- |T|\Upsilon(4S)\rangle|^2 = \frac{|A_l|^2}{4\Gamma}\Gamma[\overline{B^0}(t_-) \to f].$$

(29.2)

Here we have assumed that $t_- > 0$. The expressions for $t_- < 0$ are found by replacing t_- by $|t_-|$ in the exponentials in eqns (28.6)—see eqns (9.69) and (9.71).

In all other cases we must take 'mistags' into account, i.e., we must consider cases in which looking at a flavour-specific decay of one meson does not determine the flavour of the other meson. This is due to the oscillations of the meson before it decays into the flavour-specific final state. The resulting mistags lead to a dilution of the signal, which is related to the dilution factor $D_M \equiv (1 - y^2)/(1 + x^2) = 1 - 2\chi_0$, see Chapter 9.

29.4 CP asymmetries at the $\Upsilon(4S)$

We emphasize the conclusion of the previous subsection: for $t_- < 0$, the decay rate for a tagged decay originating in the $\Upsilon(4S)$ is *not* given simply by eqns (28.6). This has important implications for the CP asymmetries at the $\Upsilon(4S)$.

Using eqns (29.2), we see that the time-dependent asymmetry has the same form as in eqn (28.8),

$$
\begin{aligned}
A_{CP}^{\Upsilon(4S)}(t_-) &\equiv \frac{|\langle f; l^-; t_- |T|\Upsilon(4S)\rangle|^2 - |\langle f; l^+; t_- |T|\Upsilon(4S)\rangle|^2}{|\langle f; l^-; t_- |T|\Upsilon(4S)\rangle|^2 + |\langle f; l^+; t_- |T|\Upsilon(4S)\rangle|^2} \\
&= A_{CP}(t_-).
\end{aligned}
$$

(29.3)

However, one must remember that the interference-CP-violation terms appear multiplied by $e^{-\Gamma|t_-|}\sin \Delta m t_-$. This function is odd under $t_1 \leftrightarrow t_2$, and therefore it disappears when we integrate over time from $t_- = -\infty$ to $t_- = +\infty$. To see this explicitly, let us start again from eqns (29.2) and compute the time-integrated decay rates for $t_- > 0$ and for $t_- < 0$:

$$\int_0^{+\infty} dt_- \, |\langle f; l^-; t_- |T|\Upsilon(4S)\rangle|^2 = \frac{|A_l A_f|^2}{8\Gamma^2}\left[1 + |\lambda_f|^2 + \frac{1 - |\lambda_f|^2 + 2x\mathrm{Im}\,\lambda_f}{1 + x^2}\right],$$

$$\int_{-\infty}^0 dt_- \, |\langle f; l^-; t_- |T|\Upsilon(4S)\rangle|^2 = \frac{|A_l A_f|^2}{8\Gamma^2}\left[1 + |\lambda_f|^2 + \frac{1 - |\lambda_f|^2 - 2x\mathrm{Im}\,\lambda_f}{1 + x^2}\right],$$

$$\int_0^{+\infty} dt_- \, |\langle f; l^+; t_- |T|\Upsilon(4S)\rangle|^2 = \frac{|A_l A_f|^2}{8\Gamma^2}\left[1 + |\lambda_f|^2 - \frac{1 - |\lambda_f|^2 + 2x\mathrm{Im}\,\lambda_f}{1 + x^2}\right],$$

$$\int_{-\infty}^0 dt_- \, |\langle f; l^+; t_- |T|\Upsilon(4S)\rangle|^2 = \frac{|A_l A_f|^2}{8\Gamma^2}\left[1 + |\lambda_f|^2 - \frac{1 - |\lambda_f|^2 - 2x\mathrm{Im}\,\lambda_f}{1 + x^2}\right].$$

(29.4)

The terms $\mathrm{Im}\,\lambda_f$ disappear from the fully time-integrated decay rates:

$$|\langle f; l^{-}|T|\Upsilon(4S)\rangle|^{2} = \frac{|A_{l}A_{f}|^{2}}{4\Gamma^{2}}\left[1 + |\lambda_{f}|^{2} + \frac{1 - |\lambda_{f}|^{2}}{1 + x^{2}}\right],$$

$$|\langle f; l^{+}|T|\Upsilon(4S)\rangle|^{2} = \frac{|A_{l}A_{f}|^{2}}{4\Gamma^{2}}\left[1 + |\lambda_{f}|^{2} - \frac{1 - |\lambda_{f}|^{2}}{1 + x^{2}}\right].$$

$$(29.5)$$

Therefore, the fully time-integrated CP asymmetry is

$$A_{CP}^{\Upsilon(4S)} \equiv \frac{|\langle f; l^{-}|T|\Upsilon(4S)\rangle|^{2} - |\langle f; l^{+}|T|\Upsilon(4S)\rangle|^{2}}{|\langle f; l^{-}|T|\Upsilon(4S)\rangle|^{2} + |\langle f; l^{+}|T|\Upsilon(4S)\rangle|^{2}}$$

$$= \frac{a^{\mathrm{dir}}}{1 + x^{2}}. \qquad (29.6)$$

Comparing eqn (29.6) with eqn (28.11) we see that a^{int} is not present in the *tagged*, fully time-integrated asymmetries at the $\Upsilon(4S)$; these can only be used to measure direct CP violation (Deshpande and He 1996).

On the other hand, if we keep track of the time ordering of the decays, we may construct data samples in which one collects all the events with a given sign of t_{-}. These may then be used to construct the CP asymmetry

$$\frac{\int_{0}^{+\infty} dt_{-}\, |\langle f; l^{-}; t_{-}|T|\Upsilon(4S)\rangle|^{2} - \int_{0}^{+\infty} dt_{-}\, |\langle f; l^{+}; t_{-}|T|\Upsilon(4S)\rangle|^{2}}{\int_{0}^{+\infty} dt_{-}\, |\langle f; l^{-}; t_{-}|T|\Upsilon(4S)\rangle|^{2} + \int_{0}^{+\infty} dt_{-}\, |\langle f; l^{+}; t_{-}|T|\Upsilon(4S)\rangle|^{2}} = A_{CP}.$$

$$(29.7)$$

The quantity A_{CP} is the same as in eqn (28.11). Thus, if we retain the time-ordering, we avoid discarding the term that interests us most, i.e., a^{int}.

The result in eqn (29.6) does not mean that it is impossible to measure interference CP violation in fully time-integrated measurements at the $\Upsilon(4S)$; this no-go theorem is only valid for *tagged* decays. In fact, it has long been known that one can, in principle, detect interference CP violation in the decays into two CP eigenstates: $\Upsilon(4S) \rightarrow B_{d}^{0}\overline{B_{d}^{0}} \rightarrow f_{\mathrm{cp}}f_{\mathrm{cp}}$ (Wolfenstein 1984; Gavela *et al.* 1985*b*; Bigi and Sanda 1987). This is also true for time-integrated decays into two non-CP eigenstates (Silva 1998). However, the decays of B_{d}^{0} into CP eigenstates have branching ratios which are typically smaller than 10^{-3}. For example, $\mathrm{BR}[B_{d}^{0} \rightarrow J/\Psi K_{S}] = (3.7 \pm 1.1) \times 10^{-4}$ and $\mathrm{BR}[B_{d}^{0} \rightarrow \pi^{+}\pi^{-}] < 2.0 \times 10^{-5}$ (Particle Data Group 1996). The semileptonic decays have much larger branching ratios, *e.g.*, $\mathrm{BR}[B_{d}^{0} \rightarrow l^{+}\nu_{l}X] = (1.03 \pm 0.10) \times 10^{-1}$, and $\mathrm{BR}[B_{d}^{0} \rightarrow D^{-}l^{+}\nu_{l}] = (1.9 \pm 0.5) \times 10^{-2}$ (Particle Data Group 1996). Therefore, the decays into $f_{\mathrm{cp}}f_{\mathrm{cp}}$ have branching ratios that are, typically, orders of magnitude smaller than the tagged decays of the type $l^{\pm}f_{\mathrm{cp}}$. This is why one focuses on tagged decays.

29.5 Consequences of the limits on vertexing

29.5.1 *The need for asymmetric $\Upsilon(4S)$ factories*

We want to comment on the need to construct *asymmetric B* factories at the $\Upsilon(4S)$, rather than symmetric ones. The objective is to detect CP violation with theoretically clean, easily measurable observables. The ideal experiment should:

- have clean sources of B^0–$\overline{B^0}$ pairs;
- look for decays into CP eigenstates—in the hope that they might be dominated by a single weak phase;
- produce large numbers of events.

The first requirement leads naturally to $\Upsilon(4S)$, since this resonance decays predominantly into B_d^0–$\overline{B_d^0}$ and B^+–B^- pairs. The second requirement is satisfied by two important classes of experiments: those comparing the decays into $l^- f_{cp}$ with decays into $l^+ f_{cp}$; and those looking for final states $f_{cp} f_{cp}$. But, as discussed above, the branching ratios of $\Upsilon(4S) \to f_{cp} f_{cp}$ are many orders of magnitude smaller than the branching ratios of $\Upsilon(4S) \to l^{\pm} f_{cp}$. Therefore, one is led into the standard experimental set-up: production of $\Upsilon(4S)$, and search for tagged decays into CP eigenstates.

Now, as pointed out before, due to the antisymmetric nature of the wavefunction, the interference-CP-violation terms drop out of the tagged time-integrated decay rates into $l^{\pm} f_{cp}$. This feature is common to all methods based on tagged decays of the $\Upsilon(4S)$: one must follow the time dependence—or, at least, the time ordering—if one wants to measure interference CP violation. Towards this end, one must look for the distance between the primary vertex and the position at which the neutral B meson decays into f_{cp}. This can be done with silicon vertex detectors, as long as the B meson travels far enough before decaying. The limits on vertexing vary from experiment to experiment, and depend on whether the measurements are made along the beam or perpendicularly to the beam. A rule of thumb for current vertexing limits is that they preclude measurements of flight paths $\lesssim 100\,\mu$m.[78]

In symmetric $e^+ e^-$ colliders working at the $\Upsilon(4S)$ this resonance is produced at rest. The small difference between the mass of $\Upsilon(4S)$ and the mass of the $B_d^0 \overline{B_d^0}$ pair implies that those mesons move very slowly, travelling $\sim 30\,\mu$m in one lifetime. This inhibits the time-dependent measurements needed to extract a^{int} from tagged decays. The ingenious solution is to build asymmetric $e^+ e^-$ colliders; the boost of the $\Upsilon(4S)$ also boosts the B mesons. Thus, the B mesons will travel farther in the laboratory than they do when the $\Upsilon(4S)$ is produced at rest. This is the principle behind the Babar and Belle experiments.

29.5.2 Tracing the time dependence in the B_s^0–$\overline{B_s^0}$ system

This is a good point to comment on another consequence of the limits of vertexing technology. We have seen in Chapter 9 that all time-integrated rates appear multiplied by a dilution factor $1 - 2\chi_0 = 1/(1 + x^2)$. The current limit $x > 9.5$ for the B_s^0–$\overline{B_s^0}$ system implies that this is a suppression factor $\lesssim 10^{-2}$. Hence, we need to study time-dependent rates in order to probe the B_s^0–$\overline{B_s^0}$ system.

However, the term $\mathrm{Im}\,\lambda_f$ in the time-dependent tagged decay rates appears multiplied by $\sin \Delta m t$. The same limit on x puts an upper bound on the period

---

[78] As an example, the HERA-B Collaboration (1995, p. 12) will be able to measure flight paths down to $\sim 500\,\mu$m in the beam direction, and $\sim 25\,\mu$m in the direction perpendicular to the beam.

of a complete oscillation: $t = 2\pi/\Delta m_s < 1.1\,\text{ps}$, in the reference frame of the B_s^0 meson. The flight path during one oscillation ($l_{1\text{osc}}$), as measured in the laboratory frame, depends on the momentum p of the B_s^0 meson:

$$l_{1\text{osc}} = \frac{2\pi c \tau_{B_s^0}}{x_s} \frac{p}{m_{B_s^0}} \sim 300\,\mu\text{m} \left(\frac{10}{x_s}\right) \left(\frac{p}{5.37\,\text{GeV}/c}\right). \tag{29.8}$$

For example, taking $x_s = 17$, a B_s^0 with momentum $p = 121\,\text{GeV}/c$—the average momentum expected by the HERA-B Collaboration (1994, p. 10)—will travel $\sim 4\,\text{mm}$ in one oscillation. Let us assume that one needs to measure lengths down to $l_{1\text{osc}}/(2\pi)$ in order to trace the oscillation adequately. In our example, this requires measurements down to around $600\,\mu\text{m}$, which is similar to the vertex resolution for HERA-B. This means that HERA-B will be able to measure x_s and to perform time-dependent B_s^0 decay measurements, up to $x_s \sim 17$. If x_s is much larger than this value, the HERA-B Collaboration (1994, 1995) will neither measure it nor be able to measure $\text{Im}\,\lambda_f$ in B_s^0 decays. The precise numerical value varies from one experiment to another, but it is clear that, as the limit on x_s increases, the B_s^0–$\overline{B_s^0}$ oscillations may become too fast to be traced, i.e., it becomes impossible for detectors to follow the oscillations in time. Fortunately, if y_s is large enough, untagged decays allow for a new class of experiments (Dunietz 1995)—see § 37.5.

29.6   Decays into lighter neutral-meson systems

Many decays of interest involve a light neutral meson in the final state, such as D^0 or K_S. Decays of this type, known as 'cascade decays', will be analysed in Chapter 34. For the moment, we only want to address the following question: which combinations of P^0 and $\overline{P^0}$ can one detect experimentally? For definiteness, let us concentrate on the decays $B_d^0 \to J/\psi + \text{kaon}$. We wish to find out which combinations of K^0 and $\overline{K^0}$ can be identified experimentally in a straightforward way. The issue is complicated by the fact that there are two time variables playing a role in the decay chain: the time t_B it took for the initial B meson to decay, and the time interval t_K in between the creation and the decay of the light neutral meson.

We can certainly identify flavour eigenstates and mass eigenstates. In order to detect the flavour of the intermediate kaon we can use its subsequent decay into a flavour-specific final state. For instance, we can look for $J/\psi + (\pi^- l^+ \nu_l)$ to determine the flavour of the kaon. As for the mass eigenstates, we can separate out the K_L component of the beam by choosing events in which the kaon lives much longer than the K_S lifetime: $t_K \gg \tau_{K_S}$.

A more subtle question is whether one can detect CP-eigenstate combinations of the lighter neutral-meson systems. In our example, the question is whether or not we can identify final states $J/\psi K_+$ and $J/\psi K_-$, where

$$|K_\pm\rangle \equiv \tfrac{1}{\sqrt{2}} \left(|K^0\rangle \pm e^{i\xi_K}|\overline{K^0}\rangle\right). \tag{29.9}$$

If one neglects CP violation in the neutral-kaon system the answer is affirmative. There are good reasons to neglect CP violation in that system when looking for CP violation in B decays. The reason is that the CP violation in the mixing of neutral kaons is $\sim 10^{-3}$, and the direct CP violation in kaon decays is even smaller. Both are negligible in comparison to the CP-violating effects that we are looking for in B decays. Therefore, the final state $J/\psi K_S$ can be treated as a CP eigenstate—which, strictly speaking, it is not.

On the other hand, if we do not neglect the CP violation in the neutral-kaon system we must look for the subsequent decays of the neutral kaon into a CP eigenstate. For instance, we may look for $J/\psi\,(\pi^+\pi^-)_K$, where $(\pi^+\pi^-)_K$ stands for a pion pair with invariant mass equal to the mass of the neutral kaon. The pions can arise either from the CP-allowed decay of K_+, or from the CP-forbidden decay of K_-. In a correct calculation, both contributions must be taken into account, as will be discussed in Chapter 34.

The analysis is similar for decays into neutral D mesons, except that, in this case, both the mass and the lifetime differences, Δm and $\Delta\Gamma$, are very small. As a result, one usually neglects D^0–$\overline{D^0}$ oscillations. Therefore, we consider decays into D^0, $\overline{D^0}$, and $D_{f_{cp}}$—where $D_{f_{cp}}$ stands for the linear combination of D^0 and $\overline{D^0}$ which decays into the CP eigenstate f_{cp}. Note that $D_{f_{cp}}$ should not be confused with the CP eigenstates

$$|D_\pm\rangle = \tfrac{1}{\sqrt{2}}\left(|D^0\rangle \pm e^{i\xi_D}|\overline{D^0}\rangle\right). \tag{29.10}$$

The difference is that, in our notation, $D_{f_{cp}}$ refers to *a specific final state* f_{cp}; thus, any phase in the decay of the D meson to f_{cp} must be taken into account in the computation of the overall CP-violating observable. This point will be emphasized again in Chapter 34.

30

THE MIXING PARAMETERS

30.1 Introduction

In this chapter we review the determination of the mixing parameters q/p for the various neutral-meson systems. We need them because they enter in the rephasing-invariant quantities λ_f required for the calculation of the CP-violating asymmetries. We identify the conclusions which follow directly and (almost) model-independently from experiment, as opposed to expressions characteristic of the standard model (SM). Extensive use is made of the formulae in Chapter 6.

The analysis presented in this chapter is based on a very simple observation about eqn (6.58),

$$\frac{q}{p} = \sqrt{\frac{M_{12}^* - \frac{i}{2}\Gamma_{12}^*}{M_{12} - \frac{i}{2}\Gamma_{12}}} = \frac{2(M_{12}^* - \frac{i}{2}\Gamma_{12}^*)}{\Delta m - \frac{i}{2}\Delta\Gamma}. \tag{30.1}$$

That observation is the following. If $|M_{12}| \gg |\Gamma_{12}|$, or else if $\varpi \equiv \arg\left(M_{12}^*\Gamma_{12}\right)$ is very close either to 0 or to π, then we can approximate eqn (30.1) by

$$\frac{q}{p} \approx \frac{M_{12}^*}{|M_{12}|}. \tag{30.2}$$

This equation uses our choice of $\Delta m > 0$ for all neutral-meson systems. On the other hand, if $|M_{12}| \ll |\Gamma_{12}|$, or else if ϖ is very close either to 0 or to π, then we can approximate eqn (30.1) by

$$\frac{q}{p} \approx \text{sign}(\Delta\Gamma)\frac{\Gamma_{12}^*}{|\Gamma_{12}|}. \tag{30.3}$$

If either of these conditions hold, then q/p is a pure phase, which is determined either by the phase of M_{12} or by that of Γ_{12}, respectively. The validity of these approximations can be tested either experimentally—see § 30.1.1—or using theoretical arguments—see § 30.1.2.

30.1.1 Extracting experimental information on q/p

The magnitude $|q/p| = \sqrt{(1-\delta)/(1+\delta)}$ is a function of δ. This quantity has only been measured for the neutral-kaon system. For the B_d^0–$\overline{B_d^0}$ system there is a direct bound on δ, while for the B_s^0–$\overline{B_s^0}$ system that bound arises indirectly, through the bound on the mixing parameter χ.

The current experimental bounds may also be used in order to place constraints on the deviation of the argument of q/p from that of M_{12}^*, without recourse to theoretical assumptions. One starts from eqn (6.71):

$$\frac{q}{p} = \sqrt{\frac{1-\delta}{1+\delta}} \exp\left(i \arg M_{12}^*\right) \frac{1+i\delta u}{\sqrt{1+\delta^2 u^2}}, \tag{30.4}$$

which is *exact*, and where we have used our convention that Δm is positive. It follows that

$$\left| \tan \arg \left[\frac{q}{p} \exp\left(i \arg M_{12}\right)\right]\right| \leq |\delta|_{\max} |u|_{\max}. \tag{30.5}$$

Analogously, we may use eqn (6.70),

$$\frac{q}{p} = -\sqrt{\frac{1-\delta}{1+\delta}} \exp\left(i \arg \Gamma_{12}^*\right) \frac{u - i\delta}{\sqrt{u^2 + \delta^2}}, \tag{30.6}$$

to show that

$$\left| \tan \arg \left[-\frac{q}{p} \exp\left(i \arg \Gamma_{12}\right)\right]\right| \leq \frac{|\delta|_{\max}}{|u|_{\min}}. \tag{30.7}$$

The bounds in eqns (30.5) and (30.7) are very useful in the case of the neutral-kaon system. For the B systems they are not that good. There, it is customary to use instead the following theoretical argument.

30.1.2   A theoretical argument

Suppose that one has an upper bound on the quantity

$$t \equiv \left|\frac{\Gamma_{12}}{2M_{12}}\right|. \tag{30.8}$$

This quantity is real and non-negative by definition. Let us assume that one can find some argument to show that it is small, $t \ll 1$. Now, it follows from eqn (30.1) that

$$\frac{q^2}{p^2} = \frac{2M_{12}^* - i\Gamma_{12}^*}{2M_{12} - i\Gamma_{12}} = \exp\left(2i \arg M_{12}^*\right) \frac{1 - it\cos\varpi - t\sin\varpi}{1 - it\cos\varpi + t\sin\varpi}. \tag{30.9}$$

From this equation one easily derives

$$\frac{1-t}{1+t} \leq \left|\frac{q}{p}\right|^2 \leq \frac{1+t}{1-t}. \tag{30.10}$$

Thus, if we know that t is very small, we learn that q/p has modulus very close to 1, and, therefore, that it is almost a pure phase. From eqn (30.9) one may also derive

$$\left| \tan \arg \left[\frac{q^2}{p^2} \exp\left(2i \arg M_{12}\right)\right]\right| \leq \frac{t^2}{\sqrt{1-t^4}}. \tag{30.11}$$

Thus, if we know that t is very small, we learn that the phase of q/p is very close to that of M_{12}^*.

It is interesting to observe that, while the deviation of $|q/p|$ from unity is at most $\sim t$, the deviation of $\arg(q/p)$ from $\arg M_{12}^*$ is $\sim t^2$, and therefore much smaller. This is readily checked by a glance at eqn (30.9): both $t \cos \varpi$ and $t \sin \varpi$ must be non-zero in order that $(1 - it \cos \varpi - t \sin \varpi)(1 - it \cos \varpi + t \sin \varpi)$ be non-real.

30.1.3 An assumption about the decay amplitudes

We shall *assume* that the decay amplitudes corresponding to the dominant decay channels are determined by the SM tree-level diagrams. If there is new physics, then it should affect primarily the mixing in the neutral-meson systems and/or the decays which are suppressed in the SM either by CKM factors, by loop factors, or otherwise. This assumption is useful in order to get an estimate of

$$\Gamma_{12} = \sum_f \langle f|T|P^0\rangle^* \langle f|T|\overline{P^0}\rangle. \tag{30.12}$$

The assumption will be used, in the B_d^0–$\overline{B_d^0}$ and B_s^0–$\overline{B_s^0}$ systems, to place limits on $|\Gamma_{12}|$, and thus on t. Equations (30.10) and (30.11) are then used to determine q/p through the argument of M_{12}.

30.2 q_K/p_K

In the K^0–$\overline{K^0}$ system all the necessary ingredients are experimentally known and model-independent statements are possible. The determination of q/p follows most easily from eqn (30.6). Remember that in the derivation of this equation, which is *exact*, we have used our convention that Δm is positive. Experimentally, $u \equiv -\Delta\Gamma/(2\Delta m) = 1.054 \pm 0.004$, while $\delta = (3.27 \pm 0.12) \times 10^{-3}$. It follows that

$$\left|\frac{q}{p}\right| = \sqrt{\frac{1-\delta}{1+\delta}} \approx 1 - \delta \approx 1, \tag{30.13}$$

and therefore q/p is almost a pure phase, which is given by

$$\arg\frac{q}{p} \approx \pi + \arg\Gamma_{12}^* - \delta. \tag{30.14}$$

In the calculation of $\arg\Gamma_{12}^*$ one first uses the result (Lavoura 1992a)[79]

$$\left|\arg\left(\Gamma_{12}A_0\bar{A}_0^*\right)\right| \leq 5 \times 10^{-5} \tag{30.15}$$

to trade the phase of Γ_{12}^* for that of $A_0\bar{A}_0^*$. One then assumes that, even if there is physics beyond the SM, the $|\Delta S| = 1$ decay amplitudes are dominated by the usual charged-current weak interaction. Then, as in eqn (17.5),

$$\frac{A_0}{\bar{A}_0} = e^{i(\xi_K + \xi_d - \xi_s)} \frac{V_{us}^* V_{ud}}{V_{us} V_{ud}^*}. \tag{30.16}$$

---

[79]This follows from the fact that the sum in eqn (30.12) is dominated by the channel $2\pi, I = 0$.

One concludes that

$$\frac{q_K}{p_K} \approx -e^{i(\xi_K+\xi_d-\xi_s)}\frac{V_{us}^*V_{ud}}{V_{us}V_{ud}^*}. \tag{30.17}$$

This result hinges on experimental facts and on the assumption that the kaon decays to $2\pi, I = 0$ are dominated by the tree-level diagram with an intermediate W boson. Equation (30.17) neglects a term $\delta \sim 10^{-3}$ in the phase of q_K/p_K (see eqn 30.14).

Instead of eqn (30.17), one may write

$$\frac{q_K}{p_K} = -e^{i(\xi_K+\xi_d-\xi_s)}\frac{V_{cs}^*V_{cd}}{V_{cs}V_{cd}^*}e^{-2i\epsilon'}, \tag{30.18}$$

making use of the phase $\epsilon' \equiv \arg(-V_{us}V_{cd}V_{ud}^*V_{cs}^*)$ introduced by Aleksan et al. (1994). The equality between the right-hand sides of eqns (30.17) and (30.18) is exact: it holds in any model, because it follows from the definition of ϵ' (Cohen et al. 1997).

In the SM, $\epsilon' \sim 10^{-3}$ may be safely neglected. In most models beyond the SM ϵ' is still very small—see § 28.5.2. Moreover, models in which ϵ' is altered usually produce a large effect on B^0–$\overline{B^0}$ mixing—see for instance Nir and Silverman (1990). For this reason, many authors set $\epsilon' = 0$ in eqn (30.18). Indeed, if $\epsilon' \sim 10^{-3}$, then neglecting ϵ' is just as good an approximation as neglecting δ, as we did when writing down both eqns (30.17) and (30.18). On the other hand, if in a model beyond the SM ϵ' turns out to be large, then it makes sense (Kurimoto and Tomita 1997) to use eqn (30.18) while neglecting δ, because the latter quantity is *experimentally* small. This is why we prefer to keep ϵ' explicit in our formulae.

30.2.1 The phase of M_{12}^*

Instead of starting from eqn (30.6) as we did, we might start from eqn (30.4). As δ is small and $u \approx 1$ one may write

$$\arg\frac{q}{p} \approx \arg M_{12}^* + \delta. \tag{30.19}$$

Comparing eqns (30.14) and (30.19) we see that $\varpi \equiv \arg(M_{12}^*\Gamma_{12}) \approx \pi - 2\delta$, cf. eqn (8.15).

At this juncture, we should stress the important constraints placed by the experimental value of δ on model-building. We have discussed the standard-model computation of M_{12}^* in Chapter 17. The short-distance contribution comes from the box diagram. If we omit the spurious phase $\xi_K+\xi_d-\xi_s$, that contribution may be written as the sum of three pieces, proportional to λ_u^2, $\lambda_u\lambda_c$, and λ_c^2, respectively, where $\lambda_\alpha = V_{\alpha s}^*V_{\alpha d}$. There is also a long-distance contribution to M_{12}^*; its phase is that of λ_u^2.

Thus, the calculation of M_{12}^* involves three terms with different phases. By contrast, the calculation of $-\Gamma_{12}^*$ involves a single weak phase, that of λ_u^2. But the phases of M_{12}^* and of $-\Gamma_{12}^*$ have to coincide up to corrections $\propto \delta \sim 10^{-3}$. In the SM this constraint is satisfied, because the phases of λ_u and of $-\lambda_c$ coincide up to $\epsilon' \sim 10^{-3}$.

The same argument should hold if there is physics beyond the SM. Let us consider a model with no new contributions to Γ_{12}^*, but with an effective $|\Delta S| = 2$ contribution to M_{12}^*. It is an *experimental requirement* that this extra contribution to M_{12}^* should retain a phase close to that of $-\Gamma_{12}^*$. One also has to be careful that the phases of $-\lambda_c$ and of λ_u do not get too different from each other; otherwise the box diagram produces M_{12}^* with a phase very different from that of $-\Gamma_{12}^*$—barring large cancellations between the box diagram and the new-physics contributions to M_{12}^*. This is another argument for the smallness of ϵ' even when there is physics beyond the SM.

30.3 Relating $\arg\left(q_{B_q}/p_{B_q}\right)$ to $\arg M_{12}^*$

30.3.1 *The B_d case*

The CLEO Collaboration (1993c) has established that $|\delta| < 0.09$ in the B_d^0–$\overline{B_d^0}$ system. We also know that Δm must be greater than $|\Delta\Gamma|$; this arises from the combination of the observed oscillation in time-dependent measurements, with the time-independent determination of χ, as discussed in Chapter 10. The Particle Data Group (1996) finds $x = 0.73 \pm 0.05$, while, based on the data, one may conclude that $|y| < 0.3$ at 95% confidence level (Moser, personal communication). It follows that $|u| = |y/x| < 0.47$, and $|u\delta| < 0.043$. From eqn (30.4) one then knows that q/p is almost a pure phase, and, using eqn (30.5), we find that that phase does not deviate from that of M_{12}^* by more that 2.5°.

One may do much better than this by using the theoretical argument in § 30.1.2. For this we need a bound on t. The experimental bound is rather poor. Indeed, as $|\delta|$ is very small and $|\Delta\Gamma| < \Delta m$, one learns from eqn (6.64) that $|M_{12}| \approx \Delta m/2$. Combining this with $|\Gamma_{12}| \leq \Gamma$ we find the constraint

$$\left|\frac{\Gamma_{12}}{M_{12}}\right| \leq \frac{\Gamma}{|M_{12}|} \approx \frac{2\Gamma}{\Delta m} = \frac{2}{x} \leq 3. \tag{30.20}$$

This is rather useless.

But, let us assume that the contributions to Γ_{12} from decay channels common to both B_d^0 and $\overline{B_d^0}$ do not differ much from the standard-model expectations. This is likely to be the case, since these decays should be dominated by SM tree-level diagrams (Nir 1993). The branching ratios to channels common to B_d^0 and $\overline{B_d^0}$ are expected to be no larger than 10^{-3} in the SM. Moreover, in the determination of Γ_{12} via eqn (30.12) these contributions should appear with different signs. Hence, even if we allow for an enhancement by a factor of 10 due to the adding up of different contributions, one finds $|\Gamma_{12}|/\Gamma < 10^{-2}$ (Nir 1993). Using this result instead of $|\Gamma_{12}| \leq \Gamma$ in eqn (30.20), one obtains

$$\left|\frac{\Gamma_{12}}{M_{12}}\right| \leq 3 \times 10^{-2}. \tag{30.21}$$

Thus, $t < 0.015$. This guarantees that the phase of q/p deviates from that of M_{12}^* by no more than 0.007°, which is a much better result than those in the previous paragraphs. In addition, using eqn (6.28), we also find that $|\delta| \leq t \leq 0.015$.

30.3.2 The B_s case

In the B_s^0–$\overline{B_s^0}$ system we take $x \geq 9.5$ and $|y| \leq 0.3$, cf. Chapter 10. Then, using eqn (6.54),

$$\delta^2 \leq \frac{1 - y^2}{1 + x^2} = 1 - 2\chi_0 \tag{30.22}$$

we find

$$|\delta| \leq 0.1, \tag{30.23}$$

while

$$|u| = |y|/x \leq 0.03. \tag{30.24}$$

Therefore, $|\delta u| < 0.003$, and the phase of q/p deviates from that of M_{12}^* by no more than $0.18°$.

As $|\delta|$ is small and $|y| \ll x$ one may, just as in the B_d^0–$\overline{B_d^0}$ system, write $|M_{12}| = \Delta m/2$. We then find

$$\left|\frac{\Gamma_{12}}{M_{12}}\right| \leq \frac{2}{x} \leq 0.2. \tag{30.25}$$

As we did in § 30.3.1, we may find a better bound on t if we assume that the decay amplitudes to channels common to B_s^0 and $\overline{B_s^0}$ do not differ much from the standard-model expectations. The dominant decays common to B_s^0 and $\overline{B_s^0}$ are due to the tree-level transitions $b \to c\bar{c}s$; all other decays are CKM-suppressed relative to this one and should play a minor role. Therefore, Γ_{12}/Γ is expected to be large. Indeed, Beneke *et al.* (1996) found, in the SM,

$$\frac{\Gamma_{12}}{\Gamma} = -\frac{1}{2}\left(0.16^{+0.11}_{-0.09}\right), \tag{30.26}$$

in a convention in which no phases are introduced by the CP transformations.[80] Since the analysis involves primarily tree-level decays, it is likely to hold in many models beyond the SM. Using the highest value in eqn (30.26), $|\Gamma_{12}| \sim 0.14\Gamma$, one obtains

$$\left|\frac{\Gamma_{12}}{M_{12}}\right| \sim \frac{0.28}{x} \leq 3 \times 10^{-2}. \tag{30.27}$$

By coincidence, one arrives at the same final estimate as for the B_d^0–$\overline{B_d^0}$ system, cf. eqn (30.21). In the B_s^0–$\overline{B_s^0}$ system the bound on $|\Gamma_{12}|/\Gamma$ is worse, but that is compensated by the larger value of x. Therefore, the phase of q/p deviates from that of M_{12}^* by no more than $0.007°$, improving upon the experimental result. Again, using eqn (6.28), we find that $|\delta| \leq t \leq 0.015$.

---

[80]The negative sign in eqn (30.26) means that it is *predicted* that the heaviest eigenstate lives longer, i.e., that $\Delta\Gamma < 0$ (Beneke *et al.* 1996).

30.4 q_{B_d}/p_{B_d} and q_{B_s}/p_{B_s}

We have shown in the previous section, using only experimental results, that

$$\frac{q_{B_q}}{p_{B_q}} \approx \exp\left(i\arg M_{12}^*\right) \tag{30.28}$$

holds up to a reasonable accuracy. We have then followed Nir (1993) and used a mild assumption about the decay amplitudes to argue that eqn (30.28) should, in fact, be valid to extremely good accuracy. We must now determine the phase of M_{12}^*. This is readily done in the SM.

30.4.1 q/p in the SM

In the SM, an analysis of the box diagram yields

$$\left|\frac{\Gamma_{12}}{M_{12}}\right| \sim \frac{m_b^2}{m_t^2}, \quad \arg\left(M_{12}^*\Gamma_{12}\right) \sim \frac{m_c^2 - m_u^2}{m_b^2}, \tag{30.29}$$

thus vindicating eqn (30.28). Using the determination of M_{12} in eqn (18.1), we find

$$\frac{q_{B_q}}{p_{B_q}} \approx -\mathrm{sign}\left(B_{B_q}\right)\frac{V_{tb}^*V_{tq}}{V_{tb}V_{tq}^*}e^{i\left(\xi_{B_q}+\xi_q-\xi_b\right)}. \tag{30.30}$$

The CKM combination $V_{tb}^*V_{tq}$ appears because, as explained in Chapter 18, the SM box diagram for M_{12} is dominated by the contribution with two internal top quarks. Long-distance contributions to M_{12} should be negligible. The 'bag parameter' B_{B_q} is expected to be positive. However, if the vacuum-insertion approximation fails badly, B_{B_q} might be negative (Grossman et al. 1997b) in either the B_d^0–$\overline{B_d^0}$ system or the B_s^0–$\overline{B_s^0}$ system, or in both of them.

30.4.2 q_{B_q}/p_{B_q} in the presence of new physics

New physics may change the phase of q_{B_q}/p_{B_q}, or even lead to $\left|q_{B_q}/p_{B_q}\right| \neq 1$. In some models, new physics appears predominantly in the mixing of the neutral mesons, while their decays remain governed by SM physics—examples are given by Nir and Silverman (1990), Dib et al. (1991), and Branco et al. (1993). This is especially likely when the decays occur at tree level in the SM. Then, the effects of new physics on CP asymmetries may be parametrized by parameters r and θ as

$$M_{12} = M_{12}^{\mathrm{SM}}\left(re^{-i\theta}\right)^2. \tag{30.31}$$

It follows that

$$\frac{q}{p} \approx \exp\left(i\arg M_{12}^*\right) = \left(\frac{q}{p}\right)^{\mathrm{SM}}e^{2i\theta}. \tag{30.32}$$

(We take r real and positive by definition. Of course, r and θ are in principle different in the B_d^0–$\overline{B_d^0}$ and B_s^0–$\overline{B_s^0}$ systems.) We thus have

$$\frac{q_{B_q}}{p_{B_q}} = -\mathrm{sign}\left(B_{B_q}\right)\frac{V_{tb}^*V_{tq}}{V_{tb}V_{tq}^*}e^{i\left(\xi_{B_q}+\xi_q-\xi_b\right)}e^{2i\theta_q}. \tag{30.33}$$

This treatment of physics beyond the SM is originally due to Soares and Wolfenstein (1993) and to Branco et al. (1993), and has also been used by Deshpande et *al.* (1996), Cohen *et al.* (1997), Grossman *et al.* (1997a), and Silva and Wolfenstein (1997).[81]

30.5 Main conclusions

We have argued in this chapter that q/p is given, to an excellent approximation, by a pure phase. In the kaon system this is a direct result of the experimental determination of δ. In the B_d and B_s systems, we may use the limited experimental results to ascertain that the phase of q/p differs from the pure phase $M_{12}^*/|M_{12}|$ by no more than 2.5° and 0.2°, respectively. Alternatively, we may follow Nir (1993) and assume that the dominant contributions to $|\Gamma_{12}|$ arise from SM diagrams. We then conclude that the deviation is at most 0.007°, in both the B_d and B_s systems.

In the case of q_K/p_K, we have argued that its phase should retain the expression that it has in the SM, even in the presence of new physics effects. On the other hand, the phases q_{B_q}/p_{B_q} could differ significantly from their SM values. We parametrize this difference by phases θ_q, which arise from possible new-physics effects.

We found

$$\frac{q_K}{p_K} = -e^{i(\xi_K + \xi_d - \xi_s)}\frac{V_{cs}^* V_{cd}}{V_{cs}V_{cd}^*}e^{-2i\epsilon'}$$
$$= \left[-e^{i(\xi_K+\xi_d-\xi_s)}e^{-2i\epsilon'}\right], \tag{30.34}$$

$$\frac{q_{B_d}}{p_{B_d}} = -\text{sign}\,(B_{B_d})\,e^{i\left(\xi_{B_d}+\xi_d-\xi_b\right)}\frac{V_{tb}^* V_{td}}{V_{tb}V_{td}^*}e^{2i\theta_d}$$
$$= \left[-\text{sign}\,(B_{B_d})\,e^{i\left(\xi_{B_d}+\xi_d-\xi_b\right)}e^{-2i\beta}\right]e^{2i\theta_d}, \tag{30.35}$$

$$\frac{q_{B_s}}{p_{B_s}} = -\text{sign}\,(B_{B_s})\,e^{i(\xi_{B_s}+\xi_s-\xi_b)}\frac{V_{tb}^* V_{ts}}{V_{tb}V_{ts}^*}e^{2i\theta_s}$$
$$= \left[-\text{sign}\,(B_{B_s})\,e^{i(\xi_{B_s}+\xi_s-\xi_b)}e^{2i\epsilon}\right]e^{2i\theta_s}. \tag{30.36}$$

In each equation, the second line contains in between squared brackets the SM result, with the phases in the CKM matrix parametrized following the *convention* of eqn (28.24). We see that, if the mixing is only due to SM diagrams, then q/p may be parametrized exclusively in terms of β, ϵ, and ϵ', even if the CKM matrix is not unitary.

---

[81] In some of these papers θ_q has been defined differently, but the basic idea has always been the same.

DECAY AMPLITUDES: DIAGRAMMATICS

31.1 Introduction

In this chapter we work out the amplitudes governing B decays in the standard model (SM). We use a quark-diagram description which has the advantage of being very intuitive. This description may be combined with a classification of the diagrams according to their flavour-SU(3) properties (Zeppenfeld 1981; Chau *et al.* 1991; Gronau *et al.* 1994*a,b*, 1995), providing a tool to find approximate relations between different decay channels. However, the quark-diagram description does not yield exact results. In order to obtain exact results one must first derive the effective Hamiltonian governing the transitions, and then one must evaluate its matrix elements exactly. Unfortunately, the nonperturbative nature of quark confinement means that those matrix elements must be computed within some model, such as the BSW model (Bauer *et al.* 1985, 1987), and the results thus obtained are model-dependent. The effective-Hamiltonian approach will be discussed in the next chapter.

In this chapter we start by studying the dominant tree-level and gluonic-penguin amplitudes, within the spectator approximation. In this approximation, one assumes that the B meson decays through the decay of its b quark (or $\bar{b}$ antiquark), with no intervention from the other quark in the meson—hence called the 'spectator' quark. Later, we discuss briefly the amplitudes involving the spectator quark, and the electroweak-penguin amplitudes. These become important when the dominant amplitudes are either CKM-suppressed or absent.

In principle, no non-trivial model-independent statement can be made about the decay amplitudes; they must always be computed within a particular model. Here, we shall concentrate on the decay amplitudes arising from the SM. Moreover, we shall consider only the hadronic decays into two mesons. Throughout, our notation for the diagrams will be the one put forward by Gronau *et al.* (1994*a,b*, 1995).

Hadronic uncertainties arise in any computation of the decay amplitudes and decay rates. They are due to uncertain operator matrix elements, and also to unknown CP-even phases from the final-state interactions. Thus, the decay amplitudes that we need, A_f and $\bar{A}_f$, typically end up being unknown. Still, one can overcome this problem in special situations where several amplitudes can be related to each other. The simplest examples occur in the decays into CP eigenstates in which the decay amplitudes are dominated by a single weak phase. Then, the CP symmetry of the strong interaction relates A_f with $\bar{A}_f$, and the hadronic uncertainties cancel out in the ratio $\bar{A}_f/A_f$ relevant for λ_f. Other examples discussed in the literature include relations between the amplitudes for decays into final states with D^0, $\overline{D^0}$, and $D_{f_{cp}}$ (the combination of D^0 and $\overline{D^0}$ that decays into the CP-eigenstate f_{cp}); and relations obtained by using the approximate isospin or flavour-SU(3) symmetries of the strong interaction. These relations involve a comparison among several distinct decay channels, and are typically more difficult to implement experimentally.

The diagrammatic approach discussed in this chapter is commonly used in connection with additional assumptions about the relative magnitudes of the various diagrams. These assumptions are based on estimates for the diagrams in which one neglects rescattering effects. Such final-state interactions will both alter the magnitude of the contributions from individual diagrams and mix different quark-level diagrams. Here, we shall follow the usual practice and neglect those effects. We will, however, mention which parts of our analysis are more likely to be affected by rescattering effects.

31.2 SU(3) classification of mesons

The u, d, and s quarks form a triplet of flavour SU(3). The antiquarks $\bar{d}$, $-\bar{u}$, and $\bar{s}$ form an antitriplet. The mesons are defined so that they form isospin multiplets without extra minus signs (Gronau et al. 1994a,b; Grinstein and Lebed 1996). The flavour-SU(3) octet of light pseudoscalars includes the following SU(2) multiplets:

$$K^0 = d\bar{s}, \quad K^+ = u\bar{s},$$
$$\pi^- = -d\bar{u}, \quad \pi^0 = \tfrac{1}{\sqrt{2}}(-u\bar{u} + d\bar{d}), \quad \pi^+ = u\bar{d}, \tag{31.1}$$
$$K^- = -s\bar{u}, \quad \overline{K^0} = s\bar{d},$$

and

$$\eta_8 = \tfrac{1}{\sqrt{6}}(-u\bar{u} - d\bar{d} + 2s\bar{s}). \tag{31.2}$$

There is also a light pseudoscalar which is a flavour-SU(3) singlet,

$$\eta_1 = \tfrac{1}{\sqrt{3}}(u\bar{u} + d\bar{d} + s\bar{s}), \tag{31.3}$$

which mixes with η_8. Indeed, the mass eigenstates η and η' are orthogonal combinations of η_1 and η_8:

$$\eta = \eta_8 \cos\theta - \eta_1 \sin\theta,$$
$$\eta' = \eta_8 \sin\theta + \eta_1 \cos\theta. \tag{31.4}$$

The angle θ was found by Gilman and Kauffmann (1987) to be around $20°$. A good approximation (Chau et al. 1991; Gronau et al. 1994a,b) consists in taking $\theta = \theta_p \equiv \arcsin(1/3) \approx 19.5°$. This approximation gives

$$\eta = \eta_p \equiv \tfrac{1}{\sqrt{3}}(-u\bar{u} - d\bar{d} + s\bar{s}),$$
$$\eta' = \eta'_p \equiv \tfrac{1}{\sqrt{6}}(u\bar{u} + d\bar{d} + 2s\bar{s}). \tag{31.5}$$

The heavier nine vector mesons have the same quark content. One adopts the following notation, as compared to the lighter mesons: $\pi \to \rho$, $K \to K^*$, and $\eta \to \phi$. Analogously to η_1–η_8 mixing there is ϕ_1–ϕ_8 mixing; in this case the mixing angle is closer to $-40°$. Here, the mass eigenstates are designated by ϕ and ω. Sometimes (Close 1979) one uses $\theta = \theta_v \equiv \arcsin(-1/\sqrt{3}) \approx -35°$, corresponding to the eigenstates

$$\phi = \phi_v \equiv s\bar{s},$$
$$\omega = \omega_v \equiv \tfrac{1}{\sqrt{2}}(u\bar{u} + d\bar{d}). \tag{31.6}$$

One must keep in mind that this is an approximate result; in particular, ϕ also has $u\bar{u}$ and $d\bar{d}$ components. We shall not need the heavier resonances in this book.

As for the mesons with charm or bottom flavour, we shall be particularly interested in

$$
\begin{aligned}
D^0 &= -c\bar{u}, & D^+ &= c\bar{d}, & D_s^+ &= c\bar{s}, \\
\overline{D^0} &= \bar{c}u, & D^- &= \bar{c}d, & D_s^- &= \bar{c}s. \\
B_u^+ &= \bar{b}u, & B_d^0 &= \bar{b}d, & B_s^0 &= \bar{b}s, \\
B_u^- &= -b\bar{u}, & \overline{B_d^0} &= b\bar{d}, & \overline{B_s^0} &= b\bar{s}.
\end{aligned}
\tag{31.7}
$$

In addition, there are charmonium $c\bar{c}$ states. These include the states η_c (with $J^{PC} = 0^{-+}$), J/ψ, and ψ (with $J^{PC} = 1^{--}$). Similarly, we have the bottomonium $b\bar{b}$ states Υ (with $J^{PC} = 1^{--}$).

31.2.1 Final states with several possible quark-flavour contents

In the decays of one B meson to two lighter mesons, one often has to deal with final states in which one or both mesons are admixtures of several possible flavour contents. We are now ready to list the most important possibilities:

- K_S and K_L. These are linear combinations of $K^0 = d\bar{s}$ and $\overline{K^0} = \bar{d}s$.
- $D_{f_{cp}}$. By this we mean that the neutral D meson in the final state is observed through its decay into a CP eigenstate. Such CP eigenstates can be reached from both $D^0 = -c\bar{u}$ and $\overline{D^0} = u\bar{c}$.
- π^0 or ρ^0 (or heavier resonances). These particles are linear combinations of $u\bar{u}$ and $d\bar{d}$.
- η or η'. These are linear combinations of $u\bar{u}$, $d\bar{d}$, and $s\bar{s}$.
- ϕ. This meson, although it is mainly $s\bar{s}$, has $u\bar{u}$ and $d\bar{d}$ components.

These cases are special in that there may be quark diagrams with different quark content in the final state that lead into the same hadronic final state.

31.3   Tree-level diagrams

In the spectator approximation, the simplest diagrams for the decay of a $\bar{b}$ anti-quark involve only one intermediate $W^\pm$, as shown in Fig. 31.1. The letters u_α and d_k represent a generic up-type and down-type quark, respectively. The final quark contents of the various possible decays of $\bar{b}$, together with the respective CKM factors $V_{\alpha b}^* V_{\beta k}$, in the parametrization of eqn (16.38), have been listed in Table 31.1. The decays of b involve the complex-conjugate combination of CKM

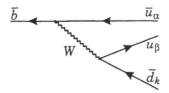

FIG. 31.1. Tree-level diagram.

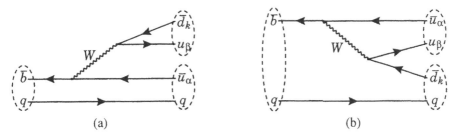

FIG. 31.2. (a) T diagram, with two charged mesons in the final state, and (b) C diagram, with two neutral mesons in the final state.

matrix elements.

When using Table 31.1, it is important to bear in mind that the CKM parameter $R_b \equiv |V_{ud}V_{ub}|/|V_{cd}V_{cb}|$ entails a suppression, since $0.24 \leq R_b \leq 0.47$ (Buras and Fleischer 1998). This is due to the fact that $|V_{ub}/V_{cb}|$ is smaller than λ, being probably closer to $\lambda/2$ or $\lambda/3$. Amplitudes with R_b thus suffer from an extra CKM suppression, over and above the usual λ suppression.

31.3.1 C and T diagrams

For an initial B_q^0 meson, the final state has quark content $\bar{u}_\alpha u_\beta \bar{d}_k q$. These quarks can hadronize into two mesons in two different ways. Either there will be two charged mesons with quark content $q\bar{u}_\alpha$ and $u_\beta \bar{d}_k$, as in Fig. 31.2 (a), or there will be two neutral mesons with quark content $q\bar{d}_k$ and $u_\beta \bar{u}_\alpha$, as in Fig. 31.2 (b). The two cases have a different colour structure. In the first case, u_β and

Table 31.1 *Classification of tree-level decays.*

Decay number	$b \to \bar{u}_\alpha u_\beta d_k$	$b \to u_\alpha \bar{u}_\beta d_k$	$V_{\alpha b}^* V_{\beta k}$
1	$\bar{c}c\bar{s}$	$c\bar{c}s$	$A\lambda^2$
2	$\bar{c}u\bar{s}$	$c\bar{u}s$	$A\lambda^3$
3	$\bar{u}c\bar{s}$	$u\bar{c}s$	$A\lambda^3 R_b e^{i\gamma}$
4	$\bar{u}u\bar{s}$	$u\bar{u}s$	$A\lambda^4 R_b e^{i\gamma}$
5	$\bar{c}c\bar{d}$	$c\bar{c}d$	$-A\lambda^3$
6	$\bar{c}u\bar{d}$	$c\bar{u}d$	$A\lambda^2$
7	$\bar{u}c\bar{d}$	$u\bar{c}d$	$-A\lambda^4 R_b e^{i\gamma}$
8	$\bar{u}u\bar{d}$	$u\bar{u}d$	$A\lambda^3 R_b e^{i\gamma}$

$\bar{d}_k$, which arise in a colour singlet from the charged current, form a—necessarily colour singlet—meson. In the second case, those two quarks are separated into different mesons. In that case, u_β and $\bar{u}_\alpha$, whose colours are initially independent, bind into a colour singlet, the same thing happening with $\bar{d}_k$ and q. Barring rescattering effects, this results in a suppression of the second option relative to the first one. One says that Fig. 31.2 (a) is colour-allowed while Fig 31.2 (b) is colour-suppressed. We shall follow Gronau *et al.* (1994a,b, 1995) and denote the two possibilities in Figs 31.2 (a) and 31.2(b) by T (for tree-level) and C (for colour-suppressed), respectively. We shall also use the notation $\bar{b} \to \bar{u}_\alpha[u_\beta \bar{d}_k]$ for T and $\bar{b} \to [\bar{u}_\alpha u_\beta]\bar{d}_k$ for C.

Since C is colour-suppressed, we would naively expect $|c/t| \sim 1/3$. Calculations within specific models yield values which are numerically closer to λ (Gronau *et al.* 1995),[82]

$$\left|\frac{c}{t}\right| \sim 0.2 \sim \lambda. \qquad (31.8)$$

The lower-case letters mean that the (potentially different) CKM factors in C and T have been explicitly factored out.[83]

Naturally, the same quark-decay schemes apply to the decays of the charged B mesons. The final quark content is the one in the first column of Table 31.1, with the addition of u and c in the decays of B_u^+ and B_c^+, respectively.

31.3.2 *Classification of the decays at tree level*

The final states common to B_q^0 and $\overline{B_q^0}$ (which we shall designate 'common final states') are especially interesting, in particular when they are CP eigenstates. Common final states that are not CP eigenstates are known as 'non-CP eigenstates' or 'CP non-eigenstates'. Of course, there are final states which are not CP eigenstates and are not common to B_q^0 and $\overline{B_q^0}$; these are the flavour-specific final states (sometimes known also as 'self-tagging final states'). In general, the final states that are not common to B_q^0 and $\overline{B_q^0}$ can only exhibit direct CP violation.

We want to classify the decays according to their flavour-changing properties, and to see which tree-level diagrams contribute to each type of decay. We may classify the decays according to whether they do or do not change the up-type quantum numbers of the first two families, and according to the net strangeness and down-flavour content of the final state—denoted by S_f and D_f, respectively.

We find that there are four classes of processes, which we list below, together with the tree-level diagrams contributing to each class. The diagrams are identified by their number in the first column of Table 31.1.

1. *Decays with $S_f = D_f = 0$ and $\Delta C = \Delta U = 0$*
   - Tree-level diagrams: $B_s^0 - 1, 4$;
   $$B_d^0 - 5, 8.$$

---

[82]The coincidence of the suppression factor $|c/t|$ in eqn (31.8) with λ is, of course, just an accident.

[83]Gronau *et al.* (1994a,b, 1995)—who introduced the diagrammatic notation that we use in this book—use only upper-case letters because they always factor out the CKM coefficients explicitly; we shall use lower-case letters when we do this.

- All final states are common to B_q^0 and $\overline{B_q^0}$.
- All diagrams can lead into CP eigenstates.
- Each amplitude only gets one tree-level contribution.

2. *Decays with $S_f = D_f = 0$ and $\Delta C \neq 0 \neq \Delta U$*
   - Tree-level diagrams: $B_s^0 - 2, 3$;
     $$B_d^0 - 6, 7.$$
   - All final states are common to B_q^0 and $\overline{B_q^0}$.
   - Final states that are CP eigenstates must contain $D_{f_{cp}}$. Examples: $B_s^0 \to D_{f_{cp}}\phi$, $B_d^0 \to D_{f_{cp}}\pi^0$.
   - Final states that are CP eigenstates receive two tree-level contributions: one from the decay into $c\bar{u}$, and another one from the decay into $\bar{c}u$.

3. *Decays with $S_f \neq 0 \neq D_f$ and $\Delta C = \Delta U = 0$*
   - Tree-level diagrams: $B_s^0 - 5, 8$;
     $$B_d^0 - 1, 4.$$
   - Common final states are the ones with K_L or K_S. Examples: $B_d^0 \to J/\psi K_S$, $B_d^0 \to \pi^0 K_S$.
   - Common final states are CP eigenstates too.
   - They have only one tree-level diagram contributing to the decay.

4. *Decays with $S_f \neq 0 \neq D_f$ and $\Delta C \neq 0 \neq \Delta U$*
   - Tree-level diagrams: $B_s^0 - 6, 7$;
     $$B_d^0 - 2, 3.$$
   - Common final states are the ones with K_L or K_S. Examples: $B_d^0 \to D^0 K_S$, $B_s^0 \to \overline{D^0} K_S$.
   - The common final states which are CP eigenstates are the ones which have simultaneously $K_{L,S}$ and $D_{f_{cp}}$.
   - The common final states which are CP eigenstates receive two tree-level contributions.

31.3.3 *Decay channels common to B^0 and $\overline{B^0}$*

The final states in the upper part of Table 31.1, with quark content $\bar{c}c\bar{s}s$, $\bar{c}u\bar{s}s$, $\bar{u}c\bar{s}s$, and $\bar{u}u\bar{s}s$, are common to B_s^0 and $\overline{B_s^0}$. Similarly, the final states in the lower part of the table, with quark content $\bar{c}c\bar{d}d$, $\bar{c}u\bar{d}d$, $\bar{u}c\bar{d}d$, and $\bar{u}u\bar{d}d$, are common to B_d^0 and $\overline{B_d^0}$. They account for most of the decay channels common to both neutrals.

But, as we have seen in the previous subsection, there are also final states common to B_s^0 and $\overline{B_s^0}$ that can be reached through the quark decays in the lower part of Table 31.1. These common final states must include either K_S or K_L, and they occur because K_L and K_S are combinations of $K^0 = d\bar{s}$ and $\overline{K^0} = \bar{d}s$. Similarly, B_d^0 and $\overline{B_d^0}$ have common final states to be reached from the quark decays in the upper part of Table 31.1; an example is $J/\psi K_S$—which is almost

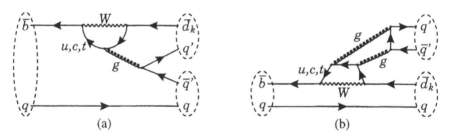

FIG. 31.3. (a) Gluonic penguin, and (b) singlet penguin.

a CP eigenstate, and may be reached by both B_d^0 and $\overline{B}_d^0$ through the quark decays $\bar{b} \to \bar{c}c\bar{s}$ and $b \to c\bar{c}s$, respectively.

Common final states are important because they are the ones that may exhibit interference CP violation; moreover, they are the ones contributing to Γ_{12} through $\Gamma_{12} = \sum_f A_f^* \bar{A}_f$. The largest terms $A_f^* \bar{A}_f$ are proportional to λ^4, and they occur for the decays in the first line of Table 31.1; they contribute to Γ_{12} in the B_s^0–$\overline{B}_s^0$ system, but not to Γ_{12} in the B_d^0–$\overline{B}_d^0$ system. This is because the decays $B_d^0 \to \overline{K}^0$ and $\overline{B}_d^0 \to K^0$ do not occur.[84] Some decays in the sixth line of Table 31.1 also give $A_f^* \bar{A}_f \sim \lambda^4$, but these must always include $D_{f_{\rm cp}}$ in the final state. Again, in the flavour basis it is easy to see that they do not contribute. As a consequence, we expect Γ_{12} to be larger in the B_s^0–$\overline{B}_s^0$ than in the B_d^0–$\overline{B}_d^0$ system and, therefore, $y_d \ll y_s$ in the SM. This is the reason why one neglects y in the B_d^0–$\overline{B}_d^0$ system, while it may be large in the B_s^0–$\overline{B}_s^0$ system (Beneke et $al.$ 1996). Moreover, the $\bar{b} \to \bar{c}c\bar{s}$ decays constitute the dominant portion of all B_s^0 decays that are common to B_s^0 and $\overline{B}_s^0$.

31.4   Gluonic penguins

We now turn to the one-loop diagrams involving one $W^\pm$ and one gluon. They are known as gluonic penguins and are shown in Fig 31.3 (a). The diagrams in which the gluons attach to the external legs—cf. Fig 20.2—must be included in the computation of the gluonic penguins. The $q'\bar{q}'$ pairs in the final state may be $u\bar{u}$, $d\bar{d}$, $s\bar{s}$, or $c\bar{c}$; the CKM factor is always the same.

In Fig 31.3 (a) we have depicted only one gluon, but there may be any number of gluon lines. As the $q'\bar{q}'$ pair created by one gluon must be in a colour-SU(3) octet state, in diagrams with only one gluon the final state from the initial B_q^0 can only have the mesons $q\bar{q}'$ and $q'\bar{d}_k$. We may denote this by $\bar{b} \to [\bar{d}_k q']\bar{q}'$. One may draw other penguin diagrams, leading to mesons $q\bar{d}_k$ and $q'\bar{q}'$ in the final state. However, such diagrams require more than one gluon; for instance, one has the diagram in Fig 31.3 (b), which has been nicknamed the 'vacuum cleaner

---

[84]It is instructive to reproduce this result in the K_S, K_L basis. The decays that are common to B_d^0 and $\overline{B}_d^0$ are those with a K_L or a K_S in the final state. However, in those cases, the values of $A_f^* \bar{A}_f$ are opposite for final states containing a K_S and for the analogous final state containing a K_L, so that the total contribution to Γ_{12} vanishes.

Table 31.2 *CKM factors for the gluonic penguins.*

$\bar{b} \to \bar{d}_k q' \bar{q}'$	Intermediate up-type quark u_γ	$V_{\gamma b}^* V_{\gamma k}$
	t	$-A\lambda^2$
$\bar{s}q'\bar{q}'$	c	$A\lambda^2$
	u	$A\lambda^4 R_b e^{i\gamma}$
	t	$A\lambda^3 R_t e^{-i\beta}$
$\bar{d}q'\bar{q}'$	c	$-A\lambda^3$
	u	$A\lambda^3 R_b e^{i\gamma}$

diagram' by Gronau (1997). Furthermore, assuming that flavour SU(3) is a good symmetry, the $q'\bar{q}'$ pair must bind into a flavour-SU(3) singlet.[85] Hence, one refers to $\bar{b} \to \bar{d}_k[q'\bar{q}']$ decays as singlet penguins (Dighe *et al.* 1996b), Thus, the diagram in Fig. 31.3 (b) generates the decay $\bar{b} \to \bar{d}_k \eta_1$ or $\bar{b} \to \bar{d}_k \phi_1$. A similar vacuum cleaner diagram, with $q'\bar{q}' = c\bar{c}$, contributes to final states with a $c\bar{c}$ meson; for instance, $B_d^0 \to J/\psi K_S$.

The CKM factors for gluonic penguins are listed in Table 31.2. There are three different gluonic penguins, corresponding to the three intermediate up-type quarks running in the loop. We may denote those three amplitudes P_t, P_c, and P_u for intermediate top, charm, and up quark, respectively. Each amplitude is separately divergent; only the sum of the three gluonic penguins,

$$P_t + P_c + P_u = V_{tb}^* V_{tk} p_t + V_{cb}^* V_{ck} p_c + V_{ub}^* V_{uk} p_u, \qquad (31.9)$$

is finite.

Within the SM, we may use the unitarity of the CKM matrix to rewrite eqn (31.9) in either of the following ways:

$$\begin{aligned}
P_t + P_c + P_u &= V_{tb}^* V_{tk} (p_t - p_u) + V_{cb}^* V_{ck} (p_c - p_u) \\
&= V_{tb}^* V_{tk} (p_t - p_c) + V_{ub}^* V_{uk} (p_u - p_c) \\
&= V_{cb}^* V_{ck} (p_c - p_t) + V_{ub}^* V_{uk} (p_u - p_t). \qquad (31.10)
\end{aligned}$$

The differences $p_\alpha - p_\beta$ are finite.

Independently of the hadronic matrix elements and of the CKM factors, gluonic penguins are suppressed by a loop factor with respect to tree-level diagrams. On the other hand, gluonic penguins with a top quark in the loop get a GIM enhancement due to the high mass of the top quark. Therefore, a rough estimate for the ratio between the gluonic penguin and the tree-level diagram is (Gronau 1993)

$$\left| \frac{p_t}{t} \right| \sim \frac{\alpha_s(m_b)}{12\pi} \ln \frac{m_t^2}{m_b^2} \sim 0.04 \sim \lambda^2. \qquad (31.11)$$

We have used $\alpha_s(m_b) = 0.2$, and assumed that the ratio of hadronic matrix elements is close to unity. The estimate in eqn (31.11) is used extensively in

---

[85]Of course, there will be flavour-SU(3)-breaking corrections.

the literature. Nevertheless, by computing in detail penguin-dominated decays, Kramer and Palmer (1995) have obtained ratios that are sometimes closer to

$$\left|\frac{p_t}{t}\right| \sim \lambda. \tag{31.12}$$

This is the outcome of calculations using the effective Hamiltonian, factorization, and the BSW model (Bauer *et al.* 1985, 1987). An experimental indication for a larger-than-expected importance of the gluonic penguins is the outcome of the measurements by the CLEO Collaboration (1998*a*) of large branching ratios for the penguin-dominated decays $B^{\pm} \to K^{\pm}\eta'$. This highlights the fact that eqns (31.11) and (31.12) are crude estimates, not to be taken too literally.

Two other important points should be made concerning the estimates in eqns (31.11) and (31.12). The first observation concerns our use of p_t in these equations. What we really mean is either $p_t - p_u$ or $p_t - p_c$, according to whether one decomposes the gluonic penguin following the first or the second line in eqn (31.10), respectively. However, we do not want to obscure the fact that those estimates are based on the GIM enhancement of the gluonic penguin, which is due entirely to the high mass of the top quark. The second observation concerns the validity of this top-quark dominance. Indeed, one *assumes* that there is no enhancement of either p_c or p_u. However, this assumption may be false for gluonic penguins—nevertheless, Buras and Fleischer (1998) argue that it is a good approximation for electroweak penguins. Indeed, gluonic penguins with internal up or charm quarks may be larger than what one usually assumes. This subject has been tackled with various strategies: see, for example, Bander *et al.* (1979), Kamal (1992*a,b*), Buras and Fleischer (1995, 1998), Atwood *et al.* (1996), Ciuchini *et al.* (1997*a,b*), Neubert (1998), and the references therein.

31.5   Classification of decays

We now compare Tables 31.1 and 31.2. We classify the various possible decays in different classes, building upon the work of Grossman and Worah (1997).

- **(c, t)** The $|\Delta C| = 1$ decays numbered 2, 3, 6, and 7 in Table 31.1 do not get contributions from gluonic penguins. We subdivide these four decays into eight classes, according to which quark pairs hadronize into mesons. The four colour-suppressed decays (C) have two neutral mesons in the final state:

  (c1) $\bar{b} \to [\bar{c}u]\bar{s}$,   (c2) $\bar{b} \to [\bar{u}c]\bar{s}$,   (c3) $\bar{b} \to [\bar{c}u]\bar{d}$,   (c4) $\bar{b} \to [\bar{u}c]\bar{d}$.

  The four colour-allowed decays (T) have two charged mesons in the final state:

  (t1) $\bar{b} \to \bar{c}[u\bar{s}]$,   (t2) $\bar{b} \to \bar{u}[c\bar{s}]$,   (t3) $\bar{b} \to \bar{c}[u\bar{d}]$,   (t4) $\bar{b} \to \bar{u}[c\bar{d}]$.

- **(p)** There are four quark decays which involve only down-type quarks and get no contributions from tree-level diagrams. We denote them $\bar{b} \to [\bar{d}_k q']\bar{q}'$. The order of the quarks in this notation is important: the last two positions correspond to the quark–antiquark pair arising out of the gluon. One has (p1) $\bar{b} \rightarrow [\bar{s}s]\bar{s}$,   (p2) $\bar{b} \rightarrow [\bar{d}s]\bar{s}$,   (p3) $\bar{b} \rightarrow [\bar{s}d]\bar{d}$,   (p4) $\bar{b} \rightarrow [\bar{d}d]\bar{d}$.

- **(pt, pc, c)** Other quark decays get contributions from both tree-level diagrams and gluonic penguins. If the two mesons in the final state are charged, one gets contributions from T and P:

    (pt1) $\bar{b} \rightarrow [\bar{s}c]\bar{c}$,   (pt2) $\bar{b} \rightarrow [\bar{d}c]\bar{c}$,   (pt3) $\bar{b} \rightarrow [\bar{s}u]\bar{u}$,   (pt4) $\bar{b} \rightarrow [\bar{d}u]\bar{u}$.

    If the final-state mesons are neutral there are two possibilities. In the case of a $c\bar{c}$ meson, there is a colour-suppressed tree-level diagram and a singlet-penguin contribution:

    (pc1) $\bar{b} \rightarrow \bar{s}[c\bar{c}]$,   (pc2) $\bar{b} \rightarrow \bar{d}[c\bar{c}]$.

    One might think that the decays $\bar{b} \rightarrow \bar{d}_k[u\bar{u}]$ only have a C contribution. However, as we shall see in the next section, this is not true for the physical states, which are combinations of $u\bar{u}$ and $d\bar{d}$ pairs:

    (c5) $\bar{b} \rightarrow \bar{s}[u\bar{u}]$,   (c6) $\bar{b} \rightarrow \bar{d}[u\bar{u}]$.

In addition to the diagrams discussed up to now in this section, there are also the singlet penguins $\bar{b} \rightarrow \bar{d}_k \eta_1$ and $\bar{b} \rightarrow \bar{d}_k \phi_1$. These multi-gluon penguins—and those appearing in (pc1) and (pc2)—cannot be estimated reliably, but they are usually taken to be smaller than diagrams (p) and (pt), because of the (colour-matching) extra gluons required. However, rescattering effects may play an important role in all decays for which singlet penguins make a significant contribution.

31.6 Decays into CP eigenstates

It looks as though the best channels to search for CP-violating phases would be those with only one tree-level contribution: (t) and (c). This is not the case for two reasons. Firstly, in the naive quark analysis just described one does not take into account the fact that some mesons in the final state have several quark contents. This occurs, for example, with π^0, ϕ and K_S. Secondly, the cleanest measurements of CKM phases arise from CP-eigenstate final states, because the hadronic uncertainties cancel in the ratio $\bar{A}_f/A_f$ when the decay is dominated by a single diagram. This severely restricts the number of final states of interest.

Most final states which are CP eigenstates consist of two neutral mesons. Final states with charged mesons can only be CP eigenstates if the charged mesons form a meson–antimeson pair. It is clear that one cannot reach any such CP eigenstate, with two charged mesons in the final state, using any of the quark decays classified as (t). But this is possible in the case of the quark decays with both tree and penguin contributions (pt). The relevant channels are those in Table 31.3, or those in which the mesons in the final state are substituted by the corresponding resonances. Unfortunately, these channels do not have only a tree-level contribution, they also have a penguin contribution. The latter may spoil the extraction of the CKM phases of the former. For the $\bar{b}s \rightarrow [\bar{s}c][\bar{c}s]$ processes there is no problem, because the gluonic penguin has the same phase as the tree-level diagram. The $\bar{b}s \rightarrow [\bar{s}u][\bar{u}s]$ processes are problematic, because the tree-level diagrams are CKM-suppressed (a factor λ^2) relative to the gluonic penguin,

Table 31.3 *Decays into CP eigenstates with two charged mesons.*

Type	Quark process	Tree-level CKM factor	Dominant-penguin CKM factor	Sample channels
(pt1)	$bs \to [\bar{s}c][\bar{c}s]$	$A\lambda^2$	$-A\lambda^2$	$B_s \to D_s^+ D_s^-$
(pt3)	$\bar{b}s \to [\bar{s}u][\bar{u}s]$	$A\lambda^4 R_b e^{i\gamma}$	$-A\lambda^2$	$B_s \to K^+ K^-$
(pt2)	$\bar{b}d \to [\bar{d}c][\bar{c}d]$	$-A\lambda^3$	$A\lambda^3 R_t e^{-i\beta}$	$B_d \to D^+ D^-$
(pt4)	$\bar{b}d \to [\bar{d}u][\bar{u}d]$	$A\lambda^3 R_b e^{i\gamma}$	$A\lambda^3 R_t e^{-i\beta}$	$B_d \to \pi^+ \pi^-$

Table 31.4 *Decays into final states with $D_{f_{cp}}$.*

Type	Quark process	CKM factor	Sample B_d decay	Sample B_s decay
(c1)	$b \to [\bar{c}u]\bar{s}$	$A\lambda^3$		
(c2)	$\bar{b} \to [\bar{u}c]\bar{s}$	$A\lambda^3 R_b e^{i\gamma}$	$D_{f_{cp}} K_S$	$D_{f_{cp}} \phi$
(c3)	$b \to [\bar{c}u]d$	$A\lambda^2$		
(c4)	$\bar{b} \to [\bar{u}c]\bar{d}$	$-A\lambda^4 R_b e^{i\gamma}$	$D_{f_{cp}} \pi^0$	$D_{f_{cp}} K_S$

offsetting the usual $|p_t/t|$ suppression of the penguin. These decays are even likely to be dominated by the penguin amplitudes, with the tree-level amplitudes acting here as the pollutant. The other two processes also have different tree-level and gluonic-penguin CKM phases, and the size of the gluonic-penguin amplitude may well be larger than 10% of the tree-level amplitude.

We turn to the decays (c1) through (c4). Observing a CP eigenstate with two neutral mesons in the final state originating from any of these decays is possible, but requires that $D_{f_{cp}}$ be present. As a result, both the $\bar{b} \to \bar{c}u\bar{s}$ and the $\bar{b} \to \bar{u}c\bar{s}$ (or both the $\bar{b} \to \bar{c}ud$ and the $\bar{b} \to \bar{u}c\bar{d}$) tree-level diagrams contribute. Some sample channels and the respective CKM factors are given in Table 31.4. For the decays in the first part of Table 31.4, one has two tree-level diagrams of similar magnitude but different weak phase. Extraction of the weak phases may be done by exploiting the relation of these decay channels with the flavour-specific decays into D^0 and $\bar{D^0}$. Those methods will be presented in Chapters 36 and 37, but they require accurate measurements of several branching ratios. For the decays in the second part of Table 31.4, there is a doubly-Cabbibo-suppressed amplitude (in the fourth line) which should not be much larger than 5% of the dominant amplitude (in the third line). This may allow the extraction of the relevant weak phase—see § 34.5.

The next-best candidates for having the decay dominated by a single diagram seem to be (c5) and (c6). However, these channels correspond to final states having a $u\bar{u}$ component, and the corresponding physical mesons also have a $d\bar{d}$ component. Then, the decays $\bar{b}q \to [\bar{s}q][u\bar{u}]$ (c5) and $\bar{b}q \to [\bar{d}q][u\bar{u}]$ (c6) may also be reached by gluonic penguins $\bar{b}q \to [\bar{s}d][\bar{d}q]$ (p3) and $\bar{b}q \to [\bar{d}d][\bar{d}q]$ (p4), respectively. It is easy to see that all the (c5) and (c6) decays have penguin contributions, with the exception of the (c5) decay of B_s^0, $\bar{b}s \to [\bar{s}s][u\bar{u}]$.

Conversely, one can show by inspection that all the final states reached by the penguin decays $\bar{b} \to [\bar{s}d]\bar{d}$ (p3) and $\bar{b} \to [\bar{d}d]\bar{d}$ (p4) have a $d\bar{d}$ meson in the final state—the exception being the $B_s^0 \to KK$ decays, which are dominated by (p3). Hence, with that exception, all the corresponding final states can also be reached by $\bar{b}q \to [\bar{s}q][u\bar{u}]$ (c5) and $\bar{b}q \to [\bar{d}q][u\bar{u}]$ (c6), respectively.

To summarize: in general, the colour-suppressed (c5) and (c6) amplitudes interfere with the gluonic-penguin (p3) and (p4) amplitudes, respectively. The exceptions occur for the $\bar{b} \to \bar{s}$ decays of B_s^0. The transitions $B_s^0 \to [\bar{s}d][s\bar{d}]$—such as $B_s^0 \to K_S K_S$ and $B_s^0 \to K^0 \overline{K^0}$—only have penguin (p3) amplitudes. The decays $B_s^0 \to [\bar{s}s][u\bar{u}]$—such as $B_s^0 \to \phi\pi^0$—only have colour-suppressed tree-level (c5) amplitudes. The last exception occurs because, in the gluonic penguin, the $q\bar{q}$ pair is created by the gluon in a zero-isospin state; thus, it cannot form a π^0, which has isospin 1.

In Table 31.5 we give the relevant CKM factors. One sees that the gluonic penguin has a weak phase different from that of the tree-level diagram. Moreover, for the processes in the first part of Table 31.5 the tree-level amplitudes are CKM-suppressed with respect to the penguin ones. This suggests that the gluonic penguin may dominate the tree-level diagram in these decays. For the processes in the second part of Table 31.5 the situation is the opposite, with those decays being dominated by tree-level diagrams, but subject to a sizeable penguin pollution. The penguin pollution may be removed for the $B_d^0 \to \pi\pi$ channels, by combining all $\pi\pi$ final states via an isospin analysis due to Gronau and London (1990)—see § 35.2.

We next consider the decays (pc1) and (pc2), which are listed in Table 31.6. The dominant contributions come from tree-level, colour-suppressed diagrams. These are corrected by (multi-gluon) singlet-penguin contributions. For the (pc1) processes in the first line the penguins have the same weak phase as the dominant tree-level contribution—these should be the cleanest channels to extract weak phases. The (pc2) processes of the second line get different phases from the tree-level and penguin amplitudes. The penguins should in principle be 5% of the

Table 31.5 *Interfering (c5) and (p3) decays, or (c6) and (p4) decays. In the first line, the decays of B_s^0 do not present interference: either they have only (c5) contributions—e.g. $B_s^0 \to \phi\pi^0$—or they have only (p3) contributions—e.g. $B_s^0 \to K_S K_S$.*

Type	Quark process	Tree-level CKM factor	Dominant-P CKM factor	Sample B_d^0 decays	Sample B_s^0 decays
(c5)	$b \to \bar{s}[u\bar{u}]$	$A\lambda^4 R_b e^{i\gamma}$	—		
				$K_S \pi^0$	see text
(p3)	$\bar{b} \to [\bar{s}d]\bar{d}$	—	$-A\lambda^2$		
(c6)	$b \to d[u\bar{u}]$	$A\lambda^3 R_b e^{i\gamma}$	—		
				$\pi^0\pi^0$	$K_S\pi^0$
(p4)	$\bar{b} \to [\bar{d}d]\bar{d}$	—	$A\lambda^3 R_t e^{-i\beta}$		

Table 31.6 *Tree-level colour-suppressed decays which also get contributions from singlet penguins.*

Type	Quark process	Tree-level CKM factor	Dominant-penguin CKM factor	Sample B_d^0 decays	Sample B_s^0 decays
(pc1)	$b \to \bar{s}[c\bar{c}]$	$A\lambda^2$	$-A\lambda^2$	$J/\psi K_S$	$J/\psi \phi$
(pc2)	$\bar{b} \to \bar{d}[c\bar{c}]$	$-A\lambda^3$	$A\lambda^3 R_t e^{-i\beta}$	$J/\psi \pi^0$	$J/\psi K_S$

Table 31.7 *Pure-penguin decays.*

Type	Quark process	P_t CKM factor	P_c CKM factor	Sample B_d^0 decays	Sample B_s^0 decays
(p1)	$b \to [\bar{s}s]\bar{s}$	$-A\lambda^2$	$A\lambda^2$	ϕK_S	$\phi\phi$
(p2)	$\bar{b} \to [\bar{d}s]\bar{s}$	$A\lambda^3 R_t e^{-i\beta}$	$-A\lambda^3$	$K_S K_S, K^0\overline{K^0}$	ϕK_S

tree-level amplitudes, but the multi-gluon structure of the singlet penguins may entail a further suppression.

We finally consider the two pure-penguin processes (p1) and (p2) in Table 31.7. For the processes in the first line all penguin diagrams carry the same phase, to leading order in the λ expansion. These processes also receive contributions from the singlet penguin $\bar{b} \to \bar{s}\phi_1$, which carries the same weak phase. However, one must keep in mind that ϕ also has small $u\bar{u}$ and $d\bar{d}$ components.[86] Hence, these decays can also be mediated by $\bar{b} \to [\bar{u}u]\bar{s}$ tree-level C diagrams, though this effect is estimated to be small—about 1% of the penguin amplitude (Grossman and Worah 1997).

For the processes in the second line of Table 31.7 the penguins P_t and P_c carry different phases. Fleischer (1994c) has shown that the presence of the subdominant contributions may lead to CP-violating asymmetries as large as 50%.

A special note must be made about the $\bar{b} \to \bar{d}$ decays of B_d^0. Barring rescattering effects, the decays $B_d^0 \to KK$—in the second line of Table 31.7—do not receive singlet-penguin contributions. Conversely, the decay $B_d^0 \to \phi\pi^0$ is mediated mostly by the singlet penguins $\bar{b} \to \bar{d}\phi_1$, with a (possibly small) correction comming from the $\bar{b} \to [\bar{u}u]\bar{d}$ tree-level diagram because of the small $u\bar{u}$ component of ϕ.

Final states with two vector mesons, such as $J/\psi\phi$ and $\phi\phi$, are in general admixtures of CP-even and CP-odd components, because they may have various orbital angular momenta. The CP-violating asymmetries in such cases suffer from partial cancellations among components with different orbital angular momentum. A study of the angular correlations should in principle allow a separation of the different angular-momentum components, at the cost of statistics (Aguila and Nelson 1986; Kayser *et al.* 1990; Dunietz *et al.* 1991).

---

[86]The possible impact of this effect must be investigated for each particular decay with a ϕ in the final state, and not just for pure-penguin decays.

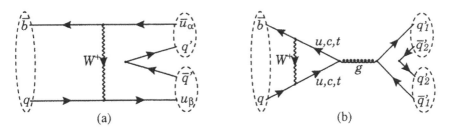

FIG. 31.4. (a) W-exchange diagram, and (b) penguin-annihilation diagram.

Tables 31.3–31.7 allow us to search for our Holy Grail: decays into CP eigenstates dominated by a single weak phase. The best channels are those in which the dominant and the leading sub-dominant amplitude have the same weak phase or, else, those in which the sub-dominant amplitudes are very suppressed. Out of the B_d^0 decays, $B_d^0 \to J/\psi K_S$ is definitely the best candidate for a clear signal. Decays in which the dominant and the leading sub-dominant amplitude have different phases and in which there is no CKM suppression of the latter, could suffer from pollution of up to 10% to 20%, should $|p_t/t| \sim \lambda$. Worse off are all the decays in which there are several phases at play, with the dominant diagram being CKM suppressed, unless the suppression is large enough for the decays to become penguin dominated—in which case, it is the measurement of the weak phase of the penguin diagram which is subject to a tree-level pollution. Such decays can even be affected by other diagrams, like those of § 31.7, involving the spectator quark, and the electroweak penguin diagrams to be presented in § 31.8. They are also more likely to be affected by rescattering effects.

31.7 Diagrams involving the spectator quark

Besides the tree-level diagrams and gluonic penguins discussed in the previous sections, B decays can also be mediated by diagrams involving the spectator quark. The W-exchange (denoted E) and penguin-annihilation (denoted PA) diagrams are shown in Figs. 31.4 (a) and 31.4 (b), respectively. They only contribute to neutral-meson decays. The W-annihilation diagram (denoted A) in Fig. 31.5 only contributes to the decays of charged-B mesons.

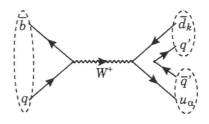

FIG. 31.5. W-annihilation diagram.

It is usually assumed that these diagrams are suppressed by f_B/m_B relative to the T, C, and P diagrams (Gronau et al. 1994a,b, 1995). This is because of the smallness of the B-meson wave function at the origin. One may therefore write

$$\left|\frac{e}{t}\right| \sim \left|\frac{a}{t}\right| \sim \left|\frac{pa}{p}\right| \sim \frac{f_B}{m_B} \sim 0.05 \sim \lambda^2, \tag{31.13}$$

where the lower-case letters remind us that the CKM matrix elements have been factored out. However, rescattering effects may affect the estimate in eqn (31.13), leading to a λ rather than λ^2 suppression (Blok et al. 1997).[87] Rescattering effects (final-state interactions) may also bring considerable T, C, and P contributions into processes where one would naively expect only A, E, or PA contributions (Blok et al. 1997). This point has been emphasized by Neubert (1998), Gérard and Weyers (1999), and others.

31.7.1 Quark diagrams and SU(3) symmetry

We are now in a position to discuss the connection between flavour-SU(3) symmetry and the quark diagrams presented above. The aim is to use this symmetry to relate different decay channels (Zeppenfeld 1981; Chau et al. 1991; Gronau et al. 1994a,b, 1995). Consider the decays $B \to PP'$ of a B meson into two light pseudoscalars belonging to the octet of SU(3), and the decomposition of these decays in terms of SU(3) representations. Zeppenfeld (1981) has shown explicitly that, if one ignores electroweak penguins, then there is a relation between this classification and the six diagrams T, C, P, E, A, and PA. Indeed, the effective Hamiltonian for these decays contains operators $\bar{b} \to \bar{k}u\bar{u}$ and $\bar{b} \to \bar{k}$ ($k = d, s$), which transform as a 3^*, 6, or 15^* of SU(3). When combined with the triplet formed by $B^+ = \bar{b}u$, $B_d^0 = \bar{b}d$, and $B_s^0 = \bar{b}s$, this leads to one singlet of SU(3), three octets, one 10, one 10^*, and one 27. This is to be contracted with the state made out of two (properly symmetrized) pseudoscalars, with representations 1, 8, and 27. As a result, there are five possible amplitudes for such decays: a singlet, three octets, and one 27-plet. Specifically, the five independent reduced matrix elements are $\langle 1||3^*||3\rangle$, $\langle 8||3^*||3\rangle$, $\langle 8||6||3\rangle$, $\langle 8||15^*||3\rangle$, and $\langle 27||15^*||3\rangle$. These, in turn, can be related to five independent combinations of the diagrams T, C, P, E, A, and PA (Zeppenfeld 1981; Chau et al. 1991; Gronau et al. 1994a,b, 1995).

The main interest of the flavour-symmetry relations in the study of CP violation lies in the possibility of extracting weak phases from the measured asymmetries, even when the decay is not dominated by a single weak phase. The first examples of this idea have appeared in connection with the penguin pollution present in the extraction of α from the time-dependent asymmetry in $B_d^0 \to \pi^+\pi^-$. Isospin was first used in this context by relating this decay to the decays $B_d^0 \to \pi^0\pi^0$, $B^+ \to \pi^+\pi^0$, and the CP-conjugate modes (Gronau and London 1990). With the same aim, Silva and Wolfenstein (1994) have used flavour

---

[87] A comparable suppression $2\pi f_B/m_B \sim \lambda$ characterizes the *inclusive* annihilation amplitude (Khoze et al. 1987). The reason is that the E, A, and PA diagrams have two quarks in the final state, and therefore enjoy a 2π phase-space enhancement with respect to the three-quark final states in T, C, and P.

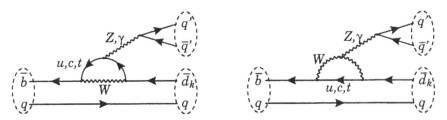

Fig. 31.6. Colour-allowed electroweak penguins.

SU(3) to relate the rates of $B_d^0 \to K^+\pi^-$ and $B_d^0 \to \pi^+\pi^-$, thereby removing the penguin pollution. A complete analysis of SU(3) relations and their application to the determination of CP-violating quantities has since been developed, starting with the articles by Gronau *et al.* (1994*a,b*, 1995).

31.8 Electroweak penguins

Besides the diagrams already discussed, one must also consider the penguin diagrams in which the gluon is substituted by a Z boson or by a photon. They are known as electroweak (EW) penguins. These diagrams always appear together with W^+W^- box diagrams, cf. Figs. 20.3 and 20.4; it is the sum of the box, Z penguin, and photon penguin which is gauge-invariant. We stress that the expression 'electroweak penguins' should always be taken to include the box diagrams too.

The importance of EW penguins in B decays was explored by Fleischer (1994*a,b*), by Deshpande and He (1994, 1995*a*), and by Deshpande *et al.* (1995). In Fig. 31.6 we display the $b \to \bar{d}_k[q'\bar{q}']$ colour-allowed electroweak penguins, denoted P_{EW}. They contribute to decays with a neutral $q'\bar{q}'$ meson in the final state. There will be two neutral mesons in the final state, for neutral-B decays; or one neutral and one charged meson, for charged-B decays. Contrary to what happens with gluonic penguins, for EW penguins the $q'\bar{q}'$ pair is not a flavour singlet, because the photon and the Z boson couple differently to up-type and to down-type quarks; the $q'\bar{q}'$ pair may be the neutral component of an isovector, such as π^0 or ρ^0.

The colour-suppressed electroweak penguins contribute to the decays $\bar{b} \to [\bar{d}_k q']\bar{q}'$ and are displayed in Fig. 31.7. They are denoted P_{EW}^C. The difference in

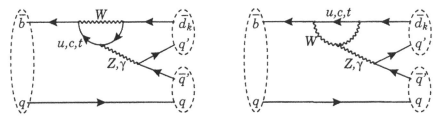

Fig. 31.7. Colour-suppressed electroweak penguins.

colour structures is expected to yield

$$\left| \frac{p_{EW}^C}{p_{EW}} \right| \sim \lambda. \tag{31.14}$$

Electroweak penguins involve the same CKM factors as gluonic penguins, namely, $V_{\gamma b}^* V_{\gamma k}$ for an up-type quark u_γ in the loop. The essential difference between gluonic penguins and EW penguins is that the former couple to all $q' \bar{q}'$ pairs equally, while the latter couple differently to up-type quark–antiquark pairs and to down-type quark–antiquark pairs. Still, quark–antiquark pairs with the same charge have identical couplings. This is obvious for Z-boson and photon penguins; it is also true for the dominant box contribution, though some highly suppressed box contributions actually differ between quark–antiquark pairs of the same electric charge. As a result, electroweak penguins violate isospin, but not the SU(3) U-spin rotation $s \leftrightarrow d$.

Just as gluonic penguins, electroweak penguins enjoy a GIM enhancement, i.e., they increase with increasing top-quark mass. The ratio of EW penguins to gluonic penguins may be roughly estimated to be

$$\left| \frac{p_{EW}}{p} \right| \sim \frac{\alpha_{\text{weak}}(m_b)}{\alpha_{\text{strong}}(m_b)} \frac{m_t^2/m_Z^2}{\ln(m_t^2/m_b^2)} \sim 0.1, \tag{31.15}$$

with CKM factors and operator matrix elements factored out.

31.9 Conclusion

In this chapter we have discussed the various diagrams involved in B decays. For the most part, we have assumed that the largest contributions to any decay are those from tree-level diagrams and from gluonic penguins. In particular, we have neglected diagrams involving the spectator quark: E, A, and PA. Moreover, we have assumed that there were no rescattering effects feeding these topologies. We will keep with standard practice; we mention these effects as the need arises, but we do not include them systematically in the analysis. We have also neglected electroweak penguins. These are expected to be important in some penguin-dominated decays—see, for example, the articles by Fleischer (1994a,b), Deshpande and He (1994, 1995a), and Deshpande et $al.$ (1995). Their effects will be mentioned whenever relevant.

We have considered how the dominant diagrams—tree-level diagrams and gluonic penguins—enter in the decays into CP eigenstates. We stress that the diagrammatic language which has been introduced applies to any other B decay as well, and not only to the very special case of decays into CP eigenstates. Decays of charged-B mesons and decays of neutral-B mesons into non-CP eigenstates involve the same diagrams and the same CKM factors.

The problem with these other channels is that one generally needs to worry about two separate amplitudes. For decays into CP eigenstates with more than one relevant weak phase, this is due to the appearance of a cosine time dependence, proportional to the difference between the squared decay amplitudes

$|A_f|^2 - |\bar{A}_{\bar{f}}|^2$, in addition to the usual sine time-dependent term, proportional to $\mathrm{Im}\,\lambda_f$. That new term measures direct CP violation. A similar direct CP violation factor is involved in the decays of charged B mesons. For the case of decays of neutral B mesons into final states that are not CP eigenstates but which are still dominated by a single weak phase, the problem is that the two amplitudes A_f and $\bar{A}_f$ are no longer related. In this case, λ_f does not depend only on a weak phase, but also on the difference between the strong phases of $\bar{A}_f$ and of A_f.

DECAY AMPLITUDES: EFFECTIVE HAMILTONIAN

32.1 Effective Hamiltonian for B decays

The diagrammatic approach discussed in the previous chapter is useful to get an order-of-magnitude estimate of the importance of the various contributions to each amplitude, and to find approximate relations among the amplitudes for different decays. However, rigorous computations require a correct treatment of the QCD corrections, and this is best accomplished using the operator product expansion (OPE) already discussed in the context of kaon decays—see Chapter 20. The OPE allows the construction of a low-energy effective Hamiltonian as a sum of local operators multiplied by Wilson coefficients. The effective Hamiltonian is the sum of pieces, each of them of the type

$$\frac{G_F}{\sqrt{2}} V_{CKM} C_n(\mu) Q_n, \tag{32.1}$$

with a definite CKM factor V_{CKM}. The Wilson coefficients $C_n(\mu)$ depend on the renormalization scale μ and on the renormalization scheme. These dependences should cancel in the physical amplitude

$$A(i \to f) = \frac{G_F}{\sqrt{2}} V_{CKM} \sum_n C_n(\mu) \langle f | Q_n | i \rangle (\mu) \tag{32.2}$$

with the renormalization-scale and renormalization-scheme dependences of the operator matrix elements $\langle f | Q_n | i \rangle (\mu)$. The cancellations may involve simultaneously many Wilson coefficients and matrix elements. The scale $\mu \sim 5 \, \text{GeV}$ separates the high-energy QCD corrections, which are treated perturbatively and included in the Wilson coefficients, from the low-energy non-perturbative confinement effects implicitly contained in the operator matrix elements. However, upon change of μ or of the renormalization scheme, contributions previously included in the matrix elements of some operator may move into the Wilson coefficients or into the matrix elements of other operators. One uses the renormalization group to run the Wilson coefficients from high to low energies. We refer the reader to the review articles by Buchalla *et al.* (1996) and by Buras and Fleischer (1998), where detailed discussions of the calculations can be found, together with extensive references to the original literature.

   The relevant diagrams are analogous to those depicted in the figures of Chapters 20 and 31. The operators that one needs are

- **Current-current operators:**

$$Q_1^{\alpha\beta k} \equiv (\bar{b}_x \alpha_y)_{V-A} (\bar{\beta}_y k_x)_{V-A},$$
$$Q_2^{\alpha\beta k} \equiv (\bar{b}\alpha)_{V-A} (\bar{\beta} k)_{V-A};$$

(32.3)

- **Gluonic-penguin operators:**

$$Q_3^k \equiv (\bar{b}k)_{V-A} \sum_q (\bar{q}q)_{V-A},$$
$$Q_4^k \equiv (\bar{b}_x k_y)_{V-A} \sum_q (\bar{q}_y q_x)_{V-A},$$
$$Q_5^k \equiv (\bar{b}k)_{V-A} \sum_q (\bar{q}q)_{V+A},$$
$$Q_6^k \equiv (\bar{b}_x k_y)_{V-A} \sum_q (\bar{q}_y q_x)_{V+A};$$

(32.4)

- **Electroweak-penguin operators:**

$$Q_7^k \equiv \tfrac{3}{2}(\bar{b}k)_{V-A} \sum_q e_q (\bar{q}q)_{V+A},$$
$$Q_8^k \equiv \tfrac{3}{2}(\bar{b}_x k_y)_{V-A} \sum_q e_q (\bar{q}_y q_x)_{V+A},$$
$$Q_9^k \equiv \tfrac{3}{2}(\bar{b}k)_{V-A} \sum_q e_q (\bar{q}q)_{V-A},$$
$$Q_{10}^k \equiv \tfrac{3}{2}(\bar{b}_x k_y)_{V-A} \sum_q e_q (\bar{q}_y q_x)_{V-A};$$

(32.5)

- **Magnetic-penguin operators** (see Fig. 32.1):

$$Q_{7\gamma}^k \equiv \frac{e}{8\pi^2} m_b \left[\bar{b}\sigma^{\mu\nu}(1+\gamma_5)k \right] F_{\mu\nu},$$
$$Q_{8G}^k \equiv \frac{g_s}{16\pi^2} m_b \left[\bar{b}_x \sigma^{\mu\nu}(1+\gamma_5)\lambda_a^{xy} k_y \right] G_{\mu\nu}^a.$$

(32.6)

The operators in eqns (32.3)–(32.5) are entirely analogous to the ones introduced in Chapter 20 for the decays of the kaons. Still, as the decays of $\bar{b}$ may involve

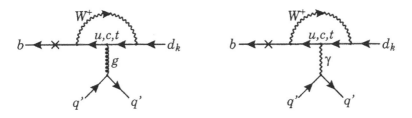

FIG. 32.1. Diagrams which generate the magnetic-penguin operators.

quarks with different flavours in the final state, one needs to introduce more operators, and a more complex notation. Thus, k stands for a light down-type quark, either d or s; while α and β stand for either u or c (out of the current-current operators in eqn 32.3, those with $\alpha = \beta$ correspond to $\Delta C = \Delta U = 0$ decays, while the $|\Delta C| = |\Delta U| = 1$ decays involve the operators with $\alpha \neq \beta$). Furthermore, x and y are colour indices, just as in Chapter 20. In the sums, q runs over the four quarks u, d, s, and c; the electromagnetic-penguin operators involve the electric charges of the quarks: $e_q = 2/3$ for $q = u$ and $q = c$, and $e_q = -1/3$ for $q = d$ and $q = s$. The Dirac structures V and A refer to the vector and axial-vector currents, respectively; $V \pm A$ stands for $\gamma^\mu (1 \pm \gamma_5)$.

The magnetic-penguin operators in eqn (32.6) correspond to the effect of the non-negligible mass of the bottom quark in penguin diagrams (see Fig. 32.1); in practice, only the external-mass effects on gluonic and photonic penguins need to be considered. These operators are important, for example, for the decay $b \to s\gamma$. This topic is beyond the scope of this book, and we shall drop these operators henceforth. We refer the reader to the reviews by Buchalla $et\ al.$ (1996) and by Buras and Fleischer (1998).

An important feature of the Wilson coefficients is that they do not depend on the flavour k of the down-type quark; that is, the Wilson coefficients are the same for $k = d$ and $k = s$.

As an example, consider the effective Hamiltonian responsible for $\Delta C = \Delta U = 0$ decays,

$$\mathcal{H}_{\text{eff}} = \frac{G_F}{\sqrt{2}} \sum_{k=d,s} \left[\sum_{\alpha=u,c} \lambda_\alpha^k \sum_{n=1}^{2} C_n(\mu) Q_n^{\alpha\alpha k} + \lambda_t^k \sum_{n=3}^{10} C_n(\mu) Q_n^k \right] + \text{H.c.}. \quad (32.7)$$

We have used the definition

$$\lambda_\alpha^k = V_{\alpha b}^* V_{\alpha k}. \quad (32.8)$$

Unitarity implies $\lambda_u^k + \lambda_c^k + \lambda_t^k = 0$.

The Wilson coefficients are renormalization-scheme- and renormalization-scale-dependent. For instance, Buras and Fleischer (1998) give, in the $\overline{MS}$ renormalization scheme with anti-commuting γ_5, for $\mu = \overline{m}_b(m_b)$, $m_t = 170\,\text{GeV}$, and with $\Lambda_{\text{QCD}}^{(5)} = 225\,\text{MeV}$,

$$
\begin{aligned}
C_1 &= -0.185, & C_2 &= 1.082, & C_3 &= 0.014, \\
C_4 &= -0.035, & C_5 &= 0.009, & C_6 &= -0.041, \\
C_7 &= -0.002\,\alpha, & C_8 &= 0.0054\,\alpha, & C_9 &= -1.292\,\alpha, \\
C_{10} &= 0.263\,\alpha,
\end{aligned}
\quad (32.9)
$$

in the next-to-leading-logarithmic approximation. Here, α is the electromagnetic fine-structure constant. Due to the large value of C_9, electroweak-penguin operators need to be considered in certain decays where tree-level diagrams are either heavily CKM-suppressed or altogether absent.

Let us define (Buras and Fleischer 1998)

$$Q^{\alpha k} \equiv \sum_{n=1}^{2} C_n(\mu) Q_n^{\alpha \alpha k} - \sum_{n=3}^{10} C_n(\mu) Q_n^{k}. \tag{32.10}$$

We then rewrite the effective Hamiltonian in eqn (32.7) as

$$\mathcal{H}_{\text{eff}} = \frac{G_F}{\sqrt{2}} \sum_{k=d,s} \sum_{\alpha=u,c} \lambda_\alpha^k Q^{\alpha k} + \text{H.c.}. \tag{32.11}$$

This expression shows explicitly that, in the standard model, one may use the unitarity of the CKM matrix to rewrite the decay amplitudes in terms of only two independent weak phases.

32.2 On the evaluation of hadronic matrix elements

Although the computations of the Wilson coefficients have become very elaborate, one still faces huge uncertainties in the estimates of the matrix elements of the operators. In B decays, the standard procedure has been to assume factorization, splitting the matrix element of each four-fermion operator into the product of matrix elements of two quark bilinears. This is achieved by inserting the vacuum state in all possible ways. The resulting matrix elements of quark bilinears are parametrized with form factors, one for each momentum structure consistent with Lorentz invariance and parity. These form factors are directly determined from experiment whenever possible; most often, they must be calculated within a model. Many recent articles use the relativistic BSW model (Bauer et al. 1985, 1987) to evaluate the matrix elements. In this approach, the matrix elements are written in terms of physical quantities and are, therefore, renormalization-scheme- and renormalization-scale-independent. Thus, they do not satisfy the requirements in the previous section.

Admittedly, the computation of the matrix elements is an awkward, unreliable step of the computation of the amplitudes, and any results must be taken as mere indications. Should the matrix elements be determined from the lattice, one will be able to get definite predictions.

32.3 CP transformations

In this section we study how the effective Hamiltonian transforms under CP, and the corresponding relations among the decay amplitudes that may be derived.

32.3.1 $\Delta C = 0 = \Delta U$ decays

We use the CP transformation of the quark field operators

$$\begin{aligned}
(\mathcal{CP})\,\psi\,(\mathcal{CP})^\dagger &= \exp\left(i\xi_\psi\right)\gamma^0 C\bar{\psi}^T, \\
(\mathcal{CP})\,\overline{\psi}\,(\mathcal{CP})^\dagger &= -\exp\left(-i\xi_\psi\right)\psi^T C^{-1}\gamma^0,
\end{aligned} \tag{32.12}$$

or else we may use directly the results in Table 3.1 for the transformation properties of quark bilinears. We find

$$(CP)\, Q_n^{\alpha\beta k}\, (CP)^\dagger = \exp\left[i\,(\xi_k - \xi_b + \xi_\alpha - \xi_\beta)\right] Q_n^{\alpha\beta k\,\dagger} \ \text{ for } n = 1, 2;$$
$$(CP)\, Q_n^{k}\, (CP)^\dagger = \exp\left[i\,(\xi_k - \xi_b)\right] Q_n^{k\,\dagger} \ \text{ for } n = 3, \ldots, 10. \tag{32.13}$$

Therefore, as the Wilson coefficients are real,

$$(CP)\, Q_n^{\alpha k}\, (CP)^\dagger = \exp\left[i\,(\xi_k - \xi_b)\right] Q_n^{\alpha k\,\dagger}. \tag{32.14}$$

Let us assume that the final states that we are considering in these $\Delta C = 0 = \Delta U$ decays only include particles of definite flavour—thus excluding the presence of mesons like K_S. In that case, the $\bar{b}q \to q\bar{a}\alpha\bar{k}$ decay leads into a final state f whose quark content is given exclusively by $q\bar{a}\alpha\bar{k}$, with no other component. Similarly, the $b\bar{q} \to \bar{q}a\bar{\alpha}k$ transition leads into a final state $\bar{f}$ whose quark content is given exclusively by $\bar{q}a\bar{\alpha}k$. Under these circumstances, $\langle f|T|B_q^0\rangle$ is obtained directly from the effective Hamiltonian. It is important to note that this is not always the case. If the final state contains particles that are not in a flavour eigenstate, such as K_S or $D_{f_{\rm cp}}$, then one must include in the calculation the transformation from the flavour-eigenstate basis into the basis of the experimentally observed particles. We shall encounter this problem when we calculate the CP-violating asymmetry in $B_d^0 \to J/\psi K_S$. For the moment, we ignore that possibility.

Using the CP transformations of the final and initial states,

$$CP|f\rangle = e^{i\xi_f}|\bar{f}\rangle,$$
$$CP|B_q^0\rangle = e^{i\xi_{B_q}}|\overline{B_q^0}\rangle, \tag{32.15}$$

we find

$$A_f = \frac{G_F}{\sqrt{2}} \sum_{k=d,s} \sum_{\alpha=u,c} \lambda_\alpha^k \langle f|Q^{\alpha k}|B_q^0\rangle$$
$$= e^{i(\xi_{B_q} - \xi_f + \xi_k - \xi_b)} \frac{G_F}{\sqrt{2}} \sum_{k=d,s} \sum_{\alpha=u,c} \lambda_\alpha^k \langle \bar{f}|Q^{\alpha k\,\dagger}|\overline{B_q^0}\rangle, \tag{32.16}$$

to be compared with

$$\bar{A}_{\bar{f}} = \frac{G_F}{\sqrt{2}} \sum_{k=d,s} \sum_{\alpha=u,c} \lambda_\alpha^{k\,*} \langle \bar{f}|Q^{\alpha k\,\dagger}|\overline{B_q^0}\rangle. \tag{32.17}$$

When f is a CP eigenstate, $\bar{f} = f$, $e^{i\xi_f} = \eta_f = \pm 1$, and eqns (32.16) and (32.17) yield

$$\frac{\bar{A}_f}{A_f} = \eta_f e^{-i(\xi_{B_q} + \xi_k - \xi_b)} \frac{\sum_{k=d,s} \sum_{\alpha=u,c} \lambda_\alpha^{k\,*} \langle f|Q^{\alpha k\,\dagger}|\overline{B_q^0}\rangle}{\sum_{k=d,s} \sum_{\alpha=u,c} \lambda_\alpha^k \langle f|Q^{\alpha k\,\dagger}|\overline{B_q^0}\rangle}. \tag{32.18}$$

We see that the ratio reduces to a pure phase if the decay is such that, in the sums in eqn (32.18), only the operators with a certain up-type quark α and a certain down-type quark k contribute.

32.3.2 $\Delta C \neq 0 \neq \Delta U$ decays

The effective Hamiltonian relevant for these decays is

$$
\mathcal{H}_{\text{eff}} = \frac{G_F}{\sqrt{2}} \sum_{k=d,s} \sum_{n=1}^{2} \left[V_{cb}^* V_{uk} C_n(\mu) Q_n^{cuk} + V_{ub}^* V_{ck} C_n(\mu) Q_n^{uck} \right.
$$
$$
\left. + V_{cb} V_{uk}^* C_n(\mu) Q_n^{cuk\,\dagger} + V_{ub} V_{ck}^* C_n(\mu) Q_n^{uck\,\dagger} \right]. \tag{32.19}
$$

This Hamiltonian leads to four amplitudes,

$$
\frac{G_F}{\sqrt{2}} V_{cb}^* V_{uk} C_n(\mu) \langle q\bar{c}u\bar{k}|Q_n^{cuk}|B_q^0\rangle,
$$
$$
\frac{G_F}{\sqrt{2}} V_{ub}^* V_{ck} C_n(\mu) \langle q\bar{u}c\bar{k}|Q_n^{uck}|B_q^0\rangle,
$$
$$
\frac{G_F}{\sqrt{2}} V_{cb} V_{uk}^* C_n(\mu) \langle \bar{q}c\bar{u}k|Q_n^{cuk\,\dagger}|\overline{B_q^0}\rangle, \tag{32.20}
$$
$$
\frac{G_F}{\sqrt{2}} V_{ub} V_{ck}^* C_n(\mu) \langle \bar{q}u\bar{c}k|Q_n^{uck\,\dagger}|\overline{B_q^0}\rangle.
$$

As we have seen in § 31.3.2, for the $S_f \neq 0 \neq D_f$ decays, i.e., for $q \neq k$, only the final states with either a K_S or a K_L are common to B_q^0 and $\overline{B_q^0}$. On the contrary, all final states reached by $S_f = 0 = D_f$ decays—when $q = k$—are common. In this case the amplitudes in eqn (32.20) describe the decays into two final states: f with quark content $q\bar{c}u\bar{q}$; and $\bar{f}$ with quark content $q\bar{u}c\bar{q}$. Those amplitudes therefore correspond to A_f, $A_{\bar{f}}$, $\bar{A}_{\bar{f}}$, and $\bar{A}_f$, respectively. Using the CP-transformation properties of the operators and kets, we find

$$
A_f = e^{i(\xi_{B_q} - \xi_f)} e^{i(\xi_q - \xi_b + \xi_c - \xi_u)} \frac{V_{cb}^* V_{uq}}{V_{cb} V_{uq}^*} \bar{A}_{\bar{f}},
$$
$$
A_{\bar{f}} = e^{i(\xi_{B_q} + \xi_f)} e^{i(\xi_q - \xi_b + \xi_u - \xi_c)} \frac{V_{ub}^* V_{cq}}{V_{ub} V_{cq}^*} \bar{A}_f. \tag{32.21}
$$

However, for decays into non-CP eigenstates this does not give us any information useful for the calculation of λ_f; indeed, $\lambda_f = (q/p)(\bar{A}_f/A_f)$ while, for instance, the first eqn (32.21) relates A_f with $\bar{A}_{\bar{f}}$.

As for CP eigenstates, the results of § 31.3.2 show that all $\Delta C \neq 0 \neq \Delta U$ decays into CP eigenstates involve $D_{f_{cp}}$. (Also, the CP eigenstates to be reached by $\Delta C \neq 0 \neq \Delta U$ and $S_f \neq 0 \neq D_f$ decays are $D_{f_{cp}} K_{L,S}$.) These are defined as the combination of D^0 and $\overline{D^0}$ which decays into the CP eigenstate f_{cp}. Then, the decay amplitudes $\langle f_{cp}|T|D^0\rangle$ and $\langle f_{cp}|T|\overline{D^0}\rangle$ must also be included in the computation—see § 34.4 and 34.5.

CKM PHASES AND INTERFERENCE CP VIOLATION

33.1 Parameters λ_f at tree level

The parameter relevant for interference CP violation in the transitions $B_q^0 \to f$ and $\overline{B_q^0} \to f$ is

$$\lambda_f = \frac{q_{B_q}}{p_{B_q}} \frac{\bar{A}_f}{A_f}. \tag{33.1}$$

In order to find its value in a given model one must compute both q_{B_q}/p_{B_q} and $\bar{A}_f/A_f$ in that model. In this chapter we assume that q_{B_q}/p_{B_q} is given by a phase, but allow this phase to differ from the standard-model (SM) one—cf. eqn (30.33).

As for the decay amplitudes, we assume that they are dominated by SM diagrams, but allow non-unitarity of the CKM matrix to alter the corresponding phases from their allowed values in the SM. We start by assuming that the decay amplitudes are dominated by a single weak phase. In that case, $\bar{A}_f/A_f$ reduces to a phase and, with our assumption that $|q/p| = 1$, so does λ_f. Later, we shall study the effect of subdominant contributions to the decay, still within the SM.

33.1.1 $B_d^0 \to \pi^+ \pi^-$

At quark level, the decay $B_d^0 \to \pi^+ \pi^-$ is $\bar{b} \to \bar{u} u \bar{d}$, while $\overline{B_d^0} \to \pi^+ \pi^-$ is $b \to \bar{u} u d$. Let us assume that the amplitudes are dominated by the tree-level diagrams of the SM. Then, their ratio is

$$\frac{\bar{A}_{\pi^+\pi^-}}{A_{\pi^+\pi^-}} = \frac{V_{ub} V_{ud}^* \langle \pi^+ \pi^- | (\bar{u}b)_{V-A} (\bar{d}u)_{V-A} | \overline{B_d^0} \rangle}{V_{ub}^* V_{ud} \langle \pi^+ \pi^- | (\bar{b}u)_{V-A} (\bar{u}d)_{V-A} | B_d^0 \rangle}. \tag{33.2}$$

We use CP symmetry of the strong interactions to relate the two matrix elements to each other. As in the decays of the neutral kaons to two pions, the state $\pi^+ \pi^-$ is CP-even. Therefore,

$$\frac{\bar{A}_{\pi^+\pi^-}}{A_{\pi^+\pi^-}} = e^{i(\xi_b - \xi_d - \xi_{B_d})} \frac{V_{ub} V_{ud}^*}{V_{ub}^* V_{ud}}. \tag{33.3}$$

Then, using eqn (30.33),

$$\lambda_{B_d^0 \to \pi^+\pi^-} = -\text{sign}\,(B_{B_d}) \frac{V_{tb}^* V_{td}}{V_{tb} V_{td}^*} \frac{V_{ub} V_{ud}^*}{V_{ub}^* V_{ud}}$$

$$= -\text{sign}\,(B_{B_d})\, e^{2i\alpha}. \tag{33.4}$$

(The spurious phases brought about by CP transformations plague the unphysical q_{B_d}/p_{B_d} and $\bar{A}_{\pi^+\pi^-}/A_{\pi^+\pi^-}$ but drop out in their product—as they should, because $\lambda_{B_d^0 \to \pi^+ \pi^-}$ is a physical quantity.) Thus, if there were no penguin diagrams, the rate asymmetries would measure the phase α.

33.1.2 $B_d^0 \to J/\psi K_S$

Let us now consider the $B_d^0 \to J/\psi K_S$ decays. J/ψ has spin one and is CP-even, while K_S has spin zero and is almost CP-even. Since B_d^0 has spin zero, the J/ψ and the K_S of the final state must have a relative $l = 1$ angular momentum. As a result, the $J/\psi K_S$ final state must be CP-odd.

The final state $J/\psi K_S$ is common to B_d^0 and $\overline{B_d^0}$. But, in the spectator-quark approximation, $B_d^0 = \bar{b}d$ only decays to $K^0 = \bar{s}d$ and not to $\overline{K^0} = s\bar{d}$; conversely, $\overline{B_d^0}$ decays to $\overline{K^0}$ but not to K^0. Using the reciprocal basis of § 6.8,

$$\langle \tilde{K}_S | = \tfrac{1}{2}\left(\langle K^0 | p_K^{-1} - \langle \overline{K^0} | q_K^{-1} \right). \tag{33.5}$$

Therefore,

$$\langle J/\psi K_S | T | B_d^0 \rangle = \frac{1}{2p_K} \langle J/\psi K^0 | T | B_d^0 \rangle,$$
$$\langle J/\psi K_S | T | \overline{B_d^0} \rangle = -\frac{1}{2q_K} \langle J/\psi \overline{K^0} | T | \overline{B_d^0} \rangle. \tag{33.6}$$

Thus,

$$\frac{\bar{A}_{J/\psi K_S}}{A_{J/\psi K_S}} = -\frac{p_K}{q_K} \frac{\langle J/\psi \overline{K^0} | T | \overline{B_d^0} \rangle}{\langle J/\psi K^0 | T | B_d^0 \rangle}. \tag{33.7}$$

Whenever B_d^0 decays into a K_S one must include a factor $-p_K/q_K$ in λ. Similarly, a K_L in the final state requires a factor $+p_K/q_K$. The reason is simple: A_f refers to the amplitude of B^0 into a given final state f, and not only to the diagram from which it originates. For decays with K_S or K_L in the final state, the overall amplitude will involve the amplitudes into the flavour eigenstates, *and* the transformation from the flavour into the mass eigenstates.

We now assume that the decay is dominated by the tree-level amplitude. Then,

$$\frac{\bar{A}_{J/\psi K_S}}{A_{J/\psi K_S}} = -\frac{p_K}{q_K} \frac{V_{cb}V_{cs}^* \langle J/\psi \overline{K^0} | (\bar{c}b)_{V-A}(\bar{s}c)_{V-A} | \overline{B_d^0} \rangle}{V_{cb}^* V_{cs} \langle J/\psi K^0 | (\bar{b}c)_{V-A}(\bar{c}s)_{V-A} | B_d^0 \rangle}. \tag{33.8}$$

Using CP symmetry to relate the matrix elements, one obtains

$$\frac{\bar{A}_{J/\psi K_S}}{A_{J/\psi K_S}} = \frac{p_K}{q_K} e^{i(\xi_K - \xi_{B_d} + \xi_b - \xi_s)} \frac{V_{cb}V_{cs}^*}{V_{cb}^* V_{cs}}, \tag{33.9}$$

where a minus sign has been included to take into account the fact that J/ψ and K_S are in a relative $l = 1$ state. Using the q/p factors for the K^0–$\overline{K^0}$ and B_d^0–$\overline{B_d^0}$ systems in eqns (30.18) and (30.33), respectively, we arrive at the final result:

$$\lambda_{B_d^0 \to J/\psi K_S} = \text{sign}\,(B_{B_d}) \frac{V_{cs}V_{cd}^*}{V_{cs}^* V_{cd}} e^{2i\epsilon'} \frac{V_{tb}^* V_{td}}{V_{tb}V_{td}^*} \frac{V_{cb}V_{cs}^*}{V_{cb}^* V_{cs}}$$

$$= \text{sign}\,(B_{B_d})\,e^{-2i\left(\beta-\epsilon'\right)}. \tag{33.10}$$

The analysis is the same for a final state $J/\psi K_L$, except that the extra factor due to K^0–$\overline{K^0}$ mixing is $+p_K/q_K$ instead of $-p_K/q_K$. Hence,

$$\lambda_{B_d^0\to J/\psi K_L} = -\lambda_{B_d^0\to J/\psi K_S} = -\text{sign}\,(B_{B_d})\,e^{-2i\left(\beta-\epsilon'\right)}. \tag{33.11}$$

The difference in sign might be expected from the fact that $J/\psi K_L$ is CP-even, while $J/\psi K_S$ is CP-odd. (In both cases, J/ψ and the kaon are in a p wave.)

33.1.3 $B_s^0 \to \rho K_S$

The computation of $\lambda_{B_s^0\to\rho K_S}$ follows a similar route. The meson ρ has spin one and is CP-even. In the spectator approximation B_s^0 decays into $\overline{K^0}$ but not into K^0. As a result,

$$\langle \rho K_S|T|B_s^0\rangle = -\frac{1}{2q_K}\langle\rho\overline{K^0}|T|B_s^0\rangle,$$
$$\langle \rho K_S|T|\overline{B_s^0}\rangle = \frac{1}{2p_K}\langle\rho K^0|T|\overline{B_s^0}\rangle, \tag{33.12}$$

and

$$\lambda_{B_s^0\to\rho K_S} = -\frac{q_K}{p_K}\frac{q_{B_s}}{p_{B_s}}\frac{\langle\rho K^0|T|\overline{B_s^0}\rangle}{\langle\rho\overline{K^0}|T|B_s^0\rangle}. \tag{33.13}$$

Therefore, when B_s^0 decays into a K_S, or K_L, we introduce an extra $-q_K/p_K$, or $+q_K/p_K$, respectively. Assuming that the decay is dominated by the current–current operators, we find

$$\frac{\langle\rho K^0|T|\overline{B_s^0}\rangle}{\langle\rho\overline{K^0}|T|B_s^0\rangle} = \frac{V_{ub}V_{ud}^*\langle\rho K^0|(\bar u b)_{V-A}(\bar d u)_{V-A}|\overline{B_s^0}\rangle}{V_{ub}^*V_{ud}\langle\rho\overline{K^0}|(\bar b u)_{V-A}(\bar u d)_{V-A}|B_s^0\rangle}$$
$$= -e^{i(\xi_b-\xi_d-\xi_{B_s}-\xi_K)}\frac{V_{ub}V_{ud}^*}{V_{ub}^*V_{ud}}. \tag{33.14}$$

We thus arrive at

$$\lambda_{B_s^0\to\rho K_S} = \text{sign}\,(B_{B_s})\,\frac{V_{cs}^*V_{cd}}{V_{cs}V_{cd}^*}\,e^{-2i\epsilon'}\,\frac{V_{tb}^*V_{ts}}{V_{tb}V_{ts}^*}\frac{V_{ub}V_{ud}^*}{V_{ub}^*V_{ud}}$$
$$= \text{sign}\,(B_{B_s})\,e^{-2i\left(\gamma-\epsilon+\epsilon'\right)}. \tag{33.15}$$

As ϵ and ϵ' are very small, the CP-violating rate asymmetry in $B_s^0\to\rho K_S$ would measure the angle γ, if the decay were dominated by the tree-level diagram. Similarly,

$$\lambda_{B_s^0\to\rho K_L} = -\lambda_{B_s^0\to\rho K_S} = -\text{sign}\,(B_{B_s})\,e^{-2i\left(\gamma-\epsilon+\epsilon'\right)}. \tag{33.16}$$

The examples above illustrate two important features of the determination of CP-violating asymmetries in a given model. Firstly, when the calculations are careful and consistent, the spurious phases in the CP transformation cancel out amongst the various factors. Secondly, the value of λ_f can, in some cases at least, be *predicted*. Both its magnitude and its *sign* are determined by the model.[88]

33.1.4 *Consequences of single-phase dominance*

Let us assume an idealized situation in which the decays $\bar{b} \to \bar{\alpha}\alpha\bar{s}$ and $\bar{b} \to \bar{\alpha}\alpha\bar{d}$ (α is an up-type quark) are dominated by tree-level diagrams, while the decays $\bar{b} \to \bar{s}s\bar{k}$ (k is a down-type quark) are dominated by the gluonic penguin with intermediate top quark.[89]

Following Soares (unpublished thesis, 1993), we define $\tilde{a}$ by

$$\eta_f \tilde{a} = a^{\text{int}} = \frac{2\text{Im}\,\lambda_f}{1 + |\lambda_f|^2}. \tag{33.17}$$

We may table the predictions of the SM for $\tilde{a}$ without specifying the CP eigenvalue η_f. Since the CP-transformation phases cancel out in λ_f, we may drop them altogether. We find that the correct results may be reproduced with[90]

$$\tilde{a} = \sin 2(\phi_{\text{diagram}} - \phi_K - \phi_M), \tag{33.18}$$

where

$$\phi_{\text{diagram}} = \begin{cases} \arg\left(V_{\alpha b}^* V_{\alpha k}\right) & \text{for decays } \bar{b} \to \bar{\alpha}\alpha\bar{k}, \\ \arg\left(V_{tb}^* V_{tk}\right) & \text{for decays } \bar{b} \to \bar{k}s\bar{s}; \end{cases} \tag{33.19}$$

$$\phi_K = \begin{cases} \arg\left(V_{cs} V_{cd}^*\right) + \epsilon' & \text{for an unpaired } K_{S,L} \text{ in } B_d^0 \text{ decays,} \\ \arg\left(V_{cs}^* V_{cd}\right) - \epsilon' & \text{for an unpaired } K_{S,L} \text{ in } B_s^0 \text{ decays,} \\ 0 & \text{otherwise;} \end{cases} \tag{33.20}$$

$$\phi_M = \begin{cases} \arg\left(V_{tb}^* V_{td}\right) & \text{for } B_d^0 \text{ decays,} \\ \arg\left(V_{tb}^* V_{ts}\right) & \text{for } B_s^0 \text{ decays.} \end{cases} \tag{33.21}$$

Equation (33.18) should be compared with eqn (28.14), where $\phi_A = \phi_{\text{diagram}} - \phi_K$. Notice that, from now on, we assume the 'bag parameters' B_{B_q} to be positive. As we have seen in this section, all λ_f change sign when the bag parameters change sign, and therefore the SM predictions for the CP asymmetries also change sign in that case (Grossman *et al.* 1997b).

---

[88] Strictly speaking, it is only after we have chosen by convention $\Delta m > 0$ that the sign of λ_f acquires physical significance. The convention $\Delta m > 0$ is implicit in our computation of q/p for all neutral-meson systems.

[89] CP-eigenstate final states arising from the $\Delta U \neq 0 \neq \Delta C$ decays of the type $\bar{b} \to \bar{\alpha}\beta\bar{k}$ with $\alpha \neq \beta$ must involve a state $D_{f_{\text{cp}}}$ in the final state and can be reached by two distinct tree-level diagrams. We shall treat these channels after we discuss cascade decays in the next chapter.

[90] This is equivalent to writing $\eta_f \lambda_f = -\exp\left[-2i\left(\phi_{\text{diagram}} - \phi_K - \phi_M\right)\right]$. But notice that, depending on the phase convention used for the CKM matrix elements, the ϕ_K and ϕ_M defined in eqns (33.20) and (33.21) may differ by π from the similar parameters defined in § 28.3 through $q/p = -e^{2i\phi_M}$. Naturally, working consistently with either definition leads to the same results.

Table 33.1 *CP-violating asymmetries in B_d^0 decays. It is assumed that the decay is dominated by a single weak phase and that the only manifestation of new physics is in B_d^0–$\overline{B_d^0}$ mixing.*

Quark process	Sample decay mode	$\tilde{a}$
$b \to \bar{c}c\bar{s}$	$J/\psi K_S$	$\sin 2(\beta - \epsilon' - \theta_d)$
$\bar{b} \to \bar{c}c\bar{d}$	D^+D^-	$\sin 2(\beta - \theta_d)$
$\bar{b} \to \bar{u}u\bar{d}$	$\pi^+\pi^-$	$-\sin 2(\alpha + \theta_d)$
$\bar{b} \to \bar{s}s\bar{s}$	ϕK_S	$\sin 2(\beta + \epsilon - \epsilon' - \theta_d)$

33.2   New physics in the mixing and the parameters λ_f

The four phases β, γ, ϵ, and ϵ' determine the phase structure of the 3×3 submatrix of an extended CKM matrix. Assuming that the decays are dominated by SM diagrams, all new-physics effects appear in the mixing of neutral mesons, and in deviations of those four angles from their allowed ranges in the SM. New-physics effects in the mixing will in general involve new phases. We have introduced phases θ_d and θ_s—see eqn (30.32)—measuring the deviations of the phases of q_{B_d}/p_{B_d} and q_{B_s}/p_{B_s}, respectively, from their SM values. As a consequence, eqn (33.21) gets substituted by

$$\phi_M = \begin{cases} \arg\left(V_{tb}^* V_{td}\right) + \theta_d & \text{for } B_d^0 \text{ decays}, \\ \arg\left(V_{tb}^* V_{ts}\right) + \theta_s & \text{for } B_s^0 \text{ decays}. \end{cases} \tag{33.22}$$

The resulting asymmetries in B_d^0 decays are listed in Table 33.1. Analogously, the $\tilde{a}$ parameters for B_s^0 decays are given in Table 33.2 (Silva and Wolfenstein 1997).

Recalling that $\alpha = \pi - \beta - \gamma$, we see that θ_d always appears in the combination $\beta - \theta_d$. Thus, we need some further input to disentangle β from θ_d. Clearly, ϵ can be extracted independently of β and γ. Now, using the unitarity of the 3×3 CKM matrix, Aleksan *et al.* (1994) have proved that

$$\sin \epsilon \approx \left| \frac{V_{us}}{V_{ud}} \right|^2 \frac{\sin \beta \sin \gamma}{\sin(\beta + \gamma)}. \tag{33.23}$$

The power of this relation lies in the fact that the ratio $|V_{us}/V_{ud}|$ is known to high precision. If there is no violation of unitarity, we may use this to disentangle β from θ_d. However, it is rather difficult to distinguish θ_d from non-unitarity of the

Table 33.2 *The same as Table 33.1, but for B_s^0 decays.*

Quark process	Sample decay mode	$\tilde{a}$
$b \to \bar{c}c\bar{s}$	$D_s^+D_s^-$	$-\sin 2(\epsilon + \theta_s)$
$\bar{b} \to \bar{c}c\bar{d}$	$J/\psi K_S$	$-\sin 2(\epsilon - \epsilon' + \theta_s)$
$\bar{b} \to \bar{u}u\bar{d}$	ρK_S	$\sin 2(\gamma - \epsilon + \epsilon' - \theta_s)$
$\bar{b} \to \bar{s}s\bar{s}$	$\eta'\eta'$	$-\sin 2\theta_s$

CKM matrix in a completely model-independent fashion (Silva and Wolfenstein 1997).

Notice that the phase ϵ' always appears in connection with the formula for q_K/p_K (Cohen *et al.* 1997). But, as discussed in § 28.5.2, its inclusion is irrelevant for most realistic models.

We have not included the decays $\bar{b} \to \bar{s}s\bar{d}$ in our tables because, as we know from Chapter 31, these are likely to be affected by penguin diagrams with virtual charm and up quarks. The decays $\bar{b} \to \bar{s}u\bar{u}$ are also not shown. These have a large penguin contribution and might even be dominated by penguin amplitudes—in which case, they will effectively measure the weak phase due to the gluonic penguin diagrams with intermediate top quark. On the other hand, we did include in the tables the decays $\bar{b} \to \bar{c}c\bar{d}$ and $\bar{b} \to \bar{u}u\bar{d}$, and used only the phase from the dominant amplitude. These decays are expected to be affected by gluonic penguins but, to the extent that p_t/t is suppressed, the penguin amplitude will not be larger than the tree-level one. The case is worse for those decays originating in colour-suppressed tree-level diagrams, such as $B_s^0 \to \rho K_S$. The effect of a second phase, coming from subdominant diagrams, will be analysed in detail in § 33.4.

33.3 $\alpha + \beta + \gamma = \pi$

In our framework the relation $\alpha + \beta + \gamma = \pi$ (mod 2π) is true by definition. Some authors refer to 'tests' of this relation, because they adopt a different definition for α, β, and γ. Their definition is the following: α is what one measures in $B_d^0 \to \pi^+\pi^-$, β is what one measures in $B_d^0 \to J/\psi K_S$, and γ is what one measures in $B_s^0 \to \rho K_S$.[91] In our language,

$$\alpha_{B_d^0 \to \pi^+\pi^-} = \alpha + \theta_d,$$
$$\beta_{B_d^0 \to J/\psi K_S} = \beta - \epsilon' - \theta_d, \qquad (33.24)$$
$$\gamma_{B_s^0 \to \rho K_S} = \gamma - \epsilon + \epsilon' - \theta_s.$$

Those authors want to test whether the sum

$$\Sigma \equiv \alpha_{B_d^0 \to \pi^+\pi^-} + \beta_{B_d^0 \to J/\psi K_S} + \gamma_{B_s^0 \to \rho K_S} \qquad (33.25)$$

is equal to π.

As one gathers from eqns (33.24), in models with new physics in the mixing one has

$$\Sigma = \alpha + \beta + \gamma - \epsilon - \theta_s$$
$$= \pi - (\epsilon + \theta_s). \qquad (33.26)$$

Nir and Silverman (1990) have pointed out that a much simpler and equivalent test would be to measure directly the phase $\epsilon_{B_s^0 \to D_s^+ D_s^-}$ in the decay asymmetry of $B_s^0 \to D_s^+ D_s^-$, which is

---

[91]It is assumed that the penguin pollution in $B_d^0 \to \pi^+\pi^-$ and in $B_s^0 \to \rho K_S$ has been removed by some method.

$$\epsilon_{B_s^0 \to D_s^+ D_s^-} = \epsilon + \theta_s, \tag{33.27}$$

as seen in Table 33.2.

In the SM, $\theta_s = 0$ and $\epsilon \sim \lambda^2$. Therefore, $\Sigma = \pi$ should be satisfied to excelent accuracy. If one finds that Σ differs from π by a large amount, then there is new physics in the mixing ($\theta_s \neq 0$) and/or ϵ must be larger than in the SM, thus signalling non-unitarity of the CKM matrix.

33.4   Corrections induced by the subleading amplitudes

33.4.1   General discussion

We now use the diagrammatic analysis of Chapter 31 to try and get an estimate of the corrections imposed by subleading diagrams. The purpose of this section is to find out whether the results in Tables 33.1 and 33.2 are really meaningful, or whether the subleading amplitudes are likely to distort the picture conveyed in those tables.

When one goes beyond the approximation of single-phase dominance there is some arbitrariness in the choice of the second phase. For decays which have a tree-level contribution, it is customary to write the decay amplitude in terms of the phase of the tree-level diagram and of the phase of the penguin diagram with a virtual top quark running in the loop, as we have done in § 31.6. However, due to the unitarity of the CKM matrix, the decay amplitudes may be written in terms of any pair of parameters λ_α. This arbitrariness may lead into erroneous estimates of the corrections introduced by the subleading diagram.

As an example, take the decay $\bar{b} \to \bar{s}c\bar{c}$. Its amplitude may be written as

$$A_{\bar{b} \to \bar{s}c\bar{c}} = V_{cb}^* V_{cs} t + V_{tb}^* V_{ts} p_t, \tag{33.28}$$

where t and p_t are essentially the tree-level amplitude and the top-quark gluonic-penguin amplitude,[92] respectively, with the CKM factors explicitly factored out. Since $|V_{cb} V_{cs}| \approx |V_{tb} V_{ts}|$, we might be tempted to say that the correction introduced by the gluonic penguin is $\sim |p_t/t|$. On the other hand, we may rewrite eqn (33.28) as

$$A_{\bar{b} \to \bar{s}c\bar{c}} = V_{cb}^* V_{cs} (t - p_t) - V_{ub}^* V_{us} p_t. \tag{33.29}$$

In this case, we would estimate the correction to be of order

$$\left| \frac{V_{ub} V_{us} p_t}{V_{cb} V_{cs} (t - p_t)} \right| \sim \lambda^2 \left| \frac{p_t}{t} \right|. \tag{33.30}$$

One thus obtains two very different estimates for the same quantity. Which of them is correct?

---

[92] In all rigour, we should use $t + p_c - p_u$ instead of t, and $p_t - p_u$ instead of p_t. We implicitly assume that t dominates the first term, while the m_t-enhanced portion of the gluonic penguin dominates the second term.

One must recall that the quantity that one wants to compute is $\tilde{a}$. In the presence of only one weak phase ϕ_1 one has $\tilde{a} = \sin 2\phi_1$. But, in the presence of two contributions with different weak phases, one gets from eqn (28.18),

$$\tilde{a} \approx \sin 2\phi_1 - 2r \sin (\phi_1 - \phi_2) \cos 2\phi_1 \cos \Delta, \qquad (33.31)$$

when the ratio of amplitudes, r, is small. Thus, the important quantity is not r itself, but rather the product $r \sin(\phi_1 - \phi_2)$. For instance, in eqn (33.28) both terms have the same phase but for a correction ϵ, and therefore $\sin(\phi_1 - \phi_2) = \sin \epsilon \sim \lambda^2$. This should be multiplied by $r \sim |p_t/t|$. Similarly, in eqn (33.29) the two terms have phase difference γ, which is large and does not introduce any further suppression. But, this appears multiplied by $r \sim \lambda^2 |p_t/t|$. If one takes this into account, evaluates the matrix elements involved, and does not do any undue approximations, the results obtained using eqn (33.28) or using eqn (33.29) are the same, as they must be.

It should be stressed that the same $r \sin (\phi_1 - \phi_2)$ combination shows up in the direct-CP-violating term a^{dir} of eqn (28.17). Indeed,

$$a^{\mathrm{dir}} \approx -2r \sin (\phi_1 - \phi_2) \sin \Delta. \qquad (33.32)$$

It will be convenient to define the deviation that the a^{int} suffers due to the presence of a second weak phase:

$$\Delta \tilde{a} \equiv \tilde{a} - \sin 2\phi_1$$
$$\approx -2r \sin (\phi_1 - \phi_2) \cos 2\phi_1 \cos \Delta. \qquad (33.33)$$

We have listed in Tables 31.3–31.7 the various types of decays into CP eigenstates. We shall now go through those tables, use the phase convention for the CKM matrix elements introduced in eqn (28.24), and list the values of ϕ_1, ϕ_2, and r for some decays. We shall build upon work done by Grossman and Worah (1997).

33.4.2   Decays with a tree-level contribution

- $\bar{b} \rightarrow \bar{c}c\bar{s}$. Examples: $B^0_s \rightarrow D^+_s D^-_s$ and $B^0_s \overset{c}{\rightarrow} J/\psi\phi$ (both with $\phi_1 = -\epsilon + \pi$),[93] and $B^0_d \overset{c}{\rightarrow} J/\psi K_S$ (with $\phi_1 = \beta - \epsilon' + \pi$). The symbol $\overset{c}{\rightarrow}$ means that these decays proceed at tree level through colour-suppressed diagrams, if we neglect final-state-interaction rescattering effects. Since the leading contribution has CKM factor $V^*_{cb} V_{cs}$ and the main gluonic-penguin amplitude has weak phase $V^*_{tb} V_{ts}$, the correction $\Delta \tilde{a} \propto r \sin (\phi_1 - \phi_2)$ is of order

$$\left| \frac{p}{t} \right| \sin \epsilon \quad \mathrm{or} \quad \left| \frac{p}{c} \right| \sin \epsilon. \qquad (33.34)$$

The second value holds for decays proceeding at tree level through colour-suppressed diagrams. As $\epsilon \sim \lambda^2$, the corrections are rather small, and therefore these decays should be dominated by a single weak phase.

---

[93] We include here phases of π, which are relevant for the caculation of $\phi_1 - \phi_2$.

- $b \to \bar{c}c\bar{d}$. Examples: $B_d^0 \to D^+D^-$ and $B_d^0 \xrightarrow{c} J/\psi\pi^0$ (both with $\phi_1 = \beta+\pi$), and $B_s^0 \xrightarrow{c} J/\psi K_S$ (with $\phi_1 = -\epsilon + \epsilon' + \pi$). In this case, $\Delta\tilde{a}$ is estimated to be

$$\left|\frac{p}{t}\right| R_t \sin\beta \quad \text{or} \quad \left|\frac{p}{c}\right| R_t \sin\beta. \tag{33.35}$$

Since β is unsuppressed, the gluonic-penguin contribution is not negligible. If one trusts the usual λ^2 estimate for $|p/t|$, the effect in $B_d^0 \to D^+D^-$ is smaller than 5%, but it may well turn out to be larger, should $|p/t|$ be of order λ. The situation is worse in colour-suppressed decays, since there the tree-level diagrams are smaller.

- $b \to \bar{u}u\bar{d}$. Examples: $B_d \to \pi^+\pi^-$ and $B_d \xrightarrow{c} \pi^0\pi^0$ (both with $\phi_1 = -\alpha+\pi$), and $B_s^0 \xrightarrow{c} \rho K_S$ (with $\phi_1 = \gamma - \epsilon + \epsilon'$). Now $\Delta\tilde{a}$ is proportional to

$$\left|\frac{p}{t}\right| \frac{R_t}{R_b} \sin\alpha \quad \text{or} \quad \left|\frac{p}{c}\right| \frac{R_t}{R_b} \sin\alpha. \tag{33.36}$$

This is worse than the previous case because R_b is smaller than unity. As a consequence, the measurement of α in the process $B_d \to \pi^+\pi^-$ suffers from large uncertainties. The extraction of γ from the decay $B_s^0 \xrightarrow{c} \rho K_S$ is even worse, due to the colour suppression of the tree-level amplitude.

- $b \to \bar{u}u\bar{s}$. Examples: $B_s^0 \to K^+K^-$, with $\phi_1 = \gamma - \epsilon + \epsilon' + \pi$, and $B_d^0 \xrightarrow{c} K_S\pi^0$, with $\phi_1 = -\alpha$. Here $\Delta\tilde{a}$ should be of order

$$\left|\frac{p}{t}\right| \frac{1}{\lambda^2 R_b} \sin\gamma \quad \text{or} \quad \left|\frac{p}{c}\right| \frac{1}{\lambda^2 R_b} \sin\gamma. \tag{33.37}$$

Here, the suppression of the tree-level diagrams is so effective, that it is probably better to think of these decays as measuring the weak phase of the top-mediated gluonic penguins, with the tree-level diagrams acting as the pollutant (Ciuchini et al. 1997b). Then, the CP-violating phase in the decay $B_s^0 \to K^+K^-$ would be close to $\phi_2 = 0$, while $B_d^0 \xrightarrow{c} K_S\pi^0$ would be dominated by the phase $\phi_2 = \beta + \epsilon - \epsilon'$.

There is one exception to this analysis of decays $b \to \bar{s}u\bar{u}$. The transition $B_s^0 \xrightarrow{c} \phi\pi^0$ only has a colour-suppressed tree-level contribution, with angle $\phi_1 = \gamma - \epsilon + \epsilon' + \pi$, since it is not affected by gluonic penguins. As explained before, this is due to the fact that, in the gluonic penguin, the $q\bar{q}$ pair arises out of the gluon in an $I = 0$ state, while π^0 belongs to an isospin triplet. However, this decay is affected by the electroweak penguins, which are even expected to be dominant (Buras and Fleischer 1997).

33.4.3 Pure penguin decays

Naively, one would think that the phase of a pure-penguin decay amplitude would be equal that of the penguin diagram with intermediate top quark. However, the subleading penguin diagrams with intermediate up and charm quarks may be relevant. The importance of the subleading penguin effects in pure penguin decays was first pointed out by Gérard and Hou (1991a,b) and by Simma et *al.* (1991), in the context of direct CP violation. These subleading effects also generate a correction of the interference CP violation term.

- $\bar{b} \to \bar{s}d\bar{d}$. Example: $B_s^0 \to K^0\overline{K^0}$. This decay cannot proceed through a tree-level diagram. For an intermediate top quark in the gluonic penguin we find $\phi_1 = 0$. This gets a correction proportional to $|p_c/p_t|\sin\epsilon$ from the charm-quark gluonic penguin. Since this is very small, the decay $B_s^0 \to K^0\overline{K^0}$ is often proposed as ideal to look for new physics.

- $\bar{b} \to \bar{s}s\bar{s}$. Examples: $B_d^0 \to \phi K_S$, with $\phi_1 = \beta + \epsilon - \epsilon'$, and $B_s^0 \to \eta'\eta'$, with $\phi_1 = 0$. The gluonic penguin with an intermediate top quark has CKM factor $V_{tb}^* V_{ts}$; the gluonic penguin with an intermediate charm quark gives a correction of order $|p_c/p_t|\sin\epsilon$. For $B_d \to \phi K_S$, this correction must be added to a second one which is due to the small $u\bar{u}$ and $d\bar{d}$ components of ϕ. As a consequence, that decay may also proceed through the tree-level diagrams $\bar{b} \to \bar{u}u\bar{s}$. Grossman and Worah (1997) have estimated the second uncertainty to be of order 1%, leading to a combined correction of order 4%.

- $\bar{b} \to \bar{d}s\bar{s}$: Examples: $B_d^0 \to K^0\overline{K^0}$ and $B_d^0 \to \phi\pi^0$ (both with $\phi_1 = 0$), and $B_s^0 \to \phi K_S$ (with $\phi_1 = -\beta - \epsilon + \epsilon'$). Here, the correction is proportional to $|p_c/p_t|\sin\beta/R_t$. Fleischer (1994c) has estimated that the corrections could lead to a CP-violating asymmetry as large as 50%. Particularly interesting is the decay $B_d^0 \to \phi\pi^0$, which proceeds via a singlet penguin. This gets a small contribution from the $\bar{b} \to \bar{u}u\bar{d}$ tree-level diagram, due to the small $u\bar{u}$ component of ϕ. However, this decay is likely to be affected by rescattering effects.

33.5 $B_d^0 \to J/\psi K_S$ is the gold-plated decay

We have concentrated our efforts in trying to identify decay channels which are dominated, in the SM, by a single weak phase. In such cases one can extract that CKM phase, unless there are new-physics contributions either to the mixing or to the decay amplitudes. Unfortunately, only a few cases satisfy these conditions. We have come in this chapter to the following conclusions (Grossman and Worah 1997). The quark decay $\bar{b} \to \bar{c}c\bar{s}$ should provide clean measurements of CKM phases. The decay $\bar{b} \to \bar{s}s\bar{s}$ should have only small uncertainties from other SM contributions. We will show in § 34.5, in which we discuss the decays with $\Delta C \neq 0 \neq \Delta U$, that this is also the case for the decay $\bar{b} \to \bar{c}u\bar{d}$. Most other decays are likely to suffer from pollution due to a subdominant amplitude with a different weak phase.

Therefore, it appears that experimentalists should devote special attention to decays which at quark level are $\bar{b} \to \bar{c}c\bar{s}$. These include $B_d^0 \to J/\psi K_S$, which measures $\sin 2(\beta - \epsilon' - \theta_d)$, and $B_s^0 \to D_s^+ D_s^-$, which measures $\sin 2(\epsilon + \theta_d)$. In the SM, $\theta_d = 0$ and $\sin 2\beta$ is positive, see eqn (18.30). On the other hand, $\epsilon \sim \lambda^2$ is very small, and therefore the CP asymmetry in $B_s^0 \to D_s^+ D_s^-$ is predicted to be small.

The decay $B_d^0 \to J/\psi K_S$ has other advantages. Machines working at the $\Upsilon(4S)$ provide a very clean source of mesons B_d^0 and $\overline{B_d^0}$; this is to be contrasted with the situation for mesons B_s^0 and $\overline{B_s^0}$ which, in the near future, will only be produced at hadron machines, where the background constitutes a severe challenge. Moreover, both the J/ψ and the K_S are easy to detect, the J/ψ through its decay into two muons, with a branching ratio $\sim 6\%$ (the decay into two electrons has the same branching ratio), and the K_S through its decay into $\pi^+\pi^-$; the final state is then composed entirely of charged particles which are easy to detect. For these reasons, $B_d^0 \to J/\psi K_S$ has been termed the gold-plated decay: it should provide a very clean measurement of $\sin 2\beta$ in the SM. This determines β up to a fourfold ambiguity.

The existence of two different amplitudes with different weak phases brings into play hadronic uncertainties, due both to uncancelled operator matrix elements and to the presence of unknown final-state-interaction CP-even phases. In Chapters 35, 36, and 37 we shall discuss methods that can be applied to some of these more challenging cases. The general idea is that one may be able to relate several decay channels in order to circumvent the hadronic uncertainties and extract the weak phases.

34

CASCADE DECAYS

34.1 Introduction

In this chapter we discuss decay chains with a neutral-meson system in an intermediate state. These are known as cascade decays. We want to study cases where the decays of B mesons proceed through intermediate states containing neutral kaons or neutral D mesons. Examples are $B_d \rightarrow J/\psi K \rightarrow J/\psi f_K$ and $B_d \rightarrow \pi^0 D \rightarrow \pi^0 f_D$. In this notation, f_K designates a set of particles which one identifies experimentally, in particular through their invariant mass, as coming from the decay of a neutral kaon; and, similarly, f_D is a set of particles originating in the decay of either D^0 or $\overline{D^0}$.

A correct understanding of cascade decays is important because the CP violation present in the intermediate-meson system will in general show up in the calculation of CP-violating observables for the overall decay chain. In particular, in a cascade decay in which a neutral-meson system decays into a lighter neutral-meson system, it may be possible to use the CP properties of the latter as an analyser for the CP properties of the former. This is now fully appreciated, following work by Azimov (1989, 1990) and by Kayser and Stodolsky (1996).

We shall discuss decay chains with intermediate neutral kaons and neutral D-mesons separately. We start with a pedagogical discussion of the $B_d \rightarrow J/\psi K \rightarrow J/\psi f_K$ decay chain. This case is simplified by the fact that the decays $B_d^0 \rightarrow \overline{K^0}$ and $\overline{B_d^0} \rightarrow K^0$ are forbidden (Azimov 1989, 1990; Dass and Sarma 1992; Kayser and Stodolsky 1996; Azimov et al. 1997; Kayser 1997). A similar situation occurs with the $B_s \rightarrow J/\psi K \rightarrow J/\psi f_K$ decays considered by Azimov and Dunietz (1997). Next, we discuss the $B \rightarrow XD \rightarrow X f_D$ decay chains, under the assumption that the mixing in the D^0–$\overline{D^0}$ system is negligible. Finally, we include an elementary introduction to those decays in which the initial state can decay into both flavour eigenstates of the neutral-meson system in the intermediate state. These cases have been introduced by Meca and Silva (1998) and by Amorim et al. (1999). Examples include $D^+ \rightarrow \pi^+ K \rightarrow \pi^+ f_K$, and also, if we allow for sizeable new-physics contributions to D^0–$\overline{D^0}$ mixing, $B^+ \rightarrow K^+ D \rightarrow K^+ f_D$. In these cases, there is CP violation in the interference between the decays from the initial state and the mixing in the intermediate neutral-meson system. This involves new CP-violating parameters, beyond the ones discussed thus far in this book (Meca and Silva 1998; Amorim et al. 1999).

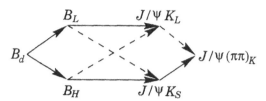

FIG. 34.1. The cascade decay $B_d^0 \to J/\psi K \to J/\psi(\pi^+\pi^-)_K$.

34.2   The decay chain $B_d^0 \to J/\psi K \to J/\psi(\pi^+\pi^-)_K$

When discussing cascade decays it is important to take into account the specific experimental conditions. We illustrate this point by considering the process $B_d \to J/\psi K \to J/\psi(\pi^+\pi^-)_K$, where $(\pi^+\pi^-)_K$ refers to two pions with an invariant mass equal to the one of the neutral kaon. Our discussion follows closely the presentation in recent reviews by Kayser—see for example Kayser (1997).[94]

The two kaon masses, m_L and m_S, differ very little and are not resolved in this experiment. Thus, the mass eigenstates of the intermediate neutral kaon may only be resolved through the lifetimes ($\tau_S \ll \tau_L$).

There are two decay times involved. There is the time t_B between the production of B_d and its decay into $J/\psi K$, and the time t_K between this event and the decay of the neutral kaon into two pions. The time interval t_B is measured in the rest frame of B_d, while t_K is measured in the rest frame of the neutral kaons.

This decay is represented in Fig. 34.1. There are four paths leading from B_d^0 to the final state. By means of a suitable choice of the two decay times, the interference among these four paths can be studied in order to determine different combinations of CKM phases (Azimov 1989, 1990; Kayser and Stodolsky 1996).

The $\pi^+\pi^-$ pair is CP-even. If CP was conserved, then the mass eigenstates would coincide with the CP eigenstates in both neutral-meson systems. We would then have $CP(B_H) = CP(J/\psi K_S) = CP(J/\psi\pi^+\pi^-) = -1$, and $CP(B_L) = CP(J/\psi K_L) = +1$.[95] Then, the decays represented by solid lines conserve CP, while the decays represented by dashed lines violate CP.

Most analyses in the literature neglect CP violation in the kaon decays. Then, K_L cannot decay into two pions and there are only two paths; $B_H \to J/\psi K_S \to J/\psi(\pi^+\pi^-)_K$ and $B_L \to J/\psi K_S \to J/\psi(\pi^+\pi^-)_K$, which interfere. Both B_d^0 and $\overline{B_d^0}$ have B_H and B_L components, so both contain interference terms. The interference involves the parameter $\lambda_{B_d^0 \to J/\psi K_S}$ computed in § 33.1.2. Neglecting ϵ' and using eqns (28.6), we then have

---

[94] This section and some of the following ones are also based on a series of tutorials given by Kayser at the *Centro de Física Nuclear da Universidade de Lisboa* in December 1997. We are indebted to Kayser for discussing the details of his work with us.

[95] We have assumed that the heaviest B_d mass eigenstate is mainly CP-odd, as predicted, for example, by Dunietz (1995) in the context of the SM. The situation may be the opposite one, in which case the roles of B_H and of B_L are interchanged.

$$\Gamma[B_d^0 \to J/\psi K_S] \propto e^{-\Gamma_B t_B} \left[1 - \sin(2\beta)\sin(\Delta m_B t_B)\right],$$
$$\Gamma[\overline{B_d^0} \to J/\psi K_S] \propto e^{-\Gamma_B t_B} \left[1 + \sin(2\beta)\sin(\Delta m_B t_B)\right]. \tag{34.1}$$

These equations display a term proportional to $\sin 2\beta$, which arises from the interference between the two competing paths.

However, this cannot be the whole picture for the decay chain $B_d^0 \to J/\psi K \to J/\psi(\pi^+\pi^-)_K$. In fact, if t_K is much larger than the lifetime of K_S, the K_S component will have completely decayed away, and the two pions must necessarily have come from a K_L. (Of course, the rate will be exceedingly small, but this is a useful thought experiment.) For this case, we combine eqns (28.6) and (33.11) to get

$$\Gamma[B_d^0 \to J/\psi K_L] \propto e^{-\Gamma_B t_B} \left[1 + \sin(2\beta)\sin(\Delta m_B t_B)\right],$$
$$\Gamma[\overline{B_d^0} \to J/\psi K_L] \propto e^{-\Gamma_B t_B} \left[1 - \sin(2\beta)\sin(\Delta m_B t_B)\right]. \tag{34.2}$$

Again, there is an interference term proportional to $\sin 2\beta$, but with the opposite sign. The reason is that $\lambda_{B_d^0 \to J/\psi K_L} = -\lambda_{B_d^0 \to J/\psi K_S}$.

It is true that $A(K_S \to \pi^+\pi^-)$ is three orders of magnitude larger than $A(K_L \to \pi^+\pi^-)$. But, as this thought experiment illustrates, when discussing observables concerning cascade decays, one must take into account *the specific experimental conditions*. The experiments on $B \to J/\psi K \to J/\psi(\pi^+\pi^-)_K$ will be looking at small times t_K. Therefore, the $J/\psi(\pi^+\pi^-)_K$ events detected are overwhelmingly due to $J/\psi K_S$, due both to the huge ratio $A(K_S \to \pi^+\pi^-)/(K_L \to \pi^+\pi^-)$ and to the t_K interval probed.

In this section we have considered two limiting cases of the decay chain $B_d^0 \to J/\psi K \to J/\psi(\pi^+\pi^-)_K$: the case of times $t_K \sim \tau_S$, in which all decays go through K_S, and the case of times $t_K \gg \tau_S$, in which all decays go through K_L. The fact that both these decays can occur is a consequence of CP violation in the kaon sector (which allows both K_S and K_L to decay into the same CP eigenstate $\pi^+\pi^-$). In the next section we shall develop formulas valid for all times t_K.

34.3   The $B_d^0 \to J/\psi K \to J/\psi(f)_K$ cascade decays

34.3.1   *The kaon state at time t_B*

Let us consider a meson B_d^0 produced at $t = 0$. At a later time t_B, it will be a linear combination of B_d^0 and $\overline{B_d^0}$ given by eqn (9.1):

$$|B_d^0(t_B)\rangle = g_+(t_B)|B_d^0\rangle + \frac{q_{B_d}}{p_{B_d}} g_-(t_B)|\overline{B_d^0}\rangle, \tag{34.3}$$

where

$$g_\pm(t_B) \equiv \tfrac{1}{2}\left(e^{-i\mu_H^B t_B} \pm e^{-i\mu_L^B t_B}\right). \tag{34.4}$$

We shall neglect the width difference in the B_d^0–$\overline{B_d^0}$ system and write

$$\mu_{H,L}^B = m_B \pm \Delta m_B/2 - i\Gamma_B/2. \tag{34.5}$$

To lowest order in the weak interaction, the decays $B_d^0 \to \overline{K^0}$ and $\overline{B_d^0} \to K^0$ are forbidden. Therefore, the decay of $B_d^0(t_B)$ into $J/\psi K$ produces a kaon state given by (Azimov 1989, 1990)

$$|K_{\text{from } B_d^0(t_B)}\rangle \propto g_+(t_B)\langle J/\psi K^0|T|B_d^0\rangle|K^0\rangle$$
$$+ \frac{q_{B_d}}{p_{B_d}} g_-(t_B)\langle J/\psi \overline{K^0}|T|\overline{B_d^0}\rangle|\overline{K^0}\rangle, \qquad (34.6)$$

Using eqns (8.6) and the fact that

$$\lambda_{B_d^0 \to J/\psi K_S} = -\frac{q_{B_d}}{p_{B_d}} \frac{p_K}{q_K} \frac{\langle J/\psi \overline{K^0}|T|\overline{B_d^0}\rangle}{\langle J/\psi K^0|T|B_d^0\rangle}, \qquad (34.7)$$

we find

$$|K_{\text{from } B_d^0(t_B)}\rangle \propto g_+(t_B)\,(|K_S\rangle + |K_L\rangle) + \lambda_{B_d^0 \to J/\psi K_S} g_-(t_B)\,(|K_S\rangle - |K_L\rangle). \qquad (34.8)$$

When ϵ' is neglected one has $\lambda_{B_d^0 \to J/\psi K_S} = \exp(-2i\beta)$ in the SM. Thus (Kayser 1997)

$$|K_{\text{from } B_d^0(t_B)}\rangle \propto \left(e^{-\frac{i}{2}\Delta m_B t_B} + i \tan\beta\; e^{+\frac{i}{2}\Delta m_B t_B}\right)|K_S\rangle$$
$$+ \left(e^{+\frac{i}{2}\Delta m_B t_B} + i \tan\beta\; e^{-\frac{i}{2}\Delta m_B t_B}\right)|K_L\rangle. \qquad (34.9)$$

A similar analysis leads to

$$|K_{\text{from } \overline{B_d^0}(t_B)}\rangle \propto \left(e^{-\frac{i}{2}\Delta m_B t_B} - i \tan\beta\; e^{+\frac{i}{2}\Delta m_B t_B}\right)|K_S\rangle$$
$$+ \left(-e^{+\frac{i}{2}\Delta m_B t_B} + i \tan\beta\; e^{-\frac{i}{2}\Delta m_B t_B}\right)|K_L\rangle. \qquad (34.10)$$

By collecting events with a particular t_B we may tune the composition of the kaon state, much as we do with a regenerator (Azimov 1989, 1990; Kayser 1997).

One may use our knowledge of the kaon system to learn about the B systems by using cascade decays. Azimov (1989, 1990) has pointed out that, in particular, we may determine whether the heaviest B_d mass eigenstate in mostly CP-even or mostly CP-odd, as well as the sign of $\Delta\Gamma$ for the B_d^0–$\overline{B_d^0}$ system, because we know this information for the kaon system. This strategy has also been explored by Dass and Sarma (1992), Kayser and Stodolsky (1996), Azimov et al. (1997), and Kayser (1997). A similar technique may be used to measure Δm in the B_s^0–$\overline{B_s^0}$ system through the cascade decay $B_s^0 \to J/\psi K \to J/\psi f_K$, even when $x_s \gg 1$ (Azimov and Dunietz 1997).

34.3.2 Cascade decay rate: t_K-dependence

Once created, the states $|K_{\text{from } B_d^0(t_B)}\rangle$ and $|K_{\text{from } \overline{B_d^0}(t_B)}\rangle$ will evolve in time, and decay into the final state f_K at time t_K. After a tedious but straightforward calculation, we find

$$e^{\Gamma_B t_B}\Gamma\left[\begin{array}{c} \overline{B_d^0} \\ B_d^0 \end{array} \xrightarrow{t_B} J/\psi K \xrightarrow{t_K} J/\psi f_K\right]$$

$$\propto e^{-\Gamma_S t_K} |A_{K_S \to f}|^2 [1 \pm \sin(2\beta) \sin(\Delta m_B t_B)]$$
$$+ e^{-\Gamma_L t_K} |A_{K_L \to f}|^2 [1 \mp \sin(2\beta) \sin(\Delta m_B t_B)]$$
$$\mp 2 e^{-\Gamma t_K} |A_{K_S \to f} A_{K_L \to f}| [\cos(\Delta m_B t_B) \cos(\Delta m_K t_K - \phi_f)$$
$$+ \cos(2\beta) \sin(\Delta m_B t_B) \sin(\Delta m_K t_K - \phi_f)], \qquad (34.11)$$

where Γ_S (Γ_L) is the width of K_S (K_L), and we have defined $\Gamma \equiv (\Gamma_S + \Gamma_L)/2$ and

$$\phi_f = -\arg\left[A_{K_S \to f} A^*_{K_L \to f}\right] \qquad (34.12)$$

as in eqn (8.25).

The behaviour of the expression in eqn (34.11) varies according to whether $|\eta_f| = |A_{K_L \to f}/A_{K_S \to f}|$ is larger or smaller than 1. If $|\eta_f| \gg 1$ the decay is dominated by K_L, for all times t_K. If $|\eta_f| \ll 1$ the structure of the decay changes with time. For $t_K \sim \tau_S$ that expression is dominated by the decays of K_S in the first line, and we reproduce eqn (34.1). For $t_K \gg \tau_S$ any event is due to the decays of K_L in the second line, and we reproduce eqn (34.2). At intermediate times one is sensitive to the interference between the decays of K_S and of K_L. Instead of depending on $\sin 2\beta$, as do eqns (34.1) and (34.2), the interference term probes $\cos 2\beta$. This is interesting, because it may allow us to determine a different trigonometric function of the CP-violating phase in $B_d^0 \to J/\psi K$ (Kayser 1997).

Let us consider a CP-allowed decay of K_S, such as $\pi^+\pi^-$. The interference term becomes relevant when

$$e^{-\Gamma_S t_K} |A_{K_S \to f}|^2 \sim 2 e^{-\Gamma t_K} |A_{K_S \to f}||A_{K_L \to f}|, \qquad (34.13)$$

i.e., for

$$t_K \sim -2\tau_S \ln(2|\eta_f|). \qquad (34.14)$$

For $f = \pi^+\pi^-$ we have $|\eta_{+-}| \sim 2.3 \times 10^{-3}$ and the interference term is important for $t_K \sim 11\tau_S$—see Fig. 8.2. Unfortunately, by then the decay rate is suppressed by a factor $e^{-11} \sim 10^{-5}$ compared to what it was at $t_K = 0$. Therefore, the interference should be easiest to detect for decays with $|\eta_f| \approx 1$, such as the semileptonic decays—see § 34.3.4.

34.3.3   Cascade decay rate: t_K-integrated

Let us integrate the rate in eqn (34.11) over t_K. We obtain

$$e^{\Gamma_B t_B} \int_0^\infty dt_K \, \Gamma \left[\begin{matrix} \overline{B_d^0} \\ B_d^0 \end{matrix} \xrightarrow{t_B} J/\psi K \xrightarrow{t_K} J/\psi f_K \right]$$
$$\propto \mathrm{BR}(K_S \to f)[1 \pm \sin(2\beta)\sin(\Delta m_B t_B)]$$
$$+ \mathrm{BR}(K_L \to f)[1 \mp \sin(2\beta)\sin(\Delta m_B t_B)]$$
$$\mp 4 \frac{\sqrt{\mathrm{BR}(K_S \to f)\,\mathrm{BR}(K_L \to f)}}{1 + x_K^2} \frac{\sqrt{\Gamma_S \Gamma_L}}{\Gamma_S + \Gamma_L} [\cos(\Delta m_B t_B)(\cos\phi_f + x_K \sin\phi_f)]$$

$$+ \cos\left(2\beta\right) \sin\left(\Delta m_B t_B\right) \left(x_K \cos\phi_f - \sin\phi_f\right)], \tag{34.15}$$

where $x_K = \Delta m_K/\Gamma \approx 1$. The interference term has a piece proportional to $\cos(\Delta m_B t_B)$, which fakes a (very small) direct CP violation in $B_d^0 \to J/\psi K_S$. The other piece is proportional to $\sin(\Delta m_B t_B)$ and to $\cos 2\beta$. Its coefficient, $x_K \cos\phi_f - \sin\phi_f$, almost vanishes if f is a two-pion state, because ϕ_{+-} and ϕ_{00} are very close to $45°$. The effect of the interference terms becomes even smaller once a cut in t_K is imposed in the integration (keeping t_K smaller than a few times τ_S) because that helps in identifying the K_S.

34.3.4   The Kayser method to measure $\cos 2\beta$

We see from eqns (34.11) and (34.15) that the cascade decay $B_d^0 \to J/\psi K \to J/\psi f_K$ may be used to determine $\cos 2\beta$ (Kayser 1997). In order to do this we must maximize the interference term. This happens when the decay amplitudes of K_L and K_S into the final state f are of similar magnitude, so that all terms are comparable at small times, when the exponentials are close to one.

Explicitly, Kayser (1997) has proposed to use the decays $B_d^0 \to J/\psi K \to J/\psi(\pi l \nu_l)_K$. In this case, $|A_{K_L \to f}| \approx |A_{K_S \to f}|$ and the amplitudes factor out. Moreover, this cascade decay has a rate comparable to that of $B_d^0 \to J/\psi K \to J/\psi(\pi^+\pi^-)_K$, because

$$\mathrm{BR}\left(K_S \to \pi^+\pi^-\right) \approx \mathrm{BR}\left(K_L \to \pi e \nu_e\right) + \mathrm{BR}\left(K_L \to \pi \mu \nu_\mu\right), \tag{34.16}$$

see eqns (8.8) and (8.9). However, while the K_S term dominates in the cascade decay $B_d^0 \to J/\psi K \to J/\psi(\pi^+\pi^-)_K$, the interference term only accounts for approximately

$$2\sqrt{\frac{\mathrm{BR}(K_S \to \pi l \nu)}{\mathrm{BR}(K_L \to \pi l \nu)}}\sqrt{\frac{\Gamma_L}{\Gamma_S}} \sim 0.003 \tag{34.17}$$

of the cascade decay $B_d^0 \to J/\psi K \to J/\psi(\pi l \nu_l)_K$ (Kayser 1997).[96] Therefore, measuring $\cos 2\beta$ will require a few hundred times more events $B_d^0 \to J/\pi K$ than measuring $\sin 2\beta$.

Actually, the need for large data samples may not be as acute as it seems. Indeed, once $\sin 2\beta$ is measured one only needs to determine the sign of $\cos 2\beta$. This measurement will reduce the fourfold ambiguity in β to the twofold ambiguity $\beta \to \beta + \pi$ (see Chapter 38).

34.4   Decay chains with intermediate neutral-D mesons

Consider a decay chain of the type $B \to XD \to Xf_D$. In most cases of interest, the state f may be reached from both D^0 and $\overline{D^0}$, and both contributions must be included. A simple algorithm to calculate such decay amplitudes arises from the observation that the linear combination of D^0 and $\overline{D^0}$ that decays into f is

$$|D_{\mathrm{into}\ f}\rangle = c_f^*|D^0\rangle + \bar{c}_f^*|\overline{D^0}\rangle$$

---

[96] In these estimates we neglect possible cancellations due to the value of β.

$$\propto \langle f|T|D^0\rangle^*|D^0\rangle + \langle f|T|\overline{D^0}\rangle^*|\overline{D^0}\rangle. \tag{34.18}$$

Here, $|c_f|^2 + |\bar{c}_f|^2 = 1$ and

$$\frac{\bar{c}_f}{c_f} = \frac{\langle f|T|\overline{D^0}\rangle}{\langle f|T|D^0\rangle}. \tag{34.19}$$

Indeed, this state is orthogonal to

$$|D_{\text{not into } f}\rangle = \bar{c}_f|D^0\rangle - c_f|\overline{D^0}\rangle$$
$$\propto \langle f|T|\overline{D^0}\rangle|D^0\rangle - \langle f|T|D^0\rangle|\overline{D^0}\rangle, \tag{34.20}$$

which clearly cannot decay into f.

We frequently use the linear combination of D mesons which decays into a CP eigenstate f_{cp}, designated by $D_{f_{\text{cp}}}$:

$$D_{f_{\text{cp}}} \equiv D_{\text{into } f_{\text{cp}}}. \tag{34.21}$$

In general, the states $D_{f_{\text{cp}}}$ do *not* coincide with the CP eigenstates $D_\pm$. If the decay of D^0 into f_{cp} is dominated by a single weak phase, then CP relates the numerator and the denominator in eqn (34.19), there is no direct CP violation in the decay, and $|c_f| = |\bar{c}_f| = 1/\sqrt{2}$.

We shall also be interested in the decays of D mesons into flavour-specific final states. In this case, either c_f or $\bar{c}_f$ vanish, implying that the decaying meson was a flavour eigenstate, either D^0 or $\overline{D^0}$. Clearly, by suitably choosing the final state f we may pick up different combinations of D^0 and $\overline{D^0}$ in the decay chain $B \to XD \to Xf_D$.

The crucial physical input in the usual analysis of cascade decays involving an intermediate neutral-D meson is the following: the mixing parameters x and y for the D^0–$\overline{D^0}$ system are very small. (In Appendix E we give both the experimental results which justify this assertion, and the theoretical expectations for those mixing parameters, in the context of the SM.) Because of the smallness of x and y *the D mesons suffer almost no oscillation before they decay*. This means that the linear combination of $|D^0\rangle$ and $|\overline{D^0}\rangle$ created by the decay of the B meson is identical to the linear combination of $|D^0\rangle$ and $|\overline{D^0}\rangle$ that later decays. This implies that we do not have to compute the full decay chain, expressed in terms of a time t_B and a time t_D, as we did in the previous section; the t_D-dependence just factors out as $\exp(-\Gamma_D t_D)$, where Γ_D is the (mean) decay width of D^0 and $\overline{D^0}$.

There are in the literature a number of proposals to determine CKM phases by comparing the decays $B \to XD^0$, $B \to X\overline{D^0}$, and $B \to Xf_D$. Such methods exploit the existence of a linear equation among the decay amplitudes,

$$\langle Xf_D|T|B\rangle = \langle XD_{\text{into } f}|T|B\rangle$$
$$= c_f\langle XD^0|T|B\rangle + \bar{c}_f\langle X\overline{D^0}|T|B\rangle. \tag{34.22}$$

Such triangular relations lie at the heart of several methods to determine the angle γ. The simplest methods of this type use a CP eigenstate, $f = f_{\text{cp}}$. That is, they compare $B \to X D^0$ and $B \to X \overline{D^0}$ with $B \to X D_{f_{cp}}$. Gronau and London (1991) studied $B^0_d \to D K_S$ and $B^0_s \to D \phi$, while Gronau and Wyler (1991) proposed $B^\pm \to D K^\pm$. As a result, such proposals are sometimes known collectively as 'the Gronau–London–Wyler method'.

34.5 λ_f for decays with $\Delta C \neq 0 \neq \Delta U$

So far, we have not yet calculated the parameter λ_f for any $\Delta C \neq 0 \neq \Delta U$ decay. Clearly, all the $\Delta C \neq 0 \neq \Delta U$ decays into CP eigenstates must contain $D_{f_{cp}}$ in the final state. This final state is identified through its subsequent decay into a CP eigenstate f_{cp} such as $\pi\pi$, $K^+ K^-$, or $\pi^0 K_S$. Therefore, that decay enters into the correct calculation of any CP-violation parameter for the overall decay chain. As seen in the previous section, this requires a knowledge of $c_f/\bar{c}_f$.

We shall *assume* that the decays of D mesons are dominated by the tree-level diagrams of the SM. If $f_{cp} = \pi^+ \pi^-$, then the auxiliary quantity required is

$$
\begin{aligned}
\frac{c_{\pi^+ \pi^-}}{\bar{c}_{\pi^+ \pi^-}} &= \frac{V_{ud} V_{cd}^* \langle \pi^+ \pi^- | (\bar{u} \Gamma^\mu d)(\bar{d} \Gamma_\mu c) | D^0 \rangle}{V_{ud}^* V_{cd} \langle \pi^+ \pi^- | (\bar{d} \Gamma_\mu u)(\bar{c} \Gamma^\mu d) | \overline{D^0} \rangle} \\
&= \frac{V_{ud} V_{cd}^*}{V_{ud}^* V_{cd}} e^{i(\xi_c - \xi_u + \xi_D)},
\end{aligned}
\tag{34.23}
$$

where $\Gamma^\mu \equiv \gamma^\mu (1 - \gamma_5)$, and $CP | D^0 \rangle = e^{i \xi_D} | \overline{D^0} \rangle$. Similarly, if $f_{cp} = K^+ K^-$, we need

$$
\begin{aligned}
\frac{c_{K^+ K^-}}{\bar{c}_{K^+ K^-}} &= \frac{V_{us} V_{cs}^* \langle K^+ K^- | (\bar{u} \Gamma^\mu s)(\bar{s} \Gamma_\mu c) | D^0 \rangle}{V_{us}^* V_{cs} \langle K^+ K^- | (\bar{s} \Gamma_\mu u)(\bar{c} \Gamma^\mu s) | \overline{D^0} \rangle} \\
&= \frac{V_{us} V_{cs}^*}{V_{us}^* V_{cs}} e^{i(\xi_c - \xi_u + \xi_D)}.
\end{aligned}
\tag{34.24}
$$

Theoretically, the practical difference between using $f_{cp} = \pi^+ \pi^-$ and using $f_{cp} = K^+ K^-$ is very small because

$$
\frac{c_{K^+ K^-}}{\bar{c}_{K^+ K^-}} = \frac{c_{\pi^+ \pi^-}}{\bar{c}_{\pi^+ \pi^-}} e^{2i\epsilon'},
\tag{34.25}
$$

and ϵ' is of order λ^4 in the SM. Many authors neglect ϵ' and use $\pi^+ \pi^-$ or $K^+ K^-$ indifferently.

The distinctive feature of these $\Delta C \neq 0 \neq \Delta U$ decays, which we have listed in Table 31.4, is that they cannot proceed through gluonic penguins. Two tree-level diagrams, $\bar{b} \to \bar{c} u \bar{s}$ and $\bar{b} \to \bar{u} c \bar{s}$, contribute almost equally to the decays $B^0_d \to D_{f_{cp}} K_S$ and $B^0_s \to D_{f_{cp}} \phi$. These diagrams have comparable magnitudes and different weak phases and, therefore, λ_f is not a pure phase and we do not measure a CKM phase directly. Fortunately, one may still compare these decays with the corresponding decays in which $D_{f_{cp}}$ is substituted by either D^0 or $\overline{D^0}$. We will show in § 36.3 how this can be used to extract the CKM phase γ.

The situation with the decays $B^0_d \to D_{f_{cp}} \pi^0$ and $B^0_s \to D_{f_{cp}} K_S$ is rather different. These decays get their main contribution from the tree-level diagram

$\bar{b} \to \bar{c}u\bar{d}$. The diagram $\bar{b} \to \bar{u}c\bar{d}$ also contributes, but it is suppressed by $|V_{ub}^* V_{cd}|/|V_{cb}^* V_{ud}|$, which is of order $\lambda^2 R_b$. In the factorization approximation, the hadronic matrix elements are the same and we find

$$\Delta \tilde{a} \propto \lambda^2 R_b \sin \gamma. \qquad (34.26)$$

Therefore, the rate asymmetry for these decays is approximately given by a single weak phase. To find it, we neglect the doubly Cabbibo-suppressed decays $B^0 \to D^0$ and $\overline{B^0} \to \overline{D^0}$. Using eqn (34.22) we get

$$\lambda_{B_d^0 \to (\pi^+ \pi^-)_D \pi^0} = \frac{q_{B_d}}{p_{B_d}} \frac{c_{\pi^+ \pi^-}}{\bar{c}_{\pi^+ \pi^-}} \frac{\langle D^0 \pi^0 | T | \overline{B_d^0} \rangle}{\langle \overline{D^0} \pi^0 | T | B_d^0 \rangle}. \qquad (34.27)$$

Now,

$$\frac{\langle D^0 \pi^0 | T | \overline{B_d^0} \rangle}{\langle \overline{D^0} \pi^0 | T | B_d^0 \rangle} = \frac{V_{cb} V_{ud}^*}{V_{cb}^* V_{ud}} \frac{\sum_{n=1}^{2} C_n \langle D^0 \pi^0 | Q_n^{cud \dagger} | \overline{B_d^0} \rangle}{\sum_{n=1}^{2} C_n \langle \overline{D^0} \pi^0 | Q_n^{cud} | B_d^0 \rangle}$$

$$= \frac{V_{cb} V_{ud}^*}{V_{cb}^* V_{ud}} e^{i(\xi_b - \xi_d + \xi_u - \xi_c - \xi_{B_d} - \xi_D)}, \qquad (34.28)$$

where we have used eqn (32.21) on the last step. Combining eqn (34.28) with eqns (30.33) and (34.23), we get

$$\lambda_{B_d^0 \to (\pi^+ \pi^-)_D \pi^0} = -\text{sign}\,(B_{B_d})\, e^{2i(\theta_d - \beta)}. \qquad (34.29)$$

Analogously,

$$\lambda_{B_d^0 \to (K^+ K^-)_D \pi^0} = -\text{sign}\,(B_{B_d})\, e^{2i(\theta_d + \epsilon' - \beta)}. \qquad (34.30)$$

34.6 Generalized cascade decays

In this section we wish to show that a new CP-violating parameter arises in those cascade decay chains for which both of the following conditions hold:

- the initial state can decay into both flavours of the intermediate neutral-meson system;
- the intermediate neutral mesons mix.

This situation generalizes the cases discussed above and is applicable to the decays $\{D^\pm, D_s^\pm\} \to X^\pm K \to X^\pm f_K$ and $\{D^0, \overline{D^0}\} \to X^0 K \to X^0 f_K$ (Amorim *et al.* 1999). This is also needed in order to study the decays $B^\pm \to X^\pm D \to X^\pm f_D$ (Meca and Silva 1998) and $\{B_q^0, \overline{B_q^0}\} \to X^0 D \to X^0 f_D$ (Amorim *et al.* 1999), whenever we go beyond the approximation used in § 34.4 and 34.5, and allow the mesons D^0 and $\overline{D^0}$ to mix.

Here we will follow Meca and Silva (1998) and concentrate on the decay chain $B^\pm \to K^\pm D \to K^\pm f_D$. We shall assume that $x_D \sim 10^{-2}$, $y_D = 0$, $|q_D/p_D| = 1$, and we shall also allow for the presence of a new CP-violating phase in D^0–$\overline{D^0}$ mixing—a variety of models of new physics which verify this scenario were presented, for example, by Nir (1996).

We have already studied several sources of CP violation which may be present in this chain:

1. Direct CP violation in the decays $B^+ \to K^+ D^0$ and $B^+ \to K^+ \overline{D^0}$. This source of CP violation would be detected through the differences

$$
\begin{aligned}
|A(B^+ \to K^+ D^0)| - |A(B^- \to K^- \overline{D^0})|, \\
|A(B^+ \to K^+ \overline{D^0})| - |A(B^- \to K^- D^0)|.
\end{aligned}
\tag{34.31}
$$

In the SM there is only one tree-level diagram contributing to each of these decays, and this source of CP violation is not present.

2. Direct CP violation in the decays $D^0 \to f$ and $D^0 \to \bar{f}$, which is measured by

$$
\begin{aligned}
|A(D^0 \to f)| - |A(\overline{D^0} \to \bar{f})|, \\
|A(D^0 \to \bar{f})| - |A(\overline{D^0} \to f)|.
\end{aligned}
\tag{34.32}
$$

3. CP violation in the interference between the mixing in the D^0–$\overline{D^0}$ system and the decay $D \to f$. This source of CP violation is related with

$$
\lambda_f = \frac{q_D}{p_D} \frac{A(\overline{D^0} \to f)}{A(D^0 \to f)},
\tag{34.33}
$$

and with $\lambda_{\bar{f}}$. Namely, CP is violated if $\lambda_f \lambda_{\bar{f}} \neq 1$.

4. CP violation in D^0–$\overline{D^0}$ mixing, probed by

$$
|q_D/p_D| - 1.
\tag{34.34}
$$

However, the sources of CP violation in 2. and 3. can be eliminated by choosing the flavour-specific final state $f_D = (K^- l^+ \nu_l)_D$, which identifies the D meson at the time of decay as D^0. The sources of CP violation in 1. and 4. are absent in the SM and are extremely small in most other models of interest (Nir 1996). We thus conclude that, when comparing the decay chain $B^+ \to K^+ D \to K^+ (K^- l^+ \nu_l)_D$ with the decay chain $B^- \to K^- D \to K^- (K^+ l^- \bar{\nu}_l)_D$, none of the sources of CP violation discussed thus far is present. We might then believe that no CP violation may be present in this case. However, Meca and Silva (1998) have shown that CP violation may still show up in this decay chain.

Indeed, a new source of CP violation exists because there are two decay paths connecting the initial state B^+ with the final state $K^+ (K^- l^+ \nu_l)_D$: the unmixed decay path $B^+ \to K^+ D^0 \to K^+ (K^- l^+ \nu_l)_D$, and also the mixed decay path $B^+ \to K^+ \overline{D^0} \to K^+ D^0 \to K^+ (K^- l^+ \nu_l)_D$. We may then have CP violation arising from the interference between the D^0–$\overline{D^0}$ mixing and the decays *from the initial state into the D^0–$\overline{D^0}$ system*. This interference is related to new parameters:

$$
\xi_+ \equiv \frac{A(B^+ \to K^+ \overline{D^0})}{A(B^+ \to K^+ D^0)} \frac{p_D}{q_D} \quad \text{and} \quad \xi_- \equiv \frac{A(B^- \to K^- \overline{D^0})}{A(B^- \to K^- D^0)} \frac{p_D}{q_D}.
\tag{34.35}
$$

If $\xi_+ \xi_- \neq 1$, then this new source of CP violation is present.

The definitions in eqns (34.35) can be generalized to include the decays from an arbitrary initial state i into a generic intermediate neutral-meson system P^0–$\overline{P^0}$ (Amorim et al. 1999):

$$\xi_i \equiv \frac{A(i \to X\overline{P^0})}{A(i \to XP^0)} \frac{p_P}{q_P} \tag{34.36}$$

We may use the techniques introduced in Chapter 7 to show that these quantities are explicitly rephasing-invariant.[97]

Notice that the parameters

$$\lambda_f \equiv \frac{q_P}{p_P} \frac{A(\overline{P^0} \to f)}{A(P^0 \to f)} \tag{34.37}$$

and ξ_i describe two completely different sources of interference. They both involve the interference between the mixing in the P^0–$\overline{P^0}$ system and some decays. But, in the parameters λ_f, the interference takes place with the decays *from the P^0–$\overline{P^0}$ system into the final state f*. On the other hand, the interference probed by the parameters ξ_i is the one occurring between *the decays into the P^0–$\overline{P^0}$ system and the mixing in that system*. The distinction is clear when we look back at the $B^\pm \to K^\pm D \to K^\pm f_D$ decay chain. The crucial difference is that, while the phase of $\lambda_f \lambda_{\bar{f}}$ is small in the SM,[98] the phase of $\xi_+ \xi_-$ is related with the phase γ, which is necessarily large, even within the SM.

Meca and Silva (1998) have noted that the presence of such a large CP-violating phase may allow for measurements of $x_D \sim 10^{-2}$ in the decay chains $B^\pm \to K^\pm D^0 \to K^\pm f_D$. This effect is similar to the one discussed by Liu (1995) and by Wolfenstein (1995), except that, in their case, x_D appears multiplied by Im λ_f—see Appendix E. This is related with ϵ' in the SM, and will be large only if the new physics that brings x_D close to 10^{-2} also produces a large new phase in D^0–$\overline{D^0}$ mixing. On the other hand, in the cascade decays studied by Meca and Silva (1998), the effect is related to the phase of $\xi_+ \xi_-$, which is proportional to the large phase γ.

We stress that the interference effects described by the parameters ξ_i are just as important as the ones described by λ_f. Amorim et al. (1999) have shown that, with the new parameters described in this section, we have all the necessary and sufficient parameters needed in order to describe the various sources of CP violation present in the most general cascade decay chain. Namely, there may be direct CP violation, CP violation in the mixing, CP violation in the interference between the mixing in the intermediate neutral meson system and the decay *from* that system (probed by λ_f), and CP violation in the interference between

---

[97]The parameters ξ_i used here have nothing to do with the spurious CP-transformation phases, for which we have used in this book the same letter ξ.

[98]The decays of the D mesons involve the first two quark families. From eqn (28.24), we know that ϵ' is the rephasing-invariant CP-odd phase present in that sector, which is extremely small in the SM.

the decay *into* the neutral meson system and the mixing in that system (probed by ξ_i).

SOME METHODS TO EXTRACT α

35.1 Introduction

In the preceding chapters we have concentrated our efforts on identifying decay channels for which, in the SM, a single weak phase dominates the amplitude. In those cases one can extract that CKM phase, barring new-physics contributions to the mixing or to the decay amplitudes. In this chapter we discuss some approaches that have been proposed to extract α, even when there are several SM diagrams contributing to the amplitude. Due to space limitations we only consider a few examples.

It is customary to distinguish between cases for which we need to know only the branching ratio—denoted $B^0 \to f$—from cases in which we also require the time dependence—denoted $B^0(t) \to f$. It will be useful to recall what one can learn at the $\Upsilon(4S)$ from tagged, fully time-integrated decays into CP eigenstates. Using eqns (29.5) and remembering that $|q/p| = 1$, we find

$$
\begin{aligned}
\left|\langle f; l^- |T| \Upsilon(4S)\rangle\right|^2 + \left|\langle f; l^+ |T| \Upsilon(4S)\rangle\right|^2 &= \frac{|A_l|^2}{2\Gamma^2} \left(|A_f|^2 + |\bar{A}_f|^2 \right), \\
\left|\langle f; l^- |T| \Upsilon(4S)\rangle\right|^2 - \left|\langle f; l^+ |T| \Upsilon(4S)\rangle\right|^2 &= \frac{|A_l|^2}{2\Gamma^2} \frac{|A_f|^2 - |\bar{A}_f|^2}{1 + x^2}.
\end{aligned}
\tag{35.1}
$$

Thus, at the $\Upsilon(4S)$, one can determine $|A_f|$ and $|\bar{A}_f|$ from tagged, fully time-integrated measurements alone. On the other hand, as we have seen in § 29.4, under the same conditions one cannot detect $\text{Im}\,\lambda_f$. All we would need in order to extract $\text{Im}\,\lambda_f$ would be the time ordering. By time ordering we mean that, after following the time dependence, one collects all the events with a given sign of t_-. See eqns (29.4).

35.2 The Gronau–London method

As we have seen, the decay $B_d^0 \to \pi^+\pi^-$ has a penguin contribution which may be sizeable (Gavela *et al.* 1985*a*; Chau and Cheng 1987; Grinstein 1989; Gronau 1989; London and Peccei 1989). Since there are two interfering amplitudes with different weak phases, the CP asymmetry will exhibit a cosine time dependence (proportional to a^{dir}) and a sine time dependence (proportional to a^{int}), instead of only a sine time dependence—cf. eqn (28.8). If the two amplitudes contributing to the decay have the same final-state, CP-even phase, then $a^{\text{dir}} = 0$. Still, the presence of two different weak phases implies that the interference-CP-violation coefficient of the time-dependent sine term, a^{int}, does not provide a clean measurement of a CKM phase (Gronau 1993). As a result, the extraction of $\sin 2\alpha$

from this decay is hindered by penguin pollution. The solution is to compare this decay with other decays related to this one by some symmetry. The first such method was proposed by Gronau and London (1990). The basic idea consists in the observation that the process $B^+ \to \pi^+\pi^0$ has a $\Delta I = 3/2$ change in isospin, which is due exclusively to the tree-level diagram. In this method one requires a measurement of the time dependence of the decay $B_d^0 \to \pi^+\pi^-$ and of the branching ratios for $B_d^0 \to \pi^0\pi^0$ and for $B^+ \to \pi^+\pi^0$, as well as the analogous quantities for the CP-conjugated decays. Symbolically:

$$B_d^0(t) \to \pi^+\pi^- , \quad B_d^0 \to \pi^0\pi^0 , \quad B^+ \to \pi^+\pi^0 , \quad \text{and CP conjugated.} \quad (35.2)$$

As we shall see, this still leaves a fourfold ambiguity in the determination of $\sin 2\alpha$. One may get rid of this ambiguity by also measuring the time dependence of $B_d^0 \to \pi^0\pi^0$. Symbolically

$$B_d^0(t) \to \pi^+\pi^- , \quad B_d^0(t) \to \pi^0\pi^0 , \quad B^+ \to \pi^+\pi^0 , \quad \text{and CP conjugated.} \quad (35.3)$$

The isospin decomposition of the decays $B \to \pi\pi$ is exactly the same as for the decays $K \to \pi\pi$, studied in Chapter 8. We include it here for completeness. Since the pions are spinless they must be in a symmetric state. For an s wave this implies a symmetric isospin configuration. Therefore, the relevant final states are

$$\begin{aligned}
\langle \pi^0\pi^0| &= \sqrt{\tfrac{2}{3}}\langle 2,0| - \sqrt{\tfrac{1}{3}}\langle 0,0|, \\
\langle \pi^+\pi^-| &\equiv \sqrt{\tfrac{1}{2}}\left(\langle \pi_1^+\pi_2^-| + \langle \pi_1^-\pi_2^+|\right) = \sqrt{\tfrac{1}{3}}\langle 2,0| + \sqrt{\tfrac{2}{3}}\langle 0,0|, \\
\langle \pi^+\pi^0| &\equiv \sqrt{\tfrac{1}{2}}\left(\langle \pi_1^+\pi_2^0| + \langle \pi_1^0\pi_2^+|\right) = \langle 2,1|.
\end{aligned} \quad (35.4)$$

The first two channels and the last channel are reached by $|B_d^0\rangle = |1/2, -1/2\rangle$ and by $|B^+\rangle = |1/2, 1/2\rangle$, respectively. In general, the transition matrix has $\Delta I = 1/2$, $\Delta I = 3/2$, and $\Delta I = 5/2$ pieces. The Wigner–Eckart theorem gives

$$\begin{aligned}
A^{+-} &\equiv \langle \pi^+\pi^-|T|B_d^0\rangle = -\sqrt{\tfrac{1}{3}}A_{1/2} + \sqrt{\tfrac{1}{6}}A_{3/2} - \sqrt{\tfrac{1}{6}}A_{5/2}, \\
A^{00} &\equiv \langle \pi^0\pi^0|T|B_d^0\rangle = \sqrt{\tfrac{1}{6}}A_{1/2} + \sqrt{\tfrac{1}{3}}A_{3/2} - \sqrt{\tfrac{1}{3}}A_{5/2}, \\
A^{+0} &\equiv \langle \pi^+\pi^0|T|B^+\rangle = \tfrac{\sqrt{3}}{2}A_{3/2} + \sqrt{\tfrac{1}{3}}A_{5/2},
\end{aligned} \quad (35.5)$$

where A_k are the relevant reduced matrix elements.

In the SM there are short-distance contributions to the $\Delta I = 1/2$ and to the $\Delta I = 3/2$ matrix elements. The tree-level diagram and its QCD corrections yield left-handed four-fermion interactions which contribute to both $A_{1/2}$ and $A_{3/2}$. On the other hand, the gluonic penguin only contributes to two-pion states with zero isospin and, therefore, they show up exclusively in $A_{1/2}$. Electroweak penguins show up in both $A_{1/2}$ and $A_{3/2}$; worse, they introduce isospin breaking, because the couplings of the photon and of the Z boson distinguish between up-type and down-type quarks (Deshpande and He 1995a). However, the EW-penguin contributions to these decays are expected to be much smaller than the other contributions, and are usually neglected in the analysis (Gronau *et al.* 1995).

In the spectator approximation, there are no short-distance diagrams leading to $A_{5/2}$. Such contributions arise, in the SM, from the $A_{1/2}$ amplitudes together with the $\Delta I = 2$ electromagnetic rescattering of the two pions in the final state (Donoghue *et al.* 1992). This contribution is naively estimated to be $A_{5/2} \sim \alpha A_{1/2}$. In the kaon sector that effect could be seen in the comparison between the decays of the charged and neutral kaons because the $\Delta I = 1/2$ rule implies a large hierarchy between the $A_{1/2}$ and $A_{3/2}$ terms. As a result, $A_{5/2}$ could be of order $0.11 \times A_{3/2}$—cf. eqn (8.72)—thus influencing the decay $K^+ \to \pi^+\pi^0$ by 11%. The situation in the decays of B_d^0 is very different because, using model calculations of $B \to \pi\pi$ decays, one expects the $A_{1/2}$ and $A_{3/2}$ contributions to be comparable (Kramer and Palmer 1995). Therefore, the corrections due to the $A_{5/2}$ term are smaller than 1% in all decays, and this term can be dropped.

Defining

$$A_0 \equiv \tfrac{1}{\sqrt{6}} A_{1/2},$$
$$A_2 \equiv \tfrac{1}{2\sqrt{3}} A_{3/2},$$

(35.6)

one obtains from eqns (35.5)

$$A^{+-} = \sqrt{2}\,(A_2 - A_0),$$
$$A^{00} = 2A_2 + A_0,$$
$$A^{+0} = 3A_2.$$

(35.7)

These three amplitudes satisfy

$$A^{+-} + \sqrt{2}A^{00} = \sqrt{2}A^{+0}.$$

(35.8)

Likewise, the amplitudes for the CP-conjugated processes satisfy

$$\bar{A}^{+-} + \sqrt{2}\bar{A}^{00} = \sqrt{2}\bar{A}^{+0}.$$

(35.9)

Equations (35.8) and (35.9) are depicted as two triangles in Fig. 35.1. They are at the root of the Gronau–London (1990) method to determine α.

Neglecting the small electroweak penguin effects, the only contribution to A_2 comes from the tree-level diagram. Therefore A_2 has a single weak phase, that of the tree-level diagram. As a consequence $|A_2| = |\bar{A}_2|$ and

$$\frac{q_{B_d}}{p_{B_d}} \frac{\bar{A}_2}{A_2} = -e^{2i\alpha},$$

(35.10)

cf. eqn (33.4). The $B^+ \to \pi^+\pi^0$ branching ratio determines $|A^{+0}| = |\bar{A}^{-0}|$ and, thus, $|A_2|$. Similarly, one may determine $|A^{+-}|$, $|\bar{A}^{+-}|$, $|A^{00}|$, and $|\bar{A}^{00}|$ from the time-integrated decay rates for $B_d^0 \to \pi^+\pi^-$, $B_d^0 \to \pi^0\pi^0$, and the CP-conjugated decays. At this stage we know the sides of both triangles in Fig. 35.1.

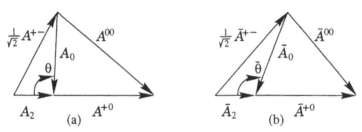

FIG. 35.1. Representation of eqns (35.8) and (35.9) as two triangles, in (a) and (b), respectively.

Let us define

$$\theta \equiv \arg\left(A_0 A_2^*\right),$$
$$\bar{\theta} \equiv \arg\left(\bar{A}_0 \bar{A}_2^*\right). \tag{35.11}$$

These angles may be determined, up to their sign, from Fig. 35.1. The ambiguity in the sign of the angles corresponds to the possibility of each of the triangles being drawn as in the figure, or upside-down.

From the time dependence of the process $B_d^0(t) \to \pi^+\pi^-$ one measures $\operatorname{Im}\lambda_{+-}$, which is given by

$$\operatorname{Im}\lambda_{+-} = \operatorname{Im}\frac{q_{B_d}}{p_{B_d}}\frac{\bar{A}^{+-}}{A^{+-}} = \operatorname{Im}\left(-e^{2i\alpha}\frac{|\bar{A}_2| - |\bar{A}_0|\,e^{i\bar{\theta}}}{|A_2| - |A_0|\,e^{i\theta}}\right). \tag{35.12}$$

The fraction in eqn (35.12) is known up to a fourfold ambiguity, which is due to the two possible signs for θ and $\bar{\theta}$. We may thus determine $\sin 2\alpha$ up to a fourfold ambiguity.

In general, this ambiguity can be removed if we look also for the time dependence of the decay $B_d^0(t) \to \pi^0\pi^0$, thereby determining

$$\operatorname{Im}\lambda_{00} = \operatorname{Im}\frac{q_{B_d}}{p_{B_d}}\frac{\bar{A}^{00}}{A^{00}} = \operatorname{Im}\left(-e^{2i\alpha}\frac{2|\bar{A}_2| + |\bar{A}_0|\,e^{i\bar{\theta}}}{2|A_2| + |A_0|\,e^{i\theta}}\right). \tag{35.13}$$

Only in some singular cases does there remain a twofold ambiguity in the determination of $\sin 2\alpha$ (Gronau and London 1990).

Models with new physics exclusively in the mixing do not affect the analysis, since they only change q_{B_d}/p_{B_d}. Thus, in models with no new contributions to the decay amplitudes this method may be used and determines $\sin 2(\alpha + \theta_d)$. The presence of new contributions to A_2 may be tested by looking for a difference between the branching ratios of $B^+ \to \pi^+\pi^0$ and of $B^- \to \pi^-\pi^0$. Of course, a null result might simply be due to very small final-state interaction phase differences, but a positive result, well above that expected from the electroweak penguins, would be a clear signal of new physics.

The application of the method of Gronau and London faces a number of experimental difficulties. The decay $B_d^0 \to \pi^0\pi^0$ is colour-suppressed and its branching ratio might be smaller than 10^{-6} (Kramer and Palmer 1995), though this estimate has come under closer scrutiny (Ciuchini *et al.* 1997*a*). In any case, the identification of two neutral pions in the final state is experimentally quite challenging.

Should the $B_d^0 \to \pi^0 \pi^0$ branching ratio be very small, one can take $|A^{00}| = |2A_2 + A_0| \sim 0$, meaning that the decay $B_d^0 \to \pi^+ \pi^-$ really measures $\sin 2\alpha$ (Sanda and Xing 1997). Grossman and Quinn (1998) have given a quantitative expression to this result. Using the definition in eqn (28.20) and eqn (28.21),

$$a_{\pi^+\pi^-}^{\text{int}} = -\sin 2\left(\alpha + \delta_\alpha\right), \tag{35.14}$$

they have proved that

$$\sin^2 \delta_\alpha \le \frac{\text{BR}(B_d^0 \to \pi^0 \pi^0) + \text{BR}(\overline{B_d^0} \to \pi^0 \pi^0)}{\text{BR}(B^+ \to \pi^+ \pi^0) + \text{BR}(B^- \to \pi^- \pi^0)}. \tag{35.15}$$

This is a nice constraint on δ_α because it does not require one to tag the initial flavour of the neutral B mesons.

35.3   The Silva–Wolfenstein method

Silva and Wolfenstein (1994) have noted that one may use flavour-SU(3) rather than isospin to remove the penguin pollution from the decays $B_d^0 \to \pi^+ \pi^-$. In this method one uses the spectator approximation and SU(3) to relate the tree-dominated process $B_d^0 \to \pi^+ \pi^-$ to the penguin-dominated $B_d^0 \to K^+ \pi^-$. Experimentally, one needs to look for the time dependence of the processes $B_d^0 \to J/\psi K_S$ and $B_d^0 \to \pi^+ \pi^-$ and for the branching ratio of $B_d^0 \to K^+ \pi^-$, as well as for the CP-conjugated processes:

$$B_d^0(t) \to J/\psi K_S, \quad B_d^0(t) \to \pi^+ \pi^-, \quad B_d^0 \to K^+ \pi^-, \quad \text{and CP conjugated.} \tag{35.16}$$

The advantage of the Silva–Wolfenstein method is that all the required observables are easy to measure: the determination of $\sin 2\beta$ from $B_d^0(t) \to J/\psi K_S$ will be the first outcome of B-factories; and the CLEO Collaboration (1998*b*) has already placed limits on the *untagged* rate ratio

$$R \equiv \frac{\Gamma[B_d^0 \to K^+ \pi^-] + \Gamma[\overline{B_d^0} \to K^- \pi^+]}{\Gamma[B_d^0 \to \pi^+ \pi^-] + \Gamma[\overline{B_d^0} \to \pi^+ \pi^-]}. \tag{35.17}$$

This quantity lies roughly between 1 and 5, at the 1σ level (CLEO Collaboration 1998*b*). Notice that the final state only contains charged particles, which are easier to detect than neutral particles.

The decay amplitudes may be written as

$$A(B_d^0 \to \pi^+ \pi^-) = \frac{V_{ub}^* V_{ud}}{|V_{ub} V_{ud}|} T + \frac{V_{tb}^* V_{td}}{|V_{tb} V_{td}|} P e^{i\Delta}, \tag{35.18}$$

$$A(B_d^0 \to K^+\pi^-) = \frac{V_{ub}^* V_{us}}{|V_{ub}V_{us}|}T' + \frac{V_{tb}^* V_{ts}}{|V_{tb}V_{ts}|}P'e^{i\Delta'}, \tag{35.19}$$

where Δ and Δ' are strong phases. These expressions are completely general, they are valid even in the presence of new physics. Any new contribution to $B_d^0 \to \pi^+\pi^-$, with an arbitrary phase, may be rewritten in terms of the two terms in eqn (35.18). In particular, within the SM, any contribution with the phase of $V_{cb}^* V_{cd} = -V_{ub}^* V_{ud} - V_{tb}^* V_{td}$ is clearly included in eqn (35.18). What such new contributions do is to alter the relative magnitudes of T and P. In the Silva–Wolfenstein method, the relative magnitude P/T is to be *experimentally* determined from the ratio of BR($B_d^0 \to K^+\pi^-$) and BR($B_d^0 \to \pi^+\pi^-$).

The decay rates are

$$\begin{aligned}
\Gamma[B_d^0 \to \pi^+\pi^-] &= T^2 + 2TP\cos(\beta + \gamma - \Delta) + P^2, \\
\Gamma[B_d^0 \to K^+\pi^-] &= P'^2 - 2T'P'\cos(\gamma - \epsilon + \epsilon' - \Delta') + T'^2.
\end{aligned} \tag{35.20}$$

The decay rates for $\overline{B_d^0} \to \pi^+\pi^-$ and for $\overline{B_d^0} \to K^-\pi^+$ are obtained from eqns (35.20) by changing the signs of Δ and of Δ', respectively. Therefore, the ratio of *untagged* decay rates in eqn (35.17) is

$$R = \frac{T'^2}{T^2} \frac{r'^2 + 2r'\cos(\alpha + \beta)\cos\Delta' + 1}{1 - 2r\cos\alpha\cos\Delta + r^2}, \tag{35.21}$$

where $r \equiv P/T$, $r' \equiv P'/T'$, and we have neglected the phases ϵ and ϵ', which are small in the SM.

The impact on R of non-vanishing strong phase shifts appears first at order Δ^2 and Δ'^2. Similarly, the change in a^{int} due to the final-state interactions also shows up first at order Δ^2, cf. eqn (28.18). Since these phases are expected to be small (Kramer and Palmer 1995), we neglect them, and use (Silva and Wolfenstein 1994)

$$R = \frac{T'^2}{T^2} \frac{r'^2 + 2r'\cos(\alpha + \beta) + 1}{1 - 2r\cos\alpha + r^2}. \tag{35.22}$$

From eqns (28.20) and (28.21), the CP violating asymmetry in $B_d^0 \to \pi^+\pi^-$ becomes

$$a_{\pi^+\pi^-}^{\text{int}} = -\sin 2(\alpha + \delta_\alpha), \tag{35.23}$$

where

$$\tan\delta_\alpha = \frac{r\sin\alpha}{1 - r\cos\alpha}. \tag{35.24}$$

The angle β will be determined from $B_d^0 \to J/\psi K_S$. The observables R and $a_{\pi^+\pi^-}^{\text{int}}$ should be measured soon thereafter. If one relates, as we will do shortly, T to T' and r to r', then eqns (35.22), (35.23), and (35.24) determine the three unknowns r, δ_α, and α.

The approximate flavour-SU(3) symmetry has long been used to relate the rates of different B decays (Zeppenfeld 1981; Savage and Wise 1989; Chau *et al.* 1991). The idea of Silva and Wolfenstein (1994) was to use an SU(3) transformation (a U-spin rotation interchanging d and s quarks), together with the spectator approximation, in the context of CP-violating asymmetries: thus, they relate T to T' and r to r'. This provides a clean determination of α, in spite of the penguin pollution in the $B_d^0 \to \pi^+\pi^-$ rate asymmetry. They use (Silva and Wolfenstein 1994)

$$\frac{T'}{T} = \left|\frac{V_{us}}{V_{ud}}\right|\frac{f_K}{f_\pi} \approx \lambda\frac{f_K}{f_\pi}, \tag{35.25}$$

$$\frac{P'}{P} = \left|\frac{V_{ts}}{V_{td}}\right|\frac{f_K}{f_\pi} \approx \frac{1}{\lambda}\left|\frac{\sin\alpha}{\sin(\alpha+\beta)}\right|\frac{f_K}{f_\pi}. \tag{35.26}$$

In these expressions the ratio of the matrix elements in the $B_d^0 \to K^+\pi^-$ to the $B_d^0 \to \pi^+\pi^-$ decays was estimated to be equal to the ratio of decay constants f_K/f_π. This is the result obtained using factorization, and provides a first order estimate of the SU(3)-breaking effects due to hadronization. Notice that we have used the unitarity of the CKM matrix in the SM in order to relate the magnitudes $|V_{ts}/V_{td}|$ with the angles α and β, on the last step of eqn (35.26).

This method has two important qualities: the fact that all observables are easy to measure and should be detected at the early stages of B-factories; and the fact that $r'/r \sim 1/\lambda^2$. As a result of the latter, a mild penguin contribution for $B_d^0 \to \pi^+\pi^-$ decays, corresponds to a dominant penguin effect in the $B_d^0 \to K^+\pi^-$ decay, where it will be easy to determine. The method has a twofold ambiguity in the determination of $\sin 2\alpha$. This is due to the fact that eqn (35.22) is quadratic in r. This ambiguity may be removed by noting that r is positive in the factorization approximation.

It should be stressed that, as presented here, this method applies within the SM. If new physics contributes to $B_d^0 - \overline{B_d^0}$ mixing with a new phase θ_d, that phase will correspond to a $\alpha \to \alpha + \theta_d$ replacement in the interference CP violation of eqn (35.23), but leaves α unchanged on the observables in eqn (35.22). One can disentangle the effect of P/T from that of θ_d by measuring, in addition, the $B^+ \to \pi^+K^0$ decay rate (Wolfenstein 1997*a*).

The precision of this method is limited by the SU(3)-breaking effects. Although f_K/f_π should be the dominant SU(3) breaking correction, and factorization provides a reasonable approximation for the matrix elements of the current–current operators, factorization of the matrix elements of penguin operators is questionable. The ratio P'/P can be left as a free parameter if one also knows the $B^+ \to \pi^+K^0$ decay rate (Gronau and Rosner 1996; Dighe *et al.* 1996*a*; Wolfenstein 1997*a*). Alternatively, this extra piece of information might be used to get at a deviation of the phase of the mixing from its SM value (Wolfenstein 1997*a*).

The precision might also be affected by the approximation $\Delta = \Delta' = 0$. Although they only show up at order Δ^2 and Δ'^2, these final-state phases have an important impact if they are large. In that case, their effects can be detected elsewhere. Under the assumption that $\Delta = \Delta'$, the same final-state phase shows up as direct CP violation in $\Gamma[B_d^0 \to K^+\pi^-] - \Gamma[\overline{B_d^0} \to K^-\pi^+]$, as well as in the

$\cos(\Delta mt)$ coefficient of the $B_d^0 \to \pi^+\pi^-$ time-dependent rate asymmetry (Deshpande and He 1995b; Gronau and Rosner 1996). This can be used to determine it.

In a series of articles, Gronau *et al.* (1994a,b, 1995) have extended this SU(3) analysis to include all B decays into two light pseudoscalars. The final states with η and η' were discussed by Dighe (1996) and by Dighe *et al.* (1996b).

35.4 The Snyder–Quinn method

Aleksan *et al.* (1991) noted that, if the decay amplitudes are dominated by a single weak phase, then one may combine the decays $B_d^0 \to \rho^\pm\pi^\mp$ with $B_d^0 \to \rho^0\pi^0$ in order to extract α, even though the final states are not CP eigenstates. However, the penguin diagrams are expected to be important, as happens in the decays $B_d^0 \to \pi\pi$. Lipkin *et al.* (1991) proposed a method to take the penguin pollution into account by combining $B_d^0 \to \rho^\pm\pi^\mp$, $B_d^0 \to \rho^0\pi^0$, $B^\pm \to \rho^\pm\pi^0$, and $B^\pm \to \rho^0\pi^\pm$ decays into an isospin analysis. One could thus determine α up to discrete ambiguities. Unfortunately, the detection of $B^\pm \to \rho^\pm\pi^0 \to \pi^\pm\pi^0\pi^0$ should be rather difficult. Besides, Gronau (1991) and Lavoura (1992b) have shown that the isospin analysis suggested by Lipkin *et al.* (1991) suffers from far less trivial ambiguities than the analysis in § 35.2, including in particular cases of continuous, rather than discrete, ambiguities.

In this section we discuss a method proposed by Snyder and Quinn (1993) to determine the angle α. The Snyder–Quinn method is based on the decays $B_d^0 \to \{\rho^+\pi^-, \rho^0\pi^0, \rho^-\pi^+\} \to \pi^+\pi^-\pi^0$ and on their CP conjugates. Thus, six possible routes to the final state are considered. The resonances ρ are described by a Breit–Wigner function of the type

$$f(s) = \frac{1}{s - m_\rho^2 + i\Pi(s)} \tag{35.27}$$

and, therefore, involve a CP-even phase which interferes with the CP-odd phases and with the final-state interaction, CP-even phases in the $B_d^0 \to \rho\pi$ decay amplitudes. The complicated interference patterns that result are the hallmark of this method. They may be used to determine various combinations of weak and strong phases.

The idea is to construct the Dalitz plot for the three-pion final state (Dalitz 1953; Fabri 1954). This is then fitted to the expression for the rate as a function of all amplitudes, weak phases and strong phases in the problem. The required multi-variable analysis contributes to the limitations of this method. Let us define

$$\begin{aligned}
a_{+-} &\equiv a(B_d^0 \to \rho^+\pi^-), \\
a_{-+} &\equiv a(B_d^0 \to \rho^-\pi^+), \\
a_{00} &\equiv a(B_d^0 \to \rho^0\pi^0), \\
\bar{a}_{+-} &\equiv a(\overline{B_d^0} \to \rho^+\pi^-), \\
\bar{a}_{-+} &\equiv a(\overline{B_d^0} \to \rho^-\pi^+), \\
\bar{a}_{00} &\equiv a(\overline{B_d^0} \to \rho^0\pi^0).
\end{aligned} \tag{35.28}$$

The CP-conjugated decay amplitude of a_{+-} is $\bar{a}_{-+}$. The amplitudes are denoted by a lower-case a in order to represent the fact that we are *not* looking into rates integrated over the whole phase space. We may then write (Snyder and Quinn 1993)

$$
\begin{aligned}
a(B_d^0 \to \pi^+\pi^-\pi^0) &= f_+ a_{+-} + f_- a_{-+} + f_0 a_{00}, \\
a(\overline{B_d^0} \to \pi^+\pi^-\pi^0) &= f_+ \bar{a}_{+-} + f_- \bar{a}_{-+} + f_0 \bar{a}_{00},
\end{aligned}
\tag{35.29}
$$

where

$$
\begin{aligned}
f_+ &\equiv f(M_{+0}^2)\cos\theta_+, \\
f_- &\equiv f(M_{-0}^2)\cos\theta_-, \\
f_0 &\equiv f(M_{+-}^2)\cos\theta_0.
\end{aligned}
\tag{35.30}
$$

Here, M_{+0}, M_{-0}, and M_{+-} are the invariant masses of the $\pi^+\pi^0$, $\pi^-\pi^0$, and $\pi^+\pi^-$ pairs, respectively. Since B_d^0 and π are spinless, ρ must have helicity zero. Thus, the dependence on the helicity angle θ is proportional to $\cos\theta$. One draws the M_{+0}^2 and M_{-0}^2 axis; each event is identified by a point in this two-dimensional Dalitz plot.

Now, using the unitarity of the CKM matrix, we may write all decay amplitudes as a sum of two terms. One of them is proportional to $V_{ub}^* V_{ud}$ and receives contributions from both tree-level and penguin diagrams. The other term is proportional to $V_{tb}^* V_{td}$ and receives contributions from penguin diagrams only. Combining this with the isospin decomposition of the decay amplitudes one may write (Lipkin *et al.* 1991; Snyder and Quinn 1993),

$$
\begin{aligned}
a_{+-} &= \frac{V_{ub}^* V_{ud}}{|V_{ub}V_{ud}|}T_{+-} + \frac{V_{tb}^* V_{td}}{|V_{tb}V_{td}|}\left(P_1 + P_0\right), \\
a_{-+} &= \frac{V_{ub}^* V_{ud}}{|V_{ub}V_{ud}|}T_{-+} + \frac{V_{tb}^* V_{td}}{|V_{tb}V_{td}|}\left(-P_1 + P_0\right), \\
a_{00} &= \frac{V_{ub}^* V_{ud}}{|V_{ub}V_{ud}|}T_{00} + \frac{V_{tb}^* V_{td}}{|V_{tb}V_{td}|}\left(-P_0\right).
\end{aligned}
\tag{35.31}
$$

There are only two pure penguin terms, P_0 and P_1, corresponding to the penguin contributions to isospin 0 and to isospin 1, respectively. Notice that the weak phases have been explicitly factored out; the T and P only contain magnitudes and strong phases. Therefore,

$$
\begin{aligned}
\bar{a}_{-+} &= \frac{V_{ub}V_{ud}^*}{|V_{ub}V_{ud}|}T_{+-} + \frac{V_{tb}V_{td}^*}{|V_{tb}V_{td}|}\left(P_1 + P_0\right), \\
\bar{a}_{+-} &= \frac{V_{ub}V_{ud}^*}{|V_{ub}V_{ud}|}T_{-+} + \frac{V_{tb}V_{td}^*}{|V_{tb}V_{td}|}\left(-P_1 + P_0\right), \\
\bar{a}_{00} &= \frac{V_{ub}V_{ud}^*}{|V_{ub}V_{ud}|}T_{00} + \frac{V_{tb}V_{td}^*}{|V_{tb}V_{td}|}\left(-P_0\right),
\end{aligned}
\tag{35.32}
$$

where we have dropped spurious phases brought about by the CP transformation. We may eliminate the terms proportional to $V_{tb}^* V_{td}$ and to $V_{tb} V_{td}^*$ in eqns (35.31) and (35.32), respectively, by considering the linear combinations[99]

$$a_{\text{sum}} \equiv a_{+-} + a_{-+} + 2a_{00}$$
$$= \frac{V_{ub}^* V_{ud}}{|V_{ub} V_{ud}|} (T_{+-} + T_{-+} + 2T_{00}),$$
$$\bar{a}_{\text{sum}} \equiv \bar{a}_{+-} + \bar{a}_{-+} + 2\bar{a}_{00}$$
$$= \frac{V_{ub} V_{ud}^*}{|V_{ub} V_{ud}|} (T_{+-} + T_{-+} + 2T_{00}). \tag{35.33}$$

As a result,

$$\text{Im} \frac{q_{B_d}}{p_{B_d}} \frac{\bar{a}_{\text{sum}}}{a_{\text{sum}}} = -\sin 2(\alpha + \theta_d). \tag{35.34}$$

We will now show how this combination can be extracted from the Dalitz plot.

As usual, the expression for the tagged, time-dependent decay rate has terms proportional to

$$\left| a(B_d^0 \to \pi^+ \pi^- \pi^0) \right|^2 \quad \text{and} \quad \left| a(\overline{B_d^0} \to \pi^+ \pi^- \pi^0) \right|^2, \tag{35.35}$$

affecting the constant and $\cos \Delta mt$ terms, and terms proportional to

$$\text{Im} \left\{ q_{B_d} p_{B_d}^* a(\overline{B_d^0} \to \pi^+ \pi^- \pi^0) a(B_d^0 \to \pi^+ \pi^- \pi^0)^* \right\} \tag{35.36}$$

in the $\sin \Delta mt$ terms. We can see from eqn (35.29) that each of these terms involves many phase combinations. Substituting eqn (35.29) in the first eqn (35.35) we get

$$\left| a(B_d^0 \to \pi^+ \pi^- \pi^0) \right|^2 = \left| f_+ a_{+-} + f_- a_{-+} + f_0 a_{00} \right|^2. \tag{35.37}$$

We stress that the form of the functions f_+, f_-, and f_0 is part of the input in the multi-variable fit. Therefore, we can extract the individual terms and recombine them into $|a_{\text{sum}}|^2$. This is obvious for terms such as $|f_+ a_{+-}|^2$. It is less obvious for the crossed terms. For example, the right-hand side of eqn (35.37) contains a term

$$f_+ f_-^* a_{+-} a_{-+}^* + f_- f_+^* a_{-+} a_{+-}^* = \text{Re} \left(f_+ f_-^* \right) \left(a_{+-} a_{-+}^* + a_{-+} a_{+-}^* \right)$$
$$+ i\text{Im} \left(f_+ f_-^* \right) \left(a_{+-} a_{-+}^* - a_{-+} a_{+-}^* \right). \tag{35.38}$$

The term multiplying $\text{Re} \left(f_+ f_-^* \right)$ is the one needed for $|a_{\text{sum}}|^2$. Similarly, substituting eqn (35.29) in eqn (35.36) we get

$$\text{Im} \left[q_{B_d} p_{B_d}^* a(\overline{B_d^0} \to \pi^+ \pi^- \pi^0) a(B_d^0 \to \pi^+ \pi^- \pi^0)^* \right]$$
$$= \text{Im} \left[q_{B_d} p_{B_d}^* \left(f_+ \bar{a}_{+-} + f_- \bar{a}_{-+} + f_0 \bar{a}_{00} \right) \left(f_+ a_{+-} + f_- a_{-+} + f_0 a_{00} \right)^* \right]. \tag{35.39}$$

We may extract the terms multiplying each individual $f f^*$ combination of functions, and thus get $\text{Im} \left(q_{B_d} p_{B_d}^* \bar{a}_{\text{sum}} a_{\text{sum}}^* \right)$. Combining this with $|a_{\text{sum}}|^2$ and using

---

[99]This presentation follows the one by Grossman and Quinn (1997).

$|q/p| = 1$, we can determine the combination in eqn (35.34), and thus measure $\sin 2\,(\alpha + \theta_d)$, despite the presence of gluonic penguins.

There are a few difficulties with this method. As we have seen, the Dalitz plot is parametrized in terms of a decay rate containing all six amplitudes in eqns (35.28). The large number of parameters involved in the fit leads to a need for large statistics. As in any other method based on isospin symmetry, one must worry about the corrections due to electroweak penguins. These are expected to be small in these channels (Grossman and Quinn 1997). The method presented here leaves the usual fourfold ambiguity in the determination of $\alpha + \theta_d$. In the absence of penguin amplitudes, there are several observables that allow us to measure also $\cos 2\,(\alpha + \theta_d)$. Although penguin amplitudes are expected to be sizeable, Grossman and Quinn (1997) argue that the interference pattern is rich enough for the determination of the sign of $\cos 2\,(\alpha + \theta_d)$, so that only the twofold discrete ambiguity $\alpha \to \alpha + \pi$ remains.

36

SOME METHODS TO EXTRACT γ

36.1 Introduction

As we have seen, the angle γ is hard to measure: the decays of B_d^0 into CP eigen-states do not measure this angle directly; and the decay $B_s^0 \to \rho^0 K_S$, which would measure $\sin 2\gamma$ if there were only tree-level amplitudes, is colour-suppressed and suffers from a significant penguin pollution. As a result, its branching ratio is very small. Aleksan *et al.* (1992) have estimated that

$$\frac{\text{BR}[B_s^0 \to \rho^0 K_S]}{\text{BR}[B_d^0 \to \pi^+ \pi^-]} \sim 2 \times 10^{-2}. \tag{36.1}$$

Using the Particle Data Group (1996) bound $\text{BR}[B_d^0 \to \pi^+ \pi^-] < 2.0 \times 10^{-5}$, one concludes that $\text{BR}[B_s^0 \to \rho^0 K_S]$ is at most $\sim 10^{-7}$.

In this chapter, we discuss a few special methods to determine γ that have been proposed in the literature. Many of them hinge on the following simple trigonometric exercise. Let us suppose that one has two complex amplitudes,

$$\begin{aligned} Z_+ &= M_+ e^{i(\varphi + \Delta)}, \\ Z_- &= M_- e^{i(\varphi - \Delta)}, \end{aligned} \tag{36.2}$$

which interfere.[100] Typically, the interference term depends on the phases

$$\begin{aligned} \varphi &= \tfrac{1}{2} \arg \left(Z_+ Z_- \right), \\ \Delta &= \tfrac{1}{2} \arg \left(Z_+ Z_-^* \right). \end{aligned} \tag{36.3}$$

The phase φ will later be related to the weak phase γ; Δ will correspond to a strong-phase difference. We define

$$\begin{aligned} c_\pm &\equiv \frac{\text{Re}\, Z_\pm}{|Z_\pm|} = \cos \left(\varphi \pm \Delta \right), \\ s_\pm &\equiv \frac{\text{Im}\, Z_\pm}{|Z_\pm|} = \sin \left(\varphi \pm \Delta \right). \end{aligned} \tag{36.4}$$

Suppose that we only know s_+ and s_-. Then, $\sin \varphi$ can be obtained from

$$\begin{aligned} \sin^2 \varphi &= \tfrac{1}{2} \left(1 + s_+ s_- - c_+ c_- \right) \\ &= \tfrac{1}{2} \left[1 + s_+ s_- \mp \sqrt{\left(1 - s_+^2 \right) \left(1 - s_-^2 \right)} \right] \end{aligned} \tag{36.5}$$

up to a fourfold ambiguity. Indeed, the sign of $\sin \varphi$ cannot be determined from eqn (36.5), which, besides, gives two possible solutions for $\sin^2 \varphi$. The latter

---

[100]The magnitudes M_+ and M_- are generally determined from branching ratios.

ambiguity corresponds to the possible interchange between $\sin^2 \varphi$ and $\cos^2 \Delta$. Indeed,

$$\cos^2 \Delta = \tfrac{1}{2} \left(1 + s_+ s_- + c_+ c_-\right)$$

$$= \tfrac{1}{2} \left[1 + s_+ s_- \pm \sqrt{\left(1 - s_+^2\right)\left(1 - s_-^2\right)}\right] \tag{36.6}$$

cannot be distinguished from eqn (36.5) when we do not know the sign of $c_+ c_-$.

Let us now suppose instead that we only know c_+ and c_-. Then, $\sin \varphi$ can be obtained from

$$\sin^2 \varphi = \tfrac{1}{2} \left[1 - c_+ c_- \pm \sqrt{\left(1 - c_+^2\right)\left(1 - c_-^2\right)}\right] \tag{36.7}$$

up to a fourfold ambiguity. This ambiguity occurs due to the undetermined sign of $\sin \varphi$, and due to the possible interchange between $\sin^2 \varphi$ and $\sin^2 \Delta$, arising because $\sin^2 \Delta$ admits the same values as eqn (36.7).

We shall present several cases in which $\gamma \approx \varphi$ can be obtained by this simple trigonometric trick.

Besides penguin pollution and the small branching ratio, determining γ from $B_s^0 \to \rho^0 K_S$ has a further disadvantage as compared to the present method: the current SM bound on γ—see § 18.5—implies that $\sin^2 \gamma$ must be non-zero, while $\sin 2\gamma$ can vanish, because $\gamma = \pi/2$ is allowed.[101] For this reason, it is probably more advantageous to try and determine $\sin^2 \gamma$, instead of trying to look for $\sin 2\gamma$, as in $B_s^0 \to \rho^0 K_S$.

36.2   The Gronau–London triangle relations $\{D^0, \overline{D^0}, D_{f_{cp}}\}$

Gronau and London (1991) have noted that one can determine CKM phases by comparing decays of the type $B \to X D^0$ and $B \to X \overline{D^0}$ with $B \to X D_{f_{cp}}$. Their observation is at the root of several experiments proposed in the literature. Gronau and London (1991) have taken $D_{f_{cp}}$ to coincide with one of the CP eigenstates of the D^0–$\overline{D^0}$ system. However, the triangle relation that they have developed can be generalized to allow also for the presence of large weak phases in the D^0–$\overline{D^0}$ system, which are possible in many models of new physics (Nir 1996). One does not even need to specify that the D meson in the third decay is observed through its decay into a CP eigenstate. We may equally well use

$$B \to X D_{\text{into } f} \to X f_D, \tag{36.8}$$

where f_D is *any* final state with an invariant mass equal to the mass of the neutral D mesons. This identifies f_D as originating in the adequate linear combination of D^0 and $\overline{D^0}$.

---

[101] This possibility has been questioned by Fleischer and Mannel (1998). See, however, the work by Neubert (1998), Gérard and Weyers (1999).

The main assumption used in the standard analysis is the following: D^0 and $\overline{D^0}$ do not oscillate, i.e., x and y are zero in the D^0–$\overline{D^0}$ system.[102] This approximation is experimentally known to hold down to $x_D \lesssim 10^{-1}$—see Appendix E. The importance of this assumption is that it guarantees that the decay chain is completely determined by $D_{\text{into } f}$, with no contribution from $D_{\text{not into } f}$, see § 34.4. The amplitudes for decays of the type in eqn (36.8) are given by

$$\langle X f_D | T | B \rangle = \langle X D_{\text{into } f} | T | B \rangle$$
$$= c_f \langle X D^0 | T | B \rangle + \bar{c}_f \langle X \overline{D^0} | T | B \rangle, \qquad (36.9)$$

where

$$\frac{\bar{c}_f}{c_f} = \frac{\langle f | T | \overline{D^0} \rangle}{\langle f | T | D^0 \rangle} \qquad (36.10)$$

and $|c_f|^2 + |\bar{c}_f|^2 = 1$. Triangular relations like the one in eqn (36.9) lie at the heart of many proposals to extract weak phases. They were first introduced by Gronau and London (1991), in the context of the decays $B_d^0 \to D K_S$ and $B_s^0 \to D \phi$; in those cases, f was chosen to be a CP eigenstate. However, we insist, in general f does not need to be a CP eigenstate.

36.3 The Gronau–London–Wyler method

36.3.1 *Procedure*

Gronau and Wyler (1991) have proposed a method to extract γ from the decays $B^{\pm} \to K^{\pm} D$, using the Gronau–London triangle relations. As a result, this is known as the Gronau–London–Wyler (GLW) method. Experimentally, it consists in the measurement of the *branching ratios* for the processes

$$B^+ \to K^+ D^0, \quad B^+ \to K^+ \overline{D^0}, \quad \text{and} \quad B^{\pm} \to K^{\pm} (f_{\text{cp}})_D, \qquad (36.11)$$

where $(f_{\text{cp}})_D$ is a CP eigenstate which is identified as coming from the decay of a neutral D meson. For example, f_{cp} may be $\pi^+ \pi^-$, $K^+ K^-$, or the CP-odd $\pi^0 K_S$. The D^0 and $\overline{D^0}$ mesons in the first two decays in eqn (36.11) are detected through their flavour-specific (for instance, semileptonic) decays. Notice that no time-dependent measurements are required in the GLW method, as we are looking at charged-meson decays.

In the GLW method one assumes that

$$|A(B^+ \to K^+ D^0)| = |A(B^- \to K^- \overline{D^0})|,$$
$$|A(B^+ \to K^+ \overline{D^0})| = |A(B^- \to K^- D^0)|, \qquad (36.12)$$

i.e., that there is no direct CP violation in these decay modes. This is the case in the SM, due to the absence of penguin diagrams for these decays.

---

[102]The analysis presented in this chapter has been generalized by Meca and Silva (1998) and by Amorim *et al.* (1998) to include non-zero D^0–$\overline{D^0}$ mixing. This brings into play new sources of CP violation—see § 34.6. In particular, Meca and Silva (1998) have shown that a value of x_D of order 10^{-2} will affect the decay rates in the Gronau–London–Wyler method (§ 36.3) and in the Atwood–Dunietz–Soni method (§ 36.5) by as much as 10%.

Applying eqn (36.9) to this case, we have

$$
\begin{aligned}
A\left[B^+ \to K^+(f_{\text{cp}})_D\right] &= c_{f_{\text{cp}}} A(B^+ \to K^+ D^0) + \bar{c}_{f_{\text{cp}}} A(B^+ \to K^+ \overline{D^0}), \\
A\left[B^- \to K^-(f_{\text{cp}})_D\right] &= c_{f_{\text{cp}}} A(B^- \to K^- D^0) + \bar{c}_{f_{\text{cp}}} A(B^- \to K^- \overline{D^0}).
\end{aligned}
\tag{36.13}
$$

This may be written

$$
\begin{aligned}
z &= z_1 + z_2, \\
\bar{z} &= z_1' + z_2',
\end{aligned}
\tag{36.14}
$$

after introducing the definitions

$$
\begin{aligned}
z_1 &\equiv \bar{c}_{f_{\text{cp}}} A(B^+ \to K^+ \overline{D^0}), \\
z_1' &\equiv c_{f_{\text{cp}}} A(B^- \to K^- D^0), \\
z_2 &\equiv c_{f_{\text{cp}}} A(B^+ \to K^+ D^0), \\
z_2' &\equiv \bar{c}_{f_{\text{cp}}} A(B^- \to K^- \overline{D^0}), \\
z &\equiv A\left[B^+ \to K^+(f_{\text{cp}})_D\right], \\
\bar{z} &\equiv A\left[B^- \to K^-(f_{\text{cp}})_D\right].
\end{aligned}
\tag{36.15}
$$

In the GLW method one measures

$$
\begin{aligned}
|z|^2 &= |z_1|^2 + |z_2|^2 + 2\text{Re}\,(z_2 z_1^*), \\
|\bar{z}|^2 &= |z_1'|^2 + |z_2'|^2 + 2\text{Re}\,(z_1' z_2'^*).
\end{aligned}
\tag{36.16}
$$

However, $|z_1|$ cannot be determined from the measured $|A(B^+ \to K^+ \overline{D^0})|$ unless the modulus of $\bar{c}_{f_{\text{cp}}}$ is known. The same holds for $|z_2|$, $|z_1'|$, and $|z_2'|$.

We now assume that the only diagrams contributing to the various decays are the tree-level diagrams of the SM, i.e., we assume that *all decay amplitudes are given solely by the W-mediated tree-level diagram*. Then, CP symmetry implies that $\left|c_{f_{\text{cp}}}\right| = \left|\bar{c}_{f_{\text{cp}}}\right| = 1/\sqrt{2}$, and

$$
\begin{aligned}
|z_1| = |z_1'| &= \frac{|A(B^+ \to K^+ \overline{D^0})|}{\sqrt{2}}, \\
|z_2| = |z_2'| &= \frac{|A(B^+ \to K^+ D^0)|}{\sqrt{2}}.
\end{aligned}
\tag{36.17}
$$

From eqns (36.16) we may determine

$$
\frac{\text{Re}\,(z_2 z_1^*)}{|z_2 z_1|} \quad \text{and} \quad \frac{\text{Re}\,(z_1' z_2'^*)}{|z_1' z_2'|},
\tag{36.18}
$$

which are to be identified as c_+ and c_- of the first eqn (36.4). Then, from eqn (36.7) we may determine the squared sine of

$$
\tfrac{1}{2}\arg\frac{z_2 z_1'}{z_1 z_2'} = \tfrac{1}{2}\arg\frac{c_{f_{\text{cp}}}^2 A(B^+ \to K^+ D^0) A(B^- \to K^- D^0)}{\bar{c}_{f_{\text{cp}}}^2 A(B^- \to K^- \overline{D^0}) A(B^+ \to K^+ \overline{D^0})}.
\tag{36.19}
$$

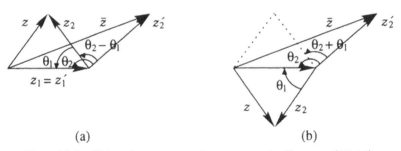

FIG. 36.1. Triangles representing geometrically eqns (36.14).

This is determined up to discrete ambiguities, including the confusion with the strong-phase difference, given by

$$\Delta_B \equiv \tfrac{1}{2} \arg \frac{z_2 z_2'}{z_1 z_1'} = \tfrac{1}{2} \arg \frac{A(B^+ \to K^+ D^0) A(B^- \to K^- \overline{D^0})}{A(B^+ \to K^+ \overline{D^0}) A(B^- \to K^- D^0)}. \tag{36.20}$$

36.3.2 *Geometrical interpretation*

Equations (36.14) may be represented in the complex plane as two triangles, the first triangle having sides with length $|z_1|$, $|z_2|$, and $|z|$, while the second triangle has sides with length $|z_1'|$, $|z_2'|$, and $|\bar{z}|$. If there was direct CP violation in the decay $D^0 \to f_{\mathrm{cp}}$, there would be no relation between the two triangles. However, our assumption that D decays are dominated by tree-level diagrams guarantees that $|z_1| = |z_1'|$ and $|z_2| = |z_2'|$. Then, the two triangles become related, as shown in Fig. 36.1 (a). Without loss of generality, we may rephase the amplitudes in such a way as to make $z_1 = z_1'$, as we have depicted in Fig. 36.1 (a). Although z_1 and z_1' have arbitrary phases, which depend on the phases of the state vectors in them, the observables that we measure do not depend on this choice.

Simple trigonometry yields

$$\cos\theta_1 = \frac{|z_1|^2 + |z_2|^2 - |z|^2}{2|z_1 z_2|},$$
$$\cos\theta_2 = \frac{|z_1| + |z_2|^2 - |\bar{z}|^2}{2|z_1 z_2|}, \tag{36.21}$$

where

$$\theta_1 \equiv \arg\left(-\frac{z_1}{z_2}\right),$$
$$\theta_2 \equiv \arg\left(-\frac{z_1'}{z_2'}\right). \tag{36.22}$$

If the physical situation were as depicted in Fig. 36.1 (a). we would have succeeded in extracting

$$\theta_2 - \theta_1 = \arg \frac{z_2 z_1'}{z_1 z_2'}, \tag{36.23}$$

identified as a weak phase in eqn (36.19). However, the physical situation could be the one depicted in Fig. 36.1 (b). In that case, when drawing Fig. 36.1 (a) we would have been making a mistake: we thought that we were calculating $\theta_2 - \theta_1$, while we were in fact calculating

$$\theta_2 + \theta_1 = -\arg \frac{z_2 z_2'}{z_1 z_1'}, \tag{36.24}$$

which is the difference of strong phases, as seen in eqn (36.20). Therefore, the two solutions for

$$\sin^2 \frac{\theta_2 - \theta_1}{2} = \tfrac{1}{2} \left[1 - \cos\theta_1 \cos\theta_2 \pm \sqrt{(1 - \cos^2\theta_1)(1 - \cos^2\theta_2)} \right] \tag{36.25}$$

correspond to the fact that this method cannot distinguish the weak phase from the difference of strong phases.

Since $\overline{D^0}K^+$ is pure isospin 1, while $D^0 K^+$ is a superposition of isospin 0 and 1, the final-state phase difference is likely to be nonzero (Gronau and Wyler 1991). Still, it should be noted that this analysis can be performed even if the final-state phase difference does vanish. In that case, the two triangles are equal: one of them points upwards, the other one points downwards, and the angle between z_2 and z_2' measures γ. As mentioned above, this situation cannot be distinguished from the one in which it is γ (the difference of weak phases) that vanishes, while the angle between z_2 and z_2' measures the strong-phase difference. In spite of this discrete ambiguity, which is inherent in the GLW method, it is remarkable that we can extract the weak phase (up to discrete ambiguities) even when the CP-violating decay-rate asymmetry vanishes.

36.3.3 The weak phase in the SM

The analysis described above can be performed using any final state $K^\pm(f_{\text{cp}})_D$. Exactly which CKM phase is measured depends on the final state considered. Still, the result of eqn (36.25) is always given approximately by $\sin^2\gamma$, in the SM.

Let us next see how the phase γ shows up. From the assumption that the decays are dominated by SM tree-level diagrams, we find

$$\frac{A(B^- \to K^- D^0)}{A(B^+ \to K^+ \overline{D^0})} = \frac{V_{cb} V_{us}^* \sum_{n=1}^{2} C_n(\mu) \langle K^- D^0 | Q_n^{cus\dagger} | B^- \rangle}{V_{cb}^* V_{us} \sum_{n=1}^{2} C_n(\mu) \langle K^+ \overline{D^0} | Q_n^{cus} | B^+ \rangle}. \tag{36.26}$$

Notice that the final-state-interaction phases for the two amplitudes are the same, and therefore they cancel in the ratio. In the same way,

$$\frac{A(B^+ \to K^+ D^0)}{A(B^- \to K^- \overline{D^0})} = \frac{V_{ub}^* V_{cs} \sum_{n=1}^{2} C_n(\mu) \langle K^+ D^0 | Q_n^{ucs} | B^+ \rangle}{V_{ub} V_{cs}^* \sum_{n=1}^{2} C_n(\mu) \langle K^- \overline{D^0} | Q_n^{ucs\dagger} | B^- \rangle}. \tag{36.27}$$

Using eqns (36.26) and (36.27), we find

$$\frac{A(B^+ \to K^+ D^0) A(B^- \to K^- D^0)}{A(B^- \to K^- \overline{D^0}) A(B^+ \to K^+ \overline{D^0})}$$

$$= \frac{V_{ub}^* V_{cs}}{V_{ub} V_{cs}^*} e^{i\left(\xi_u - \xi_b + \xi_s - \xi_c + \xi_B + -\xi_K + -\xi_D\right)} \frac{V_{cb} V_{us}^*}{V_{cb}^* V_{us}} e^{i\left(\xi_b - \xi_c + \xi_u - \xi_s - \xi_B + +\xi_K + -\xi_D\right)}$$

$$= \frac{V_{ub}^* V_{cs} V_{cb} V_{us}^*}{V_{ub} V_{cs}^* V_{cb}^* V_{us}} e^{2i\left(\xi_u - \xi_c - \xi_D\right)} \qquad (36.28)$$

We have used the CP transformations of the operators in eqns (32.13), and defined the CP transformation of the state vectors by

$$\begin{aligned}
CP|B^+\rangle &= e^{i\xi_{B^+}} |B^-\rangle, \\
CP|K^+\rangle &= e^{i\xi_{K^+}} |K^-\rangle, \\
CP|D^0\rangle &= e^{i\xi_D} |\overline{D^0}\rangle.
\end{aligned} \qquad (36.29)$$

Let us next consider the decays of D^0 and $\overline{D^0}$ to $\pi^+ \pi^-$. From our assumption and from eqn (36.10) it follows that

$$\frac{c_{\pi^+\pi^-}}{\bar{c}_{\pi^+\pi^-}} = \frac{V_{ud} V_{cd}^* \langle \pi^+ \pi^- | (\bar{u}\Gamma^\mu d)(\bar{d}\Gamma_\mu c)|D^0\rangle}{V_{ud}^* V_{cd} \langle \pi^+ \pi^- | (\bar{d}\Gamma_\mu u)(\bar{c}\Gamma^\mu d)|\overline{D^0}\rangle} = \frac{V_{ud} V_{cd}^*}{V_{ud}^* V_{cd}} e^{i(\xi_c - \xi_u + \xi_D)}, \qquad (36.30)$$

where $\Gamma^\mu \equiv \gamma^\mu \gamma_L$, and we have taken into account that the CP-parity of $\pi^+ \pi^-$ is positive.

We may now compute the ratio of amplitudes relevant for eqn (36.19). Using eqns (36.28) and (36.30), we find

$$\begin{aligned}
\frac{1}{2} \arg \frac{z_2 z_1'}{z_1 z_2'} &= \frac{1}{2} \arg \frac{c_{\pi^+\pi^-}^2}{\bar{c}_{\pi^+\pi^-}^2} \frac{A(B^+ \to K^+ D^0) A(B^- \to K^- D^0)}{A(B^- \to K^- \overline{D^0}) A(B^+ \to K^+ \overline{D^0})} \\
&= \frac{1}{2} \arg \frac{V_{ud} V_{cb} V_{ub}^* V_{cd}^*}{V_{ud}^* V_{cb}^* V_{ub} V_{cd}} \frac{V_{ud} V_{cs} V_{us}^* V_{cd}^*}{V_{ud}^* V_{cs}^* V_{us} V_{cd}} \\
&= \gamma - \epsilon'.
\end{aligned} \qquad (36.31)$$

One concludes that, when one uses final states $K^\pm(\pi^+\pi^-)_D$, one is measuring $\sin^2(\gamma - \epsilon')$.

If one uses instead final states $K^\pm(K^+K^-)_D$, then the relevant parameter is

$$\frac{c_{K^+K^-}}{\bar{c}_{K^+K^-}} = \frac{V_{us} V_{cs}^* \langle K^+ K^- | (\bar{u}\Gamma^\mu s)(\bar{s}\Gamma_\mu c)|D^0\rangle}{V_{us}^* V_{cs} \langle K^+ K^- | (\bar{s}\Gamma_\mu u)(\bar{c}\Gamma^\mu s)|\overline{D^0}\rangle} = \frac{V_{us} V_{cs}^*}{V_{us}^* V_{cs}} e^{i(\xi_c - \xi_u + \xi_D)}. \qquad (36.32)$$

With the same steps as before we get

$$\begin{aligned}
\frac{1}{2} \arg \frac{z_2 z_1'}{z_1 z_2'} &= \frac{1}{2} \arg \frac{c_{K^+K^-}^2}{\bar{c}_{K^+K^-}^2} \frac{A(B^+ \to K^+ D^0) A(B^- \to K^- D^0)}{A(B^- \to K^- \overline{D^0}) A(B^+ \to K^+ \overline{D^0})} \\
&= \gamma + \epsilon';
\end{aligned} \qquad (36.33)$$

thus, in this case one would be measuring $\sin^2(\gamma + \epsilon')$.

Obviously, these results depend only on the tree-level quark decays being used. All $\Delta S = \Delta D = 0$ decays of the neutral D mesons yield the weak phase

$$\phi = \phi_{\bar{b}\to\bar{u}c\bar{s}} + \phi_{c\to k\bar{k}u} - \phi_{\bar{b}\to\bar{c}u\bar{s}} - \phi_{\bar{c}\to\bar{k}k\bar{u}} = \begin{cases} \gamma - \epsilon' & \text{for } k = d, \\ \gamma + \epsilon' & \text{for } k = s. \end{cases} \quad (36.34)$$

In most models ϵ' is very small, and the $\Delta S = \Delta D = 0$ decays of neutral D mesons yield $\sin\gamma$, irrespectively of the final states utilized, up to a fourfold ambiguity. One may reduce the ambiguity by repeating the analysis for various distinct final states $X^\pm(f_{\text{cp}})_D$, with $X^+ = K^+$, $K^+\pi^0$, $K^0\pi^+$, and so on. Indeed, whereas the weak phase is always γ, the final-state phase difference should differ from one channel to the next. This would allow the extraction of γ up to its sign. (In the SM, the sign of γ is known, once the sign of the bag parameter B_K is fixed.)

If one uses the $\Delta S \neq 0 \neq \Delta D$ decays of the neutral D mesons, one obtains K_S or K_L in the final state, such as in $D \to K_S\pi^0$. This case is complicated by the fact that two amplitudes contribute to $D^0 \to K_S\pi^0$; the Cabbibo-allowed $D^0 \to \overline{K^0}\pi^0$, and the doubly Cabbibo-suppressed $D^0 \to K^0\pi^0$.

This method works because we have assumed that there is no CP violation in the D^0–$\overline{D^0}$ system, nor in the decays $D^0 \to f_{\text{cp}}$ and $\overline{D^0} \to f_{\text{cp}}$. Then, the state that decays into f_{cp} ($D_{f_{\text{cp}}}$) coincides with one of the eigenstates of CP ($D_\pm$). As a result, many authors perform the analysis by considering decays into CP eigenstates such as $B^+ \to K^+D_+$.[103] The difference between this analysis and the one performed by looking at $B^+ \to K^+D_{\text{into }\pi^+\pi^-} \to K^+(\pi^+\pi^-)_D$ is immaterial under the assumption above. However, in models of new physics with considerable CP violation in the D^0–$\overline{D^0}$ system and its decays, one must analyse the decay chain $B^+ \to K^+D_{\text{into }\pi^+\pi^-} \to K^+(\pi^+\pi^-)_D$ in detail.

In any case, the calculations presented in this subsection show that, under the no-oscillation requirement, the GLW method is not affected by any new phase appearing in D^0–$\overline{D^0}$ mixing. This will no longer be the case if $\Delta m/\Gamma$ in D^0–$\overline{D^0}$ mixing is $\sim 10^{-2}$ (Meca and Silva 1998).

36.3.4 *Experimental difficulties with the GLW method*

As mentioned, the measurement of $\sin 2\gamma$ with $B_s^0 \to \rho^0 K_S$ is subject to considerable penguin pollution. Moreover, one must tag the neutral meson and follow the time dependence of the decay. The Gronau–London–Wyler method does not have these shortcomings. In addition, the current bounds on γ allow $\sin 2\gamma$ to vanish, while $\sin^2\gamma$ does not.

Unfortunately, the GLW method suffers from experimental difficulties, mostly due to the fact that $|z_2| = |z_2'|$ should be much smaller than the other sides of the triangles. Indeed (Dunietz 1991; Buras and Fleischer 1998),

$$\left|\frac{A(B^+ \to K^+D^0)}{A(B^+ \to K^+\overline{D^0})}\right| \sim \left|\frac{V_{ub}^*V_{cs}}{V_{cb}^*V_{us}}\right|\left|\frac{a_2}{a_1}\right| \sim 0.26R_b, \quad (36.35)$$

---

[103]Kurimoto (1997) has presented a rephasing invariant analysis of the Gronau–London–Wyler method, using this alternative route. Naturally, his results coincide with ours.

where $|a_2/a_1| \sim 0.26$ is a phenomenological colour-suppression factor (Browder et al. 1997). Therefore, $|z_2|$ is much smaller than $|z_1|$, $|z|$, and $|\bar{z}|$. The triangles in Figs. 36.1 are very flat and $\theta_2 - \theta_1$ becomes a sensitive function of $|z_2|$ while, simultaneously, $|z_2|$ is very hard to measure.

Detecting the D^0 in the final state is rather difficult. In principle, its flavour could be tagged through the semileptonic decay $D^0 \to l^+\nu_l X_s$. However, this is subject to a huge background from the direct decay $B^+ \to l^+\nu_l X_{\bar{s}}$, which is approximately 10^6 times larger (Atwood et al. 1997). Although there are several features distinguishing the two decays, reducing the background by the required amount is a demanding task (Atwood et al. 1997). One might also attempt to identify D^0 through its decay $D^0 \to K^-\pi^+$, because $\overline{D^0} \to K^-\pi^+$ is doubly Cabibbo-suppressed (cf. Appendix E). However, as pointed out by Atwood et al. (1997), the D^0-tagging chain

$$B^+ \to K^+D^0 \to K^+(K^-\pi^+)_D \qquad (36.36)$$

cannot be distinguished from

$$B^+ \to K^+\overline{D^0} \to K^+(K^-\pi^+)_D. \qquad (36.37)$$

These two decay chains interfere at order ~ 1, because the suppression of $\overline{D^0} \to K^-\pi^+$ is compensated by eqn (36.35):

$$\left| \frac{A(B^+ \to K^+D^0)A(D^0 \to K^-\pi^+)}{A(B^+ \to K^+\overline{D^0})A(\overline{D^0} \to K^-\pi^+)} \right| \sim \frac{0.26R_b}{\sqrt{0.0077}} \sim 1. \qquad (36.38)$$

We have used $R_b \approx 0.36$ (Buras and Fleischer 1998) and the central value 0.0077 given by the CLEO Collaboration (1994) for BR$(D^0 \to K^+\pi^-)$/BR$(\overline{D^0} \to K^+\pi^-)$.

As an alternative, Atwood et al. (1997) have pointed out that some excited D^0 resonances have additional flavour-specific decay modes, such as $D^{(*)+}\pi^-$. In principle, one could measure $|z_2|$ for such modes, and proceed with the GLW method.

On the other hand, detecting $\overline{D^0}$ is much easier because

$$\left| \frac{A(B^+ \to K^+\overline{D^0})\, A(\overline{D^0} \to K^+\pi^-)}{A(B^+ \to K^+D^0)\, A(D^0 \to K^+\pi^-)} \right| \sim \frac{1}{0.26R_b\sqrt{0.0077}} \sim 10^2; \qquad (36.39)$$

the total decay chain $B^+ \to K^+D \to K^+(K^+\pi^-)_D$ is dominated by the process with an intermediate $\overline{D^0}$ to an accuracy of 1% in the amplitude (Atwood et al. 1997).

An additional difficulty with the GLW method is due to the need to measure the decay of the neutral D meson into a CP eigenstate. The product (branching ratio)$\times$(detection efficiency) for those decays lies at the 1% level (Gronau and Wyler 1991). Still, one can increase statistics by combining the analysis of several final states.

36.4 Simple extensions of the GLW method

36.4.1 *Extracting γ with self-tagging B_d^0 modes*

Dunietz (1991) has pointed out that the Gronau–London–Wyler method can also be applied to self-tagging B_d^0 decays. In this method one measures (Dunietz 1991)

$$B_d^0 \to K^{*0}D^0, \quad B_d^0 \to K^{*0}\overline{D^0}, \quad B_d^0 \to K^{*0}(f_{\mathrm{cp}})_D, \quad \overline{B_d^0} \to \overline{K^{*0}}(f_{\mathrm{cp}})_D,$$
$$(36.40)$$

where K^{*0} is detected through its decay $K^+\pi^-$, which has a branching ratio of around 2/3. The sign of the charged kaon tags the B flavour. As before, one constructs two triangles

$$A\left[B_d^0 \to K^{*0}(f_{\mathrm{cp}})_D\right] = c_{f_{\mathrm{cp}}}A(B_d^0 \to K^{*0}D^0) + \bar{c}_{f_{\mathrm{cp}}}A(B_d^0 \to K^{*0}\overline{D^0}),$$
$$A\left[\overline{B_d^0} \to \overline{K^{*0}}(f_{\mathrm{cp}})_D\right] = c_{f_{\mathrm{cp}}}A(\overline{B_d^0} \to \overline{K^{*0}}D^0) + \bar{c}_{f_{\mathrm{cp}}}A(\overline{B_d^0} \to \overline{K^{*0}}\ \overline{D^0}),$$
$$(36.41)$$

obtaining γ via a simple trigonometric exercise.

An advantage of these neutral-meson decays is that all the sides involved are expected to be of similar size,

$$\left|\frac{A(B_d^0 \to K^{*0}D^0)}{A(B_d^0 \to K^{*0}\overline{D^0})}\right| \sim \left|\frac{V_{ub}^*V_{cs}}{V_{cb}^*V_{us}}\right| \sim R_b, \qquad (36.42)$$

because both amplitudes are colour-suppressed.

The simple fact that these new decays can be studied is already useful, because it is a further aid in disentangling γ from the strong-phase differences (Dunietz 1991). Although the experimental difficulties with the identification of D^0 are ameliorated when using $B_d^0 \to K^{*0}D^0$, the problems with the detection of the decays into $D_{f_{\mathrm{cp}}}$ remain.

36.4.2 *Extracting γ from D decays into non-CP eigenstates*

The Gronau–London–Wyler method can also be applied using the branching ratios of

$$B^+ \to K^+D^0, \quad B^+ \to K^+\overline{D^0}, \quad B^+ \to K^+f_D, \quad B^- \to K^-\bar{f}_D, \quad (36.43)$$

even when f is a not a CP eigenstate. As shown by Atwood *et al.* (1997), this is possible provided that one also knows the branching ratios for the decays of D into f:

$$D^0 \to f_D, \quad \overline{D^0} \to \bar{f}_D. \qquad (36.44)$$

Of course, one still assumes that those decays are dominated by a single weak phase.

The procedure is the same as before. Defining

$$
\begin{aligned}
z_1 &\equiv \bar{c}_f A(B^+ \to K^+ \overline{D^0}), \\
z_1' &\equiv c_f A(B^- \to K^- D^0), \\
z_2 &\equiv c_f A(B^+ \to K^+ D^0), \\
z_2' &\equiv \bar{c}_f A(B^- \to K^- \overline{D^0}), \\
z &\equiv A(B^+ \to K^+ f_D), \\
\bar{z} &\equiv A(B^- \to K^- \bar{f}_D),
\end{aligned}
\tag{36.45}
$$

one may write the triangular relations

$$
\begin{aligned}
z &= z_1 + z_2, \\
\bar{z} &= z_1' + z_2'.
\end{aligned}
\tag{36.46}
$$

Assuming that there is no direct CP violation in D decays, one gets $|c_f| = |\bar{c}_f|$ and $|c_f| = |\bar{c}_f|$. Therefore,

$$
\begin{aligned}
|z_1| &= |z_1'| = |c_{\bar{f}} A(B^+ \to K^+ \overline{D^0})|, \\
|z_2| &= |z_2'| = |c_f A(B^+ \to K^+ D^0)|.
\end{aligned}
\tag{36.47}
$$

Simple trigonometry allows one to find the squared sine of

$$
\tfrac{1}{2} \arg \frac{z_2 z_1'}{z_1 z_2'} = \tfrac{1}{2} \arg \frac{c_{\bar{f}} c_f A(B^+ \to K^+ D^0) A(B^- \to K^- D^0)}{\bar{c}_{\bar{f}} \bar{c}_f A(B^- \to K^- \overline{D^0}) A(B^+ \to K^+ \overline{D^0})}.
\tag{36.48}
$$

Assuming that the weak phases dominating the decays of the D mesons are those in the tree-level SM diagrams, we obtain again the weak phase in eqn (36.34). This is determined up to the confusion with the strong phase, given by

$$
\Delta_{Df} + \Delta_B = \tfrac{1}{2} \arg \frac{z_2 z_2'}{z_1 z_1'} = \tfrac{1}{2} \arg \frac{c_f \bar{c}_{\bar{f}} A(B^+ \to K^+ D^0) A(B^- \to K^- \overline{D^0})}{\bar{c}_f c_{\bar{f}} A(B^+ \to K^+ \overline{D^0}) A(B^- \to K^- D^0)},
\tag{36.49}
$$

where we have used eqn (36.20) and

$$
\Delta_{Df} \equiv \tfrac{1}{2} \arg \frac{c_f \bar{c}_{\bar{f}}}{\bar{c}_f c_{\bar{f}}}
\tag{36.50}
$$

is the strong-phase difference in the D decays. When f is a CP eigenstate, as in the original GLW proposal, $\Delta_{Df} = 0$. The advantage of using non-CP eigenstates lies in the fact that in eqn (36.49) the difference of strong phases in B decays appears together with the difference of strong phases in D decays, and the latter is likely to be large in some channels. For these channels, the CP-violating asymmetry between the decay widths of $B^+ \to K^+ f_D$ and $B^- \to K^- \bar{f}_D$ can be large (Atwood et al. 1997). In addition, some non-CP eigenstates might be easier to detect than the CP eigenstates.

36.5 The Atwood–Dunietz–Soni method

The experimental problems discussed in connection with the Gronau–London–Wyler method may be side-stepped with a related method proposed by Atwood *et al.* (1997). The idea is to perform the analysis with non-CP eigenstates, as in § 36.4.2, but now assuming that $|A(B^+ \to K^+D^0)|$ is not measured. One can make progress without this piece of experimental information by comparing the results for two different chains:

$$B^+ \to K^+D \to K^+(f_1)_D, \quad B^- \to K^-D \to K^-(\bar{f}_1)_D, \qquad (36.51)$$

and

$$B^+ \to K^+D \to K^+(f_2)_D, \quad B^- \to K^-D \to K^-(\bar{f}_2)_D, \qquad (36.52)$$

corresponding to two different final states f_1 and f_2. In this method, the branching ratios of

$$D^0 \to f_i, \quad \overline{D^0} \to f_i, \quad \text{and} \quad B^+ \to K^+\overline{D^0}, \qquad (36.53)$$

must also be known. We continue to assume that the decays of D are dominated by a single weak phase. Due to eqn (36.39), the branching ratio for $B^+ \to K^+\overline{D^0}$ may be determined by using the decay chain $B^+ \to K^+D \to K^+(K^+\pi^-)_D$.

The final states f_i are chosen such that the decay $D^0 \to f_i$ is doubly Cabbibo-suppressed, while $\overline{D^0} \to f_i$ is Cabbibo allowed. For example, those final states might be $K^+\pi^-$ or $K^+\rho^-$. Thus, we fall under the conditions of eqn (36.38): the two amplitudes are of comparable magnitude, and the CP-violating effects are maximal. It is ironic that this method uses precisely those final states that cannot be used for a direct determination of $|A(B^+ \to K^+D^0)|$, as needed in the GLW method. The beauty of the Atwood–Dunietz–Soni method is that $|A(B^+ \to K^+D^0)|$ is not an input, but rather an output.

In the Atwood–Dunietz–Soni method one measures

$$
\begin{aligned}
\left|A\left[B^+ \to K^+(f_i)_D\right]\right|^2 &= \left|c_{f_i}A(B^+ \to K^+D^0)\right|^2 + \left|\bar{c}_{f_i}A(B^+ \to K^+\overline{D^0})\right|^2 \\
&\quad + 2\mathrm{Re}\left[c_{f_i}\bar{c}_{f_i}^*A(B^+ \to K^+D^0)A(B^+ \to K^+\overline{D^0})^*\right], \\
\left|A\left[B^- \to K^-(\bar{f}_i)_D\right]\right|^2 &= \left|c_{\bar{f}_i}A(B^- \to K^-D^0)\right|^2 + \left|\bar{c}_{\bar{f}_i}A(B^- \to K^-\overline{D^0})\right|^2 \\
&\quad + 2\mathrm{Re}\left[c_{\bar{f}_i}\bar{c}_{\bar{f}_i}^*A(B^- \to K^-D^0)A(B^- \to K^-\overline{D^0})^*\right],
\end{aligned}
$$
$$(36.54)$$

for both final states f_1 and f_2. One assumes that $|A(B^+ \to K^+\overline{D^0})|$ is known, as are the $|c_{f_i}|$ and $|\bar{c}_{f_i}|$, which can be extracted from the branching ratios of the decays of the D mesons. Therefore, the system of four eqns (36.54) may be solved for the four unknowns $|A(B^+ \to K^+D^0)|$, the strong phases $\Delta_B + \Delta_{Df_1}$ and $\Delta_B + \Delta_{Df_2}$—see eqn (36.49)—and the weak phase

$$\frac{1}{2}\arg\frac{c_{\bar{f}_i}c_{f_i}A(B^+ \to K^+D^0)A(B^- \to K^-D^0)}{\bar{c}_{\bar{f}_i}\bar{c}_{f_i}A(B^- \to K^-\overline{D^0})A(B^+ \to K^+\overline{D^0})}. \qquad (36.55)$$

Strictly speaking, the two final states should arise from the same quark-level decay, so that the weak phase is the same in both channels, cf. eqn (36.34).

However, a small ϵ' implies that the difference is irrelevant and any two final states may be used.

36.6   The Gronau–London $B_d^0 \to DK_S$ method

All the methods presented in this chapter are rooted in the triangle construction $B \to \{D^0, \overline{D^0}, D_{f_{\mathrm{cp}}}\}X$. This construction was first introduced by Gronau and London (1991) to analyse the decays $B_d^0 \to DK_S$. Their article can be viewed as containing, in fact, two distinct proposals:

1. to determine γ using the decays $B_d^0 \to \{D^0, \overline{D^0}, D_{f_{\mathrm{cp}}}\}K_S$;
2. to determine $2\beta + \gamma$ with the decays into non-CP eigenstates $B_d^0 \to D^0 K_S$ and $B_d^0 \to \overline{D^0}K_S$.

In order to use these decays one needs to tag the initial meson.

36.6.1   Extracting γ from $B_d^0 \to \{D^0, \overline{D^0}, D_{f_{\mathrm{cp}}}\}K_S$

The idea is to look for the tagged decays

$$B_d^0 \to D^0 K_S, \quad B_d^0 \to \overline{D^0}K_S, \quad B_d^0 \to (f_{\mathrm{cp}})_D K_S, \text{ and CP conjugated.} \quad (36.56)$$

Everything follows as in § 36.3, with

$$
\begin{aligned}
z_1 &\equiv \bar{c}_{f_{\mathrm{cp}}} A(B_d^0 \to K_S \overline{D^0}) = \frac{1}{2p_K}\, \bar{c}_{f_{\mathrm{cp}}} A(B_d^0 \to K^0 \overline{D^0}), \\
z_1' &\equiv c_{f_{\mathrm{cp}}} A(\overline{B_d^0} \to K_S D^0) = -\frac{1}{2q_K}\, c_{f_{\mathrm{cp}}} A(\overline{B_d^0} \to \overline{K^0} D^0), \\
z_2 &\equiv c_{f_{\mathrm{cp}}} A(B_d^0 \to K_S D^0) = \frac{1}{2p_K}\, c_{f_{\mathrm{cp}}} A(B_d^0 \to K^0 D^0), \\
z_2' &\equiv \bar{c}_{f_{\mathrm{cp}}} A(\overline{B_d^0} \to K_S \overline{D^0}) = -\frac{1}{2q_K}\, \bar{c}_{f_{\mathrm{cp}}} A(\overline{B_d^0} \to \overline{K^0}\ \overline{D^0}), \\
z &\equiv A[B_d^0 \to K_S(f_{\mathrm{cp}})_D], \\
\bar{z} &\equiv A[\overline{B_d^0} \to K_S(f_{\mathrm{cp}})_D],
\end{aligned}
\qquad (36.57)
$$

where we have used the fact that B_d^0 only decays into K^0 while $\overline{B_d^0}$ only decays into $\overline{K^0}$.

Experimentally, one determines $|z_1|$, $|z_2|$, $|z|$, and $|\bar{z}|$. The big difference between this case and the ones presented earlier lies in the fact that the determination of these magnitudes requires the tagging of the initial meson; the final state by itself does not identify the flavour of the decaying meson. Although there is interference CP violation, it drops out in the tagged time-integrated rates at the $\Upsilon(4S)$—see eqns (35.1), which are valid for any final state f. Therefore, we only need the branching ratios for these decays of B_d^0 and $\overline{B_d^0}$; we do not need to follow the time dependence.[104]

---

[104]This important consequence of the experiments performed at the $\Upsilon(4S)$, is often ignored in the literature.

The angle determined is

$$\frac{1}{2}\arg\frac{z_2 z_1'}{z_1 z_2'} = \frac{1}{2}\arg\frac{c_{f_{cp}}^2 A(B_d^0 \to K^0 D^0)A(\overline{B_d^0} \to \overline{K^0}D^0)}{\bar{c}_{f_{cp}}^2 A(\overline{B_d^0} \to \overline{K^0}\,\overline{D^0})A(B_d^0 \to K^0\overline{D^0})}. \tag{36.58}$$

This is exactly the same angle as in eqn (36.19). Its expression, which depends weakly on the final state f_{cp} used, was given in eqn (36.34). Of course, one must still face the problem that $|z_2|$ is much smaller than $|z_1|$—see eqn (36.35)—and is therefore very difficult to measure.

Notice that the B_d^0–$\overline{B_d^0}$ and K^0–$\overline{K^0}$ mixing factors have no influence on the result. This is due to the fact that we have only used the knowledge of the magnitudes of the decay amplitudes. We did not use any information about interference CP violation. As usual, the weak phase is determined up to discrete ambiguities, including the confusion with the strong phase.

36.6.2   Extracting $2\beta + \gamma$ from $B_d^0 \to \{D^0, \overline{D^0}\}K_S$

As a by-product of the experiment just described, one may also extract, from the time dependence of the decays

$$B_d^0(t) \to D^0 K_S \quad \text{and} \quad B_d^0(t) \to \overline{D^0} K_S, \tag{36.59}$$

the parameters $\mathrm{Im}\,\lambda_{B_d^0 \to D^0 K_S}$ and $\mathrm{Im}\,\lambda_{B_d^0 \to \overline{D^0} K_S}$. Decays into non-CP eigenstates were discussed by Aleksan et al. (1991), and were used by Aleksan et al. (1992) to determine γ with B_s^0 decays into $D_s^{\pm} K^{\mp}$. The method is discussed in detail in § 9.4.3 and also in § 37.3. What one measures is half the (weak) phase of $\lambda_{B_d^0 \to D^0 K_S}\lambda_{B_d^0 \to \overline{D^0} K_S}$, up to discrete ambiguities.

It is easy to show that

$$\lambda_{B_d^0 \to D^0 K_S}\lambda_{B_d^0 \to \overline{D^0} K_S} = \left(\frac{p_K\, q_{B_d}}{q_K\, p_{B_d}}\right)^2 \frac{\langle D^0 \overline{K^0}|T|\overline{B_d^0}\rangle\langle \overline{D^0}\ \overline{K^0}|T|\overline{B_d^0}\rangle}{\langle \overline{D^0} K^0|T|B_d^0\rangle\langle D^0 K^0|T|B_d^0\rangle}. \tag{36.60}$$

Using

$$\begin{aligned}
\mathcal{CP}|B_d^0\rangle &= e^{i\xi_{B_d}}|\overline{B_d^0}\rangle, \\
\mathcal{CP}|K^0\rangle &= e^{i\xi_K}|\overline{K^0}\rangle, \\
\mathcal{CP}|D^0\rangle &= e^{i\xi_D}|\overline{D^0}\rangle,
\end{aligned} \tag{36.61}$$

and the CP transformation of the operators in eqns (32.13), we find

$$\begin{aligned}
\frac{\langle D^0 \overline{K^0}|Q_n^{cus\dagger}|\overline{B_d^0}\rangle}{\langle \overline{D^0} K^0|Q_n^{cus}|B_d^0\rangle} &= e^{i(-\xi_{B_d}+\xi_K-\xi_D)}e^{i(\xi_b-\xi_s+\xi_u-\xi_c)}, \\[2mm]
\frac{\langle \overline{D^0} \overline{K^0}|Q_n^{ucs\dagger}|\overline{B_d^0}\rangle}{\langle D^0 K^0|Q_n^{ucs}|B_d^0\rangle} &= e^{i(-\xi_{B_d}+\xi_K+\xi_D)}e^{i(\xi_b-\xi_s+\xi_c-\xi_u)}.
\end{aligned} \tag{36.62}$$

Recall that the operators Q_n^{cus} and Q_n^{ucs} appear in the effective Hamiltonian with CKM coefficients $V_{cb}^* V_{us}$ and $V_{ub}^* V_{cs}$, respectively. Combining this with the phases from the mixing, we find

$$\lambda_{B_d^0 \to D^0 K_S} \lambda_{B_d^0 \to \overline{D^0} K_S} = \left(\frac{V_{cd}^* V_{cs}}{V_{cd} V_{cs}^*} e^{2i\epsilon'} \frac{V_{tb}^* V_{td}}{V_{tb} V_{td}^*} e^{2i\theta_d} \right)^2 \frac{V_{cb} V_{us}^*}{V_{cb}^* V_{us}} \frac{V_{ub} V_{cs}^*}{V_{ub}^* V_{cs}}$$

$$= e^{-2i(2\beta + \gamma - 2\theta_d - \epsilon')}. \tag{36.63}$$

Thus, this method measures the phase $2\beta + \gamma - 2\theta_d - \epsilon'$ up to discrete ambiguities. This is very interesting, because there are very few methods that probe the combination $2\beta + \gamma$.

EXTRACTING CKM PHASES WITH B_s^0 DECAYS

37.1 Introduction

In this chapter we shall discuss some methods available to extract CKM phases with the help of B_s^0 decays. In the near future, B_s^0–$\overline{B_s^0}$ pairs will only be produced at DESY, Fermilab, and LHC. In all these cases, the B_s^0–$\overline{B_s^0}$ pairs are created in uncorrelated initial states. Moreover, they are produced together with a large number of other states originating in $b\bar{b}$ quark pairs, such as $B_s^0 B^- X^+$, $B_d^0 B^- X^+$, $B_d^0 \overline{B_d^0} X_{\bar{s}d}^0$, etc. This means that one can study both B_d^0 and B_s^0 at these facilities, but that such studies must face a daunting background.

As happens with B_d^0 studies at the $\Upsilon(4S)$, full reconstruction of the events is extremely inefficient, and one usually looks for the semileptonic decay of one b-hadron (the tagging decay), only tracing the time dependence of the other b-hadron. For exclusive decays, such as $B_s^0 \rightarrow f_{\mathrm{cp}}$ or $B_d^0 \rightarrow f_{\mathrm{cp}}$, this procedure still requires the reconstruction of the exclusive channel of interest. There is, however, a very big difference with respect to experiments performed at the $\Upsilon(4S)$. There, the anti-correlated nature of the initial state guarantees that the semileptonic decay in one side of the detector does tag the flavour of the meson on the opposite side at the same instant. For uncorrelated initial states of neutral B mesons there is no such effect, and one must take the mistags into account through a dilution factor $D_M = 1 - 2\chi_0 = (1 - y^2)/(1 + x^2)$, cf. § 9.10.2. Since the semileptonic decay may come from any of the b-hadrons on the tagging side, what is of interest is the dilution factor averaged over all the b-hadron species that can decay into the channels used for tagging.

Besides these differences, that have to do with the initial states produced in the experiments, there are a number of important differences between the B_d^0–$\overline{B_d^0}$ and B_s^0–$\overline{B_s^0}$ systems. In the latter system

- the mass difference is large, $x_s > 9.5$. This has two important consequences:
  1. as we can see in eqns (28.7), the relevant terms in time-integrated rates appear with $1 + x_s^2 > 100$ in the denominator. As a result, one must trace the time-dependence of the decays, or the CP-violating asymmetries will be too small;
  2. since Δm_s is large, the oscillations of the Δm_s-dependent terms in the time-dependent decay rates in eqns (28.6) are fast, and they may even be so fast as to go beyond the experimental vertexing capabilities. That is, these oscillations may be too fast to be traced.
- the width difference may also be rather large. In fact, Beneke *et al.* (1996) have shown that $|y_s|$ lies around 0.08 in the SM, and can even be as large

as 0.15. If $|\Delta\Gamma|$ indeed turns out to be large in the B_s^0–$\overline{B_s^0}$ system, we must use the full decay rates presented in Chapter 9 rather than the simplified version in Chapter 28. This may turn out to be a benefit, because

1. we may then measure $\mathrm{Re}\,\lambda_f$ in addition to $|\lambda_f|$ and $\mathrm{Im}\,\lambda_f$, as can be seen from eqns (9.9) and (9.10). This removes the sign ambiguity of $\arg\lambda_f$, which exists when only $|\lambda_f|$ and $\mathrm{Im}\,\lambda_f$ are measured (Aleksan et al. 1992);

2. several new methods become available for the study of CP violation. Of particular importance is the possibility, pointed out by Dunietz (1995), of using untagged data samples. Untagged decays have already been discussed briefly in § 9.6.

- one may have large branching ratios for the decays into final states which, although not CP eigenstates, are common to both B_s^0 and $\overline{B_s^0}$. These are known as 'decays into non-CP eigenstates'. Looking at Table 31.1, it is easy to understand why such decays may be more relevant in the B_s^0–$\overline{B_s^0}$ than in the B_d^0–$\overline{B_d^0}$ system. The tree-level decays with larger CKM factors are those governed by $\bar{b} \to \bar{c}c\bar{s}$ and by $\bar{b} \to \bar{c}u\bar{d}$, whose amplitudes appear at order λ^2. However, the latter must be compared with $\bar{b} \to \bar{u}c\bar{d}$, which is of order λ^4. Now, the $\bar{b} \to \bar{c}c\bar{s}$ decays common to B_d^0 and $\overline{B_d^0}$ must involve a K_L or K_S in the final state. They are of the type $[c\bar{c}]K_{L,S}$, and therefore they must be CP eigenstates. The best-known example is the gold-plated decay $B_d^0 \to J/\psi K_S$. On the other hand, B_s^0 can decay through the quark process $\bar{b} \to \bar{c}c\bar{s}$ into final states which, being common to $\overline{B_s^0}$, are either CP-eigenstates—such as $B_s^0 \to D_s^+ D_s^-$—or not—some examples include $B_s^0 \to D_s^{*+} D_s^-$.[105]
The phenomenological analysis of decays into non-CP eigenstates has been discussed in § 9.4.3.

These characteristics will be exploited in the following sections in order to determine CP-violating phases in the CKM matrix.

37.2 $B_s^0 \to D_s^+ D_s^-$: the silver-plated decay

We have seen that the decays $\bar{b} \to \bar{c}c\bar{s}$ are the only ones where there is virtually no trace of a second CKM phase in the decay amplitude. This led naturally to the gold-plated decay $B_d^0 \to J/\psi K_S$ in § 33.5, which measures $\sin 2\,(\beta - \epsilon' - \theta_d)$. Its B_s^0 counterpart is the decay $B_s^0 \to D_s^+ D_s^-$, which measures $\sin 2\,(\epsilon + \theta_s)$. In the SM, $\epsilon \sim \lambda^2$ and this CP-violating asymmetry is down by an order of magnitude with respect to the one in $B_d^0 \to J/\psi K_S$. Moreover, in the near future B_s^0 will only be produced in the harsh hadronic environment, and $D_s^\pm$ mesons may be hard to detect. Therefore, although this is dominated by a single weak phase, the measurement of ϵ will not be immediate. The asymmetry in $B_s^0 \to J/\psi\phi$

---

[105] However, it turns out that this particular final state, $D_s^{*+} D_s^-$, is expected to be dominantly CP-even (Aleksan et al. 1993, Dunietz 1995). Moreover, the corresponding decay measures the small angle ϵ.

also measures $\sin 2\,(\epsilon + \theta_d)$, but one must study the angular distributions to disentangle the CP-even and CP-odd components of the final state.

However, precisely because the SM predicts that this asymmetry should be small, this is a perfect channel to look for new physics in B_s^0–$\overline{B_s^0}$ mixing. In particular, there could be a significant new phase θ_s in that mixing (Nir and Silverman 1990).

37.3 The Aleksan–Dunietz–Kayser method

Aleksan *et al.* (1992) proposed a method to measure the angle γ using B_s^0 decays into non-CP eigenstates such as $D_s^{\pm} K^{\mp}$, $D_s^{\pm} K^{*\mp}$, and $D_s^{*\pm} K^{\mp}$. For example, one would look for the time-dependent decays

$$B_s^0(t) \to D_s^+ K^-, \quad B_s^0(t) \to D_s^- K^+, \quad \overline{B_s^0}(t) \to D_s^+ K^-, \quad \text{and} \quad \overline{B_s^0}(t) \to D_s^- K^+.$$
$$(37.1)$$

These final states can be reached by colour-allowed tree-level (T) diagrams $\bar{b} \to \bar{c}u\bar{s}$ and $\bar{b} \to \bar{u}c\bar{s}$, but also by W-exchange diagrams (E). The latter, however, have the same CKM phase as the T diagrams. Moreover, there are no penguin amplitudes contributing to these decays, and rescattering effects cannot bring in a new weak phase. The rates for these decays are estimated to be $\sim 10^{-4}$, while the decay rate for $B_s^0 \to \rho^0 K_S$ is estimated to be $\sim 10^{-7}$ (Aleksan *et al.* 1992).

Thus, extracting γ with $B_s^0 \to D^{\pm} K^{\mp}$ has two advantages over $B_s^0 \to \rho^0 K_S$: the decay rates are much larger and there is only one weak phase at play. The disadvantage of working with non-CP eigenstates is that one must trace the time-dependence of four decay rates in order to extract λ_f and $\bar{\lambda}_{\bar f}$ and recover the weak phase from them.

The four amplitudes relevant for the $B_s^0 \to D_s^{\pm} K^{\mp}$ analysis are

$$
\begin{aligned}
A_f &= \langle D_s^- K^+ | T | B_s^0 \rangle \propto V_{cb}^* V_{us} \sum_{n=1}^{2} C_n(\mu) \langle D_s^- K^+ | Q_n^{cus} | B_s^0 \rangle, \\
A_{\bar f} &= \langle D_s^+ K^- | T | B_s^0 \rangle \propto V_{ub}^* V_{cs} \sum_{n=1}^{2} C_n(\mu) \langle D_s^+ K^- | Q_n^{ucs} | B_s^0 \rangle, \\
\bar{A}_{\bar f} &= \langle D_s^+ K^- | T | \overline{B_s^0} \rangle \propto V_{cb} V_{us}^* \sum_{n=1}^{2} C_n(\mu) \langle D_s^+ K^- | Q_n^{cus\dagger} | \overline{B_s^0} \rangle, \\
\bar{A}_f &= \langle D_s^- K^+ | T | \overline{B_s^0} \rangle \propto V_{ub} V_{cs}^* \sum_{n=1}^{2} C_n(\mu) \langle D_s^- K^+ | Q_n^{ucs\dagger} | \overline{B_s^0} \rangle,
\end{aligned}
\qquad (37.2)
$$

with the operators defined in eqns (32.3). Therefore,

$$
\begin{aligned}
\lambda_f &= \frac{q_{B_s}}{p_{B_s}} \frac{\bar{A}_f}{A_f} \\
&= -e^{i(2\theta_s + \xi_{B_s} + \xi_s - \xi_b)} \frac{V_{tb}^* V_{ts} V_{ub} V_{cs}^*}{V_{tb} V_{ts}^* V_{cb}^* V_{us}} \frac{\sum_{n=1}^{2} C_n(\mu) \langle D_s^- K^+ | Q_n^{ucs\dagger} | \overline{B_s^0} \rangle}{\sum_{n=1}^{2} C_n(\mu) \langle D_s^- K^+ | Q_n^{cus} | B_s^0 \rangle}, \\
\bar{\lambda}_{\bar f} &= \frac{p_{B_s}}{q_{B_s}} \frac{A_{\bar f}}{\bar{A}_{\bar f}} \\
&= -e^{-i(2\theta_s + \xi_{B_s} + \xi_s - \xi_b)} \frac{V_{tb} V_{ts}^* V_{ub}^* V_{cs}}{V_{tb}^* V_{ts} V_{cb} V_{us}^*} \frac{\sum_{n=1}^{2} C_n(\mu) \langle D_s^+ K^- | Q_n^{ucs} | B_s^0 \rangle}{\sum_{n=1}^{2} C_n(\mu) \langle D_s^+ K^- | Q_n^{cus\dagger} | \overline{B_s^0} \rangle}.
\end{aligned}
\qquad (37.3)
$$

Using the CP transformation of the operators in eqns (32.13) it is easy to prove that

$$A_f = e^{i(\xi_{B_s} - \xi_f + \xi_s - \xi_b + \xi_c - \xi_u)} \frac{V_{cb}^* V_{us}}{V_{cb} V_{us}^*} \bar{A}_{\bar{f}},$$

$$A_{\bar{f}} = e^{i(\xi_{B_s} + \xi_f + \xi_s - \xi_b + \xi_u - \xi_c)} \frac{V_{ub}^* V_{cs}}{V_{ub} V_{cs}^*} \bar{A}_f. \tag{37.4}$$

The big difference with respect to decays into CP eigenstates is that, here, the CP transformation does not relate $\bar{A}_f$, in the numerator of λ_f, with the amplitude in the denominator, A_f, but rather with $A_{\bar{f}}$. Therefore, one cannot proceed with the calculation of λ_f without either making some phase choice, or else using some symmetry to relate the numerator with the denominator. Still, we may assert that

$$\frac{\lambda_f}{\bar{\lambda}_{\bar{f}}} = \left(\frac{V_{tb}^* V_{ts}}{V_{tb} V_{ts}^*} \right)^2 \frac{V_{ub} V_{cs}^* V_{cb} V_{us}^*}{V_{ub}^* V_{cs} V_{cb}^* V_{us}} = e^{-2i(\gamma + \epsilon' - 2\epsilon - 2\theta_s)} \tag{37.5}$$

is a measurable quantity, as it should be. If we measure the real and imaginary parts of λ_f and $\bar{\lambda}_{\bar{f}}$, we can determine $\gamma + \epsilon' - 2\epsilon - 2\theta_s$. However, if $|\Delta\Gamma|$ turns out to be small in the B_s^0–$\overline{B_s^0}$ system, we can only determine the imaginary parts of λ_f and $\bar{\lambda}_{\bar{f}}$, as well as their magnitudes. Then, the sine of the relevant phase may be recovered up to a fourfold ambiguity from

$$\sin^2(\gamma + \epsilon' - 2\epsilon - 2\theta_s) = \frac{1}{2} \left[1 + s_+ s_- \pm \sqrt{(1 - s_+^2)(1 - s_-^2)} \right], \tag{37.6}$$

where

$$s_+ \equiv -\frac{\operatorname{Im} \bar{\lambda}_{\bar{f}}}{|\bar{\lambda}_{\bar{f}}|},$$

$$s_- \equiv \frac{\operatorname{Im} \lambda_f}{|\lambda_f|}. \tag{37.7}$$

This result is easier to understand if we choose a convention in which the CP transformation includes no phases. Then, we may write $q/p = -e^{2i\phi_M}$, and

$$A_f = A e^{i\phi_a} e^{i\delta_a}$$
$$\bar{A}_{\bar{f}} = A e^{-i\phi_a} e^{i\delta_a},$$
$$A_{\bar{f}} = B e^{i\phi_b} e^{i\delta_b}$$
$$\bar{A}_f = B e^{-i\phi_b} e^{i\delta_b}. \tag{37.8}$$

Therefore,

$$\lambda_f = -\frac{B}{A} e^{-i(\phi - \Delta)},$$

$$\bar{\lambda}_{\bar{f}} = -\frac{B}{A} e^{i(\phi + \Delta)}, \tag{37.9}$$

where $\phi = \phi_a + \phi_b - 2\phi_M$ and $\Delta = \delta_b - \delta_a$ are measurable weak and strong phases, respectively. In our example, $\phi = \gamma + \epsilon' - 2\epsilon - 2\theta_s$. We see that $|\lambda_f|$, $\operatorname{Im} \lambda_f$, and $\operatorname{Re} \lambda_f$ measure B/A, $B/A \sin(\phi - \Delta)$, and $-B/A \cos(\phi - \Delta)$, respectively. Similarly, $|\bar{\lambda}_{\bar{f}}|$, $\operatorname{Im} \bar{\lambda}_{\bar{f}}$, and $\operatorname{Re} \bar{\lambda}_{\bar{f}}$ measure B/A, $-B/A \sin(\phi + \Delta)$, and $-B/A \cos(\phi + \Delta)$, respectively. We can recover ϕ from these measurements.

Let us look back at eqn (37.6). One of the signs on the RHS of eqn (37.6) yields the true value of $\sin^2 \phi$; the other sign gives $\cos^2 \Delta$ instead. We stress that this trigonometric exercise is only needed if $|\Delta\Gamma|$ is small, in which case we can only measure $\text{Im } \lambda_f$, $\text{Im } \bar{\lambda}_{\bar{f}}$, $|\lambda_f|$, and $|\bar{\lambda}_{\bar{f}}|$. The ambiguity in $\sin^2 \phi$ is removed if $|\Delta\Gamma|$ is large enough to allow for the additional measurement of the real parts, $\text{Re } \lambda_f$ and $\text{Re } \bar{\lambda}_{\bar{f}}$.

The fact that we are probing $\sin^2 \gamma$ and that we do so through eqn (37.6) has three interesting consequences. Firstly, if one searches for γ with CP-violating asymmetries alone, one is really probing $\sin 2\gamma$. As pointed out before, present constraints allow this to be zero, while $\sin^2 \gamma$ is guaranteed to be nonzero. Indeed, we have seen in § 18.5 that $0.33 \leq \sin^2 \gamma \leq 1$ in the SM. Secondly, since $B_K \sin \gamma$ is positive, if we assume that B_K is positive, then eqn (37.6) only retains a twofold ambiguity. Thirdly, Aleksan $et\ al.$ (1992) argue that Δ is likely to be very small. If this turns out to be the case, then the angle γ may be extracted from eqn (37.6) without any ambiguity at all.

37.4 The Gronau–London $B_s^0 \rightarrow D\phi$ method

This method is analogous to the Gronau–London $B_d^0 \rightarrow DK_S$ method discussed in § 36.6, only with the spectator quark d being substituted by s. Both methods were suggested in the same article by Gronau and London (1991).

37.4.1 Extracting γ from $B_s^0 \rightarrow \{D^0, \overline{D^0}, D_{f_{cp}}\}\phi$

Here, one looks for the tagged decays

$$B_s^0(t) \rightarrow D^0\phi, \quad B_s^0(t) \rightarrow \overline{D^0}\phi, \quad B_s^0(t) \rightarrow (f_{cp})_D\phi, \text{ and CP conjugated.} \quad (37.10)$$

The idea is the same as in § 36.3 and 36.6.1, and one also measures the sine of the angle in eqn (36.34), $\gamma \pm \epsilon'$, up to a fourfold discrete ambiguity. Again, the hierarchy of the amplitudes makes this method difficult to implement in practice.

There is a subtle difference between this method and those other methods related to it by a change in the spectator quark: here we must follow the time dependence of the decays. In § 36.3 that was not needed because there is no mixing in charged decays. In § 36.6.1 that was also not needed, although there is B_d^0–$\overline{B_d^0}$ mixing. The point there was that, at the $\Upsilon(4S)$, the terms proportional to $\text{Im } \lambda_f$ drop out from the tagged, $time\text{-}integrated$ decay rates $\Gamma[B^0 \rightarrow f]$ and $\Gamma[\overline{B^0} \rightarrow f]$. Therefore, from these two observables we could extract the two unknowns $|A_f|$ and $|\bar{A}_f|$. Here the situation is different because the B_s^0–$\overline{B_s^0}$ pairs to be detected in experiments will not come from an odd-parity correlated state.

37.4.2 Extracting $-2\epsilon + \gamma$ from $B_s^0 \rightarrow \{D^0, \overline{D^0}\}\phi$

We have seen in § 36.6.2 that the interference terms in the time-dependent decays $B_d^0 \rightarrow \{D^0, \overline{D^0}\}K_S$ can be used to measure $2\beta + \gamma$, in the SM. Analogously, we may use the $B_s^0 \rightarrow \{D^0, \overline{D^0}\}\phi$ decays to find $-2\epsilon + \gamma$. The difference is that, in this case, we use the mixing in the B_s–$\overline{B_s}$ system, rather than the mixing in the B_d–$\overline{B_d^0}$ and K^0–$\overline{K^0}$ systems. This is just the colour-suppressed version of the same quark processes $\bar{b} \rightarrow \bar{c}u\bar{s}$ and $\bar{b} \rightarrow \bar{u}c\bar{s}$ involved in the Aleksan–Dunietz–Kayser method. Therefore, it also measures the sine of the angle $\gamma + \epsilon' - 2\epsilon - 2\theta_s$, up to a fourfold discrete ambiguity.

37.5   On the use of untagged decays

A large $|\Delta\Gamma|/\Gamma$ opens up the possibility for a new class of experiments to determine weak phases: those performed with *untagged* data samples (Dunietz 1995). Untagged data samples are interesting because they are readily produced at e^+e^- and $p\bar{p}$ machines; there is no need to tag and no extra cost in statistics (Dunietz 1995).

We have already introduced untagged decays in § 9.6. We recall that the untagged decay rates may be written as in eqn (9.50):

$$\Gamma_f(t) \equiv \Gamma[P^0(t) \rightarrow f] + \Gamma[\overline{P^0}(t) \rightarrow f] = |A_f|^2 \frac{e^{-\Gamma t}}{1 - \delta} (H - \delta I). \qquad (37.11)$$

where H and I are given by eqns (9.15):

$$\begin{aligned} H &\equiv \left(1 + |\lambda_f|^2\right) \cosh \frac{\Delta\Gamma t}{2} - 2\mathrm{Re}\,\lambda_f \sinh \frac{\Delta\Gamma t}{2}, \\ I &\equiv \left(1 - |\lambda_f|^2\right) \cos{(\Delta m t)} + 2\mathrm{Im}\,\lambda_f \sin{(\Delta m t)}. \end{aligned} \qquad (37.12)$$

The function H depends on exponentials and on $\Delta\Gamma$, while I is an oscillatory function of $\Delta m t$. Clearly, if δ is small, then the oscillatory function I drops out of the untagged decay rates. This is an advantage of untagged decays because, if Δm is very large, the time oscillations in I are too fast to be observed anyway. However, one must keep in mind that the corrections due to δ should be included whenever untagged decays are being used to look for very small effects in H.

Henceforth we take $\delta = 0$. Then,

$$\begin{aligned} \Gamma_f(t) &= e^{-\Gamma t} |A_f|^2 \left[\left(1 + |\lambda_f|^2\right) \cosh \frac{\Delta\Gamma t}{2} - 2\mathrm{Re}\,\lambda_f \sinh \frac{\Delta\Gamma t}{2}\right], \\ \Gamma_{\bar{f}}(t) &= e^{-\Gamma t} |\bar{A}_{\bar{f}}|^2 \left[\left(1 + |\bar{\lambda}_{\bar{f}}|^2\right) \cosh \frac{\Delta\Gamma t}{2} - 2\mathrm{Re}\,\bar{\lambda}_{\bar{f}} \sinh \frac{\Delta\Gamma t}{2}\right]. \end{aligned} \qquad (37.13)$$

A difference between these two untagged decays violates CP. If the decays are dominated by a single weak phase, then there is no direct CP violation, $|A_f| = |\bar{A}_{\bar{f}}|$, $|\lambda_f| = |\bar{\lambda}_{\bar{f}}|$, and the only CP violation arises from $\mathrm{Re}\,\lambda_f \neq \mathrm{Re}\,\bar{\lambda}_{\bar{f}}$.

If f is a CP eigenstate, then H is CP-conserving, and I is CP-violating. In this case, $\Gamma_f(t) = \Gamma_{\bar{f}}(t)$. If there is no direct or mixing CP violation, then $|\lambda_f| = 1$, and H only measures CP violation to the extent that $|\mathrm{Re}\,\lambda_f|$ differs from unity.

We are now in a position to appreciate the systematic analysis performed by Dunietz (1995). Firstly we consider decays into CP eigenstates such as $B_s \rightarrow D_s^+ D_s^-$. In § 37.2 we have seen that tagged decays in this channel measure

$\sin 2 \left(\epsilon + \theta_s \right)$. Here, we notice that untagged decays allow us to measure the cosine of the same angle, $\cos 2 \left(\epsilon + \theta_s \right)$. Instead of $D_s^+ D_s^-$ we may use $J/\psi \phi$ or $D_s^{*+} D_s^{*-}$. These are combinations of CP-even and (small) CP-odd components, which can be resolved by means of an analysis of the angular distributions (Dunietz *et al.* 1991; Dighe *et al.* 1996c), even in the case of untagged data samples (Dunietz and Fleischer 1997). In the SM, $\theta_s = 0$, $\epsilon \sim \lambda^2$, and this experiment looks extremely difficult, since we will be looking at the difference $\cos 2\epsilon - 1 \sim \lambda^4$. However, in some models beyond the SM there may be a large θ_s. Untagged decays can be used to look for it.

Secondly, Dunietz (1995) has proposed a version of the Aleksan–Dunietz–Kayser method of § 37.3 in which one uses instead untagged decays. This requires the value of $|A_f|$ as an additional input. In fact, only if we know this magnitude from theory can we extract $\operatorname{Re} \lambda_f$ and $\operatorname{Re} \bar{\lambda}_f$ from the fit to the time dependences in eqns (37.13). To see how to extract the weak phase from these observables, we use eqn (37.9), from which

$$
\begin{aligned}
c_+ &\equiv \frac{-\operatorname{Re} \bar{\lambda}_{\bar{f}}}{|\bar{\lambda}_{\bar{f}}|} = \cos \left(\phi + \Delta \right), \\
c_- &\equiv \frac{-\operatorname{Re} \lambda_f}{|\lambda_f|} = \cos \left(\phi - \Delta \right),
\end{aligned}
\tag{37.14}
$$

where $\phi = \gamma + \epsilon' - 2\epsilon - 2\theta_s$. We are now left with the trivial trigonometric exercise in eqn (36.7). The usual remarks about the fourfold ambiguity in $\sin \gamma$ apply. Notice that the CP-violating quantity $c_+ - c_- = -2 \sin \phi \sin \Delta$ is only different from zero if there is a non-negligible final-state phase difference Δ. Dunietz (1995) pointed out that Δ is likely to be very small for colour-allowed decays, in which case $c_+ = c_-$ would measure $\cos \phi$ directly.

Thirdly, one could perform a $B_s^0 \rightarrow \{D^0, \overline{D^0}, D_{f_{cp}}\} \phi$ analysis with untagged decays (Dunietz 1995). The idea is that one can extract from the time dependences of the untagged decays the observables

$$
\begin{aligned}
&\frac{\operatorname{Re} \lambda_f}{1 + |\lambda_f|^2}, \\
&\frac{\operatorname{Re} \bar{\lambda}_{\bar{f}}}{1 + |\bar{\lambda}_{\bar{f}}|^2}, \\
&\frac{\operatorname{Re} \lambda_{f_{cp}}}{1 + |\lambda_{f_{cp}}|^2},
\end{aligned}
\tag{37.15}
$$

with $f = D^0 \phi$, $\bar{f} = \overline{D^0} \phi$, and $f_{cp} = D_{f_{cp}} \phi$. This allows the extraction of the difference of weak phases, the difference of strong phases, and the common magnitude $|\lambda_f| = |\bar{\lambda}_{\bar{f}}|$. Contrary to the previous case, here there is no need for a theoretical input concerning amplitudes.

38

DISCRETE AMBIGUITIES

38.1 Statement of the problem

In this chapter we follow closely the analysis of Grossman *et al.* (1997a), Grossman and Quinn (1997), and Wolfenstein (1997b) on the removal of the discrete ambiguities associated with the extraction of the CKM angles from the corresponding CP-asymmetry measurements. There has been renewed interest on this subject and further developments are likely. We neglect the phases ϵ and ϵ' throughout.

Clean measurements of CKM phases must be unhindered by hadronic uncertainties. This is the case of CP asymmetries in B decays into CP eigenstates in which the amplitude is dominated by a single weak phase. These cases involve the ratio of an amplitude and its complex conjugate and, hence, twice the relevant CKM phase. In the standard model (SM), the decay $B_d^0 \to J/\psi K_S$ measures $\sin 2\beta$ and, neglecting penguin pollution, the decays $B_d^0 \to \pi^+\pi^-$ and $B_s^0 \to \rho^0 K_S$ measure $\sin 2\alpha$ and $\sin 2\gamma$, respectively.

However, a measurement of $\sin 2\phi$ can only determine the phase ϕ up to a fourfold ambiguity: if ϕ is any of the solutions between 0 and 2π, then the other three solutions are obtained from this one as $\pi/2 - \phi$, $\pi + \phi$, and $3\pi/2 - \phi$, defined mod 2π in such a way that they also lie between 0 and 2π. For $\sin 2\phi > 0$ two solutions are in the first quadrant and the other two are in the third quadrant; for $\sin 2\phi < 0$ two solutions are in the second quadrant and the other two are in the fourth quadrant. In this section we discuss ways to eliminate these ambiguities.

As pointed out before, with our definitions the angles α, β, and γ are not independent, rather

$$\alpha + \beta + \gamma = \pi \ (\text{mod } 2\pi). \tag{38.1}$$

For instance, one may use this relation to determine γ; again, all possibilities for γ should be transformed into angles lying between 0 and 2π. It depends on the model whether a specific asymmetry measures the angles defined above, or not. For example, if the dominant new-physics contributions appear in the mixing, then the asymmetries in $B_d^0 \to J/\psi K_S$ and $B_d^0 \to \pi^+\pi^-$ measure $\sin 2\tilde{\beta}$ and $\sin 2\tilde{\alpha}$, respectively, where $\tilde{\beta} = \beta - \theta_d$ and $\tilde{\alpha} = \alpha + \theta_d$, and θ_d is the new phase in the mixing. In this case it is still true (Nir and Silverman 1990) that

$$\tilde{\alpha} + \tilde{\beta} + \gamma = \pi \ (\text{mod } 2\pi), \tag{38.2}$$

and this can still be used to find out γ.

Let us suppose that $\sin 2\tilde{\beta}$ and $\sin 2\tilde{\alpha}$ have been measured. The values of $\{\tilde{\alpha}, \tilde{\beta}\}$ are then determined up to a $4 \times 4 = 16$-fold ambiguity. Equation (38.2) then yields eight solutions for γ (Grossman *et al.* 1997a):

$$
\begin{aligned}
&\pm \left(\tilde{\alpha} + \tilde{\beta}\right) \; (\mathrm{mod}\, 2\pi), \\
&\pi \pm \left(\tilde{\alpha} + \tilde{\beta}\right) \; (\mathrm{mod}\, 2\pi), \\
&\pi/2 \pm \left(\tilde{\alpha} - \tilde{\beta}\right) \; (\mathrm{mod}\, 2\pi), \\
&3\pi/2 \pm \left(\tilde{\alpha} - \tilde{\beta}\right) \; (\mathrm{mod}\, 2\pi).
\end{aligned}
\tag{38.3}
$$

These eight possibilities correspond to two possible values of $\cos 2\gamma$, and to four possible values of $\sin 2\gamma$:

$$
\begin{aligned}
\cos 2\gamma &= \cos 2\left(\tilde{\alpha} + \tilde{\beta}\right) \quad \text{or} \quad \cos 2\gamma = -\cos 2\left(\tilde{\alpha} - \tilde{\beta}\right), \\
\sin 2\gamma &= \pm \sin 2\left(\tilde{\alpha} + \tilde{\beta}\right) \quad \text{or} \quad \sin 2\gamma = \pm \sin 2\left(\tilde{\alpha} - \tilde{\beta}\right).
\end{aligned}
\tag{38.4}
$$

These, however, come in only four combinations:

$$
\left\{\cos 2\gamma, \sin 2\gamma\right\} = \left\{\cos 2\left(\tilde{\alpha} + \tilde{\beta}\right), \pm \sin 2\left(\tilde{\alpha} + \tilde{\beta}\right)\right\}
$$

or
$$
\left\{\cos 2\gamma, \sin 2\gamma\right\} = \left\{-\cos 2\left(\tilde{\alpha} - \tilde{\beta}\right), \pm \sin 2\left(\tilde{\alpha} - \tilde{\beta}\right)\right\}.
\tag{38.5}
$$

As stated above, we end up with 16 possibilities for the values of $\{\tilde{\alpha}, \tilde{\beta}, \gamma\}$. Within certain models, some of these ambiguities may be eliminated. In any model where the three complex numbers which enter into the definition of the phases α, β, and γ form a triangle, one must have (Grossman *et al.* 1997a)

$$
\tilde{\alpha} + \tilde{\beta} + \gamma = \pi \text{ or } 5\pi,
\tag{38.6}
$$

but not $\tilde{\alpha} + \tilde{\beta} + \gamma = 3\pi$. Indeed, if the angles are interior to the triangle they all lie in the range $[0, \pi]$ and they add up to π; if the angles are exterior to the triangle they all lie in the range $[\pi, 2\pi]$ and they add up to 5π. As a result, in such models there are only four possibilities for $\{\tilde{\alpha}, \tilde{\beta}, \gamma\}$ (Grossman *et al.* 1997a)—a further measurement of $\sin 2\gamma$ would remove this ambiguity completely, unless $\tilde{\alpha}$ or $\tilde{\beta}$ equal $\pi/2$ or $3\pi/2$ (Nir and Quinn 1990). In the SM, the ambiguity is further reduced because

$$
R_b \equiv \left| \frac{V_{ud} V_{ub}}{V_{cd} V_{cb}} \right| = \frac{\sin \beta}{\sin \alpha}
\tag{38.7}
$$

(see eqns 13.33 and 13.35) is smaller than $1/\sqrt{2}$. This implies that β must lie either in $[0, \pi/4]$ or in $[7\pi/4, 2\pi]$; the ranges $[3\pi/4, \pi]$ and $[\pi, 5\pi/4]$ are excluded by combination with the constraint in eqn (38.6). In addition, the SM fit of the CP-violating parameter ϵ_K forces $B_K \sin \beta$ to be positive (Nir and Quinn 1990).

Thus, if we assume the bag parameter B_K to be positive, $0 < \beta = \tilde{\beta} < \pi/4$. This leaves open only two possible solutions for $\{\alpha, \beta, \gamma\}$; in some cases, only one of them is allowed.

Grossman *et al.* (1997*a*) have given the following example to illustrate these points. Assume that $\sin 2\tilde{\alpha} = 1/2$ and $\sin 2\tilde{\beta} = \sqrt{3}/2$. Then, there are four possible values for $\tilde{\alpha}$:

$$\frac{\pi}{12}, \frac{5\pi}{12}, \frac{13\pi}{12}, \text{ and } \frac{17\pi}{12}.$$

Similarly, there are four possible values for $\tilde{\beta}$:

$$\frac{\pi}{6}, \frac{\pi}{3}, \frac{7\pi}{6}, \text{ and } \frac{4\pi}{3}.$$

The eight solutions for γ are

$$\frac{\pi}{4}, \frac{5\pi}{12}, \frac{7\pi}{12}, \frac{3\pi}{4}, \frac{5\pi}{4}, \frac{17\pi}{12}, \frac{19\pi}{12}, \text{ and } \frac{7\pi}{4}.$$

There are two possible combinations of $\tilde{\alpha}$ and $\tilde{\beta}$ leading to each value of γ. If one assumes that $\tilde{\alpha}$, $\tilde{\beta}$, and γ add up to either π or 5π, then we are left with the possible values

$$\left\{ \frac{\pi}{12}, \frac{\pi}{6}, \frac{3\pi}{4} \right\}, \quad \left\{ \frac{\pi}{12}, \frac{\pi}{3}, \frac{7\pi}{12} \right\}, \quad \left\{ \frac{5\pi}{12}, \frac{\pi}{6}, \frac{5\pi}{12} \right\}, \quad \text{and} \quad \left\{ \frac{5\pi}{12}, \frac{\pi}{3}, \frac{\pi}{4} \right\}$$

for $\{\tilde{\alpha}, \tilde{\beta}, \gamma\}$. In the SM $0 < \beta < \pi/4$, leaving only the first and third solutions. Actually, in this case the first solution is forbidden, because it leads to

$$R_t = \frac{\sin \gamma}{\sin(\beta + \gamma)} \sim 2.73, \tag{38.8}$$

while the experimental bound in eqn (18.28) is $0.79 < R_t < 1.18$.

38.2   Removing the ambiguities

Let us proceed in the assumption that $\sin 2\tilde{\alpha}$ and $\sin 2\tilde{\beta}$ have been determined. The 16-fold ambiguity in $\{\tilde{\alpha}, \tilde{\beta}\}$ can be removed by measuring in addition the *signs* of $\cos 2\tilde{\alpha}$, $\cos 2\tilde{\beta}$, $\sin \tilde{\alpha}$, and $\sin \tilde{\beta}$. These four signs resolve the ambiguities completely:

- sign($\cos 2\phi$) removes the ambiguity $\phi \to \pi/2 - \phi$;
- sign($\sin \phi$) resolves the ambiguity $\phi \to \pi + \phi$.

38.2.1   *Determining the sign of* $\cos 2\phi$

A determination of sign($\cos 2\phi$) removes the ambiguity $\phi \to \pi/2 - \phi$. Information on sign($\cos 2\alpha$) and on sign($\cos 2\beta$) can be obtained either directly or by measuring some functions of γ. Several methods include:

- A determination of $\text{sign}(\cos 2\tilde{\alpha})$ from the analysis of the Dalitz-plot distribution of the decays $B_d^0 \to \rho^{\pm} \pi^{\mp}$ and $B_d^0 \to \rho^0 \pi^0$ (Snyder and Quinn 1993). Grossman and Quinn (1997) argue that this method allows for an extraction of $\text{sign}(\cos 2\tilde{\alpha})$, even in the presence of large penguin contributions.

- A determination of the *value* of $\cos 2\gamma = 1 - 2\sin^2 \gamma$. This can be done, for instance, through the Gronau–London–Wyler method based on the decays $B^{\pm} \to K^{\pm} D$ (Gronau and Wyler 1991).[106] Recall that, given a measurement of $\sin 2\tilde{\alpha}$ and of $\sin 2\tilde{\beta}$ the constraint in eqn (38.2) guarantees that $\cos 2\gamma$ can only take the two values in eqn (38.4). In other words, once $\sin 2\tilde{\alpha}$, $\sin 2\tilde{\beta}$, and $\cos 2\gamma$ are determined, the equality

$$\cos 2\gamma = \cos 2\tilde{\alpha} \cos 2\tilde{\beta} - \sin 2\tilde{\alpha} \sin 2\tilde{\beta}, \qquad (38.9)$$

fixes the magnitude and sign of $\cos 2\tilde{\alpha} \cos 2\tilde{\beta}$. This reduces the ambiguity by a factor of two. There is a slight complication due to the fact that most simple methods determine $\sin^2 \gamma$ only up to a twofold ambiguity. However, comparing the two possible 'experimental' values for $\sin^2 \gamma$ with the two values allowed for $\{\cos 2\tilde{\alpha}, \cos 2\tilde{\beta}\}$ should be enough to remove this ambiguity (Wolfenstein 1997b).

- A determination of $\cos 2\tilde{\beta}$ from the decay chain $B_d^0 \to J/\psi K \to J/\psi f_K$ (Kayser 1997). This reduces the ambiguity by a factor of two. Once this is combined with $\cos 2\tilde{\alpha} \cos 2\tilde{\beta}$, as determined from the Gronau–London–Wyler method, we also obtain $\cos 2\tilde{\alpha}$; the only ambiguities remaining are then those of the type $\phi \to \pi + \phi$.

- A determination of $\sin 2(\gamma - \theta_s)$ from $B_s^0 \to \rho^0 K_S$. Although this observable involves θ_s and is subject to large penguin corrections, the fact that we only need to distinguish among four solutions, together with information from related channels that may be available once this measurement is performed, should alleviate the problem (Wolfenstein 1997b; Grossman and Quinn 1997). Given measurements of $\sin 2\tilde{\alpha}$ and of $\sin 2\tilde{\beta}$, eqn (38.4) guarantees that $\sin 2\gamma$ can only take four values. Once $\sin 2\tilde{\alpha}$, $\sin 2\tilde{\beta}$, and $\sin 2\gamma$ are determined, the equality

$$- \sin 2\gamma = \sin 2\tilde{\alpha} \cos 2\tilde{\beta} + \cos 2\tilde{\alpha} \sin 2\tilde{\beta} \qquad (38.10)$$

fixes the signs of both $\cos 2\tilde{\alpha}$ and $\cos 2\tilde{\beta}$. This by itself would reduce the ambiguity by a factor of four. Due to the large penguin pollution in $B_s^0 \to \rho^0 K_S$, it will be best to use it in connection with $\sin^2 \gamma$ and with the measurement of θ_s from $B_s^0 \to D_s^+ D_s^-$.

---

[106]The Aleksan–Dunietz–Kayser (1992) method uses the decays $B_s^0 \to D_s^{\pm} K^{\mp}$ to measure $\cos 2(\gamma - 2\theta_s)$. It can only be used to determine $\cos 2\gamma$ if we know, for instance from $B_s^0 \to D_s^+ D_s^-$, that θ_s is small.

38.2.2 *Determining the sign of* $\sin \phi$

A determination of the signs of $\sin \tilde{\alpha}$ and $\sin \tilde{\beta}$ would eliminate the ambiguity $\phi \to \pi + \phi$. A first attempt in this direction was made by Grossman and Quinn (1997). The idea is to compare two hadronic decays driven by the same quark-level decay: one which is dominated by a single weak phase, the other one having a substantial penguin pollution. The interference CP violation in the latter channel is given by eqn (28.18). The CP violation in the former channel is given by $\sin 2\phi_1$. This allows the determination of

$$\Delta \tilde{a} \equiv \tilde{a} - \sin 2\phi_1 \sim -2r \cos 2\phi_1 \sin (\phi_1 - \phi_2) \cos \Delta, \qquad (38.11)$$

cf. eqn (33.33). Here, $r = A_P/A_T$ is the ratio of the moduli of the penguin and tree-level amplitudes, and Δ is the difference between the strong phases of those amplitudes. One can extract $\cos 2\phi_1 \sin (\phi_1 - \phi_2)$ once the hadronic factor $r \cos \Delta$ is known. The point made by Grossman and Quinn (1997) was that the *sign* of $\cos 2\phi_1 \sin (\phi_1 - \phi_2)$ would already facilitate the resolution of the discrete ambiguities, and this only requires information about the *sign* of $r \cos \Delta$. However, one must recall that $\phi_1 - \phi_2$ is the difference between the phases of two diagrams, while ϕ_1 includes both the phase of the mixing and that of the tree-level decay amplitude. Under the assumption that the decay is given by the SM, the former is directly related to a phase in the CKM matrix—such as α or β—while the latter also contains information about possible new phases in the mixing—as in $\tilde{\alpha}$ or $\tilde{\beta}$.

Some examples of this analysis include (Grossman and Quinn 1997)

- comparing the clean measurement of $\sin 2\tilde{\beta}$ from $B_d^0 \to J/\psi K_S$ with the CP-asymmetry in $B_d^0 \to D^+ D^-$. This determines the sign of $\cos 2\tilde{\beta} \sin \beta$, once the sign of $r_{DD} \cos \Delta_{DD}$ is known;

- comparing the Dalitz-plot determination of $\sin 2\alpha$ from $B_d^0 \to \rho \pi$ (Snyder and Quinn 1993) with the asymmetry in $B_d^0 \to \pi^+ \pi^-$. This yields the sign of $\cos 2\tilde{\alpha} \sin \alpha$ once the sign of $r_{\pi \pi} \cos \Delta_{\pi \pi}$ is known.

Grossman and Quinn (1997) argue that our future knowledge of these signs is likely to be robust. Although precise calculations of the relevant ratios of matrix elements cannot be performed at present, some of the signs are common to the different computational methods (as occurs, for instance, with the sign of B_K). This makes the determination of the $\text{sign}(\sin \phi)$ conceivable. Moreover, a contradiction between several such determinations of the sign of $\sin \phi$ would be very informative. Either there would be new physics at play, or a revision of the calculations of the hadronic matrix elements would be necessary.

APPENDIX A

TWO NOTES ON CP-TRANSFORMATION PHASES

A.1 The CP-transformation phases of the quarks in q/p

When discussing interference CP violation in heavy-meson decays it is customary to assume that there is no CP violation in the mixing of P^0 and $\overline{P^0}$. In this case,

$$\frac{q}{p} = \pm e^{i\xi}, \tag{A.1}$$

where the phase ξ is the one that appears in the CP transformation

$$\begin{aligned} CP|P^0\rangle &= e^{i\xi}|\overline{P^0}\rangle, \\ CP|\overline{P^0}\rangle &= e^{-i\xi}|P^0\rangle. \end{aligned} \tag{A.2}$$

Indeed, if there is no CP violation in the mixing, then the eigenstates of mass must also be eigenstates of CP. In eqns (A.2) the $+$ sign corresponds to the heavy state being CP-even while the light state is CP-odd; the $-$ sign corresponds to the opposite situation.

Most authors choose the phase ξ in eqns (A.2) to be either 0 or π. They then compute q/p and obtain that it is equal to a nontrivial phase; however, they also assert that there is no CP violation in the mixing. This is a contradiction: if $\xi = 0$ and CP is conserved in the mixing, then q/p cannot be equal to $-\exp(2i\phi_M)$, with $2\phi_M \neq 0$ and $2\phi_M \neq \pi$. We want to resolve this contradiction.

Let us take the concrete example of the B_q^0–$\overline{B_q^0}$ systems. There, it is assumed that CP is conserved in the mixing because $\Gamma_{12} = 0$. Then,

$$\frac{q}{p} = \sqrt{\frac{M_{12}^*}{M_{12}}}. \tag{A.3}$$

The quantity M_{12} is calculated from an effective Hamiltonian having a weak (CP-odd) phase $-2\phi_M$, and a $\Delta B = 2$ operator $\mathcal{O}$:

$$\begin{aligned} M_{12} &= e^{-2i\phi_M} \langle B_q^0|\mathcal{O}|\overline{B_q^0}\rangle, \\ M_{12}^* &= e^{2i\phi_M} \langle \overline{B_q^0}|\mathcal{O}^\dagger|B_q^0\rangle. \end{aligned} \tag{A.4}$$

The operator $\mathcal{O}$ and its Hermitian conjugate are related by the CP transformation:

$$(CP)\,\mathcal{O}^\dagger\,(CP)^\dagger = e^{2i\theta_M}\mathcal{O}. \tag{A.5}$$

Also, as in eqns (A.2),

$$\mathcal{CP}|B_q^0\rangle = e^{i\xi_{B_q}}|\overline{B_q^0}\rangle,$$
$$\mathcal{CP}|\overline{B_q^0}\rangle = e^{-i\xi_{B_q}}|B_q^0\rangle. \tag{A.6}$$

We may use two insertions of $(\mathcal{CP})^\dagger(\mathcal{CP}) = 1$ in the second eqn (A.4) to derive

$$\begin{aligned}
M_{12}^* &= e^{2i\phi_M}\langle\overline{B_q^0}|(\mathcal{CP})^\dagger(\mathcal{CP})\,\mathcal{O}^\dagger(\mathcal{CP})^\dagger(\mathcal{CP})|B_q^0\rangle \\
&= e^{2i(\phi_M+\xi_{B_q}+\theta_M)}\langle B_q^0|\mathcal{O}|\overline{B_q^0}\rangle \\
&= e^{2i(2\phi_M+\xi_{B_q}+\theta_M)}M_{12}.
\end{aligned} \tag{A.7}$$

Then, from eqn (A.3),

$$\frac{q}{p} = \pm e^{i(2\phi_M+\xi_{B_q}+\theta_M)}. \tag{A.8}$$

This should be equal to $\pm e^{i\xi_{B_q}}$, as in eqn (A.1). The CP-transformation phase θ_M must therefore be chosen such that $2\phi_M + \theta_M = 0$.

However, most authors omit the phase θ_M in eqn (A.5), implicitly setting θ_M to zero.[107] They then reach the conclusion that

$$\frac{q}{p} = \pm e^{i(2\phi_M+\xi_{B_q})}. \tag{A.9}$$

If ϕ_M is non-zero, this means that the eigenstates of the Hamiltonian are not eigenstates of CP, in spite of CP being conserved in the mixing! The point is that one is not allowed to set $\theta_M = 0$. This phase must be chosen in order to reproduce CP invariance of the theory; it cannot be *a priori* given a fixed value.

Let us illustrate this point with the calculation of q/p within the SM. There,

$$\mathcal{O} \propto [\bar{q}\gamma^\mu(1-\gamma_5)b][\bar{q}\gamma_\mu(1-\gamma_5)b], \tag{A.10}$$

and

$$e^{-2i\phi_M} = \frac{V_{tb}V_{tq}^*}{V_{tb}^*V_{tq}}. \tag{A.11}$$

Now, the most general CP transformation of the quark fields b and q is

$$(\mathcal{CP})\,b\,(\mathcal{CP})^\dagger = e^{i\xi_b}\gamma^0 C\bar{b}^T,$$
$$(\mathcal{CP})\,\bar{q}\,(\mathcal{CP})^\dagger = -e^{-i\xi_q}q^T C^{-1}\gamma^0. \tag{A.12}$$

Then, from eqns (A.10) and (A.5), $\theta_M = \xi_q - \xi_b$ and

$$\frac{q}{p} = \pm e^{i(\xi_{B_q}+\xi_q-\xi_b)}\frac{V_{tb}^*V_{tq}}{V_{tb}V_{tq}^*}. \tag{A.13}$$

The requirement that $2\phi_M + \theta_M = 0$ is equivalent to

$$V_{tb}V_{tq}^* = e^{i(\xi_q-\xi_b)}V_{tb}^*V_{tq}. \tag{A.14}$$

It is clear that we may always choose ξ_q and ξ_b such that eqn (A.14) be verified, thus obtaining CP invariance.

---

[107] Most authors also set ξ_{B_q} to be either 0 or π. However, this is not the root of the problem that we wish to address.

We may check that eqn (A.14) is the correct CP-invariance condition in another way. Let us consider the charged-current Lagrangian

$$\frac{g}{2\sqrt{2}} \sum_{\alpha=u,c,t} \sum_{k=d,s,b} \left[W_\mu^+ V_{\alpha k} \overline{u_\alpha} \gamma^\mu \left(1 - \gamma_5\right) d_k + W_\mu^- V_{\alpha k}^* \overline{d_k} \gamma^\mu \left(1 - \gamma_5\right) u_\alpha \right].$$

(A.15)

The most general CP transformation is

$$\begin{aligned}
(\mathcal{CP}) \, W_\mu^+ \, (\mathcal{CP})^\dagger &= -e^{i\xi_W} W^{\mu-}, \\
(\mathcal{CP}) \, \overline{u_\alpha} \, (\mathcal{CP})^\dagger &= -e^{-i\xi_\alpha} u_\alpha^T C^{-1} \gamma^0, \\
(\mathcal{CP}) \, d_k \, (\mathcal{CP})^\dagger &= e^{i\xi_k} \gamma^0 C \overline{d_k}^T.
\end{aligned}$$

(A.16)

If there is to be CP invariance, the CP-transformation phases must be chosen such that

$$V_{\alpha k} = e^{i(-\xi_W + \xi_\alpha - \xi_k)} V_{\alpha k}^*.$$

(A.17)

It is clear that eqn (A.14) follows from eqn (A.17).

To conclude, we have shown in a particular example that one should not discard the free phases in the CP transformation of the quark fields; otherwise, one may run into contradictions. Specifically, if one sets $\xi_q = \xi_b = 0$, one finds that $q/p = \pm \exp\left[i\left(2\phi_M + \xi_{B_q}\right)\right]$, contradicting the quantum-mechanical rule that, when CP is conserved, the eigenstates of the Hamiltonian should coincide with the eigenstates of CP.

A.2   Cancellation of the CP-transformation phases in λ_f

Let us consider the decays of B_q^0 and $\overline{B_q^0}$ into a CP eigenstate f:

$$\mathcal{CP}|f\rangle = \eta_f |f\rangle,$$

(A.18)

with $\eta_f = \pm 1$. We assume that the decay amplitudes have only one weak phase ϕ_A, with an operator $\mathcal{O}'$ controlling the decay:

$$\begin{aligned}
A_f &= e^{i\phi_A} \langle f | \mathcal{O}' | B_q^0 \rangle, \\
\bar{A}_f &= e^{-i\phi_A} \langle f | \mathcal{O}'^\dagger | \overline{B_q^0} \rangle.
\end{aligned}$$

(A.19)

The CP-transformation rule for $\mathcal{O}'$ is

$$(\mathcal{CP}) \, \mathcal{O}'^\dagger \, (\mathcal{CP})^\dagger = e^{-i\theta_D} \mathcal{O}'.$$

(A.20)

Then,

$$\begin{aligned}
\bar{A}_f &= e^{-i\phi_A} \langle f| (\mathcal{CP})^\dagger (\mathcal{CP}) \, \mathcal{O}'^\dagger (\mathcal{CP})^\dagger (\mathcal{CP}) |\overline{B_q^0}\rangle \\
&= \eta_f e^{-i(\phi_A + \xi_{B_q} + \theta_D)} \langle f| \mathcal{O}' | B_q^0 \rangle
\end{aligned}$$

$$= \eta_f e^{-i(2\phi_A + \xi_{B_q} + \theta_D)} A_f. \qquad (A.21)$$

Combining eqns (A.8) and (A.21), we obtain

$$\lambda_f \equiv \frac{q}{p} \frac{\bar{A}_f}{A_f} = \pm \eta_f e^{2i(\phi_M - \phi_A)} e^{i(\theta_M - \theta_D)}. \qquad (A.22)$$

We now state the following: if the calculation has been correctly done, then the phases θ_M and θ_D, which arise in the CP transformation of the mixing and decay operators, are equal and cancel out. This cancellation is due to the mixing and decay operators involving the same quark fields. Thus,

$$\lambda_f = \pm \eta_f e^{2i(\phi_M - \phi_A)}. \qquad (A.23)$$

An explicit example of the cancellation of the CP-transformation phases occurs in the standard-model computation of the parameter λ for $B_d^0 \to \pi^+ \pi^-$, as shown in § 33.1.1.

There are two important points to note in connection with eqn (A.23):

- If we had set $\theta_M = \theta_D = 0$ from the very beginning we would have obtained the correct result for λ_f. This is what most authors do. The price to pay is, as pointed out above, an inconsistency between eqns (A.1) and (A.9).
- The $\pm$ sign in eqn (A.23) is important. That sign comes from q/p in eqn (A.8). Recalling that the sign of q/p is significant only when compared with the sign of either Δm or $\Delta \Gamma$, it is not surprising to find that λ_f always appears multiplied by an odd function of either Δm or $\Delta \Gamma$ in any experimental observable.[108]

---

[108] It is sometimes stated that the sign of $\mathrm{Im}\,\lambda_f$ can be predicted. The meaning of that statement should be clearly understood. What can be predicted is the sign of $\Delta m\,\mathrm{Im}\,\lambda_f$. Indeed, the interchange $P_H \leftrightarrow P_L$ makes Δm, $\Delta \Gamma$, q/p, and λ_f change sign. If one chooses, as we do, $\Delta m > 0$, then the sign of $\mathrm{Im}\,\lambda_f$ becomes well defined and can indeed be predicted, at least in some models.

APPENDIX B

EFFECTIVE HAMILTONIAN FOR $|\Delta S| = 2$ PROCESSES

B.1 The box diagram of the standard model

B.1.1 *Introduction*

If we interpret K^0 as $\bar{s}d$ and $\overline{K^0}$ as $s\bar{d}$, in the standard model there are two one-loop diagrams which accomplish the transition $K^0 \to \overline{K^0}$. They are box diagrams, which we have depicted in Figs. 17.2 and 17.3. The diagram of Fig. 17.3 is identical with that of Fig. 17.2 after the interchange $d_1 \leftrightarrow d_2$ has been made. The diagrams have two $W^\pm$ bosons and two up-type quarks, α and β, in the loop. The quarks α and β may be either u, c, or t. Any of the two gauge bosons $W^\pm$ may be substituted by the corresponding Goldstone bosons $\varphi^\pm$; we have not depicted the diagrams involving the $\varphi^\pm$, but we include them in the computation of the box diagrams.

In order to find the effective $|\Delta S| = 2$ interaction we integrate out the 'heavy' degrees of freedom: the $W^\pm$ bosons and the up-type quarks. When doing this we are not interested in the 'light' degrees of freedom, i.e., in the masses and momenta of the external particles in the diagram. We therefore use an approximation in which the external particles are massless ($m_s = m_d = 0$) and their four-momenta are zero. In consequence, all internal lines carry the same four-momentum k^μ, which we have to integrate over.

We compute the diagram of Fig. 17.2 in an arbitrary 't Hooft gauge in order to check its gauge-independence. We use the following shorthands for the denominators of the various propagators:

$$\begin{aligned}
D_\alpha &\equiv k^2 - m_\alpha^2, \\
D_\beta &\equiv k^2 - m_\beta^2, \\
D_W &\equiv k^2 - m_W^2, \\
D_g &\equiv k^2 - \xi_W m_W^2.
\end{aligned} \tag{B.1}$$

Gauge-independence means that the final result should be independent of ξ_W.

We use the shorthands in eqn (13.50) for the relevant combinations of CKM matrix elements; unitarity of the CKM matrix implies eqn (13.53). Indeed, the existence of an unitary CKM matrix was first suggested by Glashow, Iliopoulos, and Maiani (1970) in the context of a computation of the box diagram. For this reason, the use of eqn (13.53) is usually referred to as the GIM mechanism.

B.1.2 *Writing down the diagram*

Taking into account that each boson $W^\pm$ may be substituted by the corresponding Goldstone boson $\varphi^\pm$, we have

$$\text{Fig. 17.2} = \left(i \frac{g}{\sqrt{2}} \right)^4 i^4 \sum_{\alpha=u,c,t} \sum_{\beta=u,c,t} \lambda_\alpha \lambda_\beta \int \frac{d^4 k}{(2\pi)^4} \frac{1}{D_\alpha D_\beta}$$

$$\times \left\{ \left[\frac{-g_{\psi\chi}}{D_W} + \frac{k_\psi k_\chi}{m_W^2} \left(\frac{1}{D_W} - \frac{1}{D_g} \right) \right] \left[\frac{-g_{\eta\theta}}{D_W} + \frac{k_\eta k_\theta}{m_W^2} \left(\frac{1}{D_W} - \frac{1}{D_g} \right) \right] \right.$$

$$\times \left[\bar{s_1} \gamma^\psi \gamma_L \left(\slashed{k} + m_\alpha \right) \gamma^\eta \gamma_L d_1 \right] \left[\bar{s_2} \gamma^\theta \gamma_L \left(\slashed{k} + m_\beta \right) \gamma^\chi \gamma_L d_2 \right]$$

$$+ \frac{1}{D_g} \left[\frac{-g_{\eta\theta}}{D_W} + \frac{k_\eta k_\theta}{m_W^2} \left(\frac{1}{D_W} - \frac{1}{D_g} \right) \right]$$

$$\times \left[\bar{s_1} \frac{m_\alpha}{m_W} \gamma_R \left(\slashed{k} + m_\alpha \right) \gamma^\eta \gamma_L d_1 \right] \left[\bar{s_2} \gamma^\theta \gamma_L \left(\slashed{k} + m_\beta \right) \frac{m_\beta}{m_W} \gamma_L d_2 \right]$$

$$+ \left[\frac{-g_{\psi\chi}}{D_W} + \frac{k_\psi k_\chi}{m_W^2} \left(\frac{1}{D_W} - \frac{1}{D_g} \right) \right] \frac{1}{D_g}$$

$$\times \left[\bar{s_1} \gamma^\psi \gamma_L \left(\slashed{k} + m_\alpha \right) \frac{m_\alpha}{m_W} \gamma_L d_1 \right] \left[\bar{s_2} \frac{m_\beta}{m_W} \gamma_R \left(\slashed{k} + m_\beta \right) \gamma^\chi \gamma_L d_2 \right]$$

$$+ \frac{1}{D_g^2} \left[\bar{s_1} \frac{m_\alpha}{m_W} \gamma_R \left(\slashed{k} + m_\alpha \right) \frac{m_\alpha}{m_W} \gamma_L d_1 \right]$$

$$\times \left. \left[\bar{s_2} \frac{m_\beta}{m_W} \gamma_R \left(\slashed{k} + m_\beta \right) \frac{m_\beta}{m_W} \gamma_L d_2 \right] \right\}.$$

$$= \frac{g^4}{4} \sum_{\alpha=u,c,t} \sum_{\beta=u,c,t} \lambda_\alpha \lambda_\beta \int \frac{d^4 k}{(2\pi)^4} \frac{1}{D_\alpha D_\beta}$$

$$\times \left\{ \left[\frac{k^4 - m_\alpha^2 m_\beta^2}{m_W^4} \left(\frac{1}{D_g^2} - \frac{2}{D_W D_g} \right) + \frac{k^4}{m_W^4 D_W^2} \right] \right.$$

$$\times \left(\bar{s_1} \slashed{k} \gamma_L d_1 \right) \left(\bar{s_2} \slashed{k} \gamma_L d_2 \right) + \frac{1}{D_W^2} \left(\bar{s_1} \gamma^\psi \slashed{k} \gamma^\eta \gamma_L d_1 \right) \left(\bar{s_2} \gamma_\eta \slashed{k} \gamma_\psi \gamma_L d_2 \right)$$

$$\left. + 2 \left(\frac{k^4 - m_\alpha^2 m_\beta^2}{m_W^2 D_W D_g} - \frac{k^4}{m_W^2 D_W^2} \right) \left(\bar{s_1} \gamma^\mu \gamma_L d_1 \right) \left(\bar{s_2} \gamma_\mu \gamma_L d_2 \right) \right\}. \tag{B.2}$$

B.1.3   *Gauge independence*

We notice that

$$k^4 - m_\alpha^2 m_\beta^2 = -D_\alpha D_\beta + k^2 \left(D_\alpha + D_\beta \right). \tag{B.3}$$

As a result, the terms proportional to $k^4 - m_\alpha^2 m_\beta^2$ in eqn (B.2) yield momentum integrals whose integrand is independent either of m_α or of m_β. Equation (13.53) implies that those contributions vanish upon summation either over α or over β, respectively. We thus find that all terms with denominator D_g yield vanishing contributions, and the box diagram is gauge-independent:

$$\text{Fig. 17.2} = \frac{g^4}{4} \sum_{\alpha=u,c,t} \sum_{\beta=u,c,t} \lambda_\alpha \lambda_\beta \int \frac{d^4 k}{(2\pi)^4} \frac{1}{D_\alpha D_\beta D_W^2}$$

$$\times \left[\frac{m_\alpha^2 m_\beta^2}{m_W^4} \left(\overline{s_1} \not{k} \gamma_L d_1 \right) \left(\overline{s_2} \not{k} \gamma_L d_2 \right) \right.$$

$$+ \left(\overline{s_1} \gamma^\psi \not{k} \gamma^\eta \gamma_L d_1 \right) \left(\overline{s_2} \gamma_\eta \not{k} \gamma_\psi \gamma_L d_2 \right)$$

$$\left. - \frac{2 m_\alpha^2 m_\beta^2}{m_W^2} \left(\overline{s_1} \gamma^\mu \gamma_L d_1 \right) \left(\overline{s_2} \gamma_\mu \gamma_L d_2 \right) \right]. \tag{B.4}$$

It is now evident that the momentum integral is finite.

B.1.4   Effective operator

As the denominator of the integrand depends on k only through k^2, the integral of $k^\varsigma k^\eta$ is equivalent to $g^{\varsigma\eta}/4$ times the integral of k^2. We also use the Dirac-matrix identity

$$\gamma^\varsigma \gamma^\eta \gamma^\theta = g^{\varsigma\eta}\gamma^\theta - g^{\varsigma\theta}\gamma^\eta + g^{\eta\theta}\gamma^\varsigma + i\epsilon^{\varsigma\eta\theta\psi}\gamma_\psi \gamma_5 \tag{B.5}$$

to derive

$$\left(\gamma^\varsigma \gamma^\eta \gamma^\theta \gamma_L \right) \otimes \left(\gamma_\theta \gamma_\eta \gamma_\varsigma \gamma_L \right) = 4\Gamma^\mu \otimes \Gamma_\mu, \tag{B.6}$$

where we have used the notation in eqn (17.1). We thus obtain

$$\text{Fig. } 17.2 = \frac{g^4}{4} \left(\overline{s_1} \Gamma^\mu d_1 \right) \left(\overline{s_2} \Gamma_\mu d_2 \right) \sum_{\alpha=u,c,t} \sum_{\beta=u,c,t} \lambda_\alpha \lambda_\beta \int \frac{d^4 k}{(2\pi)^4} \frac{1}{D_\alpha D_\beta D_W^2}$$

$$\times \left(\frac{k^2 m_\alpha^2 m_\beta^2}{4 m_W^4} + k^2 - \frac{2 m_\alpha^2 m_\beta^2}{m_W^2} \right). \tag{B.7}$$

B.1.5   Integration

We introduce Feynman parameters and perform the momentum integral, obtaining

$$\text{Fig. } 17.2 = \frac{-ig^4}{64\pi^2} \left(\overline{s_1} \Gamma^\mu d_1 \right) \left(\overline{s_2} \Gamma_\mu d_2 \right) \sum_{\alpha=u,c,t} \sum_{\beta=u,c,t} \lambda_\alpha \lambda_\beta \int_0^1 dx \int_0^x dy$$

$$\times \left(\frac{m_\alpha^2 m_\beta^2}{2 m_W^4 M^2} + \frac{2}{M^2} + \frac{2 m_\alpha^2 m_\beta^2}{m_W^2 M^4} \right) y, \tag{B.8}$$

where

$$M^2 \equiv m_\alpha^2 + \left(m_\beta^2 - m_\alpha^2 \right) x + \left(m_W^2 - m_\beta^2 \right) y. \tag{B.9}$$

Defining

$$x_\alpha \equiv \frac{m_\alpha^2}{m_W^2}, \tag{B.10}$$

we have, after integrating over the Feynman parameters x and y,

$$\text{Fig. } 17.2 = \frac{ig^4}{64\pi^2 m_W^2} \left(\overline{s_1} \Gamma^\mu d_1 \right) \left(\overline{s_2} \Gamma_\mu d_2 \right) \sum_{\alpha=u,c,t} \sum_{\beta=u,c,t} \lambda_\alpha \lambda_\beta F \left(x_\alpha, x_\beta \right). \tag{B.11}$$

Here,

$$F(x_\alpha, x_\beta) = \frac{1}{(1 - x_\alpha)(1 - x_\beta)}\left(\frac{7x_\alpha x_\beta}{4} - 1\right)$$
$$+ \frac{x_\alpha^2 \ln x_\alpha}{(x_\beta - x_\alpha)(1 - x_\alpha)^2}\left(1 - 2x_\beta + \frac{x_\alpha x_\beta}{4}\right)$$
$$+ \frac{x_\beta^2 \ln x_\beta}{(x_\alpha - x_\beta)(1 - x_\beta)^2}\left(1 - 2x_\alpha + \frac{x_\alpha x_\beta}{4}\right) \tag{B.12}$$

for $\beta \neq \alpha$. For $\beta = \alpha$ one should use the limit when $m_\beta \to m_\alpha$ of the function in eqn (B.12).

B.1.6   GIM mechanism

We use eqn (13.53) to rewrite eqn (B.11) in such a way that the sums only run over the quarks c and t. We use the definition of the Fermi constant $G_F = g^2/(4\sqrt{2}m_W^2)$ to trade g^2 by G_F. Finally, the approximation $m_u = 0$ is convenient, in order to obtain expressions which do not depend on m_u. We get

$$\text{Fig. 17.2} = \frac{-iG_F^2 m_W^2}{2\pi^2}(\overline{s_1}\Gamma^\mu d_1)(\overline{s_2}\Gamma_\mu d_2)\mathcal{F}_0, \tag{B.13}$$

where

$$\mathcal{F}_0 = \lambda_c^2 S_0(x_c) + \lambda_t^2 S_0(x_t) + 2\lambda_c\lambda_t S_0(x_c, x_t). \tag{B.14}$$

Here,

$$S_0(x_c, x_t) = x_c x_t \left[-\frac{3}{4(1 - x_c)(1 - x_t)} \right.$$
$$+ \frac{\ln x_t}{(x_t - x_c)(1 - x_t)^2}\left(1 - 2x_t + \frac{x_t^2}{4}\right)$$
$$+ \left. \frac{\ln x_c}{(x_c - x_t)(1 - x_c)^2}\left(1 - 2x_c + \frac{x_c^2}{4}\right)\right], \tag{B.15}$$

and the function $S_0(x)$ is the limit when $y \to x$ of $S_0(x, y)$:

$$S_0(x) = \frac{x}{(1 - x)^2}\left[1 - \frac{11x}{4} + \frac{x^2}{4} - \frac{3x^2 \ln x}{2(1 - x)}\right]. \tag{B.16}$$

B.1.7   Effective Hamiltonian

We interpret the box diagram as resulting from an effective Hamiltonian. Following our conventions, the Lagrangian must be equal to the Feynman diagram divided by a factor i, just as the Feynman rule for each vertex is i times the

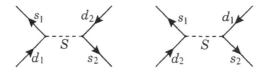

FIG. B.1. K^0-$\overline{K^0}$ mixing at tree level originating in a flavour-changing neutral Yukawa interaction.

corresponding term in the Lagrangian. Moreover, the interaction Hamiltonian is minus the interaction Lagrangian. Therefore,

$$\mathcal{H}_{\text{eff}} = \frac{G_F^2 m_W^2}{4\pi^2} \left(\bar{s}\Gamma^\mu d\right) \left(\bar{s}\Gamma_\mu d\right) \mathcal{F}_0 + \text{H.c.}. \tag{B.17}$$

We have divided $\mathcal{H}_{\text{eff}}$ by an extra 2, taking into account that $\mathcal{H}_{\text{eff}}$ is the product of two identical operators, and therefore the Feynman rule for the vertex is $-2i\mathcal{H}_{\text{eff}}$.

The effective Hamiltonian in eqn (B.17) yields the result

$$\text{Fig. } 17.3 = \frac{-iG_F^2 m_W^2}{2\pi^2} \left(\bar{s_1}\Gamma^\mu d_2\right) \left(\bar{s_2}\Gamma_\mu d_1\right) \mathcal{F}_0 \tag{B.18}$$

for the second box diagram effecting the $K^0 \to \overline{K^0}$ transition. This means that $\mathcal{H}_{\text{eff}}$, although derived from the computation of the diagram in Fig. 17.2 only, really gives rise to the two diagrams in Figs. 17.2 and 17.3, the values of those diagrams being given by eqns (B.13) and (B.18), respectively.

B.2 Scalars and K^0-$\overline{K^0}$ mixing at tree level

We now consider the possibility that, in some extension of the standard model, there is a physical scalar particle S which has flavour-changing neutral Yukawa interactions—see § 22.10—with the s and d quarks:

$$\mathcal{L} = \cdots + S\bar{s}\left(a + b\gamma_5\right)d + S^\dagger \bar{d}\left(a^* - b^*\gamma_5\right)s, \tag{B.19}$$

where a and b are dimensionless coupling constants. The interaction in eqn (B.19) leads to $K^0 \to \overline{K^0}$ transitions via the tree-level diagrams in Fig. B.1.

Just as in the computation of the box diagram, we assume all the external momenta to vanish. Then, the momentum of the propagator of S is zero. Denoting by m the mass of S, we obtain for the effective Hamiltonian which gives rise to the diagrams in Fig. B.1 the result

$$\mathcal{H}_{\text{eff}} = -\frac{1}{2m^2} \left[\bar{s}\left(a + b\gamma_5\right)d\right]^2 + \text{H.c.}. \tag{B.20}$$

Because of parity symmetry the matrix element of the operator $\left(\bar{s}d\right)\left(\bar{s}\gamma_5 d\right)$ between $\langle\overline{K^0}|$ and $|K^0\rangle$ is zero. The other matrix elements may be estimated

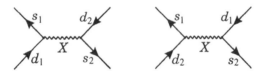

FIG. B.2. K^0–$\overline{K^0}$ mixing induced by a flavour-changing neutral current.

using the vacuum-insertion approximation in Appendix C. One obtains the following result for the contribution of the interaction in eqn (B.19) to M_{21}:

$$M_{21} = e^{i(\xi_K + \xi_d - \xi_s)} \frac{f_K^2 m_K}{24m^2} \left\{ b^2 \left[-1 - \frac{11m_K^2}{(m_s + m_d)^2} \right] + a^2 \left[1 + \frac{m_K^2}{(m_s + m_d)^2} \right] \right\}. \tag{B.21}$$

B.3   Vector bosons and K^0–$\overline{K^0}$ mixing at tree level

Suppose there is a vector boson X^μ which couples to a flavour-changing neutral current between the s and d quarks:

$$\mathcal{L} = \cdots + X^\mu \left[\bar{s}\gamma_\mu \left(a + b\gamma_5 \right) d + \bar{d}\gamma_\mu \left(a^* + b^*\gamma_5 \right) s \right], \tag{B.22}$$

where a and b are dimensionless coupling constants. This interaction leads to $K^0 \to \overline{K^0}$ transitions via the tree-level diagrams depicted in Fig. B.2. The corresponding effective Hamiltonian is

$$\mathcal{H}_{\text{eff}} = \frac{1}{2m^2} \left[\bar{s}\gamma^\mu \left(a + b\gamma_5 \right) d \right] \left[\bar{s}\gamma_\mu \left(a + b\gamma_5 \right) d \right] + \text{H.c.}, \tag{B.23}$$

where m now is the mass of the vector boson X. The contribution to M_{21}, with the relevant matrix elements computed in the vacuum-insertion approximation, is

$$M_{21} = e^{i(\xi_K + \xi_d - \xi_s)} \frac{f_K^2 m_K}{12m^2} \left\{ b^2 \left[-7 - \frac{2m_K^2}{(m_s + m_d)^2} \right] - a^2 \left[1 - \frac{2m_K^2}{(m_s + m_d)^2} \right] \right\}. \tag{B.24}$$

APPENDIX C

THE VACUUM-INSERTION APPROXIMATION IN K^0–$\overline{K^0}$ MIXING

C.1 Introduction

When we consider processes involving hadrons we often face the difficult task of evaluating matrix elements of operators between hadronic states. The difficulty is associated with the fact that the operators are written in terms of quarks while the physical asymptotic states are hadrons. Unfortunately, there is no reliable, first-principles calculation of the hadronization mechanism and, therefore, of the matrix elements of operators. For some leptonic or semileptonic decays, approximate symmetries allow trustworthy matrix-element computations to be performed. This is the case of chiral symmetry for hadrons with light quarks, in flavour-SU(3) multiplets; and heavy-quark symmetry for hadrons with a b or a c quark—the corrections being significant in the latter case. This is not possible however in the case of matrix elements for nonleptonic decays. Those matrix elements are usually evaluated using specific phenomenological models such as the BSW model (Bauer *et al.* 1985). Ultimately, they should be computed using lattice methods.

However, we may get an estimate of some matrix elements of operators quartic in the quark fields by using the vacuum-insertion approximation (VIA). The idea of the VIA is to separate the matrix element of a quartic operator into the product of two matrix elements of operators bilinear in the quark fields. The separation is not performed through a complete set of intermediate states—a partition of unity; rather, only the vacuum state is inserted. The assumption of the VIA is that a reasonable estimate, correct at least in order of magnitude, may be achieved by using only the vacuum as intermediate state. Of course, the corrections to the VIA must be estimated by using some other method for computing the matrix element.

Here we shall illustrate the VIA by evaluating a number of matrix elements relevant for neutral-kaon mixing (McWilliams and Shanker 1980). The matrix elements that we shall compute are $\langle \overline{K^0}| (\bar{s}d)^2 |K^0\rangle$, $\langle \overline{K^0}| (\bar{s}\gamma_5 d)^2 |K^0\rangle$, $\langle \overline{K^0}| (\bar{s}\gamma^\mu d)(\bar{s}\gamma_\mu d) |K^0\rangle$, and $\langle \overline{K^0}| (\bar{s}\gamma^\mu\gamma_5 d)(\bar{s}\gamma_\mu\gamma_5 d) |K^0\rangle$. We shall also compute $\langle \overline{K^0}| (\bar{s}\Gamma^\mu d)(\bar{s}\Gamma_\mu d) |K^0\rangle$, where $\Gamma^\mu \equiv \gamma^\mu (1 - \gamma_5)/2$. The operators in these matrix elements are always the product of two colour-singlet quark bilinears, with definite transformation properties under the Lorentz group.

C.2   Creation and destruction operators

The quark fields are linear combinations of destruction operators and creation operators. Those operators act on the initial and final hadron states in all possible ways. Destruction operators act to the right and creation operators act to the left. We must therefore reorder the operators, bringing the creation operators to the left and the destruction operators to the right. When we do this, there is a minus sign for each transposition, because fermionic operators anticommute.[109]

Without loss of generality, we consider d to be the sum of a single operator d^- which destroys d quarks and a single operator d^+ which creates $\bar{d}$ antiquarks. Similarly, $\bar{s}$ is the sum of an operator $\bar{s}^-$ which destroys $\bar{s}$ antiquarks and an operator $\bar{s}^+$ which creates s quarks. In the initial state K^0 one has a d quark and an $\bar{s}$ antiquark. In the final state $\overline{K^0}$ one has a $\bar{d}$ antiquark and an s quark. In the operator which effects the transition one must therefore have one destruction and one creation operator for each flavour. Thus,

$$\langle \overline{K^0} | (\bar{s}d)(\bar{s}d) | K^0 \rangle = \langle \overline{K^0} | (\bar{s}^+d^+)(\bar{s}^-d^-) + (\bar{s}^+d^-)(\bar{s}^-d^+)$$
$$+ (\bar{s}^-d^+)(\bar{s}^+d^-) + (\bar{s}^-d^-)(\bar{s}^+d^+) | K^0 \rangle$$
$$= 2\langle \overline{K^0} | (\bar{s}^+d^+)(\bar{s}^-d^-) + (\bar{s}^+d^-)(\bar{s}^-d^+) | K^0 \rangle. \quad \text{(C.1)}$$

In the same way,

$$\langle \overline{K^0} | (\bar{s}\gamma_5 d)(\bar{s}\gamma_5 d) | K^0 \rangle = 2\langle \overline{K^0} | (\bar{s}^+\gamma_5 d^+)(\bar{s}^-\gamma_5 d^-) | K^0 \rangle$$
$$+ 2\langle \overline{K^0} | (\bar{s}^+\gamma_5 d^-)(\bar{s}^-\gamma_5 d^+) | K^0 \rangle, \quad \text{(C.2)}$$

$$\langle \overline{K^0} | (\bar{s}\gamma^\mu d)(\bar{s}\gamma_\mu d) | K^0 \rangle = 2\langle \overline{K^0} | (\bar{s}^+\gamma^\mu d^+)(\bar{s}^-\gamma_\mu d^-) | K^0 \rangle$$
$$+ 2\langle \overline{K^0} | (\bar{s}^+\gamma^\mu d^-)(\bar{s}^-\gamma_\mu d^+) | K^0 \rangle, \quad \text{(C.3)}$$

$$\langle \overline{K^0} | (\bar{s}\gamma^\mu\gamma_5 d)(\bar{s}\gamma_\mu\gamma_5 d) | K^0 \rangle = 2\langle \overline{K^0} | (\bar{s}^+\gamma^\mu\gamma_5 d^+)(\bar{s}^-\gamma_\mu\gamma_5 d^-) | K^0 \rangle$$
$$+ 2\langle \overline{K^0} | (\bar{s}^+\gamma^\mu\gamma_5 d^-)(\bar{s}^-\gamma_\mu\gamma_5 d^+) | K^0 \rangle, \quad \text{(C.4)}$$

$$\langle \overline{K^0} | (\bar{s}\Gamma^\mu d)(\bar{s}\Gamma_\mu d) | K^0 \rangle = 2\langle \overline{K^0} | (\bar{s}^+\Gamma^\mu d^+)(\bar{s}^-\Gamma_\mu d^-) | K^0 \rangle$$
$$+ 2\langle \overline{K^0} | (\bar{s}^+\Gamma^\mu d^-)(\bar{s}^-\Gamma_\mu d^+) | K^0 \rangle. \quad \text{(C.5)}$$

C.3   Fierz transformations

C.3.1   *Dirac matrices*

The Fierz reshuffling theorem for the Dirac matrices states that

$$\left(\Gamma^i\right)_{\alpha\beta} \left(\Gamma_i\right)_{\gamma\xi} = \sum_j F_{ij} \left(\Gamma^j\right)_{\alpha\xi} \left(\Gamma_j\right)_{\gamma\beta}, \quad \text{(C.6)}$$

where the covariant Dirac matrices Γ^i are 1, γ^μ, $\sigma^{\mu\nu}$, $\gamma^\mu\gamma_5$, and γ_5, respectively, and the matrix of indices F is

---

[109]This is so even when the fermionic operators have different quantum numbers, like creation and destruction operators for the s and d quarks and their respective antiquarks.

$$F = \tfrac{1}{8} \begin{pmatrix} 2 & 2 & 1 & -2 & 2 \\ 8 & -4 & 0 & -4 & -8 \\ 24 & 0 & -4 & 0 & 24 \\ -8 & -4 & 0 & -4 & 8 \\ 2 & -2 & 1 & 2 & 2 \end{pmatrix}. \tag{C.7}$$

From the basic Fierz identity

$$\delta_{\alpha\beta}\delta_{\gamma\xi} = \tfrac{1}{4}\delta_{\alpha\xi}\delta_{\gamma\beta} + \tfrac{1}{4}(\gamma_5)_{\alpha\xi}(\gamma_5)_{\gamma\beta} + \tfrac{1}{8}(\sigma^{\mu\nu})_{\alpha\xi}(\sigma_{\mu\nu})_{\gamma\beta}$$
$$+ \tfrac{1}{4}(\gamma^\mu)_{\alpha\xi}(\gamma_\mu)_{\gamma\beta} - \tfrac{1}{4}(\gamma^\mu\gamma_5)_{\alpha\xi}(\gamma_\mu\gamma_5)_{\gamma\beta} \tag{C.8}$$

one may derive various other Fierz identities for Dirac matrices, in particular

$$(\Gamma^\mu)_{\alpha\beta}(\Gamma_\mu)_{\gamma\xi} = -(\Gamma^\mu)_{\alpha\xi}(\Gamma_\mu)_{\gamma\beta}. \tag{C.9}$$

C.3.2   Gell-Mann matrices

The basic Fierz identity for the Gell-Mann matrices λ^a is

$$\delta_{wx}\delta_{yz} = \tfrac{1}{3}\delta_{wz}\delta_{yx} + \tfrac{1}{2}\sum_{a=1}^{8}\lambda^a_{wz}\lambda^a_{yx}. \tag{C.10}$$

C.3.3   Applications in the VIA

The Fierz transformations are needed in the context of the VIA because the second terms in the right-hand sides of eqns (C.1)–(C.5) contain creation operators paired with destruction operators in a manner which is inappropriate to insert the vacuum state. One must therefore change the order of the operators.

Our aim is to reorder a combination of the type $(\bar{s}^+\Gamma^i d^-)(\bar{s}^-\Gamma_i d^+)$. Let us display explicitly the colour indices w, x, y, and z, and the Dirac indices α, β, γ, and ξ. We have

$$\left(\bar{s}^+_{wa}d^-_{x\beta}\bar{s}^-_{y\gamma}d^+_{z\xi}\right)\left[(\Gamma^i)_{\alpha\beta}(\Gamma_i)_{\gamma\xi}\right](\delta_{wx}\delta_{yz})$$

$$= \left(-\bar{s}^+_{wa}d^+_{z\xi}\bar{s}^-_{y\gamma}d^-_{x\beta}\right)\left[\sum_j F_{ij}(\Gamma^j)_{\alpha\xi}(\Gamma_j)_{\gamma\beta}\right]\left(\tfrac{1}{3}\delta_{wz}\delta_{yx} + \tfrac{1}{2}\sum_{a=1}^{8}\lambda^a_{wz}\lambda^a_{yx}\right). \tag{C.11}$$

The colour structure of the $\lambda^a \otimes \lambda^a$ term in eqn (C.11) forbids the insertion of a colour-singlet state like the vacuum. One assumes that this term may be neglected, and one obtains

$$(\bar{s}^+\Gamma^i d^-)(\bar{s}^-\Gamma_i d^+) = -\tfrac{1}{3}\sum_j F_{ij}(\bar{s}^+\Gamma^j d^+)(\bar{s}^-\Gamma_j d^-). \tag{C.12}$$

Applying this formula to eqns (C.1)–(C.5), one gets

$$\langle\overline{K^0}|(\bar{s}d)(\bar{s}d)|K^0\rangle = \tfrac{11}{6}\langle\overline{K^0}|(\bar{s}^+d^+)(\bar{s}^-d^-)|K^0\rangle$$

$$-\tfrac{1}{6}\langle\overline{K^0}\,|\,(\bar{s}^+\gamma_5 d^+)\,(\bar{s}^-\gamma_5 d^-)\,|\,K^0\rangle$$

$$-\tfrac{1}{12}\langle\overline{K^0}\,|\,(\bar{s}^+\sigma^{\mu\nu} d^+)\,(\bar{s}^-\sigma_{\mu\nu} d^-)\,|\,K^0\rangle$$

$$-\tfrac{1}{6}\langle\overline{K^0}\,|\,(\bar{s}^+\gamma^{\mu} d^+)\,(\bar{s}^-\gamma_{\mu} d^-)\,|\,K^0\rangle$$

$$+\tfrac{1}{6}\langle\overline{K^0}\,|\,(\bar{s}^+\gamma^{\mu}\gamma_5 d^+)\,(\bar{s}^-\gamma_{\mu}\gamma_5 d^-)\,|\,K^0\rangle, \quad\text{(C.13)}$$

$$\langle\overline{K^0}\,|\,(\bar{s}\gamma_5 d)\,(\bar{s}\gamma_5 d)\,|\,K^0\rangle = -\tfrac{1}{6}\langle\overline{K^0}\,|\,(\bar{s}^+ d^+)\,(\bar{s}^- d^-)\,|\,K^0\rangle$$

$$+\tfrac{11}{6}\langle\overline{K^0}\,|\,(\bar{s}^+\gamma_5 d^+)\,(\bar{s}^-\gamma_5 d^-)\,|\,K^0\rangle$$

$$-\tfrac{1}{12}\langle\overline{K^0}\,|\,(\bar{s}^+\sigma^{\mu\nu} d^+)\,(\bar{s}^-\sigma_{\mu\nu} d^-)\,|\,K^0\rangle$$

$$+\tfrac{1}{6}\langle\overline{K^0}\,|\,(\bar{s}^+\gamma^{\mu} d^+)\,(\bar{s}^-\gamma_{\mu} d^-)\,|\,K^0\rangle$$

$$-\tfrac{1}{6}\langle\overline{K^0}\,|\,(\bar{s}^+\gamma^{\mu}\gamma_5 d^+)\,(\bar{s}^-\gamma_{\mu}\gamma_5 d^-)\,|\,K^0\rangle, \quad\text{(C.14)}$$

$$\langle\overline{K^0}\,|\,(\bar{s}\gamma^{\mu} d)\,(\bar{s}\gamma_{\mu} d)\,|\,K^0\rangle = -\tfrac{2}{3}\langle\overline{K^0}\,|\,(\bar{s}^+ d^+)\,(\bar{s}^- d^-)\,|\,K^0\rangle$$

$$+\tfrac{2}{3}\langle\overline{K^0}\,|\,(\bar{s}^+\gamma_5 d^+)\,(\bar{s}^-\gamma_5 d^-)\,|\,K^0\rangle$$

$$+\tfrac{7}{3}\langle\overline{K^0}\,|\,(\bar{s}^+\gamma^{\mu} d^+)\,(\bar{s}^-\gamma_{\mu} d^-)\,|\,K^0\rangle$$

$$+\tfrac{1}{3}\langle\overline{K^0}\,|\,(\bar{s}^+\gamma^{\mu}\gamma_5 d^+)\,(\bar{s}^-\gamma_{\mu}\gamma_5 d^-)\,|\,K^0\rangle, \quad\text{(C.15)}$$

$$\langle\overline{K^0}\,|\,(\bar{s}\gamma^{\mu}\gamma_5 d)\,(\bar{s}\gamma_{\mu}\gamma_5 d)\,|\,K^0\rangle = \tfrac{2}{3}\langle\overline{K^0}\,|\,(\bar{s}^+ d^+)\,(\bar{s}^- d^-)\,|\,K^0\rangle$$

$$-\tfrac{2}{3}\langle\overline{K^0}\,|\,(\bar{s}^+\gamma_5 d^+)\,(\bar{s}^-\gamma_5 d^-)\,|\,K^0\rangle$$

$$+\tfrac{1}{3}\langle\overline{K^0}\,|\,(\bar{s}^+\gamma^{\mu} d^+)\,(\bar{s}^-\gamma_{\mu} d^-)\,|\,K^0\rangle$$

$$+\tfrac{7}{3}\langle\overline{K^0}\,|\,(\bar{s}^+\gamma^{\mu}\gamma_5 d^+)\,(\bar{s}^-\gamma_{\mu}\gamma_5 d^-)\,|\,K^0\rangle, \quad\text{(C.16)}$$

$$\langle\overline{K^0}\,|\,(\bar{s}\Gamma^{\mu} d)\,(\bar{s}\Gamma_{\mu} d)\,|\,K^0\rangle = \tfrac{8}{3}\langle\overline{K^0}\,|\,(\bar{s}^+\Gamma^{\mu} d^+)\,(\bar{s}^-\Gamma_{\mu} d^-)\,|\,K^0\rangle. \quad\text{(C.17)}$$

C.4 Vacuum insertion

We are now ready to insert the vacuum state, assuming for instance that

$$\langle\overline{K^0}\,|\,(\bar{s}^+ d^+)\,(\bar{s}^- d^-)\,|\,K^0\rangle = \langle\overline{K^0}\,|\,(\bar{s}^+ d^+)\,|\,0\rangle\langle 0\,|\,(\bar{s}^- d^-)\,|\,K^0\rangle$$

$$= \langle\overline{K^0}\,|\,(\bar{s} d)\,|\,0\rangle\langle 0\,|\,(\bar{s} d)\,|\,K^0\rangle. \quad\text{(C.18)}$$

This is the crucial approximation, together with that of neglecting the $\lambda^a \otimes \lambda^a$ term in the right-hand side of eqn (C.11).

At this juncture one may use a few simplifications. Firstly, the kaon state vectors carry four-momentum $p^{\mu} = (E, \vec{p})$. As it is impossible to construct an antisymmetric tensor with two indices out of $g^{\mu\nu}$, $\epsilon^{\mu\nu\rho\sigma}$, and p^{μ} alone, one concludes that

$$\langle 0|\bar{s}\sigma^{\mu\nu} d|K^0(E,\vec{p})\rangle = \langle\overline{K^0}(E,\vec{p})\,|\bar{s}\sigma^{\mu\nu} d|0\rangle = 0. \quad\text{(C.19)}$$

Secondly, we have seen in § 4.6 that, in the convention in which the s and d quarks transform under parity[110] with the same phase β_p,

---

[110] We are allowed to use parity symmetry—and later we shall be using CP symmetry and isospin symmetry too—because we are computing strong-interaction matrix elements. Weak

$$\mathcal{P}d\,(t,\vec{r})\,\mathcal{P}^\dagger = e^{i\beta_P}\gamma_0 d\,(t,-\vec{r})\,,$$
$$\mathcal{P}\bar{s}\,(t,\vec{r})\,\mathcal{P}^\dagger = e^{-i\beta_P}\bar{s}\,(t,-\vec{r})\,\gamma_0,$$

(C.20)

the neutral kaons have negative parity:

$$\mathcal{P}|K^0\rangle = -|K^0\rangle,$$
$$\langle\overline{K^0}|\mathcal{P}^\dagger = -\langle\overline{K^0}|.$$

(C.21)

The four-momentum p^μ transforms to p_μ under parity. Because of eqns (C.20), the operator $\bar{s}\gamma^\mu d$ transforms to $\bar{s}\gamma_\mu d$. Therefore, taking into account eqns (C.21),

$$\langle 0|\bar{s}\gamma^\mu d|K^0(E,\vec{p})\rangle = \langle\overline{K^0}(E,\vec{p})\,|\bar{s}\gamma^\mu d|0\rangle = 0.$$

(C.22)

In the same way,

$$\langle 0|\bar{s}d|K^0(E,\vec{p})\rangle = \langle\overline{K^0}(E,\vec{p})\,|\bar{s}d|0\rangle = 0.$$

(C.23)

On the other hand, $\bar{s}\gamma^\mu\gamma_5 d \to -\bar{s}\gamma_\mu\gamma_5 d$ under parity, and therefore

$$\langle 0|\bar{s}\gamma^\mu\gamma_5 d|K^0(E,\vec{p})\rangle = -e^{i\varphi}p^\mu f_K,$$

(C.24)

where f_K is a real positive parameter and $\exp(i\varphi)$ is an arbitrary phase, which is usually set to be either $+i$ or $-i$. Similarly, the matrix elements of $\bar{s}\gamma_5 d$ do not vanish.

Thus, if one defines (McWilliams and Shanker 1980)

$$V_1 \equiv \langle\overline{K^0}|\bar{s}\gamma_5 d|0\rangle\langle 0|\bar{s}\gamma_5 d|K^0\rangle,$$
$$V_2 \equiv \langle\overline{K^0}|\bar{s}\gamma^\mu\gamma_5 d|0\rangle\langle 0|\bar{s}\gamma_\mu\gamma_5 d|K^0\rangle,$$

(C.25)

one directly obtains from eqns (C.13)–(C.17) the following results:

$$\langle\overline{K^0}\,|(\bar{s}d)\,(\bar{s}d)|\,K^0\rangle = -\tfrac{1}{6}V_1 + \tfrac{1}{6}V_2,$$

(C.26)

$$\langle\overline{K^0}\,|(\bar{s}\gamma_5 d)\,(\bar{s}\gamma_5 d)|\,K^0\rangle = \tfrac{11}{6}V_1 - \tfrac{1}{6}V_2,$$

(C.27)

$$\langle\overline{K^0}\,|(\bar{s}\gamma^\mu d)\,(\bar{s}\gamma_\mu d)|\,K^0\rangle = \tfrac{2}{3}V_1 + \tfrac{1}{3}V_2,$$

(C.28)

$$\langle\overline{K^0}\,|(\bar{s}\gamma^\mu\gamma_5 d)\,(\bar{s}\gamma_\mu\gamma_5 d)|\,K^0\rangle = -\tfrac{2}{3}V_1 + \tfrac{7}{3}V_2,$$

(C.29)

$$\langle\overline{K^0}|\,(\bar{s}\Gamma^\mu d)\,(\bar{s}\Gamma_\mu d)\,|K^0\rangle = \tfrac{2}{3}V_2.$$

(C.30)

C.5 Use of symmetries

Let us now use the CP symmetry of the strong interaction to relate the matrix element in eqn (C.24) to $\langle\overline{K^0}(E,\vec{p})\,|\bar{s}\gamma^\mu\gamma_5 d|0\rangle$. CP acts in the following way:

interactions, which are not invariant under those symmetries, have been taken care of in the process of computing Feynman diagrams and thereby deriving an effective Hamiltonian, as was done for instance in Appendix B.

$$\langle 0| \,(CP)^\dagger = \langle 0|,$$
$$CP|K^0(E,\vec{p})\rangle = e^{i\xi_K}|\overline{K^0}(E,-\vec{p})\rangle,$$
$$(CP)\,\bar{s}\,(CP)^\dagger = -e^{-i\xi_s}s^T C^{-1}\gamma^0, \tag{C.31}$$
$$(CP)\,d\,(CP)^\dagger = e^{i\xi_d}\gamma^0 C\bar{d}^T.$$

Therefore,

$$-e^{i\varphi}p^\mu f_K = \langle 0|\bar{s}\gamma^\mu\gamma_5 d|K^0(E,\vec{p})\rangle$$
$$= -e^{i(\xi_K+\xi_d-\xi_s)}\langle 0|\bar{d}\gamma_\mu\gamma_5 s|\overline{K^0}(E,-\vec{p})\rangle. \tag{C.32}$$

Complex-conjugating eqn (C.32) and changing $\vec{p}$ into $-\vec{p}$ leads to

$$e^{i(\xi_K+\xi_d-\xi_s)}e^{-i\varphi}p_\mu f_K = \langle\overline{K^0}(E,\vec{p})\,|\bar{s}\gamma_\mu\gamma_5 d|0\rangle. \tag{C.33}$$

Therefore,

$$V_2 = -e^{i(\xi_K+\xi_d-\xi_s)}\frac{f_K^2 m_K^2}{2m_K}. \tag{C.34}$$

In this equation we have inserted a denominator $2m_K$ to correct for the relativistic normalization of states, in which there are $2E$ particles per unit of volume. (A similar factor shall be used in the next subsection when we extract the value of f_K from experiment.) As we consider the kaons to be in their rest frame we use $E = m_K$.

Contracting eqn (C.33) with p^μ and using the equations of motion for the s and d quarks, one obtains

$$e^{i(\xi_K+\xi_d-\xi_s)}e^{-i\varphi}m_K^2 f_K = (m_s + m_d)\,\langle\overline{K^0}(E,\vec{p})\,|\bar{s}\gamma_5 d|0\rangle, \tag{C.35}$$

where m_s and m_d are the current masses of the strange quark and down quark, respectively. Similarly, from eqn (C.24)

$$(m_s + m_d)\,\langle 0|\bar{s}\gamma_5 d|K^0(E,\vec{p})\rangle = +e^{i\varphi}m_K^2 f_K. \tag{C.36}$$

Therefore,

$$V_1 = +e^{i(\xi_K+\xi_d-\xi_s)}\frac{f_K^2 m_K^4}{2m_K\,(m_s+m_d)^2}. \tag{C.37}$$

We are now able to write down the final results for the matrix elements in the VIA:

$$\langle\overline{K^0}\,|(\bar{s}d)\,(\bar{s}d)|\,K^0\rangle = e^{i(\xi_K+\xi_d-\xi_s)}\frac{f_K^2 m_K}{12}\left[-1-\frac{m_K^2}{(m_s+m_d)^2}\right],$$

$$\langle\overline{K^0}\,|(\bar{s}\gamma_5 d)\,(\bar{s}\gamma_5 d)|\,K^0\rangle = e^{i(\xi_K+\xi_d-\xi_s)}\frac{f_K^2 m_K}{12}\left[1+11\frac{m_K^2}{(m_s+m_d)^2}\right],$$

$$\langle\overline{K^0}\,|(\bar{s}\gamma^\mu d)\,(\bar{s}\gamma_\mu d)|\,K^0\rangle = e^{i(\xi_K+\xi_d-\xi_s)}\frac{f_K^2 m_K}{6}\left[-1+2\frac{m_K^2}{(m_s+m_d)^2}\right],$$

$$\langle \overline{K^0} |(\bar{s}\gamma^\mu\gamma_5 d)\,(\bar{s}\gamma_\mu\gamma_5 d)| K^0\rangle = e^{i(\xi_K+\xi_d-\xi_s)}\frac{f_K^2 m_K}{6}\left[-7-2\frac{m_K^2}{(m_s+m_d)^2}\right],$$

$$\langle \overline{K^0} |(\bar{s}\Gamma^\mu d)\,(\bar{s}\Gamma_\mu d)| K^0\rangle = -e^{i(\xi_K+\xi_d-\xi_s)}\frac{f_K^2 m_K}{3}. \tag{C.38}$$

Usually, a CP transformation in which $\xi_d = \xi_s = 0$ is assumed, while ξ_K is chosen to be either 0 or π. We display all the phases explicitly so that overall rephasing-invariance becomes evident.

C.6 f_K

Sometimes one finds in the literature the quantity $\tilde{f}_K = f_K/\sqrt{2}$ being denoted by f_K. In order to avoid confusion, we explicitly give the value of f_K for the notation we use. From

$$\langle 0|\bar{s}\gamma^\mu\gamma_L d|K^0(E,\vec{p})\rangle = e^{i\varphi}p^\mu\frac{f_K}{2} \tag{C.39}$$

one derives, by means of isospin symmetry,

$$\langle 0|\bar{s}\gamma^\mu\gamma_L u|K^+(E,\vec{p})\rangle = e^{i\varphi}p^\mu\frac{f_K}{2}. \tag{C.40}$$

We normalize the ket $|K^+\rangle$ according to the conventional relativistic normalization of states $\langle K^+(E,\vec{p})|K^+(E,\vec{p})\rangle = 2EV$, which means that there are $2E$ particles per unit of volume. Using eqn (C.40) one gets

$$\Gamma\left(K^+\to\mu^+\nu\right)+\Gamma\left(K^+\to\mu^+\nu\gamma\right)=\frac{G_F^2|V_{us}|^2}{8\pi}f_K^2 m_\mu^2 m_K\left(1-\frac{m_\mu^2}{m_K^2}\right)^2$$
$$\times\left[1+O\left(\alpha\right)\right]. \tag{C.41}$$

The measured decay widths then yield (Particle Data Group 1996, p. 319) $f_K = 159.8 \pm 1.4 \pm 0.44$ MeV, where the two error bars arise, respectively, from $|V_{us}|$ and from the $O(\alpha)$ corrections. Thus, $f_K \approx 160$ MeV.

APPENDIX D

THE EFFECTIVE HAMILTONIAN FOR $K_L \to \pi^0 \nu \bar{\nu}$

D.1 Introduction

In this appendix we compute the effective $|\Delta S| = 1$ Hamiltonian responsible for the decay $K_L \to \pi^0 \nu \bar{\nu}$ in the standard model. At quark level the relevant process is $\bar{s}d \to \nu\bar{\nu}$. The one-loop diagrams for this transition are those in Figs. D.1 and D.2. We did not draw the diagrams in which the gauge bosons $W^{\pm}$ are substituted by Goldstone bosons $\varphi^{\pm}$. The quark α may be either the up, the charm, or the top quark.

Just as in the computation of the $|\Delta S| = 2$ process $\bar{s}d \to s\bar{d}$ in Appendix B, we set the four-momenta and the masses of the external fermions to zero. (The momenta p_1 and p_2 are needed in an intermediate step of the computation of Fig. D.2, but we let them tend to zero at the end—see § D.3.) We also set the mass of the charged lepton in the box diagram of Fig. D.1 to zero. This is not a very good approximation in the case of the decay $K_L \to \pi^0 \nu_\tau \bar{\nu}_\tau$, because the mass of the τ lepton, $m_\tau = 1.8\,\text{GeV}$, is relatively large—see § 19.5. In any case, the corrections to this approximation may be easily computed if needed.

We evaluate the diagrams in an arbitrary 't Hooft gauge. We use the notation in eqns (B.1) for the denominators of the propagators. We also use the notation in eqn (13.50) for the relevant combinations of CKM matrix elements. It is important to keep in mind eqn (13.53), a consequence of the unitarity of the CKM matrix. Finally, $x_\alpha \equiv m_\alpha^2 / m_W^2$.

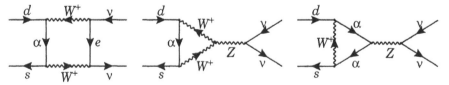

FIG. D.1. Standard-model diagrams for the process $\bar{s}d \to \nu\bar{\nu}$.

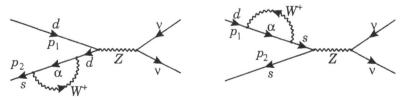

FIG. D.2. Diagrams for $\bar{s}d \to \nu\bar{\nu}$ which include flavour-changing self-energies.

Some individual diagrams are divergent. We compute them using dimensional regularization. The dimension of space–time is $d = 4 - \epsilon$. In the regularization procedure an arbitrary mass scale μ is also needed.

D.2   The diagrams in Fig. D.1

In the box diagram there cannot be internal Goldstone bosons. This is because the Yukawa couplings of $\varphi^\pm$ are proportional to the masses of the fermions, and the lepton masses are assumed to vanish in the present computation.[111] In the other two diagrams of Fig. D.1 $W^\pm$ may be substituted by $\varphi^\pm$. We use as before the shorthand $\Gamma^\mu \equiv \gamma^\mu \gamma_L$. We denote by Q the electric charge of the up-type quarks; Q enters in the evaluation of the third diagram in Fig. D.1. As we shall see, all Q-dependent terms cancel out in the final result, so that we do not need to specify that $Q = 2/3$. We use eqn (13.53) to eliminate terms which yield a vanishing contribution upon summation over α; most gauge-dependent terms disappear in this way. We obtain that Fig. D.1 is the sum of a gauge-independent part,

$$
A_{\text{gauge-independent}} = \sum_\alpha \frac{g^4 \lambda_\alpha}{4 m_W^2} \, (\bar{\nu} \Gamma_\mu \nu) \, \mu^\epsilon \int \frac{d^d k}{(2\pi)^d}
$$

$$
\times \left\{ -\frac{Q s_w^2}{D_\alpha^2 D_W} \left[2 m_\alpha^2 \left(\bar{s} \Gamma^\mu d \right) + x_\alpha \left(\bar{s} \slashed{k} \gamma^\mu \slashed{k} \gamma_L d \right) \right] \right.
$$

$$
+ \frac{\frac{1}{2} - Q s_w^2}{D_\alpha^2 D_W} \left[(2 - \epsilon) \left(\bar{s} \slashed{k} \gamma^\mu \slashed{k} \gamma_L d \right) + x_\alpha \left(2 k^2 - m_\alpha^2 \right) \left(\bar{s} \Gamma^\mu d \right) \right]
$$

$$
+ \frac{1}{D_\alpha D_W^2} \left[-4 m_W^2 + 2 m_\alpha^2 + 2 c_w^2 m_\alpha^2 \left(1 - \frac{k^2}{m_W^2} \right) \right] \left(\bar{s} \Gamma^\mu d \right)
$$

$$
+ \frac{1}{D_\alpha D_W^2} \left[-x_\alpha + 2 c_w^2 \left(2 - \epsilon + x_\alpha \right) \right] k^\mu \left(\bar{s} \slashed{k} \gamma_L d \right) \right\}
$$

$$
= (\bar{\nu} \Gamma_\mu \nu)(\bar{s} \Gamma^\mu d) \sum_\alpha \frac{i g^4 \lambda_\alpha x_\alpha}{64 \pi^2 m_W^2}
$$

$$
\times \left\{ \left(\frac{2}{\epsilon} - \gamma - \ln \frac{m_W^2}{4 \pi \mu^2} \right) \left(-\tfrac{3}{4} + \tfrac{3}{2} s_w^2 - \tfrac{3}{2} Q s_w^2 \right) \right.
$$

$$
+ \frac{1}{x_\alpha - 1} \left[\tfrac{3}{8} - \tfrac{9}{8} x_\alpha - \tfrac{11}{4} s_w^2 + \tfrac{5}{4} s_w^2 x_\alpha + Q s_w^2 \left(\tfrac{11}{4} - \tfrac{5}{4} x_\alpha \right) \right]
$$

$$
+ \frac{\ln x_\alpha}{(x_\alpha - 1)^2} \left[3 + \tfrac{3}{4} x_\alpha^2 - 3 c_w^2 x_\alpha - \tfrac{3}{2} s_w^2 x_\alpha^2 \right.
$$

$$
\left. \left. - Q s_w^2 x_\alpha \left(3 - \tfrac{3}{2} x_\alpha \right) \right] \right\},
\tag{D.1}
$$

where γ is the Euler constant; and a gauge-dependent part,

---

[111] For the same reason, the Z boson in the diagrams of Figs. D.1 and D.2 cannot be substituted by χ.

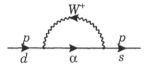

FIG. D.3. Flavour-changing self-energy diagram.

$$A_{\text{gauge-dependent}} = \sum_\alpha \frac{g^4 \lambda_\alpha}{4 m_W^2} \left[\tfrac{1}{2} + (Q-1) s_w^2 \right] (\bar{\nu} \Gamma_\mu \nu) (\bar{s} \Gamma^\mu d) \, \mu^\epsilon \int \frac{d^d k}{(2\pi)^d} \frac{2 x_\alpha}{D_\alpha D_g}. \tag{D.2}$$

D.3   The diagrams in Fig. D.2

These diagrams pose us a problem: namely, if we set both the masses and the four-momenta of the external quarks to zero then the propagator of the down-type quark becomes badly defined. We must therefore be more careful. We first consider the self-energy graph of Fig. D.3, in which p is the momentum of the quark. That graph may be written in the form, if we omit the spinors $\bar{s}$ and d,

$$\text{Fig. D.3} = \sum_\alpha \frac{g^2 \lambda_\alpha}{2} \left[A \left(p^2 \right) \not{p} \gamma_L + B \left(p^2 \right) m_s m_d \not{p} \gamma_R - C \left(p^2 \right) (m_s \gamma_L + m_d \gamma_R) \right], \tag{D.3}$$

where the functions A, B, and C of p^2 depend neither on m_s nor on m_d. In particular,

$$A \left(p^2 \right) \not{p} = \mu^\epsilon \int \frac{d^d k}{(2\pi)^d} \frac{1}{(k+p)^2 - m_\alpha^2} \left\{ \frac{1}{m_W^2 D_g} \left[\left(m_\alpha^2 + k^2 \right) \not{p} + p^2 \not{k} \right] \right.$$
$$\left. + \frac{1}{D_W} \left[(2-\epsilon)(\not{k}+\not{p}) + \frac{m_\alpha^2 - p^2}{m_W^2} \not{k} - \frac{k^2}{m_W^2} \not{p} \right] \right\}, \tag{D.4}$$

from which one obtains

$$A(0) = \frac{i}{16\pi^2} \left[\left(\frac{2}{\epsilon} - \gamma - \ln \frac{m_\alpha^2}{4\pi \mu^2} \right) \left(\xi_W + \frac{x_\alpha}{2} \right) \right.$$
$$\left. + \xi_W + \frac{3}{4} \frac{x_\alpha^2 - x_\alpha + 2}{x_\alpha - 1} - \xi_W \frac{x_\alpha + \xi_W}{x_\alpha - \xi_W} \ln \frac{x_\alpha}{\xi_W} - \frac{3}{2} \frac{x_\alpha \ln x_\alpha}{(x_\alpha - 1)^2} \right]. \tag{D.5}$$

One uses the self-energy formula of eqn (D.3) in computing Fig. D.2. One applies the on-shell conditions for the momenta of the external quarks, $p_2^2 = m_s^2$ and $p_1^2 = m_d^2$. Also, $\not{p}_2$ acting on the left should be put equal to m_s and $\not{p}_1$ acting on the right should be put equal to m_d, as a consequence of the Dirac equation for the quark spinors. Indeed, gauge-invariance in general only holds for physical processes, in particular for processes in which external particles are on mass shell. One obtains

$$\text{Fig. D.2} = \sum_\alpha \frac{g^4 \lambda_\alpha}{4 m_W^2} (\bar{\nu} \Gamma_\mu \nu) \frac{1}{m_s^2 - m_d^2} \left\{ \left[-\tfrac{1}{2} - (Q-1) s_w^2 \right] \gamma^\mu \gamma_L \right.$$

$$\times \left[A\left(m_s^2\right) m_s^2 - A\left(m_d^2\right) m_d^2 + B\left(m_s^2\right) m_s^2 m_d^2 - B\left(m_d^2\right) m_s^2 m_d^2 \right.$$
$$-C\left(m_s^2\right)\left(m_s^2 + m_d^2\right) + C\left(m_d^2\right)\left(m_s^2 + m_d^2\right)\right]$$
$$- (Q-1) s_w^2 m_s m_d \gamma^\mu \gamma_R \left[A\left(m_s^2\right) - A\left(m_d^2\right) \right.$$
$$\left. \left. + B\left(m_s^2\right) m_s^2 - B\left(m_d^2\right) m_d^2 - 2C\left(m_s^2\right) + 2C\left(m_d^2\right)\right]\right\}. \tag{D.6}$$

We now let m_s and m_d tend to zero. The expression in eqn (D.6) becomes

$$\text{Fig. D.2} = \sum_\alpha \frac{g^4 \lambda_\alpha}{4 m_W^2} \left(\bar{\nu} \Gamma_\mu \nu\right) \left\{ \left[-\tfrac{1}{2} - (Q-1) s_w^2 \right] \gamma^\mu \gamma_L \left[A(0) + B'(0) m_s^2 m_d^2 \right.\right.$$
$$\left. - C'(0)(m_s^2 + m_d^2) \right] - (Q-1) s_w^2 m_s m_d \gamma^\mu \gamma_R \left[A'(0) + B(0) - 2C'(0) \right]$$
$$+\text{terms of higher order in } m_s \text{ and } m_d \}, \tag{D.7}$$

where the prime denotes the derivative relative to p^2. Clearly, when m_s and m_d are zero only $A(0)$ is relevant. Thus, in that limit

$$\text{Fig. D.2} = \sum_\alpha \frac{g^4 \lambda_\alpha}{4 m_W^2} \left[-\tfrac{1}{2} - (Q-1) s_w^2 \right] \left(\bar{\nu} \Gamma_\mu \nu\right) \left(\bar{s} \Gamma^\mu d\right) A(0). \tag{D.8}$$

This contribution cancels out the gauge-dependence in $A_{\text{gauge-dependent}}$. Indeed,

$$A_{\text{gauge-dependent}} + \text{Fig. D.2} = \left[-\tfrac{1}{2} - (Q-1) s_w^2 \right] \left(\bar{\nu} \Gamma_\mu \nu\right) \left(\bar{s} \Gamma^\mu d\right) \sum_\alpha \frac{i g^4 \lambda_\alpha x_\alpha}{64 \pi^2 m_W^2}$$
$$\times \left[-\frac{3}{2} \left(\frac{2}{\epsilon} - \gamma - \ln \frac{m_W^2}{4\pi\mu^2} \right) \right.$$
$$\left. + \frac{11 - 5 x_\alpha}{4 (x_\alpha - 1)} + \frac{3 x_\alpha (x_\alpha - 2) \ln x_\alpha}{2 (x_\alpha - 1)^2} \right]. \tag{D.9}$$

D.4 Final result

One adds the contributions in eqns (D.1) and (D.9) and obtains the finite and gauge-invariant result

$$\text{Fig. D.1} + \text{Fig. D.2} = \sum_\alpha \frac{-i g^4 \lambda_\alpha}{16 \pi^2 m_W^2} \left(\bar{\nu} \Gamma_\mu \nu\right) \left(\bar{s} \Gamma^\mu d\right) X\left(x_\alpha\right), \tag{D.10}$$

where

$$X(x) = \frac{x}{8 (x-1)} \left(x + 2 + \frac{3x - 6}{x - 1} \ln x \right). \tag{D.11}$$

The final result turns out to be independent both of s_w and of the charge Q of the up-type quarks.

The effective Hamiltonian for the process $\bar{s} d \to \nu \bar{\nu}$ therefore is

$$\mathcal{H}_{\text{eff}} = \frac{2 G_F^2 m_W^2}{\pi^2} \left(\bar{\nu} \Gamma_\mu \nu\right) \left(\bar{s} \Gamma^\mu d\right) \left[\lambda_c X\left(x_c\right) + \lambda_t X\left(x_t\right)\right] + \text{H.c.}. \tag{D.12}$$

We have set the mass of the up quark to zero, just as in Appendix B.

APPENDIX E

D^0–$\overline{D^0}$ MIXING

E.1 Theoretical expectations

The mesons $D^0 \sim c\bar{u}$ and $\overline{D^0} \sim \bar{c}u$ have lifetime $(4.15 \pm 0.04) \times 10^{-13}$ s, corresponding to

$$\Gamma = (1.586 \pm 0.015) \times 10^{-12}\,\text{GeV}. \tag{E.1}$$

They decay faster than B_d^0 and $\overline{B_d^0}$, in spite of their mass 1.8645 ± 0.0005 GeV being much smaller than the mass of the latter mesons; this is because the decay of the charm quark is not suppressed by small CKM-matrix elements, contrary to the decay of the bottom quark, which is suppressed by $|V_{cb}| \ll 1$ and $|V_{ub}| \ll 1$.

In the standard model (SM) D^0–$\overline{D^0}$ mixing should be tiny: one expects $x \ll 1$ and $|y| \ll 1$. If x and $|y|$ are very small, as predicted and as, to some extent, experimentally confirmed, then D^0 and $\overline{D^0}$ practically do not oscillate into and from each other while decaying; the linear superposition of D^0 and $\overline{D^0}$ which is created at production time is identical with the one to be found at decay time. This has important consequences in the theoretical analysis of some decays of the B^0–$\overline{B^0}$ systems, as seen in particular in Chapters 36 and 37.

A good measure of mixing is $F = \left(x^2 + y^2\right) / \left(2 + x^2 - y^2\right)$; this quantity tends to 1 when either $x \to \infty$ or $|y| \to 1$.

In the SM D^0–$\overline{D^0}$ mixing receives three main contributions: from box diagrams, from dipenguin diagrams, and from long-distance effects. In this section we review briefly each of these contributions and its expected size, and turn afterwards to D^0–$\overline{D^0}$ mixing in extensions of the SM.

E.1.1 The box diagrams

The box diagrams for D^0–$\overline{D^0}$ mixing are analogous to the ones for K^0–$\overline{K^0}$ mixing, which were analysed in Appendix B. One has charm and up quarks in the external lines, and any of the three down-type quarks in the internal fermion lines. One might expect the bottom quark to dominate, because the function $S_0(x)$ (see eqn B.16) grows with x. However, the loops with bottom quarks end up being negligible, for two reasons: firstly, the bottom quark is not that heavy; secondly, its contribution is very much suppressed by the small CKM-matrix elements V_{cb} and V_{ub} in the vertices. Hence, only the strange and down quarks contribute effectively and, due to the GIM suppression, the effective Hamiltonian is proportional to (Datta and Kumbhakar 1985; Donoghue *et al.* 1986c)

$$\frac{\left(m_s^2 - m_d^2\right)^2}{m_W^2 m_c^2}, \tag{E.2}$$

and thus very small. In the computation in Appendix B one could neglect the masses and momenta of the external s and d quarks. In the box diagrams for D^0–$\overline{D^0}$ mixing, on the other hand, the masses of the internal s and d quarks are small compared to the mass of the external charm quarks, and the latter cannot be neglected. An extra operator $(\bar{c}\gamma_R u)(\bar{c}\gamma_R u)$ then appears in the effective Hamiltonian, and its matrix element must be evaluated.

One may use the vacuum-insertion approximation to estimate the matrix elements. One obtains $|M_{12}| \sim 10^{-17}$ to $10^{-16}\,\mathrm{GeV}$, corresponding to $F \sim 10^{-10}$ to 10^{-8}, which is extremely small.

E.1.2 *Dipenguin diagrams*

The contributions of dipenguin diagrams to K^0–$\overline{K^0}$ mixing and to B^0–$\overline{B^0}$ mixing are negligible (Donoghue *et al.* 1986d; Eeg and Picek 1987, 1988), and one may be tempted to neglect them in D^0–$\overline{D^0}$ mixing too. However, Petrov (1997) has claimed that in the latter case they yield a short-distance contribution to M_{12} not much smaller than, and with the opposite sign to, the contribution from the box diagrams.

E.1.3 *Long-distance contributions*

The exact evaluation of the box diagrams and dipenguin diagrams is not so important because D^0–$\overline{D^0}$ mixing is probably (Wolfenstein 1985) dominated by long-distance effects, i.e., by intermediate hadronic states—not quarks—in the $D^0 \leftrightarrow \overline{D^0}$ transitions. In order to understand why this is so, one may make a comparison with the K^0–$\overline{K^0}$ and B_d^0–$\overline{B_d^0}$ systems. In the latter system the three up-type quarks α couple with CKM-matrix factors $V_{\alpha b}V_{\alpha d}^*$ of the same order of magnitude λ^3; the top quark being very heavy, it overwhelms the contributions from the light quarks; then, the low-energy physics of hadrons, i.e., the long-distance effects, are irrelevant. In K^0–$\overline{K^0}$ mixing the charm quark competes with the top quark because, in spite of being much lighter, it couples with $V_{cs}V_{cd}^* \sim \lambda$, while the coupling of the top quark $V_{ts}V_{td}^* \sim \lambda^5$. Thus, the couplings of light hadrons to K^0 and $\overline{K^0}$ are relatively strong, and one therefore expects relevant long-distance contributions to M_{12}, cf. § 17.6. In $D^0 \to \overline{D^0}$ transitions the important intermediate quarks are the light s and d quarks; it can then be expected that light hadrons couple strongly to D^0 and $\overline{D^0}$, from which large long-distance contributions should follow.

The long-distance contributions are non-perturbative and we cannot compute them from first principles. Donoghue *et al.* (1986c) have evaluated the contributions of intermediate states with two charged pseudoscalar mesons—$\pi^+\pi^-$, K^+K^-, π^+K^-, and $K^+\pi^-$, for which some experimental data are available; they have obtained $F \sim 10^{-8}$. However, there are other intermediate states— with two vector mesons, or one pseudoscalar and one vector meson, as well as with one, three, four, ... mesons. It is likely that these intermediate states yield contributions of the same order of magnitude as the one studied by Donoghue *et al.* (1986c), and moreover it is likely that all those contributions have different signs and partially cancel each other, in such a way that one may guess that the sum of all of them ends up giving $F \sim 10^{-8}$.

A different approach to the long-distance contributions is based on heavy-quark effective theory. This approach was pioneered by Georgi (1992) and followed by Ohl *et al.* (1993); they obtained $F \sim 10^{-10}$. This is much smaller than the estimate by Donoghue *et al.* (1986c).

It should be pointed out that both these approaches concentrate on the long-distance contribution to the dispersive part of the $\overline{D^0} \to D^0$ transition amplitude, M_{12}. The absorptive part, Γ_{12}, remains unchecked, and might be larger than M_{12}. The original estimate of Wolfenstein (1985) was that F might be as large as 10^{-4} due to the long-distance contributions; this estimate seems to stay on firm ground for Γ_{12} (Le Yaouanc *et al.* 1995).

Golowich and Petrov (1998) have suggested that the rich spectrum of resonances with masses between 1.6 and 2.1 GeV may give important contributions to D^0–$\overline{D^0}$ mixing. In a partly phenomenological analysis they obtained $F \sim 10^{-8}$, and found that $|\Gamma_{12}/M_{12}|$ might be larger than unity.

E.1.4 D^0–$\overline{D^0}$ *mixing beyond the standard model*

The fact that the SM predicts D^0–$\overline{D^0}$ mixing to be so small means that there is a large window of opportunity to check extensions of the SM via a possible large D^0–$\overline{D^0}$ mixing. Various extensions of the SM may lead to large mixing (Burdman 1995; Nir 1996). In particular,

- A fourth generation would contribute to M_{12} through box diagrams with intermediate b' quarks. With $|V_{ub'}V_{cb'}| \sim \lambda^3$ and $m_{b'} \sim 100\,\text{GeV}$ the current experimental limit on mixing is saturated (Burdman 1995).

- Vector-like singlet quarks of charge 2/3 lead to flavour-changing couplings of the Z boson with the up-type quarks. These couplings generate a potentially large D^0–$\overline{D^0}$ mixing at tree level.

- Multi-Higgs-doublet models without flavour conservation similarly lead to D^0–$\overline{D^0}$ mixing at tree level. A neutral scalar with mass $\sim 100\,\text{GeV}$ and coupling to $\bar{c}u$ with strength $\sim \sqrt{m_c m_u}/v$ saturates the experimental bound (Burdman 1995).

- Multi-Higgs-doublet models include charged scalars, which enter box diagrams for D^0–$\overline{D^0}$ mixing similar to the SM boxes but with one or both $W^\pm$ replaced by charged scalars. This, too, may easily saturate the experimental bound.

One or more of these mechanisms may be simultaneously operative. Thus, various viable theoretical ideas lead to values of F within reach of current or planned experiments.

E.2   Experimental results

Using eqns (9.14) and (9.15) one has

$$\Gamma[P^0(t) \to f] = e^{-\Gamma t} \left|\frac{q}{p}\right|^2 |\bar{A}_f|^2 \left\{ |\bar{\lambda}_f|^2 - \Gamma t \left(x \operatorname{Im} \bar{\lambda}_f + y \operatorname{Re} \bar{\lambda}_f\right) \right.$$
$$\left. + \frac{\Gamma^2 t^2}{4} \left[|\bar{\lambda}_f|^2 \left(y^2 - x^2\right) + y^2 + x^2 \right] + O\left(\Gamma^3 t^3\right) \right\},$$
$$\Gamma[\overline{P^0}(t) \to \bar{f}] = e^{-\Gamma t} \left|\frac{p}{q}\right|^2 |A_{\bar{f}}|^2 \left\{ |\lambda_{\bar{f}}|^2 - \Gamma t \left(x \operatorname{Im} \lambda_{\bar{f}} + y \operatorname{Re} \lambda_{\bar{f}}\right) \right. \tag{E.3}$$
$$\left. + \frac{\Gamma^2 t^2}{4} \left[|\lambda_{\bar{f}}|^2 \left(y^2 - x^2\right) + y^2 + x^2 \right] + O\left(\Gamma^3 t^3\right) \right\}.$$

Therefore, when x and y are small,[112]

$$\Gamma[P^0(t) \to f] \approx e^{-\Gamma t} \left|\frac{q}{p}\right|^2 |\bar{A}_f|^2 \left\{ |\bar{\lambda}_f|^2 - \Gamma t \left(x \operatorname{Im} \bar{\lambda}_f + y \operatorname{Re} \bar{\lambda}_f\right) \right.$$
$$\left. + \frac{\Gamma^2 t^2}{4} \left(y^2 + x^2\right) \right\} \qquad \Leftarrow |\bar{A}_f| \gg |A_f|,$$
$$\Gamma[\overline{P^0}(t) \to \bar{f}] \approx e^{-\Gamma t} \left|\frac{p}{q}\right|^2 |A_{\bar{f}}|^2 \left\{ |\lambda_{\bar{f}}|^2 - \Gamma t \left(x \operatorname{Im} \lambda_{\bar{f}} + y \operatorname{Re} \lambda_{\bar{f}}\right) \right. \tag{E.4}$$
$$\left. + \frac{\Gamma^2 t^2}{4} \left(y^2 + x^2\right) \right\} \qquad \Leftarrow |A_{\bar{f}}| \gg |\bar{A}_{\bar{f}}|.$$

In particular, for flavour-specific decay modes,

$$\Gamma[P^0(t) \to \bar{o}] \approx e^{-\Gamma t} \left|\frac{q}{p}\right|^2 |\bar{A}_{\bar{o}}|^2 \frac{\Gamma^2 t^2}{4} \left(y^2 + x^2\right),$$
$$\Gamma[\overline{P^0}(t) \to o] \approx e^{-\Gamma t} \left|\frac{p}{q}\right|^2 |A_o|^2 \frac{\Gamma^2 t^2}{4} \left(y^2 + x^2\right). \tag{E.5}$$

When mixing is small $F \approx \left(x^2 + y^2\right)/2$. This is the quantity r_{mix} that experimentalists strive to measure.

Equations (E.5) have been used by the E791 Collaboration (1996) to set an experimental limit on D^0–$\overline{D^0}$ mixing. The E791 Collaboration (1996, 1998) has used the decays $D^{*+} \to \pi^+ D^0$ and $D^{*-} \to \pi^- \overline{D^0}$ to identify the flavour of the neutral-D meson at production time. They have then compared the 'right-sign' decays

$$D^0 \to K^- l^+ \nu_l \text{ and } \overline{D^0} \to K^+ l^- \bar{\nu}_l \tag{E.6}$$

with the 'wrong-sign' decays

$$D^0 \to K^+ l^- \bar{\nu}_l \text{ and } \overline{D^0} \to K^- l^+ \nu_l, \tag{E.7}$$

where l may be either e or μ. They have used the fact that, according to eqns (E.5), the time-evolution of the wrong-sign decays should be given, when x

---

[112]The product Γt must remain of order 1 lest the factor $\exp(-\Gamma t)$ renders the decays unobservable.

and y are small, by $\Gamma^2 t^2 \exp(-\Gamma t)$. They have obtained the 90%-confidence-limit $r_{\text{mix}} < 5.0 \times 10^{-3}$.

Later, the E791 Collaboration (1998) has observed the 'wrong-sign' decays

$$D^0 \to K^+ \pi^- \left(\pi^+ \pi^-\right) \quad \text{and} \quad \overline{D^0} \to K^- \pi^+ \left(\pi^+ \pi^-\right), \tag{E.8}$$

and has compared them to the 'right-sign' decays

$$D^0 \to K^- \pi^+ \left(\pi^+ \pi^-\right) \quad \text{and} \quad \overline{D^0} \to K^+ \pi^- \left(\pi^+ \pi^-\right), \tag{E.9}$$

where the notation $(\pi^+ \pi^-)$ indicates the possible presence of an extra pair of charged pions in the final state. The right-sign decays are proportional to $|V_{ud} V_{cs}|^2 \sim 1$, while the wrong-sign decays are proportional to $|V_{us} V_{cd}|^4 \sim \lambda^4 \sim 2.5 \times 10^{-3}$. Thus, in this case the wrong-sign decays are not really forbidden, rather they are 'doubly Cabibbo-suppressed', i.e., their decay amplitudes are suppressed by two powers of the Cabibbo angle. Then, in eqns (E.4), with $P^0 \equiv D^0$, $\overline{P^0} \equiv \overline{D^0}$, $f \equiv K^+ \pi^-$ or $K^+ \pi^- \pi^+ \pi^-$, and $\bar{f} \equiv K^- \pi^+$ or $K^- \pi^+ \pi^+ \pi^-$, one expects $|\bar{\lambda}_f|^2$ and $|\lambda_{\bar{f}}|^2$ to be $\sim 2.5 \times 10^{-3}$.

One must be careful to distinguish the different decay-time dependences:

- Mixing-induced decays—$(\Gamma t)^2 \exp(-\Gamma t)$;
- Doubly-Cabibbo-suppressed decays—$\exp(-\Gamma t)$;
- Interference terms—$(\Gamma t) \exp(-\Gamma t)$.

Carefully taking this into account,[113] the E791 Collaboration (1998) obtained the 90%-confidence-limit $r_{\text{mix}} < 8.5 \times 10^{-3}$, comparable to the bound extracted from semileptonic decays.

The Particle Data Group (1996) refers to the pre-1996 limits on D^0-$\overline{D^0}$ mixing. Those experimental searches too have used either the semileptonic or the $K^{\pm} \pi^{\mp}$ (together with $K^{\pm} \pi^{\mp} \pi^+ \pi^-$) decays of the neutral-$D$ mesons. They have obtained results which were either weaker or less general than the ones by the E791 Collaboration (1996, 1998).

E.3   Conclusions

- The SM predicts $r_{\text{mix}} \lesssim 10^{-8}$.
- Experiment can at present only guarantee that $r_{\text{mix}} < 10^{-2}$.
- Various models beyond the SM could saturate or even exceed the experimental bound.

---

[113] The importance of the interference terms was emphasized by Blaylock *et al.* (1995); Wolfenstein (1995); Browder and Pakvasa (1996); Liu (1996). Earlier it had been usual to neglect terms with time-dependence $(\Gamma t) \exp(-\Gamma t)$.

REFERENCES

Abers, E. and Lee, B. (1973). *Physics Reports* **9**, 1.

Ademollo, M. and Gatto, R. (1964). *Physics Letters* **13**, 264.

Adler, S. L. (1969). *Physical Review* **177**, 2426.

Aguila, F. del and Cortés, J. (1985). *Physics Letters* **156B**, 243.

Aguila, J. R. del and Nelson, C. A. (1986). *Physical Review* **D33**, 101.

Aguila, F. del, Aguilar-Saavedra, J. A., and Branco, G. C. (1998). *Nuclear Physics* **510B**, 39.

Akundov, A., Bardin, D., and Riemann, T. (1986). *Nuclear Physics* **276B**, 1.

Aleksan, R., Dunietz, I., Kayser, B., and Le Diberder, F. (1991). *Nuclear Physics* **361B**, 141.

Aleksan, R., Dunietz, I., and Kayser, B. (1992). *Zeitschrift für Physik* **C54**, 653.

Aleksan, R., Le Yaouanc, A., Oliver, L., Pène, O., and Raynal, J.-C. (1993). *Physics Letters* **B316**, 567.

Aleksan, R., Kayser, B., and London, L. (1994). *Physical Review Letters* **73**, 18.

Alfimenkov, V. P. *et al.* (1990). *JETP Letters* **52**, 373.

Ali, A. and London, D. (1997). *Nuclear Physics B (Proceedings Supplements)* **54A**, 297.

Aliev, T. M., Eletskii, V. L., and Kogan, Ya. I. (1984). *Soviet Journal of Nuclear Physics* **40**, 527.

Allton, C. R., Ciuchini, M., Crisafulli, M., Lubicz, V., and Martinelli, G. (1994). *Nuclear Physics* **431B**, 667.

Altarelli, G., Cabibbo, N., Corbo, G., Maiani, L., and Martinelli, G. (1982). *Nuclear Physics* **208B**, 365.

Altarev, I. S. *et al.* (1992). *Physics Letters* **B276**, 242.

Altomari, T., Wolfenstein, L., and Bjorken, J. D. (1988). *Physical Review* **D37**, 1860.

Alvarez-Gaumé, L., Kounnas, C., Lola, S., and Pavlopoulos, P. (1998). CERN report CERN-TH/98-392 (hep-ph/9812326).

Amorim, A., Santos, M. G., and Silva, J. P. (1999). *Physical Review* **D59**, 056001.

Amundson, J. F., Rosner, J. L., Kelly, M. A., Horwitz, N., and Stone, S. L. (1993). *Physical Review* **D47**, 3059.

Anderson, C. D. (1933). *Physical Review* **43**, 491.

Anderson, G. W. and Hall, L. J. (1992). *Physical Review* **D45**, 2685.

Anselm, A. A., Chkareuli, J. L., Uraltsev, N. G., and Zhukovskaya, T. A. (1985). *Physics Letters* **156B**, 102.

Antaramian, A., Hall, L. J., and Rašin, A. (1992). *Physical Review Letters* **69**, 1871.

ARGUS Collaboration (Albrecht, H. *et. al.*) (1990). *Physics Letters* **B234**, 409.

ARGUS Collaboration (Albrecht, H. *et. al.*) (1991). *Physics Letters* **B255**, 297.

ARGUS Collaboration (Albrecht, H. *et. al.*) (1993). *Zeitschrift für Physik* **C57**, 533.

Atiyah, M. F. and Ward, R. S. (1977). *Communications in Mathematical Physics* **55**, 117.

Atwood, D., Blok, B., and Soni, A. (1996). *International Journal of Modern Physics* **A11**, 3743.

Atwood, D., Dunietz, I., and Soni, A. (1997). *Physical Review Letters* **78**, 3257.

Azimov, Ya. I. (1989). *JETP Letters* **50**, 447.

Azimov, Ya. I. (1990). *Physical Review* **D42**, 3705.

Azimov, Ya. I. and Dunietz, I. (1997). *Physics Letters* **B395**, 334.

Azimov, Ya. I., Rappoport, V. L., and Sarantsev, V. V. (1997). *Zeitschrift für Physik* **A356**, 437

Babu, K. S. and Mohapatra, R. N. (1990). *Physical Review* **D15**, 1958.

Bahcall, J. N. (1989). *Neutrino astrophysics.* Cambridge University Press.

Bahcall, J. N. and Pinsonneault, M. H. (1992). *Review of Modern Physics* **64**, 885.

Bahcall, J. N. and Pinsonneault, M. H. (1995). *Review of Modern Physics* **67**, 781.

Bahcall, J. N., Basu, S., and Pinsonneault, M. H. (1998). *Physics Letters* **B433**, 1.

Ball, P., Beneke, M., and Braun, V. M. (1995). *Physical Review* **D52**, 3929.

Baluni, V. (1979). *Physical Review* **D19**, 2227.

Bamert, P., Burgess, C. P., Cline, J. M., London, D., and Nardi, E. (1996). *Physical Review* **D54**, 4275.

Bander, M., Silverman, D., and Soni, A. (1979). *Physical Review Letters* **43**, 242.

Bardeen, W. A. (1969). *Physical Review* **184**, 1848.

Bardeen, W. A. and Tye, S.-H. H. (1978). *Physics Letters* **74B**, 229.

Barenboim, G. and Botella, F. J. (1998). *Physics Letters* **B433**, 385.

Barenboim, G., Botella, F. J., Branco, G. C., and Vives, O. (1998). *Physics Letters* **B422**, 277.

Barger, V., Phillips, R. J. N., and Whisnant, K. (1986). *Physical Review Letters* **57**, 48.

Barker, F. C. (1992). *Nuclear Physics* **540A**, 501.

Barker, F. C. (1994). *Nuclear Physics* **579A**, 62.

Barmin, V. V. *et al.* (1984). *Nuclear Physics* **247B**, 293.

Barr, S. M. (1984). *Physical Review* **D30**, 1805.

Barr, S. and Langacker, P. (1979). *Physical Review Letters* **42**, 1654.

Barr, S. M., Chang, D., and Senjanović, G. (1991). *Physical Review Letters* **67**, 2765.

Basdevant, J. L., Froggatt, C. D., and Petersen, J. L. (1974). *Nuclear Physics* **72B**, 413.

Basdevant, J. L., Chapelle, P., Lopez, C., and Sigelle, M. (1975). *Nuclear Physics* **98B**, 285.

Bauer, M., Stech, B., and Wirbel, M. (1985). *Zeitschrift für Physik* **C29**, 637.

Bauer, M., Stech, B., and Wirbel, M. (1987). *Zeitschrift für Physik* **C34**, 103.

Beall, G., Bender, M., and Soni, A. (1982). *Physical Review Letters* **48**, 848.

Beenakker, W. and Hollik, W. (1988). *Zeitschrift für Physik* **C40**, 141.

Bég, M. A. B. and Tsao, H.-S. (1978). *Physical Review Letters* **41**, 278.

Bég, M. A. B., Maloof, D. J., and Tsao, H.-S. (1979). *Physical Review* **D19**, 2221.

Belavin, A. A., Polyakov, A. M., Schwartz, A. S., and Tyupkin, Yu. S. (1975). *Physics Letters* **59B**, 85.

Bell, J. S. and Jackiw, R. (1969). *Nuovo Cimento* **60A**, 47.

Bell, J. S. and Steinberger, J. (1966). In *Proceedings of the Oxford international conference on elementary particles, 1965* (ed. R. G. Moorhouse, A. E. Taylor, and T. R. Walsh). Rutherford Laboratory, Chilton.

Beneke, M., Buchalla, G., and Dunietz, I. (1996). *Physical Review* **D54**, 4419.

Beneke, M., Buchalla, G., and Dunietz, I. (1997). *Physics Letters* **B393**, 132.

Bento, L., Branco, G. C., and Parada, P. A. (1991). *Physics Letters* **B267**, 95.

Bernabéu, J., Branco, G. C., and Gronau, M. (1986a). *Physics Letters* **169B**, 243.

Bernabéu, J., Pich, A., and Santamaría, A. (1986b). *Zeitschrift für Physik* **C30**, 213.

Bernabéu, J., Pich, A., and Santamaría, A. (1988). *Physics Letters* **B200**, 569.

Bernard, C. and Soni, A. (1989). *Nuclear Physics B (Proceedings Supplements)* **9**, 155.

Bertolini, S., Eeg, J. O., and Fabbrichesi, M. (1996). *Nuclear Physics* **476B**, 225.

Bertolini, S., Eeg, J. O., Fabbrichesi, M., and Lashin, E. I. (1998a). *Nuclear Physics* **514B**, 93.

Bertolini, S., Fabbrichesi, M., and Eeg, J. O. (1998b). SISSA (Trieste) report SISSA-19-98-EP (hep-ph/9802405).

Bethe, H. A. (1986). *Physical Review Letters* **56**, 1305.

Bigi, I. I. and Sanda, A. I. (1981). *Nuclear Physics* **193B**, 85.

Bigi, I. I. and Sanda, A. I. (1987). *Nuclear Physics* **281B**, 41.

Bigi, I. I., Khoze, V., Uraltsev, N. G., and Sanda, A. I. (1989). In *CP violation* (ed. C. Jarlskog). World Scientific, Singapore.

Bijnens, J. and Wise, M. B. (1984). *Physics Letters* **137B**, 245.

Bilenky, S. M. and Petcov, S. T. (1987). *Review of Modern Physics* **59**, 671.

Bjorken, J. D. and Drell, S. D. (1964). *Relativistic quantum mechanics*. Mc-Graw-Hill, New York.

Bjorken, J. D. and Dunietz, I. (1987). *Physical Review* **D36**, 2109.

Blaylock, G., Seiden, A., and Nir, Y. (1995). *Physics Letters* **B355**, 555.

Blok, B., Gronau, M., and Rosner, J. L. (1997). *Physical Review Letters* **78**, 3999.

Boehm, F. and Vogel, P. (1992). *Physics of massive neutrinos*. Cambridge University Press.

Bopp, P. *et al.* (1989). *Physical Review Letters* **56**, 919; erratum *ibid.* **57**, 1192.

Botella, F. J. and Silva, J. P. (1995). *Physical Review* **D51**, 3870.

Bowser-Chao, D., Chang, D., and Keung, W.-Y. (1998). *Physical Review Letters* **81**, 2028.

Branco, G. C. (1980*a*). *Physical Review Letters* **44**, 504.

Branco, G. C. (1980*b*). *Physical Review* **D22**, 2901.

Branco, G. C. and Lavoura, L. (1985). *Physics Letters* **B165**, 327.

Branco, G. C. and Lavoura, L. (1986). *Nuclear Physics* **278B**, 738.

Branco, G. C. and Lavoura, L. (1988*a*). *Physics Letters* **B208**, 123.

Branco, G. C. and Lavoura, L. (1988*b*). *Physical Review* **D38**, 2295.

Branco, G. C. and Rebelo, M. N. (1985). *Physics Letters* **B160**, 117.

Branco, G. C., Frère, J.-M., and Gérard, J.-M. (1983). *Nuclear Physics* **221B**, 317.

Branco, G. C., Gérard, J.-M., and Grimus, W. (1984). *Physics Letters* **136B**, 383.

Branco, G. C., Lavoura, L., and Rebelo, M. N. (1986). *Physics Letters* **180B**, 264.

Branco, G. C., Grimus, W., and Lavoura, L. (1989). *Nuclear Physics* **312B**, 492.

Branco, G. C., Morozumi, T., Parada, P. A., and Rebelo, M. N. (1993). *Physical Review* **D48**, 1167.

Branco, G. C., Grimus, W., and Lavoura, L. (1996). *Physics Letters* **B380**, 119.

Branco, G. C., Délepine, D., Emmanuel-Costa, D., and González-Felipe, R. (1998). *Physics Letters* **B442**, 229.

Branco, G. C., Rebelo, M. N., and Silva-Marcos, J. I. (1999). *Physical Review Letters* **82**, 683.

Browder, T. E. and Pakvasa, S. (1996). *Physics Letters* **B383**, 475.

Browder, T. E., Honscheid, K., and Pedrini, D. (1997). *Annual Review of Nuclear and Particle Science* **46**, 395.

Brown, B. A. and Ormand, W. E. (1989). *Physical Review Letters* **62**, 866.

Buchalla, G. (1997). *Nuclear Physics B (Proceedings Supplements)* **59**, 130.

Buchalla, G. and Buras, A. J. (1993*a*). *Nuclear Physics* **398B**, 285.

Buchalla, G. and Buras, A. J. (1993*b*). *Nuclear Physics* **400B**, 225.

Buchalla, G. and Buras, A. J. (1994). *Nuclear Physics* **412B**, 106.

Buchalla, G. and Isidori, G. (1998). *Physics Letters* **B440**, 170.

Buchalla, G., Buras, A. J., and Harlander, M. K. (1990). *Nuclear Physics* **337B**, 313.

Buchalla, G., Buras, A. J., and Harlander, M. K. (1991). *Nuclear Physics* **349B**, 1.

Buchalla, G., Buras, A. J., and Lautenbacher, M. (1996). *Review of Modern Physics* **68**, 1125.

Buchanan, C. D., Cousins, R., Dib, C., Peccei, R. D., and Quackenbush, J. (1992). *Physical Review* **D45**, 4088.

Buchmüller, W., Fodor, Z., Helbig, T., and Walliser, D. (1994). *Annals of Physics* **234**, 260.

Buras, A. J. (1997). In *Proceedings of the 28th international conference on high-energy physics, Warsaw, Poland, 25–31 July 1996* (ed. Z. Ajduk and A. K. Wroblewski). World Scientific, Singapore.

Buras, A. J. and Fleischer, R. (1995). *Physics Letters* **B341**, 379.

Buras, A. J. and Fleischer, R. (1998). In *Heavy flavours II* (ed. A. J. Buras and M. Lindner). World Scientific, Singapore.

Buras, A. J. and Gérard, J.-M. (1987). *Physics Letters* **192B**, 156.

Buras, A. J., Jamin, M., and Weisz, P. H. (1990). *Nuclear Physics* **347B**, 491.

Buras, A. J., Jamin, M., Lautenbacher, M. E., and Weisz, P. H. (1992). *Nuclear Physics* **370B**, 69.

Buras, A. J., Jamin, M., Lautenbacher, M. E., and Weisz, P. H. (1993a). *Nuclear Physics* **400B**, 37.

Buras, A. J., Jamin, M., and Lautenbacher, M. E. (1993b). *Nuclear Physics* **400B**, 75.

Buras, A. J., Jamin, M., and Lautenbacher, M. E. (1993c). *Nuclear Physics* **408B**, 209.

Buras, A. J., Lautenbacher, M. E., and Ostermaier, G. (1994). *Physical Review* **D50**, 3433.

Buras, A. J., Jamin, M., and Lautenbacher, M. E. (1996). *Physics Letters* **B389**, 749.

Burcham, W. E. and Jobes, M. (1995). *Nuclear and particle physics*. Longman Scientific & Technical, Harlow.

Burdman, G. (1995). Fermi National Accelerator Laboratory report FERMI-LAB-CONF-95/281-T (hep-ph/9508349).

Byrne, J. *et al.* (1990). *Physical Review Letters* **65**, 289.

Cabibbo, N. (1963). *Physical Review Letters* **10**, 531.

Callan Jr., C. G., Dashen, R. F., and Gross, D. J. (1976). *Physics Letters* **63B**, 334.

Carter, A. B. and Sanda, A. I. (1980). *Physical Review Letters* **45**, 952.

Carter, A. B. and Sanda, A. I. (1981). *Physical Review* **D23**, 1567.

Cartwright, W. F., Richman, C., Whitehead, M., and Wilcox, H. (1953). *Physical Review* **91**, 677.

CCFR Collaboration (Rabinowitz, S. A. *et. al.*) (1993). *Physical Review Letters* **70**, 134.

CCFR Collaboration (Bazarko, A. O. *et al.*) (1995a). *Zeitschrift für Physik* **C65**, 189.

CCFR Collaboration (Barish, B. *et al.*) (1995b). *Physical Review* **D51**, 1014

CDHS Collaboration (Abramowicz, H. *et. al.*) (1982). *Zeitschrift für Physik* **C15**, 19.

Chamberlain, O., Segrè, E., Wiegand, C., and Ypsilantis, T. J. (1955). *Physical Review* **100**, 947.

Chang, D. (1983*a*). Unpublished Ph.D. thesis. Carnegie-Mellon University.

Chang, D. (1983*b*). *Nuclear Physics* **214B**, 435.

Chang, D., Mohapatra, R. N., and Parida, M. K. (1984). *Physical Review Letters* **52**, 1072.

Chau, L.-L. (1983). *Physics Reports* **95**, 1.

Chau, L.-L. and Cheng, H.-Y. (1987). *Physical Review Letters* **59**, 958.

Chau, L.-L. and Keung, W.-Y. (1984). *Physical Review Letters* **53**, 1802.

Chau, L.-L., Cheng, H.-Y., Sze, W. K., Yao, H., and Tseng, B. (1991). *Physical Review* **D43**, 2176.

Cheng, H.-Y. (1988). *Physics Reports* **158**, 1.

Cheng, T.-P. and Li, L.-F. (1994). *Gauge theory of elementary particle physics*. Oxford University Press.

Cheston, W. B. (1951). *Physical Review* **83**, 1118.

Chetyrkin, K. G., Dominguez, C. A., Pirjol, D., and Schilcher, K. (1995). *Physical Review* **D51**, 5090.

Chinowsky, W. and Steinberger, J. (1954). *Physical Review* **95**, 1561.

Choi, K., Kim, C. W., and Sze, W. K. (1988). *Physical Review Letters* **61**, 794.

Christenson, J. H., Cronin, J. W., Fitch, V. L., and Turlay, R. (1964). *Physical Review Letters* **13**, 138.

Christos, G. A. (1984). *Physics Reports* **116**, 751.

Ciuchini, M. (1997). *Nuclear Physics B (Proceedings Supplements)* **59**, 149.

Ciuchini, M., Franco, E., Martinelli, G., and Reina, L. (1993). *Physics Letters* **B301**, 263.

Ciuchini, M., Franco, E., Martinelli, G., and Reina, L. (1994). *Nuclear Physics* **415B**, 403.

Ciuchini, M., Franco, E., Martinelli, G., and Silvestrini, L. (1997*a*). *Nuclear Physics* **501B**, 271.

Ciuchini, M., Franco, E., Martinelli, G., Masiero, A., and Silvestrini, L. (1997*b*). *Physical Review Letters* **79**, 978.

Clark, D. L., Roberts, A., and Wilson, R. (1951). *Physical Review* **83**, 649.

Clark, D. L., Roberts, A., and Wilson, R. (1952). *Physical Review* **85**, 523.

CLEO Collaboration (Bartelt, F. *et. al.*) (1993*a*). *Physical Review Letters* **71**, 4111.

CLEO Collaboration (Bean, A. *et. al.*) (1993*b*). *Physics Letters* **B317**, 647.

CLEO Collaboration (Bartelt, F. *et. al.*) (1993*c*). *Physical Review Letters* **71**, 1680.

CLEO Collaboration (Cinabro, D. *et. al.*) (1994). *Physical Review Letters* **72**, 1406.

CLEO Collaboration (Alexander, J. P. *et. al.*) (1996). *Physical Review Letters* **77**, 5000.

CLEO Collaboration (Behrens, B. H. *et. al.*) (1998*a*). *Physical Review Letters* **80**, 3710.

CLEO Collaboration (Godang, R. *et. al.*) (1998*b*). *Physical Review Letters* **80**, 3456.

Cleveland, B. T. *et al.* (1995). *Nuclear Physics B (Proceedings Supplements)* **38**, 47.

Close, F. E. (1979). *An introduction to quarks and partons.* Academic Press Limited, London.

Cohen, A. G. (1994). In *The building blocks of creation: from microfermis to megaparsecs* (ed. S. Raby and T. Walker). World Scientific, Singapore.

Cohen, A. G., Kaplan, D. B., and Nelson, A. E. (1993). *Annual Review of Nuclear and Particle Science* **43**, 27.

Cohen, A. G., Kaplan, D. B., Lepeintre, F., and Nelson, A. E. (1997). *Physical Review Letters* **78**, 2300.

Coleman, S. (1978). In *The whys of subnuclear physics* (ed. A. Zichichi). Plenum Press, New York.

Crewther, R. J. (1978). In *Facts and prospects of gauge theories* (ed. P. Urban). Springer-Verlag, Heidelberg.

Crewther, R. J., di Vecchia, P., Veneziano, G., and Witten, E. (1979). *Physics Letters* **88B**, 123; erratum *ibid.* **91B**, 487.

Dalitz, R. H. (1953). *Phil. Mag.* **44**, 1068.

Dar, A. and Shaviv, G. (1996). *Astrophysical Journal* **468**, 933.

Dass, G. V. and Sarma, K. V. L. (1992). *International Journal of Modern Physics* **A7**, 6081; erratum *ibid.* **8**, 1183.

Dass, G. V. and Sarma, K. V. L. (1994). *Physical Review Letters* **72**, 191; erratum *ibid.* **72**, 1573.

Dass, G. V. and Sarma, K. V. L. (1996a). *Physical Review* **D54**, 5880.

Dass, G. V. and Sarma, K. V. L. (1996b). Tata Institute of Fundamental Research (Mumbai) report TIFR-TH-96-57 (hep-ph/9610466).

Datta, A. and Kumbhakar, D. (1985). *Zeitschrift für Physik* **C27**, 515.

Davidson, S. and Ellis, J. (1997). *Physics Letters* **B390**, 210.

Deshpande, N. G. and He, X.-G. (1994). *Physics Letters* **B336**, 471.

Deshpande, N. G. and He, X.-G. (1995a). *Physical Review Letters* **74**, 26; erratum *ibid.* **74**, 4099.

Deshpande, N. G. and He, X.-G. (1995b). *Physical Review Letters* **75**, 1703.

Deshpande, N. G. and He, X.-G. (1996). *Physical Review Letters* **76**, 360.

Deshpande, N. G., He, X.-G., and Trampetić, J. (1995). *Physics Letters* **B345**, 547.

Deshpande, N. G., Dutta, B., and Oh, S. (1996). *Physical Review Letters* **77**, 4499.

Deutsch, M. (1953). *Progress of Theoretical Physics* **3**, 131.

Dib, C. O., London, D., and Nir, Y. (1991). *International Journal of Modern Physics* **A6**, 1253.

Dighe, A. S. (1996). *Physical Review* **D54**, 2067.

Dighe, A. S., Gronau, M., and Rosner, J. L. (1996a). *Physical Review* **D54**, 3309.

Dighe, A. S., Gronau, M., and Rosner, J. L. (1996b). *Physics Letters* **B367**, 357.

Dighe, A. S., Dunietz, I., Lipkin, H. J., and Rosner, J. L. (1996c). *Physics Letters* **B369**, 144.

Dine, M., Fischler, W., and Srednicki, M. (1981). *Physics Letters* **104B**, 199.

Dirac, P. A. M. (1958). *The principles of quantum mechanics*. Oxford University Press.

Donoghue, J. F., Golowich, E., and Holstein, B. R. (1986a). *Physics Reports* **131**, 319.

Donoghue, J. F., Golowich, E., Holstein, B. R., and Trampetić, J. (1986b). *Physics Letters* **179B**, 361.

Donoghue, J. F., Golowich, E., Holstein, B. R., and Trampetić, J. (1986c). *Physical Review* **D33**, 179.

Donoghue, J. F., Golowich, E., and Valencia, G. (1986d). *Physical Review* **D33**, 1387.

Donoghue, J. F., Holstein, B. R., and Klimt, S. W. (1987). *Physical Review* **D35**, 934.

Donoghue, J. F., Golowich, E., and Holstein, B. R. (1992). *Dynamics of the Standard Model*. Cambridge University Press.

Dunietz, I. (1991). *Physics Letters* **B270**, 75.

Dunietz, I. (1994). In *B decays* (ed. S. Stone). World Scientific, Singapore.

Dunietz, I. (1995). *Physical Review* **D52**, 3048.

Dunietz, I. and Fleischer, R. (1997). *Physical Review* **D55**, 259.

Dunietz, I., Greenberg, O. W., and Wu, D.-D. (1985). *Physical Review Letters* **55**, 2935.

Dunietz, I., Quinn, H. R., Snyder, A., Toki, W., and Lipkin, H. J. (1991). *Physical Review* **D43**, 2193.

Durbin, R., Loar, H., and Steinberger, J. (1951). *Physical Review* **84**, 581.

E731 Collaboration (Gibbons, L. K. *et. al.*) (1997). *Physical Review* **D55**, 6625.

E791 Collaboration (Aitala, E. M. *et al.*) (1996). *Physical Review Letters* **77**, 2384.

E791 Collaboration (Aitala, E. M. *et al.*) (1998). *Physical Review* **D57**, 13.

E799 Collaboration (Weaver, M. *et al.*) (1994). *Physical Review Letters* **72**, 3758.

Ecker, G. and Grimus, W. (1985). *Nuclear Physics* **258B**, 328.

Ecker, G., Grimus, W., and Konetschny, W. (1981a). *Nuclear Physics* **177B**, 489.

Ecker, G., Grimus, W., and Konetschny, W. (1981b). *Nuclear Physics* **191B**, 465.

Ecker, G., Grimus, W., and Neufeld, H. (1988). *International Journal of Modern Physics* **A3**, 603.

Eeg, J. O. and Picek, I. (1987). *Nuclear Physics* **292B**, 745.

Eeg, J. O. and Picek, I. (1988). *Zeitschrift für Physik* **C39**, 521.

Ellis, J. and Gaillard, M. K. (1979). *Nuclear Physics* **150B**, 141.

Enqvist, K., Maalampi, J., and Roos, M. (1986). *Physics Letters* **B176**, 396.

Enz, C. P. and Lewis, R. R. (1965). *Helvetica Physica Acta* **38**, 860.

Erozolimskii, B. G., Kuznetsov, I. A., Kujda, I. A., Mostovoi, Yu. A., and Stepanenko, I. V. (1991). *Physics Letters* **B263**, 33.

Fabri, E. (1954). *Nuovo Cimento* **11**, 479.

Feynman, R. P. (1949). *Physical Review* **76**, 749.

Fishbane, P., Meshkov, S., Norton, R. E., and Ramond, P. (1985). *Physical Review* **D31**, 1119.

Fishbane, P., Norton, R. E., and Rivard, M. J. (1986). *Physical Review* **D33**, 2632.

Fleischer, R. (1994*a*). *Zeitschrift für Physik* **C62**, 81.

Fleischer, R. (1994*b*). *Physics Letters* **B321**, 259 and *Physics Letters* **B332**, 419.

Fleischer, R. (1994*c*). *Physics Letters* **B341**, 205.

Fleischer, R. and Mannel, T. (1998). *Physical Review* **D57**, 2752.

Fogli, G. L., Lisi, E., Marrone, A., and Scioscia, G. (1999). *Physical Review* **D59**, 033001.

Flynn, J. M. (1997). In *Proceedings of the 28th international conference on high-energy physics, Warsaw, Poland, 25–31 July 1996* (ed. Z. Ajduk and A. K. Wroblewski). World Scientific, Singapore.

Flynn, J. M. and Randall, L. (1989) *Physics Letters* **B224**, 221; erratum *ibid.* **B235**, 412.

Franco, E., Maiani, L., Martinelli, G., and Morelli, A. (1989). *Nuclear Physics* **317B**, 63.

Fridman, A. (1988). CERN report CERN-EP-88-123.

Friedman, J. I. and Telegdi, V. L. (1957). *Physical Review* **105**, 1681.

Fritzsch, H. and Minkowski, P. (1981). *Physics Reports* **73**, 67.

Froggatt, C. D. and Petersen, J. L. (1977). *Nuclear Physics* **129B**, 89.

Furry, W. F. (1937). *Physical Review* **51**, 125.

García, A., Huerta, R., and Kielanowski, P. (1992). *Physical Review* **D45**, 879.

Garwin, R. L., Lederman, L. M., and Weinrich, M. (1957). *Physical Review* **105**, 1415.

Gasser, J. and Leutwyler, H. (1982). *Physics Reports* **87**, 77.

Gavela, M. B. *et al.* (1985*a*). *Physics Letters* **154B**, 425.

Gavela, M. B. *et al.* (1985*b*). *Physics Letters* **162B**, 197.

Gavela, M. B., Hernández, P., Orloff, J., Pène, O., and Quimbay, C. (1994). *Nuclear Physics* **430B**, 382.

Gell-Mann, M., Ramond, P., and Slansky, R. (1979). In *Supergravity* (ed. D. Z. Freedman and P. van Nieuwenhuizen). North-Holland, Amsterdam.

Gelmini, G. B. and Roncadelli, M. (1981). *Physics Letters* **99B**, 411.

Georgi, H. (1981). *Hadronic Journal* **1**, 155.

Georgi, H. (1992). *Physics Letters* **B297**, 353.

Georgi, H. and Glashow, S. L. (1988). Harvard University report HUTP-98-A048 (hep-ph/9807399).

Gérard, J.-M. and Hou, W.-S. (1988). Max-Planck Institute (Munich) report MPI-PAE/PTh-26/88.

Gérard, J.-M. and Hou, W.-S. (1989). *Physical Review Letters* **62**, 855.

Gérard, J.-M. and Hou, W.-S. (1991a). *Physical Review* **D43**, 2909.

Gérard, J.-M. and Hou, W.-S. (1991b). *Physics Letters* **B253**, 478.

Gérard, J.-M. and Weyers, J. (1999). *European Physical Journal* **C7**, 1.

Gilman, F. J. and Kauffmann, R. (1987). *Physical Review* **D36**, 2761; erratum *ibid.* **D37**, 3348(E).

Gilman, F. J. and Wise, M. B. (1979). *Physical Review* **D20**, 2392.

Glashow, S. L. (1961). *Nuclear Physics* **22B**, 579.

Glashow, S. L. and Weinberg, S. (1977). *Physical Review* **D15**, 1958.

Glashow, S. L., Iliopoulos, J., and Maiani, L. (1970). *Physical Review* **D2**, 1285.

Goldhaber, M., Grodzins, L., and Sunyar, A. W. (1958). *Physical Review* **109**, 1015.

Golowich, E. and Petrov, A. A. (1998). *Physics Letters* **B427**, 172

Gottfried, K. (1966). *Quantum mechanics.* Benjamin, New York.

Grimus, W. and Rebelo, M. N. (1997). *Physics Reports* **281**, 239.

Grinstein, B. (1989). *Physics Letters* **B229**, 280.

Grinstein, B. and Lebed, R. F. (1996). *Physical Review* **D53**, 6344.

Grinstein, B., Isgur, N., and Wise, M. B. (1986). *Physical Review Letters* **56**, 298.

Grinstein, B., Isgur, N., Scora, D., and Wise, M. B. (1989). *Physical Review* **D39**, 799.

Gronau, M. (1989). *Physical Review Letters* **63**, 1451.

Gronau, M. (1991). *Physics Letters* **B265**, 389.

Gronau, M. (1993). *Physics Letters* **B300**, 163.

Gronau, M. (1997). In *Flavour-changing neutral currents: present and future studies* (ed. D. B. Cline). World Scientific, Singapore.

Gronau, M. and London, D. (1990). *Physical Review Letters* **65**, 3381.

Gronau, M. and London, D. (1991). *Physics Letters* **B253**, 483.

Gronau, M. and Rosner, J. L. (1996). *Physical Review Letters* **76**, 1200.

Gronau, M. and Wyler, D. (1991). *Physics Letters* **B265**, 172.

Gronau, M., Kfir, A., and Loewy, R. (1986). *Physical Review Letters* **56**, 1538.

Gronau, M., Rosner, J. L., and London, D. (1994a). *Physical Review Letters* **73**, 21.

Gronau, M., Hernández, O. F., London, D., and Rosner, J. L. (1994b). *Physical Review* **D50**, 4529.

Gronau, M., Hernández, O. F., London, D., and Rosner, J. L. (1995). *Physical Review* **D52**, 6374.

Grossman, Y. and Nir, Y. (1997). *Physics Letters* **B398**, 163.

Grossman, Y. and Quinn, H. R. (1997). *Physical Review* **D56**, 7259.

Grossman, Y. and Quinn, H. R. (1998). *Physical Review* **D58**, 017504.

Grossman, Y. and Worah, M. P. (1997). *Physics Letters* **B395**, 241.

Grossman, Y., Nir, Y., and Worah, M. P. (1997a). *Physics Letters* **B407**, 307.

Grossman, Y., Kayser, B., and Nir, Y. (1997b). *Physics Letters* **B415**, 90.

Gunion, J. F., Haber, H. E., Kane, G. L., and Dawson, S. (1990). *The Higgs hunter's guide.* Addison-Wesley, Redwood City (California).

Gupta, R. (1998). Los Alamos report LAUR-98-271 (hep-ph/9801412).

Gupta, R. and Bhattacharya, T. (1997). *Nuclear Physics B (Proceedings Supplements)* **53**, 292.

Hagberg, E., Hardy, J. C., Koslowsky, V. T., Savard, G., and Towner, I. S. (1997). In *Non-nucleonic degrees of freedom detected in nuclei NNDF'96* (ed. T. Minamisono, Y. Nojiri, T. Sato, and K. Matsuta). World Scientific, Singapore.

Hagelin, J. (1979). *Physical Review* **D20**, 2893.

Hagelin, J. S. (1981). *Nuclear Physics* **193B**, 123.

Hall, L. J. and Weinberg, S. (1993). *Physical Review* **D48**, 979.

Hardy, J. C., Towner, I. S., Koslowsky, V. T., Hagberg, E., and Schmeing, H. (1990). *Nuclear Physics* **509A**, 429.

Heinrich, J., Paschos, E. A., Schwarz, J.-M., and Wu, Y.-L. (1992). *Physics Letters* **B279**, 140.

HERA-B Collaboration (Lohse, T. *et al.*) (1994). DESY report DESY-PRC 94/02.

HERA-B Collaboration (Hartouni, E. *et al.*) (1995). DESY report DESY-PRC 95/01.

Hooft, G. 't (1976). *Physical Review Letters* **37**, 8 and *Physical Review* **D14**, 3432.

Hooft, G. 't (1980). In *Recent developments in gauge theories, Cargèse 1979* (ed. G. 't Hooft *et al.*). Plenum, New York.

Hooft, G. 't (1986). *Physics Reports* **142**, 357.

Huet, P. and Sather, E. (1995). *Physical Review* **D51**, 379.

Iconomidou-Fayard, L. (ed.) (1997). *Proceedings of the workshop on K physics, Orsay, France*. Editions Frontières, Gif-sur-Yvette.

Inami, T. and Lim, C. S. (1981). *Progress of Theoretical Physics* **65**, 297.

Jackiw, R. and Rebbi, C. (1976). *Physical Review Letters* **37**, 172.

Jamin, M. and Münz, M. (1995). *Zeitschrift für Physik* **C66**, 633.

Jarlskog, C. (1985a). *Physical Review Letters* **55**, 1039.

Jarlskog, C. (1985b). *Zeitschrift für Physik* **C29**, 491.

Jaus, W. and Rasche, G. (1987). *Physical Review* **D35**, 3420.

Joshipura, A. S. (1992). In *Particle phenomenology in the 90's* (ed. A. Datta, P. Ghose, and A. Raychaudhuri). World Scientific, Singapore.

Joshipura, A. S. and Rindani, S. D. (1991). *Physics Letters* **B260**, 149.

Jost, R. (1957). *Helvetica Physica Acta* **30**, 409.

Jost, R. (1963). *Helvetica Physica Acta* **36**, 77.

Kabir, P. K. (1970). *Physical Review* **D2**, 540.

Kajantie, K., Laine, M., Rummukainen, K., and Shaposhnikov, M. E. (1996). *Nuclear Physics* **466B**, 189.

Kamal, A. N. (1992a). *International Journal of Modern Physics* **A7**, 3515.

Kamal, A. N. (1992b). *Zeitschrift für Physik* **C54**, 411.

Kandaswamy, J., Salomonson, P., and Schechter, J. (1978). *Physical Review* **D17**, 3051.

Kaplan, D. B. and Manohar, A. V. (1986). *Physical Review Letters* **56**, 2004.

Kayser, B. (1982). *Physical Review* **D26**, 1662.

Kayser, B. (1996). Fermi National Accelerator Laboratory report FERMILAB-CONF-96-429-T (hep-ph/9702264).

Kayser, B. (1997). National Science Foundation report NSF-PT-97-2 (hep-ph/9709382).

Kayser, B. and Shrock, R. E. (1982). *Physics Letters* **B112**, 137.

Kayser, B. and Stodolsky, L. (1996). Max-Planck-Institut (Munich) report MPI-PhT/96-112 (hep-ph/9610522).

Kayser, B., Gibrat-Debu, F., and Perrier, F. (1989). *The physics of massive neutrinos*. World Scientific, Singapore.

Kayser, B., Kuroda, M., Peccei, R. D., and Sanda, A. I. (1990). *Physics Letters* **B237**, 508.

Khoze, V. A., Shifman, M. A., Uraltsev, N. G., and Voloshin, M. B. (1987). *Soviet Journal of Nuclear Physics* **46**, 112.

Khriplovich, I. B. and Lamoreaux, S. K. (1997). *CP violation without strangeness: electric dipole moments of particles, atoms, and molecules*. Springer-Verlag, Berlin.

Kilcup, G. W. (1991). *Nuclear Physics B (Proceedings Supplements)* **20**, 417.

Kim, J. E. (1979). *Physical Review Letters* **43**, 103.

Kim, J. E. (1987). *Physics Reports* **150**, 1.

Kim, C. W. and Pevsner, A. (1993). *Neutrinos in physics and astrophysics*. Harwood Academic Publishers, Chur.

Klemt, E. *et al.* (1988). *Zeitschrift für Physik* **C37**, 179.

Kobayashi, M. and Maskawa, T. (1973). *Progress of Theoretical Physics* **49**, 652.

Kostelecký, V. A. (1998). Indiana University report IUHET 384 (hep-ph/9809584).

Kostelecký, V. A. and Potting, R. (1991). *Nuclear Physics* **359B**, 545.

Kramer, G. and Palmer, W. F. (1995). *Physical Review* **D52**, 6411.

Kroll, N. M. and Wada, W. W. (1955). *Physical Review* **98**, 1355.

Kurimoto, T. (1997). In *CP violation and its origin* (ed. K. Hagiwara). KEK, Tsukuba.

Kurimoto, T. and Tomita, A. (1997). *Progress of Theoretical Physics* **98**, 967.

Landau, L. (1957). *Nuclear Physics* **3**, 127.

Langacker, P. and London, D. (1988). *Physical Review* **D38**, 886.

Lavoura, L. (1989*a*). *Physical Review* **D40**, 2440.

Lavoura, L. (1989*b*). *Physics Letters* **B223**, 97.

Lavoura, L. (1991). *Annals of Physics* **207**, 428.

Lavoura, L. (1992*a*). *Modern Physics Letters* **A7**, 1367.

Lavoura, L. (1992*b*). *Modern Physics Letters* **A7**, 1553.

Lavoura, L. (1994). *International Journal of Modern Physics* **A9**, 1873.

Lavoura, L. (1997). *Physics Letters* **B391**, 441.

Lavoura, L. and Silva, J. P. (1993*a*). *Physical Review* **D47**, 1117.

Lavoura, L. and Silva, J. P. (1993*b*). *Physical Review* **D47**, 2046.

Lavoura, L. and Silva, J. P. (1994). *Physical Review* **D50**, 4619.

Leader, E. and Predazzi, E. (1982). *Gauge theories and the 'new physics'*. Cambridge University Press.

Lee, T. D. (1973). *Physical Review* **D8**, 1226.

Lee, T. D. (1990). *Particle physics and introduction to field theory*. Harwood Academic Publishers, Chur.

Lee, T. D. and Wick, G. C. (1966). *Physical Review* **148**, 1385.

Lee, T. D. and Yang, C. N. (1956). *Physical Review* **104**, 254.

Leutwyler, H. (1996). CERN report CERN-TH-96-25 (hep-ph/9602255).

Leutwyler, H. and Roos, M. (1984). *Zeitschrift für Physik* **C25**, 91.

Le Yaouanc, A., Oliver, L., Pène, O., and Raynal, J.-C. (1995). Laboratoire de Physique Théorique et Hautes Energies (Orsäy) report LPTHE-Orsay 95/15 (hep-ph/9504270).

Lipkin, H. J., Nir, Y., Quinn, H. R., and Snyder, A. E. (1991). *Physical Review* **D44**, 1454.

Liu, T. (1996). In *Workshop on the tau charm factory* (ed. J. Repond). American Institute of Physics, New York.

London, D. and Peccei, R. D. (1989). *Physics Letters* **B223**, 257.

Löwdin, P. -O. (1998). *Linear Algebra for Quantum Theory*. John Wiley and Sons.

Lueders, G. (1954). *Kgl. Danske Videnskab. Selskab Mat. Fys. Medd.* **28**, No. 5.

Lusignoli, M. (1989). *Nuclear Physics* **325B**, 33.

Lusignoli, M., Maiani, L., Martinelli, G., and Reina, L. (1992). *Nuclear Physics* **369B**, 139.

McFarlane, W. K. *et al.* (1985). *Physical Review* **D32**, 547.

McWilliams, B. and Shanker, O. (1980). *Physical Review* **D22**, 2853.

Mampe, W., Ageron, P., Bates, C., Pendlebury, J. M., and Steyerl, A. (1989). *Physical Review Letters* **63**, 593.

Marciano, W. J. (1991). *Annual Review of Nuclear and Particle Science* **41**, 469.

Marciano, W. J. and Parsa, Z. (1986). *Annual Review of Nuclear and Particle Science* **36**, 171.

Marciano, W. J. and Parsa, Z. (1996). *Physical Review* **D53**, R1.

Marciano, W. J. and Sirlin, A. (1986). *Physical Review Letters* **56**, 22.

Mark-III Collaboration (Adler, J. *et al.*) (1989). *Physical Review Letters* **62**, 1821.

Mark-III Collaboration (Bai, Z. *et al.*) (1991). *Physical Review Letters* **66**, 1011.

Marshak, R. E. (1951). *Physical Review* **82**, 313.

Meca, C. C. and Silva, J. P. (1998). *Physical Review Letters* **81**, 1377.

Méndez, A. and Pomarol, A. (1991). *Physics Letters* **B272**, 313.

Mikheev, S. P. and Smirnov, A. Yu. (1985). *Soviet Journal of Nuclear Physics* **42**, 913.

Mohapatra, R. N. (1985). *Physics Letters* **159B**, 374.

Mohapatra, R. N. and Pal, P. B. (1991). *Massive neutrinos in physics and astrophysics*. World Scientific, Singapore.

Mohapatra, R. N. and Pati, J. C. (1975). *Physical Review* **D11**, 566 and *Physical Review* **D11**, 2558.

Mohapatra, R. N. and Senjanović, G. (1978). *Physics Letters* **79B**, 283.

Mohapatra, R. N. and Senjanović, G. (1980). *Physical Review Letters* **44**, 912.

NA31 Collaboration (Barr, G. D. *et al.*) (1993). *Physics Letters* **B317**, 233.

Nardi, E., Roulet, E., and Tommasini, D. (1995). *Physics Letters* **B344**, 225.

Narison, S. (1995). *Physics Letters* **B358**, 113.

Nelson, A. (1984). *Physics Letters* **136B**, 387 and *Physics Letters* **143B**, 165.

Neubert, M. (1995). In *'95 Electroweak interactions and unified theories* (ed. J. Tran Thanh Van). Editions Frontières, Paris.

Neubert, M. (1998). *Physics Letters* **B424**, 152.

Nieves, J. F. (1982). *Physical Review* **D26**, 3152.

Nilles, H.-P. (1984). *Physics Reports* **110**, 1.

Nir, Y. (1993). In *The third family and the physics of flavor* (ed. L. Vassilian). Stanford Linear Accelerator Center, Stanford (California).

Nir, Y. (1996). *Nuovo Cimento* **109A**, 991.

Nir, Y. and Quinn, H. R. (1990). *Physical Review* **D42**, 1473.

Nir, Y. and Silverman, D. (1990). *Nuclear Physics* **345B**, 301.

Ohl, T., Ricciardi, G., and Simmons, E. H. (1993). *Nuclear Physics* **403B**, 605.

Onogi, T. *et al.* (1997). *Nuclear Physics B (Proceedings Supplements)* **53**, 289.

Pais, A. and Treiman, S. B. (1975). *Physical Review* **D12**, 2744.

Panofsky, W. K. H., Aamodt, R. L., and Hadley, J. (1951). *Physical Review* **81**, 565.

Parada, P. A. (1996). Unpublished Ph.D. thesis. Instituto Superior Técnico. (In Portuguese.)

Particle Data Group (Barnett, R. M. *et al.*) (1996). *Physical Review* **D54**, 1.

Particle Data Group (Caso, C. *et al.*) (1998). *European Physical Journal* **C3**, 1.

Paschos, E. A. (1977). *Physical Review* **D15**, 1966.

Paschos, E. A. and Wu, Y.-L. (1991). *Modern Physics Letters* **A6**, 93.

Pati, J. C. and Salam, A. (1974). *Physical Review* **D10**, 275.

Patterson, J. R. (1995). In *Proceedings of the XXVII international conference on high-energy physics, Glasgow, Scotland, July 20-27, 1994* (ed. P. J. Bussey and I. G. Knowles). Institute of Physics, Bristol.

Pauli, W. (1955). In *Niels Bohr and the development of physics*. Pergamon, London.

Peccei, R. D. (1989). In *CP violation* (ed. C. Jarlskog). World Scientific, Singapore.

Peccei, R. D. and Quinn, H. R. (1977). *Physical Review Letters* **38**, 1440 and *Physical Review* **D16**, 1791.

Perkins, D. H. (1987). *Introduction to high-energy physics*. Addison-Wesley, Menlo Park.

Petrov, A. A. (1997). *Physical Review* **D56**, 1685.

Plano, R. *et al.* (1959). *Physical Review Letters* **3**, 525.

Raffelt, G. G. (1996). *Stars as laboratories for fundamental physics: the astrophysics of neutrinos, axions, and other weakly interacting particles*. University of Chicago Press.

Rajaraman, R. (1982). *Solitons and instantons*. North-Holland Publishing Company, Amsterdam.

Rindani, S. D. (1995). *Pramana* **45**, S263.

Roesch, L. P., Telegdi, V. L., Truttmann, P., Zehnder, A., Grenacs, L., and Palffy, L. (1982). *American Jounal of Physics* **50**, 931.

Roldán, J. (1991). Unpublished Ph.D. thesis. University of València. (In Castilian.)

Rosner, J. L. (1990). *Physical Review* **D42**, 3732.

Rosner, J. L. (1994). In *B decays* (ed. S. Stone). World Scientific, Singapore.

Rubakov, V. A. and Shaposhnikov, M. E. (1996). *Phys. Usp.* **39**, 461; erratum *ibid.* **39**, 1276.

Sachs, R. G. (1963). *Physical Review* **129**, 2280.

Sachs, R. G. (1964). *Annals of Physics* **22**, 239.

Sachs, R. G. (1987). *The physics of time reversal*. University of Chicago Press.

Sakharov, A. D. (1967). *JETP Letters* **5**, 24.

Sakurai, J. J. (1964). *Invariance principles and elementary particles*. Princeton University Press.

Sakurai, J. J. (1994). *Modern quantum mechanics*. Addison-Wesley, New York.

Salam, A. (1968). In *Elementary particle theory: relativistic groups and analyticity (Nobel symposium No. 8)* (ed. N. Svartholm). Almquist and Wiksell, Stockholm.

Sanda, A. I. and Xing, Z.-Z. (1997). Nagoya University report DPNU-97-43 (hep-ph/9709491).

Savage, M. and Wise, M. (1989). *Physical Review* **D39**, 3346; erratum *ibid.* **D40**, 3127(E).

Schröder, H. (1994). In *B decays* (ed. S. Stone). World Scientific, Singapore.

Scott, I. J. *et al.* (1995). In *Proceedings of the XXVII international conference on high-energy physics, Glasgow, Scotland, July 20-27, 1994* (ed. P. J. Bussey and I. G. Knowles). Institute of Physics, Bristol.

Segré, G. and Weldon, H. A. (1979). *Physical Review Letters* **42**, 1191.

Senjanović, G. (1979). *Nuclear Physics* **153B**, 334.

Senjanović, G. and Mohapatra, R. N. (1975). *Physical Review* **D12**, 1502.

Sharpe, S. R. (1991). *Nuclear Physics B (Proceedings Supplements)* **20**, 429.

Shifman, M. A., Vainshtein, A. I., and Zakharov, V. I. (1980). *Nuclear Physics* **166B**, 493.

Shrock, R. E. and Wang, L. L. (1978). *Physical Review Letters* **41**, 1692.

Silva, J. P. (1998). *Physical Review* **D58**, 016004.

Silva, J. P. and Wolfenstein, L. (1994). *Physical Review* **D49**, R1151.

Silva, J. P. and Wolfenstein, L. (1997). *Physical Review* **D55**, 5331.

Silverman, D. (1996). *International Journal of Modern Physics* **A11**, 2253.

Simma, H., Eilam, G., and Wyler, D. (1991). *Nuclear Physics* **352B**, 367.

Sirlin, A. (1987). *Physical Review* **D35**, 3423.

Sirlin, A. and Zucchini, R. (1986). *Physical Review Letters* **57**, 1994.

Snyder, A. E. and Quinn, H. R. (1993). *Physical Review* **D48**, 2139.

Soares, J. M. (1992). *Physical Review Letters* **68**, 2102.

Soares, J. M. (1993). Unpublished Ph.D. thesis. Carnegie Mellon University.

Soares, J. M. and Wolfenstein, L. (1992). *Physical Review* **D46**, 256.

Soares, J. M. and Wolfenstein, L. (1993). *Physical Review* **D47**, 1021.

Sohnius, M. F. (1985). *Physics Reports* **128**, 39.

Streater, R. F. and Wightman, A. S. (1964). *CPT, spin, statistics and all that.* Benjamin, New York.

Super-Kamiokande Collaboration (Fukuda, Y. *et al.*) (1998). *Physical Review Letters* **81**, 1562.

Swain, J. and Taylor, L. (1997). Hep-ph/9712421.

Swain, J. and Taylor, L. (1998). *Physical Review* **D58**, 093006.

Towner, I. S. (1992). *Nuclear Physics* **540A**, 478.

Towner, I. S. and Hardy, J. C. (1998). Talk delivered at WEIN98 (nucl-th/9809 087).

Trodden, M. (1998). Case Western Reserve University report CWRU-P6-98 (hep-ph/9803479).

Tsao, H.-S. (1980). In *Proceedings, 1980 Guangzhou conference on theoretical particle physics* (ed. H. Ning and T. Hung-Yuan). Science Press, Beijing.

Turck-Chièze, S., Däppen, W., Fossat, E., Provost, J., Schatzman, E., and Vignaud, D. (1993). *Physics Reports* **230**, 57.

Turok, N. (1993). In *Perspectives on Higgs physics* (ed. G. L. Kane). World Scientific, Singapore.

Vainshtein, A. I., Zakharov, V. I., and Shifman, M. A. (1975). *JETP Letters* **22**, 55.

Vainshtein, A. I., Zakharov, V. I., and Shifman, M. A. (1977). *Soviet Physics JETP* **45**, 670.

WA2 Collaboration (Bourquin, M. *et al.*) (1983). *Zeitschrift für Physik* **C21**, 27.

Weinberg, S. (1967). *Physical Review Letters* **19**, 1264.

Weinberg, S. (1975). *Physical Review* **D10**, 3583.

Weinberg, S. (1976). *Physical Review Letters* **37**, 657.

Weinberg, S. (1978). *Physical Review Letters* **40**, 223.

Weinberg, S. (1990). *Physical Review* **D42**, 860.

Weinberg, S. (1995). *The quantum theory of fields.* Cambridge University Press.

Weisskopf, V. F. and Wigner, E. P. (1930*a*). *Zeitschrift für Physik* **63**, 54. (In German.)

Weisskopf, V. F. and Wigner, E. P. (1930*b*). *Zeitschrift für Physik* **65**, 18. (In German.)

Wick, G. C., Wightman, A. S., and Wigner, E. P. (1952). *Physical Review* **88**, 101.

Wigner, E. P. (1932). *Nachr. Ges. Wiss. Göttingen, Math-Physik K1* **32**, 546. (In German.)

Wilczek, F. (1978). *Physical Review Letters* **40**, 271.

Wilkinson, D. H. (1982). *Nuclear Physics* **377A**, 474.

Winstein, B. (1992). *Physical Review Letters* **68**, 1271.

Winstein, B. and Wolfenstein, L. (1993). *Review of Modern Physics* **65**, 113.

Wolfenstein, L. (1964). *Physical Review Letters* **13**, 562.

Wolfenstein, L. (1978). *Physical Review* **D17**, 2369.

Wolfenstein, L. (1979). *Physical Review* **D20**, 2634.

Wolfenstein, L. (1981). *Physics Letters* **B107**, 77.

Wolfenstein, L. (1983). *Physical Review Letters* **51**, 1945.

Wolfenstein, L. (1984). *Nuclear Physics* **246B**, 45.

Wolfenstein, L. (1985). *Physics Letters* **B164**, 170.

Wolfenstein, L. (1991). *Physical Review* **D43**, 151.

Wolfenstein, L. (1995). *Physical Review Letters* **75**, 2460.

Wolfenstein, L. (1997*a*). *Physical Review* **D56**, 7469.

Wolfenstein, L. (1997*b*). Carnegie Mellon University report 'Searching for new physics from CP violation in B decays', to be published in the Proceedings of the B-physics conference, Honolulu, Hawaii, 1997.

Wu, C. S. and Shaknov, I. (1950). *Physical Review* **77**, 136.

Wu, C. S., Ambler, E., Haywood, R. W., Hoppes, D. D., and Hudson, R. P. (1957). *Physical Review* **105**, 1413.

Xing, Z.-Z. (1996). *Physical Review* **D53**, 204.

Yamamoto, H. (1997*a*). *Physics Letters* **B401**, 91.

Yamamoto, H. (1997*b*). *Physical Review Letters* **79**, 2402.

Yanagida, T. (1979). In *Proceedings of the workshop on unified theory and baryon number in the universe* (ed. O. Sawata and A. Sugamoto). KEK, Tsukuba.

Yang, C. N. (1950). *Physical Review* **77**, 242.

Yang, C. N. and Mills, R. L. (1954). *Physical Review* **96**, 191.

Yuan, C.-P. (1995). *Modern Physics Letters* **A10**, 627.

Zeppenfeld, D. (1981). *Zeitschrift für Physik* **C8**, 77.

Zhitnitsky, A. R. (1980). *Soviet Journal of Nuclear Physics* **31**, 260.

INDEX